MASTERS THESES IN THE PURE AND APPLIED SCIENCES

ACCEPTED BY COLLEGES AND UNIVERSITIES
OF THE UNITED STATES AND CANADA

VOLUME 33

MASTERS THESES IN THE PURE AND APPLIED SCIENCES

ACCEPTED BY COLLEGES AND UNIVERSITIES OF THE UNITED STATES AND CANADA

VOLUME 33

Edited by

Wade H. Shafer

Center for Information and Numerical Data Analysis and Synthesis

PLENUM PRESS • NEW YORK AND LONDON

The Library of Congress cataloged the first volume of this title as follows:

Masters theses in the pure and applied sciences accepted by colleges and universities
of the United States. v. [1]—1955–56 —
Lafayette, Ind., School of Mechanical Engineering, Purdue University.
v. 29 cm.

Title varies: 1955–56 – 1957, Master's theses accepted by U. S. colleges and
universities in the fields of chemical engineering, chemistry, mechanical engineering,
metallurgical engineering, and physics.—1958, Master's theses and doctoral dissertations in the pure and applied sciences accepted by colleges and universities of the
United States.

Vols. for 1955–56 —   issued by Purdue University's Thermophysical
Properties Research Center (1955–56 as its TPRC publication)
QC251.P8

1. Science—Bibl. 2. Engineering—Bibl. 3. Dissertations, Academic—U.S.—Bibl.
I. Purdue University, Lafayette, Ind. Thermophysical Properties Research Center.
(Series: Purdue University, Lafayette, Ind. Thermophysical Properties Research
Center. TPRC publication)
Z7401.M35   58-62673

Library of Congress Catalog Card Number 58-62673

ISBN 0-306-43732-5

© 1990 Plenum Press, New York
A Division of Plenum Publishing Corporation
233 Spring Street, New York, N.Y. 10013

Printed in the United States of America

PREFACE

Masters Theses in the Pure and Applied Sciences was first conceived, published, and disseminated by the *Center for Information and Numerical Data Analysis and Synthesis (CINDAS)* * at Purdue University in 1957, starting its coverage of theses with the academic year 1955. Beginning with Volume 13, the printing and dissemination phases of the activity were transferred to University Microfilms/Xerox of Ann Arbor, Michigan, with the thought that such an arrangement would be more beneficial to the academic and general scientific and technical community. After five years of this joint undertaking we had concluded that it was in the interest of all concerned if the printing and distribution of the volumes were handled by an international publishing house to assure improved service and broader dissemination.

Hence, starting with Volume 18, *Masters Theses in the Pure and Applied Sciences* has been disseminated on a worldwide basis by Plenum Publishing Corporation of New York, and in the same year the coverage was broadened to include Canadian universities. All back issues can also be ordered from Plenum.

We have reported in Volume 33 (thesis year 1988) a total of 13,273 theses titles from 23 Canadian and 185 United States universities. We are sure that this broader base for these titles reported will greatly enhance the value of this important annual reference work.

While Volume 33 reports theses submitted in 1988, on occasion, certain universities do report theses submitted in previous years but not reported at the time. This fact is indicated for each thesis citation by the thesis year shown in parentheses.

The organization of Volume 33 is identical to that of past years. It consists of theses titles arranged by discipline and by university within each discipline. Any accredited university or college with a graduate program in the pure and applied sciences (excluding mathematics and the life sciences) is invited to contribute titles in the fields presently covered. Inquiries concerning interlibrary loan or purchase of any title mentioned in this volume should be sent directly to the institution responsible for the thesis.

Some theses may be shifted in their listing to best reflect their contents with respect to the 44 different study disciplines. For example, a thesis in civil engineering may be listed in this publication under disciplines 09—Civil Engineering, 23—Irrigation Engineering, 42—Sanitary Engineering, Water Pollution, and Water Resources, or 44—Transportation Engineering.

The editor gratefully acknowledges the cooperation of the librarians and graduate school officials whose continued interest in this work makes its publication possible. Also, *CINDAS* has played a key part in the publication of this reference work since thesis year 1955.

*Formerly the *Thermophysical Properties Research Center (TPRC)*

CONTENTS

*Mathematics and most life sciences have been excluded from this publication, on a purely arbitrary basis, simply to limit the scope of the work. Biochemistry, biophysics, and bioengineering are included in the coverage when titles in these areas are reported together with chemistry, physics, and engineering and not as a separate discipline.

COMPARATIVE DATA ON THE SERIES

Volume	Publication Date	Thesis Year	Contributing Institutions	Titles Reported
1	Oct. 1957	1955	93	1,002
		1956	93	1,027
2	Aug. 1958	1957	154	1,727
3*	Oct. 1959	1958	139	3,736
4	Dec. 1960	1959	162	4,984
5	Dec. 1961	1960	183	5,708
6	Dec. 1966	1961	186	5,911
7	Aug. 1966	1962	186	6,321
8	Aug. 1966	1963	175	6,505
9	Jan. 1968	1964	174	6,940
10	Jan. 1968	1965	170	7,310
11	Jan. 1968	1966	173	7,099
12	Jul. 1968	1967	167	6,909
13	Jul. 1969	1968	174	7,802
14	Jan. 1971	1969	175	7,160
15	Jul. 1971	1970	183	7,437
16	Jul. 1972	1971	182	7,170
17	Jul. 1973	1972	250	8,513
18†	Dec. 1974	1973	251	10,381
19	Dec. 1975	1974	229	10,045
20	Dec. 1976	1975	267	10,374
21	Dec. 1977	1976	244	10,586
22	Oct. 1978	1977	255	10,658
23	Nov. 1979	1978	247	10,432
24	Nov. 1980	1979	241	10,033
25	Dec. 1981	1980	241	10,308
26	Dec. 1982	1981	242	11,048
27	Dec. 1983	1982	227	10,918
28	Dec. 1984	1983	223	10,611
29	Dec. 1985	1984	225	12,637
30	Aug. 1987	1985	212	12,400
31	Aug. 1988	1986	206	11,480
32	Dec. 1989	1987	198	12,483
33	Jul. 1990	1988	208	13,273

*Part Two of Volume 3 included doctoral dissertations from the 1956–1957 academic year, citing 2,846 titles from 103 universities.

†Effective with Volume 18, the coverage was extended to include Canadian universities.

PARTICIPATING UNIVERSITIES

1. ABILENE CHRISTIAN COLLEGE
   ABILENE, TEXAS 79699-0001

2. ACADIA UNIVERSITY
   WOLFVILLE, NOVA SCOTIA
   CANADA B0P IXO

3. AKRON, UNIVERSITY OF
   AKRON, OHIO 44304

4. ALABAMA A - M UNIVERSITY
   NORMAL, ALABAMA 35762

5. ALABAMA, UNIVERSITY OF
   HUNTSVILLE, ALABAMA 35899

6. ALABAMA, UNIVERSITY OF
   UNIVERSITY, ALABAMA 35486

7. ALBERTA, UNIVERSITY OF
   EDMONTON, ALBERTA
   CANADA T6G 2E8

8. ALFRED UNIVERSITY
   ALFRED, NEW YORK 14802

9. AMERICAN UNIVERSITY
   WASHINGTON, D.C. 20016

10. ARIZONA STATE UNIVERSITY
    TEMPE, ARIZONA 85287

11. ARKANSAS, UNIVERSITY OF
    FAYETTESVILLE, ARKANSAS 72701

12. AUBURN UNIVERSITY
    AUBURN, ALABAMA 36849

13. BALL STATE UNIVERSITY
    MUNCIE, INDIANA 47306-0160

14. BAYLOR UNIVERSITY
    WACO, TEXAS 76798-7264

15. BOSTON COLLEGE
    CHESTNUT HILL, MASSACHUSETTS 02167

16. BOWLING GREEN STATE UNIVERSITY
    BOWLING GREEN, OHIO 43403

17. BRIGHAM YOUNG UNIVERSITY
    PROVO, UTAH 84602

18. BROCK UNIVERSITY
    ST. CATHARINES, ONTARIO
    CANADA L2S 3A1

19. BROOKLYN COLLEGE
    BROOKLYN, NEW YORK 11210

20. BROOKLYN, POLYTECHNIC INSTITUTE OF
    BROOKLYN, NEW YORK 11201

21. BROWN UNIVERSITY
    PROVIDENCE, RHODE ISLAND 02912

22. BUCKNELL UNIVERSITY
    LEWISBURG, PENNSYLVANIA 17837

23. CALGARY, UNIVERSITY OF
    CALGARY, ALBERTA, CANADA T2N 1N4

24. CALIFORNIA STATE UNIVERSITY
    FRESNO, CALIFORNIA 93740-0034

25. CALIFORNIA STATE UNIVERSITY
    FULLERTON, CALIFORNIA 92634-4080

26. CALIFORNIA STATE UNIVERSITY
    LONG BEACH, CALIFORNIA 90840

27. CALIFORNIA STATE UNIVERSITY
    SACRAMENTO, CALIFORNIA 95819-2695

28. CALIFORNIA, UNIVERSITY OF
    DAVIS, CALIFORNIA 95616

29. CALIFORNIA, UNIVERSITY OF
    IRWINE, CALIFORNIA 92717

30. CALIFORNIA, UNIVERSITY OF
    LA JOLLA, CALIFORNIA 92093

31. CARLETON UNIVERSITY
    OTTAWA, CANADA K1S 5B6

32. CARNEGIE-MELLON UNIVERSITY
    PITTSBURG, PENNSYLVANIA 15213

33. CASE WESTERN RESERVE UNIVERSITY
    CLEVELAND, OHIO 44106

34. CENTRAL FLORIDA, UNIVERSITY OF
    ORLANDO, FLORIDA 32816-0450

35. CITY COLLEGE OF NEW YORK
    NEW YORK, NEW YORK 10031

36. CLARK ATLANTA UNIVERSITY
    ATLANTA, GEORGIA 30314

37. CLARKSON COLLEGE OF TECHNOLOGY
    POTSDAM, NEW YORK 13676

38. CLEMSON UNIVERSITY
    CLEMSON, SOUTH CAROLINA 29631

39. CLEVELAND STATE UNIVERSITY
    CLEVELAND, OHIO 44115

40. COLORADO SCHOOL OF MINES
    GOLDEN, COLORADO 80401

41. COLORADO STATE UNIVERSITY
    FORT COLLINS, COLORADO 80523

42. COLORADO, UNIVERSITY OF
    BOULDER, COLORADO 80309-0424

43. CONCORDIA UNIVERSITY
    MONTREAL, QUEBEC,
    CANADA H3G IMB

44. CONNECTICUT, UNIVERSITY OF
    STORRS, CONNECTICUT 06268-1005

45. CORNELL UNIVERSITY
    ITHACA, NEW YORK 14853

46. DALHOUSIE UNIVERSITY
    HALIFAX, NOVA SCOTIA
    CANADA B3H 4H6

47. DELAWARE, UNIVERSITY OF
    NEWARK, DELAWARE 19711

48. DREXEL UNIVERSITY
    PHILADELPHIA, PENNSYLVANIA 19104

49. DUKE UNIVERSITY
    DURHAM, NORTH CAROLINA 27706

50. EAST CAROLINA UNIVERSITY
    GREENVILLE, NORTH CAROLINA 27834

51. EAST TENNESSEE STATE UNIVERSITY
    JOHNSON CITY, TENNESSEE 37614-0002

52. EAST TEXAS STATE UNIVERSITY
COMMERCE, TEXAS 75428

53. EASTERN ILLINOIS UNIVERSITY
CHARLESTON, ILLINOIS 61920

54. EASTERN KENTUCKY UNIVERSITY
RICHMOND, KENTUCKY 40475-3101

55. EASTERN NEW MEXICO UNIVERSITY
PORTALES, NEW MEXICO 88130

56. EMORY UNIVERSITY
ATLANTA, GEORGIA 30322

57. EMPORIA STATE UNIVERSITY
EMPORIA, KANSAS 66801-5087

58. FLORIDA ATLANTIC UNIVERSITY
BOCA RATON, FLORIDA 33431-0991

59. FLORIDA STATE UNIVERSITY
TALLAHASSEE, FLORIDA 32306

60. FLORIDA, UNIVERSITY OF
GAINESVILLE, FLORIDA 32611

61. FORT HAYS KANSAS STATE UNIVERSITY
HAYS, KANSAS 67601-4099

62. GEORGETOWN UNIVERSITY
WASHINGTON, D.C. 20057

63. GEORGIA INSTITUTE OF TECHNOLOGY
ATLANTA, GEORGIA 30332

64. GEORGIA, UNIVERSITY OF
ATHENS, GEORGIA 30602

65. GUAM, UNIVERSITY OF
GUAM 96923

66. GUELPH, UNIVERSITY OF
GUELPH, ONTARIO, CANADA N1G 2W1

67. HAWAII, UNIVERSITY OF
HONOLULU, HAWAII 96822

68. HOUSTON, UNIVERSITY OF
HOUSTON, TEXAS 77004

69. HOWARD UNIVERSITY
WASHINGTON, D.C. 20059

70. IDAHO STATE UNIVERSITY
POCATELLO, IDAHO 83209-0009

71. IDAHO, UNIVERSITY OF
MOSCOW, IDAHO 83843

72. ILLINOIS INSTITUTE OF TECHNOLOGY
CHICAGO, ILLINOIS 60616

73. ILLINOIS STATE UNIVERSITY
NORMAL, ILLINOIS 61761-6901

74. INDIANA STATE UNIVERSITY
TERRE HAUTE, INDIANA 47809

75. INDIANA UNIVERSITY
BLOOMINGTON, INDIANA 47405

76. IOWA STATE UNIVERSITY
AMES, IOWA 50011-2140

77. JOHN CARROLL UNIVERSITY
CLEVELAND, OHIO 44118

78. JOHNS HOPKINS UNIVERSITY
BALTIMORE, MARYLAND 21218

79. KANSAS, UNIVERSITY OF
LAWRENCE, KANSAS 66045-2800

80. KENTUCKY, UNIVERSITY OF
LEXINGTON, KENTUCKY 40506

81. LAKE HEAD UNIVERSITY
THUNDER BAY, ONTARIO,
CANADA P7B 5E1

82. LAMAR UNIVERSITY
BEAUMONT, TEXAS 77710

83. LAURENTIEN UNIVERSITY
SUDBURY, ONTARIO
CANADA P3E 2C6

84. LEHIGH UNIVERSITY
BETHLEHEM, PENNSYLVANIA 18015

85. LONG ISLAND UNIVERSITY
BROOKLYN, NEW YORK 11201-5372

86. LOUISIANA STATE UNIVERSITY
BATON ROUGE, LOUISIANA 70803-3300

87. LOUISIANA TECH UNIVERSITY
RUSTON, LOUISIANA 71272-0046

88. LOUISVILLE, UNIVERSITY OF
LOUISVILLE, KENTUCKY 40292

89. LOWELL, UNIVERSITY OF
LOWELL, MASSACHUSETTS 01854

90. LOYOLA UNIVERSITY
CHICAGO, ILLINOIS 60611

91. MAINE, UNIVERSITY OF
ORONO, MAINE 04469

92. MANITOBA, UNIVERSITY OF
WINNIPEG, MANITOBA
CANADA R3T 2N2

93. MARQUETTE UNIVERSITY
MILWAUKEE, WISCONSIN 53233

94. MARSHALL UNIVERSITY
HUNTINGTON, WEST VIRGINIA 25755-2100

95. MASSACHUSETTS INSTITUTE OF
TECHNOLOGY
CAMBRIDGE, MASSACHUSETTS 02139

96. MASSACHUSETTS, UNIVERSITY OF
AMHERST, MASSACHUSETTS 010003

97. MC NEESE STATE UNIVERSITY
LAKE CHARLES, LOUISIANA 70609-2180

98. MIAMI, UNIVERSITY OF
CORAL GABLES, FLORIDA 33124

99. MICHIGAN TECHNOLOGICAL UNIVERSITY
HOUGHTON, MICHIGAN 49931

100. MINNESOTA, UNIVERSITY OF
MINNEAPOLIS, MINNESOTA 55455

101. MISSISSIPPI STATE UNIVERSITY
STARKVILLE, MISSISSIPPI 39762

102. MISSISSIPPI, UNIVERSITY OF
UNIVERSITY, MISSISSIPPI 38677

103. MISSOURI, UNIVERSITY OF
COLUMBIA, MISSOURI 65201

104. MISSOURI, UNIVERSITY OF
ROLLA, MISSOURI 65401

105. MONTANA COLLEGE OF MINERAL
SCIENCE AND TECHNOLOGY
BUTTE, MONTANA 59701

106. MONTANA STATE UNIVERSITY
BOZEMAN, MONTANA 59717-0022

107. MONTANA, UNIVERSITY OF
MISSOULA, MONTANA 59812

108. MONTREAL, UNIVERSITY OF
MONTREAL, QUEBEC
CANADA H3C 3J7

109. MURRAY STATE UNIVERSITY
MURRAY, KENTUCKY 42071-3308

110. NEBRASKA, UNIVERSITY OF
LINCOLN, NEBRASKA 68588-0410

111. NEW BRUNSWICK, UNIVERSITY OF
FREDERICTON, NEW BRUNSWICK
CANADA E3B 5H5

112. NEW JERSEY INSTITUTE OF TECHNOLOGY
NEWARK, NEW JERSEY 07102

113. NEW MEXICO INSTITUTE OF MINING
AND TECHNOLOGY
SOCORRO, NEW MEXICO 87801

114. NEW MEXICO STATE UNIVERSITY
LAS CRUCES, NEW MEXICO 88003-0006

115. NEW MEXICO, UNIVERSITY OF
ALBUQUERQUE, NEW MEXICO 87131

116. NEW YORK, STATE UNIVERSITY OF
ALBANY, NEW YORK 12222

117. NEW YORK, STATE UNIVERSITY OF
BUFFALO, NEW YORK 14260

118. NEW YORK, STATE UNIVERSITY OF
FREDONIA, NEW YORK 14063

119. NORTH CAROLINA STATE UNIVERSITY
RALEIGH, NORTH CAROLINA 27695-7102

120. NORTH CAROLINA, UNIVERSITY OF
CHAPEL HILL, NORTH CAROLINA 27599-3022

121. NORTH DAKOTA STATE UNIVERSITY
FARGO, NORTH DAKOTA 58105-5599

122. NORTH DAKOTA, UNIVERSITY OF
GRAND FORKS, NORTH DAKOTA 58202

123. NORTH TEXAS, UNIVERSITY OF
DENTON, TEXAS 76203

124. NORTHEASTERN LOUISIANA UNIVERSITY
MONROE, LOUISIANA 71209-0600

125. NORTHERN ARIZONA UNIVERSITY
FLAGSTAFF, ARIZONA 86011

126. NORTHERN ILLINOIS UNIVERSITY
DEKALB, ILLINOIS 60115-2854

127. NORTHERN IOWA, UNIVERSITY OF
CEDAR FALLS, IOWA 50613

128. NOTRE DAME, UNIVERSITY OF
NOTRE DAME, INDIANA 46556

129. NOVA SCOTIA, TECH UNIVERSITY OF
HALIFAX, NOVA SCOTIA
CANADA B3J 2X4

130. OAKLAND UNIVERSITY
ROCHESTER, MICHIGAN 48309-4401

131. OKLAHOMA, UNIVERSITY OF
NORMAN, OKLAHOMA 73069

132. OLD DOMINION UNIVERSITY
NORFOLK, VIRGINIA 23508-8506

133. OREGON STATE UNIVERSITY
CORVALLIS, OREGON 97331

134. OREGON, UNIVERSITY OF
EUGENE, OREGON 97403-1219

135. PACIFIC, UNIVERSITY OF THE
STOCKTON, CALIFORNIA 95211

136. PENNSYLVANIA STATE UNIVERSITY
UNIVERSITY PARK, PENNSYLVANIA
16802

137. PENNSYLVANIA, UNIVERSITY OF
PHILADELPHIA, PENNSYLVANNIA 19104-6206

138. PITTSBURGH, UNIVERSITY OF
PITTSBURGH, PENNSYLVANIA 15261

139. PRINCETON UNIVERSITY
PRINCETON, NEW JERSEY 08540

140. PUERTO RICO, UNIVERSITY OF
RIO PIEDRAS, PUERTO RICO 00931-3346

141. PURDUE UNIVERSITY
WEST LAFAYETTE, INDIANA 47907

142. QUEENS COLLEGE
FLUSHING, NEW YORK 11367

143. QUEENS UNIVERSITY
KINGSTON, ONTARIO
CANADA K7L 3N6

144. REGINA, UNIVERSITY OF
REGINA, CANADA S4S OA2

145. RENSSELAER POLYTECHNIC INSTITUTE
TROY, NEW YORK 12180-3590

146. RHODE ISLAND, UNIVERSITY OF
KINGSTON, RHODE ISLAND 02881-0803

147. ROCHESTER, UNIVERSITY OF
ROCHESTER, NEW YORK 14627

148. ROSE-HULMAN INSTITUTE OF
TECHNOLGY
TERRE HAUTE, INDIANA 47803-3999

149. SAINT LOUIS UNIVERSITY
SAINT LOUIS, MISSOURI 63108

150. SAN DIEGO STATE UNIVERSITY
SAN DIEGO, CALIFORNIA 92182-0511

151. SAN FRANCISCO STATE UNIVERSITY
SAN FRANCISCO, CALIFORNIA 94132

152. SASKATCHEWAN, UNIVERSITY OF
SASKATOON, CANADA S7N OWO

153. SHERBROOKE, UNIVERSITY OF
SHERBROOKE, QUEBEC,
CANADA J1K 2R1

154. SOUTH CAROLINA, UNIVERSITY OF
COLUMBIA, SOUTH CAROLINA 29208

155. SOUTH DAKOTA SCHOOL OF MINES AND
TECHNOLOGY
RAPID CITY, SOUTH DAKOTA 57701-3995

156. SOUTH DAKOTA STATE UNIVERSITY
BROOKINGS, SOUTH DAKOTA 57007-1098

157. SOUTH DAKOTA, UNIVERSITY OF
VERMILLION, SOUTH DAKOTA 57069-2390

158. SOUTH FLORIDA, UNIVERSITY OF
TAMPA, FLORIDA 33620-5350

159. SOUTHERN ILLINOIS UNIVERSITY
EDWARDSVILLE, ILLINOIS 62026-1046

160. SOUTHERN METHODIST UNIVERSITY
DALLAS, TEXAS 75275

161. SOUTHWEST TEXAS STATE UNIVERSITY
SAN MARCOS, TEXAS 78666-4604

162. SOUTHWESTERN LOUISIANA, UNIV. OF
LAFAYETTE, LOUISIANA 70503-2038

163. STEPHAN F. AUSTIN STATE UNIVERSITY
NACOGDOCHES, TEXAS 75962-3024

164. STEVENS INSTITUTE OF TECHNOLOGY
HOBOKEN, NEW JERSEY 07030

165. TEMPLE UNIVERSITY
PHILADELPHIA, PENNSYLVANIA 19122

166. TENNESSEE TECHNOLOGICAL
UNIVERSITY
COOKEVILLE, TENNESSEE 38501

167. TENNESSEE, UNIVERSITY OF
KNOXVILLE, TENNESSEE 37996-0220

168. TENNESSEE, UNIVERSITY OF
TULLAHOMA, TENNESSEE 37388

169. TEXAS A AND M UNIVERSITY
COLLEGE STATION, TEXAS 77843

170. TEXAS CHRISTIAN UNIVERSITY
FORT WORTH, TEXAS 76129

171. TEXAS SOUTHERN UNIVERSITY
HOUSTON, TEXAS 77004-4814

172. TEXAS WOMEN S UNIVERSITY
DENTON, TEXAS 76204

173. TEXAS, UNIVERSITY OF
ARLINGTON, TEXAS 76019

174. TEXAS, UNIVERSITY OF
AUSTIN, TEXAS 78713-7330

175. TEXAS, UNIVERSITY OF
EL PASO, TEXAS 79968-0566

176. TOLEDO, UNIVERSITY OF
TOLEDO, OHIO 43606

177. TORONTO, UNIVERSITY OF
TORONTO, ONTARIO
CANADA M5S 1A4

178. TRENT UNIVERSITY
PETERBOROUGH, ONTARIO
CANADA K9J 7B8

179. TULSA, UNIVERSITY OF
TULSA, OKLAHOMA 74104-3189

180. TUSKEGEE INSTITUTE
TUSKEGEE INSTITUTE, ALABAMA 36088

181. U.S. NAVAL POSTGRADUATE SCHOOL
MONTEREY, CALIFORNIA 93943-5002

182. U.S.A.F. INSTITUTE OF TECHNOLOGY
WRIGHT PATTERSON AFB, OHIO 45433-6583

183. UTAH STATE UNIVERSITY
LOGAN, UTAH 84322

184. UTAH, UNIVERSITY OF
SALT LAKE CITY, UTAH 84112

185. VERMONT, UNIVERSITY OF
BURLINGTON, VERMONT 05405

186. VICTORIA, UNIVERSITY OF
VICTORIA, BRITISH COLUMBIA
CANADA V8W 2Y2

187. VILLANOVA UNIVERSITY
VILLANOVA, PENNSYLVANIA 19085

188. VIRGINIA STATE COLLEGE
PETERSBURG, VIRGINIA 23803

189. VIRGINIA, UNIVERSITY OF
CHARLOTTESVILLE, VIRGINIA 22906

190. WAKE FOREST UNIVERSITY
WINSTON-SALEM, NORTH CAROLINA 27109

191. WASHINGTON STATE UNIVERSITY
PULLMAN, WASHINGTON 99164-3200

192. WASHINGTON UNIVERSITY
SAINT LOUIS, MISSOURI 63130

193. WASHINGTON, UNIVERSITY OF
SEATTLE, WASHINGTON 98195

194. WESLEYAN UNIVERSITY
MIDDLETOWN, CONNECTICUT 06457

195. WEST FLORIDA, UNIVERSITY OF
PENSACOLA, FLORIDA 32514-5750

196. WEST TEXAS STATE UNIVERSITY
CANYON, TEXAS 79016

197. WEST VIRGINIA UNIVERSITY
MORGANTOWN, WEST VIRGINIA 26506-6069

198. WESTERN CAROLINA UNIVERSITY
CULLOWHEE, NORTH CAROLINA 28723

199. WESTERN KENTUCKY UNIVERSITY
BOWLING GREEN, KENTUCKY 42101

200. WESTERN MICHIGAN UNIVERSITY
KALAMAZOO, MICHIGAN 49008-3899

201. WESTERN ONTARIO, UNIVERSITY OF
LONDON, ONTARIO
CANADA N6A 5B8

202. WESTERN WASHINGTON UNIVERSITY
BELLINGHAM, WASHINGTON 98225

203. WICHITA STATE UNIVERSITY
WICHITA, KANSAS 67208

204. WINDSOR, UNIVERSITY OF
WINDSOR, ONTARIO
CANADA N9B 3P4

205. WORCESTER POLYTECHNIC INSTITUTE
WORCESTER, MASSACHUSETTS 01609

206. WRIGHT STATE UNIVERSITY
DAYTON, OHIO 45435

207. WYOMING, UNIVERSITY OF
LARAMIE, WYOMING 82071

208. YOUNGSTOWN STATE UNIVERSITY
YOUNGSTOWN, OHIO 44555-0001

MASTERS THESES

ALABAMA, UNIVERSITY OF (UNIVERSITY)

Numerical Simulation of Unsteady Viscous
Flow Past a Circular Cylinder
(1988) / Fang J

A Study of the Use of Fast Fourier
Transforms for Solving the Elliptic Grid
Generation Equations
(1988) / Garriz J A

Optimal Wind Shear Control Using an Energy
Equivalent Performance Index
(1988) / Jea S J

ALBERTA, UNIVERSITY OF

The Design and Construction of an Outdoor
Icing Wind Tunnel, and Some Preliminary
Sea-Spray Icing Experiments
(1988) / Nowak A

AUBURN UNIVERSITY

An Aerodynamic Analysis of Several Grid Fin
Configurations Using Vortex Lattice Method
(1987) / Brooks R A

Two-Dimensional Analysis of Steady Compressible
Fluid Flow in a Channel with Friction and Heat
Transfer
(1987) / Chen C G

An Investigation of a Regenerative Cooling
System for a Laser-Heated Thruster
(1987) / Craig T L

Application of Lifting Surface Theory to
Non-Planar Wings
(1987) / Gills P F

A Direct Optimization Method for Estimation of
Supersonic Flow Turbine Stator Profiles
(1987) / Hatfield J A

Behavior of Particulate Clouds Ejected from a
Rigid, Rotating Platform in Earth Orbit
(1987) / Phelps K A

CALIFORNIA STATE UNIV. (LONG BEACH)

Panel Model Calculation on Flow About
Fan-In-Wing Aircraft
(1988) / Hauser D L

Calculation of Unsteady Potential Flow
About an Arbitrary Airfoil
(1988) / MacInnes H D

CALIFORNIA, UNIVERSITY OF (DAVIS)

A Manned Simulator Investigation of Vehicle
and Display Dynamics in Hovering Flight
(1988) / Gorder P J

Wind-Tunnel Measurements of Mass Transport
Rates and Velocity Profiles in Saltating
Turbulent Boundary Layers
(1988) / Mounla H

Measurement of Human Pilot Dynamic
Characteristics in Flight Simulation
(1988) / Reddy J T

CALIFORNIA, UNIVERSITY OF (LA JOLLA)

An Assessment of New Space Transportation
Technologies Applied in an Infrastructure
for Transporting Mass from the Moon's
Surface to Low Earth Orbit
(1988) / Henley M W

CARLETON UNIVERSITY

Low Speed Testing of Propellers
Incorporating a Helicoidal Surface and
Swept Blades
(1988) / MacKenzie S C

CORNELL UNIVERSITY

An Experimental Study of the Shearless, Grid
Generated Turbulence Mixing Layer
(1988) / Ikeda E

Implicit Multigrid Solution of the Euler
Equations for Two-Dimensional Inlets
(1988) / Iyer R K

Solution Adaptive Mesh Calculations of
Conical Potential Flows
(1988) / Solomon G M

FLORIDA, UNIVERSITY OF

A Programming System for Optimum Design of
Gust Loaded Structures
(1988) / Bach C

Optimization of Isotope Separation in
Laminar Oscillatory Gas Flow
(1988) / Demas C

Aeroelastic Synthesis of the Oblique Wing
Configuration
(1988) / Soepardi T

IOWA STATE UNIVERSITY

Geometric Modeling for Flowfield
Applications Using Interpolating B-Spline
Surfaces
(1988) / Atwood C A

Comparison of TVD Schemes Applied to the
Navier-Stokes Equations
(1988) / Buelow P E

Minimum-Time Retrieval of Tethered
Subsatellite Through Traction Control
(1988) / Chu S

Development of Atmospheric Guidance Schemes
for the Aeroassist Flight Experiment
(1988) / Dukeman G A

Design, Analysis, and Optimization of a
Curved Anisotropic Beam Undergoing Large
Nonlinear Dynamic Motion
(1988) / Martin M D

KANSAS, UNIVERSITY OF

Calculation of Aerodynamic Characteristics
of Airfoils with Laminar Separation
(1988) / Chou I K

Results of a Preliminary Design Study for a
Family of Commuter Airplanes
(1988) / Creighton T R

KANSAS, UNIVERSITY OF
(continued)

An Integrated Study of Structures,
Aerodynamics, and Controls on the Forward
Swept Wing X-29A and the Oblique Wing
Research Aircraft
(1988) / Dawson K S

Investigation of Empennage Buffeting
(1988) / Lee I G

A Technique for Creating and Modifying a
Preliminary Airplane Configuration Using
a Three-Dimensional Computer-Aided Design
System
(1988) / Morgan L K

Calculation of Aerodynamic Characteristics
of Airplane Configurations at High Angles
of Attack
(1988) / Tseng J A

MASSACHUSETTS INSTITUTE OF TECHNOLOGY

Human Perception and Indication of the
Vertical: Experiments and Models
(1988) / An B

Supervisory Control Algorithms for
Telerobotic Space Structure Assembly
(1988) / Anderson D E

Competition and Collaboration in the
Commercial Aircraft Industry
(1988) / Baldwin M R

Chemical Kinetics of Scramjet Propulsion
(1988) / Biasca R J

Experimental Component Mode Synthesis of
Structures with Joint Freeplay
(1988) / Blackwood G H

Energy Absorption Methods for Fluid Quality
Gauging in Low Gravity
(1988) / Bokhour E B

An Optical Phase Locked Loop
(1988) / Boyd R L

An Experimental Investigation of the Effects
of Inlet Radial Temperature Profiles on the
Aerodynamic Performance of a Transonic
Turbine Stage
(1988) / Cattafesta L N

A Comparison of Multivariable Design
Methodologies for a Two Degree of Freedom
Gryo Torque-Rebalance Loop
(1988) / Coffee J R

Enhancing the Orbital Utilization of
Expendable Launch Vehicles
(1988) / Comstock D A

Ascent Guidance for a Winged Boost Vehicle
(1988) / Corvin M A

A Design Study of Onboard Navigation and
Guidance During Aerocapture at Mars
(1988) / Fuhry D P

Manipulator Control with Obstacle Avoidance
Using Local Guidance
(1988) / Funaya K

Characteristics of Wall Pressure and Near
Wall Velocity in a Flat Plate Turbulent
Boundary Layer
(1988) / Gresko L S

Development and Experimental Verification of
Damping Enhancement Methodologies for Space
Structures
(1988) / Hagood N W

Nonlinear Dynamics of a Rotating, Extending
Spacecraft Appendage
(1988) / Hain R M

A Multiple Scales Approach to the Stability
and Control of a Hypersonic Re-Entry Glider
(1988) / Jafry Y R

A Numerical Investigation of Hot Streaks in
Turbines
(1988) / Krouthen B B

A Multivariable Controller for an
Electromagnetic Bearing/Shaft System
(1988) / LaRocca P J

A Comparison of Numerical Schemes on
Triangular and Quadrilateral Meshes
(1988) / Lindquist D R

An Elegant Lambert Algorithm for Multiple
Revolution Orbits
(1988) / Loechler L A

Fabrication Methods and Costs for Thermoset
and Thermoplastic Composite Processing for
Aerospace Application
(1988) / Masi B A

An Implicit Algorithm for Analysis of an
Airframe-Integrated Scramjet Propulsion
Cycle
(1988) / Mattson E A

Active Control of Tiltroter In-Plane Loads
During Maneuvers
(1988) / Miller D G

Load Relief Optimization of a First Stage
Boost Vehicle Using an Automated Computer
Design Tool
(1988) / Moreno C S

Optimization of Manual Control Dynamics in
Space Telemanipulation: Impedance Control
of a Force Reflecting Hand Controller
(1987) / Paines J D

Control System Design Using H(Infinity)
Optimization
(1988) / Playter R R

Implementation of a Separated Flow Panel
Method for Wall Effects on Finite Swept
Wings
(1988) / Robinson D E

Analysis of a Heavy Lift Launch Vehicle
Design Using Small Liquid Rocket Engines
(1988) / Russ D P

Neuromorphic Regulation of Dynamic Systems
Using Back Propagation Networks
(1988) / Sanner R M

Design Optimization of a Transatmospheric
Launch Vehicle
(1987) / Sansoy S A

MASSACHUSETTS INSTITUTE OF TECHNOLOGY
(continued)

The Effect of a Gap Nonlinearity on
Recursive Parameter Identification
Algorithms
(1988) / Schaffer S E

Aircraft Performance Enhancement with Active
Compressor Stabilization
(1988) / Seymour J C

Control Applications Using Neural Networks
(1988) / Showalter B E

Analysis of the Altitude Tracking
Performance of Aircraft-Autopilot Systems in
the Presence of Atmospheric Disturbances
(1988) / Sturdy J L

A Systematic Approach to Modeling for
Stability-Robustness in Control System
Designs
(1988) / Tilley J A

High Performance Control of the
Supermaneuverable F18 HARV Aircraft
(1988) / Voulgaris P G

MASSACHUSETTS, UNIVERSITY OF

The Design and Construction of a Calibration
Wind Tunnel
(1988) / Shreder C F

MINNESOTA, UNIVERSITY OF

On the Compatibility of Eigenstructure
Assignment and Reduced Order Modeling
(1988) / Ekblad M K

Aerodynamics of Two-Dimensional Slotted
Bluff Bodies
(1988) / Takahashi F

MISSISSIPPI STATE UNIVERSITY

Correlations for the Drag Coefficient of Blunt
Circular Cylinders in Transonic Axial Flow
(1986) / Arabshahi A

A Feasibility Study of the Use of a Control Line
Flying Model Aircraft for Wing Tip Vortex
Visualization
(1986) / Brownlee J A

An Investigation of a Navier-Stokes Solution for
Quasi-Three-Dimensional Flow
(1986) / Chae Y S

An Investigation of Digital Control System with
Transport and Time Delay
(1986) / Iles R W

The Development of a Three Dimensional Split
Flux Vector Euler Solver with Dynamic Grid
Applications
(1986) / Janus J M

On Some Computer Programs to Generate Grids About
Three-Dimensional Bodies for Euler Equation Solvers
(1986) / Kumar De S

A Comparison of an Optimal Control Law and a
Classical Control Law for a Lateral Stability
Augmentation Device on NAVION Airplane
(1986) / Lee Y D

Performance Analysis of Beech T-34B Navy Trainer
Using Three Steady State Methods
(1986) / Moosaviasi M

A Numerical Method for Analyzing Performance
Flight Test Data
(1986) / Sabzehparvar M

An Investigation of the Drag of Multiple
Aligned Bluff Bodies
(1986) / Takahei N

Generation of Surface Coordinates in Curved Surfaces
and in Plane Regions
(1986) / Tiarn W D

MISSOURI, UNIVERSITY OF (COLUMBIA)

A Knowledge Base for Corrosion Prevention
(1988) / Fei Y F

Acceleration/Deceleration Testing for
Diagnostics of Spark Ignition Engines
(1988) / Mueller G P

MISSOURI, UNIVERSITY OF (ROLLA)

Numerical Study of Hydrogen-Air Mixing
Flowfield in a Ramjet Combustor
(1987) / Trinh H P

NEW YORK, STATE UNIVERSITY OF (BUFFALO)

Vibrations of Non-Proportionally Damped
Structures
(1988) / Bellos J

Calculations of the Stokes-Flow Resistance
Matrices for a Helical Particle, Using
Slender Body Theory
(1988) / Kim Y J

Digital Control of a Two Degree of Freedom,
Lightly Damped, Low Natural Frequency
Oscillatory System
(1988) / Lochocki R

Magnetohydrodynamic Flow in a Cylindrical
Channel
(1988) / Miller K

Correcting Finite Element Models with
Measured Modal Results Using Eigenstructure
Assignment Methods
(1988) / Minas C

Generalized Thermoelastic Plane Stress and
Displacement by Finite Differences
(1988) / Mule J

Enhanced Tracking of Aircraft Utilizing
Nonlinear Force Moment, and Control
Estimation
(1988) / Shyu I M

A Numerical Study on a Plane Wall Jet
(1988) / Yang C

NORTH CAROLINA STATE UNIVERSITY

Theoretical Evaluation of Engine Auxiliary
Inlet Design for Supersonic V/STOL Aircraft
(1988) / Heavner R L

Experimental Verification and a Parametric
Study of the Variational Modal
Identification Method
(1988) / Hedgecock C E

NOTRE DAME, UNIVERSITY OF

Control of Vortex Breakdown on Delta Wings
(1988) / Bhobe S M

Experimental Studies of the Transitional
Separation Bubble on the NACA 663-018
Airfoil at Low Reynolds Numbers
(1988) / Fitzgerald E J

An Investigation of the Vortex Flow Over a
Delta Wing with and Without External Jet
Blowing
(1988) / Iwanski K P

Leading Edge Vortex Dynamics on a Pitching
Delta Wing
(1988) / LeMay S P

An Investigation of the Effects of an
External Jet on the Performance of a Highly
Swept Delta Wing
(1988) / Visser K D

PENNSYLVANIA STATE UNIVERSITY

Simulation and Parametric Study of a Direct
Expansion Energy Recovery System Utilizing
100 Percent Supply and Exhaust Air
(1987) / Richichi C P

Prediction of the Downwash Associated with a
Jet-Flapped, Wing-Body Combination by Means
of a First Order Panel Method
(1987) / Semenza J T

PURDUE UNIVERSITY

Orbit Determination of a Spacecraft with
Undetermined Control Inputs
(1988) / Allan T A

Identification of Equivalent Short Period
Dynamics of the X-29A
(1988) / Bauer J E

An Experimental Investigation of Wind Tunnel
Quality
(1988) / Behrens R D

Nonlinear Finite Element Analysis of Plates
and Shells of Composite Laminates
(1988) / Chen G

Reducing Free Edge Effect on Laminate
Strength by Edge Modification
(1988) / Chu G D

Search Problems in Mission Plannaing and
Navigation of Autonomous Aircraft
(1988) / Krozel J A

A Computational Investigation of
Propeller/Wing Interaction
(1988) / Lee F K

Sun-Synchronous Trajectory Design Using
Consecutive Lunar Gravity Assists
(1988) / Marsh S M

Sensitivity Reducing Controller Design
Method for Uncertain Parameter Systems
(1988) / Okada K

Tracking Fixed Wing Aircraft and Helicopters
Using Attitude and Rotor Angle Information
(1988) / Schierman J D

Earth-Mars Trajectories Using a Lunar Flyby
(1988) / Schlossberg V E

An Experimental Study of Swirling Flows as
Applied to Annular Combustors
(1988) / Seal M D

Modeling Human Perception and Estimation of
Kinematic Responses During Aircraft Landing
(1988) / Silk A B

The Evaluation of Three Control Laws for the
(H-4) Helicopter
(1988) / Townsend B K

Microcomputer-Based Flight Simulation
(1988) / Williamson R Z

An Experimental Investigation of Propeller
Wing Interaction
(1988) / Witkowski D P

Dynamic Aeroelastic Stability of a Generic
x-Wing Aircraft Configuration with Design
Parameter Variations
(1988) / Woods J A

RENSSELAER POLYTECHNIC INSTITUTE

Evaporation Waves in Superheated Liquids
(1988) / Cho J

Stagnation Line Heat Transfer on a Highly
Swept Wing at a Mach Number of 10
(1988) / D'Orsi N C

An Experimental Investigation of Heat
Transfer to a Flat Plate and Flap at
Mach 10.5
(1988) / Donahue D J

Applications of the Snapthrough Airfoil
as a Higher Harmonic Control Device
(1988) / Dubben J A

Hydraulic Damping of an Injection Piston
(1988) / Spiewak D R

Analysis of a Three Lifting Surface Aircraft
for Minimization of Induced Drag
(1988) / Stavetski D

Hinge Moments on a Wing-Flap Configuration
at Zero Incidence at Mach 10.5
(1988) / Terzis P B

Passive Shock Wave/Boundary Layer Control of
a Complete NACA 0012 Airfoil Model Using a
Contoured Transonic Wind Tunnel
(1988) / Trilling T W

SAN DIEGO STATE UNIVERSITY

Finite Element Programs for Orthotropic
Axisymmetric Structures
(1987) / Cerkovnik M E

Acoustic Radiation from an Infinite,
Fluid-Loaded, Inhomogeneous, Thin Plate
(1987) / Cole T M

Flow Induced Tones in a Side Dump Ramjet
Combustor
(1988) / Hildebrand G C

Delta-V Savings Using the Synergistic
Maneuver as Applied to a Class of
Transatmospheric Vehicle
(1988) / Huynh C T

SAN DIEGO STATE UNIVERSITY
(continued)

Effects of Lifting-Ejector Jet on the
Longitudinal Aerodynamics of Delta Wings
(1988) / Kern D N

Visualization of Acoustic Scattering from
Flexible Surfaces
(1988) / Rockholt H M

TENNESSEE, UNIVERSITY OF (KNOXVILLE)

Computation of Isothermal Laminar Swirling
Flow in a Model Arc Heater
(1987) / Chiang A S

Surface Temperature Prediction of a Bridge
for Tactical Decision Aide Modelling
(1987) / Cross P T

Liquid Cooled Engine Installations in
General Aviation Aircraft: Design
Considerations and Analysis of Coolant Heat
Exchangers and Their Installation
(1987) / Groff R D

An Experimental Investigation of the Air
Drag on Monofilament Fibers and Flowfield
Behavior Resulting from Two Convergent High
Velocity Jets: Melt Blowing Application
(1987) / Haynes B D

An Improved Technique for Normalizing
Engine-Out Minimum-Control Airspeed Data
(1987) / McCarthy B G

Fluid Streamlines Flow in an Annular Cavity
(1987) / Sanford C S

A Library of Computer Subprograms for Use in
Air-Breathing Propulsion Analysis Computer
Programs
(1987) / Saunders J L

TENNESSEE, UNIVERSITY OF (TULLAHOMA)

Alternative Flow Diverting Devices for the
Four Foot Transonic Wind Tunnel Arnold
Engineering Development Center
(1988) / Abel R J

Effects of Coefficient Errors on the
Performance Predictions of a Typical
Fighter at Medium Altitude
(1988) / Cloyd J D

Selection of an Aircraft Type to Sevice the
Flexible Air Transportation System of the
Tennessee Valley Aerospace Region Air
Transportation Market Area
(1988) / Haney R W

Calibration of Flow Angularity Probes
Using a Numerical Search Algorithm
(1988) / Hodges D A

Flying Qualities and Performance Evaluation
Testing of the AH 64A Helicopter
(1988) / Joseph S E

Transionspheric Propagation of Broadband
Wave Forms
(1988) / Miller J A

Correction of Aerodynamic Coefficients
from Flight Test Data
(1988) / O'Donnell K

Shadow Spectrometer Measurements of Particle
Sizes from Air-Assisted Atomizing Spray
Nozzles
(1988) / Qian X

Rocket Engine Test Facility Run Time
Limitations for Four Typical Liquified
Propellant Combinations
(1988) / Quinn R W

Use of a Personal Computer as a Data
Acquisition System for Flight Testing
(1988) / Sims J

A Study of Flows Behind Bluff Bodies
Immersed in a Sheared Flow
(1988) / Tirres L

TEXAS A AND M UNIVERSITY

Analysis and Optimization of Incompressible
Separated Flow Around an Airfoil with Two
Finite-Gap Flaps
(1988) / Anderson M B

A Comparative Investigation of Laminar
Separation Bubbles at Low Reynolds Numbers
in Wind Tunnels and Free Flight Environments
(1988) / Blohowiak J R

A Comparative Study of Reaction Rate, Species, and
Vibration-Dissociation Coupling Models for an AOTV
Flowfield
(1988) / Bobskill G J

A Finite Element Model for Laminated Composite
Plates with Matrix Cracks and Delaminations
(1988) / Buie K D

An Experimental and Theoretical Acoustic
Investigation of Single Disc Propellers
(1988) / Bumann E A

Flow Field Characteristics of a Radial Jet
Reattaching on a Flat Plate
(1988) / Carbone J S

The Effect of Freestream Variations on the
Propagation of Detonation and Combustion Waves
(1988) / Clark M L

Numerical Computation of Aerodynamic Sensitivity
Coefficients in the Transonic and Supersonic Regimes
(1988) / El Banna H M

Aeroassisted Orbital Transfer Vehicle Guidance
Performance in the Presence of Density Dispersions
(1988) / Fitzgerald S M

Matrix Cracking and Bending Stiffness Reduction
in Composite Laminates
(1988) / Frailey J A

A Parametric Study of Shock Jump Chemistry, Electron
Temperature, and Radiative Heat Transfer Models in
Hypersonic Flows
(1988) / Greendyke R B

Simulation of a Stol Airlifter in Wind Shear, Using
Total Energy and Glideslope Angular Error Methods
for Glidepath Control
(1988) / Johnson E W

Feedback Control and Steering Laws for Spacecraft
Using Single Gimbal Control Moment Gyros
(1988) / Oh H S

TEXAS A AND M UNIVERSITY
(continued)

An Experimental Investigation of Damage-Dependent
Material Damping of Laminated Composites
(1988) / Smith S A

A Kalman Filtering Technique for Spacecraft Attitude
Determination and Control Using Gibbs Vector
(1988) / Tallant G S

Trajectory Simulation for Design of Ram-Air
Inflated Parachutes
(1988) / Tallant S U

TEXAS, UNIVERSITY OF (ARLINGTON)

Computation of the Interaction of a
Transverse Jet with Supersonic External Flow
(1988) / Amaya Guzman M A

Hybrid Control of an Intelligent Compliant
Wrist
(1988) / Barber K S

Automatic Vibration Controller for a
Variable Rotor Speed Helicopter
(1988) / Fisher M R

A Coupled Aerodynamic and Structural
Analysis of a Flexible Pitchihng Airfoil
(1988) / Flansburg B D

A Study of Flow in Shock Tubes
(1988) / Huang S J

An Expert System for Aircraft Conceptual
Design
(1988) / Logan M J

A Full Potential Scheme Applied to Two
Dimensional Steady Transonic Nozzle Flows
(1988) / Lu H

The Propulsive Characteristics of a Cross
Flow Fan Installed in an Airfoil
(1988) / Nieb T W

Application of the Frequency Domain Testing
Technique to the Analysis of the Open-Loop
Dynamics of a Conventional Helicopter
(1988) / Oetting A M

Response of Stagnation Pressure Probe System
to a Suddenly Applied Pressure
(1988) / Pereira V C

Application of an Adaptive Grid Scheme to a
Shock Capturing Code
(1988) / Ross J L

TEXAS, UNIVERSITY OF (AUSTIN)

Pulse-Modulated Controller Synthesis for a
Flexible Spacecraft
(1988) / Anthony T C

An Orbit Simulation Study of a Geopotential
Research Mission Including
Satellite-to-Satellite Tracking and
Disturbance Compensation Systems
(1988) / Antreasian P

Boundary Element Method:: Vectorized
Solutions and Inverse Design
(1988) / Batista J C

Minimum Time Control of Large Space
Structures
(1988) / Boucher R L

On the Effect of Transverse Shear on the
Stability of Long Composite Cylindrical
Shells Under External Pressure
(1988) / Chan K S

Departure Windows for Earth-Moon
Transportation
(1988) / Chesley B C

Autonomous Navigation Using an Orbital
Sextant Configuration: A Near Earth
Application
(1988) / Ericson N L

Long-Arc Orbit Determination Improvements
Using Encke's Method
(1988) / Fields R K

The Development of a Parabolized
Navier-Stokes Code and Its Application to
Planar and Axisymmetric Flows
(1988) / Fritz D A

Short Baseline Determinations Using Global
Positioning Satellite Differenced Phase
Measurements
(1988) / Guinn J R

Analysis of Least Squares Orbit
Determination Accumulation Algorithms
(1988) / Hasan D A

Pure Shear Testing of Adhesives in Situ
(1988) / Hayashi T

Trajectory Determination and
Characterization of CisLunar Low-Thrust
Spacecraft
(1988) / Korsmeyer D J

Preliminary Space Mission Planning Utilizing
a Mission Planning Subroutine Library
(1988) / Mahoney R J

Minimum-Time Rescue Trajectories for the
Manned Maneuvering Unit
(1988) / Neff J M

A Comparison of GPS Orbits Estimated
Using Observations from North America
and Europe
(1988) / Pollmeier V M

Identification of Nonclassical Aeroelastic
Responses Using Bispectral Analysis
(1988) / Schwarz J D

Rapid Numerical Calculations of Reentry
Aerodynamics
(1988) / Sehlke B D

Variable Thrust Modeling for Rocket Booster
Trajectory and Parameter Estimation
(1988) / Shelton J R

Parameter Identification Applied to the
Adaptive Control of a Hypersonic Flight
Vehicle
(1988) / Smith E C

Precision Orbit Determination for the
GEOSAT Exact Repeat Mission
(1988) / Smith J C

TEXAS, UNIVERSITY OF (AUSTIN)
(continued)

An Adaptive Flight Control System Based on a
New Explicit Parameter Identification
Technique
(1988) / Smith R H

Flow Fluids for an Underexpanded,
Single-Nozzle Missile Exhausting Into a
Vertical Launching System
(1988) / Sterk T M

Synthesis of a Periodic-Disturbance
Accommodating Controller for Space Station
Momentum Management
(1988) / Warren V W

Detection Filters for Aircraft Fault
Detection
(1988) / Wilbers D M

U.S. NAVAL POSTGRADUATE SCHOOL

Hot-Wire Measurements of Compressor Blade
Wakes in a Cascade Wind Tunnel
(1988) / Baydar A

Microcomputer Implementation and Use for the
Aviation Logistic Support Functions of the
Third Marine Aircraft Wing: A Case Study
(1987) / Buranosky J F

An Independent Evaluation of the CH-46
Safety Reliability and Maintainability
Program (SR&M)
(1987) / Conklin M J

Dynamic Stall Computations Using a Zonal
Navier-Stokes Model
(1988) / Conroyd J H

High Reynolds Number, Low Mach Number,
Steady Flow Field Calculations Over a
NACA 0012 Airfoil Using Navier-Stokes and
Interactive Boundary Layer Theory
(1987) / Cowles L J

An Aerodynamic Performance Evaluation of
the NASA/AMES Research Center Advanced
Concepts Flight Simulator
(1987) / Donohue P F

A Study of the Effect of Design Parameter
Variation on Predicted Tilt-Rotor Aircraft
Performance
(1988) / Dunston M C

A Mapping of the Viscous Flow Behavior in a
Controlled Diffusion Compressor Cascade
Using Laser Doppler Velocimetry and
Preliminary Evaluation of Codes for the
Prediction of Stall
(1988) / Elazar Y

Interaction of a Vortex Pair with a Free
Surface
(1987) / Elnitsky J

Cost Analysis for Aircraft System Test and
Evaluation: Empirical Survey Data
Structuring and Parametric Modeling
(1987) / Fonnesbeck R W

The Effects of Torque Response and Time
Delay on Rotorcraft Vertical Axis Handling
Qualities
(1987) / Fyles P A

Aerodynamic Performance of Wings of
Arbitrary Planform in Inviscid,
Incompressible, Irrotational Flow
(1988) / Holm C I

Cost Estimating Relationships for Fighter
Aircraft
(1987) / Hong W P

Effect of Vapor Velocity During Condensation
on Horizontal Finned Tubes
(1988) / Hopkins C R

Experimental Design of a Piloted Helicopter
Off-Axis-Tracking Simulation Using a Helmet
Mounted Display
(1987) / Hopkins G J

Microcomputer Implementation and Use for the
Aviation Logistic Support Functions of the
Third Marine Aircraft Wing: A Case Study
(1987) / Howes C W

The Importance of Aircraft Performance and
Signature Reduction Upon Combat
Survivability
(1988) / Langford J D

Cost Analysis for Aircraft System Test and
Evaluation: Empirical Survey Data
Structuring and Parametric Modeling
(1987) / Lee D J

Parametric Analysis of Airland Combat Model
in High Resolution
(1988) / Lee J Y

A Computer Graphics Simulation of the
Endgame in Aircraft Combat Survivability
(1987) / Lee Y M

An Experimental Investigation of a Fighter
Aircraft Model at High Angles of Attack
(1988) / Leedy D H

An Autopilot Design for the United States
Marine Corps' Airborne Remotely Operated
Device
(1987) / Lloyd S D

Unmanned Air Vehicles - Real Time
Intelligence Without the Risk
(1988) / Miller J B

A Comparative Analysis of Tilt Rotor
Aircraft Versus Helicopters Using Simulator
Results
(1988) / Mislick G K

Aerothermodynamics of a Jet Cell Facility
(1988) / Nicolaus E A

Dynamic Stall Analysis Utilizing
Interactive Computer Graphics
(1988) / Pagenkope E L

The Effects of Flight Hours and Sorties on
Failure Rates
(1987) / Phillips S J

Analysis of Aircraft Combat Sustainability
Using a Markov Chain
(1988) / Reuss G C

A Study of the Department of Defense
Configuration Management Policies and
Procedures as Applied to the FA-18
Strike/Fighter Program
(1987) / Roum C J

U.S. NAVAL POSTGRADUATE SCHOOL
(continued)

A Prototype of Pilot Knowledge Evaluation by
an Intelligent CAI (Computer-Aided
Instruction) System Using a Bayesian
Diagnostic Model
(1987) / Rugsumruad Y

Investigation of Dynamic Stall Using LDV
(Laser Doppler Velocimetry): Mean Flow
Studies
(1987) / Ryles R R

Graphic Simulation of an Air Defense
Scenario
(1987) / Schuster K

Control of Embedded Vortices Using Wall Jets
(1988) / Schwartz G E

Pilot Study on the Applicability of Variance
Reduction Techniques to the Simulation of a
Stochastic Combat Model
(1987) / Smith C

An Inexpensive Real-Time Interactive Three
Dimensional Flight Simulation System
(1987) / Smith D B

An Inexpensive Real-Time Interactive Three
Dimensional Flight Simulation System
(1987) / Streyle D G

Viscous/Inviscid Interaction Analysis of the
Aerodynamic Performance of the NACA 65-213
Airfoil
(1987) / Subroto P H

The Development of a Computer Code (U2DIIF)
for the Numerical Solution of Unsteady,
Inviscid and Incompressible Flow Over an
Airfoil
(1987) / Teng N H

A Case Study of a Combat Helicopter's
Single Unit Vulnerability
(1987) / Trueblood J W

Computer Program for Conceptual Tandem
Rotor Helicopter Design
(1987) / Vandenbos B A

A Piloted Simulation Investigating Handling
Qualities and Performance Requirements of a
Single-Pilot Helicopter in Air Combat
Employing a Helmet-Driven Turreted Gun
(1987) / Williams J N

U.S.A.F. INSTITUTE OF TECHNOLOGY

QFT Design of a Robust Landing Longitudinal
Digital Controller for AFTI/F-16 Using
Matrix
(1988) / Adams J M

Hypervelocity Orbital Intercept Guidance
(1988) / Alfano S

Evaluation of Flight Test Bed
(1988) / Bagley R

Multiple-Model Parameter-Adaptive Control
for In-Flight Simulation
(1988) / Berens T J

Navier-Stokes Solution for NACA Airfoil
(1988) / Boyles P D

Volumetric Rendering Techniques for the
Display of Three-Dimensional Aerodynamic
Flow Field Data
(1988) / Bridges D J

Numerical Simulation of Flow Over Iced
Airfoils
(1988) / Coleman L A

Internal Debonding of Solid Propellant
Fillers
(1988) / Copp P

Structural and Materials Aspects of Turbine
Technology
(1988) / Copp P

An Analysis of Laterial-Directional Handling
Qualities and Eigen-Structure of High
Performance Aircraft
(1988) / Costigan M J

A Real-Time Simulator for Man-in-the-Loop
Testing of Aircraft Control Systems
(SIMTACS-RT)
(1988) / Dameron G G

Real Time Conflict Resolution in Automated
Guided Vehicle Scheduling
(1988) / Daniels S C

Water Tunnel Investigation of Lift
Augmentation Using Dynamic Stall on
Various Wing Platforms
(1988) / David M

A Wind Tunnel and Computer Investigation of
the Low Speed Characteristics of the Prone
Ejection System (PRESS)
(1988) / Dillon L R

Investigation of the Flowfield Created by
the Interaction of a Sonic Jet and
Co-Flowing Supersonic Stream
(1988) / Durand B J

Robotic Compliant Motion Control for
Aircraft Refueling Applications
(1988) / Duvall D J

Hot Section Turbine Technology
(1988) / Elrod W

Steady State Dynamic Behavior of an
Axisymmetric Confined Jet Thrust Vector
Control Nozzle
(1988) / Fridell J H

Multivariable Control Law Design for the
Control Reconfigurable Combat Aircraft
(CRCA)
(1988) / Hammond D I

Control of Large Scale Space Structures
(1988) / Hinnerichs T

Controlled Degradation of Resolution of High
Quality Flight Simulation Images for
Training Effectiveness Evaluation
(1988) / Kaip D D

Flight Controller Design with Nonlinear
Aerodynamics, Large Parameter Uncertainty,
and Pilot Compensation
(1988) / Kobylarz T J

Special Purpose Vehicle Research
(1988) / Kramer S

U.S.A.F. INSTITUTE OF TECHNOLOGY
(continued)

An Experimental Investigation of Crack
Closure Mechanism During Thermal Mechanical
Fatigue Crack Growth
(1988) / Mall D

Simulation and Evaluation of Control Systems
at High Angle of Attack on YF-16
(1988) / McCormack L B

Experimental Investigation of Stability and
Controllability of a Low Aspect Ratio
Fighter Aircraft with Variable Tail Dihedral
(1988) / McDonald D C

Experimental Investigation of Flow Mixing
on Thrust Ejector Efficieny
(1988) / Morfitt D J

Early Jet Engines and the Transition from
Centrifugal to Axial Compressors: A Case
in Technology Change
(1988) / Nichelson B J

Multiple Model Adaptive Controller for the
STOL F-15 with Sensor/Actuator Failures
(1988) / Pogoda D L

Response Equivocation Analysis for the
Smart Stick Controller
(1988) / Pracher J M

Computer Aided Reliability Design (CARD) of
a Positive Pressure Oxygen Mask and New
Night Vision Goggles
(1988) / Robinson D

Application of Linearized Kalman Filter
Smoother to Aircraft Trajectory Estimation
(1988) / Ruley J M

Ejector Effects on a Supersonic Nozzle at
Low Altitude Mach Number
(1988) / Seaver C A

Aircraft Performance Enhancement with Active
Compressor Stabilization
(1988) / Seymour J C

Numerical Grid Generation and Potential
Airfoil Analysis and Design
(1988) / Siladic M F

Analysis of Space-Based Lidar for Aircraft
(1988) / Simmons S P

Thin-Layer Navier-Stokes Solutions for a
Cranked Delta Wing
(1988) / Smith F R

Improvements to the AFIT Tactical Mission
Planner (TMP)
(1988) / Spear J L

Research in Supersonic Combustion
(1988) / Tanis F

VIRGINIA, UNIVERSITY OF

Automated Helicopter Design
(1988) / Bruneau B Y

WASHINGTON STATE UNIVERSITY

A Wind Tunnel Facility for Use in Shear Flow
Related Research
(1988) / Hanle R N

WASHINGTON, UNIVERSITY OF

Studies of CW Lasing Action in CO2-CO,
N20-CO, CO2-H2O and N20-H2O Mixtures Pumped
by Blackbody Radiation
(1987) / Abel R W

A Mean Flow Field Study of Coaxial Jets with
Swirl Using Five-Hole Pressure Probe and
Hot-Wire Anemometry
(1987) / Dang P M

Mathematical Techniques for the Mechanical
Characterization of Composite Materials
(1987) / Fadale T D

A Study of Aerodynamic and Displacement
Components for a Brayton-Cycle Engine
(1987) / Fietier N

A Comparison of Glauert's Wind Tunnel Wall
Corrections to Deflected Wake Corrections
(1987) / Hadi K

Numerical Studies of Impact-Fusion Target
Dynamics
(1987) / Sankaran L

A General Flight Simulation for Use in
Engineering Analysis
(1987) / Slocum D C

A Study of Inviscid Supersonic Mixing Layers
Using a Second-Order TVD Scheme
(1987) / Soetrisno M

Application of an Adaptive Mesh Algorithm to
a Finite Element Method Solution of the
Two-Dimensional Burger's Equations
(1987) / Torgerson G L

Laminar Heat Transfer to an Airfoil in
Incompressible Flow
(1987) / Wang H

Time-Dependent Behavior of Composite
Laminates in Hygrothermal Environments
(1987) / Yi S

WEST VIRGINIA UNIVERSITY

Development of Flight Test Data Acquisition
Software and Applications to Stability
Derivative Analysis
(1988) / Arrington E A

Numerical Design Study of Winglets on Low
Aspect Ratio Wings Using Supercritical
Airfoil Technology
(1988) / Cerney M J

WICHITA STATE UNIVERSITY

Flight Performacne of a Dual-Cycle
Interceptor Engine Used for Reconnaissance
Missions
(1987) / Shih W

Practical Optimal Control Theory for Reduced
Static Stability Aircraft
(1987) / Vian J L

Predicting Transient Skin Pressures in
Electro-Impulse De-Icing of Airfoils
(1987) / Walsh J L

ALABAMA A - M UNIVERSITY

Nutritional Improvement of Sorghum Based
Kisra
(1988) / Ahmed A M

Effect of Row Spacings and Weed Control
Methods on Non-Till and Conventionally
Planted Soybeans
(1988) / Anaele A O

The Influence of Corn, Sorghum and Triticale
Based Diets on the Abdominal Fat Pad
Deposition in Broilers
(1988) / Bello G A

Evaluation of Soil and Nutrient Loss Due to
Runoff Under Different Cropping and
Management Systems
(1988) / Eke A U

In Vitro Dry Matter and Organic Matter
Digestibility of Triticale Forages and
Nutritive Value of Triticale Grains in
Rations for Feeder Pigs
(1988) / Fearon A L

Effect of Blanching and Packaging Materials
on Sensory Quality and Stability of Salted
and Unsalted Roasted Peanuts
(1988) / Hashim I B

The Influence of Different Strip Crop
Widths on Soil Conservation and Crop
Productivity
(1988) / Ide H B

Quality Attributes of Faba Bean Flours
Prepared by High and Low-Moisture Processing
(1988) / Idowu A O

Uptake of 15N by Algae and Their Evaluation
of Plant Available Nitrogen Sources
(1988) / Jaggernauth M

Participation of Alabama Landowners in the
Conservation Reserve Program of the 1985
Farm Bill
(1988) / Kairumba J N

Evaluation of Feeding Value of Kudzu to
New Zealand White Rabbits
(1988) / Kalu G O

Thematic Mapper and Digital Vector Data
Integration for Improving Management of
Natural Resources
(1988) / Massay L E

The Development of a Fermented Soy Beverage
(1988) / McRee C J

Phosphorus Status of Some North Alabama
Soils
(1988) / Mehadi A A

A Study on Proximat Composition, Amino Acids
and Fatty Acids of Peanuts Grown in Jamaica
(1988) / Miller E A

Evaluation of Four Commercial Egg Producing
Strain of Chickens for Egg Production and
Performance from Day-Old to 72 Weeks of Age
(1988) / Okechukwu C A

Effects of Low Dose Irradiation on the
Sensory Qualaities of Fresh Nonfrozen
Poultry Meat
(1988) / Stevens T C

The Inhibitory Effect of Streptococcus
Thermophilus and Pencillium Roqueforti on
the Growth and Toxin Production of
Aflatoxigenic Molds
(1988) / Thomas B A

Extractable Micronutrient Cations,
Exchangeable Cations and Cation Exchange
Capacity of Acid Forest Soils
(1988) / Ubi M W

Characteristics of Proteins and Amino Acids
in Seeds and Mill-Fedd Fractions of Some
Advanced Lines of Triticale
(1988) / Watson H

ALBERTA, UNIVERSITY OF

An Exploration of the Economic
Responsiveness of Emerging Commercial Beam
Farmers in the Colombian Andes
(1988) / Anderson Medellin M

Risk and Returns on Alberta Grain Farms
(1988) / Boyda A C

The Influence of Community Learning Centers
on the Development of Yukon Communities
(1988) / Bruners D A

Impacts of Futures Prices on Canola Street
Prices, 1983 to 1986
(1988) / Burden R E

Effect of Beta-Glucans and Processing Method
on the Utilization of Barley by Cattle
(1988) / Engstrom D F

Effects of Somatotropin Injection and
Feeding Frequency on Plasma Hormone
Concentrations, Rumen Parameters, Milk
Yield and Milk Composition in Holstein
Cows
(1988) / French N J

The Potential for Sino-Canadian Joint
Ventures in Beef Production
(1988) / Gietz R E

Cultivation and Dehydration of Oyster
Mushrooms
(1988) / Li-Shing-Tat B

Effect of Isopropyl-N-Chlorophenyl Carbamate
Application and Storage Variables on Potato
Chip Color
(1988) / Mahmood G

Is Canadian Farmland a Growth Stock? The
Financial Risks and Returns from Investing
Capital in Canadian and Alberta Farmland,
1970-1986
(1988) / Main Freeman C A

A Bioeconomic Analysis of the Potential for
Game Ranching in Northern Alberta
(1988) / Matthews L C

Structure-Trade Links in Japanese Cereal
Agriculture
(1988) / Nakamura M L

Structure and Technology on Alberta Grain
Farms
(1988) / Ruud L E

Studies on an Actin-Like Protein from
Chicken Pectoralis Muscle
(1988) / Swallow K

ALBERTA, UNIVERSITY OF
(continued)

Variation in Mycena Citricolor on Coffee
in Costa Rica
(1988) / Wang A

The Effects of Thermal Environment on the
Metabolism of Mature Sheep
(1988) / Whitmore W T

Agricultural Land Tenure in Alberta: A
Socio-Economic Analysis
(1988) / Wilde D L

ARIZONA STATE UNIVERSITY

Highway Landscape Maintenance Water
Conservation
(1988) / Goy M

ARKANSAS, UNIVERSITY OF

A Study of Agricultural Production
Technology in East Arkansas Using a Multiple
Input Output Profit Function
(1988) / Ajmera S

In-Vitro Plant Regeneration of Zoysiagrass
from Embryo-Derived Callus
(1988) / Al-Khayri J M

Maximization of Present Value of Net Return,
a Model for Optimal Sustained-Yield Aquifer
Management
(1988) / Asghari K

Bean Production and the Relationship with
Recommendation Domains in the Rwandan
Highlands
(1988) / Brewster M L

Effect of Variations in Parboiling on Yield,
Color, and Texture of Recooked Rice
(1988) / Carrol R J

Factors Affecting Lipoxygenase,
Hydroperoxide Lyase and Oxidation in
Pickling Cucumbers and Pickel Products
(1988) / Cho M

A Predictive Model of Head Yield Using
Selected Factors and Near Infrared
Reflectance of Long Grain Rough Rice
(1988) / Crawley G C

A Preliminary Investigation Into the
Efficacy of Dimethyldithiocarbamate as a
Therapeutic Agent for Use in Poultry
(1988) / Delap S K

Employment Opportunities and Educational
Needs of Arkansas' Level One Agricultural
Industry
(1988) / Ellibee M A

Calcium Binding Capacity of Natural Phytic
Acid
(1988) / Farkvam D O

Modeling Heat and Moisture Changes in
Bunker-Stored Rice
(1988) / Freer M W

Soft Wheat Yield as Affected by Row Spacing,
Seeding Rate, and Genotype
(1988) / Freeze D M

Protein Hydrolysis of Milled Rice Flour for
Use in Infant Formula
(1988) / Garrett C L

The Effect of Foliar and Postharvest
Application Cytokinin and Gibberellic Acid
on Yield and Quality of Snap Beans
(1988) / Giese J H

Influence of Scion and Rootstock on Growth
Characteristics and Nodulation of Soybean
(1988) / Goldmon D L

Effects of Fructose and Gonadtropins on
Rabbit Reproduction and Performance
(1988) / Heird C E

The Effect of Feeding Mineral Oil and
Butylated Hydroxy Anisole of Heptachlor
and Chlordane to Dairy Cows
(1988) / Highnight S D

A Cross Sectional Analysis of the
Relationship Between Socio-Economic
Variables and Income Measures by Type of
County for Seven Southern States, 1979
and 1984
(1988) / Hogan T G

An Economic Analysis of Fresh Market Peach
Production in Arkansas
(1988) / Kirchner D

The Effects of Different Body Weights of
Broiler Breeder Females on Their Subsequent
Performance
(1988) / Nnyagu C C

Growth Rate, Body Measurements, and Feed
Efficiency in New Zealand White, Japanese
White, F1, and F2 Hybrid Rabbits
(1988) / Oetting B C

The Feasibility of Hedging Soybean Oil and
Cross-Hedging Cottonseed, Corn, Coconut, and
Palm Oil
(1988) / Ooi C S

A Re-Evaluatioon of Nutritional Factors
Reported to be Growth Promotants for
Broilers
(1988) / Rydell J A

Control and Seedhead Suppression of Red Rice
in Soybean
(1988) / Salzman F P

Response of Barrows and Gilts to Various
Dietary Energy Levels During Different
Seasons of the Year
(1988) / Smith J F

Fractionation and Characterization of
Rice Glutelin
(1988) / Snow S D

Dietary Energy Levels Utilizing Different
Low Energy Feedstuffs for Laying Hens
(1988) / Stilborn H L

Effects of Postruminal Antibiotic
Administration, Protein Supplementation and
Ruminal Fill on Intake, Digestion and
Digesta Characteristics in Cattle Fed
Native Grass Hay
(1988) / Stokes S R

ARKANSAS, UNIVERSITY OF
(continued)

The Effects of Hedging on Farm Loan Default
Risks for Given Leverage Ratios
(1988) / Stutte G W

Determination of Vernalization Requirements
of Common Arkansas Wheat Cultivars
(1988) / Sutton R L

Hedging Fort Smith Feeder Cattle
(1988) / Thompson S L

Energy and Amino Acid Programs for Male and
Female Broilers Grown Separately
(1988) / Tidwell N M

Phenological Development and Nutritive
Changes in Small Grains for Forage
(1988) / Walker D W

Arginine Vasotocin and Mesotocin Response to
Acute Heat Stress in Domestic Fowl
(1988) / Wang S

The Economics of Alternative Production and
Marketing Strategies for Beef-Forage
Producers: A Simulation Analysis Using
"Graze"
(1988) / Watts W M

The Efficacy of Clomazone (COMMAND) Under
Arkansas Conditions
(1988) / Westberg D E

Growth and Reproduction of Nutrient
Restricted New Zealand Rabbits in a Heat
Stressed Environment
(1988) / Wittorff E K

AUBURN UNIVERSITY

Secretory Patterns and Metabolic Clearance Rate of
Porcine Growth Hormone, and Somatomedin-C
Concentrations in Swine Selected for Growth
(1987) / Arbona J R

The Impact of Increased Price Variability on Farm
and Retail Egg Prices
(1987) / Brand A H

Effect of Urea-Molasses Supplement on Forage
Intake, Digestibility and Ruminal Turnover in
Steers Grazing Bermudagrass
(1987) / Dawkins R A

Feasibility of Centralized Menu in Choice Menu,
Offer vs Serve Lunch Program
(1987) / Dillon M L

Effect of Form and Level of Dietary Selenium
on Murine Mammary Tumorigenesis
(1987) / Dukes J R

Measuring Neighborhood Cohesion in
Residential Settings
(1987) / Graham C R

Using Volatile Fatty Acid Relationships as
Indicators of Failure in Anaerobic Digestion
of Swine Waste
(1987) / Holmberg R D

Recreational Use and Preservation Values of Free-
Flowing River Segments in Alabama
(1987) / Malone J W

Government Farm Program Participation Under
Selected Policy Provisions for Typical Alabama
Cotton Farms
(1987) / Mims A M

Economic Analysis of Public Investment in
Agricultural Biotechnology Research
(1987) / Mulla M A

Concentrations of Metabolic Hormones and Metabolites
Between Higher and Lower Producing Dairy Cows
During Early Lactation
(1987) / Pierce A C

Modeling Soil-Metal Sliding Resistance
(1987) / Robbins D H

Effect of Cooking on Selenium Retention in Chicken
(1987) / Tanticharoenkiat O

BRIGHAM YOUNG UNIVERSITY

Reproductive Tract Stimulation on Cows
Injected with Prostaglandin
(1988) / Betts V W

Oxidative Stability of Cholesterol-Free
Egg Lipid Fractions Prepared with Food
Grade Solvents
(1988) / Blackwelder J A

Gallium(II) Does Not Actively Substitute
for Iron(III) in Iron/Gallium Competition
Studies
(1988) / Blaylock M J

A Small-Scale Agricultural Strategy for
Developing Nations
(1988) / Flores M Q

Shelf-Life of Refrigerated Turkeys Varying
in Basting Quantity
(1988) / LeFevre M J

Petiole Nitrate Profiles and N Uptake: A
Comparison Among Seven Potato Cultivars
(1988) / Lewis R J

Spring Wheat Response to Nitrogen and Plant
Growth Regulators
(1988) / Lutz P H

Sorghum Growth Response to Polyacrylamide
Treated Soil
(1988) / Stimpson D C

CALIFORNIA STATE UNIV. (FRESNO)

The Influence of Zeranol on Biologic and
Economic Efficiency of the Intact Male
Bovine
(1987) / Borelli C L

Reverse-Phase Paired-Ion Liquid
Chromatographic Analysis: Niacin in
Mushroom
(1987) / Chen M H

Lignin Adsorbent Effects on the Performance
of Metribuzin and Hexazinone
(1987) / Gerecke T J

In-Vitro Seed Germination of Early-Ripening
Genotypes of Vitis Vinifera, L
(1987) / Massoudi M

CALIFORNIA STATE UNIV. (FRESNO)
(continued)

Effect of Nitrogen Fertilization on Growth,
Reproductive Development, and Seed
Production of 'Anaheim' Chili Pepper
(1987) / Payero J O

A Study of Grape Harvest Mechanization
(1988) / Shaghaghi B

Nitrate Reductase Activity in Grapevine
Leaves and Roots
(1987) / Tarpley L

Evaluation of Laboratory Vigor Tests for the
Prediction of Pepper Seedling Emergence in
the Field
(1987) / Trawatha S E

The Effects of Delinting Whole Cottonseed
on Its Value in Supporting Lactation of
Dairy Cattle
(1988) / Wagner E P

CALIFORNIA, UNIVERSITY OF (DAVIS)

Tomato Seed Vigor: Relationship to Priming,
Deterioration and Tolerance of Herbicides
(1988) / Argerich C A

Competitive Survival of Split and
Nonmanipulated Porcine Embroys
(1988) / Ash K L

Nitrogen Nutrition and Susceptibility of
Sweet Corn to Drought
(1988) / Awono J P

Effect of Hedging Upon Herbaceous Character
of Cabernet Sauvignon Wine
(1988) / Bengard S

Interactions of Ascorbic Acid and Sulfur
Dioxide in Grape Juice and Wine
(1988) / Bertheau R J

Effect of Blanching on the Stability of
Undermilled Calrose Rice
(1988) / Bruenner B A

The Effects of Premature Defoliation on
Nitrogen Remobilization and Vegetative
Regrowth of Deciduous Fruit Trees
(1988) / Castagnoli S P

The Nutritional Value of Amaranth Seed In
Growing-Finishing Swine
(1988) / Chi F

The Role of Ethylene in the Senescence of
Detached Broccoli Florets
(1988) / Ching T W

Effects of Prenatal Zinc Deficiency on
Tubulin Polymerization
(1988) / Cuellar S

The Effects of the Nutrition Assistance
Program on the Production and Consumption
of Local Foods in the Commonwealth of the
Northern Mariana Islands
(1988) / Denman V M

Ovulation Rate and Pre and Post-Implantation
Survival in mice with a Major Gene for
Postweaning Growth
(1988) / Dilts R B

Factors Involved in the Consumers' Decision
to Use Calcium Supplements
(1988) / Dreith A Z

A Simulation Model for Winery Operations
(1988) / Eckstein M S

Copper Metabolism and Antioxidant Status in
the Dahl-S Rat: Effects of Varying Dietary
Copper and Zinc
(1988) / Garrow T A

Women, Agricultural Cooperatives and the
Economic Crisis in Costa Rica
(1988) / Gilliland L A

Use of Carbohydrate Fragments as Precursors
to Pyrazines
(1988) / Hardy R N

Fostering Lambs by Odor Transfer: Two-Lamb
Substitution as a Means of Increasing the
Number of Lambs Reared by Ewes
(1988) / Holmes L N

The Effects of Trellis System, Pruning Level
and Row Spacing on Yield, Yield Components,
and Fruit Composition of Cabernet Sauvignon
Grapes
(1988) / Johnson R A

Pyrazine Formation in Lactose/Casein
Derived Model Systems
(1988) / Kao C F

Susceptibility of 'Haas' Avocado Fruit to
Mechanical Injury During Postharvest
Handling
(1988) / Katz P M

A Farming System in the Mandara Mountains
of North Cameroon
(1988) / Kerr A T

Community Land Management in the Andean
Context: The Sectoral Following System
(1988) / Kraft K E

The Residual Effects of Controlled
Atmospheres on Postharvest Physiology and
Quality of Strawberries
(1988) / Li C

Investigation on the Mutagenecity of
N-Nitrosothioproline
(1988) / Liu I C

Energy Utilization of Alfalfa
Hay. I. Digestibility and Utilization
by Steers and Sheep
(1988) / Miller P S

The Effectiveness of Chemical Preservative
Systems, Singularly or in Combination, on
the Shelf Life Extension of Vacuum Packaged
Rockfish Fillets
(1988) / Miller S A

Effects of Water and Salt Stress on Field
Grown Processing Tomato: Changes in Yield
and Fruit Quality
(1988) / Mitchell J P

Welfare Effects of Technical Change in the
Mexican Egg and Poultry Industry
(1988) / Perali C F

CALIFORNIA, UNIVERSITY OF (DAVIS)
(continued)

Magnetic Resonance Imaging to Obtain
Moisture Profiles and Moisture Dependent
Diffusion Coefficient During Apple Drying
(1988) / Perez E

Assessing Dairy Potential of Western
Whiteface Ewes
(1988) / Reynolds L L

The Effects of Nitrogen on the Growth and
Development of California Rice Varieties
(1988) / Roberts S R

Production of Sheep and Goat Hybrid/Sheep
Chimeras by Inner Cell Mass Transplantation
(1988) / Roth T L

Development Implementation: An
Interdisciplinary Assessment of the Village
Fish Pond Project in Northeast Thailand
(1988) / Sai-Ngarm J

Salt-Tolerant Triticum and Lophopyrum
Derivatives Limit the Accumulation of Sodium
and Chloride Ions under Saline-Stress
(1988) / Schachtman D P

The Viability of Targhee Lambs at Birth from
Growth Selected and Unselected Lines
(1988) / Schwartz M R

Changes in Water Soluble, Chelator Soluble,
and NaOH Soluble Pectins in 'Cabernet
Sauvignon' Grape Berries During Maturation
(1988) / Silacci M W

The Effects of Leaf Removal on Canopy
Microclimate, Crop Yield, and Composition
of Juice and Wine of Several Grape Cultivars
(1988) / Smit S P

The Effects of Selection for Gain in Mice
on the Direct-Maternal Genetic Covariance
(1988) / Swartz A R

Effect of Width of Cordon Separation and
Shoot Positioning on Growth, Yield and Fruit
Quality of Sauvignon Blanc, Chardonnay and
Cabernet Sauvignon Grapevines
(1988) / Tizio-Mayer C R

Sequential Sensitivity Analysis:
Refinements to the Model Using Predictions
for the Triangle Test
(1988) / Vie A

The Use of Monoclonal Antibodies to Detect
Embryonic H-Y Antigen
(1988) / Waltemyer C A

Rheological Characterization of Fluid Dairy
Products
(1988) / Wayne J E

A Comparative Study of Political and
Bureaucratic Aspects of Emergency Famine
Relief
(1988) / Westcott J A

Genetic Differentiation Amond Cutthroat
Trout Populations
(1988) / Xu R

CLEMSON UNIVERSITY

Competitive Position of United States
Soybeans in a Dynamic World Economy: A
Sectoral Sensitivty Analysis
(1988) / Arburn G W

The Economic Feasibility of Centralized
Packing Houses in the Tri-State Region
(1988) / Bundrick A E

Electric Field Levels Lethal to Gambusia
(1988) / Christopher D E

Financial Analysis of Investment in Poultry
and Swine Facilities in South Carolina
(1988) / Collins C B

Policy Impacts on Structural Change in South
Carolina: The Case of Feedgrain and Oilseed
Producers
(1988) / Doherty M J

Estimation of the Contributory Value of
Riparian Rights to Real Property
(1988) / Evans A A

The Importance of Price and Nonprice
Incentives on Supplier Selection in the
Fresh Produce Industry
(1988) / Harrison R W

An Econometric Analysis of Corn and Soybean
Yields in South Carolina
(1988) / Justice S T

Financial Condition of South Carolina Farms
(1988) / McKenzie J G

COLORADO STATE UNIVERSITY

A Simulation Model for the Grazing Behavior
of Beef Cattle
(1988) / Baker B B

Temperature and Moisture Content Control of a
Solid Substrate Fermentation on Extruded Corn
(1988) / Barstow L M

Growth and Destruction of Listeria Monocytogenes
in Meat Plant Environments and Products
(1988) / Boyle D L

Monitoring the Acetone-Butanol Fermentation
by High Pressure Liquid Chromatography
(1988) / Buday Z

Production of Phenolics by Immobilized and
Encapsulated Cells of Nicotiana Tabacum
(1988) / Haigh J R

The Influence of Water Treatment Sludge on
the Growth of Sorghum-Sudangrass
(1988) / Heil D M

Pore Size Distributions from Soil Water Retention
Characteristics for Studying Soil Organisms
(1988) / Heil J W

Economic Implications of a Geographic
Information System for the Uncompahgre Valley
(1988) / Hermann K A

Water Induced Soil Erosion Processes and Modeling
(1988) / King J P

Economies of Size and Technical Efficiency on
Farms of the Zona Mixta Cerealera in Argentina
(1988) / Manchado J C

COLORADO STATE UNIVERSITY
(continued)

Melengesterol Acetate and Prostaglandin
Fluoride-Alpha Combinations for Estrus
Synchronization in Beef Heifers
(1988) / Mauck H S

Evaluation of Hard Red Winter Wheat Seed Quality
(1988) / Miller D R

Factors Effecting Alfalfa Production in the
Southern San Luis Valley
(1988) / Moerdyk T

Morphological Characterization of Six Dry Bean
Genotypes Grown Under Non-Irrigated Conditions
in Colorado
(1988) / Ortega P J

Utilization of Brewers Dried Grains in
Diets of Pheasant Breeder Hens
(1988) / Pfaff W K

Phenotypic and Genetic Responses to
Selection in Beef Cattle
(1988) / Posegate R P

Energetic Efficiency and Body Characteristics of
Lambs Fed Cimaterol
(1988) / Rikhardsson G

Economic Evaluation of Reproductive Management
Alternatives in Beef Cattle
(1988) / Snelling W M

Nutrient Status in Cyclists Consuming a Buffered
Electrolyte or a Sodium Bicarbonate Solution
(1988) / Tamis B

Contributions of Autotrophic and Heterotrophic
Nitrifiers to Soil NO and N(2)O Emissions
(1988) / Tortoso A C

The Effects of Stress, E. Coli and Dietary Edta
on Trace Element Kinetics in Chicks
(1988) / Tufft L S

CONNECTICUT, UNIVERSITY OF

Markmap: A Geographic Information System
for Market Area Analysis
(1988) / Annitto R N

The Effect of Unilateral Castration on
Leydig Cell Testosterone Production
(1988) / Cracchiolo S A

The Orchard House Landscape Concord,
Massachusetts 1857-1868
(1988) / Davis H A

A Study of Italian Immigrant Gardens in the
Northeast United States, 1900-1945
(1988) / Deshefy-Longhi T

Interspecific Hybridization in the Genus
Impatiens
(1988) / Gager C F

The Control of Estrus and Ovulation in Dairy
Heifers with Alfaprostol and Gonadotropin
Releasing Hormone
(1988) / Leers S C

A Multi-Objective Approach to Evaluating
Line Generalization
(1988) / Morse S J

Follicular Atresia in Cattle:
Interrelationship of Follicular Fluid
Estradiol and Progesterone with Plasma
Vitamin A and Beta-Carotene
(1988) / Nichols D M

An Empirical Analysis of Financial
Performance Based on a Sample of New England
Milk Producers
(1988) / Wadsworth J J

The MacDowell Colony: An Historic Survey of
the Designed Landscape 1908-1938
(1988) / Wertheimer S M

The Influence of Boar Exposure and Outdoor
Relocation on Puberty Attainment of
Confinement Raised Gilts
(1988) / Wojtusik R M

CORNELL UNIVERSITY

The Physiological Relationship Between
Milking Induced Changes in Oxytocin and
Prolactin Secretion and Mammary Blood Flow
During the Estrous Cycle
(1988) / Aromando M C

Partial Chemical Characterization of the
Aroma of Cortland Apples
(1988) / Bartik K F

Broadleaf Weed and Grass Weed Responses to
Dinoseb and Fluazifop-P or Sethoxydim
Combinations When Applied as Tank-Mixes or
Sequential Treatments
(1988) / Boring R A

Early Season, Thematic Mapper-Based Crop
Separability as Influenced by Regional
Variation in New York State
(1988) / Buechel S W

Growth and Morphology of the Maize Root
System
(1988) / Cahn M D

Determination of the Presence of Geotrichum
Candidum in Fermented Dairy Products and
the Use of Acidity and Starter Cultures as
Control Mechanisms of Its Growth
(1988) / Conklin D M

An Analysis of the Impact of the Federal
Tax Reform Act of 1986 on New York Dairy
Farmers
(1988) / Conrad J A

Evaluation of the Nitrogen Supplying
Potential of Legumes Placed on the Surface
or Incorporated in a Brazilian Oxisol
(1988) / Costa F

Inhibition of the Enzyme Polyphenol Oxidase
by the Amino Acid L-Cysteine
(1988) / Dudley E D

Effect of Fiber Level and Source on Growth
Rate and Feed Intake in Neonatal Dairy
Calves Receiving Two Different Milk Feeding
Programs
(1988) / Giner-Chavez B I

The Influence of Computer Information
Systems on Dairy Farm Management
(1988) / Giraudo E J

CORNELL UNIVERSITY
(continued)

Alternative Methods of Forecasting Japanese
Wheat and Corn Imports
(1988) / Gomersall C N

Evaluation of Progress Breeding Alfalfa
Populations with the Capability to
Regenerate Roots After Taproot Severing
(1988) / Hansen J L

Delipidation of Fish Muscle with
Supercritical Fluid Extraction
(1988) / Hardardottir I

The Design of a Semicontinuous Supercritical
Fluid Extractor
(1988) / Hohner A S

Texture Changes of Egg White and Frozen
Stored Minced Fish
(1988) / Hsieh Y L

Influence of N Source, RTZ, and Mg Level on
Nutritional Quality of Winter Wheat Forage
(1988) / Huang J

Estimation of Calcium Status in Selected
Food Systems by an In Vitro Method
(1988) / Keane L A

Behavioral Enhancement of Ewe Lamb
Reproduction Through Use of the Ram Effect
(1988) / Lai S A

Permeation of High Barrier Films by Ethyl
Esters Effect of Permeant Molecular Weight
and Relative Humidity
(1988) / Landois-Garza J

Evaluation of Pork Carcass Composition
with a New Instrumental Technique
(1988) / Liu Y

The Effects of Herbicide Treatment and
Tillage on Soil Moisture and Yield of Corn
(1988) / Malek R S

Diversification of the Cheddar Cheese
Industry Through Specialty Cheese
Production: An Economic Assessment
(1988) / Martin J C

Root Distribution of Dry Beans and Red
Fescue in a Living Mulch System
(1988) / Maynard E T

Changes in United Kingdom Cereal
Production: An Econometric Investigation
of Area and Yield Response in the Eastern
Counties of England, 1970-1985
(1988) / Melencovitch A

A Segmentation Analysis of Consumer
Attitudes and Usage of Grocery Coupons
(1988) / Meloy M G

Factors Explaining the Difference in Winter
Wheat Yields Between New York State and
England 1965-1985
(1988) / Mendoza J B

A Profitability and Loan Pricing Analysis of
a Commercial Bank's Agricultural Loan
Portfolio: A Case Study
(1988) / Miller L H

Effect of the Pore Size of Ultrafiltration
Membranes on the Quality of Apple Juice
(1988) / Padilla O I

Kinetic Phase Separation of Thermophilic
Anaerobic Digestion of Sorghum
(1988) / Phillips D E

A Study of the Foaming Properties of
Proteins
(1988) / Phillips L G

The Effect of Soil Temperature, Alachlor and
Metotachlor on the Growth of Field Corn
(1988) / Pier J W

Effect of Dietary Energy Dilution and
Restriction on Pre- and Post-Molt
Performance of Laying Hens
(1988) / Rangel-Lugo M

Effects of Flooding on Transpiration and
Root Hydraulic Conductance of Conifer
Transplants
(1988) / Reece C F

A Study on the Residence Time Distribution
in the Holding Tube of a Pasteurizer
(1988) / Sancho M F

The Effects of Packaging Materials and
Methods on the Shelf-Life Stability of
Frozen Fish
(1988) / Santos E E

Labile and Total Zinc Concentrations in the
Rhizosphere and Bulk Soils of Oats and Rice
(1988) / Sarong L Q

Design and Performance of an Automatic
Electro-Hydraulic Unit to Control the Depth
of Seed Placement as a Function of Soil
Moisture
(1988) / Shahin M A

Estimating Volatilities for Setting Margins
for Soybean Futures
(1988) / Shane J G

Use of Organic Waste Products for Improving
Onion Production on Organic Soils with High
Levels of MARL
(1988) / Shepard M T

Evaluation of an Extrinsic Tracer Method
for the Measurement of Calcium
Bioavailability from Dairy Products
(1988) / Sowizral K C

Operational Typologies for Gravity
Irrigation Systems
(1988) / Steiner R A

Gaseous Atmosphere Effects on the Growth
and Differentiation of Tomato Hypocotyl
Sections in Culture
(1988) / Thome C R

Cholesterol Reduction in Anhydrous Milk Fat
Through Supercritical Fluid Extraction
(1988) / Tse B L

Shade Tolerance of Vegetables: Experiments
in New York and a Survey in Costa Rica
(1988) / VanDusen K C

Gelation Characteristics of Whey Protein
Isolates
(1988) / Zirbel F I

CORNELL UNIVERSITY
(continued)

Dietary Fish Oil: Effects on Fatty Acid
Composition and Enzymatic Activity of Rat
and Mouse Liver Plasma Membranes
(1988) / Zuniga Saldierna M E

Hormones, Libido and Sperm Output in
Holstein Bulls
(1988) / von Brackel A J

DELAWARE, UNIVERSITY OF

Regulation of Hepatic Thyroxine 5'
Monodeiodination Growing Chicken (Gallus
Domesticus)
(1987) / Alfonso C P

Economic and Financial Analysis of Food
Supply and Input Demand in Mauritania
(1987) / Ba D

Pathotypic Differences Among Infectious
Bursal Disease Viruses
(1988) / Carr M Y

The Economic Impact of a Public Garden on
Its Community: A Case Study of Phipps
Conservatory, Pittsburgh, Pennsylvania
(1987) / Dolinar E M

The Effects of Ionophore Drugs in Two
Protein Diets on Broiler Chickens' Growth
Performance
(1987) / Fan W H

The Feasibility of Utilizing an Electronic
Marketing System Among Delaware and New
Jersey Produce Growers
(1987) / Houser M B

A Stochastic Analysis of Broiler Farm
Financial Conditions and Farmer Expectations
(1988) / Hunter B S

Factors Affecting the Micropropagation of
Artemisia Dracunculs L. Var Sativa, French
Tarragon, Using Shoot Tips as the Explant
Source
(1987) / Mac Kay W A

Immunization Against Infectious Bursal
Disease Virus in Chicken
(1987) / Meza G G

A History of the American Association of
Botanical Gardens and Arboreta, Inc.
(1987) / Nace G A

Effects of Cow Sequence and Related Factors
on the Accuracy of Milk Yield Estimates
(1987) / Quartuiccio J S

Heat Production in Broiler Chickens Fed
Three Levels of Dietary Energy, Given Daily
Injections of Growth Hormone, or Fed
Thyroid-Active Substances
(1987) / Rand A L

Aspects of the Host Range of the Soybean
Cyst Nematode Heterodera Glycines
(1987) / Rivera L M

Chemistry and Availability of Potassium
Silicates in Soils
(1988) / Sadusky M C

Cholecystokinin Receptor Binding in Sheep
Brain as Affected by Feeding and Fasting
(1987) / Solomons R N

Endogenous Abscisic Acid Levels in Asexual
Embryos of Carrot
(1987) / Spencer T M

Risk-Returns Analysis of Delaware
Agriculture: A Capital Asset Pricing
Model Approach
(1988) / Tambe A M

William Penn's Estate at Pennsbury and the
Plants of Its Kitchen Garden
(1987) / Thomforde C

A Dual Form to Retail Multiproduct Analysis:
an Application to the Northeast Food
Industry
(1987) / Wobus P

Relationship of the Southern Cornstalk Borer
to Stalk Rot of Corn in Delaware
(1987) / Yocum J A

Social Perspectives in Horticulture
(1988) / Zadik M H

Grounds Maintenance Time Requirements
(1987) / Zelonis M E

FLORIDA, UNIVERSITY OF

U. S. Demand for Avocados, Bananas, and
Tropical Fruits: Tobit Analysis and
Projections
(1988) / Acon W

An Econometric Analysis of Media Effect of
Generic Advertising on Fluid Milk Demand
(1988) / Armotrading G P

Assessment of Competencies Possessed by
Students Enrolled in Ornamental Horticulture
(1988) / Aviles-Rodriguez L N

Mineral Status of Beef Cattle in the Eastern
Chaco Region of Argentina
(1988) / Balbuena O

Effect of Planting Data and Row Position of
Soybean and Pigeonpea Into Maturing Maize on
Light Availability, Growth, and Grain Yield
(1988) / Bidja R

Double-Cropped Soybean/Oat Root
Growth and Development Affected by Tillage
(1988) / Bruniard G

Response of Long-Juvenile Soybean Genotypes
to Sowing Date
(1988) / Bruniard J

Development of a Continuous Stormwater
Management Model for Agriculture in
Florida's Flatwoods Resource Area
(1988) / Burleson R W

The Cost Determination of Five Cold
Protection Methods Utilized on Young Citrus
Trees
(1988) / Cardin B

Mutagenicity Study of Nitrite-Treated Soy
Sauce Mixtures
(1988) / Chang F C

FLORIDA, UNIVERSITY OF
(continued)

A Differential Plating Medium for Rapid
Detection and Isolation of Prolific
Histamine-Producing Bacteria from Tuna
(1988) / Chen C

Effects of Photoperiod on Mare Gestation
Length, Postpartum Ovulation Interval, and
Foal Birth Weight
(1988) / Cheves L

Preparation, Stability, and Acceptability
of Breadfruit Chips
(1988) / Clos L

Physical Properties of Restructured Beef
Roast Formulated with Various Levels of Fat
and Connective Tissue
(1988) / Cruz Lantigua R J

Policy and Economics of Rice Production in
the Ivory Coast: Cost Minimization in the
Government Distribution Network
(1988) / Diarassouba M

An Attempt to Explain the Mode of Action of
a Synthetic Zeolite when Fed to Laying Hens
and Broiler Chicks
(1988) / Fethiere R

Assessment of Rock Shrimp Melanosis: Sulfite
Treatment and Residuals, Phenoloxidase
Activity, and Microbial Growth
(1988) / Fink K

An Improved Analysis of Formaldehyde by
Derivatization with 2-Nitrophenylhydrazine
and Packed Column Gas-Chromatography
Electron Capture Detection
(1988) / Fishkind M

The Communication Component of the
International Agricultural Research Centers
and Its Role in the Transfer of Agricultural
Technology
(1988) / Frarey L

Periparturient Endocrine Patterns of
Progesterone, Prostaglandin F(2a),
Luteinizing Hormone, Oxytocin and Relaxin
in Beef Cattle
(1988) / Ganz N

Seasonal Price Premium and Supply Response
by Florida Dairy Producers: A Stochastic
Dynamic Programming Analysis
(1988) / Gao X

Effect of Procine Conceptus Secretory
Proteins of Uterine Development,
Interestrous Interval and Uterine Secretion
of Prostaglandins
(1988) / Harney J P

A Statistical Model of Smallholder Cropping
Strategies in Northern Nigeria:
Implications for Agricultural Extension
(1988) / Hessels R

Simulation of Gum Naval Stores Production
Systems
(1988) / Hodges A

Effects of Percentage Brahman and Angus
Breeding, Age-Season of Feeding, and
Slaughter End Point on Feedlot Performance
and Carcass Characteristics
(1988) / Huffman R

Gonadotrophin-Releasing Hormone in Pig
Ovaries
(1988) / Jackson K

The Conceptual Design, Development, and
Testing of an Environmental Controller for
Tomato Greenhouses
(1988) / Jacobson B

Resource Allocation Among Limited Resource
Farmers on Sitiung 5c, West Sumatera,
Indonesia
(1988) / Kan S K

Development and Use of Techniques for
Monitoring the Biological Waste Treatment of
Pulp and Paper Industry Wastes
(1988) / Malpass J

Supermarket Produce Demand and Shelf Space
Effects
(1988) / Molina G

Using Thermal Infrared Thermometry to
Estimate Soil Water Content
(1988) / Myhre B E

Development of the Growing DDay Index for
Use in Corn Yield Weather Relations
(1988) / Ndiyoi M

Differential Responses of Maize, Peanut, and
Grain Sorghum to Water Stress
(1988) / Nzeza K

Effect of Harvest Date on Yield, Grade, and
Seed Quality of Southern Runner Peanut
(1988) / Pedelini R

Crossing Percentages Among Families and
Evaluation of Selection for Maturity in
Crimson Clover
(1988) / Reith P E

Development and Evaluation of an Adaptive
Control Procedure for Carbon Dioxide
Control in Environmental Growth Chambers
(1988) / Roy B

The Impact of Reverse Osmosis on Florida
Milk Supplies
(1988) / Schiek W

Effects of Potassium and Sodium on the
Biological Responses of Rumen Bacteria to
Ionophores
(1988) / Schwingel W R

Quantifying and Evaluating Agriculture
Chemical Drift Under Various Florida
Application and Microweather Conditions
(1988) / Semmes R H

Nitrate Movement in Sandy, High-Water-Table,
Bedded Soil in Florida
(1988) / Siaka K

Machinery System Reliabilities, Capacities
and Field Efficiencies for Sugarcane
Operations
(1988) / Steidinger G W

FLORIDA, UNIVERSITY OF
(continued)

Topping Time and Bottom Leaf Removal of
Flue-Cured Tobacco
(1988) / Stocks G

Evaluation of a Sprinkling and Fan
Evaporative Cooling System on Lactating
Dairy Cattle in Florida
(1988) / Strickland J

Tissue Carnitine Status of Colostrum
Deprived Neonatal Piglets Fed Carnitine-Free
and Carnitine-Supplemented Formula
(1988) / Sullivan M

Remote Sensing and Geographic Information
System for Studying Agricultural Land Use
Changes and Abandoned Well Identification
in St. Lucie County, Florida
(1988) / Tan Y R

Calcium Selenite as a Dietary Source of
Selenium for Poultry and Ruminants
(1988) / Tarla F

Influence of Copper Supplementation and
Pelleted Versus Extruded Concentrate on
Growth and Development of Weanling Horses
(1988) / Thomas M

Kinetic Reduction of Wastewater Constituents
in Overland Flow
(1988) / Thompson R H

Influence of Plant Density and Planting
Pattern on Peanut Yield and Market Quality
(1988) / Tshitebwa M

Wheat/Crimson Clover Intercrop Competition
Due to Season, Cropping System and N, and
K Fertilization
(1988) / Uyanik M

Interactive Effects of Nematode and Water
Stresses on Soybean Growth, Development, and
Physiology
(1988) / Widodo S E

Using Satellite Imagery for Vegetation
Classification and Land Use Change
Assessment in the Florida Everglades
(1988) / Williams L

Relative Biological Availability of
Manganese Sources for Poultry and Ruminants
(1988) / Wong-Valle J

Purification, Serology, Characterization,
Effects, and Genetic Mapping of a Potyvirus
Isolated from Alyceclover
(1988) / Zhao G

GEORGIA, UNIVERSITY OF

The Development of an Improved Serum Plate
Agglutination Assay for the Detection of
Antibodies to Mycoplasma Gallisepticum in
Poultry
(1988) / Ahmad I

Evaluation of Carcass Composition Traits in
a Four-Breed Diallel Among Simmental,
Limousin, Polled Hereford and Brahman Beef
Cattle
(1988) / Arnold J W

Whole Cottonseed and Tallow as Supplements
for Lactating Primiparous Beef Cows
(1988) / Barron J E

Effect of Alpha-Ketoisocaproic Acid on the
Cholesterol Content of Tissue and Eggs in
Broilers and Laying Hens
(1988) / Beyer R S

Effect of Light, CO2, O2, and Water
Submersion on Photorespiratory
Characteristics of C3-C4 Species
(1988) ./ Byrd G T

Effect of Naloxone on Luteinizing Hormone
Secretion in Prepubertal Beef
(1988) / Chang W J

Integrated Reproductive Management of Beef
Cattle
(1988) / Deason R

Some Effects of Dietary Silicon and Aluminum
of Broiler Chickens
(1988) / Elliot M A

Row Spacing and Plant Population Effects on
Yield of Soybean
(1988) / Ethredge W J

An Analysis of the Profitability and
Riskiness of Using Legume Cover Crops in
No-Till Feed Grain Production
(1988) / Franklin R L

A Probability Distribution Based Analysis of
Corn Programs
(1988) / Goggin B D

The Effects of Energy Source and Level Upon
Endocrine Regulation of Growth and Mammary
Tissue Development in Beef Heifers
(1988) / Houseknecht K L

Characterization of Hybrids Between
Photosynthetically Distinc Flaveria Species
(1988) / Huber W E

The effect of Different Dietary Levels of
Calcium on Bone and Egg Parameters in White
Leghorn Pullets at the Onset of Lay
(1988) / Hudson H A

Market Share Valuation of Quality
Differentiation in Certified Soybean
Seed in the Southeast
(1988) / Jeong K

Rapid Non-Subjective Test Using
Bioluminescence to Measure Lactate
Dehydrogenase as an Indicator of Bovine
and Embryo Viability
(1988) / Jordan J E

The Introduction of New Maize Technology
in the Farming System of the Kasai-Oriental
Region of Zaire
(1988) / Lukusa T M

Manipulation of Rumen Protozoal Population
to Alter Rumen Function
(1988) / Martin A C

Early Post-Mortem Chemical and Physical
Changes in Broiler Breast Muscle
(1988) / McGinnis J P

GEORGIA, UNIVERSITY OF
(continued)

The Effect of Isoacids on Performance and
Blood and Rumen Parameters of Growing
Finishing Beef Cattle
(1988) / McKissack K E

Winter Legumes as a Nitrogen Source for
No-Tillage Corn and Grain Sorghum
(1988) / McVay K A

Tolerance of Soybean to the Soybean Cyst
Nematode: A Rhizotron Study
(1988) / Miltner E D

Improving Dynamic Visual Inspection
Performance
(1988) / Myers J B

Soil and Plant Response to Phosphogypsum in
Reconstructed Acid Soil Profiles
(1988) / O'Brien L O

Expert System Based Coupling of Crop Growth
and Water Management Models
(1988) / Perry C D

Alfalfa Persistence and Yield Under
Continuous Grazing and Seed Yield of
Grazing Tolerant Germplasms
(1988) / Smith S R

The Development of an Automated Rapid
Static Ballasting System for Farm Tractors
(1988) / Vande Linde G F

Effects of Tolerance and Initial Soybean
Cyst Nematode Population Density on
Reproductive Advantage of Resistant Soybean
Genotypes
(1988) / Waller R S

GUELPH, UNIVERSITY OF

Growth Analysis of a Spring Canola Hybrid,
Its Parents and a High Yielding Cultivar
(1988) / Alam S M

The Transfer of Nitrogen from 15N-Labelled
Legume Residues to Corn and Barley Using
Two Tillage Systems
(1988) / Alder V J

Adapted Hedging Strategies for Eastern
Canadian Hog Producers
(1988) / Auger G

Feeding Programs for Broiler Breeders
(1988) / Bennett C D

Role of Lipid Soluble Antioxidants in the
Desiccation Tolerance of Germinating Soybean
(1988) / Berry T L

Nitrogen Degradation of Soybean Meal, Raw
Soybeans and Jet-Splotched Soybeans in
Holstein Steers
(1988) / Cahill L W

The Effect of T-2 Toxin on Neurochemistry
and Amino Acid Metabolism in the Rat
(1988) / Cavan K R

The Inheritance and Variation of Fatty Acids
in Spring Rapeseed Using in Vitro Techniques
and Single Seed Descent
(1988) / Chen J

Evaluation of the Waiting-Bed System for
Out-of-Season Strawberry Production in
Quebec and Ontario
(1988) / Chercuitte L

Senescence of Barley Anthers Cultured
in Vitro
(1988) / Cho U H

Analysis of a Gametoclonal Variant Resulting
from Another Culture of the Hard Red Spring
Wheat Cultivar Sinton
(1988) / Cober E R

Instructional design for Management Training
Necessitated by Rapid Technological Change
(1988) / Craig D M

The Effect of Nutrient Intake on Performance
of Laying Hens Exposed to Acute Short-Term
Heat Stress
(1988) / De Schutter A C

An Economic Assessment of Benefits Arising
from Adoption of Conservation Tillage
Practices in Southwestern Ontario
(1988) / Dickson E J

Cultivar Identification of Rapeseed Using
Isoenzyme Analysis
(1988) / Duke L H

Signals in Lymphocyte Activation
(1988) / Ebanks R O

Carcass Characteristics, Distribution of
Tissues and Predictions of Lean Content
from Carcasses in Beef Cattle
(1988) / Fan L

Inheritance of Somatic Embryogenesis in
Alfalfa
(1988) / Fernandez M H

Investigations of Alfalfa Plant and Callus
Response to Verticillium Albo-Atrum and
Fungal Filtrates
(1988) / Frame B R

A Thermal and Structural Analysis of
Amylose-Lipid Complexes
(1988) / Galloway G I

Evaluation of Two Cycles of Recurrent
Selection in Maize for Improved Forage
Performance Relative to Grain Production
Performance
(1988) / Germain C

Membrane Mediated Decolourization of
De Chaunac Cold Press Wine
(1988) / Giesbrecht R J

Performance and Inheritance of EMS Induced
Early Flowering Mutants of the Corn Imbred
Line A632
(1988) / Harwood D J

The Feasibility of Direct Broadcast
Satellite (DBS) in Delivering Agricultural
Extension Programs in Saskatchewan
(1988) / Hobin B A

Polymorphism of Palm Oil and the Effect of
Addition of Palm Oil on the Polymorphic
Properties of Hydrogenated Canola Oil
(1988) / Hong Y P

GUELPH, UNIVERSITY OF
(continued)

Returns to Canadian Federal Sheep Research
from 1968 to 1984
(1988) / Horbasz C N

A Structure-Function Study of Whey-Potato
and Whey-Pea Protein Composite Blends
(1988) / Jackman R L

Modification of Fats and Oils by Chemical
and Enzymatic Interesterification
(1988) / Jah F

The Effects of Salinity and Water Table
Depth on Maize Production
(1988) / Johnson T G

Soybean Seed Vigour Effects on Plant Growth,
Development and Yield, and Interactions with
Seed Moisture and Soil Temperature
(1988) / Jones D J

The Water Retention Characteristics and
Resistances to Water Loss and Water Uptake,
of Dead Winter Wheat Tissue
(1988) / Kalliomaki N M

An Evaluation of Factors that Influence the
Use of Recommended Farm Practices in
Nazerth Area, Ethiopia
(1988) / Kebede L

Biochemical-Physical Characteristics and
Processing Quality of Field and Storage
Type Potatoes
(1988) / Leszkowiat M J

The Effects of Salinomycin on the Production
and Absorption of Volatile Fatty Acids in
the Large Intestine of Pigs
(1988) / McMillan E G

Analysis of the Effects of Credit Terms and
Financial Assistance on the Financial
Performance of Ontario Cash Grain Farms
(1988) / Musaba E C

Towards the Production of Somatic Hybrids
Between Haploid Brassica Napus and Haploid
B. Oleracea
(1988) / Nichols R L

Simulation of Agricultural Credit Policies
of the Thai Bank for Agriculture and
Agricultural Cooperatives
(1988) / Nitsmer S

An EX ANTE Analysis of Technological
Change: Bovine Somatotropin and the
Ontario Dairy Industry
(1988) / Oxley J C

Development of a Computer-Assisted Learning
Softward Package for Teaching Agricultural
Production Economics
(1988) / Pan X

Textural Properties of Cheddar Cheese Made
from Ultrafiltered Milk
(1988) / Pasari S

Effect of Soil Treatments on Root and
Shoot Growth During the Post-Silking
Period in Maize
(1988) / Paul T

The Effect of Farmland Prices of Rural
Severances
(1988) / Pinto F R

Writing to Learn: The Impact of the Use of
Word Processing as a Learning/Writing Tool
Upon University-Level Developmental Learning
(1988) / Pletsch G D

The Development of a Method to Characterize
Soil Structural Stability and Its Use to
Measure the Effects of Bromegrass and Corn
Root Exudates
(1988) / Pojasok T

The Effect of Leanness in Pigs on Growth
Performance, Carcass Traits and Meat Quality
(1988) / Rae W A

Hedging with Commodity Options: The Effects
on Risk and Returns for Ontario Beef
Producers
(1988) / Rasmussen D L

Thiazyl and Selenazyl Radicals and Dimers
(1988) / Reed R W

Proline Metabolizing Enzymes in the
Neonatal Pig
(1988) / Samuels S E

Denitrification in Soil Aggregates of
Different Sizes
(1988) / Seech A G

The Role of Mycorrhizae in the Absorption of
Phosphorus and Zinc by Maize in Field and
Growth Chamber Experiments
(1988) / Shen L

Measurement of Rheological Properties of
Selected Cheeses by Instrumental Methods
(1988) / Shinn J M

A Comparison of Androgenic Culture Methods
and Assessment of Agronomic and Quality
Traits in Androgenic Populations
(1988) / Siebel J

A Study of Asparagus Toughening
(1988) / Smith J L

An Analysis of Farm Structural Change in
Heilongjiang, China -- A Linear Programming
Approach
(1988) / Song C

Thin-Layer Microwave Drying of Peanuts
(1988) / St. John B A

Effect of Palen of Nutrition on Growth and
Mammary Gland Development in Holstein
Replacement Dairy Heifers
(1988) / Stelwagen K

Estimation of Variance Components for
Lactation Traits of Canadian Dairy Goats
(1988) / Sullivan B P

The Impact of Subsidized Boneless
Manufacturing Beef Imports from the European
Economic Community on the Canadian Cattle
Industry
(1988) / Van Duren E H

Spatial and Temporal Dynamics of Soil-Water
Content Under a Corn Crop
(1988) / Van Wesenbeeck I J

GUELPH, UNIVERSITY OF
(continued)

Estimation of Genetic Parameters for
Production Traits in Boiler Chickens
(1988) / Wang L

Heterogeneity of Variances Among Herds and
Its Effects on Dairy Sire Evaluation
(1988) / Winkelman A M

The Inheritance of Leaf Angle, Leaf
Curvature, and Leaf Length, in Winter Wheat
(1988) / Wood D G

HAWAII, UNIVERSITY OF

Isolation of Plant Genes by Differential
Hybridization of cDNA Libraries
(1988) / Albert H H

Production of Flavor Related Amino Acids
and Nucleotides in Fish Liquefaction
(1988) / Chern T S

An Investigation Into the Use of Electric
Fields in Prawn Harvesting
(1988) / Droz T E

Effect of Oviductal Vein Resection on
Embryonic Survival in the Ewe and PGF(2)
Challenge During Mid Pregnancy in Sheep
(1988) / Gemmer K R

Evaluation of Evapotranspiration
Requirements Using Canopy Size for Selected
Tree Crops in Hawaii
(1988) / McNulty M V

Dietary History in Three Villages of Guam
Prewar and Wartime, (1925-1950), with
Special Reference to the Role of Calcium
Intake in Motor Neuron Diseases
(1988) / Parker C M

Delivery System for an Electro-Hydraulic
Vegetable Transplanter
(1988) / Rumsey M D

Soil Minerals, Plant Minerals and Plant
Oxalate: A Preliminary Investigation Into
the Calcium and Aluminum Content of Select
Root Crops Consumed by Five Guamanian
Families
(1988) / Standal G S

The Effect of Ovariectomy and Progesterone
Replacement on Uterine Secretion of
Prostaglandin F2 Alpha in Sheep and the
Effect of Progesterone on the Lifespan of
the Corpus Luteum in Heat Stressed Dairy
Heifers
(1988) / Thamotharan M

An Interactive Methodology for Determining
Sub-Optimality of Production Systems
(1988) / Tian K

IDAHO, UNIVERSITY OF

Effect of Heat Treatment on Ruminal
Degradation of Canola Meal
(1988) / Ahmadi A R

Influence of Soil Water Content and
Temperature on Emergence of Winter Rape
Seedlings
(1988) / Avery H R

The Investigation of Three Factors Important
for Wheat Anther Culture Success
(1988) / Barchowsky S L

Immunosuppression of Somatotropin-Release
Inhibiting Hormone: Effect on Serum
Somatotropin Concentrations and Steer
Performance
(1988) / Bauer C A

Factors to Consider in Evaluating and
Designing the Future Pre-Service Technical
Agriculture Curricular Component of a
B.S. Degree in Agricultural Education at
the University of Idaho
(1987) / Beitia M G

A Laboratory Study on Infiltration Into
Frozen Soils Under Simulated Rainfall
Conditions
(1988) / Boll J

Tools and Equipment Necessary to Conduct
Agricultural Mechanics Instruction in
Vocational Agriculture Programs
(1987) / Carter B L

Winter Rapeseed with Differential Levels
of Glucosinolates Evaluated as a Green
Manure Crop to Suppress Aphanomyces Root
Rot of Peas
(1988) / Davis J B

Beef Cattle Replacement Heifers and Their
Reactions to Selenium Boluses and Copper
Injections in Washington County, Idaho
(1988) / Edmiston F L

Disease Interaction of Pratylenchus Thornei
and Fusarium Roseum 'Culmorum' on Winter
Wheat with a Method for the Culture of
Pratylenchus Thornei on Carrot Disks
(1987) / Eschen D J

An Analysis of White Wheat Carryover Stocks
(1988) / Faux R B

Differential Response of Barley Cultivars
to Ethephon
(1988) / Gaiser D R

Sources of Information on New and/or
Innovative Farming Practices and How They
Are Accessed by Farmers in New Perce
County of the State of Idaho
(1988) / Gor C O

Characterization of Vegetable Oil Combustion
(1988) / Griend L V

An Assessment of the Direction of Vocational
Agriculture/FFA for the Future
(1987) / Harper E

Adult Agricultural Education Needs in
Franklin County, Idaho
(1987) / Hjorth L

The Effects of Follicle Stimulating Hormone
in Twice Daily Versus Once Daily Treatments
on Norgestomet Synchronized Ewes
(1987) / Hoffman K

Soil-Habitat Type Relationships in the Idaho
Selkirk Mouantains
(1988) / Houston K E

IDAHO, UNIVERSITY OF
(continued)

A Farm Systems Approach to the
Implementation of Agricultural
Innovations: Nikhom Non Sang, Thailand: A
Case Study
(1988) / Leacock W B

Phosphorus Effects on Corn Hybrids Differing
in Cold Tolerance
(1988) / Long L D

The Effects of Limb Spreading on Fruit Set
and Fruit Quality of 'Friar' Plum
(1987) / Longstroth M

Interference Between Triaaine-Resistant
Oilseed Rape and Wild Proso Millet
(1987) / Miller T W

Design and Development of a Low Damage
Sugarbeet Cleaner and Transloader
(1988) / Pohl J J

Screening for High Concentrations of Erucic
Acid in Mixed and Mutated Populations of
Winter Rapeseed
(1987) / Romero J E

A Dynamic Comparative Advantage Analysis of
White Wheat Production for the Palouse
Region
(1988) / Rugube L M

The Effectiveness of Cross Hedging Western
Barley Using Midwest Corn Futures: An
Empirical Analysis
(1988) / Ruhoff N F

A Study of Cropland Leasing Arrangements
in Idaho
(1988) / Seavert C

Integrated Economic Assessment of Soil
Erosion and Water Quality in Idaho's Tom
Beall Watershed
(1988) / Shi H Q

An Assessment of Community Attitudes Toward
Vocational Agriculture/FFA
(1987) / Wilder S

The Effects of Methyl Ester of Winter
Pareseed Oil on Diesel Engine Durability
(1988) / Zhang Q

ILLINOIS STATE UNIVERSITY

The Metabolic Profiling of Soybean Leaves
(1988) / Unger B C

IOWA STATE UNIVERSITY

The Effect of the Sire Breed of the Ewe on
the Birth and Weaning Weights of her
Crossbred Lambs
(1988) / Al Tamimi R M

Effect and Reliability of Sample Volume Size
in Soil Nitrate-Nitrogen and Moisture
Sampling
(1988) / Baker D G

Preliminary Evaluation of Two Recurrent
Selection Procedures in Maize
(1988) / Blackburn D J

Stress-Strain Characteristics of Soybean
Meal as Related to Flowability and Pressure
of Bulk Solids
(1988) / Bokhoven W H

Selection for Iron Efficiency of Soybean in
Nutrient Solution, Field and Tissue Culture
(1988) / Diers B W

Minimum Nitrogen Requirement for Gravid
Swine
(1988) / Dunn J M

The Effect of Sodium Bicarbonate in the
Ration on Genetic Differences of Leg
Structure in Duroc Swine
(1988) / Ernst C W

Hybrid Performance of Sorghum Parental Lines
Developed by Mass Selection and S1 Yield
Testing
(1988) / Ess K R

Optimization of Film Adhesion to Cook-in
Package Meat Products
(1988) / Goodmann T E

Effects of Gender and Castration of the Male
on the Somatotropic Axis of Cattle
(1988) / Gronowski A M

Regulation of Progesterone and Relaxin
Secretions of the Boar and Its Effects on
Sperm Motility
(1988) / Huang C

Relaxin in Blood and Seminal Secretions of
the Boar and Its Effects on Sperm Motility
(1988) / Juang H

The Effects of Substituting Mechanically
Separated Pork for Hand Deboned Pork on the
Organoleptic Qualities of a Fermented Snack
Sausage
(1988) / Kramer D G

Effects of Organic Acids on Seed Germination
and Seedling Growth in Soil
(1988) / Krogmeier M J

Fly Ash and Cement Kiln Dust Effects of
Elemental Composition of Soybean
(1988) / Labios J D

Effects of Skeletal Muscle Collagen and
Alkaline Phosphate on the Characteristics
of Finely Comminuted Cooked Sausage
(1988) / Ladwig K M

Effects of Salt and Phosphate on the
Stability of High Collagen Meat Emulsions
(1988) / Li X

Effects on Soybean Carbohydrate on Pig
Growth Performance and Nutrient
Digestibility
(1988) / Lii C K

Effects of the Addition of Glucose and
Electrolytes to the Drinking Water on the
Performance of Early Weaned Pigs
(1988) / Murphy J K

Time Course of Ethanol Evolution During the
Early Germination of Artificially Aged
Soybean Seeds
(1988) / Reedy M E

IOWA STATE UNIVERSITY
(continued)

Climate Date Retrieval and Analysis Using
an Interactive Microcomputer Based
Methodology
(1988) / Reinke B C

Pollen Fertility and Agronomic Performance
of Sorghum Hybride with Different Male
Sterility-Inducing Cytoplasms
(1988) / Secrist R E

Effects of Rosemary Oleoresin and Other
Antioxidants on the Sensory and Chemical
Properties of Precooked Pork Patties
(1988) / Sheu C

A Study of Residential Combustion Venting
Failures
(1988) / Shouse S C

Recurrent Selection for Test Weight in Oats
(1988) / Smith M A

Determination of Crop Production Risks Using
Simulation Yield Distributions and an Expert
System Shell
(1988) / Stefanski R J

Effects of Genetic Merit of Mates and Herd
Level of Daughteers on Sires' Predicted
Differences
(1988) / Welper R D

KENTUCKY, UNIVERSITY OF

The Conflicting Decision in a Lease Versus
Purchase Analysis
(1988) / Abbey A D

Identification and Antibiotic Resistance of
Gram-Negative Bacteria Isolated from Chicken
Breasts Obtained at Retail Outlets
(1988) / Aguilar-Castro D I

Effect of Canopy Structure and Availability
of Forage on Ingestive Behavior of Grazing
Ruminants
(1988) / Arias J E

The Effects of Electrical Stimulation and
Carcass Suspension on Tenderness of Several
Major Beef Muscles
(1988) / Arpi N

Acceptability and Digestibility of
Preservative Treated Hay by Horses
(1988) / Battle G H

A Simulation of Compaction as a Function of
Soil and Machinery Characteristics
(1988) / Bingner R L

The Protective Effects of Thiamine on
4-Methylimidazole Toxicity in Swiss Mice
(1988) / Bonfiglio M L

The Determination of a Relationship Between
the Copper, Zinc and Selenium Levels in Mare
and Those in Their Foals
(1988) / Breedveld L

Commercial and Exchange Rate Policies in the
Dominican Republic, 1970-1986: Their
Effects on the Price Incentives Structure
(1988) / Caamano-Valdez F

The Effect of Different Levels of Dietary
Magnesium and Potassium on Glucose Kinetics
in Sheep
(1988) / Dame E L

Soybean Filling Period: Measurement Methods
and Value for Parental Selection
(1988) / Dotti S

The Actions of Estradiol Benzoate and
Prostaglandin P2a on Parturition in Swine
(1988) / Eckerle B T

Comparison of Five Intermating Designs on a
Soybean Base Population Synthesized Prior to
Recurrent Selection
(1988) / Fatmi A

Genetic Variation in Burley Tobacco for
Bud Growth and Response to Maleic Hydrazide
(HM) Application
(1988) / Gorman D P

An Economic Analysis of Alfalfa Hay Flow
Patterns in Kentucky
(1988) / Greatbatch J L

Evaluation of the Hydraulic Mechanisms of a
Hazardous Waste Landfill Cover Under Stable
and Substistence Conditions
(1988) / Harned J A

Use of Early Maturing Soybean Cultivars to
Improve Production Efficiency of Late
Planted Soybean
(1988) / Hayati R

The Dynamic Elements of Yield and Revenue
Probability Distributions: Individual and
Group Learning in Tomato Production
(1988) / Hearne R

Effects of Age and Feed Restriction on
Intermediary Metabolism and Secrection of
Luteinizing Hormone in Ovariectomized Ewes
(1988) / Hileman S M

The Effects of Assumed Available Water, Salt
and Acetic Acid on the Stability of
Intermediate Moisture Meat
(1988) / Jose C

Studies of the Mechanisms of Cadmium and
Zinc Accomodation in Cultured Cells of
Nicotiana and Other Tissues
(1988) / Krotz R M

Studies on the Metabolism of Dievitone by
Rhizoctonia Solani
(1988) / Li D

Potassium and Ammonium Adsorption-Desorption
Characteristics of Soils with High Potassium
Fixing Capacity
(1988) / Lumbanraja J

Dynamics of Denitrifier Populations and
Deitrifier Enzyme Activity in the Soil
(1988) / Martin K J

Somatic Embryogenesis in Trifolium: Protein
Profiles Associated with High- and
Low-Frequency Regeneration
(1988) / McGee J D

Erodibility and Sediment Yield by Natural
Rainfall Events from Reconstructed Topsoil,
Subsoil, and Mine Spoil
(1988) / McIntosh J E

KENTUCKY, UNIVERSITY OF
(continued)

The Effects of Somatropin and Dietary Energy
on Growth, Puberty, and Intermediary
Metabolism in Beef Heifers
(1988) / McShane T M

Effect of Homogenization of (or) Lecithin on
Agglutination of Starter Cultures
(1988) / Milton K J

Mefluidide Effects on Quality and Growth of
Perennial and Annual Warm-Season Grasses
(1988) / Ni M

Manganese Toxicity in Nicotiana Tabacum-L
(1988) / Ogedegbe S A

Prediction of Soybean Rewetting Rates Under
Fully Exposed Conditions
(1988) / Osborn G S

Employment Instability During a Recession in
Selected Mon-Metro Kentucky Counties:
Evidence on Factors Determining Employment
Instability at the Manufacturing Firm Level
(1988) / Peters D R

Hedging with Live Hog Options in the
Kentucky Swine Industry
(1988) / Ponder J

Puffing Diced Green Peppers with Carbon
Dioxide
(1988) / Saputra D

Effect of Protease-Negative Starter Cultures
on Cheese Yield
(1988) / Sauer C J

Effect of Soil PH on Chlorimuron
Phytotoxicity and Persistence
(1988) / Schmitz G L

Effects of Nitrogen Source, Phosphorus, and
Molybdenum on the Mineral Nutrition of
Burley Tobacco, as Dependent on Soil pH and
Nutrient Placement
(1988) / Schwamberger E C

Growth, Carbohydrate Content and Yield of
Sweet Pepper Plants as Affected by Clipping,
Growth Regulators, Temperature Regimes and
Carbon Dioxide Levels
(1988) / Stamper M A

Finite Element Analysis of Air Flow Through
Alfalfa Bales and in Cracks Between Bales
(1988) / Tolzin J L

Metabolic Responses and Nitrogen Metabolism
in the Ovine Administered
Alpha-Ketoglutarate Via Abomasal or
Intravenous Infusion
(1988) / Van Der Veen J L

Variability for the Rate of Recombination in
a Diverse Soybean Population
(1988) / Vogt S

The Influence of No-Tillage on Growth and
Yield of Burley Tobacco
(1988) / Yo F

The Application of the "Techno-Corporate
Complex" Model to the Agricultural Policy
and Defense Policy Realms
(1988) / Young G M

LAKE HEAD UNIVERSITY

Growth of Wild Rice, in Flocculent Sediments
(1988) / Day W R

LOUISIANA STATE UNIVERSITY

Chemical Treatment of Lignocellulosic
Material for Use as Ruminant Feed
(1988) / Agrawal B D

Gypsum Effects on an Alligator Clay Soil
and Sugarcane in Louisiana
(1988) / Buselli E M

A Linear Programming Approach to Local
Optimization of Supplemental Irrigation
Under Limited Water Use Conditions
(1988) / Cardonacastillo H

A Tractor Database System for Dynamic
Modeling
(1988) / Hamman-Costello J

The Optimal Number, Size and Location of
Livestock Auctions in Louisiana
(1988) / Maher D S

An Analysis of Selected Post Harvest
Marketing Strategies for Louisiana Soybean
Producers
(1988) / McCann K L

A Geared Two-Link Harvester Reel
(1988) / Miller M D

Evaluating the Financial Performance of
Agricultural Cooperatives: A
Multidimensional Model
(1988) / Rambaldi A N

Bermudagrass Response to Helminthosporium
Cynodontis: Whole Plant Versus Cell Culture
(1988) / Richardson M D

MAINE, UNIVERSITY OF

An Evaluation of the Return to Agricultural
Research at the Maine Agricultural
Experiment Station
(1988) / Adams G D

The Time-Temperature Effect on Sugar
Migration and Physical Characterization in
Lowbush Blueberries
(1988) / Benner L C

The Estimation of Implicit Border Costs and
Evaluation of Their Effect on the Supply of
Round White Potato Exports from Prince
Edward Island to the United States
(1988) / Carroll J M

Forced Air Drying of Woody Biomass Fuel
Chips Under Ambient or Near Ambient Air
Conditions
(1988) / Drechsel C M

Economic Threshold Model for Postemergence
Herbicide Use in Round White Potatoes
(1988) / Gould T D

A Study of the Optimal Timing of Marketings
and Transport Schemes for a Fresh-Market
Blueberry Cooperative
(1988) / Hoelper A L

MAINE, UNIVERSITY OF
(continued)

The Evaluation of Production Traits in Pure
Population Mating and Reciprocal Crosses of
the American Oyster
(1988) / Scully K C

The Development of a New Pepperoni Using
Mutton and Spent Fowl
(1988) / Stickney M R

Evaluation of Sweet White Lupon Seed as an
Alternative Protein Source for Ruminant
Rations in Maine
(1988) / Tracy V A

Quality Improvement of Sunmi-Based Gel
Prepared from Minced Cod by Means of
Pretreatment
(1988) / Tsai P P

A Survey of Salmonella in Maine Swine
(1988) / Xuan H

MANITOBA, UNIVERSITY OF

The Effects of Selection for Growth Rate and
of Heterosis on Long-Term Reproductive
Performance in Male Mice
(1988) / Boucchart S M

Analysis of Fat, Fatty Acids, and
Alpha-Tocopherol in Mature Human Milk and
Relationships to Donor Diet
(1987) / Britten J R

Factors Affecting the Fluidity of Boar Sperm
Membranes
(1988) / Canvin A T

An Evaluation of the Tripartite Red Meat
Stabilization Program for Manitoba Hog
Procedures
(1988) / Cromwell G H

An Economic Evaluation of Winter Wheat
(1988) / Dornian B L

The Effect of Plant Growth Regulators on
Plant Height, Lodging and Yield in Barley
(1988) / Entz P J

Nutrition-Related Attitudes and Health
Practices of Elderly Women Living Alone
(1987) / Gauthier M J

The Eolf of Noncovalent Forces in the
Micellization Phenomenon Using the
Globulin, Legumin, as a Study System
(1988) / Georgiou C

Blackspot Bruising in Russet Burbank
Potatoes Subject to Impact Loads
(1988) / Ghadge A D

The Effects of Molybdenum and Sulfur on the
Flow and Solubility of Various Minerals
Along the Digestive Tract of Steers
(1988) / Golfman L S

Anthocyanins of the Saskatoon Berry:
Interactions with Physical and Chemical
Parameters and Colour Intensification of
the Pigment Extracts
(1988) / Green R C

Utilization of Whole or Extruded Canola
Seed by Sheep and Dairy Cattle
(1987) / Grumpelt B P

The Economics of Wheat Production in Zambia
(1987) / Kasalu E

Income Stabilization for Grain Farms
(1988) / Kovacs M

Effect of Moldy Mycotoxin-Free Barley and
Ochratoxin-A Contaminated Barley-Based
Diets on Young Growing Pigs
(1988) / Lippold C C

Economic Evaluation of the Use of Ammoniated
Feeds in Beef Cow Rations
(1988) / Melville G

The Economic Potential of U.S. Routes for
the Movement of Grain from Western Canada
to Export Destinations
(1988) / Miller P M

Closed-Loop Blanching of Apple Slices
(1988) / Muller A V

Effects of Enzyme Supplementation on the
Nutritional Value of Barley in Chicken
Diets
(1987) / Neskar M F

Sensory Descriptive Analysis of Guatemalan
Black Beans, and Prediction of Hedonic
Ratings Using Multiple Regression and
Principal Component Statistical Analysis
(1988) / Rios-Sierra B L

Effects of Enzyme Supplementation on the
Nutritional Value of Fababeans in Chicken
Diets as Assessed by Growth and Available
Energy Studies
(1988) / Rodriguez-Castanon J I

Surface Hydrophobicity Manipulation Through
Ammonium Sulfate Concentration
(1988) / Rogers J W

The Utilization of Canola Meal by Young
Growing Pigs
(1987) / Seddon I R

The Use of Selected Plant Growth Regulators
for the Production of Small Whole Seed
Potatoes
(1988) / Shongwe V D

The Influence of Eugenol on the Growth and
Extracellular Enzyme Production of Bacillus
Subtilis
(1988) / Thoroski J H

Market Information: Needs and Sources for
the Manitoga Grain Farmer
(1988) / Timko M L

Utilization of Manitoba Whitefish for the
Fabrication of a Texturized Seafood Analogue
Prototype
(1988) / Tonogai J R

Utilization of Wheat Bran by Growing Gilts
(1988) / Tuitoek J K

Seasonal Changes in the Odor Volatiles
Associated with Microfloral Infection and
Acarine Infestation in Bin-Stored Wheat
(1988) / Tuma D

MANITOBA, UNIVERSITY OF
(continued)

Application of New Methodology to Canola
Protein Isolation
(1988) / Welsh W

Evaluating the Photomethod: A Validity
Study
(1988) / Zacharias E

MASSACHUSETTS, UNIVERSITY OF

A Comparison of Calculated and Analyzed
Amino Acid Values for Commercial Products
(1988) / Blom K J

Flexibility in Fall Harvest Management of
Alfalfa
(1988) / Dempsey M

The Effect of Clostridium Perfringens
Enterotoxin on Growth, Sporulation, and
Macromolecular Synthesis in C. Perfringens
(1988) / Dillon M E

The Effect of Antecedent Wetness on Flow
Instability During Infiltration Into Layered
Soil
(1988) / Edelstein D M

Effects of Nitrification Inhibitors and
Nitrogen Fertilizers on Growth and
Composition of Plants
(1988) / Feng J

The Production of Sodium-Lactate from
Cheese Whey Using a Membrane Cell
(1988) / Fitzpatrick J J

Proteins Produced in Vitro by the Goat
Blastocyst Near the Time of Maternal
Recognition of Pregnancy
(1988) / Gnatek G G

An Economic Analysis of Subsistence Farming
in 18th Century Massachusetts
(1988) / Hurlbutt C L

Aroma Recovery During Evaporative
Concentration
(1988) / Iakovidis A

Effects of Cultured Bovine Uterine
Epithelial Cells on Spermatozoa in Vitro
(1988) / Krunkosky T

Rice Stickiness
(1988) / Lee S J

Extruder Texturization of Rice Flour-Fish
Flour Blends
(1988) / Miscourides D N

Nuclear Size in Relation to Cellular Size
During Oogenesis and Early Embryogenesis
in the Mouse
(1988) / Pandolf T

Effects of LH and Indomethacin on Bovine
Luteal Cells in Long- and Short-Term
Culture
(1988) / Pike A C

A Growth Analysis of the Competitive Posture
of Field Corn with Respect to Fall Panicum
and Large Crabgrass
(1988) / Podmayer S M

Co-Culture of Preimplantation Mouse Embryos
with Oviduct Epithelium
(1988) / Priestley J B

Selective Affinity and Interparticle
Migration in Food Powder Mixtures
(1988) / Sapru V

MINNESOTA, UNIVERSITY OF

Amino Acid Availability in Poultry
Feedstuffs
(1988) / Akavanichan O

The Retention and Storage Stability of
Encapsulated Orange Peel Oil as Influenced
by Spray Drier Operating Parameters
(1988) / Anker M H

West German Agricultural Policy Under the
Common Agricultural Policy of the European
Community
(1988) / Becker D V

Post-Anthesis Storage of Nitrate, Fructan,
and Water-Soluble Carbohydrates in Barley
Stems
(1988) / Benson L L

Microwave-Conventional Heating of Model Cake
Systems Containing Components of Differing
Dielectric Properties
(1988) / Brand B A

Effects of a Bacterial Inoculant on the
Fermentation of High Moisture Shelled and
Ear Corn
(1988) / Faber D A

Genetic and Phenotypic Variation in Milk
Production of Sheep
(1988) / Fecht J W

Economic Development Assistance Under
Eisenhower and Kennedy: Events and Ideas
Leading to the Development Decade of the
1960's
(1988) / Hagen J M

Avoiding the Centering Bias or Range Effect
When Determining an Optimum Level of
Sweetness in Lemonade and a Hedonic Price
Index for Chocolate Chip Cookies
(1988) / Johnson J R

A Method for Determining B-Galactosidase
Activity of Yogurt Culture in Skim Milk
(1988) / Lin W J

Fan Management Strategies for Ambient Air
Drying of Corn
(1988) / Lynch B F J

An Analysis of Links Between Foreign
Agricultural Assistance, Productivity and
International Trade
(1988) / Maginnis H J

A Comparison of Anther-Derived Double
Haploid and Single Seed Descent Lines in
Spring Wheat
(1988) / Mitchell M J

Alfalfa Establishment with Small Grain
Companion Crops: Small Grain Genotypes
with Differing Morphological Characteristics
(1988) / Nickel E

MINNESOTA, UNIVERSITY OF
(continued)

Small Grain Genotypes with Differing
Morphological Characteristics
(1988) / Nickel S E

Phase Resistance in Streptococcus
Cremoris KR4
(1988) / Olander A J

The Farm-to-Retail Price Spreads and
Linkages for Dairy Products
(1988) / Papadas C

Survival of Listeria Monocytogenes at Low
Temperatures in Dairy Systems
(1988) / Petran R L

The Recovery, Survival, and Characterization
of Growth of Listeria Monocytogenes in
Food-Related Systems
(1988) / Petran R L

An Economic Analysis of Hedging Wheat Flour
Purchases and Sales
(1988) / Poirier L A

Genetic and Phenotypic Variation in Growth
and Reproductive Performance of Mouflon
Sheep and Its Crosses
(1988) / Quiggle D W

The Occurrence of Cholesterol Oxides in
Dairy Products
(1988) / Sander B D

Whey Protein and Starch Interaction in
Model Food Systems
(1988) / Schanen P A

Genetic Characterization of a Foodborne
Multiple-Antibiotic-Resistant Salmonella
Typhimurium Strain
(1988) / Schuman J D

Mapping Chicken Chromosomes by In-Situ
Hybridization and G-Banding
(1988) / Shaw E M

An Economic Evaluation of Alternative
Tillage Systems for a Minnesota Dairy Farm
(1988) / Stommes D G

Survival of Salmonella at Low Temperatures
(1988) / Strantz A A

Dairy Cow Sensitivity to Shrot Duration
Electrical Currents
(1988) / Sun Z

Effects of Variety, Calcium Composition, and
Heating Methods on Fracturability of Pootato
Tissues Treated by Prewarming
(1988) / Taguchi M

Reorganization Performance of Local
Agricultural Cooperatives: A Minnesota
Study
(1988) / Taitt J L

The Contribution of World Bank Assistance to
Agricultural and Rural Development in
Pakistan
(1988) / Traxler G J

Developmental Changes in the Carbon Costs of
Dinitrogen Fixation in Legumes
(1988) / Twary S N

Flavor Compounds of Significance on Cheddar
Cheese
(1988) / Vandeweghe P

Aerodynamic Characteristics of Parched Wild
Rice Kernels and Hulls
(1988) / Voehl M R

MISSISSIPPI STATE UNIVERSITY

Sack Drying of Rice, Soybeans, and Corn Seed at
Various Airflow Rates
(1986) / Abdullah W M

Correlation of Titratable Acidity and pH Values at
Various Steps in Cheddar Cheese Manufacture
(1986) / Acevedo B C

Estimation of Suitable Field Working Days for
Mississippi Black Prairie
(1986) / Acharya B P

Computer Analysis System for Bovine Oviduct
Motility Data
(1986) / Adsit P D

The Effects of the Addition of Nonfat Dry Milk
Solids on the Shelf-Life of Fluid Milk
(1986) / Allen W W

The Evaluation of Soybean Meal, Roasted Whole
Soybeans or Whole Cottonseed as a Concentrate
Ingredient for Lactating Dairy Cows
(1986) / Baker J G

Study of the Effect of Fiber Modification of Sorghum
Silage Compared to Corn Silage for Milk Production
(1986) / Balogu D O

The Effects of Culture, Time of Incubation, and Heat
Treatment of Milk on the Number of Viable Cells and
Organoleptic Qualities of Unflavored Yogurt
(1986) / Baril T A

An Economic Analysis of Producing Grapes as a
Commercial Crop in Mississippi
(1986) / Benoist L A

Programming Greenhouse of Floricultural Crops
(1986) / Boyd R M

Economic Analysis of Double-Cropped Soybeans and
Wheat in the Delta Area of Mississippi
(1986) / Boykin B S

An Analysis of Equity Ownership as Related to the
Characteristics of Members of Selected Mississippi
Cooperatives
(1986) / Brown R B

An Economic Assessment of Production Alternatives
Resulting from Changes in the Machinery Complement
of Representative Farms in the Delta Area of
Mississippi
(1986) / Caillavet D F

The Effect of Proteolytic Enzymes in Raw Milk on the
Shelflife of Ultra High Temperature Processed Milk
(1986) / Cartledge M F

Prediction of Soybean Moisture Content After Maturity
(1986) / Chen W S

An Economic Analysis of the Rough Rice Futures
Contract
(1986) / Covey T P

Overhead Labor Cost in the Delta Area of Mississippi
(1986) / Cox L R

MISSISSIPPI STATE UNIVERSITY
(continued)

Programming Container-Grown Woody Ornamental Crops
(1986) / Crafton V W

Design of a Small Experimental Seed and Grain Dryer
(1986) / Davila S I

Market Potential for Farm Raised Catfish for Food
Within Current Market Area and Structure
(1986) / Dixon D A

Influence of Selected Microorganisms on Flavor
Intensity of Butter and Butter-Margarine Mixtures
(1986) / Embarek M

Deriving Use-Value Estimates for Appraisal of Farm
Land in Bolivar County
(1986) / Fitts C E

Optimum Organization of Gins and Warehouses For
Marketing Cotton in the Mississippi Delta
(1986) / Fondren T H

A Microcomputer Program to Evaluate Soybean
Marketing Alternatives
(1986) / Fuller M J

Use-Value Appraisal Analysis of Farmland in
Hinds County
(1986) / Gillis W G

Pre-harvest Variations in Seed and Pod Moisture
Content of Soybeans related to Diurnal Changes in
Relative Humidity and Temperature
(1986) / Gonzalez-Hernandez O

A Digestion Study Involving Six Species of
Wild Carnivore
(1986) / Hamor G A

Effects of Reduced Levels of Sodium Chloride and
the Addition of Potassium Chloride on the
Characteristics of Tumbled Cured Ham
(1986) / Hong D P

Evaluation of Multiple Loading Station Sweet
Potato Transplanter
(1986) / Hsieh J F

The Fiscal Equity of the Mississippi Public
School Finance System-Measurement and Analysis
(1986) / Huddleston K A

The Effect of Exogenous Recombinant or Pituitary
Extracted Bovine Growth Hormone on Performance
of Dairy Cows
(1986) / Hutchison C F

A Model for Administering Food Aid Programs
(1986) / Jabbour B G

Modeling Yield Reduction for Soybean
(1986) / Kimani P K

The Utilization and Energy Value of Wheat for Swine
(1986) / Lapjatupon W

The Structural and Operational Characteristics and
the Procurement and Marketing Practices of the
U.S. Catfish for Food Processing Industry
(1986) / Miller J S

Production and Economic Efficiency of Four Forage
Systems for Lactating Dairy Cows
(1986) / Murphey E J

An Exploration of Financial Liquidity and Solvency
Conditions, and Internal Rate of Return on
Investments at Replacement Cost of a Representative
Delta of Mississippi Farm
(1986) / Negbenebor A I

Alternative Marketing Strategies for Wintergrazing
Feeder Cattle in Mississippi
(1986) / O'Bannon J L

Evaluation of Forage Systems Using Beef Cattle in
Mississippi
(1986) / Omar S

Establishment of Soybean Drying Parameters in
Relation to Initial Moisture Levels
(1986) / Ravandi M A

An Economic Analysis of Soybean Irrigation in the
Mississippi Delta
(1986) / Rebsamen P S

Effect of Microbiological Quality of Raw Milk
on Shelf-Life of UHT Sterilized Milk
(1986) / Rodriguez I R

Influence of Temperature, Protein and Breed on
Digestion by Wether Lambs
(1986) / Rodriquez H E

Effect of Cooking on the Collagen Content of
Chicken Gizzard and Breast Meat Tissues
(1986) / Ruangtrakool B

An Economic Evaluation of the Dynamic Threshold
Concept in the Control of the Heliothis Complex
in Cotton
(1986) / Sanford S O

Viscosity, pH and Acid Development as Affected
by Culture Combinations, Hours of Incubation
and Heat Treatment of Milk
(1986) / Sasso Y

The Effect of Total Mixed Ration vs Computerized
Feeders on Performance of Dairy Cows in Early
Latation
(1986) / Schillings J G

Economic Analysis of Alternative Dairy Feeding/
Management Systems
(1986) / Sewell W H

The Evaluation of Risk Reducing Strategies on Two
Representative Sizes of Mississippi Delta Farms
(1986) / Simmons C H

Comparison of the Airflow-Static Pressure
Relationships When Air is Pushed and Pulled
Through a Mass of Grain
(1986) / Sivira O R

An Evaluation of the Relative Profitability of
Broilers of Various Weights
(1986) / Slice J K

The Influence of Raw Milk Quality on Shelf-Life of
Ultra High Temperature Processed Milk as Determined
by Acid Degree Value
(1986) / Suarez E J

The Growth-Interfering Effects of Spoilage
Microorganisms by Gram-Positive Cocci on
Broiler Carcasses
(1986) / Tanteeratarm K

Evaluation of Aquatic Plant Materials using Lambs
and Laboratory Techniques
(1986) / Tayei S R

MISSISSIPPI STATE UNIVERSITY
(continued)

The Effect of Computerized Feeders vs A Total Mixed
Ration on the Performance of Lactating Dairy Cows
(1986) / Tucker W B

Effects of Reduced Levels of Sodium Chloride and
the Addition of Potassium Chloride on the
Characteristics of Cured Hams
(1986) / Upara U

Lysine and Energy Levels in Corn-Soybean Meal
Diets for Growing-Finishing Pigs
(1986) / WaKamau N

Insemination Management for a One Injection
Prostaglandin F(2) Alpha Synchronization P
otocol: I. Single Versus Double Insemination
(1986) / Wahome J N

Response and Fertility of Dairy Heifers Following
Injection of PGF2 Alpha in Early, Mid or Late
Diestrus
(1986) / Watts T L

The Comparison of a High Urea Liquid Supplement
vs. Cottonseed Meal for Dairy Cows in Early
Lactation
(1986) / Weisenberger T P

A Comparative Study of Regional Blood Flow
Measurement Techniques in the Rabbit Oviduct
(1986) / Whatley S J

Effect of Frame Size on Growth, Performance, Carcass
Traits, Lipogenesis, Muscle Traits, and Carcass
Composition in Beef Cattle
(1986) / Williams J B

Estimated Cost Structure for New and Existing
Specialized On-the-Rail Beef Slaughter Plants
in Mississippi
(1986) / Wilson J C

The Effects of Pen Location, Age at Weaning and
Calcium and Phosphorus Levels on the Performance
of Nursery Pigs
(1986) / Wood C M

Studies on Curing, Cooking and Shelf-life of
Chicken Ham
(1986) / Yang S C

Trade Relations Between the USA and the Ivory
Coast: Potential for Improvement
(1986) / Yapo R A

A Descriptive Analysis of Venezuela's Cereal
Grain Market
(1986) / de Sanchez Y B

MISSOURI, UNIVERSITY OF (COLUMBIA)

The Effect of Iron Supplements on the Excretion
of Copper, Zinc, and Iron as Measured by ICP-MS
(1988) / Beltran M F

Comparison of the Chelation of Iron-59 by
Soyproteins and Myofibrillar Proteins, In vitro
(1987) / Blase R J

Effect of Pre-Slaughter Fasting and Transportation
on Live Hog and Carcass Characteristics
(1988) / Bryan R W

Inhibition of Warmed-Over Flavor Using
Maillard Reaction Products in Beef Roasts
(1988) / Choate G K

Crop Management Effects on Grain Sorghum Growth
and Development, Grain Yield, and Grain Dry-Down
(1988) / Diaz O H

The Effects of Quality of Diet on Labor
Productivity and Population Growth in India
(1988) / Errass C F

Alternative Marketing Strategies for Missouri
Slaughter Hog Producers
(1988) / Gure M M

Effect of Cookery Method and Endpoint Temperature
on Sensory Characteristics, Nutritive and Shear
Values of Fresh Pork Cuts
(1988) / Karrasch M A

Conversation Easements: An Analysis of Landlords'
Response to a Multi-Period Land Retirement Program
(1988) / Kula O

Effect of Extracellular Nucleotides on Inositol
Trisphosphate Formation in Bovine Pulmonary
Artery Endothelial Cells
(1988) / Landis D M

Intensive Management Practices in Soft Red
Winter Wheat in Missouri
(1988) / Lavado J M

Characterization of Cheddar-Cheese-Associated
Non-Starter Lactobacilli and Their Role in
Proteolysis During Cheddar Cheese Ripening
(1988) / Peterson S D

Zinc and Alcohol Relationships During Pregnancy
in Sows and Offspring
(1988) / Ratliff V L

Megagametophyte Development in Lotus Corniculatus, L.
Conibricensis, and their Protoplast Fusion Hybrid
(1988) / Rim Y W

Effects of the Addition of 20mM Hepes Buffer to
Freezing Media upon Physical Condition and
Culturing Survival after Freezing of Mouse Embryos
(1988) / Slapak J R

Influences of Heat Stress and Related Endocrine
and Uterine Environmental Changes on Embryonic
Survival
(1988) / Spencer-Johnson K J

Economic Factors Involved in the Importation of
Feed Grains by South Korea
(1988) / Suh M W

The Influence of Wet and Dry Feed on Egg
Production Under Heat Stress
(1988) / Tadtiyanant C

MONTANA STATE UNIVERSITY

Optimal Crop Sequences to Control
Cephalosporium Stripe in Winter Wheat
(1987) / Danielson J G

A Comparison of Alternative Production
Function Models Using Non-Nester
Hypothesis Tests
(1987) / Embleton M E

The Impact of Expenditure Category Exemption
on the Incidence of a Retail Sales Tax
(1987) / Judd F A

The Small Sample Properties of a Nonstandard
Estimator in the Context of First Order
Autocorrelation
(1987) / Siebrasse P B

MURRAY STATE UNIVERSITY

Comparison of the Effects of Two Types of
Forced Exercise and Two Stabling Practices
on the Muscular Development of Yearling
Horses
(1988) / Arrigon J R

Cracked Corn-Soybean Meal Versus Whole
Shelled Corn-Supplement Diets for Background
Beef Calves: Performance and Economics
(1988) / Asssadi-Rad A M

Comparison of Broadleaf Weed and Grass
Control of Two Dinitroanalines and Their
Safety When Applied to Kentucky Bluegrass
(1988) / DeHaven C W

Comparison of Performance of Cows/Calves
Grazing Kentucky-31 or Johnstone Fescue
(1988) / Jazi S H

A Comparison of Selected Factors Affecting
Percent Protein of Milk and Recommendations
for Pricing Milk Based on Protein Content
(1988) / Jennings M P

A Methodology for Measuring and Managing
Quality in a Food Service Activity
(1988) / King A A

Comparison of Crystalyx to Corn Silage or
Commercial Range Cubes for Supplement
Wintering Beef Cows
(1988) / Lawson G S

A Comparison of the Effects of Whole Shelled
Corn and Cracked Corn in Finishing Rations
on Beef Cattle Performance and Carcass
Characteristics
(1988) / Smith S L

A Direct Marketing Study: Henry County
Farmers' Market
(1988) / Weatherford J W

NEBRASKA, UNIVERSITY OF

Water and Nitrogen Relations of Alfalfa,
Corn, and Soybean on Farms Using Low-Input
Management Systems
(1988) / Aiken R M

The Occurrence and Properties of Xanthomonas
Compestris PV. Phaseoli (SMITH) Dye and
Other Xanthomonads on Symptomless Weeds
(1988) / Angeles-Ramos R

A Comparison of Costs of Different Turkey
Feed Purchasing Strategies
(1988) / Bachenberg T C

The Influence of Planting Data, Seeding
Rate, and Phosphorus Application Rate on
Yields and Growth of 'Brule' Wheat
(1988) / Blue E N

Perceptions of Iowa and Nebraska Legislators
Regarding Secondary and Adult Agricultural
Education Programs
(1988) / Burger B

The Effects of Alternative Forms of Risk and
Uncertainty on Regional Corn Supply
Functions
(1988) / Chuang C

Carbon Dioxide and Energy Exchanges of
Cereal Crops
(1988) / Clement R J

Sulphur Nutrition and the Effect of N/S
Ratio on Poinsettias
(1988) / Dale M E

Plant Growth Regulator Effects on Kentucky
Bluegrass
(1988) / Doyle J M

Hydraulic Roughness Coefficients as Affected
by Random Roughness
(1988) / Finkner S C

Propagation, Establishment, and Ecological
Characteristics of Penstemon Hydenii S.
Watson
(1988) / Flessner T R

Genetic and Environmental Effects on
Incidence and Causes of Lamb Mortality
(1988) / Gama L T

Grain Hybrids and Starch Utilization in
Ruminants
(1988) / Gramlich S M

Soil Water Content Measurements Using
Fiber-Optics
(1988) / Ku Y

Performance of Seedigated and
Conventionally Planted Crops
(1988) / Leander W S

Parental Perceptions of and Involvement in
Nebraska Secondary Agriculture Education
Programs
(1988) / Lechner M A

Economic Factors that Affect Groundwater
Irrigation Development in Nebraska
(1988) / Mckenzie G D

Millet-Soybean Rotation and Nitrogen
Fertilizer Effects in Millet Productivity
(1988) / Mohamed M S

The Effect of Trim Level, Cooking Method and
Chop Type on Lipid Migration and Retention,
Caloric Content and Cholesterol Level in
Cooked Pork
(1988) / Morgan J B

Identification and Evaluation of Molds
Associated with Blue-Eye Condition in
Stored Popcorn
(1988) / Nelson M J

The Effect of Socio-Political Culture on
National Development Planning in Nigeria
(1988) / Ogbenta A O

Striking the Nerve of a Theater: Design of
a Symphonic Hall
(1988) / Pierrottet T M

Agrometerological Inputs Into the Management
and Production of Rice in a Developing
Nation
(1988) / Pollonais S R

NEBRASKA, UNIVERSITY OF
(continued)

The Role of Roughage in Feedlot Diets
(1988) / Poppert G L

Crop Response to Root Plowing Along
Multi-Row Field and Border Windbreaks
(1988) / Rasmussen S D

The Relationship Between Calcium Dependent
Protease I, II, CDP Inhyibitor and Growth
Rate in Beef Cattle
(1988) / Rhynalds C D

Tillage, P Source and Method, and
Cultivation Effects in Continuous Irrigated
Corn Cropping Systems
(1988) / Roberson K G

Estimation of Corn Canopy Temperature and
Water Budget Using Automated Weather Station
Data
(1988) / Sagar R M

Implementing the Resource Protection
Planning Process (RP3) at the Nebraska State
Historic Preservation Office: An
Operational Model for Historic Contexts
(1988) / Sawyers S C

Optimal Deficit Irrigation Management
(1988) / Severin M A

Effects of Tillage, Corn and Wheat Residue,
and Application Time on Metribuzin Efficacy
and Dissipation in Soil
(1988) / Sorenson B A

Farm Level Risk Analysis Using a Discrete
Stochastic Programming Model
(1988) / Spilker M F

Mineralogical Implications of High Ratios
of 15 Bar Water Content to Percent Measured
Clay for Two Soils of Western Nebraska
(1988) / Stolpe N B

A Multiplicative Selection Index Applied to
Nine Cycles of Full-Silicon Boron Recurrent
Selection in Maize
(1988) / Stromberg L D

New Towns in Malaysia: A Study of Planned
Integration and Interaction in a Multiracial
Society
(1988) / Sulaiman S A

Farm Size and Economic Efficiency in a
Sample of Dryland Farms in Morocco: A
Profit Function Approach
(1988) / Tamehmacht Z

Hybrid Expert System of Beef-Forage Grazing
Systems
(1988) / Tao Y

Emergence, Development, and Yield of Surface
Seeded Corn Under Irrigated Conditions
(1988) / Thrailkill D J

Parameter Estimates from Two Replicated
Recurrent Selection Procedures in Maize
(1988) / Tragesser S L

High Density Sorghum Production for Late
Plantings in Eastern Nebraska
(1988) / Villar J L

Effective Cooperation Among Grain Marketing
and Farm Supply Cooperatives
(1988) / Vossler K J

NEW MEXICO STATE UNIVERSITY

Mineral Content and Nutritive Value of
Alfalfa Hay
(1988) / Ali A M

Farm Tasks, Responsibilities and
Preferences: The Case of Women in Dona
Ana County, New Mexico
(1988) / Alvarez R C

A Wheat Support Buy-Out in Guatemala
(1988) / Armas-Marroquin L

Comparison of Natural Service and Artificial
Insemination in Superovulated Ewes
Inseminated with Intrauterine Methods
(1988) / Baezz-Kohn A R

Expression of the Bacillus Thuringiensis
Var. Kurstaki HD-73 Delta-Endotoxin Gene in
Transgenic Tomato
(1988) / Bailie S E

Effects of Training on Muscle Fiber Types
and Blood Serum Parameters in Two-Year Old
Quarter Horse Mares
(1988) / Barnett B R

An Economic Comparison of Livestock
Enterprises in New Mexico, 1986
(1988) / Brockman B A

Milk Production Estimates and Cow
Productivity of Two-and Three-Breed
Rotational Crossbreeding Systems Utilizing
Brangus, Hereford, Charolais and Simmental
Cattle Under Semidesert Range Conditions
(1988) / Chabo R G

Undergrowth Response Following Manual
Control of Daniela Oliveri/Detarium
Microcarpum Woodland at the N'Dama Cattle
Ranch Unit of Madina Diassa (Republic of
Mali)
(1988) / Dakono J

Effects of a Monensin Ruminal Delivery
Device on Forage Intake and Ruminal
Fermentation in Steers Grazing Irrigated
Winter Wheat Pasture
(1988) / Davenport R W

Allium Species and Cultivar Screening for
Pink Root Resistance Across Isolates and
Inoculum Levels
(1988) / Del Cid A R

The Economics of Controlling Black Grass
Bugs in New Mexico
(1988) / Garrett K M

Economics of Big Sagebrush Control in the
Colorado Plateau
(1988) / Gordon H W

A Survey of Twenty-Four Grazed and
Nongrazed Areas in Southern New Mexico
(1988) / Herman H J

An Analysis of Collateral-Based Hedging
Strategies for Commercial Bank Loan
Portfolios
(1988) / Jones B D

NEW MEXICO STATE UNIVERSITY
(continued)

Linking Forecasting and Least Cost
Rationing for Management Option Analyses
in New Mexico Beef Cattle Industry
(1988) / Keita C S

Elimination of ^{14}C-Heptachlor from Body
Burdens of Sheep
(1988) / Khan M F

Investment Decisions and Crops
Diversification as a Risk Reduction
Strategy for a Commercial Farm
(1988) / Konchou L

Phyto-Estrogenic Effect of Alfalfa Hay and
Alfalfa Pasture on the Reproductive
Performance of Yearling Ewes
(1988) / Matanis M L

Gamma-Irradiation of Crucifer Seed: Effects
on Glucosinolates and Major Fatty Acids
(1988) / Maxwell C J

Economic Evaluation of Alfalfa Production
Under Less than Optimum Irrigation Levels
(1988) / Melaku E

Potato Yield as Affected by Soil Cover
Practice
(1988) / Mendoza-Salcido F

Cost and Returns of Producing Apples in
North Central New Mexico Based on a
Probability Disstribution of Yields: A
Microclimatic Computer Simulation Model
Approach
(1988) / Minja P K

Growth Response of Grass Species to Grazing
Intensities in a Savanna of Eastern Botswana
(1988) / Mphinyane W N

Effects of Zeranol and Ovine Growth Hormone
on Growth, Endocrine Responses and Carcass
Characteristics in Fine-Wool Lambs
(1988) / Olivares V H

Effects of Elephant Butte Irrigation
District Water Rights on Land Values in
Las Cruces, New Mexico, 1983-1987
(1988) / Pritchard R

Effect of Non-Phenolic and Phenolic Browse
Plants on the Nutritive Value of Angora Goat
Diets
(1988) / Saiwana L L

Effect of Price Support Policy on the Type
of Cotton Produced in New Mexico
(1988) / Samatana M

Rooting Effects of Four Levels of
Indole-3-Butyric Acid, Picloram and
Paclobutrazol Upon Hardwood, Softwood and
Semi-Hardwood of Hybrid 1613 and Saltcreek
Grape Rootstocks
(1988) / Santizo Flores L E

Shifting Demand for Meats in Mali
(1988) / Sidibe I

Urea Foliar Fertilization on Cotton at Two
Soil Nitrogen Fertilization Rates
(1988) / Solares-Pareja G E

Growth, Endocrine Profiles and Reproductive
Responses of Fine-Wool Ewe Lambs Treated
with Ovine Prolactin Before Breeding
(1988) / Spoon R A

Milk Production in the United States:
Interrelationship Between Allocative
Efficiency and Quality Factors
(1988) / Tandia S M

Economic Feasibility of Cotton Ginning
Technology Alternatives
(1988) / Tjirongo M T

Incidence and Impacts of Locoweed Poisoning
in New Mexico
(1988) / Tomalino L M

Reproductive Performance of Finewool Range
Ewes Treated with Syncro-Mate B
(1988) / Tubbs L W

Salinity-Enhanced Growth of Russion-Thistle
(1988) / Valenzuela-Vazquez M

Comparative Food Habits of Gila Trout and
Speckled Dace in a Southwestern Headwater
Stream
(1988) / Van Eimeren P A

A Computer Approach to Trading Commodity
Futures
(1988) / Ward J B

NEW YORK, STATE UNIVERSITY OF (FREDONIA)

Molecular Cloning and Overexpression in
E. Coli of Nif-H Gene from Anabaena
(1988) / Donahue J G

The Effects of Seed Mass and Emergence Time
on Seedling Growth in Four Aesculus Species
(1988) / Smith P

The Effects of Mechanical and Chemical
Control of Aquatic Macrophytes on the
Littoral Zone Fishes in Chautaqua Lake,
New York
(1988) / Somerville M

NORTH CAROLINA STATE UNIVERSITY

The Influence of Poultry Species, Muscle
Groups and Sodium Chloride Level on
Strength, Deformability and Water Retention
in Heat Set Muscle Gels
(1988) / Amato P M

Nonadditive Genetic Effects on First
Lactation Milk Yield of Holstein Cows
(1988) / Choi Y S

Social Coat Analysis of Alcohol Production
from Argentine Sugarcane
(1988) / Gargiulo C A

A Plant Color Scale and Its Use in a Study
of the Effects of Cultivar and Topping Stage
on Indicators of Maturity and Agronomic
Traits in Burley Tobacco
(1988) / Kelley W T

Inheritance of Fatty Acid Content in Peanut
(1988) / Mercer L C

NORTH CAROLINA STATE UNIVERSITY
(continued)

Broiler Performance Comparing Bacitracin
Methylene Disalicylate and
3-Nitro-4-Hydroxyphenylarsonic Acid with
Virginiamycin in the Presence of the
Coccidiostat Halofuginone Hydrobromide
(1988) / Nazario C A

Effects of Four Growth Regulators on the
In-Vitro Culture of Peanut Embryos
(1988) / Reece P

NORTH DAKOTA STATE UNIVERSITY

The Effect of Physiological Age, Sex Class and
Marbling Class on Cholesterol Content of Beef
Longissimus Muscle
(1988) / Berg P M

Effect of Dietary Fiber on Swine Prolificacy
(1988) / Carter D I

Intermarket Wheat Spreads
(1988) / Chan A K

Optimal Trade Patterns for Wheat, Corn and Soybeans
Under Changing Competitive Advantage
(1988) / Drennan R T

Effects of Soilborne Fungi on Germinability, Medulla
Color, and Integrity of Sclerotia of Sclerolia
Sclerotia (Lib.) de Bary
(1988) / Duval D L

Selecting Crop Production and Marketing Plans to
Minimize Risk for Farmers in Southeast Central
North Dakota
(1988) / Elhard E A

Comparative Agronomic and Quality Characteristics
of Two Seli of Wheat Hybrids
(1988) / Eriksmoen E D

Adjuvants and Grass Control Herbicides Applied
with DPX-M6316
(1988) / Feist D A

Sharp-Tailed Grouse Nesting and Brood Rearing
Habitat in Grazed and Nongrazed Treatments in
Southcentral North Dakota
(1988) / Grosz K L

Quality and Price Competition in United Kingdom
Wheat Import Market
(1987) / Heilman R G

R-nj Aleurone Color Selection in Two Maize
Synthetics and Their Reciprocal Hybrids
(1988) / Houghton R D

Foliar Applied UAN Fertilizer, Tilt Fungicide,
and Cerone Plant Growth Regulator as Related to
Intensive Management of Hard Red Spring Wheat
(1988) / Jenny R D

Models of Net Blotch Severity and Yield Loss in
Barley
(1988) / Jin Y

Forecasting Cattle Price Trends
(1988) / Kinnischtzke J N

Drying Energy Requirements and Breakage of Northern
Bred Corn
(1988) / Koeckeritz A R

The Role of Plant Hormones and Population Dynamics
of Corynebacterium Sepedonicum on the Development
of Potato Ring Rot
(1988) / Kurowski C J

Tank-Mixes of Agricultural Chemicals
(1988) / Pahl S J

Studies on the Cellular Proteins of Sporidia from
Four Races and Two Mating Type of Ustilago Hordei
(1988) / Park H

Grain Sorghum Production as Influenced by Row
Spacing and Plant Population
(1988) / Schatz B G

Use of Straw and Alfalfa to Vary Energy Level for
Confined Ewes During Periods of Low Energy
Requirements
(1987) / Schmidt J T

Response of Western Snowberry Dominated Communities
to Grazing in Southcentral North Dakota
(1987) / Sturn G M

Leafy Spurge Control and Soil Residue With
Bulfometuron
(1988) / Swenson O R

Role of Angiogenic Factors in Ovarian Function
(1988) / Taraska T

Analysis of Importers' Purchasing Behavior Toward
Major Wheat Exporters
(1988) / Tedros Y M

Clopyralid Persistence in Soil
(1987) / Thorsness K B

Effect of Sunflower Cytoplasms on Male Fertility
Restoration, Agronomic, and Morphologic
Characteristics
(1987) / Wolf S L

Effects of Bacteria on Germination and Degradation of
Sclerotia of Sclerotinia Sclerotiorum (Lib.) DeBary
(1988) / Wu H

NORTHEASTERN LOUISIANA UNIVERSITY

Soil Amendment Soybean Seedling Rhizosphere
Conducement
(1988) / Claverie E F

High-Performance Liquid Chromato-Graphic
Separation and Quantitation of Soybean
Leghemoglobins Using an Acetate Gradient
(1988) / Roberts E A

NORTHERN IOWA, UNIVERSITY OF

Permeability of Isolated Leaf Cuticles of
Lemon and Pear to Acids
(1988) / Hauser H D

NOVA SCOTIA, TECH UNIVERSITY OF

Network Model for the Selection of Dairy
Waste Management Systems
(1988) / Hengnirun S

OREGON STATE UNIVERSITY

Computer simulation of Transient Refrigeration
Load in a Cold Storage for Apples and Pears
(1987) / Adre N

OREGON STATE UNIVERSITY
(continued)

Influence of Cattle Grazing and Forage Seeding
on Establishment of Conifers in Southwest Oregon
(1987) / Alejandro-Castro M

Cordeauxia Edulis: Production and Forage Quality
in Central Somalia
(1988) / Ali H M

Balanced Structural Cross-Sections of the Central
Salt Range and Potwar Plateau of Pakistan:
Shortening and Overthrust Deformation
(1987) / Baker D M

Possible Association Between Dwarfing Genes
Rht(1) and Rht(2) and the Reaction to Septoria
Tritici Blotch in Winter Wheat
(1987) / Baltazar B

Cultural Evolution and Small-Scale Farming
(1988) / Caday P P

Cooling and Mass Transfer Rates in Fresh Fruits
(1988) / Chadwick J M

Urease Induction in Barley Leaves by Foliar
Application of Urea
(1987) / Chen Y

The Effects of Density and Proportion on Spring
Wheat and Lolium Multiflorum Lam
(1987) / Concannon J A

Effects of nitrogen and Storage Time on the
Quality of Highbush Blueberry Fruit
(1987) / DeFrancesco J T

Factors Affecting Selective Weedy Rye (Secale Sp.)
Control in Winter Wheat with the Herbicide Ethiazin
(1987) / Diener P R

Determinants of Off-Farm Labor Supply Among Farm
Households in the North Willamette Valley
(1987) / Doyle D J

Seed Depth Influence on Position of the Growing
Point and Chemical Control of Wild Proso Millet
(1987) / Fernandez Mendez R J

Response of Cuscuta Campestris and Cuscuta
Indecora to Glyphosate Applied After
Attachment to Alfalfa
(1987) / Fessehaie R

The Effect of Various Management and Policy
Options on the Financial Stress Situation
of Oregon Grain and Cattle Producers
(1987) / Hewlett J P

Demand Analysis for Surimi-Based Products in Japan
(1988) / Kim S W

The Effect of Temperature on Biochemical Activities
of Embryo and Seedling of Germinating Rice Seed
(1987) / Luo S

The Economic Impact of Nonearnings Exports on
Residentiary Sectors for Rural Oregon Counties,
1979-1984
(1987) / McLeod D M

Interrelation of the Components of Grain Yield
in a Cross Between Dwarf and Semidwarf Wheat
Cultivars
(1988) / Medina L E

Assessment of Alternative Raw Product Valuation
Methodology with Respect to Cooperatives Single
Pool Returns
(1987) / Meyersick R R

The Conservation Reserve Program and Its Impact
on the Economies of Rural Communities
(1987) / Nofziger S D

Evaluating Grain Yield as Influenced by Biological
Yield and Harvest Index in Four Winter Wheat
Crosses Involving Near-Isogenic Lines
(1988) / Rahman M M

Beans, Cabbage, and Sugar Beets in a Chemically
Suppressed Sod of Manhattan II Perennial Ryegrass
(1987) / Rinehold J W

Evaluation of Rhizobium Strains with Three
Mediterranean Forage Legumes for Biological
Nitrogen Fixation
(1987) / Streeter D J

The Use of Remote Sensing in an Integrated
Approach for Stress Detection on the Cultivated
Cranberry
(1988) / Thoma M N

Performance of Cultivars, Hybrids and Composites
of Winter Wheat Grown at Three Locations
(1987) / Verges R P

PENNSYLVANIA STATE UNIVERSITY

Nutrient Dynamics of a Southeastern
Pennsylvania Dairy Farm
(1987) / Bacon S C

Environmental Influences on Corn Stalk
Nitrate Levels Used for Tissue Testing
(1988) / Bamka W J

The Effect of Calcium Chloride Added to the
Irrigation Water on the Shelf Life and
Quality of Harvested Mushrooms
(1987) / Barden C L

Long-Term Effects of Treated Domestic
Wastewater on Growth, Histopathology, and
Swimming Performance of Brown Trout
(1987) / Benson A J

Energy Input to Small Scale Microprocessor
Controlled Mash Production Processes
(1988) / Berg M P

Effect of Soil Tillage on Residual Nitrogen
Contribution by a Red Clover Green Manure
Crop to Subsequent Corn Crops
(1987) / Craig P H

Addition of Rumen-Protected Methionine and
Lysine to the Diets of Lactating Holstein
Cows: Nitrogen Status, Milk Production and
Composition, Plasma Amino Acids and Growth
Hormone
(1987) / Donkin S S

User Charges in Pennsylvania Local
Government
(1987) / Harp A J

The Cryopreservation of Murine and Bovine
Embryos Using Trehalose in Combination with
Glycerol
(1987) / Honadel T E

PENNSYLVANIA STATE UNIVERSITY
(continued)

Effects of Concentrate Feeding on Carcass
Characteristics and Cooked Beef
Acceptability
(1987) / Kasubick S

Sulfate Retention in a Pennsylvania Soil
(1987) / Kaufman R M

The effect of Initial Soil Moisture Content
and Soil Drying on the Susceptibility of
Ammonia Volatilization Loss from Surface
Applied Urea
(1987) / Kern J

Kinetics of Shrinkage of Mushrooms During
Blanching
(1987) / Konanayakam M

Economic and Sociological Determinants of
Married Malaysian Women's Migration
Choices: The Case of Conjugal Separation
(1987) / Le Clere F B

The Effect of Heat Processing of Dried Egg
White on Iron Bioavailability and
Iron-Protein Interactions
(1988) / Leahey J M

A Feasibility Study of Central Manure
Management for the Rural Clean Water Project
of Pennsylvania
(1987) / Lengerich E J

Evaluation of Cold Acidified Milk Replacer
Fed Ad-Libitum to Dairy Calves
(1987) / Magliaro A L

Comparison of the Effects of Water-Cooled
and Air-Cooled High Pressure Sodium Lamps
on Lettuce Growth
(1987) / Mankin K R

Determinants of Household Food Packaging
Choices and Solid Waste Generation: A
Household Garbage Analysis
(1987) / Mauger P C

The Characteristics of a Stationary Free
Floating Nitrogen Discharge Generated in a
Microwave Resonant Cavity
(1988) / Maul W A

Coal-Water Fuel Combustion in Oxygen
Enriched Atmospheres
(1987) / Mc Ilvried T S

The Effect of Monensin on Growth,
Reproductive Performance, and Body
Composition of Holstein Heifers
(1988) / Meinert R A

Pennsylvania Dairy Farmers' Preferences for
Milk Quotas
(1987) / Parsons R L

An Examination of the Lackawanna County 4-H
and Family Ski Club to Evaluate the Effect
of Family Recreation on Family Relations
(1987) / Pencek W G

Chemical Growth Regulation of Tall Fescue
on Pennsylvania Roadsides
(1987) / Prinster M G

Telemetry Device for Measuring the Damage
Potential of Harvesting and Handling
Equipment
(1987) / Welling R L

Real Property Tax Equity in Pennsylvania: A
Comparison of Farm and Non-Farm Tax Burdens
(1988) / Wollover D R

Effects of Artificial Selection and Planting
Date on Flowering, Yield, and Survival in
Red Clover
(1988) / Zabalgogeazcoa I A

The Survival of Clostridium Sporogenes and
Clostridium Perfringens in Thermal-Processed
Vacuum-Packaged Turkey Breast Meat
(1988) / Zwally J K

PURDUE UNIVERSITY

Influence of Melatonin on Serum
Concentrations of Progesterone During the
Luteal Phase of the Estrous Cycle and
During Early Pregnancy of Mature Ewes
(1988) / Aguiar L M

Nutrient and Phytotoxic Contributions of
Residues to Soil in No-Till Continuous
Corn Ecosystems
(1988) / Breakwell D P

Effects of Selection for 365-Day Weight on
Carcass Characteristics in Beef Cattle
(1988) / Bryant D E

Influence of Within-Row Spacing on
Morphology and Forage Quality of Diverse
Sorghum Genotypes
(1988) / Caravetta G J

Milk Production in Serra-da-Estrela
Sheep: A Study of Environmental Factors,
Lactation Curve Characteristics and
Genetic Parameters
(1988) / Delgado F J

Nonenzymatic Browning in Skim Milk During
Dehydration
(1988) / Franzen K A

Influence of Ethephon on Stalk Strength
Components and Grain Yield in Corn
(1988) / Hall T E

Influence of Rotation and Tillage on Soybean
Infection by the Northern Stem Canker
Pathogen and Other Diaporthe/Phomopsis
Biotypes
(1988) / Holland G J

A Computerized Decision Aid for Hedging: An
Analysis of Spread and Basis Behavior
(1988) / Holzer J L

Fat Utilization by Young Pigs
(1988) / Howard K A

Methods of Examining Intrademic and
Interdemic Competition in Tribolium
Castaneum Using Group Selection
(1988) / Imel J D

Evaluation of the Agronomic Performance and
Food Quality Characteristics of Experimental
Sorghum Hybrids in Niger, West Africa
(1988) / Kapran I

PURDUE UNIVERSITY
(continued)

Establishment of Late-Summer Seeded Alfalfa
as Influenced by Seeding Date, Cultivar, and
Inclusion of Orchardgrass
(1988) / Kramer F D

Detection of 3-Methyl-Indol in Boar Fat and
Blood Serum Using High Performance Liquid
Chromatography
(1988) / Lin R S

Chalcone Synthase and Its Role in
Phytoalexin Synthesis in Sorghum
(1988) / Lue W L

Row, Broadcast and Hill or Strip Placement
of P-Fertilizer on Millet and Rice in Niger
(1988) / Mahaman M I

The Development of a Corn-Nitrogen-Weather
Model for Use in Studying Corn Producers'
Nitrogen Management Decisions
(1988) / Merz L D

Long Term Selection Response for Pupal
Weight in Tribolium Castaneum
(1988) / Miles D A

Microbial Ecology of Topdressing and
Topdressing Created Layers of Creeping
Bentgrass Golf Greens
(1988) / Robbeloth E K

Vegetative Propagation to Utilize Heterosis
in Lima Beans
(1988) / Schoneweis S D

Developmental Study of a Hard, Intermediate,
and Soft Variety of Sorghum
(1988) / Shull J M

Calculated Soil Moisture Regimes of the
United States and Northern Africa
(1988) / Toure A

Heterotic Pattern and Combining Ability for
Agronomic and Food Grain Quality Traits in
Exotic x Exotic, Exotic x Intermediate and
Exotic x Local Sorghum Hybrids in Niger
(1988) / Tyler T A

The Diversification Potential for
Traditional Grain Operations Into Vegetable
Crops in Southern Indiana
(1988) / Whipker B E

Growth and Reproductive Performance of Ewe
Lambs Supplemented with Corn or Soybean Meal
While Grazing Pasture
(1988) / Yoder R A

Elongated Fish Chromosomes Prepared from Fin
Tissue Culture
(1988) / Zhuo L

RENSSELAER POLYTECHNIC INSTITUTE

An Urban Form for the Irish Climate
(1988) / Tobin J

RHODE ISLAND, UNIVERSITY OF

Influence of Tillage Systems and Winter Cover
on Off-Site Losses of Sediment, Nutrients and
Atrazine in Silage Corn Production
(1987) / Arcieri W R

Nutrient Density of Diets of Elderly
Persons Living in Rhode Island
(1988) / Ash S A

Diepoxides of Beta-Carotene and Their Provitamin
A Activity
(1988) / Heinonen M

Effect of Pseudomonas Syringae on the Nucleation
Temperature of Food Material
(1988) / Izquierdo M P

Optimization in the Formulation of Surimi-Based
Extruded Products
(1987) / Jon-Shang C

Cell-Free Filtrate for Xanthan Gum Production
(1987) / Tabatabaie M

A Study of the Production of Cell-Free Xanthan
Gum in a Fed-Batch Fermentation System with
Tangential-Flow Filter
(1987) / Talebi F

PCB and Heavy Metal Residues Survey in Winter
Flounder from Three Selective Sites in
Narragansett Bay and Vicinity
(1988) / Wang J S

SASKATCHEWAN, UNIVERSITY OF (SASKATOON)

Heritability of Hull Peeling in Two-Row
Barley
(1988) / Aidun V L

A Study of the Green-Ampt Infiltration
Equation and Its Performance in Modelling
Intermittent Application of Water
(1988) / Aliwa N J

The International Effects of the Western
Grain Stabilization Program
(1988) / Cameron D L

A Prior Loan Classification for
Agricultural Credit in Thailand
(1988) / Chantakasorn W

Valuing Tradeoffs Between Soil Quality and
Net Returns in Saskatchewan Using Stochastic
Dynamic Programming
(1988) / Chinthammit D

A Dynamic Model of Co-Operative Growth
(1988) / Corman J P

Resistance of Bulk Lentil Seeds to Airflow
(1988) / Falacinski A A

Active Immunization Against Gonadotrophin
Releasing Hormone in Sheep and Cattle
(1988) / Goubau S M

An Economic Analysis of the Effects of
Domestic Transportation Subsidies and
Foreign Tariffs on the Canadian Flaxseed
Industry
(1988) / Jack L A

The Influence of Various Cultural Factors on
Anther Culture of Spring Wheat: Cultivars
Katepwa and Columbus
(1988) / McGregor L J

An Investigation Into Future Directions for
Electronic Identification of Livestock
(1988) / McLean J W

SASKATCHEWAN, UNIVERSITY OF (SASKATOON)
(continued)

Effect of Application Method on Herbicide
Efficacy
(1988) / Mkunga M J

An Economic Analysis of the Lesotho Daury
Industry: A Spatial Equilibrium Model
(1988) / Mochebelele M T

An Economic Analysis of the Lesotho Wool
and Mohair Marketing System
(1988) / Mokitimi N L

Protein Quality Evaluation of Indian
Cottonseed, Groundnut and Mustard Seed
Meals Using Mice
(1988) / Musonda N M

Germination and Establishment of Perennial
Grasses Under Water Stress
(1988) / Qi M Q

Analysis of Protein and Tannin Content in
Sorghum Using Near Infrared Reflectance
Spectroscopy
(1988) / Shayo N B

Utilization of Ammoniated and Inoculated
Barley Silage by Lactating Cows
(1988) / Undi M

The Incentive for Horizontal Merger in the
North American Farm Machinery Industry
(1988) / Vercammen J A

The Economics of Soil Erosion in
Saskatchewan: A Stochastic Dynamic
Programming Approach
(1988) / Weisensel W P

An Econometric Analysis of the Canadian
Sheep and Lamb Industry
(1988) / Yeung Y L

Tractive Performance and Fuel Economy Study
of a Four-Wheel Drive Tractor
(1988) / Zhang Q

SOUTH CAROLINA, UNIVERSITY OF

Aerial Insecticide Losses Over Conventional
and Nonconventional Tillaged Tomato Plots
(1988) / Carver D R

SOUTH DAKOTA STATE UNIVERSITY

Hedging Prime Interest Rate and Bank for
Cooperatives Rates by Using Financial
Futures for South Dakota Agricultural
Cooperatives
(1988) / Chan L K

Inheritance of Nine Enzyme Loci in Sunflower
and Estimates of Outcrossing Rates in
Helianthus Germplasm Pool II
(1988) / Gerdes J T

Characteristics of Milk and Butter from Cows
Fed Sunflower Seeds
(1988) / Middaugh R P

South Dakota Elevator's Winter Wheat
Marketing Practices and Seasonality of Cash
Wheat Prices
(1988) / Ober B M

Zone and Drip Cooling Comparison for
Lactating Swine
(1988) / Raap D L

Asset-Liability Management/Interest Rate
Risk of Banks in South Dakota
(1987) / Reis P

Significance of Exchangeable Soil Sodium on
Implement Draft
(1988) / Schaefer S

Residual Effects of Chloride Fertilization
on Selected Plant and Soil Parameters
(1988) / Schumacher W K

Control of Downy Brome with Combinations of
BAY-SMY-1500 and Metribuzin
(1988) / Smart J R

In Vitro Rescue of Helianthus Annuus
Embryos by Ovule Culture
(1988) / Volin J C

TENNESSEE, UNIVERSITY OF (KNOXVILLE)

A Comparison of Cooked Beef Rolls Made from
Flaked Pre- and Post-Rigor Beef
(1987) / Archer O M

Relationships Between Selected Personal and
Family Characteristics of Tennessee
Homemakers and Their Use of Clothing
Construction Practices
(1987) / Atkinson B W

Influence of Feeder Pig Body Type on
Performance and Fat Deposition
(1987) / Bacon C D

An Evaluation of Nonselective and Selective
Pricing Strategies for Finished Hog
Producers in Tennessee
(1987) / Banker K D

Selected Farm Characteristics of Tennessee
Farrow-to-Finish Swine Producers and Their
Use of Selected Productin Practices in
Relationship to Extension Contacts and Farmer
Employment Status
(1987) / Blair R E

Sensory and Chemical Evaluation of Flavor in
Ground Beef from Grass- and Grain-Fed Steers
(1987) / Bolton J C

Effect of Cultiver, Processing, and Storage
on the Quality Characteristics of Sweet
Potato Chips
(1987) / Bozorgmehr K

Selected Characteristics of Tennessee Feeder
Pig Producers and Their Use of Recommended
Swine Management Practices in Relation to
the Number of Contacts Producers Had with
County Extension Agents
(1987) / Chadwell J S

Development of a Technique for the Isolation
of Plasmids from Food-Related Pseudomonas
Fluorescens
(1987) / Chappell M C

Relationships Between Tennessee 4-H Agents
Opinions About Awards and Their Use of
Recognition in County 4-H Programs
(1987) / Cordell J E

TENNESSEE, UNIVERSITY OF (KNOXVILLE)
(continued)

An Assessment of the Tennessee Agricultural,
Forest, and Open Space Land Act in Knox
County
(1987) / Davis R C

Econometric and Time Series Models for
Predicting the Futures Market Basis for
Tennessee Feeder Cattle
(1987) / Dodd A J

Relationships Between Tennessee Homemakers
Employment Status, the Number and Types of
Contacts Made with Extension Agents and
Their Use of Clothing Cnsumer Practices
(1987) / Donaldson J E

Nonmarket Valuation of the University of
Tennessee Arboretum
(1987) / Downing M E

Development of Yogurt with Sweet Potato as
an Ingredient
(1987) / Ebah C B

An Analysis of the Effect of Bank
Management, Market Structure, and Economic
Conditions on Nonmetropolitan Bank Lending
Performance in Tennessee: Pre- and
Post-Deregulation
(1987) / Green S C

A Study of Breed Differences in Steers in
the Obion County Junior Livestock Show
During the Six-Year Period, 1973 to 1978
(1987) / Grooms C W

Comparisons Between Tennessee Dairy
Producers that Continued or Discontinued DHI
in 1983 and Analysis of Reasons for Doing so
and Relationships Between Reasons and
Selected Producer and Herd Characteristics
(1987) / Hale S J

A Comparison of Selected Groups of County
Leaders Perception of Selected Community
Problems
(1987) / Heiskell G M

Analysis of Energy Consumption on Tennessee
Farms
(1987) / Hensley K G

Recovery of Clostridium Perfringens in the
Presence of an Oxygen-Reducing Membrane
Fraction
(1987) / Hoskins C B

Relationships Between the Use of Recommended
Energy Conservation Practices in Tennessee
Homes and Characteristics of the Home, the
Homemaker and the Number of Contacts Made
with Extension Agents
(1987) / Idles M E

An Analysis of the Results of the 1985 Beef
Cattle Survey of the 21 Counties of
Extension Service District One and Some
Comparisons with 1981 Survey Results
(1987) / Kimery L D

Performance of Growing-Finishing Pigs Fed
from Conventional Self Feeders Versus a
Programmed Hog Feeding System
(1987) / Ligon J C

Characteristics of Snack Food with Added
Soy Protein
(1987) / Mancuso P

The Effects of Fungal-Infested Fescue on
Hormonal Secretions and Ovarian Development
in the Beef Heifer
(1987) / McKenzie P P

Effect of Nickel Concentration, Iron
Concentration, and Iron Form on Performance,
Hematological, Tissue and Parameters of
Young Poultry and Weaned Pigs
(1987) / Moore J R

Economic Analysis of Waste Management
Systems on Tennessee Dairy Farms
(1987) / Morgan R D

A Comparison of Parallel and Serial
Distribution of Septic Tank Effluent
(1987) / Mucke F A

Relationships Between Characteristics of
Soybean Production in Tennessee, the Number
of Contacts the Producer Had with Extension
and Their Use of Certain Production
Practices
(1987) / Officer S L

A Multi-Period Linear Programming Model for
Optimally Scheduling the Distribution of
Food-Aid in West Africa
(1987) / Ray J J

Energy Metabolite and Glucoregulatory
Hormone Concentrations and Net Fluxes Across
Portal-Drained Viscera, Hepatic, Total
Splanchnic and Peripheral Tissues in
Gestating, Lactating and Weaning Ewes
(1987) / Sensenig S C

Economic Analysis of In-Store Experiments
Regarding Sales of Locally Grown Tomatoes
to Urban Consumers in Know County, Tennessee
(1987) / Stout C L

A Plant Morphology Integrator
(1987) / Sui R

The Effect of Form and Level of Fat and
Supplemental Lysine on the Performance of
Weaning Pigs
(1987) / Tanner J W

Inhibition of Bovine Peripheral Blood
Mononuclear Cell Blastogenesis by Mammary
Secretions
(1987) / Torre P M

The Relationship Between Prepuberal Hormone
Levels and Sertoli Cell Number to Spermatid
Production in the Beef Bull
(1987) / Whaley P D

Dietary Effects on Mixed Function Oxidase
Activity in the Ruminant
(1987) / Zanzalari K P

TEXAS A AND M UNIVERSITY

Microbiology of Precooked, Uncured, Refrigerated
Vacuum-Packaged Meat Products
(1988) / Anderson M L

Effects of Inbreeding on Performance in
Afrikaner Cattle
(1988) / Beffa M L

TEXAS A AND M UNIVERSITY
(continued)

Effects of Bacitracin Methylene Disalicylate,
Bacitracin Zinc, and Virginiamycin in Combination
with New Coccidiostats on the Performance of
Commercial Broiler Chickens
(1988) / Bekoe K O

An Economic Analysis of Factors Influencing
Texas and U.S. Dry Onion Prices
(1988) / Bello H M

The Effect of Intramuscular and Subcutaneous
Fat on the Microbiological and Sensory
Characteristics of Pork Loin Chops
(1988) / Boxer M K

Performance of Technical Analysis of Short Hedging
Fed Cattle
(1988) / Brasher W E

Effects of Sodium Chloride, Sucrose, and Storage on
Rheological Parameters of Heat Induced Gels of
Liquid Egg Products
(1988) / Brough J

Proper Use of Sodium Bisulfite with Minimal
Salt Penetration during Brine Immersion
Freezing of Shrimp
(1988) / Broussard S R

Reliability Models for Finger Joint Strength
and Stiffness Properties in Douglas-Fir
Visual Laminating Grades
(1988) / Burk A G

Evaluation of the Fatty Acid Composition and Caloric
Value of Ground Beef Formulated to Contain Low
Levels of Fat
(1988) / Campbell L E

Effects of Various Factors on the Value and
Marketability of Lamb
(1988) / Cannell R C

Effects of Chronic Ingestion of the Repartitioning
Agent Clenbuterol on the Development of Adipose
Tissue in Sheep and Cattle
(1988) / Coleman M E

Effect of Dietary High-Oleic Sunflower Oil in
a Swine Diet on Properties of Raw and Cooked
Pork and Pork Products
(1988) / Davidson T L

A Mathematical Model for the Simulation of
Closed-Loop Earth-Coupled Heat Exchangers
for a Water Source Heat Pump
(1988) / DeLange K J

Evaluation of a Catalase-Based Method for
Estimating the Shelf-Life of Pasteurized
Whole Milk
(1988) / Dill S L

Cholesterol Content of Longissimus and Semimembranosus
Muscles and Associated Adipose Tissues from Beef,
Pork and Lamb
(1988) / Dohmann S S

Exercise Overloading in the Equine --
Cardiorespiratory and Metabolic Response to a
Combined Long, Slow, Distance and Interval
Training Exercise Regimen
(1988) / Drozd L F

A Portfolio Analysis of Grain Options and Futures
in a Hedging Context with a Comparison of Quadratic
Programming, Motad and Target Motad
(1988) / Ferdinandsen J A

Strategic Planning for Mixed Class Grazing
in the Edwards Plateau
(1988) / Field E H

Plot Size and Location Within a Cotton Block: Their
Effects on the Canopy Temperature Function and Crop
Water Stress Index
(1988) / Gaitan C A

Role of Yield Grade and Carcass Weight on the
Cutability of Lamb Carcasses at Multiple Trim
Levels and Fabrication Styles
(1988) / Garrett R P

Economic Implications of Irrigation and New Crop
Alternatives in Selected Regions in Texas
(1988) / Hall K D

The Influence of Quality Grade, Fat Trim Level and
Degree of Doneness on the Fatty Acid Composition
of Beef
(1988) / Harberson T J

Hitching of Cable Drawn Implements
(1988) / Harrison C R

Cellulose and Psyllium Supplementation in
10 Females: The Effect on Food Intake and
In Vitro Fermentation Variables
(1988) / Haynes S R

Performance Evaluation of PM(10) and High-Volume
Air Samplers Using a Coulter Counter Particle
Size Analyzer
(1988) / Herber D J

Co-Existence of a Perennial C(3) Graminoid in a
C(4) Dominated Grassland -- Evaluation of Gas
Exchange Characteristics
(1988) / Hicks R A

Cytology, Morphology, and Method of Reproduction of
Some Diverse Buffelgrass Morphotypes
(1988) / Hignight K W

The Determination of the Bioavailability and the
Site of Absorption of Magnesium in Lambs Fed
Different Sources of Magnesium
(1988) / Hurley L A

Impact of Different Subcutaneous Fat Trim Levels
on the Composition of Beef Retail Cuts
(1988) / Jones D K

Use of the Minhas Function in Water Deficit and
Yield Analysis of Grain Sorghum
(1988) / Karama A S

Effect of Reducing Amino Acid Excess in a Corn-
Soybean Meal Diet on Performance, Nitrogen Balance
and Nutrient Digestibilities of Growing Pigs
(1988) / Kelly K A

Evaluation of Zebu Breeds Based on Postweaning
Growth, Feedlot growth, and Carcass Characteristics
(1988) / Kerr J L

Characterization of Cattle Types to Meet
Specific Beef Targets
(1988) / Knapp R H

TEXAS A AND M UNIVERSITY
(continued)

Environmental (Season) Effects on the Maintenance
Requirements and Body Composition of Two Diverse
Biological Types of Cattle
(1988) / Laurenz J C

The Effect of Artificial Drying Temperature on
the Quality of Early Harvested Pecan Kernels
(1988) / McLean R W

Structural Impacts of the 1985 Farm Bill on
Typical Farms in the Texas Southern High Plains
and Delta Region of Mississippi
(1988) / Miller C F

Viscosity of Plant Oils as a Function of Temperature,
Fatty Acid Chain Length, and Unsaturation
(1988) / Neo T H

Responses of Lactating Holstein Cows to
Chilled Drinking Water in the Summer
(1988) / Noel D L

The Relationship Between Blood Carbon Dioxide,
Acid-Base Balance and Calcium Metabolism in the
Hyperthermic Laying Hen
(1988) / Ono Y

A Framework for Identifying Rural Agribusiness
Centers
(1988) / Outlaw J L

Comparison of Double-Sampling Techniques for
Estimating Forage Production
(1988) / Peterson M A

The Role of Microorganisms in the Production of
Volatile Sulfhydryl Compounds in Cheddar Cheese
Slurries
(1988) / Ponce-Trevino P

Country Specific Demand Estimates for U.S. Corn
and Soybean Exports
(1988) / Quintana R I

Incorporating Risk and Uncertainty Into Extension
Applications: Case Example -- The Wheat and
Stocker Cattle Grazing Evaluator
(1988) / Ralston R E

The Growth of Glossy and Nonglossy Sorghums
Subjected to Water Deficits
(1988) / Ramirez J I

Physical and Chemical Characteristics of an
Interesterified Blend of Butterfat and Cottonseed
Oil with Possible Industrial Applications
(1988) / Rashidi N

Investigation of Factors Influencing Stalk
Strength in Sorghum
(1988) / Richie W E

Characteristics of Brush Treatments Affecting
Habitat Use Patterns by Cattle in South Texas
(1988) / Rowland M E

The Effect of Heat Stress on 1,25-
Dihydroxycholecalciferol Induced Calcium-
Binding Protein in Laying Hens
(1988) / Schaeffer B H

Changes in Adipose Tissue Cellularity, Adipose
Tissue Lipogenesis and Muscle Growth in Steers
Administered the Synthetic Beta-Adrenergic
Agonist, Clenbuterol
(1988) / Schiavetta A M

Headspace Profiles of Modified Atmosphere Packaged
Fresh Red Snapper by Gas Liquid Chromatography
(1988) / Scorah C D

Growth and Nutrient Utilization by Yearling Horses
Fed Added Dietary Fat
(1988) / Scott B D

Purification and Expression of Fatty Acid Binding
Proteins in Chicken Liver and Intestine
(1988) / Sewell J E

Effects of Processing, Home Preparation and Maturity
on the Mineral Content of Cowpeas
(1988) / Shows D L

Fatty Acid Composition of Retail Cuts Trimmed to
Different External Fat Levels from Choice, Select,
and Standard Beef Grade Carcasses
(1988) / Sullivan D R

Nutritional Value of Quality Protein Maize, Food
and Feed Corn for Starter and Grower Pigs
(1988) / Sullivan J S

Effects of Cattle Grazing Systems on
Shrub-Grassland Birds in South Texas
(1988) / Swanson D W

The Effect of Antimicrobial Agents and Modified
Atmosphere Packaging on the Microbial Shelf Life
of Corn Tortillas
(1988) / Tellez-Giron A

Yields and Values of By-Products of Different
Types of Cattle
(1988) / Terry C A

Prediction of Genetic Values for Weaning Weight from
Field Data on Calves Produced by Embryo Transfer
(1988) / Thallman R M

Characteristics of Dough and Tortillas Prepared
with Composite Wheat-Sorghum Flours
(1988) / Torres P I

Separation of 23 Cynodon Turf-Type Cultivars Using
Electrophoretic Banding Patterns
(1988) / Venmeulen P H

Effects of Vesiculectomy and Bulbourethralectomy
on Stallion Spermatozoa
(1988) / Webb R L

The Fatty Acid Esterification Pathway in
Bovine Liver and Adipose Tissue
(1988) / Wilson J J

An Economic Analysis of Integrated Pest
Management Strategies for Texas Pecans
(1988) / Woods T A

Involvement of the Endogenous Opioid Peptides in
Luteinizing Hormone-Releasing Hormone Release in
vitro During the Breeding and Non-Breeding Season
in the Mature Ewe
(1988) / Wu T Y

The Relationship between Engineering Design
and Construction Costs of Aquaculture Ponds
(1988) / Yates M E

TEXAS SOUTHERN UNIVERSITY

The Development of a New Medium for the
Rapid Cultivation of Candida Albicans and
Aspergillus Fumigatus
(1988) / Madu S I

TEXAS, UNIVERSITY OF (AUSTIN)

The Effects of Pharmacological Doses of
Tryptophan on Aggressive Tendencies in
Head-Trauma Patients
(1988) / Denison R

TUSKEGEE INSTITUTE

Assessment of Protein and RNA Synthesis in
the Bursa and Spleen of Ducks and Geese
Treated with Aroclor 1254
(1987) / Gager J V

Effect of Dietary Calcium and Protein on
Young Pregnant Rats: Bioavailability of
Calcium and Folic Acid in Collard Greens
(1987) / Gandra A R

Protein and RNA Synthesis in Cardiac and
Skeletal Muscle of Geese and Ducks Treated
with Aroclor 1254
(1987) / Harvey B A

The Nutritional Composition of Smoked and
Stored Herring in Ghana
(1987) / King W M

The Effect of Selection and the Response of
Dietary Energy Level on Abdominal Fat and
Other Quantitative Traits in White Plymouth
Rock Female Chickens
(1987) / Mafeni M J

Strategies for Increasing Institutional
Credit Delivery to Small Farmers in Cross
River State of Nigeria
(1987) / Mbaba V E

Nutritive Value of Dehydrated Sweet Potato
Meal for Gestating and Lactating Swine
(1987) / Moten S J

Interrelationship of Age and Diet on the
Riboflavin Status of Low-Income Lactating
Women in Jamaica
(1987) / Severin C M

Studies on Abdominal Fat and Other
Quantitative Traits in Male White Plymouth
Rock Chickens as Influenced by Nutrition
and Selection
(1987) / Tivzenda P T

Effect of Sweet Potato Meal on Performance
and Carcass Traits of Growing Swine
(1987) / Tor-Agbidye Y

Effects of Gamma-Radiation and Storage Time
on the Physical, Chemical and Sensory
Qualities of Georgia Jet Sweet Potatoes
(1987) / Yakubu P I

UTAH STATE UNIVERSITY

An Appropriate Technology Cultivator for
Less Developed Countries
(1987) / Ahmed I

Reducing the Impact of Weeds in Ornamental
Landscapes
(1987) / Astrove S M

Ultrafiltration: Retentate-Permeate
Partitioning of Milk Constituents
(1987) / Bastian E D

A Model for Estimating Available Iron from
Total Nutrient Intakes
(1987) / Black A M

Factors Affecting Puberty, Estrus and
Ovulation in Corriedale and Criollo Sheep
of the Southern Peruvian Highlands
(1987) / Bravo P W

The Effect of Dietary Calcium and
Phosphorus Levels on Audiogenic Seizure
Susceptibility and Brain Neurotransmitters
in Magnesium Deficient Rats
(1987) / Chaistitwanich R

Effects of Sire, Ration, and Interaction of
Sire with Ration on Reproductive
Performance of Holstein Dairy Cows
(1987) / Chen J J

The Interaction of the Sulfonylurea
Herbicides with Diclofop-Methyl on Control
of Wild Oat
(1987) / Downard R W

The Effect of Utah Population Growth on
Conversion of Agricultural Land to
Residential Land
(1987) / Dyner S S

Selected Parameters of Reproduction in
Rambouillet and St. Croix Ewes
(1987) / Evans R C

Nutrition of Sheep on Grazing Crested
Wheatgrass-Shrub Versus Crested Wheatgrass
Pastures in Winter
(1987) / Gade A E

Seedling Emergence of Springwheat Under
Conservation Tillage as Affected by Planting
Date and Seeding Depth
(1987) / Ghassemzadeh M F

The Effect of Pantethine and Its Metabolites
on Serum Cholesterol Levels in Cholesterol
Fed Rabbits
(1987) / Graves C P

Effects of Incorporated Coarse-Fragments
on Erosion
(1987) / Horne C S

Effects of Micrococci on Improving Sensory
Acceptability of Motton Summer Sausage
(1987) / Jung H

Economic Aspects of Reproductive Problems in
Utah and Southeastern Idaho
(1987) / Lemrick S W

Plasmid-Linked Maltose Utilization
Isolated from Meat
(1987) / Liu M C

Influence of Land Ownership on Range
Condition in Rich County, Utah
(1987) / Loring M W

Comparative Study of Elongated Chromosomes
in Sheep and Goats and a Proposed Standard
(1987) / Mensher S H

Manpower and Training Needs for Landscape
Activities in Ornamental Horticulture in
Utah
(1987) / Mickelson T V

UTAH STATE UNIVERSITY
(continued)

Prediction of Secondary Cavity Nester
Habitat: A Test of the Forest Survey
Data Base
(1987) / O'Brien R A

The Feasibility of Artificial Insemination
of Dairy cattle Managed by Fulani Tribesmen
in Kaduna State, Nigeria
(1987) / Ojomo C O

Factors Affecting Growth of Proteinase
Positive and Proteinase Negative
Ultrafiltered Milk Retentate
(1987) / Pope B K

A No-Drill Comparison Study on Dryland
Wheat in Northern Utah
(1987) / Rich W J

A Model for Testing Draft Requirements of
Some Tillage Equipment in Utah
(1987) / Rose M E

Conversion of a Conventional Grain Drill to
Allow Spacing Using Acra-Plant Double-Disk
Drill Modules
(1987) / Sadeghmoghaddam G

Manufacture of White Soft Cheese from
Ultrafiltered Whole Milk Retentate
(1987) / Shammet K M

Revegetation of Denuded High Elevation
Watersheds in Mustang, Nepal
(1987) / Upadhya M

Physiology and Endocrinology of the
Post-Partum Interval in Lactating Ewes
(1987) / Villalta Rojas P I

Effect of Slope Length on Rill Erosion from
Phosphate Mine Embankments
(1987) / Whitson C R

Effect of Short Duration Grazing on Soil
Moisture Depletion and Plant Water Status
in a Crested Wheatgrass Pasture
(1987) / Wraith J M

Effects of Temperature During Stratification
of Apple Seeds
(1987) / del Real Laborde J I

UTAH, UNIVERSITY OF

Reactive Separation Processes
(1988) / Chang Y A

VERMONT, UNIVERSITY OF

New Techniques to Quantify Evaluation of
Bovine Oocyte Viability and Follicular
Fluid Constituents
(1988) / Ballard C S

Cotton Production in the Binga District of
Zimbabwe
(1988) / Coia S

Identification of Regional Priorities for
Economic Development Programs in the
Agricultural Sector of Panama
(1988) / De Requena C C

A Test to Detect Adulteration of Butterfat
in Certain Dairy Products with Vegetable Fat
(1988) / Fox J R

Market Failure in the Provision of Honeybee
Pollination: A Heuristic Investigation
(1988) / O'Grady J H

Atrazine and Carbofuran Losses from a Corn
Field in Vermont
(1988) / Williams J W

VIRGINIA STATE COLLEGE

A Study on Farm Labor Participation
(1989) / Watnangkura P

WASHINGTON STATE UNIVERSITY

Lipid Oxidation and Pigment Stability in
Restructured Beef, Pork and Turkey Steaks
During Frozen Storage
(1988) / Akamittath J G

Phytate-Protein Interaction in Peas
(1988) / Al-Wesali M S

The Effects of Agrarianism on Agricultural
Policy Preference
(1988) / Altman N K

Impact of Non-Corrosive Forage Stabilizers
on Chemical Composition, Storage
Characteristics, and Nutritive Value of
Alfalfa Hay for Lactating Dairy Cattle
(1988) / Deetz D A

The Economic Value of Recreational Steelhead
Fishing in Washington
(1988) / Demirelli L K

Evaluation of Three Techniques for
Determination of Nutritional Value of
Barley in Diets for Swine
(1988) / Garcia Martinez V F

No-Till and Conventionally Tilled Winter
Barley Production with Respect to Landscape
Position
(1988) / Hopkins A G

Investigation of Management Factors Which
Influence the Somatic Cell Count Status of
the Herd
(1988) / Hutton C T

Effects of Sources of Soybean Meal, Dietary
Phosphorus, Chloride and Magnesium on Tibial
Dyschondroplasia in Broiler Chicks
(1988) / Luo L

Consumer Potato Demand and Nutrition
Education
(1988) / Marotz C C

Utilization and Transport of Storage
Carbohydrates in Sweet Cherry
(1988) / McCamant T

Influence of Rugose Mosaic Disease on
Growth, Gas Exchange, and Carbohydrate
Translocation in Sweet Cherry
(1988) / Mermer S

Soluble Protein Characteristics of
Gewurztraminer and White Riesling Grapes
and Wine During Maturation
(1988) / Murphey J M

WASHINGTON STATE UNIVERSITY
(continued)

Serum Luteinizing Hormone and Testosterone
Concentrations After Administration of LHRH
Analog Microcapsules in Pony Stallions and
Synchronization of Estrus in Yearling Beef
Heifers with Melengestrol Acetate and
Prostaglandin F
(1988) / Neiberggs H L

Covering Sweet Cherries with Plastic:
Economic and Horticultural Analyses
(1988) / Opperman D A

Adipose Tissue Cellularity and Hormone
Sensitive Lipase Activity in Dairy Cattle
of Different Genetic Merit and Energy Status
(1988) / Smith T R

Evaluation of Nadh as an Indicator of
Maturity in Golden Delicious Apples
(1988) / Strecker T

Allocation of Washington Apple Supply
Between Fresh and Processed Markets
(1988) / Tengkuahmad A B

Low Dose Gamma Irradiation and the Retention
and Dynamics of Ascorbic Acid and B-6
Vitamers in Pure Solutions and Potatoes
(1988) / Walker M K

Economic Trade-Offs Between Quality
Characteristics and Crop Yield in the
Production of Concord Grapes
(1988) / Yarnell J L

WASHINGTON, UNIVERSITY OF

Analysis of Residual Stress Based on
Destructive Test Measurement
(1987) / Huang J S

Project Management for the Small Design
Firm
(1987) / Lauritsen G L

WEST TEXAS STATE UNIVERSITY

The Yield and Nutritive Value of a Native
Pasture in the Texas Panhandle
(1988) / Nsibande L I

The Effect of Antitranspirants and Water
Stress on Plant Response of Sorghum
(1988) / Robinson C A

The Role of an Agricultural Extension
Specialist and Recommendations for Increased
Effectiveness in Cameroon
(1988) / Tata R M

WEST VIRGINIA UNIVERSITY

Micropropagation of Eastern Black Walnut
(1988) / Akers R A

Using a Key Factor Indicator Approach for
Holstein Dairy Farm Management Change: A
Case Study
(1988) / Bean T G

Major Roles of Agricultural Extension Agents
in the Agricultural Technology Delivery
System in the Year 2000
(1988) / Bonanno S C

Influences on the Job Satisfaction of
Vocational Agriculture Teachers
(1988) / Cochran J E

Job Satisfaction of College Agriculture and
Forestry Teaching Faculty at West Virginia
University
(1988) / Cowie M L

Analysis of West Virginia Agricultural
Trends
(1988) / Dike L C

Major Problems Which Hinder the Increased
Use of Integrated Pest Management by Farmers
as Perceived by Pest Management Specialists
(1988) / El-Ayouby N S

An Analysis of the Changing Employment
Structure in West Virginia, 1940-1980
(1988) / Eresechinma D L

Perceived Training Needs of Cooperative
Extension Agents in West Virginia
(1988) / Howell J A

The Influence of Summer Pruning on Yield and
Fruit Quality of Apple Trees Trained to the
Lincoln Canopy Trellis System
(1988) / Keiser T A

An Economic Analysis of Alternative Meadow
Management Systems for West Virginia Beef
Cow/Calf Producers
(1988) / Maxwell E W

Animal Waste Management Practices in West
Virginia, 1985
(1988) / Miller J L

Perceived Communications and Support
Linkages of Secondary/High School Principals
and Agriculture Teachers in Schools of
Swaziland
(1988) / Nxumalo N N

Microbiology of Refrigerated, Fermented
Dairy Products with Emphasis on Shelf Life
(1988) / Nye W R

Agricultural Research Needs and Priorities
in Zanzibar as Perceived by Administrators
and Extension Workers
(1988) / Rajab K M

An Analysis of Employment Changes in Service
Industries in West Virginia, 1950-1980
(1988) / Rouhani E M

Major Problems Encountered in
Administrating/Supervising Vocational
Agriculture as Perceived by State
Vocational Agriculture Supervisors
(1988) / Smith S D

Perceptions of Agriculture Faculty at Land
Grant Institutions in the Northeastern US
Concerning the Land Grant Mission
(1988) / Snively D W

Economic Assessment of Agricultural Land-Use
Values in West Virginia
(1988) / Tall M M

Organizational Commitment of Agents in the
West Virginia Extension Service
(1988) / Wright R W

Evaluation of a Socio-Economic Impact Model
for West Virginia
(1988) / Yost P L

WESTERN KENTUCKY UNIVERSITY

Use of Soybean Meal, Raw Soybeans, and
Heat-Treated Soybeans as Protein Supplements
with and Without Niacin for Dairy Cows in
Early Lactation
(1988) / Aguilar D E

Performance and carcass Traits of Swine of
Four Different Prototypes and Three USDA
Grades
(1988) / Althaus J

Nitrogen Solubility In-Vitro and Dry Matter
and Nitrogen Disappearance of Protein
Supplements In-Situ
(1988) / Fatouhi N

Response of Three Cultivars of Bell Pepper
to Mulching and Irrigation
(1988) / Gonzalez C O

Evaluation of Omesafan for Broadleaf Weed
Control in Soybeans (Clycine Max)
(1988) / Jimenez R

WYOMING, UNIVERSITY OF

Distribution of Water Application Under Low
Pressure Center Pivot Irrigation Systems
(1989) / Baron M D

Design and Evaluation of an Irrigation
Furrow Packing System
(1989) / Larsen J F

The Effect of Mow/Fertilization and Clipping
Treatments on Leafy Spurge Control with
Herbicides
(1989) / Madsen K A

Geology of a Precambrian Igneous/Metamorphic
Sequence, Laramie Range, Wyoming
(1989) / Molnar G R

ARIZONA STATE UNIVERSITY

A Multiple Criteria Satisfying Methodology
for the Design of Energy-Efficient Buildings
(1988) / Addison M S

BALL STATE UNIVERSITY

A Comparison of the Use of Video and Slides in
Testing Landscape Scenic Preference
(1988) / Bell D J

Statistical Approach Toward Designing Expert System
(1988) / Hu Z

Adaptive Use Study of the Pennsylvania Station at
Fort Wayne, Indiana
(1988) / Leonard C

BRIGHAM YOUNG UNIVERSITY

County-Level Land Use Planning Policies
and Regulations Impacting the Pattern of
Settlement in Utah County, Utah
(1988) / Johnson E A

Earthquake Perceptions and Housing Problems
Resulting from the February 4, 1976,
Guatemalan Earthquake
(1988) / Pratt S M

BROWN UNIVERSITY

Methods of Analysis of Task Distribution and
Resource Allocation in a Class of
Distributed Control Multiple Processor
Architectures
(1988) / Leibholz D L

CALGARY, UNIVERSITY OF

An Architectural Evaluation Case Study:
Activity Settings in a Children's Hospital
Dayroom
(1988) / Johnston W A

CALIFORNIA, UNIVERSITY OF (DAVIS)

Tree Shade in Urban Parking Facilities: An
Evaluation of Parking Patterns and the Davis
Mandatory Shade Ordinance
(1988) / Elliott K

CLEMSON UNIVERSITY

A Gateway to a Great American City: Kansas
City, Missouri
(1988) / Beck T G

Towards a New Awareness of the Everyday: A
College of Popular Culture at North Carolina
State University
(1988) / Bynum R T

A New Awakening: A New Baptist Chruch for
Liberty, North Carolina
(1988) / Dawkins J D

Context and Memory: A Levee Museum for
Augusta, Georgia
(1988) / Hopkins J M

Architectural Dialogue and the Parthenon: A
Nashville Acropolis
(1988) / Kennon J E

The Institution and the Changing Face of the
American Small Town: Clemson Town Hall
(1988) / Malloy M J

The Essence of Dance: Its Exploration in
Movement and Architecture, the University
of South Carolina
(1988) / Mason K I

Man, Music, and Architecture: Charleston
Symphony Hall
(1988) / McClure F A

Work Order Analysis: A Tool for
Understanding and Improving the Maintenance
of Public Housing in Charlotte, North
Carolina
(1988) / Stroud J F

A Celebration of Life: Eden, North Carolina
(1988) / Turner R L

Olympus Space Station: Architecture for the
Final Frontier
(1988) / Walker M J

COLORADO STATE UNIVERSITY

Dispersed Forest Roadside Campers -An
Analysis of Three Types-
(1988) / Anderson G C

CORNELL UNIVERSITY

A Proposal for an Ithaca Urban Cultural Park
(1988) / Acosta P

The Rehabilitation and Adaptive Use of the
Carriage House of the 1890 House Museum and
Center for the Arts, Cortland, New York
(1988) / Carter M A

Controlling Computer Generated Motion with
Dynamics, Kinematics, and Behavior Functions
(1988) / Isaacs P M

No Direction Home: The Inability of Local
Governments to Solve the Housing
Affordability Crisis
(1988) / Iurino T J

Future Operation of the Panama
Canal: Monopoly or Competition
(1988) / Iwabuchi M

The Case for Conflict Resolution Programs
in Schools
(1988) / Kidde A D

Caught in the Wake of the Urban, Post
Industrial Crisis: A Case Study of
Household Strategies in New Brunswick,
New Jersey
(1988) / Klot J F

The History of the Development of the
Semarange: The Causes of Urban Problems
in a Colonial City, 1870-1940
(1988) / Kwanda T

The Characteristics of Income Distribution
Among Different Sizes of SMSAs in the United
States: A Study of Economic Inequality and
Size of Urban Population
(1988) / Lee H C

Curved Object Modeling and Rendering
(1988) / Lu W

Inquiry Into the Preliminary Formulation of
a Strategic Plan on Housing, Ithaca, New
York
(1988) / Lynn D J

CORNELL UNIVERSITY
(continued)

The Role of the Federal Land Development
Authority (FELDA) in Agricultural
Development and Land Settlement in
Malaysia, 1971-1990
(1988) / Maidin S L

Urbanization and Urgan Growth in
Siberia: A Comparative Analysis
(1988) / Rice D L

Socio-Economic Development and the Evolution
of the Urban System in
Malaysia: Implications for the New Economic
Policy and National Unity
(1988) / Teoh H S

A Two-Pass Solution to the Rendering
Equation: A Synthesis of Ray Tracing and
Radiosity Methods
(1988) / Wallace J R

FLORIDA, UNIVERSITY OF

Post-Rural Regional Planning: A North
Florida Case Study
(1988) / Al-Moussallie H

A Software Package for Use in Ultrasonic
Acoustical Modeling
(1988) / Battle J

A Place for Christian Worship: A Study of
Christianity and Its Roots in Florida
(1988) / Baxter C

The Changing Morphology of Rural County
Seats in North Central Florida
(1988) / Bennett S R

The Ultimate in Spa Resorts: The
Renaissance of the Alcazar, St. Augustine,
Florida
(1988) / Boykin-Hayter K

The Methodist Camp Meeting in America: An
Analysis and Contemporary Reinterpretation
(1988) / Brewington G

Environment: The Physical, Cultural, and
Social Dimensions for the Design of
Educational Facilities
(1988) / Brooks-Pilling L

The Use of Cables in the Counteraction of
Wind Forces in Tall Buildings
(1088) / Burke M J

Planning the Metropolitan Region:
Institutional Responses to Change
(1988) / Burrows T

The Jerusalem Temple: Its Symbolism and
Influence in Synagogue and Church
Architecture
(1988) / Carlton J

The Origins of the Historical Preservation
Movement in Pensacola
(1988) / Curenton M C

Compatible Design in St. Augustine, Florida,
Through Interpretation of a Visual Code
(1988) / Dodge T

The Applicability of a Geo-Facilities
Information System to the Criminal Justice
System
(1988) / Dove C

Historic Preservation Planning for
Florida's Small Communities
(1988) / Foster J

Performance Characteristics of the
Ventilated Skin Roof
(1988) / Frankel G

Capturing a Culture: Latin Quarters and
Design Review in Florida
(1988) / Gonzalez A

The Use of Computer Mapping and Property
Appraiser Tapes to Analyze Alafia River
Basin Agricultural Land Use, Past and
Present, with a Future Preservation
Alternative
(1988) / Hoover M

Continued Prominence of a Symbolic Area
Landmark: The Polk County Courthouse
(1988) / Horton S

The Mizner Style: A Synthesis of Spanish
Revival Architecture and American Country
House Architecture
(1988) / Kino D E

Designing Compatible Infill Structures in
Historic Districts: A Case Study in the
Ybor City Historic District, Tampa, Florida
(1988) / Miranda J

An Exploration Into the Revitalization of an
Orlando, Florida, Neighborhood and How the
Implementation of Tenant Management Within
a New Housing Development Will Aid This
Revitalization
(1988) / Murphy S L

Alternative Concepts in Civic Architectural
Design: A City Hall Proposal for Palatka,
Florida
(1988) / Obarowski P

Tourism Planning in Trinidad and Tobago: A
Critical Look at the Government's Draft
Tourism Policy 1987
(1988) / Ojah-Maharaj S

Analysis of Via Del Corso in Rome:
Attributes that Affect Its Pedestrian Use
(1988) / Penso M

Protecting Sources of Potable Ground Water:
A Study of a Developing Strategy
(1988) / Pierce A

Automated People Movers: An Update of the
Miami Metromover
(1988) / Ramirez E

Domestic Architecture of Southeastern
Louisiana: Continuing the Tradition
(1988) / Smith L

An Architectural Resource File of the
Greater Caribbean: A Proposal
(1988) / Stauffer M

The Application of a Geographic Facilities
Information System to a Local Fire
Protection Model
(1988) / Terry S R

FLORIDA, UNIVERSITY OF
(continued)

Florida Stream Networks and Drainage Basin
Morphology
(1988) / Tighe R

New Faces in Old Places: A Survey of
Architectural Guidelines for New
Construction Within Historic Districts
(1988) / Turner J

GEORGIA INSTITUTE OF TECHNOLOGY

Mortgage Redlining in Metropolitan Atlanta
(1988) / Fitterman S F

GUELPH, UNIVERSITY OF

An Evaluation of Retail Floorspace Expansion
as a Means of Downtown Revitalization
(1988) / Cundy C K

IDAHO, UNIVERSITY OF

Publishing Your Architectural Documents from
Database to Output Through Desktop
Publishing
(1988) / Egli C A

ILLINOIS INSTITUTE OF TECHNOLOGY

An Application of Sign Systems in
Instructional Manuals: Correlating
Information Types and Sign Systems
(1988) / Ichikawa T

Housing Development on a Suburban Site
(1988) / Jabbour S F

Low-Rise Self-Help Infill Housing for
Chicago
(1988) / Jacobs L M

Design and Analysis of Energy Efficient
Apartments for Kathmandu
(1988) / Maskey N

A Multiuse Arena with a Cable-Stayed Roof
(1988) / Oechsner T H

IOWA STATE UNIVERSITY

The Planning and Design of a New Botanical
Research Library for Iowa State University
(1988) / Li S

KANSAS, UNIVERSITY OF

Core Area Pedestrian Malls: A Study for
Outdoor Malls in the United States
(1988) / El Tabba M M

A City's Cry for Survival: A Case for
Studying the Cultural and Spatial
Organization of an Islamic City
(1988) / Ferwati M S

LONG ISLAND UNIVERSITY

Growth of a Community: A History of Bay
Ridge Brooklyn
(1988) / Bartolomeo J G

LOUISIANA STATE UNIVERSITY

Planning Activities and Characteristics
Important to Effective Campus Master
Planning - An Evaluation of Policy Statement
23 "Campus Planning Effort" at Louisiana
State University
(1988) / Canon W H

The Renaissance of the Traditional Streetcar
in the Urban Revitalization Movement
(1988) / Evans S D

An Analysis and Evaluation of Nine Landscape
Ordinances Emphasizing Information that Will
Aid in the Development of a More
Comprehensive and Unique Landscape Ordinance
(1988) / Foster S P

The Development of a Short Course to Aid in
the Presentation of Information on
Barrier-Free Site Requirements
(1988) / Hackett M B

A Computer Assisted Instruction for Drainage
Design Using Hypercard
(1988) / Kim Y

A Survey of Corporate Facility Planners
Within High Tech Companies, and Chamber of
Commerce Executives Regarding the Degree of
Importance of Quality of Life as a Site
Selection Factor
(1988) / Stovall M S

Using Mini- and Microcomputers to Satisfy
the Date Needs of the Mandeville, Louisiana,
Planning Department
(1988) / Wilkerson G W

MASSACHUSETTS INSTITUTE OF TECHNOLOGY

Traditional Building Trades and Crafts in
Changing Socio-Economic Realities and
Present Aesthetic Values: Case Studies in
Syria
(1988) / Abed J H

The Function of Repetition in Typology
(1988) / Antoniades C G

Reassessing the Role of Tradition in
Architecture
(1988) / Arshad S

Architecture as Evocation of Place:
Thoughts on an Architectural Beginning in
Bangladesh
(1988) / Ashraf K K

When Metaphors Speak: A Design for a
Theater in Boston
(1988) / Baker S D

Transformation of an Institution: The
Design of a Catholic Church for Boston's
South End
(1988) / Beliveau E M

Can the Market Create What the City Wants?
Boston's Midtown Cultural District
(1988) / Berry R L

Emotion, Myth and Meaning in Architecture:
Psyche's Journey Through a Warehouse
(1988) / Bill J A

Dudley: A Communicational Project
(1988) / Borges-Mendez R

MASSACHUSETTS INSTITUTE OF TECHNOLOGY
(continued)

A Tectonics Approach to Architecture: A New
Church Building for Buffalo
(1988) / Bovenzi C A

High School Dropouts in Boston
(1988) / Byrne G A

Building Exchange and Multiplicity Into
Housing for the Elderly: an Exercise in
Synthesizing Associative References
(1988) / Campbell J A

Back to the Crossroads of Flatbush - The
Junction: Student Housing for Brooklyn
College, Brooklyn, New York
(1988) / Campbell K A

Understanding Changes and Constants of the
Courtyard House Neighborhoods in Beijing
(1988) / Casault A

Negotiating Better Superfund
Agreements - Prospects and Protocols
(1988) / Cassel S A

The Reawakening of the Chinese Heritage
Through the Creation of a Cultural Embassy:
Transformation of the Chinese Architectural
Language
(1988) / Chin H Y

A House for all Seasons: A Suggested
Habitation Model for Great Barrington
(1988) / Davis J R

An Artists' Community in the Back Bay:
Continuity and Change
(1988) / Duckham K L

The Mall at Chapel Square: A Pivotal
Public/Private Partnership in New Haven,
Connecticut
(1988) / Edwards A I

Estimated Regional Impacts from
Rub-and-Spoke Operations at U.S. Airports
(1988) / Eisenberg E

A Pattern of State and Non-Governmental
Organizations Interaction - The Philippine
Case
(1988) / Gorospe E S

Communication Tools for Community-Based
Development Programs
(1988) / Gray S A

Image, Text and the Female Body: Rene
Magritte and the Surrealist Publication
(1988) / Greeley R A

from Sketch to Layout: Using Abstract
Descriptions and Visual Properties to
Generate Page Layouts
(1988) / Greenlee R L

The Nursing Shortage and Its Relationship to
Part Time and Temporary Employment Growth:
How Should Unions Respond?
(1988) / Greiner A C

Realities at the Fringe of the Economy:
Women Cloth Traders in Kinshasa
(1988) / Guetta I S

Increasing the Scope of Activity on the
Private Sector in Urban Development
(1988) / Gupta V K

Combining Organizing and Housing
Development: Conflictive yet Synergistic
(1988) / Hadrian R A

Illegitimate Settlements in West Beirut: A
Manipulation of Tenure Policies and Class
Struggle Over Land
(1988) / Halabi M B

Lessons in Matchmaking: Conflicts in Job
Assignments for China's Scientists and
Technicians
(1988) / Hammond J C

A New Town Hall for Norwich, Vermont
(1988) / Harboe P T

The Invented Eye Looks at Architecture
(1988) / Hazra A

Landscape and Form: Observation and
Transformation of Farm Form
(1988) / Henrich D K

Programs and Precedents: Future Prospects
of Housing Theory and Practice in Lebanon
(1988) / Jabr A Z

The Common Greenway and the Establishment
of Park Character
(1988) / Johnson J M

Design and Implementation of Decision
Support Systems for Environmental Management
(1988) / Johnson S T

Sports and Community Empowerment: Moving
the Game Into the Community
(1988) / Jones D M

In Search of a Direction in the Contemporary
Architecture of Arabia
(1988) / Khan S I

From Work to Home: Boston's Hotel Workers
and the Prospects for Union-Sponsored
Housing
(1988) / Kluver J

Skyscrapers in Context
(1988) / Kobayashi K

A Seaside Resort in an Island of the Aegean
(1988) / Kriezis C A

A Sub-Systems Approach to Small Lot
Single-Family Housing
(1988) / Kuhn H

Technology as a Strategic Variable: The
Perspective of Export-Oriented Firms in
Chile
(1988) / Laureda F S

The Rhode Island State House: The
Competition (1890-1892)
(1988) / Lewis H A

Movement and the Neighborhood Fabric as the
Generators for an Urban Elementary School
(1988) / Liberatore D L

Land for Housing the Poor Through Urban
Agriculture: The Case of Lusaka, Zambia
(1988) / Lucenet F P

MASSACHUSETTS INSTITUTE OF TECHNOLOGY
(continued)

Community Control and Neighborhood
Development: Anarchy, Pluralism and
Supports as Models for Analysis
(1988) / Ludwig K M

The Facade and the Contextual Expression of
Dynamic Use: A Culinary Arts School and
Restaurant in Paris
(1988) / Luthringshauser H E

The Time of Nature: A Retreat in Northern
California
(1988) / Lyon C M

Twentieth Century Chinese Architecture:
Examples and Their Significance in a Modern
Tradition
(1988) / Marcus K K

Architectural Theory after 1968: Analysis
of the Works of Rem Koolhaas and Bernard
Tschumi
(1988) / Martin L

A Plan for Metropolitan State Hospital:
Imagery as a Therapy for an Institution
(1988) / McMurrin S J

Community Architecture: Myth and Reality
(1988) / Mongold N J

Interpreting Territorial Structure: the
Machiya of Kyoto and the Rowhouses of
Boston's South End
(1988) / Moore J R

The Fan Piers/Pier Four Development: An
Analysis of Negotiated Development Review
in Boston
(1988) / Mugnani L C

Unauthorized Colonies and the City of
Delhi
(1988) / Mukherjee S

Housing Policies in Mexico: A Case Study
(1988) / Munoz G

Between Asylum and Independence: Toward a
System of Community Care for People with
Long-Term Mental Illness
(1988) / Murphy E T

Sei-ichi Shirai and Subjective Method of
Synthesis
(1988) / Nagaya T

The University as Mediator: A New Model for
Service
(1988) / Nash J H

Post-Nollan - Land Use Planning: How
Communities are Coping with the Supreme
Court Decision
(1988) / Natoli N S

Service Settings for an Aging Society: A
Community Ordering Principle
(1987) / Nelson G G

From the Cheney House to Taliesin: Frank
Lloyd Wright and Feminist Mamah Borthwick
(1988) / Nissen A D

A Cross Sectional Multivariate Analysis of
the Determinants of Maintenance Costs in
Boston's Public Housing
(1988) / Okundaye O O

Developing the Small, Mixed-Use Urban
Project: A Contribution to Neighborhood
Revitalization
(1988) / Olivier J M

Rationalized Structural Systems for Diverse
Applications
(1988) / Panayides F A

Towards a New High Technology Development in
the Silicon Valley: A 21st Century Urban
Design Vision
(1988) / Pang J K

Paradigm or Pariah? An Architectural
Intervention Into Daniel Burnham's Timeless
Plan of Chicago
(1988) / Pettigrew P S

The Dynamics of Needs and Interests: A
Mediator's Response to the Israeli-
Palestinian Conflict
(1988) / Podziba S L

Economic Feasibility and Impact Analysis of
an Improved Ferry Service Between Java and
Sumatra
(1988) / Poerwosoenoe A P

From the Squatter Settlement: A Program to
Build the City
(1988) / Reeves N M

Transitions: A Sanatorium on the Seaside
(1988) / Rodriguez B H

Interpreting a Contemporary Urban Vernicular
for Cities: The Case of Delhi
(1988) / Roy A

The Urban Poor Challenge: Delivering
Services to the Urban Poor - Government
Organizations Versus Non-Government
Organizations (NGOs)
(1988) / Ruiz Cepeda M R

Narrative Light: A Design of a Nonastic
Retreat
(1988) / Sama J M

A Locational Approach to Providing
HIV-Testing Services: Sao Paulo, Brazil
(1988) / Schall S M

Real Property Portfolio Management: A
Decision-Support Model
(1988) / Schcolnik A E

A Design Methodology: The Object as a
Vessel for Vitality
(1988) / Schwarz A D

The City in Change: Socio-Spatial
Dimensions of Urban Environmental Change
(1988) / Shah B

The Architectural Implications of Passive
Solar Cooling Systems in Hot Arid Climates
(1988) / Sharag-Eldin A M

The Massachusetts College of Art Disposition
Case: Evaluating State and Community Roles
(1988) / Sklar S

MASSACHUSETTS INSTITUTE OF TECHNOLOGY
(continued)

Alternative Dispute Resolution in the United
States Army Corps of Engineers:
Opportunities and Barriers
(1988) / Sobel G B

A Rationalized Building System for
Low-Income Housing as a Response to the
Issues of Flexibility and Participation
(1988) / Spinazzola A R

Industrial Structures: An Analysis and
Transformation of Their Formal
Characteristics
(1988) / Strub D

The American College and Its Architecture:
An Institutional Imperative
(1988) / Taylor R T

Beyond Dualism in Urban Development: The
Role of the Informal Sector as a Mode of
Production in the Urban Economy - Concepts,
the Indian Evidence and Policy Implications
(1988) / Tewari M

Space, Time and Acoustics
(1988) / Thompson P R

From Ground to Sky - an Exploration in
Urban Continuities
(1988) / Thornton K M

Limited Development as a Tool for
Agricultural Preservation in Massachusetts
(1988) / Tuttle W D

Designing National Identity: Recent
Capitols in the Post-Colonial World
(1988) / Vale L J

Labor Market Consequences of Immigration: A
Study of Recent Puerto Rican Migrants in
Select Occupations and Cities in the U.S.
(1988) / Valenzuela A

Issues in Agricultural Restructuring for
Ensuring Food Security in Developing Island
Economics
(1988) / Williams S D

MASSACHUSETTS, UNIVERSITY OF

Large Scale Mapping of Wetlands in Amherst,
Massachusetts: Analysis of Available Data
Sources
(1988) / Dickman A C

Municipal Recreation Agencies in
Massachusetts: The Relationship of
Community Characteristics to Recreation
and Conservation Planning Issues
(1988) / Edmonds H L

Low Income Housing in Developing
Countries: Shelter for the Urban Poor
(1988) / Extross K M

Cooperative Strategies for Economic
Development
(1988) / Johnson M R

Results of a Survey of Massachusetts'
Planning Board Chairmen Regarding
Residential Growth Management
(1988) / Matuszko T E

Urbanization and Internal Migration in the
Republic of Korea: Examined in Relation
to the Demographic and Mobility Transition
Hypotheses
(1988) / Oh P A

Managing the Changing Land Uses of Urban
Waterfronts
(1988) / Perry J R

Inland Water Transport in South and East
Asia
(1988) / Sinha S

MIAMI, UNIVERSITY OF

Future Saudi-Arabian City Planning Based
on the Use of Solar-Hydrogen Energy
(1988) / Al-Ajlan S A

Urban Design on the Basis of Islamic
Principles: An Approach Toward Designing
the Future Saudi Arabian City with Special
Reference to Military Cities
(1988) / Al-Harrbi A

MINNESOTA, UNIVERSITY OF

The Study of the Image of Downtown
Minneapolis: A Pragmatic Approach
(1988) / Bhaskaran A

An 'Affective' Approach to Nurturing
Creative Potential in Landscape Architecture
(1988) / Frost W W

A Participatory Design Process Theory
Utilizing Visualization
(1988) / Helgeson S D

Renewal of Old Urban Neighborhoods Into
Humane and User-Responsive Environments
(1988) / Mittal S

A Center for the Media Arts, Minneapolis,
Minnesota
(1988) / Neer P S

Architecture: Major Factor of the City New
York Waterfront Development
(1988) / Obayawat S

A Museum of Architecture
(1988) / Shields R S

A Retail Development in Nicollet Mall
(1988) / Sulaeman W

Architecture in the Natural Environment
(1988) / Wicks R V

MISSISSIPPI STATE UNIVERSITY

The Role of Landscape in Small Town Design
(1986) / Vander Mey G A

MONTANA, UNIVERSITY OF

Confluence Park, a Proposal for
Revitalization of the Colorado Riverfront
in Grand Junction, Colorado
(1988) / Portner K M

MONTREAL, UNIVERSITY OF

l'Espace Agricole, un Enjeu Social le Cas
de Laval, 1978-1984
(1988) / Archambault J

MONTREAL, UNIVERSITY OF
(continued)

Peut-on Batir l'Habiter un Architecte
s'Interroge
(1988) / Baggio P

Logement en Banlieue Pour Les Personnes du
Troisieme Age
(1988) / Baillard V

Les Facteurs Lies a l'Implamtation de Condos
de Luxe au Centre-Ville de Hull
(1988) / Barrette M C

Les Notions de Flexibilite et de Pseudofixe
en Architecture
(1988) / Begin C

Traitement de la Fonction Transport au Sein
de la Ville de Montreal
(1988) / Bertrand J

Etude de Design Urbain des Lieux Pietonniers
Nationaux
(1988) / Botti D

l'Efficacite d'une Etude d'Impact Visuel:
le Cas de Deux Lignes Hydro-Electriques
(1988) / Casgrain V

Enquete Sur le Design Urbain: Les Lignes
Directrices de Dix Villes d'Amerique du
Nord
(1988) / Charbonneau N

Le Processus de Consultation Publique et
l'Elaboration des Schemas d'Amenagement
(1988) / Cloutier M

La Capacite Portante du Quartier Saint-Jean
Baptiste a Montreal
(1988) / Couture M S

Vers Une Analytique Epistemologique du
Discours en Amanagement
(1988) / Daunais C N

La Mediation des Conflits en Matiere
d'Environnement
(1988) / Decoste S

La Problematique d'un Developpement
Ecologique au Niveau Municipal Etude de
Cas: Archipel Versus Laval
(1988) / Dubois S

La Reglementation Patrimoniale: le Cas de
la M. R. C. de Bellechasse
(1988) / Emond C

l'Application de Systems Prefabriques Pour
l'Auto-Construction
(1988) / Gamboa G E

Conception des Elus Locaux Concernant
l'Amenagement, le Processus de Planification
et la Loi Sur l'Amenagement et l'Urbanisme
(1988) / Groulx J

Etude de la Problematique de la Couleur en
Design Industriel
(1988) / Halle E

Deux Problematiques d'Amanagement Dans un
Etablissement d'Hebergement Prive Pour
Personnes Agees
(1988) / Landry J

La Planification des Amenagements Cyclables
(1988) / Lavoie R

La Consultation Publique: un Instrument
d'Amenagement a la Portee de Qui
(1988) / Leblanc D

Considerations Sur la Methodologie des
Etudes d'Impact Sur l'Environnement: le
Cas du Ministere du Transport et de
l'Etude d'Impact d'Amqui
(1988) / Legault R

Sistema de Transporte y Desarrollo Regional
en la Orinoquia Colombiana
(1988) / Lizarazo P

Etude Theorique et Pratique sur la
Connaissance Spatiale de l'Enfant
(1988) / Patenaude G

Reflexion sur la Participation a
l'Amenagement du Centre Ville a Partir du
Cas de Dolbeau
(1988) / Pelchat B

Planification Globale et Integree le Cas de
la Degradation Des Sols Agricoles
(1988) / Perrault H

Le Plan d'Urbanisme en Milieu Rural
(1988) / Roy L

Derogations Mineures aux Reglements
d'Urbanisme: Limites et Voies d'Avenir
(1988) / Savard D

Amenagement du Secteur du Metro Plamondon
a Montreal
(1988) / Simon P P

Amenagement Dans la MRC de la Vallee du
Richelieu: Les Questions d'Environnement
(1988) / Turenne A

Le Credit Foncier au Cameroun
(1988) / Yepmo S M

Les Corporatives d'Habitation au Quebec
(1988) / Zoanga A

NEBRASKA, UNIVERSITY OF

Design Guidelines for a City Entrance:
Study of the North 27th Street Corridor,
Lincoln, Nebraska
(1988) / Andrade M A

The Influence of Large Corporations on the
Planning Process: A Case Study
(1988) / Bauer K

Queen Anne Victorian Architectural Details
on Domestic Architecture in Red Oak, Iowa
and Mount Pleasant, Iowa
(1988) / Griffen E P

Squatters in Kuala Lumpur: Housing Problems
and Proposed Solution
(1988) / Hanafiah M K

Traffic Related Studies of Child Care
Centers: Lincoln, Nebraska
(1988) / Hayatgheybi M

Strategic Planning as a Tool for Economic
Development in Rural Nebraska
(1988) / Hendricks J J

Self-Help Housing: An Approach to the
Squatter Housing Problems in Malaysia
(1988) / Naim A

NEW JERSEY INSTITUTE OF TECHNOLOGY

Planning Design Change: Focus on Ottawa,
Illinois
(1988) / Carney D J

The Architecture of Automobile and Building
Design: Learning from 100 Years of Parallel
Processes
(1988) / Ristic V

NEW YORK, STATE UNIVERSITY OF (BUFFALO)

Economic Development in an East Asian
NIC: A Case Study of South Korea
(1988) / Choong P

Women, Development and Planning: A
Theoretical Perspective
(1988) / Ford-Ross C

The Interior City: From the Street to
the Mall
(1988) / Jerozolimski G

Criticality of Cultural Considerations in
Urban Development: The Case of Bangalore,
India
(1988) / Maruliah M

Housing Options for the Welfare Household:
Case Study of Buffalo, New York
(1988) / Morrell J

The Place of Bioregionalism in Environmental
Planning
(1988) / Wooster M

NOVA SCOTIA, TECH UNIVERSITY OF

An Historical Perspective on Tourism
Planning and Development in Nova Scotia
(1988) / Clark A R

A Comparison of the Subdivisional Approval
Processes in Metropolitan Halifax
(1988) / Fraser P D

Towards a Provincial Policy Statement on
Flood Damage Reduction in Nova Scotia
(1988) / Gallivan J V

Development Control -- the Use of the City
of Dartmouth's Municipal Development
Boundary in Practice
(1988) / Jenkins D S

A Program in Practice: An Assessment of the
Nova Scotia Mainstreet Program
(1988) / Mitchell E C

The Impact of the 1983 Planning Act on Rural
Planning in Nova Scotia
(1988) / Smith D M

OREGON STATE UNIVERSITY

The Architecture and Design of a Parallel
Sorting Engine
(1988) / Bate J G

PENNSYLVANIA STATE UNIVERSITY

The Satellite Towns in China: A Comparative
Study
(1987) / Deng D

Understanding the Significance of Computers
in Architecture
(1987) / Gamboa M T

Bid Unbalancing in Unit Price Contracts for
Public Building Construction
(1987) / Nold R B

Connection Flexibility in Single-Story
Frames
(1987) / Volz J S

Interdependence of Issues in the Work of
Andrea Palladio: A Study of Villas in
Chapters XIV and XV of His Four Books of
Architecture
(1987) / Wingert M K

The Analysis of Critical Path Method-Based
Time Versus Cost Management Model for
Reinforced Concrete Structural Frames
(1987) / Yuan R

PURDUE UNIVERSITY

Development and Evaluation of X-Band
Microwave Density Indepent Moisture
Control Model
(1988) / Nichols C J

RENSSELAER POLYTECHNIC INSTITUTE

Community Meeting-Places
(1988) / Harris B

The English and American Arts and Crafts
Movements: Appropriate Technology in the
Nineteenth Century
(1988) / Keefe-Watkinson J M

Village and Landscape: The Between
Become Manifest
(1988) / Mellor J M

Home in the City: A Way of Life for Urban
Families
(1988) / Montgomery M W

TEXAS, UNIVERSITY OF (AUSTIN)

Productivity Estimate Using the CII Model
Plant
(1988) / Arsaelsson L

Targeting Desired Businesses in City
Economic Development Programs
(1988) / Barnett E

A Generative Theory of Architectural
Surfaces: Aspects of Vision, Music, and
Architecture
(1988) / Bechtel J A

Communication and Related Work Problems in
the Construction Industry
(1988) / Berntzen P

The Academia-Practice Schism: Political
Response of Municipal Planners in Austin,
Texas
(1988) / Borchard G

TEXAS, UNIVERSITY OF (AUSTIN)
(continued)

Development of a Strategic Marketing Plan
for a Small Construction Company
(1988) / Curtis M R

An Investigation of Standard Test Procedures
for Polymer Concrete
(1988) / Harris C M

The Analysis and Behavior of Pin-Jointed
Schywedler Domes
(1988) / Kelley J D

Elder Cottages: Housing Strategy and Design
for the Aging
(1988) / Leiser N M

Noncollinear Interaction of a Tone with
Noise
(1988) / Lind S J

Instruction Set Design System with
Performance Evaluation Capability
(1988) / Massoudian K

The Role of the Planner in Zoning
Administration: A Case Study of Austin,
Texas
(1988) / Mullen G J

Construction Cost Reduction Through Improved
Design Quality
(1988) / Osei-Adubofour K

Architecture and Performance Evaluation of a
VLSI Instruction Cache for METRIC
(1988) / Patel R N

Investigation of the Behavior of Hardwood
Paneled Shear Walls Used in Manufactured
Homes
(1988) / Powell M G

A Parallel Architecture for High Quality
Graphics
(1988) / Rashid M

A Regional Analysis of the Economic and
Chemical Impacts of Agricultural
Diversification
(1988) / Rowley T D

Impact Fees: Municipal Experiments in
Fundraising
(1988) / Schulze A M

Evaluation of Logical External Memory
Architectures for Multiprocessor Systems
(1988) / Sivaramakrishnan S

An Open Systems Approach to Regional Service
Provision: Housing for the Developmentally
Disabled in Region III, New Hampshire
(1988) / Townsend L C

An Expert System for Steel Design Change
Response
(1988) / Yamaji M

U.S. NAVAL POSTGRADUATE SCHOOL

A Software Architecture for a Commander's
Display System
(1987) / Adams R M

Project Skyline: A Design Exploration
(1987) / Landers M F

Architecture and Allocation Considerations
for Group Expert Systems
(1988) / Rattigan M B

Project Skyline: A Design Exploration
(1987) / Welch W J

A Software Architecture for a Commander's
Display System
(1987) / Zyda M J

U.S.A.F. INSTITUTE OF TECHNOLOGY

Architecture and Design for a Laser
Programmable Double Precision Floating
Point Application Specific Processor
(1988) / Comtois J H

Finite Precision Arithmetic in Singular
Value Decomposition Architectures
(1988) / Duryea R A

A Performance Study of Hypercube
Architecture
(1988) / Lamanna C A

UTAH STATE UNIVERSITY

Urban Residents' Fear of Crime and the
Urban Environment
(1987) / Baker K R

The Role of Visitor Preference Information
in the Planning, Design and Management of
Archeological Resource Areas in National
Parks
(1987) / Cowley J P

Central Campus Open Spaces: An Evaluation
of Image and Use
(1987) / Woodward G K

UTAH, UNIVERSITY OF

The Laboratory Environment as an
Experimental Shell in Which Creativity is
Fostered Through Human Interaction and
Flexibility: A Medical Research Facility
of California, Irvine
(1988) / Berry G

A Senior Citizen Center and Congregate
Living Combination
(1988) / Blackner L

Exploratorium for the Physical Sciences
(1988) / Chen W

The Expression of Theology and Symbolism in
the Design of a Conservative Baptist Church
Building
(1988) / Ellis D A

A Self-Sufficient Research Colony on Mars
(1988) / Henriksen D S

The U.S.S. Johannes Kepler: A Space
Research Facility
(1988) / Hunter D R

A Youth Hostel at Cape Perpetua, Oregon
(1988) / Johnson D H

Seismic Research and Information Center,
Los Angeles, California
(1988) / Keefe C

An Asian Community Center
(1988) / Lee S L

UTAH, UNIVERSITY OF
(continued)

A Ski Hut System in the Wasatch Mountains
(1988) / Louie M

A Public Library in San Francisco
(1988) / Miller K B

VERMONT, UNIVERSITY OF

Obstacles to Land Use Planning in the
Matanuska-Susitna Borough, Alaska
(1988) / Chmielewski T M

VICTORIA, UNIVERSITY OF

Coastal Zone Conflict Resolution and the
Importance of Negotiation Preliminaries
the Case of Buckley Bay, Vancouver Island
(1988) / Carr C A

The Evaluation and Compatibility of Tourism
and Retirement in Parksville and Qualicum
Beach, British Columbis
(1988) / Foster D M

WASHINGTON STATE UNIVERSITY

Centralized Yard and Garden Waste Composting
and Alternative for Washington State
(1988) / Clemens C L

Utility of the Environmental Threshold
Concept in Managing Natural Resources: A
Groundwater Analysis
(1988) / Glasoe S D

The Effect of Temperature and Cytokinin on
Carrot Seed Variability and Germination
(1988) / Pfleging E J

WASHINGTON, UNIVERSITY OF

The Blackrock Valley Winery
(1988) / Albertin C R

Character and Townscape of Dubai Central
Area
(1987) / Bamardoof S M

A New Seattle City Hall at First Avenue and
University Street
(1987) / Brunner D P

Accomodating Increased Density Through
Transformation of the Wallingford
Residential Context
(1988) / Coleman C

Planners as Educators: Assessing the
Responsibility for Environmental Education
(1987) / Dean B A

The Roosevelt/Fairview Marketplace
(1988) / Dentoni S M

A Small Convention Center for Durango,
Colorado
(1988) / Evensen E

Understanding Place: Subjective Approaches
(1987) / Felberg K

A Design for Community Housing in East
Palo Alto, California
(1987) / Foug A

The Effects of Environmental Attitudes on
Landscape Preferences in an Urbanizing
County
(1987) / Freeman M R

Everett Waterfront for the Past 30
Years: Mapping Land Use and Analysis of
Land Use Change
(1987) / Fukushima Y

Placemaking in the Suburbs: Urban Design of
a Light Rail Suburg
(1988) / Gilbert G W

Design for the Seattle Art Museum
(1987) / Grant S D

An Achitectural Office Business Plan

Housing Families in City Neighborhoods: The
Design of Urban Housing and Open Space
(1987) / Hamer E L

Seattle Symphony Hall: D Design Proposal
and Study of Architectural Ornament
(1988) / Hirshkowitz R

Passage to Chinatown, Managing the Sense of
a Place - Chinatown - International District
Seattle
(1987) / Hsu H H

Qualitative Aspects of Daylighting in the
Design of a Church
(1987) / Hunting M L

Main Street as a Tool for Community
Development and Urban Re-Integration
(1987) / Jimerson J A

A Maritime Center for Seattle's Central
Waterfront
(1988) / Khan N J

A Secular Cemetery for the City of Seattle
(1987) / Kirkendall L A

Hugo Haering: In Search of an Organic
Architecture
(1988) / Koppe J C

Japanese Canadian Cultural Centre. A
Symbolic and Physical Focus for the Japanese
Canadian Community
(1987) / Lim W

Evolution of the Faculty Club on the
University of Washington Campus: 1909
Through Present
(1987) / Lindsay A L

A Finnish Community Center
(1988) / Lonka H

A Maritime Center for Seattle
(1988) / McIntire M L

Builder Housing: A Seattle Survey
(1988) / Mutter W E

A Japanese American Cultural Service Center
for Seattle, Washington
(1987) / Nakata R J

Population Concentration in Seoul and
Decentralization Policies in Korea
(1987) / Nam H C

WASHINGTON, UNIVERSITY OF
(continued)

Urban Residential Forms: An Analysis of
House Types and Residential Patterns in
Athens, Greece
(1987) / Porte D S

The Role of Spatial Quality and Familiarity
in Determining Countryside Landscape
Preference
(1987) / Porter M V

The Northlight Building. A Mixed-Use
Building at 6th Avenue South and South King
Street in Seattle's International District
(1988) / Riehl J

Investigations of Environmental Meaning
and Detached Housing
(1987) / Roe D S

A Filipono-American Cultural Community
Center in Seattle
(1988) / Rolluda A J

Access to Health Care for the Low Income
Population and the Issue of Uncompensated
Care in Seattle
(1987) / Sapunar K M

Architecture Hall: A Review of Its Past
and a Proposal for the Future
(1987) / Senn R P

Designing in a Divided Public
(1987) / Sharp S L

Current Practices in Land Use Plan
Monitoring in North America
(1987) / Slachowitz M A

An Identification and Analysis Procedure for
Real Property Marketability and
Developability in the Urban/Rural Interface
(1988) / Watson J L

Music and Dance at B. C. Place: A Proposed
Performing Arts Centre in Vancouver, B. C.
(1988) / Watts D L

Removing Barriers to Small Business
Expansion: A King County Case Study
(1987) / Wilson R K

The Politics of Adoptions: A Case Study of
the King County Comprehensive Plan-1985
(1987) / Wolf K

WEST VIRGINIA UNIVERSITY

The Social Production of Urban Space: A
Case Study of Pittsburgh's Hill District,
1945-1970
(1988) / Mallett W J

ARIZONA STATE UNIVERSITY

Quasars and Their Physical Association with Galaxies
(1988) / Landenburger C J

CALGARY, UNIVERSITY OF

On the Origin of Carbon Stars
(1988) / Chan S J

CORNELL UNIVERSITY

Long Term Pulsar Intensity Variations and Propagation Effects in the Interstellar Medium
(1988) / Thiering I

FLORIDA, UNIVERSITY OF

Astrographic Plate Reduction Using an Improved Weighted Least Squares Analysis
(1988) / Probst C

Statistical Studies of Photographic Data for Optically Violent Variables
(1988) / Shepherd D

INDIANA STATE UNIVERSITY

Solution of Asymmetric Light Curve of W-Ursae Majoris Binary Star System
(1988) / Wynne D

MANITOBA, UNIVERSITY OF

Spectra Associated with Commutative Rings
(1989) / Watanabe I

MASSACHUSETTS, UNIVERSITY OF

Cosmic Grain Size Evolution
(1988) / Field C A

MIAMI, UNIVERSITY OF

The Influence of Planetary Wave Instability on the Mean Circulation of the Venus Upper Atmosphere
(1988) / Turner R S

MINNESOTA, UNIVERSITY OF

Cerenkov Line Radiation from Quasars
(1988) / Chan V K

The Effects of Mass Loss on Subsequent Star Formation and the Evolution of Galaxies
(1988) / Congdon C W

Late-Type Luminous Supergiants
(1988) / Helper C E

SAN DIEGO STATE UNIVERSITY

A Photometric Analysis of the Eclipsing Binary OX Cassiopeia
(1988) / Crinklaw G R

A Photometric Study of the Ursae Majoris Star AP Leo
(1987) / Hickey J P

The Internal Geometry and Density Distribution and Their Effects on the Radio Continuum of Planetary Nebulae
(1988) / Thawadi A I

Effects of Mass Loss on the Internal Structure of White Dwarf Stars
(1988) / Webb K W

TEXAS A AND M UNIVERSITY

The Chaotic Terrain of Elysium Planitia, Mars
(1988) / Roberts S L

TEXAS, UNIVERSITY OF (AUSTIN)

The Gravitational Instability of Collisionless Particles in a Cosmological Self-Simular Shell
(1988) / Hwang J C

The Use of Narrow Band CCD Photometry for Determining the Spatial Distribution of Molecules in the Coma of Comet Halley
(1988) / Matney C G

The Search for Past Starbursts in the Milky Way
(1988) / Noh H R

U.S.A.F. INSTITUTE OF TECHNOLOGY

Three Probable Sellar Mass Black Holes
(1988) / Fry D W

VICTORIA, UNIVERSITY OF

Evolution of the Quasar Luminosity Function
(1988) / Schade D J

WESLEYAN UNIVERSITY

A Continued Photometric Survey of the H-Alpha Line in Late-K and M-Dwarf Stars with an Improved System
(1988) / Miller J R

NORTH CAROLINA, UNIV. OF (CHAPEL HILL)

A New Color Magnitude Diagram for NGC 5694
and Its Implications for the Mass of the
Milky Way Galaxy
(1987) / Luerich S A

ALFRED UNIVERSITY

Preparation and Evaluation of Silicon
Nitride Matrices for Silicon Nitride - SiC
Fiber Composites
(1988) / Axelson S R

A Property/Structure Study of Alkaline Earth
Galliosilicate Glasses
(1988) / Balewick L A

The Effect of Processing Variables on the
Electrical Properties of the System
Lead Manganese Niobium Oxide - Lead
Magnesium Tungsten Oxide - Lead Zirconium
Oxide - Lead Titanium Oxide for Low Firing
Multilayer Capacitor
(1988) / Barus A M M

The Effect of Magnesium Sulfate and Surface
Area on the Casting Rate of Brazil Sanitary
Ware Clays
(1988) / Bliss M J

Effects of A.C. Electric Current in the
Corrosion of Refractories for Electric
Glass Melting
(1988) / Breviari A

The Healing of Glass in Humid Environments
(1988) / Clark M K

Effects of Abrasion Flaws and Dislocations
on Stress-Induced Dauphine Twinning in
Alpha-Quartz
(1988) / Cocuzzi D A

The Effect of Atmosphere and Temperature on
the Sintering Behavior of Porous Glass
(1988) / DeCarr S M

Defect Structure of Calcia-Stabilized
Zirconia
(1988) / Dwivedi A

Properties and Structure of Lead
Fluoroborate Glasses
(1988) / Gressler C

Corrosion of Refractories by Metal Drilling
in Tank Bottoms
(1988) / Gupta A

Zinc Doped Lead Magnesium Niobate Relaxor
Dielectric Produced by Conventional Method
(1988) / Hsu W

Dense Cordierite Ceramics Via Sol-Gel
Coating of Powders for Electronic Packaging
Application
(1988) / Iyengar L

Automated Thermal Diffusivity Measurement
of Glasses and Their Melts
(1987) / Jain A

Electronic Defects in Zinc Oxide-Bismuth
Oxide-Manganese Oxide System
(1987) / Jiang J

An Analysis of Parameters Affecting Grinding
Rate and Powder Contamination During
Attrition Milling
(1988) / Kerr M C

Formation and Firing of Alkoxide Derived
Gels in the Zirconia-Silica System
(1988) / Kilinski B M

Intert Gas Solubility in Alkali Borate
Glasses and Their Melts
(1987) / Kohli J T

Study of Zirconium Oxide - Cerium
Oxide - Silicon Oxide - Titanium Oxide
System: Structural and Electrical Behavior
(1988) / Kudesia R

Glass Melt - Refractory Interactions
(1988) / Leaskey L

The Structure and Properties of Tin Doped
Zirconium Titanate Made by Coprecipitation
(1988) / Lee A H

Effects of Sol-Gel Coatings on Properties
of Glass
(1988) / Makni I B H

Investigation of Hardness Load Dependence
Using a Recording Microhardness Testing
Machine
(1988) / Mason W

The Effects of Process Gas Parameters on
Plasma Enhanced Chemical Vapor Deposition
of Silicon Oxynitrides Using 1.9 Percent
Silane
(1988) / Niermann D L

The Effects of the Float Glass Forming
Process on Ion Exchange Characteristics of
Soda-Lime Glass
(1988) / Pfitzenmaier R W

The Effects and Methods of the Dispersion
of Yttria in a Reaction Bonded Silicon
Nitride System
(1988) / Rabidoux C W

Effects of Deflocculation to Improve
Beneficiation and Mixing with Subsequent
Flocculation on the Filtration Rate of Two
Porcelain Bodies
(1988) / Ross D A

Processing of Electronic Ceramics Via
Sol-Gel Coating of Powders
(1987) / Selmi F

The Effect of Particle Size Distribution on
the Melting of Soda-Lime-Silica Glass
(1988) / Sheckler C A

Deep Level Transient Spectroscopy of Single
Crystal Zinc Oxide
(1988) / Simpson J C

Characterizing Guyana's Clays for
Sanitaryware Production
(1988) / Singh S

Nonstoichiometry in Alkaline Earth Excess
Alkaline Earth Titanates
(1988) / Udayakumar K R

The Effect of Flaw Origin on the
Ion-Exchange Strengthening of Glass
(1988) / Vitch S M

Investigation of Calcium Phosphate Cements
(1987) / Xie L

Effects of Green Packing Characteristics on
Solid State Sintering
(1988) / Zheng J

BROWN UNIVERSITY

Dynamic Fracture Initiation Testing of
Ceramics and Ceramic Matrix Composites
(1988) / Bopp E R

Plate Impact Experiments for Studying
Microcracking in Ceramics
(1988) / Raiser G F

CLEMSON UNIVERSITY

Preparation and Characterization of a
Composite Vanadium Dioxide Switch
(1988) / Chen C Y

Colloidal Processing and Mechanical
Properties of Alumina-Zirconia Composite
(1988) / Cherian I K

The Reduction of Nitrate Leaching in a
Concrete Waste Form by the Addition of
Waterproofing Admixtures
(1988) / Kass M D

Chemically Derived Near-Net-Shape
Ceramic-Ceramic Composites
(1988) / Park S Y

GEORGIA INSTITUTE OF TECHNOLOGY

Monolithic Structures Formed from Alumina
Hollow Microspheres
(1988) / Munne V

LOUISIANA STATE UNIVERSITY

Computer Simulation of Flow, Reaction and
Heat Transfer in Reaction Injection Molding
(1988) / Perrett F J

MARQUETTE UNIVERSITY

Chemical Vapor Deposition of
Borophosphosilicate Glass (BPSG)
(1988) / Jackson C

Piezoelectric Properties and Orthopedic
Implant Studies of Barium Titanate Ceramic
(1988) / Tai W C

MISSOURI, UNIVERSITY OF (ROLLA)

Surface Treatment and Chemical Dissolution
of Samarium-Lithium-Silicate Glasses
(1988) / Helma A E

Modelling and Optical Properties of
Composites
(1988) / Rutz H L

Evaluation of Crystallite Size and Strain
from X-Ray Diffraction Pattern
(1988) / Yau J K

PENNSYLVANIA STATE UNIVERSITY

The Role of Manganese on the Resistivity and
Stoichiometry of a Lead-Iron-Niobate Ceramic
(1987) / Gilmour P C

Slow Crack Growth in Fluorozirconate Glass
(1987) / Gonzalez A C

Zirconia-Yttria-Titania Ceramics
(1988) / Hsu C H

Thermomechanical Behavior of Oxygen in
Germanium-Antimony-Selenium Chalcogenide
Melts
(1988) / Rengan A R

Microstructure Development and Pyroelectric
Properties of Fresnoite Polar Glass-Ceramics
(1987) / Wheeler M B

WASHINGTON, UNIVERSITY OF

Mechanical and Thermomechanical Properties
of CVI Silicon Carbide Matrix Continuous
Ceramic Fiber Composites
(1987) / Eckel A J

Potential Application of Microemulsions in
Ceramic Processing
(1988) / Pak S S

AKRON, UNIVERSITY OF

Reaction Mechanisms of Chlorinated Methanes
Over Potassium Chloride Vanadium Dioxide
Catalyst
(1989) / Danals R

Supercritical Extraction of Coal
(1989) / Ghosh A

The Processing and Electrical Testing of
Filled Conductive Polystyrene Composites
(1989) / Govil A

Application of Artificial Intelligence to
Nonlinear Process Control and Information
Systems
(1989) / Karbhari P R

Integration Algorithm Development
(1988) / Natarajan L

Effect of Surfactants/Cobalt-Surfactants on
the Phase Separation Temperature and
Volatility of Methanol/Hydrocarbon/Water
Blends
(1989) / Nitiraharjo S

Effects of Heptanoic and Oleic Acid on the
Phase Separation Temperatures of Methanol
Hydrocarbon Water Mixtures
(1988) / Patel R S

Steady State Simulation of Three Phase
Recycle Reactor for Solvent Methanol Process
(1989) / Rastogi R

Thermal Stability and Activity of Mixed
Copper/Potassium Chloride Catalysts
(1989) / Saraf A

The Effects of Sulfur on the Oxygenate
Synthesis Over Silica Supported Rhodium,
Ruthenium, and Nickel
(1989) / Sze C

A Kinetic Study for the Growth of
Kluyveromyces Fragilis on Mixed Substrates
(1989) / Yun H S

ALABAMA, UNIVERSITY OF (UNIVERSITY)

Concentration and Structure of Coke on a
Zeolite Cracking Catalyst
(1988) / Andrews R M

Combustion of CWS: Maintaining a Steady
Flame and Measuring the Temperature Profile
(1988) / Haymon T D

Development of a Multipurpose Interface
Board for the Color Computer
(1988) / Hill W O

Methane Synthesis with Zeolite Catalyst
(1988) / Kim S H

The Optimization of Prehydrolysis Yield from
Hardwood Using a Semibatch Reactor
(1988) / Kuo J H

The Synthesis of Methyl Tertiary Butyl
Ether Over H-ZSM-5 Zeolite Catalyst
(1988) / Pien S I

The Prehydrolysis Study for Maximizing
Xylose Yield from Hardwood
(1988) / Wei M S

ALBERTA, UNIVERSITY OF

Multivariable Optimal Constrained Control
Algorithm (MOCCA)
(1988) / Lim K Y

Equilibrium Solubility of Hydrogen Sulphide
and Carbon Dioxide in a Mixed Solvent
(1988) / MacGregor R J

Measurement of Free Calcium Ion
Concentrations by Ion Exchange/Atomic
Absorption
(1988) / Mutucumarana V P

Near-Infrared Diffuse Reflectance Analysis
of Oil Sand
(1988) / Shaw R C

Attomole Determination of DABSYL-Amino Acids
with Capillary Zone Electrophoresis
Separation and Laser-Induced Thermo-Optical
Absorbance Detection
(1988) / Yu M

ARIZONA STATE UNIVERSITY

Continuous Ambulatory Peritoneal Treatment
of Diabetes: Design and Simulation of the
Hybrid Pancreas
(1988) / Aniuk L M

Microprocessor Based Process Monitoring and
Control for the SISO System
(1988) / Curtis M D

Estimated Knee and Hip Contact Forces During
Exercises Selected from the Arthritis
Foundation's Exercise Program
(1988) / Fuller J J

Rapid Thermal Nitridation of Thin Silicon
Dioxide Films
(1988) / Henscheid D K

In Vitro and In Vivo Stability of
Co-Immobilized Glucose Oxidase and Catalase
(1988) / Ireland T W

Evaluation of Polyetherurethane Urea,
Polytetrafluoroethylene, and Silicone
Rubber for Use in Total Pericardial
Replacement
(1988) / Rhodes J E

Mathematical Model of the Dual Origin
Circular Coronary Artery Bypass Graft
(1988) / Rice R W

The Efficacy of Oxygenated Asanguineous
Crystalloid Cardioplegia
(1988) / Shemansky F A

Experimental and Theoretical Studies on a
Cascaded Falling Film Liquid-Liquid Heat
Exchanger
(1988) / Turner T J

Pressure Drop and Flow Regimes of Concurrent
Two Phase Flow Through Small Scale Porous
Media
(1988) / Wilemon M S

BROOKLYN, POLYTECHNIC INSTITUTE OF
(continued)

An Increase in the Interfacial Bond Strength
of Kevlar 49/Epoxy Resin by the Introduction
of Pendent Chains at the Fiber Surface
(1988) / Jutis B

Parameter Estimation in
Differential/Algebraic Systems
(1988) / Lalka C J

Modeling of Chemorheology of an Emine-Epoxy
System of the Type Used in Advanced
Composites
(1988) / Lee C H

An Experimental and Theoretical
Investigation of the Reactivity and Oxygen
Chemisorption Behavior of Partially Gasified
Coal Chars
(1988) / Lizzio A A

Metal Requirement of Residual Protease in
Bacillus Subtilis DB104
(1988) / Longo A

Application of Thermodynamic Models to
Process Design and Control
(1988) / Norton C J

Modeling of Chemorheology of an Amine-Epoxy
System of the Type Used in Advanced
Composites
(1988) / Ott J D

BUCKNELL UNIVERSITY

Simulation of Heat Transfer in Tumors During
Photoradiation Therapy
(1988) / Johnson M D

CALGARY, UNIVERSITY OF

An Evaluation of the Trebble-Bishnoi
Equation-of-State with Respect to Prediction
of Phase Equilibrium in Binary and Ternary
Systems
(1988) / Da Sie W J

Evaluation of Immobilized Cell Continuous
Penicillin Fermentations
(1988) / Forestell S P

Computer Simulation of an Industrial Quench
Cooler on a Large Ethane Cracking Furnace
(1988) / Huntrods R S

Adsorption of Foam-Forming Surfactants
(1988) / Mannhardt K

Modified Shape Factors for Improved
Visicosity Predictions Using Corresponding
States
(1988) / Monnery W D

Ultrapyrolysis of n-Hexadecane in a Novel
Micro-Reactor
(1988) / Wallace J A F

CALIFORNIA, UNIVERSITY OF (LA JOLLA)

Mechanism of Electrodeposition of
Permalloy Thin Films
(1988) / Gangasingh D

Studies of Convective and Diffusive Mass
Transport Phenomena in Semiconductor
Fabrication Processes
(1988) / Heble B F

Comparison of Adsorption and Dynamic
Oxidation of Carbon Oxide on Supported
Platinum/Aluminum Dioxide Catalysts
(1988) / Racine B N

Adsorption and Desorption of Carbon
Monoxide on Supported Platinum Catalysts
(1987) / Sally M J

CARNEGIE-MELLON UNIVERSITY

Prevade: A Computer Program to Aid the
Preliminary Design of Multi-Effect
Evaporator Processes
(1988) / Dorsey M T

Diffusiphoresis of Latex Particles in a
Non-Ionic Surfactant Gradient
(1988) / Kaba M W

Effects of Preparation on the Properties of
Niobia-Alumina Binary Oxides
(1988) / Maurer S M

The Use of Total Internal Reflection
Microscopy to Examine Colloidal Forces
(1988) / Pino M A

Computer-Aided Scheduling of Multiproduct
Batch Plants
(1988) / Pradhan S

Express: An Expert System for Separations
Systems Synthesis
(1988) / Wehe R

CASE WESTERN RESERVE UNIVERSITY

The Calculation of Surface Elastic Moduli
for Langmuir-Blodgett Prelayers
Phthalocyanine Dimer Compound
(1988) / Boggs K E

The Electrochemical Fluorination of Octanoic
Acid
(1988) / Chuey S R

Electrolytic Production of Manganese
Dioxide: Rotating Cylinder Electrode and
Particulate Bed Electrode Reactor Studies
(1988) / Dietrich J J

Nonlinear Model Predictive Control of
Styrene Polymerization at Unstable Operating
Points
(1988) / Hidalgo P M

Rheological Characterization of a
Zwitterionic Suurfactant Solution with and
Without Pigment Addition
(1988) / Padolewski J P

Process Improvements in the Electrochemical
Fluorination of Octanoyl Chloride
(1988) / Prokop H W

Rehology and Casting Behavior of Aqueous
Suspensions of Modified Silicon Carbide and
Silicon Nitride Ceramic Powders
(1988) / Thuer A M

CITY COLLEGE OF NEW YORK

Study of Radial Densityt Profiles in a Fast
Fluidized Bed Using X-Ray Imaging
(1988) / Bandlamudi E K

Instabilities and Feedback Identification
of Dynamic Systems with Multiple Delays
(1988) / Pantelides G P

Mobility Enhancement in Partially Miscible
Systems with Thin-Layuer Chromatography
(1988) / Thomas D K

CLARKSON COLLEGE OF TECHNOLOGY

A Predictive Model for the Steady Shear
Viscosity of Polymer Melts
(1988) / Adams M E

Particle Contamination in Compressed Gas
System
(1988) / Alberhasky S J

Separation Process Simulation Using a
Nonequilibrium Model
(1988) / Arehole A N

Preparation and Properties of Uniformly
Coated Inorganic Collidal Particles
(1988) / Garg A K

Scale-Up of the Forced Hydrolysis
Presipitation Process
(1988) / Her Y S

Application of Thermodynamically Constrained
Hybrid Methods to Gibbs Energy Minimization
(1988) / Korchinski B

Phase Diagram of Titanium Chloride-Magnesium
Chloride System
(1988) / Kumar K A

Evaporation Control in Float Zone Refining
of Cadmium Telluride
(1988) / Kweeder J

Characterization of Oil Residuals by Means
of Hydrodynamic Measurements
(1988) / Kyriacou K

CF4/O2 Plasma Etching of Chromium and Cermet
(1988) / Malocsay E

Design and Characterization of a Chemical
Vapor Deposition Reactor
(1988) / Moravec P S

Special Non-Thesis Program
(1988) / Morton D

Multicomponent Reactive Condensation
(1988) / Pugliese R A

Particle Dispersion in a Wall Bounded
Turbulent
(1988) / Raja R

Excimer Laser-Induced Deposition of Copper
(1988) / Ritz M J

The Migration of a Large Non-Spherical
Bubble in a Rotating Liquid
(1988) / Ruggles J S

The Evolution of 3-Dimensional Structure in
Sedimenting Suspensions
(1988) / Rwei S P

Transient Behavior and the Kinetics of
Polymer Etching in Plasmas
(1988) / Scott P M

CLEMSON UNIVERSITY

Excess Properties of Ternary Lennard-Jones
Mixture
(1988) / Ashok K C

The Role of Micromixing in the Scale-Up
of Geometrically Similar Batch Reactors
(1988) / Baud R

Flue Gas Desulfurization by In-Duct Dry
Scrubbing Using Calcium Hydroxide
(1988) / Bond G A

Improvement and Scale Up of a Process to
Extract Mineralization Inhibiting Proteins
from Oyster Shells
(1988) / Kort M

Evaluation of the Perturbed Hard Chain
Equation of State Combined with Chemical
Theory for Prediction of Near Critical and
Supercritical Phase Equilibria for Mixtures
of Acetic Acid-Water-Carbon Dioxide and
Acetic Acid-Water-Propane
(1988) / McCully M A

The Analysis and Control of Commercial
Carpet Dryers
(1988) / Robertson A E

Effects of Physical Aging on Viscoelastic
Properties of Polymers and Composites
(1988) / Wang S F

CLEVELAND STATE UNIVERSITY

Modeling of a Modified Low Pressure Chemical
Vapor Deposition Reactor
(1988) / Inamdar A S

Design of Jet Reactor for Kinetic Study of
Catalytic Reactions
(1988) / Koronich E C

COLORADO SCHOOL OF MINES

Enhancement of Coal Liquefaction Reactivity
by Mild Alkylation Pretreatment
(1988) / Armstrong M E

Kinetics of Colorado Oil Shale Pyrolysis and
Hydropyrolysis
(1988) / Chu S H

A Penetration-Theory Solution for
Nonisothermal Gas Absorption Accompanied by
a Second-Order Reaction
(1988) / Evans J D

Direct Hydrogenation Reactivity of High
Volatile Bituminous Coal
(1988) / Flahive J P

Exploratory Studies in Enhanced Low Severity
Coal Liquefaction
(1988) / McHugh K J

Isochoric PVT Measurements for (0.99 Carbon
Dioxide + 0.01 Ethane) and Modelling PVT
Behavior of Carbon Dioxide Rich Mixtures
(1988) / Sherman G J

COLORADO SCHOOL OF MINES
(continued)

Extraction of Molybdenum Ore Using Gaseous
Water and Additives at Temperatures Above
the Critical Temperature of Water
(1988) / Simon M A

A Group Contribution Study for Synthetic
Fuel Constituents-Naphthenes
(1988) / Voss S F

COLORADO STATE UNIVERSITY

Cell Density Measurements in Hollow Fiber
Bioreactors
(1988) / Blute T

Mixing Requirements for Enzymatic
Hydrolysis of Cellulose
(1988) / Elander R T

Renaturation Kinetics of Recombinant
Secretory Leucoprotease Inhibitor
(1988) / Jachim S W

Modelling and Control of a Continuous Fermentation
(1988) / Kramer J F

Convective Diffusion in Catalytic Reactors of
Rectangular Duct with Non-Uniform Catalyst Activity
(1988) / Lee S Y

Performance Studies for a Solar Open-Cycle
Liquid Desiccant Cooling System
(1988) / Patnaik S

Chemical Vapor Deposition of Carbon Filaments
from Acetylene as Catalyzed by Iron and Nickel
(1988) / Schmitt C C

COLORADO, UNIVERSITY OF

Design and Simulation of Strippers for the
Removal of Volatile Organics from
Contaminated Groundwater
(1988) / Boyer J M

Selective Recycle of Flocculent Bacteria
Using Continous Bioreactors
(1988) / Henry K L

Dynamics of Packed Bed Superheaters
(1988) / Morrow A B

Use of an Expert System to Change Process
Model Structures for Control of a CFSTR in
Real-Time
(1988) / Morrow P W

The Experimental Determination of the
Permeability, Viscosity and Shear-Induced
Diffusivity of Fine Latexes and Microbial
Suspensions
(1988) / Ogden G E

Transient Isotope Studies of Iron-Catalyzed
Gasification of Carbon
(1988) / Perkins D A

Factors Affecting Yeast Flocculation
(1988) / Szlag D C

CONNECTICUT, UNIVERSITY OF

Interactions of Plasticizers with PVC
(1988) / Boo H K

Removal of Heavy Metals from Wastewaters by
Chitosan and Shellfish Waste
(1988) / Deshaies M R

Production of Penicillin-G Using Immobilized
Living Cells of Penicillium Hrysogenum
(1988) / Flanagan W F

Growth Kinetics of Environmental Crazing in
Polycarbonate
(1988) / Lee W C

Evaluation of Surface Treatment for Glass
Fibers in Composite Materials
(1988) / Lex P J

Modeling and Simulation of Electrode
Processes in Zinc Halogen Batteries
(1988) / Shenov R V

CORNELL UNIVERSITY

Mass Transfer from Flowing Emulsion in the
Microcirculation
(1988) / Tufail N

DELAWARE, UNIVERSITY OF

Process Induced Fiber Orientation and Weld
Line Studies on TMC and BMC Materials
(1988) / Eduljee R F

Partitioning of Amino Acids and Small
Peptides in Aqueous Two Phase Systems
(1987) / Prilutski S J

Modeling the Influence of Diffusion on
Polymer Fragmentation Kinetics
(1987) / Simko M J

Industrial Applications of Computerized
Tomography
(1988) / Sweany T N

DREXEL UNIVERSITY

Enhancement in Thermal Oxidation of Silicon by Ozone
(1988) / Chao S C

Enhanced Butanol Fermentation
(1987) / Zakrzewski J S

EAST CAROLINA UNIVERSITY

Single-Wire Coupled Microwave Diagnostics of
an Anisotropic Plasma
(1979) / Carawon R E

Implementation of Computer-Controlled
Electron Angular Distribution Measurements
of the ECU Accelerator
(1985) / Chen F Y

Development of a Fast Fourier Transform
System Using the Forth Language
(1984) / Colston E F

A Study of a Three Port Network and Its
Applications to Surface Acoustic Wave
Transducer
(1980) / Debnath N C

An Investigation of Light Intensity
Fluctuations During the 30 May 1984 Solar
Eclipse and Their Relationship to Lunar
Topography
(1986) / Harper M A

EAST CAROLINA UNIVERSITY
(continued)

A Matrix Transformation Between Taylor and
Fourier Series Representations of Functions
(1979) / Jones R R

The Determination of Direct Normal Isolation
Using Static Tilted Devices
(1983) / Lamm L O

Development of a Terminal Concentrator
Using an LSI-11/02 Microcomputer
(1981) / Lane A S

Trace Element Analysis with the East
Carolina University Van De Graff Accelerator
(1973) / Larkins A L

Determination of Size and Dielectric
Constant of Liquid Droplet Using Microwave
Systems
(1976) / Lee C S

An Iterative Technique for Calculating
Charged Particles Trajectories in Nonuniform
Fields
(1977) / Lin B S

A Linear Representation for the Left Hand
Branch Cut of the Scattering Amplitude
Using Chebyshev Polynomials
(1984) / Marguglio K A

Electron Spin Resonance in Single Crystal
Natural Danburite (Calcium Boron Silicon
Dioxide)
(1983) / Mazed M A

A Dedicated Computer System for the Study
and Control of Lyophilizer Operations
(1988) / Riggs C F

A Study of Various Perturbation Methods for
Solving the Time Independent Schrodinger
Equation
(1975) / Rose O J

A Spherical Electrostatic Electron Energy
Analyzer
(1980) / Sankoorikal J T

A Multifunction Local Computer Network
(1983) / Sarangi A G

Heat Transfer from an Internally Heated
Circular Cylinder to a Moving Incompressible
Medium
(1972) / Tamul J

Development of a Multiple Ion Source
(1978) / Thomas E C

Two-Dimensional Motion of a General
Relativistic Projectile in a Uniform
Gravitational Field
(1988) / White T A

Deconvolution and Characteristics of Cusp
Spectra for Electron Capture and Loss to
the Projectile's Continuum
(1986) / Yu Y C

FLORIDA, UNIVERSITY OF

Thermochemistry of Indium-Thallium and
Gallium-Thallium Liquid Alloys
(1988) / Krawczyk J R

Surface Passivation of Gallium Arsenide
(1988) / Racicot R

GEORGIA INSTITUTE OF TECHNOLOGY

Characterization of a Cobolt Molybdenum/
Aluminum (20) (3) Catalyst Exposied to a
Coke Inducing Environment
(1988) / Baumgart J W

Incinerator Ash Dissolution Model for the
System: Plutonium, Nitric Acid and
Hydrofluoric Acid
(1988) / Brown E V

High Temperature Deformation of Pan-Based
Carbon Fiber Precursors
(1988) / Daga V

Radinox Process Applied to Formaldehyde
Oxidation
(1988) / Fleming R W

Development of a Conductive Elastomeric
Matrix for Robotic Tactile Sensors
(1988) / Hammond P T

Effect of Disc Angulation on the Fluid
Dynamics of a Tilting Disc Mitral Valve
Prosthesis
(1988) / Mumpower E L

Rheological Characterization of High Density
Polyethylene with Processing Aids in
Capillary Flow and Its Implications in a
Non-Isothermal Annular Flow Process
(1988) / Nguyen K P

In Vitro Continuous Monitoring of Cardiac
Output Using Ultrasound Doppler in
Pulsatile Flow
(1988) / Ruo J B

The Kinetics of the Chlorine Dioxide
Generation Reaction
(1988) / Tenney J D

The Relation of Dynamic to Equilibrium
Values of Surfactant-Induced Interfacial
Tensions
(1988) / Wingard D B

HOUSTON, UNIVERSITY OF

Application of Quantitative Two-Dimensional
Polyacrylamide Gel Electrophoresis to Study
the Alterations in Intracellular Protein
Synthesis Due to Environmental stresses in
Mamalian Cells
(1988) / Biyani P

Finite Element Analysis of the Start-Up
Dynamics of a Deep-Well Oxidation Process
(1988) / Gelus E

Enhanced Oil Recovery Processes in Geologic
Models
(1988) / Hanauer D

Experimental and Theoretical Studies of
Gas-Liquid Two-Phase Flow at Reduced
Gravity Conditions
(1988) / Janicot A

Carbon Monoxide Rich Methanation Kinetics on
Supported Rhodium and Nickel Catalysts
(1988) / Keehan D

Sintering Kinetics of Supported Rhodium
Catalysts
(1988) / Patterson E

HOWARD UNIVERSITY

Application of a Gaussian Dispersion Model
to Predict Sulfur Dioxide Concentrations in
Washington, D.C.
(1988) / Adams T L

Graft Copolymerization of Acrytonitrile Onto
Cellulose
(1989) / Mariam H G

IDAHO, UNIVERSITY OF

Adaptive Design of a Zeolite-Water Solar
Refrigerator
(1987) / Asbe D

Design and Characterization of Microwave
Plasma System for Plasma-Solid Reaction
Studies
(1988) / Beaudry A

Design and Construction of a Supercritical
Fluid Reactor System for a Corrosion Study
of Metals in Supercritical Fluids
(1988) / Daehling K W

Pentachlorophenol Degradation in
Immobilized Cell Reactors
(1988) / Kadakia R K

Microcomputer Applications for Operator
Training in Relation to the University
of Idaho's Wood-Fired Boiler
(1987) / Lewis A

Optimization, Dynamics and Control of a
Kraft Mill Pre and Post Oxygen
Delignification Washing System
(1988) / McHugh T J

Thermocentrifugometric Analysis
(1988) / Milligan R J

Preparation and Performance of Rapeseed
Oil Fatty Esters as Diesel Fuels
(1987) / Mosgrove D

Purex Diluent Degradation Products and
Their Effect on Solvent Extraction
(1987) / Nguyen D

Development and Testing of a Rotary Disc
Pulser for Pulse Column Application
(1988) / Olson A L

Water-Steam Flow, Tube Burnout, and
Dynamics in Natural Circulation Boilers
(1988) / Pereyra E J

Prediction of Properties of
Nitrogen-Argon-Oxygen Mixtures Using
Extended Corresponding States Methods
(1988) / Rousseau M F

Development and Characterization of a
Plasma Fluidized Bed Reactor
(1988) / Wierenga C R

IOWA STATE UNIVERSITY

Enzymatic Hydrolysis of Corn Gluten Proteins
(1988) / Hardwick J E

Numerical Studies of Turbulent Reacting
Flows
(1988) / Leonard A D

JOHNS HOPKINS UNIVERSITY

Growth and Gas Production for the
Hyperthermophilic Archaebacterium:
Pyrococcus Furiosus
(1989) / Malik B

An Experimental Program to Determine the
Effect of Shear on the Phase Behavior of
High Pressure Polymer Solutions
(1989) / Watkins J J

KANSAS, UNIVERSITY OF

Modification and Application of the TORP
Stream-Tube Model to Calculate Tracer Flow
in a Single Layer, Homogeneous Reservoir
(1988) / Hwang M W

KENTUCKY, UNIVERSITY OF

Selective Separation of Organic and
Inorganic Species in Aqueous Solutions Using
Charged, Thin-Film Composite Membranes
(1988) / Adams R

Numerical Solutions of Fractional-Order
Chemical Reactions in Tubular Reactors with
Axial Diffusion
(1988) / Arnett W D

On a Petrov-Galerkin Finite Element Method
for Evaporation of Polydisperse Aerosols
(1988) / Huang L

A Second Order Closure Model Study of
Atmospheric Boundary Layer
(1988) / Pai P P

A Finite Element Algorithm to Model
Condensational Growth of Multi-Component
Aerosols
(1988) / Rao A

Coupling of Ode Solver and Statistical
Softwares for Determining Reaction Rate
Constants of Complex Kinetics
(1988) / Shah A D

Surface Areas and Pore Properties of Raw and
Biologically Desulferized Coal
(1988) / Shyam V

Characterization and Evaporation of Aerosol
Droplets
(1988) / Tilley H L

Evaporation of Single Aerosol Particles in
Continuum, Knudsen and Kinetic Regimes
(1988) / Zaveri J S

LOUISIANA STATE UNIVERSITY

The Effect of Fuel Additives on the
Formation of Soot During Benzene Pyrolysis
(1988) / Arbour G B

Laboratory Simulation and Diffusion Model
for Leaching of Organic Chemicals from
Marine Bottom Sediment
(1988) / Baron J A

Pressure Response Cell: Determination of
Non-Linear Kinetics
(1988) / Bhargava S P

LOUISIANA STATE UNIVERSITY
(continued)

Extraction of Lignin from Poplar Wood
Using Supercritical Ammonia-Water Mixtures
(1988) / Bludworth J G

Modeling Fluidized-Bed and Moving-Bed Coal
Gas Desulfurization Reactors
(1988) / Cockrill D E

Environmental Effects on the Segregative
Plasmid Instability in Eschericia Coli
(1988) / Coutinho T M

Design and Assessment of a Crossflow
Airstripper for Contaminated Groundwater
Cleanup
(1988) / Locicero L

Ion Exchange Kinetics of Dextranase on an
Agarose Based Resin
(1988) / Pethe S G

Batch Process Control of a Batch Reactor
and Its Implementation on ACS
(1988) / Singh R

An Investigation Into the Possibilities of
Using Archaelogical Data for Studying Metal
Movement in Soils Over Long Time Periods
(1988) / Thomas J

Mass Transfer and Pressure Drop for a
Cascade Cross Flow Picked Column in an
Air-Stripping Mode
(1988) / Wood F D

Kinetics of Sucrose Crystallization by
Cooling
(1988) / Youssef O

LOUISVILLE, UNIVERSITY OF

Synthesis and Characterization of Polyacetylene
(1988) / Bajikar S S

Kinetic Studies of Xylene Isomerization Over
Silica-Alumina Catalysts
(1988) / Bharati K D

Fortran Programs for Pressure Flow in Polymer
Processing
(1988) / Chiu M B

An Assessment of a Membrane Distillation Process
(1988) / Jenkins K D

Excess Volumes of Mixing: The Benzene +
Trichloroethylene System
(1988) / Pickerell D S

Kinetic Studies of Xylene Isomerization Over
Hydrogen, Iron, Ruthenium - Offretite Zeolite
Catalysis
(1988) / Shelukar S D

Thermal Analysis Comparison of Microwave Oven and
Convection Oven Cured EPON-812
(1988) / Shroff M H

LOWELL, UNIVERSITY OF

Computer Controlled Optimization of
Sheffield and Gurley Instruments
(1989) / Change K K

Microcomputer-Based Process Control System
for Batch Distillation
(1988) / Dauerman S A

Modelling of Gas Separation by Hollow Fiber
Membranes
(1989) / Giglia S

Study of the Heat Transfer Process During
the Melting and Freezing of Paraffin Wax
Particles in Wax-in-Water Emulsion
(1988) / Kalyani V J

Flash Pulse Method for the Determination of
Thermal Conductivities of Pultrusion Formed
Composite Materials
(1988) / Lee H C

Fracture Toughness of Elastomer Modified
Epoxy
(1988) / Mooney C T

Advanced Control of a Jacketed Tubular
Reactor with Multiple Reactions
(1988) / Rastogi A

Zircon from Zirconia and Silica in Neutral
Atmosphere for Use in Ceramic Inclustion
Pigments
(1988) / Trappen K M

MAINE, UNIVERSITY OF

Kinetic Modeling of Thermal Decomposition of
Lipnocellulosic Materials
(1988) / Balkan H

Physicochemical Properties of Multicomponent
Polymeric Systems
(1988) / Bhavsar C R

Effects of Shear Forces on Bacterial
Adhesion
(1988) / Colacino M G

Alternating Voltage Synthesis of Organic
Chemicals
(1988) / Hart P W

A Mass Transfer Analog for the Natural
Convection Heat Transfer in a Rectangular
Enclosure with a Constant Heat Flux at
Two Facing Vertical Walls
(1988) / Iyengar V R

Hemicellulose Retention During Kraft
Pulping
(1988) / Medhora H K

Membrane Permeate Interactions a Qualitative
View
(1988) / Morin M J

Tyhe Separation of Water/2-Propanol Mixtures
by Membrane Pervaporation
(1988) / Vachaspati J

MASSACHUSETTS INSTITUTE OF TECHNOLOGY

Morphology of Particle Size Distribution
from Fuel to Fly Ash in Coal-Water Fuel
Flames
(1988) / Paloposki T A

Packed Fiber Bed Reactor Design for Animal
Cell Culture
(1987) / Perry S D

MASSACHUSETTS INSTITUTE OF TECHNOLOGY
(continued)

Fluxes and Net Reaction Rates of High
Molecular Weight Material in a Near-Sooting
Benzene-Oxygen Flame
(1988) / Pope C J

Convection Inside a Porous Microbial
Partical as a Means of Nutrient Transport
(1988) / Tsiveriotis K G

MASSACHUSETTS, UNIVERSITY OF

Theoretical Modelling of Some Solids
Processing Equipment
(1988) / Guilloit P P

Characterization of Polyethylene Produced
at Low Yields Via Supported Chromium Oxide
Catalysts
(1988) / Naik B G

Continuous Cell and Cell Debris Filtration
in a Magnetically Stabilized Fluidized Bed
(1988) / Terranova B E

MICHIGAN TECHNOLOGICAL UNIVERSITY

The Effects of Surface Tension and
Wettability on Liquid-Film Mass Transfer
Coefficients in a Packed Column
(1988) / Au-Yeung P H

The Physical and Hydroliquefaction
Characteristics of Superclean Coals
(1988) / Everett R D

Flue Gas Desulfurization Using a Fluidized
Bed Spray Dryer
(1988) / Garlapati B R

Pressure and Velocity Correlations in a
Plane Turbulent Water Jet
(1988) / Menning E

Prediction of Effective Transport Properties
of Continuous Unidirectional Fibrous
Composites
(1988) / Mohan K K

Microencapsulation of Unsaturated Oil
(1988) / Patil R

Microwave Heating of Solids
(1988) / Patil S N

Optimization of the (CD) (EO) Bleaching
Sequence at Quinnesec
(1988) / Powell D A

Effect of Damaged Fibers on the Mechanical
Behavior of Unidirectional Short-Fiber
Composites
(1988) / Raju V K

Development of Rapid Small-Scale Column
Tests
(1988) / Reddy P S

Scale-up Studies for Non-Newtonian
Gas/Liquid Mixing
(1988) / Wierenga M

MINNESOTA, UNIVERSITY OF

Carbon-to-Nitrogen Ratio Effect on the
Anaerobic Digestion of Cheese Whey
(1988) / Backus B D

FTIR Spectroscopy of Formaldehyde Reactions
on Copper/Zinc Oxide Catalysts
(1988) / Ganapathy K A

Poly(Ethyl Acrylate) and Poly
(Alpha-Benzyl-L-Glutamate): An
Interpenetrating Polymer System
(1988) / Gibson P R

Engineering Analysis of a Continuous Flow
Fermentor with High Biomass Concentrations
of Flocculating Yeast
(1988) / Hochalter J B

Modeling of Galvanic Effects at Partially
Protected Metal Surfaces
(1988) / Kassimati A

Interfacial Mixing in Fast Polymerizations
(1988) / Machuga S C

Photopolymerization Reactor Engineering
(1988) / Martin P R H

Instrumentation for Microcarrier Cultures
(1988) / Oberg M G

Fluid Phase Equilibria from Minimization of
the Free Energy
(1988) / Quinones-Cisneros S E

Three Zone Animal Cell Bioreactors
Incorporating Contractable Fibrous Supports
(1988) / Scholz M T

Instrumentation of a Cell-Retention Reactor
for Suspension Culture of Hybridomas
(1988) / Seamans T C

Tracer Diffusion in Polymer Solutions: An
Investigation of Anomalous Behavior in
Viscous Solvents by Forced Rayleigh
Scattering
(1988) / Smeltzly M A

MISSISSIPPI STATE UNIVERSITY

The Effect of Paper Machine White Water on the
Uniform and Pitting Corrosion Rates of Selected
Steels
(1986) / Ahmad Z B

Design and Testing of a Low-Cost Small
Scale Alcohol Distillation System
(1986) / Kadir M O

Kinetic Studies of the Cyclohexene-Ozone
Reaction in Aqueous Solutions
(1986) / Keady H D

Mass Transfer and Selectivity in Ozone
Absorption Processes
(1987) / Mehta Y

Kinetics of Reactions of Aniline and Alpha-
Naphthylamine With Ozone in Water
(1986) / Peng S C

Steady State Operation of a Low-Cost Small-Scale
Alcohol Distillation System
(1986) / Silva J L

Liquid Phase Oxidation of Benzene by Ozone
(1986) / Soong H S

Design Fundamentals for Ozone Absorption
in Bubble Columns
(1987) / Stephens J W

Kinetic Studies of Ozonation of Toluene and
Naphthalene in Aqueous Solutions
(1986) / Wang P C

MISSISSIPPI, UNIVERSITY OF

Rapid Pyrolysis of Oil Shales
(1988) / Pullammanappallil C P

Nitrogen Oxide from the Fluidized Bed
Combustion of Lignite
(1988) / Sheu W

MISSOURI, UNIVERSITY OF (COLUMBIA)

PVTX of Binary Mixtures: Equations of State
and Mixing Rules
(1988) / Drame N R

Binders for Coal Log Extrudates
(1988) / Nika K A

Alginate Beads: Formation in a Two-fluid Atomizer
and Their Effect on Gas Hold-up in a Three-phase
Bubble Column
(1988) / Su H

Experimental Considerations of Chaotic
Behavior in Two CSTR's in Series
(1988) / Wu F

Numerical Analysis of the Temperature
Distribution in an Unbounded Medium
(1988) / Yeh L P

MONTANA STATE UNIVERSITY

The Reaction Kinetics and Film Morphology of
Molybdenum Films Deposited by LPCVD on a
Silicon Surface
(1987) / Flanigan E J

Acid and Enzymatic Hydrolysis of
Autohydrolyzed Lignocellulosic Substrates
(1987) / Lamar D A

Hot-Water Drying and Briquetting of Lignite
(1987) / Poor K J

Separation by Prevaporation of Para and Meta
Xylene in the Presence of Carbon
Tetrabromide
(1987) / Wytcherley R W

MONTREAL, UNIVERSITY OF

Etude de la Mesure de l'Ouverture de
Filtration des Geotextiles Non-Tisses par
Tamisage Hydrodynamique
(1988) / Belisle J

Caracterisation de la Cinetique du
Biotraitement Aerobie d'un Effluent
Reconstitue d'une Conserverie de
Betteraves de Table
(1988) / Duguay P

Etude de la Stabilisation et la Degradation
des Polyolefines Chargees
(1988) / Fetoui A

Etude de l'Hydrodynamique et du Transfert
de Masse Dans un Reacteur a Biofilm a
Fluidisation Inversee
(1987) / Garnier A

Etude de Faisabilite de la Separation de la
Vapeur d'Eau d'un Melange Gazeux Par un
Procede a Membrane
(1988) / Meyer M

Proprietes de Surface des Fibres de Verre
et Leur Effet Renforcant Dans Deux Matrices
Thermoplastiques
(1988) / Osmont E

Modelisation Rheologique de Solutions
Diluees: Macro-Molecules Semiflexibles
Versus Inflexibles
(1988) / Rollin A

Modelisation d'une Extrudeuse Monovis en
Regime Permanent
(1988) / Vincellette A

NEBRASKA, UNIVERSITY OF

An Analytical Evaluation of Future Nebraska
Bridgerail-Guardrail Transition Designs
Using Computer Simulation Model Barrier VII
(1988) / Ataullah S

Resource Recovery from Electroplating Rinse
Baths
(1988) / Bray S S

Design and Modelling of a Radiantly Heated,
Spouted Bed Biomass Pyrolysis Reactor
(1988) / Dingare N A

Thermal Analysis of a Filament-Wound
Composite During Cure
(1988) / Hsiung H

Modeling Studies of the Interaction of
Hydrogen with Polycrystalline Zirconium:
Effect of Pre-Adsorbed Oxygen
(1988) / Jahan I

A Unified Approach for Solving Nonlinear
Partial Differential Equations in Chemical
Engineering Applications
(1988) / Quan Leon J R

Dynamic Mechanical Spectroscopy: Model
Epoxy Resins
(1988) / Shih T

Performance Limits for a Flame-Tube Heater
(1988) / Yan Y

NEW BRUNSWICK, UNIVERSITY OF

Compton Scattering Nondestructive Inspection
Inspection of Concrete Structures
(1988) / Whynot T M

NEW JERSEY INSTITUTE OF TECHNOLOGY

A Dynamic Model of a Fill-and-Draw Reactor
and its Implications for Hazardous Waste
Treatment
(1988) / Chang S H

A Study on the Poly(Ethylene Terephthalate)/
Fluoropolymer Blends
(1988) / Chiu C

Respiratory Impedances Computed from the
Forced Oscillation Technique Using an
Esophageal Balloon
(1988) / Fisler R T

NEW JERSEY INSTITUTE OF TECHNOLOGY
(continued)

Oxidation of Tolueme, Benzene and p-Xylene
in a Photoreactor
(1988) / Horng S S

Determination of Aromatic Amines by
Diazonium Salt
(1988) / Hsu H H

Synergistic Fillers in Polymer Concrete:
New Composites
(1988) / Javed S H

A Dynamic Model of Fill-and-Draw Reactor
Utilizing an Inhibitory Substrate
(1988) / Ko Y F

Microbial Phenol Degradation Utilizing a
Complete-Mix Biological Reactor: The
Effects of Dissolved Oxygen Content
(1988) / Kollar K

Bead Design for Biodegradation of
2-Chlorophenol Using Micro-Organisms
Entrapped in Alginate Gel
(1988) / Lakhwala F

A Study of Human Exposure to Benzo(a)Pyrene
(BAP) Through Different Pathways. Part A:
Assessment of Benzo(a)Pyrene (BAP) Exposure
Through Urine Analysis with the Hydriodic
Acid Reduction Reaction. Part B
(1988) / Liang S K

Synthesis and Characterization of a Novel
Blocked Isocyanate Dental Adhesive Based on
Diphenylmethane - 4,4' - Diisocyanate
(1988) / Lin Y

Programming for the Prediction of
Thermodynamic Phenomena
(1988) / Rim O K

The Design, Manufacture and Mechanical
Analysis of an L4-5 Spinal Disc Prosthesis
(1988) / Siryi B A

Oxidation of N,N-Dimethylformamide and
N,N-Dimethylacetamide in a Photoreactor
(1988) / Syu S J

Agitation Requirements for Complete
Dispersion of Emulsions
(1988) / Tsai D H

Detection and Determination of
N-Nitrosohexamethyleneimine in
Benzenesulfonamide, N-[[(Hexahydro-1-H
Azepin-1-YL) Amino]-Carbonyl]-4 Methyl
by HPTLC
(1988) / Valdez A A

Degradation of Dichloromethane,
Trichloromethane and Carbon Tetrachloride
in a Photoreactor
(1988) / Wadhwa S

Use of Recirculation Reactor to Study
Biodegradation of 2-Chlorophenol
(1988) / Yang K C

Preconcentration and Direct Flame Ionization
Method for Measurement of Non-Methane
Organic Compounds in Ambient Air and
Statistical Analysis of Experimental Data
(1988) / Zuo L

NEW MEXICO STATE UNIVERSITY

Axial Migration of Spirulina Platensis in
Laminar Tube Flow
(1988) / Chappell M L

Aerobic Treatment of Wastewater by Means of
a Centrifugal Film Reactor
(1988) / Chavez J

Integrated Solar Aquafarming
(1988) / Koesoemodiprodjo C W

NEW MEXICO, UNIVERSITY OF

Static and Dynamic Prore Structure Via Nuclear
Magnetic Resonance
(1988) / Glaves C L

Characterization of Partially Sintered Silica
Compacts
(1988) / Holt T E

Effective Diffusivity Measurements of Methane and
Nitrogen in Coal
(1988) / Olague N

Modeling Acid Migration Through Soils
(1988) / Partin K

Fractal Analysis of Well-Defined Surfaces Using
Physical Absorption
(1988) / Stermer D

NEW YORK, STATE UNIVERSITY OF (BUFFALO)

Synthesis of Ceramic Fibers by Chemical
Vapor Deposition
(1988) / Arya P

A Study of the Catalytic Formation of
Carbon Filaments
(1988) / Bianchini E

Investigation on the Adsorption of Methane,
Carbon Monoxide, Carbon Dioxide, Hydrogen,
Hydrogen Sulfide and Their Mixtures on 5A
Molecular Sieves
(1988) / Chen Y

Surface Modification by Two-Liquid Process
Deposition of AB Block Copolymers
(1988) / Chung B

Fractionation of Polyethylene with
Supercritical Propane: Static Phase
Equilibrium Measurements and Calculations
(1988) / Colman E

An Experimental Study of the Effects of
Micropore Diffusion Limitations and Pressure
Gradient on Breakthrough Concentration
Profiles in Fixed Beds of Molecular Sieve
Adsorbent
(1988) / Haas O

Pulse Adsorption on Zeolite Partial
Oxidation Catalysts
(1988) / Knoerzer A

Viscous Fingering in Hele-Shaw Cells
(1988) / Manoussidis K

Improved Catalysts for Oxidation of
Dimethyl Methyl Phosphonate
(1988) / Padukone N

NEW YORK, STATE UNIVERSITY OF (BUFFALO)
(continued)

Experimental Investigation of the Dynamics
of Thermoforming
(1988) / Parisi M

Polyphenylacetylene Formed in the Presence
of Light-Activated W(CO)6: Reaction
Mechanism and Some Polymer Properties
(1988) / Park J S

The Effect of Controlled Levels of Carbon
Dioxide on the Metabolism, Morphology, and
Secondary Metabolite Formation of
P. Chrysogenum
(1988) / Shanahan J

An Investigation of the High Pressure
Kinetics of the Water-Gas Shift Reaction
Over a Sulfided Molybdenum Oxide-Alumina
Catalyst Promoted by Cobolt Oxide and an
Alkali Metal or Rare Earth Oxide
(1988) / Spillman D

Hybridoma Culture and Monoclonal Antibody
Production in a Hollow-Fiber Bioreactor
(1988) / Srinivasan V

NORTH CAROLINA STATE UNIVERSITY

A Model of Steady-State and Impedance
Measurements in Flooded Porous Electrodes
(1988) / Viner A A

NORTH DAKOTA, UNIVERSITY OF

Multiphase Flow in Vertical and Inclined
Annuli
(1988) / Patel R M

NOVA SCOTIA, TECH UNIVERSITY OF

Heat Transfer in a Circulating Fluidized
Bed Combustor
(1988) / Konuche F K

OKLAHOMA, UNIVERSITY OF

Development of Chemical Process Models for
Enhanced Oil Recovery
(1988) / Ahmed M S

Analysis of the Staged Construction of an
Integrated Gasification Combined Cycle Plant
with Methanol Synthesis and Carbon Dioxide
Recovery
(1988) / Boyd J A

The Use of Liquid-Coacervate Extraction for
the Removal of Organics
(1988) / Gullickson N D

Mathematical Modeling of Surfactant
Adsorption at the Mineral Oxide/Aqueous
Solution Interface
(1988) / Subramanian K

OREGON STATE UNIVERSITY

The Cloud Point Composition and Flory-Huggins
Interaction Parameters Polyethylene Glycol and
Sodium Lignin Sulfonate in Water-Ethanol Mixtures
(1987) / Luh S

PENNSYLVANIA STATE UNIVERSITY

Friction and Wear of Vapor-Deposited
Lubricant Films
(1987) / Chad K K

Vapor Phase Deposition Studies of Phosphate
Esters on Metal and Ceramic Surfaces
(1987) / Deckman D E

A Study of the Effect of Mixing on the
Precipitation of Barium Sulfate in a
Continuous Stirred Tank Reactor
(1988) / Fitchett D E

Ceramic Tribology: Methodology and
Mechanisms of Alumina Wear
(1987) / Gates R S

Measurement and Prediction of Critical
Temperature of Binary Hydrocarbon-
Nonhydrocarbon Mixtures
(1987) / Huang Z J

Influence of Additive Molecular Structure on
Lubricant Oxidation
(1987) / Hutter J C

Qualification and Family Analysis of
Viscosity and Thermal Conductivity
(1987) / Ibrahim S B

Development of a Microwave Hall Effect
Technique to Characterize Heterogeneous
Catalysts
(1987) / Kelly S L

The Preparation and Characterization of
Bimetallic, Silica-Supported Palladium
Copper Catalysts
(1988) / Leon C A

Algorithms for Condensed Phase Reactions
(1987) / Subramanian D

Experimental Determination and Prediction
of Critical Pressure for Nonhydrocarbon
Mixtures
(1987) / Wei G

A Study of the Effects of Pressure and/or
Shear Rate on the Viscosity of Liquid
Lubricants
(1988) / Wu C S

The Design and Construction of a
State-of-the-Art High Temperature Tribometer
(1987) / Yellets J P

Rheology and Formulation of Complex Fluids
(1987) / Zarkalis V S

PENNSYLVANIA, UNIVERSITY OF

Process Scheduling Using Artificial
Intelligence
(1988) / Weinstein M R

PITTSBURGH, UNIVERSITY OF

Oxidative Coupling at Methane Over Potassium
Doped Antimony Oxide Catalysts
(1989) / Agarwal S K

Experimental Determination of LDL Uptake and
Metabolism by the Arterial Wall: Role of
Hemodynamics
(1989) / Calvo W J

PITTSBURGH, UNIVERSITY OF
(continued)

Characterization of Silica Supported Nickel
Catalysts: Effects of Support and Alkali
(1989) / Ciocco M V

Studies in the Acceleration Zone of
Pneumatic Systems at Various Inclinations
(1989) / Dhodapkar S V

Testing of a System Designed to Monitor the
Chloride Channel of the Gaba Receptor
(1989) / Ferrance J P

Dense Phase Transport: The Vertical Plug
Flow in 2 and 4 Inch Pipes
(1989) / Gu H

Fundamental Study of the Performance of a
Laboratory-Scale Chlorine Dioxide Gas
Delivery System
(1989) / Liaw S J

Synthesis and Characterization of Poly
(Organo) Phosphazenes
(1989) / Mujumdar A N

Acid and Catalytic Properties of
Non-Stoichiometric Aluminum Borates
(1989) / Peil K P

Interfacial Property and Adsorption Study of
Lignite, Graphite Powder, Kaolin, and Pyrite
(1989) / Richardson A G

Behavior of a Laboratory-Scale Chlorine and
Chlorine Dioxide Gas-Adsorption System
(1989) / Sun L

PRINCETON UNIVERSITY

Analysis of Chemical Kinetic Systems with
Computational Singular Perturbation
(1988) / Konopka D M

PURDUE UNIVERSITY

Group Contribution Method for Prediction of
Effective Molecular Volumes in Gel
Permeation Chromatography
(1988) / Achar R

Moving-Withdrawal Liquid Chromatography of
Amino Acids
(1988) / Agosto M

Surface Tensions of Viscous and Anisotropic
Fluids
(1988) / Alexopoulos A H

Performance of a Novel Foam Practionation
Column for the Recovery of Proteins from
Dilute Waste Effluents
(1988) / Brown L K

Solubility of Synthesis Gases in a Fischer
Tropsch SASOL Wax and in Three High
Paraffins
(1988) / Chou J S

The Effect of Chemical Structure on the
Linear Viscoelastic Properties of Epoxy
Resins
(1988) / Gibson F W

Ammonia Catalyzed Extraction of Coal
(1988) / Gossen T J

Production of Colloidal and Silica Supported
Silver Clusters with a Multiple Expansion
Cluster Source (MECS)
(1986) / Jarosch T R

Unstable Steady States of Methylotrophs and
Their Potential Application to
Polysaccharide Production
(1988) / Jayakumar S

Experimental and Modeling Studies of the
Selective Epitaxial Growth of Silicon from
the Chemical Vapor Deposition of
Dichlorosilane
(1988) / Kastelic M M

Synthesis of Monodisperse Microspheroidal
Particles
(1984) / Keville K M

Use of Porous Glass Fiber and Composite
Cellulosic Matrix Column as a Support for
Glucoamylase Immobilization
(1988) / Lo W

Size Exclusion Chromatography on Corn Starch
(1988) / Long S E

Localization of Invertase in Recombinant
Saccharomyces Cerevisiae
(1988) / Marten M R

Deposition of Stabilized Metal Cluster
Films Using the Langmuir-Blodgett Technique
(1988) / Osifchin R G

Bacterial Growth on Lactose: Experiments
Versus Cybernetic Models
(1988) / Straight J V

A Study of Batch Chemical Process Capacity
Using a Batch/Semicontinuous Simulator
(1988) / Young R A

QUEENS UNIVERSITY

Turbulent Mixing of Enclosed Multiple Plane
Jets
(1988) / Caws M

XES FES and BUSI - Expert Systems for
Chemical Process Development
(1988) / Crowe C J

Fluidization Properties of Silica Sand and
Bed Material from Fluidized Bed Consumption
(1988) / Farkas N A

Culture of Hybridoma and Insect Cells in
Microcapsules of Controlled Membrane
Molecular Weight Cut-Off for the Enhanced
Production and Recovery of Monoclonal
Antibodies and Baculoviruses
(1988) / King G A

Delayed Release of Water-Soluble
Macromolecules from Polylactide Pellets
(1988) / Marcotte N

Particle-Size-Resolved Carbon Distribution
in Solid Waste of a Pilot Scale Atmospheric
Bubbling Fluidized Bed Combustor
(1988) / Poirier D J

RENSSELAER POLYTECHNIC INSTITUTE

The Impact of Household Hazardous Wastes
on Landfill Leachates
(1988) / Gapinski D P

RENSSELAER POLYTECHNIC INSTITUTE
(continued)

A Numerical Model to Describe the Kinetic
Behavior of a Substrate Amended Soil System
Under One Dimensional Unsaturated Flow with
Accompanying Biological Degradation of a
Trace Contaminant
(1988) / Janezic R T

A Study of Combustion Performance and
Dioxin Formation in a Spouted Bed Reactor
(1988) / Nugent P W

Frictional Behavior of Polymer Pellets and
Their Mixtures
(1988) / Shim W

Theoretical Calculations of Surface Tension
Gradients and Intermolecular Forces in
Thin Wetting Films
(1988) / Tee J J

Compositional Quenching of Phenolic Resins
for Impact Modification
(1988) / Tirrell S G

RHODE ISLAND, UNIVERSITY OF

Precipitation Studies of Calcium and Magnesium
Hydroxide from Aqueous Phase
(1988) / Bhandarkar S

Hot Corrosion of Sputter Coated Ceramics on
Ceramic Substrates
(1988) / Davies G B

Stability of Chemical Vapor Deposition Coated
Ceramics in Corrosive Melts
(1988) / Holmes T M

Computer Simulation of the Microstructure Developed
in Reaction-Sintered Silicon Nitride Ceramics
(1988) / Ku W

The Effect of Substrate Charge on Wettability
(1988) / Owiti C A

Protein Precipitation by Electrodialytic Salting-Out
(1988) / Poirier J A

ROCHESTER, UNIVERSITY OF

Growth Studies of Mixed Bacterial
Populations in Batch and Continuous Systems
(1987) / Bartel M E

Microspectrophotometric Method for Measuring
Oxygen Saturation in Single Red Blood Cells
(1988) / Considine L K

Purification and Characterization of
Thermotropic Liquid Crystalline Cyclic
Siloxane Oligomers
(1988) / Krishnamurthy S

ROSE-HULMAN INSTITUTE OF TECHNOLOGY

Decoupled Composition Control of a Binary
Distillation Column
(1989) / Morris D M

A Study of Aerobic Batch Baker's Yeast
Fermentation
(1989) / Wernimont C P

SASKATCHEWAN, UNIVERSITY OF (SASKATOON)

Flowmeter and Entry Length Tests with
Slurries
(1988) / Colwell J M

Direct Conversion of Plant Oils and
De-Pitched Tall Oil to Fuels and Chemicals
Using H-ZSM-5 Catalyst
(1988) / Furrer R M

Pyrolysis of Spent Coffee in a Batch Reactor
(1988) / Shankaranarayanan G V

A Generalized Correlation for Calculating
Viscosities of Mixtures of Hydrocarbons and
Related Fluids
(1988) / Vermani R

SHERBROOKE, UNIVERSITY OF

Spheroidisation des Poudres dans un Reacteur
a Plasma
(1988) / Bokhari A H

Etude de Modelisation de Plasma h.f. en
Ecoulement Turbulent
(1988) / Chehade M E

Etude du Transfert de Chaleur par Radiation
Thermique dans un Reacteur de Pyrolyse Sous
Vide des Vieux Pneumatiques
(1988) / Labrecque B

Production de Composes Aromatiques Liquides
a Partir du Gaz Naturel, Utilisant un
Reacteur a Plasma
(1988) / Laflamme C

Developpement d'un Reacteur a Niveaux
Multiples Pour la Pyrolyse Sous Vide de
la Biomasse
(1988) / Lemieux R

Modelisation Mathematique de l'Interaction
d'un Jet Avec un Ecoulement Isotherme et
Non-Isotherme
(1988) / Mjah Z

Modelisation Mathematique des Ecoulements
Plasma-Particules
(1988) / Proulx P

Contribution a l'Etude de la Purification
du Silicium par Voie Plasma Pour Obtention
de Silicium Haute Purete
(1988) / Robert W

SOUTH CAROLINA, UNIVERSITY OF

An Experimental Study of the Dynamic
Behavior of a Diaphragm-Type Chlorine
Caustic Electrolyzer
(1988) / Chin P C

Effect of Pipeline Configuration on Upstream
Mixing of Two-Phase Liquid-Liquid Flow by
Laser Image Processing
(1988) / Demetriou G D

Optimal Temperature and Initiator Policies
for Batch Solution Polymerization of Styrene
(1988) / Soots C A

An Empirical Model to Describe the Effects
of Velocity, Pipe Diameter and Viscosity on
Two-Phase Liquid-Liquid Flow Using Enhanced
Laser Image Processing
(1988) / Su H T

Simultaneous Recovery of Heavy Metal Ions
from Dilute Solution Using Flow-Through
Porous Electrodes
(1988) / Zeng S S

SOUTH DAKOTA SCHOOL OF MINES AND TECHNOL.

A Vector Algebra Geometry Based Heat
Transfer Simulation Algorithm
(1988) / Brost D M

Feedforward Control for Optimization of a
Continuous Stirred Tank Reactor
(1988) / Solaas D M

SOUTH FLORIDA, UNIVERSITY OF

Simulation Studies in Design of Flexible
Heat Exchanger Networks
(1988) / Chatchupong T

Supercritical Delignification of Wood in
Alkaline Environment
(1988) / Chen S L

Electrochemical Impedance Measurements of
the Effect of Deaeration on the Corrosion
of Steel in Concrete
(1988) / Hsieh S H

A Knowledge-Based System for the Selection
of Separation Technologies
(1988) / Netterfield T A

Study of the Dynamics of Free Radical
Polymerization
(1988) / Shetty S S

Effect of Surface Roughness on the Double
Layer Capacitance at the Steel/Alkaline
Medium Interface
(1988) / Thomas M N

SOUTHWESTERN LOUISIANA, UNIVERSITY OF

The Effect of Model Error on Optimum Control
Using the Smith Algorithm Predictor
(1988) / Kanpittaya S

Parameter Affecting the Formation of Iron
Carbonate in Aqueous Carbon Dioxide
Oxygen-Free Environment
(1988) / Mbagwu F O

Process and Disturbance Identification
(1988) / Raksakij W

The Effect of Model Error on Optimum PID
Controller Tuning
(1988) / Venable S W

STEVENS INSTITUTE OF TECHNOLOGY

Polyblends of LDPE: A Simple Rheological
Model for Phase Morphology Formation in
Melt Flow
(1987) / Bendaikha H

TENNESSEE TECHNOLOGICAL UNIVERSITY

Thermophysical Properties of Two Foam
Insulations
(1988) / Destephen M R

Analysis of Heat Flow Through Dynamic
Insulation
(1988) / Lin C

Heat Transfer Through an Assembly of Spheres
(1988) / Yang M H

TENNESSEE, UNIVERSITY OF (KNOXVILLE)

A Detailed Steady-State Control Analysis of
an Ethanol-Water Distillation Column
(1987) / Canter D L

The Effects of Matrix Scaling on the Design
and Control of Multivariable Processes
(1987) / Dyer C W

Application of Process Control Cultivation
to a Liquid-Liquid Heat Exchanger
(1987) / Farell A E

A Novel Technique for Measuring Interfacial
Tension in a Binary System
(1987) / Harris M T

A Comparison of Multivariable Controllers:
Single Value Analysis Controller and
Multivariable Tuning Regulator
(1987) / Khor G C

A Mathematical Model for Equilibrium
Distribution of Uranium(VI), Plutonium(IV),
and Nitric Acid in Purex Solvent Extraction
Processes
(1987) / Lee J H

Bioconversion of D-Xylose Into Ethanol
(1987) / Liu H S

Evaluation of Control Systems and Energy
Conservation Techniques for Distillation
Columns
(1987) / Roat S D

A Study of the Sedimentation of Bimodal
Distributions of Microspheres
(1987) / Shor J T

Evaporation of a Volatile Solute from an
Aqueous Spray Pond
(1987) / Smith G D

The Effect of Selected Environmental
Factors on the Growth of Wild Carrot Cells
in a Continuous Flow Perfusion Chamber
(1987) / White W K

TENNESSEE, UNIVERSITY OF (TULLAHOMA)

Evaluation of Improved Anion-Exchange Resin
Based Seed Regeneration Concept for Coal
Fired MHD Power System
(1988) / Modeste D C

Development of Potassium Recovery System for
MHD Spent Seed Containing Materials
(1988) / Patil A

TEXAS A AND M UNIVERSITY

Inhibition of Trichoderma Reesei Cellulase by
Cellobiose, Glucose, Ethanol and Butanol
(1988) / Cognata M B

Selective Reduction of Nitric Oxide with Ammonia
Over Vanadia on Pillared Titanium Phosphate
(1988) / Czarnecki L J

Interpretation of Data from a Pulse Reactor
(1988) / Desai N S

TEXAS A AND M UNIVERSITY
(continued)

Dependence of Viscous Properties of Dilute Drag
Reducing Solutions on Concentration and Salt
(1988) / Lackey D A

Synthesis of Ultrafine Grain Ferrites
(1988) / Livingston T W

Aged Asphalt
(1988) / Martin K L

Effects of Vapor-Liquid Equilibrium on Wetting
Efficiency in Hydrodesulfurization Trickle-Bed
Reactors
(1988) / Mills A L

A Description of the Vapor Phase in the Lithium
Thionyl Chloride Battery
(1988) / Morales R

Numerical Simulation of the Non-Isothermal
Developing Flow of a Nonlinear Viscoelastic
Fluid in a Rectangular Channel
(1988) / Nikoleris T

Identification of Toxic Components in Beechwood
and Petroleum Creosotes
(1988) / Okaygun M S

A Solid-State Nuclear Magnetic Resonance Study of the
Reactions of Propene on HY Zeolite
(1988) / Oshiro I S

Hydrodynamics of Bubble Columns with Application
to Fischer-Tropsch Synthesis
(1988) / Raphael M L

A Mathematical Model of a Gas-Fed Oxygen Reduction
Porous Electrode
(1988) / Ridge S J

Metal Loading and Reactivity of Zeolite Y
(1988) / Saenz M G

An Algebraic Model for a Zinc/Bromine Flow Cell
(1988) / Simpson G D

Synthesis and Characterization of Catalysts
Containing Nickel for Reforming Methane with
Carbon Dioxide
(1988) / Sommer M E

The Effect of WC Grain Size and Cobalt Content on
Properties and Electrical Discharge Machining of
WC-CO Composites
(1988) / Tsai W L

Regeneration of Adenosine 5'-Triphosphate Using
Atpase Immobilized in an Artificial Membrane
(1988) / VanDuker F C

TEXAS, UNIVERSITY OF (AUSTIN)

Modelling, Simulation and Stability of
Complex Crystallizers
(1988) / Chan W M

Blood Compatibility of Methacrylate
Copolymers: An Interfacial Approach
(1988) / Desai N P

Synthesis of Filler Particles for Dental
Composites
(1988) / Gu M R

Performance and Scale-Up of Laboratory
Scale Distillation Columns Containing
Structured Packing
(1988) / Hufton J R

Solvent Effects on Reactions in
Supercritical Fluids
(1988) / Mehta A J

A Light Scattering Device for Studying Human
Blood Platelet Aggregation
(1988) / Pohl P I

A Moving Phase Boundary in the Quasi-Steady
Solution of Laser Induced Droplet
Vaporization
(1988) / Slesar T J

TOLEDO, UNIVERSITY OF

Photo-Oxidation Products of Vinyl Porphyrins
(1988) / Cain S D

Mathematical Modeling of Systems Involving
Chemical Coupling
(1988) / Goetz D D

The Effect of Ethanol on the Kraft Pulping
of Poplar
(1988) / Hama B H

Development and Demonstration of a Hollow
Fiber Extractive Fermentor for the
Production of Ethanol
(1988) / Kagolanu R

Stabilization of a Hazardous Refinery Waste
with Cement Kiln Dust
(1988) / Majumdar P S

An Experimental Investigation of the Effect
of Cell Pressure on the Performance of
Resistojets
(1988) / Manzella D H

A Laboratory Study of the Catalytic
Conversion of Methane to Endothermic Gas
(1988) / Misra S

The Cooling of a Continuously Moving and
Stretching Sheet or Filament
(1988) / Murali K N

A Comparison of Numerical Methods for the
Prediction of Two-Dimensional Heat Transfer
in an Electrothermal Deicer Pad
(1988) / Wright W B

TULSA, UNIVERSITY OF

Microbial Removal of Sulfur Dioxide from a
Gas Stream
(1988) / Dasu B N

Vacuum Degasification of Water
(1987) / Hussain M A

A Method for Evaluating the Effects of
Two-Phase Flow Velocity on the Performance
of Oil Field Corrosion Inhibitors
(1987) / Sundar K

UTAH, UNIVERSITY OF

Shape Selective Hydrodewaxing of Jet Fuel
Boiling Range Distillates
(1988) / Longstaff D C

Light Scattering and Ultrafiltration of
Macromolecules in the Presence of
Submicron Particles
(1988) / Soong L F

VILLANOVA UNIVERSITY

Enhanced Direct Digital Control System
for Multi-Effect Evaporator Process
(1988) / Frandsen M

Experimental Determination of Heat-Transfer
Coefficients and Process Control Parameters
for a Helical-Flow Heat Exchanger
(1988) / Peterson K

VIRGINIA, UNIVERSITY OF

An Expert System for Process Control Loop
Tuning
(1988) / Bofinger K A

Comparison of Organic and Aqueous Two-Phase
Extraction in 2.3 - Butanediol Fermentation
(1988) / Eiteman M A

Network Models for Flow and Adsorption in
Porous Media
(1988) / George B E

The Mobilization of Penicillin Aeylase on
a Ceramic Filter
(1988) / Kelleher J M

Metabolism and Diffusion in Small Animals
(1988) / Kelly W J

Catalyzed Absorption of Nitrogen Oxide in
Nitric Acid
(1988) / Rose H P

A.C. Impedance Method for Characterization
of Deep Level States in Semiconductors
(1988) / Smolko F L

Oxygen Transfer Coefficients in Yeast
Suspensions
(1988) / Takemura Y

Ion-Exchange of Phenylalanine and
Tryosine on a Strongly Acidic Resin:
Equilibrium and Fixed Bed Studies
(1988) / Vierow J B

Molecular Dynamics of Shear Viscosity of
Supercritical Carbon Dioxide
(1988) / Wang R Y

WASHINGTON STATE UNIVERSITY

The Application of Fractals to Sintering
(1988) / Bruinsma P J

SMA and EAA Polymer Thin Films from a
Fluidized Bed Electrode Reactor
(1988) / Desai V M

Effect of Ionizable Surface Groups on the
Adsorption of Linear Alkyl Sulfates on
Polystyrene Latex Surfaces
(1988) / Gwin J L

A Feasibility Study on the Production of
Foreign Protein Products with Plant Tissue
Cultures
(1988) / Hogue R S

The Effect of Shear and Agitation on Plant
Tissue Culture
(1988) / Hooker B S

Rate Studies on Poly(Aniline) Film
Preparation in Static Cells and in Fluidized
Bed Electrode Reactors
(1988) / Segelke S

Burnout Studies of a Thick Film Copper
Conductor
(1988) / Whitehead W K

WASHINGTON UNIVERSITY

Mass Transfer Effects Involved with
Microencapsulated Hybridoma Cells
(1987) / Edmunds W W

Immobilization of Plant Cells Through
Adsorption and Entrapment
(1987) / Freidel-Summers I

Diffusion Through Microcapsules Made by the
Extrusion Process
(1987) / Scheidelaar M H C

Calcium Carbonate as a Phosphate Binder in
Uremia
(1987) / White K C

WASHINGTON, UNIVERSITY OF

Proclessing and Characterization Studies of
Poly(Ether-Ether-Ketone) (PEEK) Matrix
Composites
(1987) / Ahlstrom C

Monomer Solution Processed Polyimide Matrix
Composites
(1987) / Gallagher N B

The Aggregation Stability of a Vasicular
Drug Delivery System
(1987) / Gamon B L

Modelling the Surface Velocity Distribution
on AT-Cut Piezoelectric Quartz Crystals for
Liquid-Phase Applications
(1987) / Martin B A

Decolorization of Chitosan
(1986) / Muvundamina M

Carbon Oxide Detection by the Photo-Response
of Zinc Oxide Thin Films Deposited on
Chemical Field Effect Transistors
(1987) / Nguyen P

A Real-Time Data Acquisition and Control
System for Anaerobic Fluidized Bed
Bioreactors
Response Characterization and Dynamic
Modeling
(1987) / Slater W R

Internal Model Control with Constraints on
Process Inputs and Outputs
(1988) / Subrahmanian T

The Synthesis and Evaluation of an
Amino-s-Triazine Substituted Polyacrylamide
Dry Strength Resin for Paper
(1987) / Sullivan M P

Development of a Piezoelectric Sensor to
Monitor Fluid Properties and Flow Rates
(1987) / Wise B M

ABILENE CHRISTIAN COLLEGE

Isomerism in Heptanuclear Osmium Clusters
and Reactivity of [Re(2)Cl(7)(dto)]-Toward
Tertiary Phosphines
(1988) / Tekut T F

AKRON, UNIVERSITY OF

Formation Constants of B-Diketonato
Complexes of Copper (III)
(1988) / Alhedai A B

In Vitro Activation of Mouse Peritoneal
Macrophages by Quadrol-Bio-Gel
(1988) / Song J

Theoretical Investigation of Molecular
Behavior of Metal Dithizonate Complexes
(1988) / Sun Y

Synthesis, Characterization, and Phase
Equilibria of Water Soluble Dispersible
Polymers
(1988) / Vaidya M M

Purification and Kinetic Studies of
Aspartokinase III from Escherichia Coli K12
(1988) / Wokem A I

ALABAMA, UNIVERSITY OF (HUNTSVILLE)

Synthesis, Crystal Growth and Polymerization
of Diacetylene Monomers
(1988) / Patel D N

Synthesis and Accumulation of Myosin Heavy
Chain and Actin in Cultured Embryo Cardiac
Cells
(1988) / Zwanziger L L

ALABAMA, UNIVERSITY OF (UNIVERSITY)

Chemistry of Butenolides Derived from
D-Ribonolactone
(1988) / Blazis V J

Activation of Pyrazine Diazohydroxide,
Sodium Salt (NSC-361456) by Rat Liver
(1988) / Bridges E G

Cloning and Overexpression of the
Excherichia Coli trpB Gene and Purification
of the Tryptophan Synthase B2-Subunit
(1988) / McDaniel M L

Slurry Nebulization Inductively Coupled
Plasma Analysis of Sediments
(1988) / Palmer J L

ALBERTA, UNIVERSITY OF

The Structure of the Complex of Polypeptide
Chymotrypsin Inhibitor 1 and Streptomyces
Proteinase B
(1988) / Greenblatt H M

AMERICAN UNIVERSITY

Thermal Encapsulation of Toxic Metal Salts
(1987) / Baranek T M

The Isolation of Chlorinated Pesticides and
Polychlorinated Biphenyls in Specimens of
Human Liver and Other Biological Materials
(1987) / Kline W F

Induction of Rat Hapatic Weight Gain and
Functional Capacity by Barbiturates:
Correlation with Tumor-Promoting Ability
(1987) / Nims R W

Oxidation of Monosaccharides with Oxygen in
Alkaline Solution
(1987) / Shalaby M A

Synthesis of Chromium(III) Complexes with
Nucleotide, Amino Acid and Peptide
(1987) / Weng W

ARIZONA STATE UNIVERSITY

Elemental Abundances and Origin of the Iron
Dike
(1988) / Johnson T D

Determination of Formic Acid in Chondritic
Meteorites
(1988) / Kimball B A

Vapor Phase Sorption of Trichloroethylene on
Quartz Sand and Partially Saturated Soils
(1988) / Oja K J

Characterization of the Cell Wall and
Associated Protein Material of Seliberia
Stellata
(1988) / Santos M E

The Determination of Short-Range Nonbonded
Interactions in Alkali Halides
(1988) / Scarbrough S A

Characterization of the Protein Encoded by
DEB-A: A Gene Regulated During Drosophila
Embryogenesis
(1988) / Stilwell J L

Structure and Dynamics of DNA Hydration
Shells Studies by Light Scattering
(1988) / Tuo N

ARKANSAS, UNIVERSITY OF

Topological Studies of Monomeric and Dimeric
Cytochrome Oxidase Specifically Labeled with
a Fluorescent Probe on Subunit III
(1988) / Cho Y

Improvements in the Detection and
Identification of Antibiotic Materials
Using Laser-Based Polarimetry
(1988) / Shao Y Y

The Mechanism of Cytochrome c(2) in
Photosynthetic Electron Transport
(1988) / Zha X

AUBURN UNIVERSITY

Synthesis, Reaction Studies, and Pharmacological
Evaluation of 4-Benzylidene- and 4-Isopropylidene-1-
Aryl-3-Methyl-2-Pyrazolin-5-Ones
(1987) / Carter D A

Hyaluronidase-Purification and Inhibition by a
Gold Complex and a Steroid Derivative
(1987) / Durante G G

Loss of Propranolol to Ultrafiltration Devices
and Effects of a Perfluorochemical Emulsion on
Propranolol Binding by Alpha 1-Acid Glycoprotein
(1987) / Fan H F

AUBURN UNIVERSITY
(continued)

Computer Use in Independent Community Pharmacies
in Alabana
(1987) / Fulford M D

Effects of a Perfluorochemical Blood Substitute on
Diazepam Binding by Human Albumin
(1987) / Graben R D

A Study of the Electrocatalyzed Oxidation of
Trivalent Phosphorus Compounds with Dioxygen
(1987) / Ho H L

Studies of Sulfuranes with Oximate Apical Ligands
(1987) / Hornbuckle S F

The Synthesis and Characterization of Some
Bisphosphine Platinum Complexes
(1987) / Hsieh W C

An Index for Assessing Pharmacy Costs Against the
Medicare Standard Under the Prospective Payment
System
(1987) / Lai M M

The Syntheses and Characterization of Symmetrical
and Unsymmetrical Bidentate Phosphine Ligands
(1987) / Perry L M

Structure-Retention Relationship Studies in Reversed-
Phase Liquid Chromatography, Intramolecular Hydrogen
Bonding in Benzamides and Related Compounds
(1987) / Roura L E

An Investigation of the Inhibition of Acetaminophen
Hepatotoxicity by Chlorpromazine
(1987) / Saville J K

The Chemistry of Tris (2,6-Dialkoxyphenyl) Phosphine-
Approaches to a Novel Phosphorane and Studies of an
Ionophoric Phosphine
(1987) / Sun Y J

Synthesis and In Vitro Aldose Reductase Inhibitory
Activity of Compounds Possessing the N-Acylglycine
Pharmacophore
(1987) / Swearingen B E

Platinum Drug-DNA Binding, Studied by Restriction
Endonuclease Electrophoretic Fragmentation Patterns
(1987) / Tsai L L

BAYLOR UNIVERSITY

The Electrochemical Oxidation of Insoluble
Materials in Emulsion and Micelle Systems
(1988) / Adeniyi W K

I. Studies on the Competitive Thermal
Fragmentation of Hydroxyketones:
Dehydration Versus Formaldehyde Elimination
II. Studies on the Conjugate Addition of
Grignard Reagents to -Unsaturated Ketones:
Isolation and Oxidation of E and Z Enols
of a Ketone
(1988) / Logaraj S

The Use of Volumes of Activation to
Determine the Mechanism of Action of
Additives on Electrodeposition
(1988) / Mathew S A

BOSTON COLLEGE

Ganglioside Composition of Normal and Mutant
Mouse Embryos
(1988) / Bouvier J D

Studies Directed Towards the Synthesis of
Diazaquinomycin A
(1988) / Field J A

Tyhe Synthesis and Properties of
2-Pyrmidinone Containing
Oligodeoxynucleotides
(1988) / Gildea B D

Separation of Extracellular Endochitinase
and Exochitinase from Streptomyces Plicatus
(1988) / Hickey K A

Fluorescent Labeling of Nucleic Acids
(1988) / Hodges R R

The Role of DNA Secondary Structure on
Restriction Endonuclease Activity
(1988) / Rigby S

Modifying Catalysts by Coordination Site
Distortion
(1988) / Sanchez M Z

Sequencing and Comparison of Drosophila
Melanogaster Vitelline Membrane cDNA Clones
(1988) / White M K

BOWLING GREEN STATE UNIVERSITY

Effect of Substituents in the 13C NMR
Chemical Shifts of Gamma Carbons
(1988) / Alemayehu M

Cationic and Free Radical Reactivity of
Fluorinated Bicyclopentanes
(1988) / Amburgey J

Electrocatalytic Reduction of Oxygen on
Modified Oxide Surfaces
(1988) / Cahffins S

Flourine, Proton, and Carbon NMR
Investigations of a Rigid Molecular System:
A Halogen Substituted Series of Methyl
Tricyclo[4.2.1.0 2.5]Non-7-Ene-2-
Carboxylate and Related Compounds
(1988) / Huff A F

The Synthesis of Fluorinated Four Membered
Rings Especially 3,3-Difluorocyclobutene
(1989) / Jetter J W

A Chloride Ion Conducting Membrane for Use
in a Rechargeable Battery Using an Ambient
Temperature Molten Salt
(1989) / Lee C

Development of Cationic Rearrangement
Reaction of Bicyclo(2.1.0) Pentane
Derivatives for Synthesis
(1989) / Mu N

Electrochemical Investigations with
Microelectrodes: Rheometric Study of
Non-Netwonian Fluids and the Kinetics
of Superoxide Dismutation
(1988) / Parthasarathy A

A Stern-Volmer Analysis of Isomerization of
2-Methylcyclopentadience Using Vibrational
Overtone Activation
(1989) / Ranatunga D R

Vibrational Overtone Activation of Methyl
Isocyanide and Methylcyclopropene
(1988) / Samarasinghe S

BOWLING GREEN STATE UNIVERSITY
(continued)

Dihydrobenzoannulation of Methyl-5-Aryl-2,
4-Pentadienoates by Flash Vacuum Pyrolysis
(1988) / Senanayake C B

Synthesis and Spectral Properties of Rose
Bengal Derivatives Containing Anthracene
Chromophores
(1988) / Zhou J

BRIGHAM YOUNG UNIVERSITY

Attenuation Corrections Applied to Proton
Induced X-Ray Emission in Spherical
Particles
(1988) / Jex D G

I. Construction of a Low Temperature
Adiabatic Calorimeter. II. Calorimetric
Studies of Phase Transitions in CHI(3),
K(2)Mn(2)(SO(4))(3), and K(2)Mg(2)(SO(4))(3)
(1988) / Woodfield B F

BROCK UNIVERSITY

The Use of Diatom Assemblages to Determine
Paleotrophic Changes in Lakes
(1987) / Agbeti M D

The Identification, Distribution and
Persistence of Oxamyl and Its Degradation
Products in Planted Corn Seed, Seedling Root
and Soil from Oxamyl-Treated Corn Seeds
(1987) / Fulop G J

The Chemistry of Heterocyclic Methane Bases
and of Their Acyl and Thioacyl Derivatives
(1987) / Gupta A

Amino Acid Transport: The Special Case of a
H/L-Glutamate Co-Transport System in
Asparagus Cells
(1987) / McCutcheon S L

Biotransformation of Aromatic Hydrocarbons
and Organic Sulphides by Fungi
(1987) / Munoz B

Amplification of the DFR1 Gene in
Saccaromyces Cerevisiae
(1987) / Ondrusek N K

A Fast Atom Bombardment Mass Spectrometry
Study of Tris (3, 6-Dioxaheptyl)
Amine-Alkali Metal Halide Complexes and
Hydrogen Bonded Complexes
(1987) / Theberge R

The Fast Atom Bombardment Mass Spectrometric
Analysis of the Carbonate Pesticide Carbaryl
(1987) / Waters G J

Polytypism and Silicon Carbide: A Solid
State Nuclear Magnetic Resonance Study
(1987) / Winsborrow B G

BROWN UNIVERSITY

1,3,4-Thiadiazolidine-2,5-Diones and
Cis-Oxadiaziridines: A Photochemical
Approach to the Study of AZO Compounds
(1988) / De Felippis J

X-Ray Studies of Alkoxide Incorporation in
ENOLATE Aggregates Preparation of
Lithium OPYLAMIDE and Determination of a
Novel Ketone ENOLATE
(1988) / Orioli G J

BUCKNELL UNIVERSITY

HPLC of 9-Fluorenylmethylchloroformate
Polyamine Derivatives with Fluorescence
Detection
(1988) / Price J R

CALGARY, UNIVERSITY OF

Site-Directed Mutagenic Replacement of
GLU-461 with GLN in Beta-Galactosidase
(1988) / Bader D E

Amplification and Mutation of IMP
Dehydrogenase in Mycophenolic Acid Resistant
Neuroblastoma Cells
(1988) / Hodges S D

The Electrochemical Oxidation and Reduction
of Nickel-Cobalt Glassy Alloys in Alkaline
Solutions
(1988) / Lian K

Synthesis of Lupane Geomarkers
(1988) / Ottosen M K

An FT-IR Study of the Conversion of Methanol
to Higher Hydrocarbons Over Fluorided Zeolite
H-ZSM-5
(1988) / Villeneuve E

The Stimulation of Smooth Muscle Actomyosin
ATPase by Fodrin and Its Modulation by
Calmodulin and Protein Kinases
(1988) / Wang C

Oxidation and Reduction Processes of Lead
in pH 9 to 14 Solutions
(1988) / Waudo W

CALIFORNIA STATE UNIV. (FRESNO)

Development of Ion Chromatography for
Determinant of Selenium Species in
Irrigation Drainage Water
(1987) / Sauer N

A Test of the Reactivity-Selectivity
Principle Reactions of Amines with
Picryl-Chloride
(1987) / Tufon C

CALIFORNIA STATE UNIV. (FULLERTON)

The Interaction of Gly RS with Glys DNA
(1988) / Dittrich P

Dissociation of Horse Spleen Apoferritin
(1988) / Kokavandi H

Chiral Separation of Pharmaceutical
Compounds by High Performance Liquid
Chromatography: A Selective Review of
Recent Literature
(1988) / Kuan T

Affinity Labeling and Cross-Linking of
Glycyl-tRNA Synthetase Using Periodate
Oxidized tRNA
(1988) / Mendoza R

Copper Binding to Rat Plasma Proteins
(1988) / Vu H

An Investigation Into the Reactions Between
Phenylboron Dichloride and Grignard
Reagents
(1988) / Wildgoose R

CALIFORNIA STATE UNIV. (LONG BEACH)

Partial Purification of a Di-N-Butyl
Phthalate Metabolizing Enzyme in
Artemia Salina
(1988) / Healy P A

The Kinetics and Mechanism of Metal
Ion-Promoted Hydrolysis of Benzaldehyde
o-Ethyl s-Phenyl Acetal
(1988) / Hwang F S

A Theoretical Study of the Effect of Carbon
Monoxide on the Hemoglobin-Oxygen
Equilibrium Using the MWC Model
(1988) / Lesniewski E K

Substituent Effects in the Mercury(II)
Promoted Hydrolysis of O,S-Acetals of
Benzaldehyde
(1988) / Maynard D F

Oxidation of 4-Methyl-2-Mercaptopyrimidine
by Silver (II) Macrocycles
(1988) / Radzian R

CALIFORNIA STATE UNIV. (SACRAMENTO)

A Software Set for Laboratory Interfacing of
a Microcomputer Using PASCAL
(1988) / Hong K K

CALIFORNIA, UNIVERSITY OF (DAVIS)

Lewis Acid Adducts of Octaethylpophyrin
N-Oxide and Prophyrin Oxidation with Ozone
(1988) / Bailey D A

Total Syntheses of Regioselectively
Deuterated Protoporphyrin III and
Deuteroporphyrin XIII
(1988) / Chu S S

Gas Phase Study of Enol-Keto Tautomerism
and Nuclear Spin-Lattice Relaxation
(1988) / Folkendt M M

Covalent Binding and Enzymatic Oxidation and
Hydrolysis of Pyrrolizidine Alkaloids and
Their Metabolites
(1988) / Grasse L D

1. Synthesis of Esters of
1,2,3-Propanetricarboxylic Acid. 2. Novel
Compounds from Indian Medicinal Plants
(1988) / Sturm N S

ATP Sulfurylase from Rat Liver:
Purification, Kinetics, and Chemical
Modifications
(1988) / Yu M

CARLETON UNIVERSITY

One-Step Conversion of Cellulose to Fructose
Using Co-Immobilized Cellulase,
B-Glucosidase and Glucose Isomerase
(1988) / Chakrabarti A C

Glycolytic Enzyme Binding and Metabolic
Control
(1988) / Duncan J A

The Biosynthesis of Butenolide by HXL 1503
(1988) / Kendall J E

The Effects of Potential Energy Surface
Features on Product Vibrational
Distrubutions for the Reaction F+H HF+H
(1988) / Polowin J E

The Covalent Modification of
Phosphofructokinase and Pyruvate Kinase
in the Hypometabolic Pulmonate
(1988) / Whitwam R E

CITY COLLEGE OF NEW YORK

The Exploration of a Stereospecific
Reductive Cyclization as a Means of Creating
a Key Intermediate in the Synthesis of
Kaurenoid Diterpenes
(1988) / Nielsen C M

CLARK ATLANTA UNIVERSITY

The Study of Polymethylsilylethylenediamine and
Polydimethylsilylethylenediamine as Potential
Ceramic Precursors
(1988) / Abrahams P A

The Synthesis and Characterization of Poly[Styrene-
Co-(4-Amino-3-Methylamino-2-Oxabutyl)Styrene] and
Derivatives
(1987) / Lu Z

The Synthesis and Characterization of Chlorobutyl
Rubber-G Polyoxymethylene prepared Via Alkyl
Halide-Metal Salt Initiated Cationic
Graft Copolymerization
(1987) / Rashada Ya

Modification of Fructose 1, 6-Bisphosphatase
with Butanedione
(1988) / Simmons C M

The Photochemistry of the Azidopentaammine
Rhodium (III) Complex: Triplet State
(1987) / Zulu U

CLARKSON COLLEGE OF TECHNOLOGY

Preparation of Uniform Collidal Dispersions
by Chemical Reactions in Aerosols: A.
Titanium Dioxide: B. Titanium Dioxide
Coated with Polyureathane: C. Titanium
Dioxide Coated with Polydivinylbenzene: D.
Titanium Dioxide Coated with Silicon Dioxide
(1988) / Mayville F C

Effects od Deuterium Oxide on Locomotor
Activity and Courtship Rhythms of Drosophila
Bearing Mutations at the Period Locus
(1988) / Robinson D

CLEMSON UNIVERSITY

A Study of Organoarsenical (II) Compounds
and Biologically Significant Thiols and
Dithiols
(1988) / Adams E R

The Total Synthesis of the Alkaloid Mearsine
(1988) / Crouse J R

Oligomerization of Cytochrome c-Oxidase by
Cardiolipin in Detergent Solutions
(1988) / Kim Ahn K

Transition Metal Polyselenides and
Polysulfides
(1988) / O'Neal S C

Silyl Enol Ether Variation of Robinson
Annulation
(1988) / Potris S M

COLORADO STATE UNIVERSITY

Chloracnegenic Potential of
2,3,7,8-Tetrabromodibenzo-P-Dioxin
(1988) / Bird S

Cytochrome C Oxidase: A Step Toward
Synthetic Analogs
(1988) / Cranmer B K

Electrochemical Properties of Electrodes Modified
with Polymers Containing Metal Complexes
(1988) / Dunkle J R

Photoresponsive Polyquinolines
(1988) / Nyitray A M

CONCORDIA UNIVERSITY

Cytosine Deaminase from Saccharomyces
Cerevisiae: A Partial Purification and
Inhibition by Pyrimidines
(1988) / Alleva A A

An Investigation of the Behaviour of the
Maleonitriledithiolate Dianion of Platinum
in Heterogeneous and Homogeneous Systems
Under Visible Irradiation
(1988) / Domingue A R

Gradient Columns in the GC Analysis of Metal
Diethyldithiocarbamates
(1988) / El-Gawahergui A S

Preparation and Characterization of Leached
Asbestos Materials
(1988) / Kipkemboi P K

Cytochrome Oxidase and Metallorderivatives
of Cytochrome
(1988) / Laberge M M

Thermal Dissociation of Carbon Monoxide from
Heme Peroxidases
(1988) / Monshipouri M

The Characterization of Alkaline Phosphatase
in Rat Tests
(1988) / Nagy F S

Conversion of Aqueous Ethanol to Ethylene
Over Zeolite ZSM-5 and Chryso-Zeolite ZSM-5
Catalysts
(1988) / Nguyen T M

Influence of Fluoride Ions, Stoichiometry,
Reaction Temperature and Time on the
Composition and Symmetry of Superionic
Lead Tin Fluoride Obtained by Aqueous
Reactions Between Lead (II) Nitrate and
Stannous Fluoride
(1988) / Parris J

CONNECTICUT, UNIVERSITY OF

The Simultaneous Separation and
Quantification of Human Milk Lipids
(1988) / Colline S E

Comparison of Colorimetric and HPLC Analysis
of Vitamin A in Plasma and Liver from Rats
with Differing Initial Stores and Times of
Depletion
(1988) / Contois J H

Characterization of Two Glutamine
Synthetases from Frankia sp. Strain CPI1
(1988) / Edmands J A

Crystal Chemistry and Some Magnetic
Properties of Vanadium Oxides with the
Spinel Structure
(1988) / Jian A

Photo-Birch Reduction of Some Alkyl
Substituted Benxenes
(1988) / Kumar A

Microwave Spectroscopic Studies of Nitrous
Acid Allyl Ester and 1-Pyrrolidine
Carboxaldehyde
(1988) / Lee S G

The Development and Application of a
Reactive Labelling Technique for Monitoring
the Use of Polyimides
(1988) / Mathisen R J

Identification of Virulence Factors
Associated with Strains of Campylobacter
Jejuni Utilizing an Avian System
(1988) / Nettles C G

Electron Spin Resonance Study of Copper in
Ionomers
(1988) / Oh J J

The Mechanism of Inactivation of
Chloramphenicol by Chloramphenicol Acetyl
Transferase
(1988) / Russell M W

Interfacial Studies of Optical Fibers
(1988) / Saini A K

Proton Fluxes in Samanea Leaflet During
Dark Induced Leaflet Closure
(1988) / Xu Y

Analysis of PMR-15 by High Performance
Liquid Chromatography
(1988) / Yungk R E

CORNELL UNIVERSITY

Polymer Modified Electrodes for Trace
Metal Ion Analysis
(1988) / McCracken L L

DALHOUSIE UNIVERSITY

The Effect of an Electron Releasing
Substituent on the Reactivity of Radical
Cations
(1988) / Henseleit K M

The Photosensitized (Electron Transfer)
Tautomerization of Alkenes
(1988) / Mines S A

Cholesterol Metabolism in Cultured Human
Fibroblasts: Evidence for Heterogeneity
in the Defect in Type II Neuman-Pick Disease
(1988) / Rastogi S A

Expression of the Apolipoprotein B Gene in
the Developing Chick Embryo: Tissue
Specificity and Estrogen Responsiveness
(1988) / Wiktorowicz M

DELAWARE, UNIVERSITY OF

Effects of Temperature and Solvent on
Florescence Spectra: A Predictive Model
(1988) / Brams K F

Development and Study of Biosensors
Utilizing Intact-Chemoreceptor Structures
(1988) / Buch R M

Synthesis and Structural Characterization of
Metal Clusters Containing Main Group
Elements
(1987) / Fountain M

An Apoenzyme Electrode for the Determination
of Pyridoxal -5'- Phosphate
(1988) / Francis A J

Inactivation of Pig Kidney General Acyl-CoA
Dehydrogenase by 2-Alkynoyl-CoA Derivatives
(1987) / Freund K

DREXEL UNIVERSITY

Analysis of Blood Flow in Normal and Cancerous Breast
Tissues at Normal and Reduced Ambient Pressures
(1987) / Abraham V P

Peptidases of the Dimorphic Fungus Mucor Racemosus
(1988) / DiSanto M E

Diatoms as a Source of Omega-3 Fatty Acids
(1988) / Drewicz G A

Computer-Assisted Morphometry and Shape Analysis of
Sezary Cells in the Peripheral Circulation of
Sezary Syndrome Patients
(1988) / Hess S K

Synthesis of Alpha-Keto Esters
(1988) / Ly C Q

Three Dimensional Reconstruction of Chromatin
Fibres from Multiple Tilt Scanning Electron
Microscope Images
(1987) / Rundle D A

The Effect of Two Hypolipidemic Agents,
Cholestyramine Resin and Niacin, on the
Development, Sterol Composition and
Glyceride Content of Heliothis Zea
(1988) / Smith V B

Magnesium in Methanol: A Selective Reducing
Agent
(1988) / Suchismita

Methods for Prevention of After-Cataract Formation
Inhibition of Lens Epithelial Cell Proliferation
Using Targeted Cytotoxins
(1987) / Tyndall R

A Multi-Channel Data Acquisition System for
Biological Signals
(1988) / Walsh D F

Maternal and Infant Characteristics Associated with
Breast Versus Formula Versus Combination Feeding
(1988) / White S

Synthesis Directed at 2,3,6-Trideoxy-3-Aminohexoses
(1988) / Yuan C K

DUKE UNIVERSITY

1: An Approach to the Synthesis of
Lincosamine. 2: Study of Nucleophilic
Reactors of the Pyranose Ring Oxygen
(1988) / Date V

A Study of the Selective Reduction of Chiral
Iminium Ions
(1988) / Kaufman C R

Preparation and Chemistry of
Trimethylsilymethylarsine
(1988) / Kwag C Y

Diasteromeric Control of Two-Chain Ketone
Amphiphiles Via Hydrophobic Effect
(1988) / Lee E C

A Study of Two Reactions of Chiral
Hydroxylamines
(1988) / McFadyen R B

Design and Synthesis of a Photoreactive
Serine Proteinase Inhibitor
(1988) / Westin C D

EAST CAROLINA UNIVERSITY

Aldol Condensation: A Stereoselective
Approach to Tetra-Hydrophenanthrene
Derivatives and 3-oxa-bicyclo [3.3.1]
Nonan-6-Ones
(1989) / Biggers M S

Wavelength Selection and Nonlinear Response
in Multivariate Spectroscopic Assays
(1989) / Gregoriou V G

Application of Evolving Factor Analysis as
a Self-Modeling Curve Resolution Technique
(1987) / Hamilton J C

Kinetic Studies of Electrode Reactions by
Osteryoung Square Wave Voltammetry
(1988) / Jenq J

Liquid Chromatography/Electrochemical
Determination of Phenols Via Precolumn
Derivatization
(1987) / Kemp M W

The Selection of Ions by Solvents Confined
to Narrow Pores: Their Physical and
Biophysical Significance
(1988) / Liles T L

Raw Material Testing Using SIMCA Analysis
of Near Infrared Reflectance Spectra
(1988) / Webber L D

A Stereoselective Synthesis of d,1-16-oxa-15
Alpha-Methyl-19-Nortestosterone
(1988) / Yu B C

EAST TENNESSEE STATE UNIVERSITY

Genetic and Biochemical Characterization of
Novel Ribosomal Protein Mutant of
Escherichia Coli
(1988) / Amlinger A B

Some Interesting Observations on the
Dissolution of Copper in DMSO-CCL4 Systems
(1988) / Bunn B B

A Study of Some Synthetic Methods for the
Preparation of Pyrrolizidine Precursors
(1988) / Lin Y C

EAST TEXAS STATE UNIVERSITY

The Investigation of a Possible Correlation
Between Selenium in Spruce Trees and Fossil
Fuel Production of Power
(1988) / Colestock C N

EASTERN ILLINOIS UNIVERSITY

Isolation and Partial Characterization of
Serogroup Specific Antigens
(1988) / Carrion M E

The Structure of O-Antigen from
Liposaccharide of Rhizobium Leguminosarum
128C53 and its NOD- FIX- Mutant
(1987) / Chen T B

Coal Desulfurization Using Pyrite Conversion
Through Dielectric Heating
(1988) / Cleaveland D C

Characterization of the Lipopolysaccharides
from Symbiotic Mutants
(1988) / Garcia F

Sequential Solvent Extraction of Argone
Premium Coal Samples
(1987) / Mai W

Electrochemical Studies of Iron, Cobalt, and
Nickel Complexees of Bidentate Phosphine
Thioether Ligands
(1987) / Turner R L

EASTERN KENTUCKY UNIVERSITY

Characterization of a Commercial Grade of
Polyethylene for Extrusion of Thin Films
(1988) / Gill A M

The Synthesis of Some N,N'-Bis-(Aralkyl)-1,
4-Diaminobutanes and 1,3-Bis(Aralkyl)
2-Aryl-1,3-Perhydrodiazepines
(1988) / Halliday B D

EASTERN NEW MEXICO UNIVERSITY

Spectroscopic Properties of
Super-Exchange-Coupled Binuclear
Chromium(II) Complexes
(1988) / Chen L

Possible Effects of a Plant Lectin
(Concanavalin-A) on Gaillardia Multiceps
(Greene) Seed Germinability in the Presence
of Calcium Sulfate and Manganous Sulfate
(1988) / Upshur G

EMORY UNIVERSITY

Synthesis and Use of Polyoxometalates as
Anti-Viral Compounds
(1988) / Hartnup M A

A Database Program on the Storage, Search,
and Retrieval of Solubility Data
(1987) / Iwamoto M

Studies in Bioorganic Chemistry: 1.
Synthetic Approaches Toward Novel
Transmembrane Ion Transport Molecules.
2. Studies of the Alkylation of
Pyrimidine Nucleotides by Covalent
Catalysis
(1988) / Johnson E J

Electrophoretic Variation for
Glutamate-Oxaloacetate Transaminase Enzymes
in Talinum Mengesii
(1987) / Kemp A C

Purification and Characterization of Human
Redoxyendonuclease: A DNA Enzyme that
Recognizes Oxidative and Radiation-Induced
DNA Base Damage
(1987) / Lee K

A Practical Synthesis of Substituted
Benzocyclobutenediones
(1988) / Lescosky L J

The Functionalization of Adamantane Using
Transition Metal-Substituted
Polyoxometalates as Catalysts
(1988) / Lin C H

Recent Efforts Towards the Maximization of
Regiochemistry in Substituted
1,4-Naphthoquinone Synthesis
(1987) / McSwain C M

Inhibition of Serine Palmitoyltransferase
In Vitro and Long-Chain Base Biosynthesis
in Intact Chinese Hamster Ovary Cells by
Beta-Cl-Alanine
(1988) / Medlock K A

I. Ether Phospholipids: New Synthetic
Methods and Compounds. II. New Potential
PKC Activators
(1988) / Sisti N J

The Influence of Tert-Butyl Alcohol as a
Mobile Phase Modifier on High Performance
Liquid Chromatography Using Bonded Phase
Cyclodextrin Columns
(1988) / Tarr M A

A Specific Methodology to Facilitate the
Development of New Tc-99m Diagnostic
Imaging Agents
(1988) / Williams H L

EMPORIA STATE UNIVERSITY

The Application of the Fuoss-Onsager Conductance
Theory to Chlorides of Alkali Metals in
Ethanol-Water Mixtures
(1988) / Bastianpillai R E

Investigation of the Use of Positive Chemistry
Halogen Compounds as Oxidants of Cholesterol
(1988) / Sahli A

FLORIDA STATE UNIVERSITY

Determination of Force Constants for Group 6
Metal Carbonyls for Use in Mechanistic Studies
(1988) / Berger T G

The Role of Anisotropic Attractive Potentials
in Nematic Liquid Crystal Theory
(1988) / Eldredge C P

Electron Transfer Reactions of Nickel(II),
Copper(II) and Vanadyl Schiff Base Complexes
(1988) / Hoferkamp L A

Solid State ^{15}N NMR of Uniformly [15]N
Labeled Gramicidin
(1988) / Lograsso P V

Solid State Phosphorus Nuclear Magnetic Resonance
Study on Alignment and Characterization of
Oriented Phospholipids Containing Gramicidin D
(1988) / Moll F

FLORIDA, UNIVERSITY OF

Computational Studies of Molecular
Collisions
(1988) / Cohen J

Synthesis and Characterization of Donor
Acceptor Network Polymers
(1988) / Johnson M

The Construction and Evaluation of a Helium
Inductively Coupled Plasma
(1988) / Manning T

The Construction and Evaluation of a Pulsed
Hollow Cathode Lamp Excited Inductively
Coupled Plasma Atomic Fluorescence
Spectrometer
(1988) / Masamba W

Interaction of a Three-Element Thin Film
Detector with Selected Low Energy Heavy Ions
(1988) / Milosevich Z

Characterization of Electrochemical Probes
in CTAB/1-Butanol/Hexadecane/Water
Microemulsions
(1988) / Myers S

The Evaluation of Light Emitting Diodes as
a Source for Molecular Fluorescence
(1988) / Napier R

A Mechanistic Study of the Photolysis of
Dicarbonyl-h(5)-Cyclopetadienyl-(1-Phenyl
Cyclobutyl-1-Carbonyl) Iron
(1988) / Trace R

Attempted Synthesis of a
Hexahydro-2H-Naphtho-{1,8-bc}-Furan-2-One
Precursor to Forskolin
(1988) / Walker M

GEORGETOWN UNIVERSITY

Spectroscopic Investigations of
Molybdophosphate Heteropoly Blues
(1989) / Barrows J N

Nuclear Magnetic Resonance Studies of the
Interaction of Proton Donors with Organic
Free Radicals
(1988) / Bonesteel J K

The Administration of Folic Acid to
Epileptics with Phenytoin Gingival
Hyperplasia: A Double-Blind, Placebo
Controlled Pilot Study
(1988) / Brown R S

Magnetic Properties of Mixed-Valence
Heteropoly Blues Containing Delocalized
Electrons and Paramagnetic Transition Metal
Atoms. X-Ray Crystal Structures of
Alpha-12-Tungstocobaltate(II) and the 2E
Blue of Alpha-12-Tungstocobaltate(II)
(1988) / Casan-Pastor M N

Diversity of Algae (Chlorophyta, Phaeophyta,
Rhodophyta) Related to Depth, Light
Intensity, NO(3)-NO(2) and Water Flow Across
Two Transects of Long Reef, Grand Turk, BWI
(1989) / Goertemiller T R

Cloning and Sequencing of Compensatory DNA
Coding Mammalian Aspartyl-tRNA Synthetase:
Acquisition of Amphiphilic Helixes in High
Eukaryotes
(1989) / Jacobo-Molina A

The Study of the Electronic Structures of
One and Two Dimensional Solids by the Tight
Binding Approximation
(1988) / Lee Y S

Anisaldehyde Production Upon Irradiation of
4-Fluoroanisole and Its Analogs in the
Presence of Aqueous Potassium Cyanide and
Other Nucleophiles
(1989) / Ortega-Luoni P A

The Drosophila ETS-2' Gene from Schneider
Cell Line No. 2: Isolation, Sequence, and
Expression
(1988) / Pribyl L J

Study of Band Profiles in Non-Linear
Chromatography
(1988) / Shirazi S G

GEORGIA INSTITUTE OF TECHNOLOGY

Correlation of Antihypertensive Activity of
Novel DBM Substrates with Adrenergic
Catecholamine Levels in Spontaneously
Hypertensive
(1988) / Evans C N

Studies with Solvent Introduction in
Inductively Coupled Plasma-Atomic Emission
Spectroscopy
(1988) / Marmolejo E B

Investigation of the Lifetimes of
Electrochemically Generated Anthracene
Cation Radicals in the Presence of Selected
Nucleophiles
(1988) / Popp A E

The Determination of the Partition
Coefficients for a Variety of Pyrrolizidine
Alkaloids and the Relationship of These
Values to the Anti-Tumor Activity of the
Alkaloids
(1988) / Ryan E T

GEORGIA, UNIVERSITY OF

Biochemical and Molecular Analysis of
Drosphila Chromosomal Protein D1: A
Satellite DNA-Associated Protein Containing
a Novel Repeating Motif
(1988) / Ashley C T

Analysis of the Chromatin Structure of
Developmentally Regulated Gene Sets in
Cotton
(1988) / Borroto-Esoda K

Development of Intravacity Photothermal
Deflection Spectrometry
(1988) / Chen M

Structure of Newly Synthesized O-Linked
Oligosaccharides on Glycoproteins in Day
9 Embryonic Chick Neural Retina
(1988) / McDaniel G M

Factors Controlling the Intensity of Room
Temperature Phosphorescence in Micellar
Systems Containing Alkyl Sulfates in the
Presence of Thallium(I) and Sulfite Ions
(1988) / Nugara N E

GEORGIA, UNIVERSITY OF
(continued)

The Synthesis of Benzyloxycarbonyl-(E)
and (Z)-2,3-Methano-Aspartic Acid
(1988) / Park J A

Synthesis and Characterization of
Superconducting Oxides
(1988) / Tang H Y

Synthesis of (+-)-E-2,3-Methanotyrosine
(1988) / Turocy G

GUELPH, UNIVERSITY OF

Binding of Dichlorobenzidine to DNA in
Salmonella Typhimurium
(1988) / Debruin L S

High Temperature Deuterium NMR Studies of
Liquid Crystalline Polymers
(1988) / Facey G A

A Study of the Spin Adducts Produced During
the Respiratory Burst of Canine Neutrophils
(1988) / Jandrisits L T

Lymphocyte Plasma Membrane 5'-Nucleotidase:
Isolation, Reconstitution, and Detergent
Interactions
(1988) / Loe D W

Horseradish Peroxidale Catalysed Oxidation
of 3-Methylindole
(1988) / Martos P A

Membrane Association of Proline
Dehydrogenase in Escherichia Coli K12
(1988) / Mitchell D R

Construction of a Photoelectron Spectrometer
and the Study of reactive Molecules
(1988) / Sammynaiken R

The Synthesis of O-Nitroveratrole and the
Nucleophilic Aromatic Photosubstitution
Reactions of O-Nitroanisole and
O-Nitroveratrole
(1988) / Stephenson K D

The Electrodeposition of Cadmium Telluride
from Tri-N-Butylphosphine Telluride and the
Characterization of a Cadmium
Sulfide/Cadmium Telluride Heterojunction
(1988) / Von Windheim J A

HAWAII, UNIVERSITY OF

A Cyclic Product of Thiamin Nitrate in
Methanol: The Synthesis, Spectral and
Crystallographic Studies of 16-Pyrimidinium
Crown-4 Nitrate
(1988) / Fermin V N

The Effect of Calcium Entry Blockers on
Endotoxin-Induced Procoagulant Activity of
Mononuclear Cells
(1988) / Lee L L

The Effects of Phenyl-Methyl-Sulfonyl
Fluoride, Aprotinin, and Trypsin on
Respiratory Synctial Virus Induced
Cytopathology in Hep-2 and A549 Cell Lines
(1988) / Reese P E

A Structural Study of Two Heavy Metal
Lithium Organosulfur Clusters and a
Tetrachloromercurate Salt of Vitamin B1
(1988) / Taogoshi G J

A Study of Fluorinated, Geometric, and
Miscellaneous Bacteriohodopsin Analogues
(1988) / Tierno M E

HOWARD UNIVERSITY

Rare Earth Doping and Its Effect on the Hear
Capacity of the REBa2 Cu3 O7 (RE - Y, Pr,
Nd, Sm, Eu, Gd, La) High Temperature
Superconduction Ceramics
(1988) / Gray M C

Isocitrate Dehydrogenases of the Oyster,
Crassostrea Virginica Gmelin
(1988) / Kargbo A O

Reactions of B-Hydroxyvinylsilanes with
Base: Elimination (to Give Allenes) in
Competition with Protiodesilylation
(1988) / Kassim A M

Periodic Acid Oxidation of Methyl Vernolate
and Derivatives: Isolation and
Characterization of Products
(1989) / Nana E Y

Acid Catalyzed Synthesis of Dibasic Acids
from Vernonia Galemensis Seed Oil
(1988) / Powers F T

Synthesis of Dodecanedioic Acid from
Vernonia Galamensis Seed Oil
(1988) / Tabi D N

IDAHO STATE UNIVERSITY

The Synthesis of Some Oxazinoisoindoles and
and Attempted Pyrrolooxazole Preparation
(1989) / Xia J

IDAHO, UNIVERSITY OF

Changes in the Vitamin B-6 Content in
Potatoes During Storage
(1987) / Addo C

Kinetics and Mechanism of Iron Removal from
Human Serum Transferrin
(1988) / Bali P K

Parameter Estimation in Electrochemical
Kinetics
(1987) / Greenberg H

Synthesis and Structure of New
1,3,2-Diazaheteroles Formed by Reaction of
Antimony Chloride and Bismuth Chloride with
Diimines
(1988) / Hubler T L

In Situ, Real Time FT/IR/CIR/ATR in
Conjunction with Pre and Post XPS/AES Study
of the Biocorrosion of Copper by Gum Arabic,
Alginic Acid, Bacterial Culture Supernatant
and Pseudomonas Atlantica Exopolymer
(1988) / Jolley J G

Stability and Auxin Activity of Phenylacetic
Acid in Selected Tissue Culture Systems
(1988) / Leuba V

IDAHO, UNIVERSITY OF
(continued)

Diastereoselectivity in the Lateral
Metalation and Electrophilic Quenching of
Chiral Isoxazolyl-Oxazolines
(1988) / Mallet B

Spectroscopic Studies of TCNQ Molecular
Structure in Solution
(1987) / Merriam M J

Extracellular Peroxidase in Lentil
Suspension Cultures and the Influence of
Sodium Chloride on Its Activity
(1988) / Wright A

ILLINOIS STATE UNIVERSITY

Cardiobacterium Hominis - Susceptibility
Testing, Light and Electron Microscopy
Studies
(1988) / Carver S L

2-Substituted Pyridine N-Oxides and Their
Metal Ion Complexes
(1988) / Daraska C A

Role of Fibronectin in Lymphocyte Migration
(1988) / Davis J M

Nitrosation Reactions of 9.B-Unsaturated
Oximes
(1988) / Easter J A

The Specific Activity of Lactate
Dehydrogenase and 9-Keto-Prostaglandin
Reductase in Reproductive Tissues
(1988) / Farley K A

The Effect of Ingested Gossypol on the
Ultrastructure of Thyroid and Adrenal
Glands from the Rat
(1988) / Franco S A

The Induction of Heat Shock Proteins in
Resting and Mitogen Stimulated Lymphocytes
(1988) / Ghassemi M

Influence of Cadmium on the Ultrastructure
of Developing Chloroplasts in Soybean and
Corn
(1988) / Ghoshroy S

Role of Various Solutes in Staphlylococcus
Aureus Osmoregulation
(1988) / Graham J E

Analysis of Age-Related Changes in Protein
Phosphorylation by Rat Hepatocytes
(1988) / Heydari A R

Hemolymph Binding of Juvenile Hormone and
Related Compounds in Arthopods
(1988) / Li H

Development of Methodology to Identify
Metabolites from Maize Leaves
(1988) / Liu W S

Partial Isolation and Characterization of
Lysine and Ornithine Decarboxylases from
Serratia Marcescens
(1988) / Mathiss J P

A Revolutionary Method fo Isotopic Isomer
Separation
(1988) / Pescatore J A

The Prostaglandin Accumulation in Vitro
by Mouse Uteri, Oviducts, and Ovaries as a
Function of Reproductive States and
Maturation
(1988) / Saffaripour S

The Effect of Growth Conditions on
Cofactor-Linked Xylose Reductase Activity
in Pachysolen Tannophilus
(1988) / VanCauwenberge J E

Plasmid Deoxyribonucleic Acid Profiles of
Betw-Lactamase Producing Aeromonas
Hydrophila Strains
(1988) / Yousuf M

INDIANA STATE UNIVERSITY

Determination of the Molar Absorptivities
and Formation Constants on Iron (II)
Perchlorate with 1,10-Phenanthroline
(1988) / Rivera-Rivera J

INDIANA UNIVERSITY

Multiphoton Ionization of Formaldehyde
(1988) / Engel A I

The Effects of Acute Ethanol on the Levels
of Several Amino Acids in the CNS
(1988) / Gongwer M A

Synthesis and Cyclizations of
1-Iodo-5-Methyl-5,6-Heptadiene
(1988) / Mennan K E

An Investigation Into the Mechanism of the
Diels-Alder Reaction
(1988) / Peterson K B

A Possible Role for the Mitochondrial
Polymerase in the Nucleus of Saccharomyces
Cerevisiae
(1988) / Schultz P

Purification and Investigation of the
Anion Activation of Mitochondrial
F(1)-Atpase
(1988) / Smith J H

Nonlinear Phenomena in Precipitation and
Cascade Kinetics
(1988) / Tupper K

Mechanistic Aspects of the Electrochemical
Reduction of 1,1,4,4-Tetraphenyl-1,
2,-Butadiene
(1988) / Vincent M L

IOWA STATE UNIVERSITY

Peak Identification Using Optical Activity
in Series with Ultraviolet Absorption
Detection and Comments on Peak Response in
Liquid Chromatography
(1988) / Chan K C

Influence of Dietary Calcium and Vitamin D3
on Composition of Plasma Lipids in Young
Pigs
(1988) / Foley M K

Comparison of Various Reference Electrodes
for Pulsed Amperometric Detection of
Carbohydrates Under Gradient Chromatographic
Conditions
(1988) / Mead D A

IOWA STATE UNIVERSITY
(continued)

Collision Removal of Molecular Ions in
Inductively Coupled Plasma Mass Spectrometry
(1988) / Rowan J T

New Chemistry of Azlactones
(1988) / Yu F

Fluorescence Line-narrowing Spectrometry:
Application to the Study of Benzo[A]Pyrene
Metabolic Pathways
(1988) / Zamzow D S

JOHN CARROLL UNIVERSITY

Reduction of a-Nitro-(2)-Alkyl-(4,
4)-Dimethyl-(2)-Oxazolines
(1988) / Gelbach R V

KANSAS, UNIVERSITY OF

Catalysis of the Methanolysis of Aryl Methyl
Phosphates by Acetate Buffer and
Methoxide: Solvent Isotope Effects
(1988) / Bryan C D

Ascorbic Acid Analysis in Human Brian
(1988) / Cammack J

Role of Endogenous Opioid Peptides in
Catecholamine Secretion from the Rat
Adrenal Medulla
(1988) / Chen Y M

tRNA Processing by Spinach Chloroplast
RNase P
(1988) / David N W

In Vitro Evaluation of Cisplatin Reactivity
with Selected Amino Acid, Peptides, and
Proteins
(1988) / Deeken R A

The Evaluation of
p-Methylthiophenylisothiocyanate as an
Electrogenic Reagent for the Detection of
Primary and Secondary Amines
(1988) / Edwards D F

Complexation of Procainamide with Dextrose
Irreversibility of the Reaction in Plasma
In-Vitro
(1988) / Lacerte J A

Kinetic Deuterium Isotope Effects on
Phase-Transfer-Catalyzed (PTC) Chlorination
and Microsomal Hydroxylation of Xylenes
(1988) / Ling K H

Urokinase Plasminogen Activator: Proenzyme
Binding and Binding Protein Isolation
(1988) / McKay L D

Partial Purification and Characterization
of an Intracellular Trypsin/Thrombin
Complexing Protein
(1988) / Morgenstern K A

A Synthesis of Oxygenated Neolignans: Total
Synthesis of (+)-Kadsurenone and
(+)-Denudatin
(1988) / Ray J E

Adenylate Cyclase and the Decidual Cell
Reaction
(1988) / Rayford A

Molecular Cloning of Chicken Metallothionein
cDNA
(1988) / Wei D

KENTUCKY, UNIVERSITY OF

Hydrolysis Studies of 2-Chloroethyl Ethyl
Sulfide
(1988) / Barnes E L

Bonding in Selected First-Row Transition
Metal Diatomic Molecules: Ased Molecular
Orbital Theory
(1988) / Bucknum M J

Determination of Hydroxyl Group
Concentration in Coal Liquids by 31P NMR
(1988) / Dadey E

Electronic Contributions to Structures of
Complexes of Dithioacetals
(1988) / Kaufmann G B

Synthetic Approaches to Novel Copolymers
Based on a New Multifunctional Monomer
(1988) / Kulkarni S

A Multichannel Data Acquisition System for
Laser Spectroscopic Studies
(1988) / Rutherford M L

ESR Studies of Pharmacueticals: I.
Photolytic Decomposition of Mitotane and
Related Compounds Using the Technique of
Spin Trapping
(1988) / Umhauer S A

LAMAR UNIVERSITY

A Study of the Ethynaylation of
2-Substituted-Cyclohexanones with 2-Aryl
Substituents
(1988) / Chen J C

Copper (II), Palladium (II) and Mercury (II)
Complexes of 1,2-Diaminocyclohexane and
1,3-Cyclohexanebis (Methylamine)-Potential
Antitumor Agents
(1988) / Chiang C K

Indoles from 1-Halo-2-Nitrobenzene and
Acetylides and Substituted Acetylides
(1988) / Hoang T D

Synthesis and Biological Screening of Some
Potential Insecticides: N-(3-Chloro-2 Benzo
(b) Thienocarbonyl)-Substituted Anilines,
N-(3-(6-Chloropyriadizyl))-N'-(Substituted
Benzoyl) Hydrazines and Their
Fluoroalkoxylated Derivatives
(1988) / Hwang Y T

A Solidification/Stabilization Study for the
Disposal of Pentachlorophenol
(1988) / Lee K C

Potentiometric and Thermometric Titrations
of Sparingly Soluble Compounds of O/W
Microemulsion and Construction of a Computer
Interfaced Colorimeter to Perform
Thermometric Titrations of Such Compounds in
O/W Microemulsion
(1988) / Patel B M

The Effect of 16-Hydroxyestrone on Antibody
Production in Rats Experimentally Infected
with Trypanosome Lewisi
(1988) / Smith E H

LAURENTIEN UNIVERSITY

Electroorganic Synthesis of Octylaniline
(1988) / Esmiel F

Chemical Basis for Sub-Banding of Creatine
Kinase, M
(1988) / Marcotte B

Uptake of 226Ra by Muskrat and Beaver Near
Uranium Mine Tailings at Elliot Lake,
Ontario
(1988) / Mirka M

Copolymerization and Cyclohexene:
Initiation/Termination Mechanism by End
Group Analysis
(1988) / Su P

LONG ISLAND UNIVERSITY

A Kinetic Study of the Oxidation of
Bis(Phenanthroline)-Iron (II) by Hydrogen
Peroxide
(1988) / Dworkin S A

The Effect of Metal Ions on Lipid
Peroxidation and the Thiobarbituric Acid
Assy for Malondialdehyde
(1988) / Farhoud R

A Study of the Electrochemical and Magnetic
Properties of Ferrioxamine B
(1988) / Helman R

Development of Pre-Column Purification
Methods for Physotigmine Extraction from
Biological Fluids
(1988) / Yatim N B

LOUISIANA STATE UNIVERSITY

The Effect of Fuel Additives on Soot
Formation During Benzene Pyrolysis and
Oxidation
(1988) / Adams N P

Complexation of Deoxyribonucleic Acids by
Antitumor Metallocene (IV) Compounds and
Other Transition Metal Complexes and the
Synthesis of Fulvenes from Various Cyclic
Diones
(1988) / Cronan J M

Addition Reactions to Unsaturated
Organosilicon Compounds
(1988) / Xin Z

Mechanisms of Solidification/Stabilization
of Hazardous Wastes Studied by Nuclear
Magnetic Resonance Relaxation Times
(1988) / Yang S L

LOUISIANA TECH UNIVERSITY

Levamisole Effect on 5-Mucleotidase and
Alkaline Phosphatase of Bovine Milk Fat
Globule Membrames
(1988) / Sharifi S H

LOUISVILLE, UNIVERSITY OF

Transforming Growth Factor-Alpha: Synthesis
and Evaluation of Fragments and Its Detection
in Human Malignancies
(1988) / Franklin G A

LOWELL, UNIVERSITY OF

Purine Metabolizing Enzymes: Evaluation of
Adenylate Kinase and Non-Specific
Nucleotidase in the Cytosol
(1988) / Martin P J

Electrocatalytic Detection of Triglycerides
Via the Insitu Formed Silver Oxide
(1988) / Michalakis G M

Evaluation of Semi-Alpha-Hemoglobin as an
Intermidiate in Assembly: The
Ferroprotoporphyrin IX Binding Step
(1988) / Park R Y

LOYOLA UNIVERSITY

An Unusual Extension of the Lramero
Rearrangement
(1988) / Bacino D J

The Development of Electrochemical
Derivatization Techniques for Detection
of Prostaglandins by High Pressure Liquid
Chromatography
(1988) / Beck G M

Thermal Denaturation of Hemoglobin A
Cross-Linked with Bifunctional Imidoesters
(1988) / Corso T D

New Syntheses of Indolizidine Alkaloids
Swainsonine and Castanospermine from Chiral
Precursors
(1988) / Holms J H

Immunological Mechanism of Action of an
Experimental Nucleoside Drug, GR1784:
Effects on Hematopoiesis and Lymphocyte
Mitogenesis
(1988) / Landis R C

Intramolecular Cyclization of
Omega-Iodoepoxides: An Annulation
Methodology
(1988) / Spina K P

MAINE, UNIVERSITY OF

Isolation and Characterization of the
5-Prime Region of Genomic Sequences
Encoding the Small Subunit of Rubisco
in Larix Laricina
(1988) / Brunner A F

Cloning of Messenger RNA's Differentially
Expressed on Hormonally Induced
Tuberization in Potatoe
(1988) / Cook L

Isolation and Structure Elucidatron of
Tetranortriterpenes from Turraea
Robusta (Meliaceae)
(1988) / Gaul F E

Determination of Vitamin-C in Fruits and
Vegetables from Supermarkets Versus Farmer's
Markets Using a New HPLC Method
(1988) / Krishnan M R

Steric and Electronic Effects in the
Synthesis of 1-Benzyl-2-Benzazepinones
(1988) / Omburo G A

MANITOBA, UNIVERSITY OF

BALB/c 3T3 Fibroblasts: A Study of the
Sensitivity to Melphalan as a Function of
the Proliferative Rate
(1988) / Blosmanis R E

Studies of Ribonucleotide Reductase Activity
in a Highly Hydroxyurea Resistant Mouse Cell
Line and Inhibition of Ribonucleotide
Reductase by Gossypol
(1988) / Chan A K

Evaluation of the Rabbit as an Animal Model
for Determining the Bioavailability of
Sustained Release Chlorpheniramine Dosage
Forms
(1988) / Chen X

Some FORTRAN-77 Programs for the Study of
Large Amplitude Vibrations in Small
Molecules
(1988) / Davie K J

Heterobimetallic Dicyclohexylphosphide
Bridged Complexes
(1988) / Jenkins H A

X-Ray Crystallographic Studies of a
4-Demethoxythiodaunomycinone Derivative
and 1,2,4-Triazole
(1988) / Li N H

Characterization of Melphalan Transport in
Murine L5178Y Lymphoblasts
(1988) / Miller L

Hydrolysis and Inner-Sphere Electron
Transfer Reactions of Iron and Cobalt
Complexes Containing 2,2'-Bipyridine and
4-Fluoro-2,2,'-Bipyridine
(1988) / Nguyen T Q

Regulation of Phosphatidylcholine Metabolism
in Mammalian Tissues
(1988) / Oh K

The Use of Five and Six Bond Long-Range
1H-1H and 1H-19F Coupling Constants in the
Determination of Conformational Preferences
in Some Substituted Derivatives of Toluene
and the Determination of the Nature of the
Rotational Barrier in
a,a-Diacetoxytoluene Derivatives
(1988) / Takeuchi C S

Perturbations of (6)J(F,CH(3)) by Ring
Substituents in 4-Fluorotoluene
Derivatives: Conformational Implications
(1988) / Tseki P F

MARQUETTE UNIVERSITY

Catalytic Carbopalladation of Upsilon
Methylenebicyclo [4.1.0] Alkanes
(1988) / Brodt C

Investigation of Hemoglobin Cooperativity
Through Structural Studies with Modified
Hemes
(1988) / Oconnor E

Peak Shapes as a Function of the Difference
Between Eluite Solvent and Mobile Phase
Strength in Liquid Chromatography
(1988) / Pan S L

The Fluorination of Organic Compounds by
Fluorine and Cesium Fluoroxysulfate in
Acetonitrile and Aqueous Solution
(1988) / Suarez L

MARSHALL UNIVERSITY

Mechanistic Investigation Into the 1,2-Diol
Cleavage of 1,1,2,3-Tetraphenyl-1,
2-Ethanodiol with N,N-Dimethylformamide
Eimethylacetal
(1988) / Goolsby T L

MASSACHUSETTS INSTITUTE OF TECHNOLOGY

Thermal Decomposition of Diazetines
(1988) / Allen R M

Amino Acid Analysis by Gas Chromatography
(1988) / Block E H

Triazolinediones: Studies of Imino
Derivatives and Studies of Lewis Acid
Catalyzed Reactions with Arenes
(1987) / Ippolito T M

The Synthesis of Chorismate and Anthranilate
Analogs and Investigations of Their
Biological Activity
(1988) / Wolff J R

MASSACHUSETTS, UNIVERSITY OF

Quantum and Classical Studies of the
Dissociation Dynamics of Hydrogen and Its
Isotopes on Nitrogen
(1988) / Chiang C M

Quantitation and Toxicological Assessment
of the PCB Levels of Fourteen Sites in the
Acushnet River Estuary
(1988) / Markham P M

Oligomeric Forms of Organotin Compounds:
Synthesis and Characterization
(1988) / Nadim H

Analysis of Palm Oil Carotenoids by
High-Performance Liquid Chromatography
with UV-VIS Photodiode-Array Detection
(1988) / Ng J H

High Performance Liquid Chromatography
Separation of Tetracyclines on Silica and
Polymer Based Reversed Phase Columns
(1988) / Passaretti C L

The Ultrastructural and Molecular
Organization of the Mammalian Kinetochore
(1988) / Pikora C A

Laser Induced Two-Photon Photoionization as
a Detection Mode for High-Performance
Liquid Chromatography
(1988) / Snyder P A

The Fluorescence Phenomena of Aluminum
Fluvic Acid Complexes
(1988) / Wang A L

MIAMI, UNIVERSITY OF

Electron Spin Resonance Studies of Novel
Cyclooctatetraene-Crown-Ether Compounds
(1988) / Almirall J R

MICHIGAN TECHNOLOGICAL UNIVERSITY

X-Ray Diffraction Studies on Aqueous Purine
and Purine-Sodium Chloride Solutions as a
Function of Temperature
(1988) / Ellis T W

The Effects of 2,4,7-Trinitro-9-Fluorenone
Dopant on the Luminescence Properties of
Poly-(N-Vinylcarbazole)
(1988) / Guo X

Phosphoinositide Involvement in the Action
of the Plant Hormone, Gibberellin
(1988) / Renders J M

Synthesis and FT-IR Study of [18 O] Labeled
Gibberellic Acid
(1988) / Zhou M M

MINNESOTA, UNIVERSITY OF

Induced Lysogeny in Bradyrhizobium Japonicum
(1988) / Abebe H M

Complexation and Cocrystallization with
Triphenylphosphine Oxide
(1988) / Baures P W

Pharmacokinetics of 6-Thiouric Acid in
Rental Allograft Recipients: Evaluation for
Use in Monitoring Azathioprine Therapy
(1988) / Chan G L C

Sequence Characterization of Sterol Carrier
Protein from Rat Liver, Heart, and Kidney
Tissue
(1988) / Cooper J A

Effect of Adsorption in the Khand-Pauson
Cyclizations
(1988) / Froen D E

Calmodulin Interaction with Cardiac
Sarcoplasmic Reticulum
(1988) / Hanson T P

Synthesis and Mass Spectral Analysis of
Partially Labeled Permethylated
Anhydroalditols
(1988) / Kemper M S

Hybride Transfer from Sodium Borohydride
to NAD + Analogues
(1988) / Kim D

New Khand-Pauson Reactions
(1988) / Kim D Y

Application of the Reductive Cleavage Method
to the Structural Analysis of Carbohydrates
Containing Terminal Neuraminic Acid Groups
(1988) / Kotchevar A T

Effects of Heteroatoms and Geometry on
Allylic Carbanions
(1988) / Li J

Evaluation of a PS-DVB Stationary Phase for
Reversed-Phase HPLC Applications
(1988) / Pedigo S

Yolk Glycoprotein Complexes of
Caenorhabditis Elegans
(1988) / Sutherlin M E

A Study of the Interactions of Small Solutes
and Triazine Dyes Used in High Performance
Affinity Chromatography
(1988) / Wade T M

Novel Binding Site for L-Quisqualate
Sensitizes Nuerons to Depolarization by a
Group of Phosphonate-Containing Analogues
of Glutamate
(1988) / Whittemore E R

Assemblying Complex Molecules with Stepwise
AdE Reactions
(1988) / Wiitala K W

Application of Manganese(III) Acetate in the
Synthesis of Bioactive Lignans
(1988) / Yang F Z

Luminescence Spectra of Europium Triple
Nitrites
(1988) / Yu H

An Alphal-Adrenergic Receptor Mediated
Phosphatidylinositol Effect in Canine
Cerebral Microvessels
(1988) / Zeleznikar R J

MISSISSIPPI STATE UNIVERSITY

An Approach to Isolating the Gene Coding for
Bacterial Superoxide Dismutase Using an
E. Coli Genomic Library
(1986) / Chang P K

Synthesis and Reactions of Some Di-, Tetra-,
and Hexahydropyridazines
(1986) / Crook L C

The Effect of Paper Machine White Water Closure
on the Uniform Corrosion of 304L Stainless Steel
(1986) / Keady G D

Effects of Diet on Lipogenic Enzyme Activities in
Channel Catfish Hepatic and Adipose Tissue
(1986) / Likimani T A

Storage Proteins in Rice
(1986) / Rahman N N

Studies on Olfactory Tissue Positive Sodium -
Potassium Ions ATPhase Activity
(1986) / Smith J W

Autoimmune Antibodies in Systemic Rheumatic Disease.
Characteristics of Antigenic Nuclear Particles
(1986) / Williamson G G

Peroxidase, Luminol and Plant Chemiluminescence
 I. Evidence of Peroxidase Involvement in Plant
   Chemiluminescence
 II. Quenching of Peroxidized Luminol
   Chemiluminescence by Reductants
(1986) / Wong J K

MISSISSIPPI, UNIVERSITY OF

High Temperature Differential Thermal
Analysis Studies of Ternary and Quaternary
Phased Transition Metal Fluorides
(1988) / Du T Y

Characterization of the Binding of m-AMSA, a
Potent Antitumor Agent, to Nucleic Acids
(1988) / Elmore R H

MISSOURI, UNIVERSITY OF (COLUMBIA)

Study of the Orientation of DNA in Filamentous
Bacteriophage
(1988) / Bauer M E

Solution Chemistry of EDTMP Complexes which
are Potential Therapeutic Bone Agents
(1988) / Clark P A

Expression of an HSP70 Cognate, HSC-1, in Tomato
(1988) / Duck N B

An Examination of the Kestoses Produced in Asparagus
Roots, Fescue Leaf Blades, Banana Fruit, and Yeast
Invertase - Sucrose Reaction
(1988) / Forsythe K L

Prostacyclin Release from Cultured Pulmonary
Artery Endothelial Cells Stimulated with
Extracellular Nucleotides
(1988) / Hicks C S

Microwave Plasma Atomic Emission Spectrometry
as an Element Selective Gas Chromatographic
Detector
(1988) / McClendon E L

Reactions of 2-Methylene-1, 3-Cyclopentane-dione
with Electron-Rich Alkenes and Preparation and
Characterization of Alkoxy Ozonides
(1988) / Meyer L A

Synthesis of Deuterated Sesquiterpenes
(1988) / Mustiga M D

Chemical Characterization of Dairy Biomass and
Supercritical Fluid Extraction of Hazardous
Chemicals
(1988) / Nam K S

Development of a Cryogenic Purge - Trap
System for Analysis of Volatile Organics
(1988) / Palmer C L

Soybean Embryo Lipoxygenases: Molecular Analysis
of Null Mutants and Germination Specific Species
(1988) / Park T K

Comparison of Induced Proteins Synthesized in
Hydrogen Peroxide and Near-Ultraviolet-Treated
Bacterial Cells
(1988) / Pierceall W E

Introduction of Large Molecules into Viable 3T6
Mouse Fibroblasts by ATP-Induced Permeabilization
(1988) / Saribas A S

Reactions of the Oxodiperoxocomplexes of Mo(VI)
and W(VI) with a Variety of One-Equivalent
Reductants
(1988) / Schwane L M

Reactions of Cyclohexene Oxide with Hypochlorite
Salts
(1988) / Shabany H

Isolation and Characterization of a Soybean
Urease-Like Genomic Clone
(1988) / Torisky R S

Comparison of Radiochemical Properties of Amine
Oxime and Imine Oxime Complexes Technetium-99M
(1988) / Wang R

Effects of Increased Endogenous Catalase and
Dihydroxyacid Dehydratase on Resistance/
Sensitivity to Near-Ultraviolet Radiation in
Escherichia Coli
(1988) / Wilke A L

The Influence of DNA Adenine Methylase on the
Sensitivity of Escherichia Coli to Near-
Ultraviolet Radiation and Hydrogen Peroxide
(1988) / Yallaly P E

MISSOURI, UNIVERSITY OF (ROLLA)

Thermal, Mechanical and Interfacial
Characteization of a Fiberglass Reinforced
Plastic Composite
(1988) / Williams T J

MONTANA STATE UNIVERSITY

Grasshopper Hemagglutinin: Immunochemical
Localization in Hemocytes and Confirmation
of Non-Opsonic Properties
(1987) / Bradley R S

Carbon Monoxide-Facilitated Rearrangements
of CIS-1,2 Disubstituted Platinum(VI)
Metallacyclobutane
(1987) / Ernst G E

The Determination of Chlorsulfuron by the
Inhibition of Acetolactate Synthase
(1987) / Kagel G W

Use of Ion Clustering Equilibrium for Isomer
Identification in Electron Capture Mass
Spectrometry
(1987) / Krieger L A

MONTANA, UNIVERSITY OF

A Kinetic Study of the Oxidation of
Arsenite by Birnessite
(1988) / Hayes T H

The Chemistry of 5,6-(9,10-Anthraceno)-1,
2,4,6-Cycloheptatetraene
(1988) / McCarthy T J

MONTREAL, UNIVERSITY OF

l'Effet de la Solvation Sur la Conductivite
Ionique du Poly (Oxyethylene) Complexe par
Les Thiocyanates Alcalins
(1988) / Besner S

Developpement d'un Analyseur Multi-Canaux
Pour l'Analyse d'Impuretes Dans l'Argon
(1988) / Boudreau D

Determination de Metaux Dans le Liquide
Amniotique par Spectroscopies Atomiques
(1988) / Bussiere L

Changement de Conformation des Ribosomes de
Tissus Musculaires de Hamsters Atteints de
Dystrophie Musculaire Hereditaire
(1988) / Caron R

La Pediculose Causee Par Pediculus Humanus
Var. Capitis: Etude Epidemiologique de
Cette Infestation Chez les Enfants d'Age
Scolaire
(1988) / Champagne M

MONTREAL, UNIVERSITY OF
(continued)

Dosage des Anions Inorganiques Dans Les
Eaux Naturelles par Chromatographie
Liquide Avec Detection Spectrophotometrique
Indirecte
(1988) / Chauret N

Mise en Evidence d'un Precurseur de la
Phospholipase A2 dans le Spermatozoide
Humain: Purification et Caracterisation de
l'Enzyme et de Son Precurseur
(1988) / Guerette P

Etude Microcalorimetriques des Effets du
Chlorure de Sodium et du Chlorure de
Magnesium sur le Poly (Acrylate de Sodium)
en Solution Aqueuse
(1988) / Laberge F

Etude Structurale des Dithiobenzoates
d'Oligomethylenes
(1988) / Leblanc C

Expressionn et Secretion d'une Proteine
Prostatique Humaine (PSP94) par Escherichia
Coli
(1988) / Linard C

Etude de Quelques Composes de Nickel (II)
et de Cuivre (II) Avec des Derives de Purine
(1988) / Maheu S

Caracterisation des arn Messagers du
Tubercule de la Pomme de Terre Durant la
Reponse d'Hypersensibilite
(1988) / Marineau C

Role des Ions-Agglomerats Dans la Generation
de Spectres de Masse Obtenus Par
Bombardement d'Atomes Rapides
(1988) / Perreault H

Synthese in Vitro et Structure des Trnai Met
(1988) / Perreault J P

Reactivite des Pentachlorures de Niobium et
de Tantale Avec l'Aza-7 Indole
(1988) / Poitras J

Determination de Metaux Presents a l'Etat
de Traces Par Spectroscopie d'Emission
Atomique a Plasma Microondes
(1988) / Quellet J

Etude de l'Assemblage et du Recouvrement de
Derives du Fluorene a l'Etat Solide
(1988) / Raymond S

Construction d'un Vecteur Eucaryote SV40
Pour l'Expression de la Proopiomelanocortine
Porcine
(1988) / Rondeau S

Computer Techniques for the Classification
and Fast Retrieval of Mass Analyzed Ion
Kinetic Energy Spectra (Mikes) Obtained by
High Energy Collision Activation
(1988) / Roussis S

Cinetique d'Hydrolyse d'Esters Analogues de
l'Acide Pivalique Comme Promedicaments du
Ketoprofene
(1988) / Scherer E

Introduction de la Methode du Recuit Simule
Dynamique au Formalisme LCGTO-HF
(1988) / St-Amant A

Formal Transfers of Hybride from Carbon
Hydrogen Bonds: Attempted Intramolecular
Reductions of Substrates Bound by
2,3-Dihydro-1,3-Dimethyl-2-(2-Pyridinyl)
1H-Benzimidazo
(1988) / Tarazi M

Contribution a l'Etude de la Quantification
des Caracteristiques Morphologiques et de
la Geometrie de Surface de Particules
Solides Dans les Operations Unitaires Phar
(1988) / Thibert R

Etude du Systeme Chimique de la Butyne-2
Diol-1,4 en Solutions Aqueuse et
Methanolique
(1988) / Zauhar B

MURRAY STATE UNIVERSITY

Synthesis and Characterization of
Dimolybdenum Tetrakis-(5-Alkylthio-1,3,
4-Thiadiazole 2-Thidate) Complexes: A Most
Intriguing Set of Derivatives of Vanchem
Dmtd
(1988) / Estes M L

Proton Inventory of the Hydrolysis of Some
2- and 4-Fluoro-N-Heteroarenes
(1988) / Jialun M

Determination of Trace Metals and Ions in
Tree Ring Wood and Correlations Regarding
Acidic Precipitation and Tree Growth
(1988) / McClain K

The Effect of Physicochemical Interaction
Between the Supercritical Mobile Phase and
the Stationary Phase on the Retention
Mechanism in Supercritical Fluid
Chromatography
(1988) / Strubinger J R

NEBRASKA, UNIVERSITY OF

A Gas-Phase Study of the Diels Alder
Reaction and the Reformatsky Condensation
(1988) / Castle L W

Studies of Molybdenum Complexes Containing
Dinitrogen
(1988) / Hayes R K

Product Analysis of the Oxidation of
Methylamine, Cyanide Ion, and Cyanate Ion
by Potassium Ferrate (VI)
(1988) / Humphrey M D

The Use of Activated Zinc in the
Simmons-Smith Reactions
(1988) / Lenk B E

Synthesis of Bridged Polycarbocyclic
Natural Products
(1988) / Novak P M

Coupling Constants Study of Pyrrolidine Ring
System Using a Modified Karplus Equation
(1988) / Son P S

Construction of a High Resolution Electron
Energy Loss Spectronometer (HREELS) and
Testing Results from a Cobalt/Iron (110)
System
(1988) / Wulser K W

NEW BRUNSWICK, UNIVERSITY OF

A Study of Ion-Exchange, Equilibrium
Dialysis, and Ultrafiltration Separation
Procedures with Respect to Aluminum
Speciation
(1988) / Armstrong C M

Crystal Structure of (SN)x and Other
Compounds
(1988) / Drummond D F

Recovery of Chlorine from a Residual Gas
(1988) / Garg A

Gadolinium Poison Concentration Monitoring
in CANDU Reactors
(1988) / O'Connor D R

NEW JERSEY INSTITUTE OF TECHNOLOGY

Hydrodechlorination Reactions of
Trichloroethylene and 1,1,1, Trichloroethane
Over Alumina Supported Palladium Catalysts,
and Trichloroethylene Over Nickel Catalyst
(1988) / Shah A R

Synthesis of Strained Bridgehead Lactams
(1988) / Tu Q

Thermal Decomposition of Dichloromethane/1,
1,1-Trichloroethane Mixture in an Atmosphere
of Hydrogen
(1988) / Won Y S

NEW MEXICO INSTITUTE OF MINING & TECH.

The Correlation of Lightning Activity with
Nitrate and Nitrite Concentration in
Precipitation
(1987) / Chavez R R

Extraction and Analysis of Transdermally
Applied Nicotine from Biological Materials
(1988) / Divine K K

Elongated Flow Properties of Hydrolyzed
Polyacrylamide Solutions
(1988) / Nimir H B

The Effect of Thyroid Status on the Membrane
Bound Cyclic AMP Phosphodiesterase in the
Rat Liver
(1987) / Okmen A A

Ice-Nucleation-Active Bacteria Associated
with Lichens
(1987) / Okmen F

Synthesis, Characterization and Solution
Properties of Tri-Alkyl-Tin Fluorides
(1987) / Taylor C

Thermal Decomposition of Ammonium Nitrate
(1988) / Tewari M P

The Adsorption/Desorption Kinetics and
Characteristics of Toluene in Soil
Water/Air Systems
(1987) / Tucker A R

Variation of Michaelis-Menten Constants with
Change in Substrate in Wheat Germ Acid
Phosphatase
(1987) / Wilson K V

NEW MEXICO STATE UNIVERSITY

The Reaction of O-Alkyl Hydroxylamines with
m-Trifluoromethyl Benzenesulfonyl Peroxide
(1988) / Christophe N B

Immuno-Identification of Ribulose 1,
5-Bisphosphate Carboxylase/Oxygenase as a
Natural Substrate of Transglutaminase in
Plants
(1988) / Dharma A

Transformation of Rat Tracheal Epithelia
Cells in Culture by Combined Exposure to
Radiation and Chemicals
(1988) / Maio S M

Utilization of Mycobacterium Smegmatis
Spheroplasts as a Means for Efficient
Transfection
(1988) / Smith G B

Characterization of a Temperate
Mycobacterium Avium Bacteriophage
(1988) / Tupponce A K

NEW MEXICO, UNIVERSITY OF

The Bipolar Transistor Microelectrode-Sensor
(1988) / Paulter N G

NEW YORK, STATE UNIVERSITY OF (ALBANY)

The Synthesis, Characterization and Crystal
and Molecular Structure of (n-Bu-4-N)3
[Mo-6-O-18(NNC-6-F-5)]
(1987) / Sun X Y

NEW YORK, STATE UNIVERSITY OF (BUFFALO)

Simultaneous Electrochemistry and Electron
Paramagnetic Resonance
(1988) / Male R

The Design, Synthesis and Evaluation of
5-Alkylsulfonylsalicylanilides and
Quarternary Ammonium Salts of N,N,N,
Dimethylalkylamino Salicylanilides as
Potential Anti-Plaque Agents
(1988) / Poopisut N

Radioactive Calcium Channel Blockers as
CNS Diagnostic Imaging Agents
(1988) / Posthauer J

X-Ray Crystal Structures of the Compounds
(u-H)2Ruc(CO)9 (u3-n20MeC=COMe), [Et4N]
[Bi2Fe2Co(CO) 10] and (C9H6CH3)Rh(C2H4)2
(1988) / See R

The Analytical Utility of Cylindrical
Microelectrodes
(1988) / Singleton S

NEW YORK, STATE UNIVERSITY OF (FREDONIA)

Characterization and Thermal Analysis of a
Novel Azo-Containing Photoresponsive
Polyester
(1988) / Swanson S

NORTH CAROLINA STATE UNIVERSITY

Characterization of Steryl Ester
Synthetase from Saccharomyces Cerevisiae
(1988) / Luckin K A

NORTH CAROLINA STATE UNIVERSITY
(continued)

Physical, Thermal and Optical
Characterization of Selected Metal-Organic
Compounds
(1988) / Poston S L

Substrate Recognition Features of Juvenile
Hormone Esterase: A Survey of a New Series
of Unsaturated and Acetylenic
Trifluoromethyl Keton Inhibitors
(1988) / Upchurch L R

NORTH CAROLINA, UNIV. OF (CHAPEL HILL)

Characterization of Chemically Modified
Gamma-Carboxyglutamyl Residue Containing
Proteins
(1987) / Kushan J P

A Study of the Feasibility of Using
Molecular Mechanics Simulations to Explain
in Vivo Site-Specific Mutagenesis in the
418 Region of the E-Coli GPT Gene
(1987) / Srinivasan N

NORTH DAKOTA STATE UNIVERSITY

Evaluation of a Theophylline Pharmacokinetic Dosing
Program in Geriatric Patients
(1987) / Diffendorfer R J

Synthesis of conformationally Restrained Benzodioxans
(1987) / Gholami K

Pin Milling and Air Classification Studies of Dry
Edible Beans
(1988) / Han J

The Effects of Flour Particle Size on Baking Quality
(1988) / Kurimoto Y

Laboratory Scale Production and Properties of
Sunflower Curd
(1988) / Liu T

Synthesis of Conformationally Restrained Amino-
1,2,3,4,4a,10a- hexa-Hydrodibenzo[b,e] [1,4] Dioxin
(1988) / Liu W

Influence of Chlorpromazine on Motility and Calcium
Uptake of Boar Sperm
(1987) / Namyniuk M A

Chromatographic Determination of Amantadine in
Biological Fluids
(1987) / Pai G G

The Lignin of Hard Red Spring Wheat Bran
(1987) / Schwarz P B

NORTH TEXAS STATE UNIVERSITY

Ion Chromatography of Soluble Chromium(III)
and Chromium(VI)
(1988) / Huang J S

A Reinvestigation of the Kinetics and
Mechanism of Ligand Exchange in -(2,2,8,
8,-Tetramethyl-3, 7-Dithianonane)
Decacarbonylditungsten (0)
(1988) / Liao J

Elemental Analysis of Brainstem in Victims
of Sudden Infant Death Syndrome
(1988) / Oquendo J

Detector Comparison for Simultaneous
Determination of Organic Acids and Inorganic
Anions
(1988) / Pannell D K

Trace Elemental Analysis of Ashes in the
Combustion of the Binder Enhanced D-RFB by
Inductively Coupled Plasma Atomic Emission
Spectroscopy
(1988) / Tai C

NORTHEASTERN LOUISIANA UNIVERSITY

N-Butyl-N-(4-Hydroxybutyl) Nitrosamine (BBN)
and Selected Microflora of Rat Colon
(1988) / Rokneddini J

NORTHERN ILLINOIS UNIVERSITY

Modification of Cytochrome c Peroxidase with
Hydrogen Peroxide: Characterization of the
Endogenous Decay Products at pH 8.0
(1988) / Hernan R A

Comparison of F(1)-ATPase and
Chloroform-Released ATPase: Dependence of
Activity on Temperature, Substrate
Concentration, and Inhibitors
(1988) / O'Brien K L

Toward the Synthesis of Podophyllotoxin
(1988) / Surjasasmita I B

A Photoluminescence Study of the
Deprotonation of Tertiary Amine Local
Anesthetics: Dibucaine-HCl
(1988) / Williamson L N

NOTRE DAME, UNIVERSITY OF

A Study of the Effects of Ion Bombardment
on the (100) Surface of Cubic Silicon
Carbide
(1988) / Chan S H

Functional Molecular Weight Determination of
Rat Lung UDP-GlcNAc: Dolichol-Phosphate
N-Acetylglucosaminyl 1-Phosphate Transferase
by Radiation Inactivation
(1988) / Doody M B

The Synthesis of N(5)-Hydroxy-N(5)-Acyl-L
Ornithine Derivatives: Entry Into the
Rhodotorulic Acid Family of Hydroxamate
Siderophores
(1988) / Ciglio C A

A New Approach to Spirocyclic Compounds
(1988) / Seutet P

Studies Directed Towards the Synthesis of
the Nine-Membered Lactone of Griseoviridin
(1988) / Strohbach J W

OAKLAND UNIVERSITY

Trace Level Analysis of Environmental
Organotin Residues Using Liquid
Chromatography
(1988) / Brown K L

Electrical Conductivity of Iodine Doped
Schiff Base Oligomers
(1988) / December T S

Perturbation of Rat Hepatic Acyl-Cobalt: A
Synthetase Activity by Valproic Acid
(1988) / Papin M M

OKLAHOMA, UNIVERSITY OF

I. Carbon-13 Chemical Shift Anisotrophy of
4-Halobenzonitriles. II. A Search for
Quantitative Standards for Carbon-13 NMR
Characterization of Coal
(1988) / Sierra B D

OLD DOMINION UNIVERSITY

Low Pressure Gas Flow Analysis Through an
Effusive Inlet Using Mass Spectrometry
(1988) / Brown D R

Gas Permeability Measurements on Small
Polymer Specimens
(1987) / Burns K S

Synthesis of Potential Long Acting Calcium
Channel Antagonists: A Study of
1,4-Dihydropyridine Analogs
(1987) / Christos T E

Modulation of Queuine Uptake in Cultured
Human Fibroblasts by Phorbol Esters and
Interferons
(1988) / Crane D L

Investigation of Complex Formation by
Oligomers of Cytosine and Guanosine
(1987) / Davis S R

Use of Chromium for the Quantitative
Determination of Red Cell Volume
(1988) / Handbury M

Synthesis and Characterization of
N-Alkyl-Adenosine-5-Diphosphates
(1987) / Huang H P

A Method for the Large Scale Purification of
HCG (Human Chorionic Gonadotrophin) from the
Urine of Pregnant Women
(1988) / Kuo C S

Reactions of Organic N-Chloramines in the
Gastric Fluid of the Rat
(1987) / Mazina K E

Reactions of Aqueous Chlorine in the Gastric
Fluid of the Rat
(1988) / Nickelsen M G

Optimization of the Tungstic Acid Denuder
Technique for the Measurement of
Atmospheric Ammonia
(1988) / Roberts P D

The Synthesis and Evaluation of
5-Phenyloxazolidines as Potential
Cardiovascular Drugs
(1988) / Wang T C

OREGON STATE UNIVERSITY

The Photochemistry of Polychlorobenzenes in
Micellar Media
(1987) / Lee Y S

A Critical Comparison of Methods for the
Determination of Phytoplankton Chlorophyll
(1988) / Salinas J T

Polybenzimidazole Synthesis from Compounds with
Sulfonate Ester Linkages
(1987) / Seckin T

Synthesis of a Model of the Spiroketal Portion
of Avermectin B
(1987) / Warrier U S

Synthetic Transformations on Shikimic Acid
(1988) / White E D

OREGON, UNIVERSITY OF

Thermodynamic and Kinetic Aspects of the
Reversible Unfolding and Refolding of
Mutants and Phage T4 Lysozyme at Low
Temperature
(1988) / Chen B L

Part I: Statistical Mechanical Study of the
Profile of a Fused Salt Near a Wall.
Part II: Ultraviolet Resonance Raman
Studies of Benzene Derivatives
(1987) / Li S

Far Ultraviolet Resonance Raman Studies of
Protein Components
(1988) / Mayne L C

The Development of Cobaloxime Mediated
Radical Alkyl-Alkenyl Cross Coupling
Reactions and an Application to the
Synthesis of Ammonium 3-Deoxy-D-Manno-2
Octulosonate (KDO)
(1988) / Meier M S

Novel Molecules Useful for Biological and
Medical Applications: Part I. Toward the
Development of Free Radical Based
Photoaffinity Labels. Part II. New
Nitroxide Formulations as Contrast-Enhancing
Agents for Magnetic Resonance Imaging:
Part III. Mammalian Brain Sigma Recepter
Specific Ligands: Structures and Molecular
Mechanics Calculations
(1987) / Pou S

Modulation of Methyl Esterase Activity in
Bacterial Chemotaxis
(1987) / Russell C B

Raman Spectroscopic Study of Sequence
Dependence of Local and Global Conformations
of Nucleic Acid Duplexes
(1987) / Wang Y

PACIFIC, UNIVERSITY OF THE

Possible Interference by Common Ordoriferous
Foodstuffs in the Determination of
Breath-Alcohol Content Using the Intoxilyzer
4011AS
(1988) / Jones G E

Investigation of Solubility and Dissolution
of Famotidine from Solid Glass Dispersions
of Xylitol
(1988) / Mummaneni V

PENNSYLVANIA STATE UNIVERSITY

The Synthesis, Structure and Ring-Opening
Polymerization of
Organosilylcyclotriphosphazenes
(1987) / Dunn B S

Soluble Species of Arsenic Trisulfide in
Amine Systems
(1987) / Guiton T A

Aqueous Speciation of the Orthophosphate and
Aluminophosphate Systems: A Spectroscopic
Study
(1987) / Luhrs R C

PENNSYLVANIA, UNIVERSITY OF

Synthesis and Characterization of Tertiary
Butyl-Phthalocyaninato-Rhodium Compounds
(1988) / Schmidt B F

PRINCETON UNIVERSITY

The Analysis of Three Immunological Models
Including Differentiated T Cells and
Replicating Antigen
(1988) / Zecha A D

PUERTO RICO, UNIVERSITY OF

Analysis of High Molecular Weight Organic
Compounds by Modified Purge and Trap
(1988) / Cuevas S N

Photochemistry and Photophysics Studies of
Some Iridium(II) and Rhodium(III) Complexes
(1988) / Lasso J I

Photochemical Studies of Adenine Nucleotides
as Models for ADN and ARN
(1988) / Lopez L

Rates of Cleavage of Binuclear Complexes of
Rhodium(III)
(1988) / Santos J

PURDUE UNIVERSITY

A Qualitative Analysis of Graduate Student
Problem Solving in Organic Synthesis
(1988) / Bowen C W

Study of the Associated Risk Factors
Contributing to Aspirin-Induced Renal
Dysfunction
(1988) / Bowman L

Two-Phase Reactions of HOCl with Beta,
Gamma-Unsaturated Carboxylic Acids and with
Delta 4-Cholesten-b-Beta-ol
(1988) / Corman M L

In Vitro Modelling of Photosynthetic Light
Harvesting
(1988) / Gotch A J

A Comparative Study of Protein Isolated from
Microcystis Aeruginosa
(1988) / Harper C F

The Insertion Chemistry of Mixed Alkyl,
Acyloxide Compounds of Zirconium (IV) and
Hafnium (IV)
(1988) / Kobriger L M

Effect of Competing Anions on Adsorption of
Bile Salts by Cholestyramine
(1988) / Kos R

Computer-Assisted Mechanistic Evaluation
of Pericyclic Reactions
(1988) / MacMahon C S

Modeling the Kinetics of Ammonia-Selective
Potentiometric Electrode
(1988) / Pagan G

An Alternate Synthesis of Calixarenes and
the Attempted Synthesis of Rotund [8] Arenes
(1988) / Payne E W

Studies in Aqueous Size Exclusion
Chromatography
(1988) / Principi J M

A Study of Aluminum Compounds Used as
Adjuvants in Biologicals
(1988) / Shirodkar S

Preparation and Characterization of
Apomyoglobin Reconstituted with
Selected Hemes
(1988) / Steup M B

Natural Product Analysis by Tandem Mass
Spectrometry
(1988) / Sweatlock J D

A Study of Heterocyclic Ring Reductions by
Calcium in Amines
(1988) / Thrasher K J

Reproducibility and Correlation Studies in
Planar Chromatography
(1988) / Uhegbu C E

Calcium-Ethylenediamine Reduction of
Phenylacetic Acid and Its Methylated
Derivatives
(1988) / Williams S M

A Theoretical and Experimental Study of the
Size Exclusion Chromatography of
Polyelectrolytes
(1988) / Wu J

QUEENS UNIVERSITY

Rubber Toughening of Polystyrene Through
Reactive Blending
(1988) / Fowler M W

RENSSELAER POLYTECHNIC INSTITUTE

X-Ray Photoelectron Spectroscopy of
Trimethyl Aluminum Adsorbed on High and
Low Surface Area Materials
(1988) / Chera J J

Automatic Control of Mean Arterial Blood
Pressure Using a Self-Tuning Pole
Placement Algorithm
(1988) / Chillrud D

EPR Spin Probe Study of Poly(Ethylene
Oxide)/NaSCN Complexes
(1988) / Gault K M

Modeling the Hemodynamic Response to
Dopamine in Acute Heart Failure
(1988) / Gingrich K J

Beta Absorption Dilatometry
(1988) / Greer S K

Aggregation State of the RuBisCO Large
Subunit Binding Protein: In-Vitro and
In-Vivo
(1988) / Hubbs A E

Studies of Low Pressure Chemical Vapor
Deposition of AIN and SiC Films Using
Organometallic Precursors
(1988) / Lee W

Growth, Plasmid Isolation, and the Methane
Monoxygenase of Methylobacterium
Organophilum XX
(1988) / Trautman C R

RHODE ISLAND, UNIVERSITY OF

Freshwater Gastropods of Equatorial Africa:
Correlations Between Shell Isotope Ratios and
Environment
(1988) / Amegashitsi L L

Resolution and Signal-to-Noise Enhancement
on Real Spectra
(1988) / Chen T R

Studies of Powder Flow of Pharmaceutics
Formulations
(1988) / Chueh H R

Analysis of Marine Sediments and Soft-Shell
Clams, MYA Arenaria, for Petroleum Derived
Hydrocarbons and Metals
(1988) / Franklin F E

Formulation of an Oral Controlled Release
Dosage Form for Nifedipine
(1988) / Garg V

Modification of Polymer Morphology: Electron
Microscopy Study of Polyaniline
(1988) / Hwang J H

A Study of the Effects of Sodium Benzoate on
Pyruvate Carboxylase Activity in Isolated
Mitochondria
(1988) / Jacob A G

Environmental Change and Speciation in the
Evolution of Turkana Basin Gastropods
(1988) / Lin Y

An Improved Synthesis of Racemic Onosital
Phosphates
(1988) / Liu Y C

The Study of Lithium Transport in Rat
Cortical Synaptosomes
(1988) / Mao W Z

The Effects of Streptozotocin-Induced Diabetes
on Systolic Arterial Pressure and Aortic
Collagen Biosynthesis in the Goldblatt
Renovascular Hypertensive Rat
(1988) / Mariani M J

The Chemistry of L-Ascorbic and D-Isoascorbic Acids:
Synthesis of Chiral-1',2'-Seco-2'-Deoxynucleosides
As Potential Inhibitors of Purine and Pyrimidine
Phosphorylases
(1987) / Mikkilineni A B

SAINT LOUIS UNIVERSITY

Application of Simplex Optimization in High
Pressure Liquid Chromatography and
Continuous Flow Ion Exchange Procedure
(1988) / Biwald C E

Synthesis of Benzochroman Derivatives
(1988) / Erickson J E

16S rRNA Structure Analysis
(1988) / Lawson E A

Synthesis and Polarographic Studies of
4-Arylaminocyclohexanones
(1988) / Mannino A

Kinetic Studies of the Lanthanide (III)
Catalyzed Reaction of Ethylenediamine
with Acetonitrile
(1988) / Meyer K E

SAN DIEGO STATE UNIVERSITY

The Application of N-Hydroxymethylacetamide
in the Synthesis of S-Acetamidomethyl
Pantebutatheine 4-Phosphate
(1988) / Atshaves B P

Doppler-Free Spectroscopy Based on Optical
Phase Conjugation by Degenerate Four Wave
Mixing in a Hollow Cathode Discharge
(1988) / Chen D A

Shock Sensitivity of Silane
(1988) / Famil-Ghiriha J

FMLP Binding in Adipose and Hepatic Plasma
Membranes
(1988) / Mazis G S

Detection of Protein Factors that Recognize
the Enhancer and Promoter Sequences of the
Chicken U1 and U4 RNA Genes
(1987) / Walker R J

Determination of PCO2 in Seawater
(1987) / Wang B

SAN FRANCISCO STATE UNIVERSITY

Synthesis and Spectroscopic Studies of a
Series of Octa-Substituted Porphyrins and
Their Iron (III) Complexes
(1988) / Isaac M

Interspecific Chemical Variations Within
the Subphylum Urochdata
(1988) / Kendall T

Investigation of the Structual role of
Cysteine Residues and Chemical Modification
of Lysine Residues in Lamb Protein of
Escherichia Coli
(1988) / Malloy B

Molecular Oxygen in Bilayer Membranes: An
Electrochemical Titration
(1988) / Moy F

Nucleophilic Activation of Coordinated
Carbon Monoxide. Reactions of Tungsten
Hexacarbonyl with Methoxide and Hydroxide
(1988) / Muraoka M

A Versatile Instrument to Study Static and
Quasi-Elastic Light Scattering
(1988) / Thompson G

Acid Promoted Carbon-Carbon Cleavage of
9-Xanthyl Compounds: A Kinetic and
Mechanistic Study
(1988) / Trudell M

Dichloroketene Additions to Small Ring
Systems: Synthetic Applications
(1988) / Watson S

The Electrochemistry of Monomeric and
Dimeric Platinum Organometallic Complexes
Containing Phosphine Ligands
(1988) / Yee L

Models of the Cytochromes b. Syntheses and
Spectroscopic Studies of a Series of
Mono-Ortho-Amidotetraphenylporphinatoiron
(III)-bix (N-Methylimidazole) Complexes
(1988) / Zhang H

SASKATCHEWAN, UNIVERSITY OF (SASKATOON)

Pharmacokinetics of Single and Multiple
Doses of Ketoprofen with and Without
Cimetidine
(1988) / Corman C L

The Use of Tris-Bipyridine Ruthenium
Complex as a Photosensitizer to Activate
Cyclopentane in Aqueous Medium and at
Room Temperature
(1988) / Njapba N J

SHERBROOKE, UNIVERSITY OF

Less Recepteurs de l'Aldosterone dans les
Reins d'Embryons de Poulet
(1988) / Allard C

Synthese du N-Bromomethylcetoanimo-6-Hexyl
O-B-D-Galactophyranosyl-(1-4)-(Acetamino
2-Desoxy-2-B-D-Glucopyranosyl)-(1-2)
D-Manno-Pyranoside Poru le Marquage par
Affinite du Site de Reconnaissance des
Sucres Chez la PHA
(1988) / Ammann H

Spectroscopie Infrarouge par Transformee de
Fourier, Spectre d'Adsorption a Resolution
Elevee du Phosphure de Methine Entre 3148
et 3248 CM-1
(1988) / Boulanger B

Electrolyse de l'Eau en Milieu Alcalin sur
le Nickel et le Cobalt de Raney
(1988) / Choquette Y

Electrohydrogenation Catalytique des
Composes Polyaromatiques sur Electrodes de
Nickel de Raney
(1988) / Comtois M

Syntheses de Produits Tricycliques par
Reaction de DIELS-ALDER
Trans-Annulaire: Approche de la Synthese
de Steroides Fonctionnalises en Position 1
(1988) / Dallaire C

Etude de l'Adsorption des Hydrocarbures
Aromatiques Polycycliques sur des Materiaux
Particulaires et Fibreux
(1988) / Hassani C A

Etude de l'Adsorption d'Electrolytes Simples
et de Complexes d'Inclusion a l'Interface
MERCURE/SOLUTION
(1988) / Nadeau P

Enolisation et Alkylation de DIOXANNES-1,3
Fonctionnalises en Position 5: Evidence du
Controle Stereoelectronique: Etude de la
Reaction de DIELS-ALDER Transannulaire des
Trienes Macrocycliques a Quatorze
Membres: Vers une Synthese Generalisee de
Molecules Polycycliques
(1988) / Ndibwami A

Nouvelle Approche de Synthese aux
Triterpenes: Synthese du TRANS-1,
10-CISOIDE-CIS-2, 7-DIMETHYL-1,
2-TETRACARBOMETHOYX-5, 5,12,12-TRICYCLO
(8.4.0.0.[2,7]TETRADECENE-8
(1988) / Roberge J

Modelisation Moleculaire d'Analogues
Cyclisses de l'Enkephaline: Differentiation
des Crepteurs Opiaces Mu et Delta
(1988) / Villeneuve G

Radical Oxidation of Pyrimidine Nucleosides
(1988) / Wagner R J

SOUTH CAROLINA, UNIVERSITY OF

Three Complexes of Cadmium(II) with
2,2'-Dipyridylamine: X-Ray Crystal
Structures and Cadmium-113 Nuclear Magnetic
Resonance Chemical Shielding Tensor Elements
(1988) / Adams M T

Synthesis of a C(7)-C(13) Fragment of
Erythronolide
(1988) / Chandler A C

Fluorometric Analysis of the Unformity of
Deposition on Membrane Filter Cassettes
(1988) / Jenkins S L

Macrocyclization Methodologies for
Cembranes: Approaches Toward B-CBT
(1988) / Kaltenbach R F

The Synthesis and Characterization of a
Rhenium Triflate Complex, Dirhenium Bridging
Hydrido Complex and a Diruthenium, Cobalt
Mixed-Metal Sulfido Complex
(1988) / Kuhns J D

Isolation and Characterization of Pure cDNA
Coding for Avian Keratin Filament Associated
Polypeptide #15/16
(1988) / Lambros I P

Feasibility Studies Toward the Furano
Lactone Natural Product Kallolide A
(1988) / Nelson D J

Risk Factor Profiles of Primary Carcinoma of
the Breast Stratified by Estrogen Receptor
Status
(1988) / Watkins S

SOUTH DAKOTA STATE UNIVERSITY

Development of an Analytical Method for the
Trace Determination of Mycophenolic Acid by
High Performance Liquid Chromatography
Following Extraction of the Compound from
Feedstuffs
(1988) / Kahsal T Z

Analysis of the Herbicides Chlorsulfuron,
Sulfometuron Methyl and Metsulfuron Methyl
in Soil and Water
(1988) / Mao F W

Aroyl Phosphates: The Mechanism of
Isomerization
(1988) / Ning S X

SOUTH DAKOTA, UNIVERSITY OF

Production of Solutions of Bicarbonate Ions
and OH(n) Species in Single Crystals of the
Potassium Halides and Cesium
(1988) / Kearney W R

Analysis of DNA Photolyase Mutants of
Chlamydomonas Reinhardtii
(1988) / Munce D B

Study of Lithiomethyldimethylamine
Dimethylaminomethylborane and Its
Dimethylgallata Derivative
(1988) / Webb K M

SOUTHERN ILLINOIS UNIVERSITY

Transition-State Structure in an Enzymatic
Reaction
(1988) / Arnold R

Annealation of Imidazoles and Benzimidazoles
to Their Fused Ring Derivatives
(1988) / Custer J H

Fluorination of Estrone
(1988) / Mortezani R

SOUTHERN METHODIST UNIVERSITY

Investigation of Simple Anion Addition
Versus Tandem Cyclization Rearrangement in
Aryne Reactions: Mechanism,
Regioselectivity and Synthetic Utility
(1988) / Crenshaw L I

Synthesis and Characterization of
Polystyrenegraft-Poly
(Methylphenylphosphazene)
(1988) / Schaefer M A

SOUTHWEST TEXAS STATE UNIVERSITY

Polyimidines from 3,3',4,4'-
Benzophenonetetracarboxylic Dianhydride and 4,4'-
(Hexafluoroisopropylidene)Diphthalic Anhydride
(1988) / Mores M A

SOUTHWESTERN LOUISIANA, UNIVERSITY OF

The Effect of Disaccharides and Divalent
Metal Ions on the Thermal Stability of
Enzymes
(1988) / Ercikan E A

A Comparison of Intermolecular Potential
Approximations in Gas Phase Hot Atom Model
Calculations
(1988) / Su L

STEPHAN F. AUSTIN STATE UNIVERSITY

Vibrational Correlation of Selected
Fluoroberyllates
(1988) / Hodo J D

STEVENS INSTITUTE OF TECHNOLOGY

Simulation of the Low Pressure Methanol
Synthesis Reactor
(1987) / Betaimi K

Steroselective Transformatiions of B-Lactams
(1987) / Chiang J

Design of a Pid Controlled for the
Activated Sludge Process
(1987) / Cooper A M

Integrated Approach to the Packing of Liquid
Chromatography Microcolumns
(1987) / Edkins T J

External Effectors of Dolichol and
Cholestrol Biosynthesis
(1987) / Herald E G

LDPE/PS Polyblends: Elongational Viscosity
and Prediction of Structure and Morphology
Effects of Rheodegradation in Industrial
Mixers
(1987) / Mehta D D

Effect of Extensive Oxidative Degradation
on Isotactic Poly(Butane-1) Properties
(1987) / Mezaache D

TEMPLE UNIVERSITY

Determination of Low Molecular Weight
Dialkylamines in Blood
(1988) / Gilmartin G J

Synthesis of Biologically Active Analogues
of Dehydroepiandrosterone
(1988) / Tran P B

TENNESSEE TECHNOLOGICAL UNIVERSITY

Computational Studies of Heme Proteins
(1988) / Boles J O

The Characterization of Desorbed Species
from Weathered Charcoal
(1988) / Brown R P

The Characterization of the Kerogen in
Chattanooga Shale by Oxidation with
Perchloric Acid
(1988) / Stanton B J

TENNESSEE, UNIVERSITY OF (KNOXVILLE)

Resistance to Copper Toxicity in a
Spontaneously Generated Mutant of Salmonella
(1987) / Armstrong D A

Aerosol Direct Fluorination of Esters and
the Synthesis of Perfluorinated Olefins
(1987) / Brennan J M

The Rat Liver Genome Contains Sequences
Homologous to Sequences Within pUC and M13
DNA
(1987) / Cheatham R B

Effects of Insulin and the Phorbol Ester
Tumor Promoter TPA on Cell Growth and
Transcript Levels of the Growth-Associated
Enzyme, Ornithine Decarboxylase
(1987) / Goodman S A

A Facile Synthesis of Primary Amines Via
Organoboranes
(1987) / Henderson D A

The Role of Multivalency in Antibody
Mediated Liposome Targeting
(1987) / Houck K S

Interleukin 2 Receptor Expression by K-572
Cells
(1987) / Kouns W C

Fiber Optic-Based Time-Resolved Fluorimetry
for Immunoassays
(1987) / Petrea R D

The Addition of HI to Unsaturated
Hydrocarbons Using Iodine and Aluminum
Dioxide
(1987) / Stewart L J

TENNESSEE, UNIVERSITY OF (KNOXVILLE)
(continued)

Matrix Isolation Fourier Transform Infrared
Spectroscopy of Nitro Polycyclic Aromatic
Hydrocarbons
(1987) / Stout P J

Aeros Direct Aerosol Fluorination of
Ethers and Esters
(1987) / Taylor D R

TEXAS A AND M UNIVERSITY

The Iron(III)-Catalyzed Oxidation of DTPA in
an Aqueous Solution
(1988) / Christiansen S H

A Site-Directed Mutation in the Regulatory Subunit
of Aspartate Transcarbamoylase: Effects on
Allosteric and Catalytic Response
(1988) / Corder T S

Investigation of Silver Electrodeposition on
Polycrystalline Platinum by Iodine Chemisorption
(1988) / Harris J E

Regulation of Protein Metabolism, DNA Synthesis,
and Growth by Epidermal Growth Factor in A431
Mutants
(1988) / Miller S D

Mechanistic Studies of Coenzyme B(12) Model Systems
(1988) / Robinson J W

The Stability and Chemical Modification of
Synthetically Useful Enzymes
(1988) / Smith G L

The Binding of Spermine to SSB Protein-
Poly(dT) Complexes
(1988) / Wei T F

Investigation of High Temperature Gaseous Species
by Knudsen Cell Mass Spectrometry Above the
Condensed Systems Copper Yttrium Ruthenium Carbide,
Silver Yttrium Ruthenium Carbide, and Gold Yttrium
Ruthenium Carbide
(1988) / Wilhite D W

TEXAS CHRISTIAN UNIVERSITY

A Study of Boron-Activated Dienamies Derived
from Pyridine-n-Borane
(1988) / Park E S

The Rearrangement Reactions of
Gem-Dichlorocyclopropyl-Carbinyl Cations
(1988) / Qian C

TEXAS SOUTHERN UNIVERSITY

Determination of Sulfide in Core Samples
Using Ion-Selective Electrode
(1988) / Anyatonwu A C

Quantitative Determination of Glutamine and
the Determination of Glutiminase Activity
in Saccharomyces Cerevisiae Via Biocatalytic
Enzyme Immobilization
(1988) / Aremo S

Kinetic and Inhibition Studies of the Mixed
Function Oxidase System in Human Uteri
(1988) / Bahar A I

A Study on the Effect of Temperature and
Time on Direct Liquefaction of Coal Using
Bis[Dihydrobis(Pyrazolyl)Borate]Cobalt (II)
as a Catalyst
(1988) / Charles A L

Aromatic and Fries Rearrangement
(1988) / Ebrahimi M

A Polarographic Study of Certain Nickel
Complexes
(1988) / Hunt F J

Liquefaction of Illinois Number Six River
King Coal Using Cobalt(II) Poly(Pyrazolyl)
Borates as Catalysts
(1988) / Jackson K A

Application of Energy Dispersive X-ray
Spectrometry to the Determination of Trace
Metals in the Environment (Lake Houston)
(1988) / Lawal B K

A Study of Bis[Tetrakis(1-Pyrazolyl)
Borate]Nickel (II) as a Catalyst for
Direct Coal Liquefaction
(1988) / Lin H M

A Trace Metal Study of the Aqueous Sediments
Surrounding a Coal-Fired Electrical Facility
(1988) / Mitchell D L

A Study of Trace Metals in Human Urine
(1988) / Nassimi Z

A Study of Trace Metals from Selected Sites
Along the Houston Ship Channel
(1988) / Nwaiwu L O

Organic Geochemistry of the Ellenberger
Group in Central Texas
(1988) / Okon E A

A Study of Aqueous Sediments for Trace
Metals
(1988) / Ozurumba E O

A High Temperature and High Pressure Study
of the Effects of Alkaline Earth Metal and
Sulfate Ions on Calcium Carbonate Scaling
in Brine
(1988) / Sethuraman G

Determination of Oxylate Using Immobilized
Oxalate Oxidase Enzyme
(1088) / Shah V P

Purification and Characterization of Human
Uterine Cytochrome P-450 Reductase
(1988) / Sirossian S

Bioselective Electrode Studies of Cystine
Via Enzyme Immobilization
(1988) / Smith K L

Petroleum Geochemistry of Part of the
Western and Central Great Basin, Nevada
(1988) / Tanyingu L G

Direct Coal Liquefication Using
Bis[Dihydrobis(Pyrazolyl) Borate] Iron (II)
as a Catalyst
(1988) / Watson R W

Direct Coal Liquefaction Using
Bis(Dihydrobis(1-Pryazolyl)Borate)Nickel
(II) as a Catalyst
(1988) / Yuen P K

TEXAS WOMEN'S UNIVERSITY

Kinetics of the Reaction of
Fluoroanthraquinones with Secondary Amines
(1988) / Parker C D

TEXAS, UNIVERSITY OF (ARLINGTON)

Electrochemical Studies of
Poly(3-Methylthiophene) and Other
Heterocyclic Compounds
(1988) / Hsu S

Luminescence Probe Studies of Zinc Oxide and
Polypyrrole Surfaces
(1988) / Phan L L

TEXAS, UNIVERSITY OF (AUSTIN)

The Effects of Explosive Shock Wave
Propagation Through a Solid Molecular
Structure
(1988) / Clark D E

Synthesis of a Oxazole[2,3-c] [1,4]
Benzodiazepine with Potential Antitumor
Activity
(1988) / Davis M E

Cyclizations of Alpha Amino Radicals
(1987) / Laswell W L

Zeolite Modified Electrode
(1988) / Li Z

Equilibrium Calculations for the
Optimization of Chlorine Determination by
InCl Flame Molecular Emission Spectrometry
(1988) / Rodriguez L F

Effect of the Liquid Film Thickness on the
Corrosion of Steel
(1988) / Tebbal S

Elucidation of High Temperature Interactions
Between Chromium and Graphite During
Graphite Furnace Atomic Absorption Analysis
(1988) / Wolfe K I

TEXAS, UNIVERSITY OF (EL PASO)

Aqueous Solubilities and Partition Coefficients
of Quinones
(1988) / Bhave M P

Crown Ether Inhibition of Bacterial
Bioluminescence
(1988) / Darvish M

Affinity Chromatography of Vibrio Harvey Luciferase
(1988) / Gajiwala K S

Polynitroaromatic Compounds, Potential Precursors to
Carbon Films
(1988) / Han Y D

Hydrosilylation of Phenylacetylene with Transition
Metal Silane in the Presence of Hexachloroplatinic
Acid
(1988) / Lii J

Bicyclo (6.1.0) Nona-3.5-Diene and Ester Derivatives
(1988) / Okafor E M

Ferrocenyl-Silyl Ethers
(1988) / Pena-Cabrera E

Spin Hamiltonian Simulation of Electron Paramagnetic
Resonance Spectra of Gadolinium III in Clays and
Crystals
(1988) / Samuels A C

Soluble Protein and Enzyme Changes During the
Senescence of Cercis Canadensis
(1988) / Skorpenske R G

Synthesis of Diphenyl-Bridged Bifunctional
Intercalators Bis-Purine Analogs
(1988) / Tsai M

Oligosilane-Transitionmetal Complexes: A Dramatic
New Reformation Reaction
(1988) / Wang L J

TOLEDO, UNIVERSITY OF

The Synthesis and Characterization of
2,2,4,4-Tetramethyl-3,
4-Di-Meta-Toluyl-3-Hexene, a Sterically
Congested Alkene
(1988) / Garry P A

The Determination and Characterization of
Carbon Black in Elastomer Systems
(1988) / Manley P E

Carboxyl-Terminal Analysis of Soybean
Lipoxygenase
(1988) / McNemar T B

Observations on the Synthesis of Verdins
(1988) / Nonis P J S

U.S.A.F. INSTITUTE OF TECHNOLOGY

The Redox Chemistry of 5-Imino
7-Deoxydaunomycinione the Preparation and
Characterization of Leuco Daunomycin
(1988) / Bird D M

Synthesis of Thienylamphetamine Derivatives
Via Borane Chemistry
(1988) / Marks R C

Phosphorus-Stabilized Carbanion-Accelerated
Claisen Rearrangements: Asymmetric
Induction Via 1,3,2-Oxaza Phosphorinanes
(1988) / Marlin J E

Rate Acceleration of the Dielsalder Reaction
of Anthracene Cycloadducts by Polysiloxy
Substituents
(1988) / McKelvey T

UTAH STATE UNIVERSITY

Studies Toward the Synthesis of
Acanthicifoline, Jasminidin and
3-Substituted Pyridine Carboxaldehydes
(1987) / Herrick J J

Mode of Action of 2-Digeranylamino Ethanol,
an Experimental Species-Selective Piscicide
(1987) / Jones L A

A Study of the Conformational States of the
Plasma Membrane Adenosine Triphosphatase
from Redbeet Storage Tissue
(1987) / Kesav B V

Synthetic Studies Toward
B-Alkylthiolanthionines
(1987) / Kim H O

UTAH, UNIVERSITY OF

T Cell Profilerative Response to Mitogen
Derived from Mycoplasma Arthritidis in
Presence of 1a+Epidermal Cells as an
Accessory Cell
(1988) / Ghadessi M

A Model for Effective Monitoring of Patients
Receiving Aminoglycoside Antibiotic Therapy
(1988) / Gibbs F G

Isolation of Unique DNA Sequences from
Upopathogenic Escherchia Coli Using a
Subtraction Hybridization Technique
(1988) / Perkins M L

Absorbance Measurements of Optically
Inhomogeneous Samples Using Phase Conjugate
Thermal Lens Spectroscopy
(1988) / Plumb D M

8-Deazafolate Analog Inhibitors of
Thymidylate Synthase
(1988) / Samathanam C A

The Synthesis, Characterization and
Chemistry of Rhenium Organosilanes:
Precursors to Metal-Stabilized Silenes
(1988) / Young C S

VERMONT, UNIVERSITY OF

Asymmetric Vincadifformine
(1988) / Kearney T

VICTORIA, UNIVERSITY OF

The Synthesis and Coordination Chemistry of
Amino-Substituted Bisphosphines
(1988) / Phillips A J

Detection and Distribution of Estrogen
Receptor in Human Breast Cancer by a
Combined Biochemical/Immunohistochemical
Multiple Microsample Assay
(1988) / Thornton I G

VILLANOVA UNIVERSITY

The Isolation and Partial Purification of a
Protease from Bacteroides Intermedius
(1988) / Hopkins N

Investigation of Hetero-Dimer Formation of
Some Water-Soluble Metalloporphyrins
(1988) / Sag S

Isolation and Characterization of Protease
from Bacteroides Intermedius
(1988) / Shandilya H G

WASHINGTON STATE UNIVERSITY

X-Ray Photoelectron Spectroscopic
Characterization of the Nitrogen-Phosphorus
Bonding in Phosphazene Trimers and Polymers
(1988) / Dake L S

WASHINGTON, UNIVERSITY OF

Synthesis and Evaluation of Phospholipid
Analogue Inhibitors of Phospholipase A(2)
(1987) / Berman R J

Synthesis and Characterization of Terminal
and Bridging Nitride and
Trimethylsilylimido Complexes of Vanadium
(1988) / Lerchen M

Multilinear Regression and Curve Resolution
in Supercritical Fluid Chromatography-Mass
Spectrometry
(1987) / Moore S D

Synthesis of Substituted Pyrimidines for a
Study of DNA Sequence Recognition
(1987) / Smirnow D

An Automated Audio/Protein Recorder for Use
with Physiological Fluids
(1987) / Warner W S

WESLEYAN UNIVERSITY

A Model System for the Structure and
Reactions of L-Dehydroascorbic Acid
(1988) / Lali M D

WEST TEXAS STATE UNIVERSITY

Transformation of the Blaz Gene in
Staphylcoccus Aureus Utilizing Lysostaphin
as the Lytic Agent
(1988) / Sharp S B

WEST VIRGINIA UNIVERSITY

New Synthetic Route to Vinylallenones and
Their Reactions with Enamines to Provide
Highly Substituted Arylsilanes
(1988) / Andemichael Y W

Reactivity Studies of Isonitriles with
Electron-Deficient
1-Sila-3-Zirconacyclobutane Complexes
(1988) / Berg F

Regio- and Stereoselective Synthesis of
Tri- and Tetrasubstituted Alkenes
(1988) / Chen J H

Synthesis and Pharmacological Evaluation of
Spiro-Analogues of 5-Benzyl-5-Ethyl
Barbituric Acid
(1988) / King S B

Spectroscopic Studies of Aldehyde
Dehydrogenases and a Model Dehydrogenase
Enzyme Complex
(1988) / Meckley M M

Calcium and Taaurine Matebolism in Chicken
Blood Cells
(1988) / Porter D W

Characterization of Cytochrome P450 Ia-1 and
Vitellogenin Synthesis in Isolated Trout
Liver Cells
(1988) / Saito N

New Semiconducting Copper Phthalocyanine
Dimers
(1988) / Teng G

Drug Removal by Hemofiltration
(1988) / Yu S S

Stereoselective Synthesis of Terminal
1,3-Butadienes Via Trimethylsilyl
Substituted Allylborane
(1988) / Zhuang H

WESTERN CAROLINA UNIVERSITY

Transfer Protein Specificity for DPPC in
Rat Lung
(1988) / Childress S K

WESTERN CAROLINA UNIVERSITY
(continued)

The Derivatives of 3-Carene
(1988) / Choubal M D

Hydrogen Atom Addition to Sulfur Dioxide in
Solid Argon in 12K
(1988) / Fender M A

Studies Involving the Photocycloaddition
Reactions of (+) 3-Carene and Selected
Quinones
(1988) / Holsten G A

A New Route to 2-Carene-4-One from 3-Carene
(1988) / Pan C J

Oxidation of Manganese Gluconate Complexes
(1988) / Ramsey H C

WESTERN KENTUCKY UNIVERSITY

Spectral and Kinetic Studies of the
Interaction of Cyanide and Detergents with
Cobalt (II) Phthalocyanine in DMSO
(1988) / King H M

WESTERN MICHIGAN UNIVERSITY

Synthesis and Reactions of a-Diketones and
Preparation of Some New Benzimidazoles
(1988) / Agharahimi M

Micellar Catalyzed Hydrolysis of Hydroxamic
Acids by Perfluorooctanoic Acid
(1988) / Akhavan-Tafti M

Synthesis and Study of Novel Quinoxaline
DI-N-Oxide and 3-Acyl-2-Methylquinoxaline
DI-N-Oxides
(1988) / Ogbu C

A Synthetic Approach to Novel Heterocyclic
Herbicides
(1988) / Pathadan P

Characteristics of Binding of Interleukin-1
to YT.NCI Cells
(1988) / Speziale S

WESTERN ONTARIO, UNIVERSITY OF

Nuclear Magnetic Resonance Studies of
In-Vitro and In-Vivo Glial Tumors
(1988) / Megyesi J E

The Purification and Properties of
Monoacylglycerol Kinease from Bovine Brain
Cytosol
(1988) / Shim Y H

Approaches to the Total Synthesis of
Cycloseychellene
(1988) / Tse D W C

Examination of an Alteration in the Long
Terminal Repeat of a Highly Leukemogenic
Type-B Retrovirus
(1988) / Waight P W

WESTERN WASHINGTON UNIVERSITY

Response to Temperature, Diet, and Substrate
During Larval Development of Two Species of
the Genus Hemigrapsus
(1988) / Holzer K A

Fuel Structure and Fire in Subalpine Fir
Forests, Olympic National Park, Washington
(1988) / Taylor K L

A Survey of Biogenic and Anthropogenic
Hydrocarbons in Temperature Glacier Snow
and Ice
(1988) / Wienkers L C

WICHITA STATE UNIVERSITY

Multi-Purpose Water Analysis System
(1987) / Campbell K S

Inhibitors of Human Leukocyte Elastase
(1987) / El-Shoubary M T

Inhibitors of Human Leukocyte Elastase:
Aryl Azolides
(1987) / Gray A V

WINDSOR, UNIVERSITY OF

[1,4] 0>C Silyl Migrations of Furans and
Thiophenes: Applications to Preparation
of 3,4-Disubstituted
(1988) / Bures E J

Part I - Determination of Iron, Copper and
Zinc in Single Aliquot of Serum
Sample - Part II - Fluorometri
(1988) / Castillo G M

An Improved Kinetic Fluorometric Enzymatic
Coupled Assay for Determination of
Galactose 1-Phosphate
(1988) / Catomeris P

Preparation and Characterization of Defined
Conjugates of Glucose Oxidase and Peroxidase
(1988) / Harake B

Structure-Function Studies on the Mechanism
of Action of Staphylococcal Alpha Toxin
(1988) / Hebert T

Fine-Structure Mixing in 92D Rubidium Atoms
by Collisions with Ground-State Rubidium
and Noble Gas
(1988) / Mallory T R

Applications of Supported Enzymes in
Clinical Chemistry
(1988) / Sullivan E

WORCHESTER POLYTECHNIC INSTITUTE

Ethanol Production by Immobilized
Saccharomyces Cerevisiae: Application of
Kinetic and Dimensional Analyses
(1987) / McConville F X

The Role of Cytochrome P-450-11-B in
Biosynthesis of Aldosterone
(1986) / Shen W

Production of Sinefungin by Free and
Immobilized Cells of Streptomyces Incarnatus
(1987) / Talanian K M

WRIGHT STATE UNIVERSITY

The Effects of Biodynamic Stress on Workload
in Human Operators
(1987) / Albery W B

Synthesis of Aromatic Polymides Containing
Oxyalkylene Linkages
(1987) / Beltz M W

WRIGHT STATE UNIVERSITY
(continued)

Scanning Electron Microscopic Studies of
Skin Equivalents Maintained in Vitro and
in Vivo
(1987) / Blough R I

The Solubility of Ten Gases in the
1-Alkanols at 298.15 K
(1987) / Bo S

Synthesis, Characterization and NMR Studies
of Binuclear Nickel-Zinc and Nickel-Cadmium
Schiff Base Complexes
(1987) / Bollinger M K

The Cytotoxicity of Perfluoro-N-Decanoic
Acid on Cultured Mouse Neuroblastoma Cells
(1987) / Brickson B M

NTP-AMP Phosphotransferases: Structural and
Functional Comparisons
(1987) / Burnette B L

The Use of Portable Refreezable Head Coolers
to Increase Heat Tolerance of Personnel
Working in Hot Environments
(1987) / Bustamante M J

Preparation and Hydrolysis of Polymers
Containing 2,4-Dichlorophenoxy Acetic Acid
as a Pendent Substitute
(1987) / Cardona L F

Isotopic Dilution/Mass Spectrometry
Measurements of Deuterium Oxide in
Urine: Application to Total Body Water
Determinations
(1987) / Coldiron S J

The Role of Interferon in the Replication of
Two Mengovirus Variants in Mouse L 929 and
L (Y) Cells
(1987) / DelRaso N J

Synthesis and End-Group Analysis of
Copolymers of Hexafluoropropylene and
Vinylidene Fluoride
(1987) / Donahue W H

Radial Keratotomy: A Review of U. S.
Literature with Recommendations for Far
Pilot Qualifications
(1987) / Duncan D L

Certification of Hypertensive Pilots--A
Normative Approach
(1987) / Ferreira J M

Basic Critical Care and Anesthesiology for
Long-Duration Space Missions: Orbital
Stations to Planetary Expeditions
(1987) / Forest R

Factors Responsible for the Decline of White
Oak Regeneration: A Historical Analysis
(1987) / Hagan K A

Calculation of the Isotropic Compton X-Ray
Scattering Profile of Titanium Copper
Hydrogen
(1987) / Hsu T J

Comparison of Tolerance Differences Between
Standard and Sequentially Inflating G-Suits
During +Gz Acceleration
(1987) / Jennings R

Plasmid Control of Exoenzyme Excretion in
Pseudomonas Aeruginosa
(1987) / Johnson J

Encephalomyocarditis Virus-Induced Diabetes
Mellitus-Like Syndrome in the ICR Swiss
Mouse
(1987) / Kaptur P E

The Gas-Phase Thermal Decomposition of
Alkylaromatic PCB Substitutes
(1987) / Mazer S L

A Comparative Study of Macro and Micro
Methods to Assay for Alpha and Beta
Interferon
(1987) / McIntyre T M

Insertion of Transposon Tn7 Into Pseudomonas
Aeruginosa
(1987) / Morgenstern M R

Calcium Metabolism in Whole-Body Vibration
(1987) / Myers K J

The Synthesis and Thermal Evaluation of
Acetylene Terminated Sulfone Resins
(1987) / Picklesimer D K

Synthesis and Structural Characterization of
Polyphosphate and Polyphosphonate Esters
Derived from Biologically Active Diols
(1987) / Powers D S

Synthesis and Characterization of a Series
of Copper - Cobalt Binuclear Base Complexes
(1987) / Ridgely M M

Ethynyl End-Capped Polyimide Oligomers
Containing Oxyethylene Linkates: Synthesis
and Characterization
(1987) / Sridhar K

The Synthesis of 1,2-Bis(Aminophenoxy)
Propane and Its Polymerization with Selected
Dianhydrides
(1987) / Tjugito S

Simulator Induced Syndrome in Coast Guard
Aviators
(1987) / Ungs T J

Modification of the Phase Relationship
Between a Vestibular Stimulus and Eye
Movement
(1987) / Woodard D

WYOMING, UNIVERSITY OF

Antisense Sequences Directed Against Tobacco
Mosaic Virus
(1989) / Nelson A W

YOUNGSTOWN STATE UNIVERSITY

The Fragmentation Patterns of Nickel,
Palladium, and Platinum Dimethylglyoximes
by Electron and Chemical Ionization Mass
Spectrometry
(1988) / Besozzi L

Neutron Activation Analysis of Samples Taken
from the Lake Milton Bed and Various
Surrounding Tributaries
(1988) / Heipel K

Determination of Trace Amounts of Aqueous
Sulfide by Indirect CVAAS of Mercury
(1988) / Kurtes R A

AKRON, UNIVERSITY OF

Vibration Response to Variable Stiffness
Beams and Frames
(1989) / Chaarani W Z

Dynamic Hinge Concept for Beams and Frames
(1989) / Chidiac G E

Computational Simulation of Composite
Structures with and Without damage
(1988) / Wilt T E

ALABAMA, UNIVERSITY OF (UNIVERSITY)

Integrated Design of Concrete Flat Plate
Structures
(1988) / Delahay J M

ALBERTA, UNIVERSITY OF

Soil-Geogrid Interfacial Shear Strength
(1988) / Bobey L W

An Investigation and Analysis of a Landslide
Near Raush Valley, British Columbia
(1988) / Bradshaw M T

The Behaviour of a Reinforced Soil Slope
(1988) / Chalaturnnyk R J

Calculation of Ice Jam Profiles
(1988) / Flato G M

Pile Driveability Prediction
(1988) / Vasavithasan M

Scales for the Development of Longitudinal
Velocity in a Curved Channel
(1988) / Weng D L

The Flexural Creep Behaviour of OSB Stressed
Skin Panels
(1988) / Wong P C

ARIZONA STATE UNIVERSITY

Evaluation and Testing of Polyacrylamide
Synthetic Rubber as an Asphalt Modifier
(1988) / Aljilani M Y

Modeling Lateral Scour and Deposition in
Alluvial Channels Using HEC-6 and Stream
Tubes
(1988) / Krueger G H

A Batch-Type Testing for Estimation of
Adsorption of Gaseous Compounds Onto Soils
(1988) / Marwig R

The Development of a Compacted Earth
Anchoring System
(1988) / Perkins J R

Analysis of Spherical Lamella Domes by
Triangular Elements
(1988) / Voutsinas F

State of the Art Review and Conceptual
Analysis of Partially Saturated Flow
Modeling
(1988) / Walsh K D

An Expert System in Steel Connection Design
(1988) / Wang Q

Effects of Changing Exclusive/Permissive Left
Turn Phasing from Variable to Fixed Cycle
Length
(1988) / Warne T R

Development of Transport and Stiffness
Matrixes for an Arch Acted on by Second
Order Forces
(1988) / Wuellner P D

Parametric Evaluation of Partially
Prestressed Concrete Members by Non-Linear
Analysis
(1988) / Yamani V R

ARKANSAS, UNIVERSITY OF

Simplification of the Subgrade Resilient
Modulus Test (ASHTO T-274)
(1988) / Foo K Y

Resilient Modulus Sample Preparation for
Granular Materials
(1988) / Qedan B A

AUBURN UNIVERSITY

Computer-Aided Piping Design Interface for
Microcomputer Materials Management System
(1987) / Crispin G S

Bicycle Safety Evaluation
(1987) / Davis W J

Review, Modification and Application of
Intersection Sight Distance Models
(1987) / Famili R

Finite Element Analysis of the Space Telescope
Focal Plane Structure Joint
(1987) / Hwang D Y

Stress Intensity Factors for Cracks in Rectangular
Blocks and Round Bars
(1987) / Jung H Y

Generation of Overhead Conductor Aeolian Vibration
"Load"-N Curve Modification Factors
(1987) / Logue H R

Laboratory Investigation and Analysis of
Borehole Dilution
(1987) / Matteson K T

Unit Price Estimating Using Microcomputer
Spreadsheet Software
(1987) / Sabet R F

An Evaluation of the Factors that Impact Risk and
Profitability in the Construction Industry
(1987) / Sanders S R

Distress Evaluation of Portland Cement Concrete
Pavement
(1987) / Stiegler N D

BRIGHAM YOUNG UNIVERSITY

Development of a Strategy for Automatic
Selection of Standard I-Section in 3-D
Steel Frames
(1988) / Fonseca F S

Verification of F-4 Torque Box Structural
Analysis and Crack Growth Analysis
(1988) / Jensen R P

BRIGHAM YOUNG UNIVERSITY
(continued)

Three-Dimensional Flow Modeling of a
Perched Water Zone Underlying Waste
Disposal Ponds at the Idaho National
Engineering Laboratory
(1988) / Naylor P N

Binary Image Boundary Extraction with
Polygonal Approximation
(1988) / Oliphant P H

CALGARY, UNIVERSITY OF

The Permanent Tide and Its Effects on the
Deformable Earth
(1988) / Arden D A

Durability of Sulphur Concrete
(1988) / Czarnecki B

Effect of Vertical Holes on Concrete Column
Slab Connections
(1988) / Finnigan C

Studies in Gas-Solid Reactions
(1988) / Hassam M S

Integrated Deformation Analysis of the
Olympic Oval, Calgary
(1988) / Rauhut A C

Use of Control Specimens During Freeze-Thaw
Durability Testing
(1988) / Rutherford J H

Optimum Design Method for Concrete Flat
Plates
(1988) / Saretsky K P

Numerical Methods for Transietn Flow
Analysis of Closed Conduits
(1988) / Seneviratne P A R

Microstructure and Hydration of Cement and
Silica Fume Paste
(1988) / Yogendran V

CALIFORNIA STATE UNIV. (FULLERTON)

Study of the Deflection, Cracking and
Ultimate Strength of Partially and Fully
Prestressed Beams
(1988) / Arshad K

CALIFORNIA STATE UNIV. (LONG BEACH)

A Review of the Causes of River Bank Erosion
Including Methods of Analysis and Methods of
Prevention
(1988) / Jones P M

A Study of Sediment Deposition in Reservoirs
(1988) / Majaj N H

A Study of the Variation of the Friction
Factor in Alluvial Channels
(1988) / Thompson J R

CALIFORNIA STATE UNIV. (SACRAMENTO)

The Remedial Investigation of the McNamara
and Peepe Lumber Mill
(1988) / Gills E

Design of Gravity Loaded Frames
(1988) / Lee G Y

The Effect of Lateral Support on Corrugated
Pipe Stiffness
(1988) / Moran E T

Calculation of Effective Length for
Nonprismatic Columns
(1988) / Namazi R

Enhancing Ring Compression in Rigid Conduits
(1988) / Tokas C V

CALIFORNIA, UNIVERSITY OF (DAVIS)

Stochastic Model for the Management of a
Stream-Aquifer System
(1988) / Hantush M M

Review and Evaluation of Design Methods for
Eccentrically Braced Frames
(1988) / Regan D M

CALIFORNIA, UNIVERSITY OF (IRVINE)

Comparative Analysis of Two Numerical
Methods for Computing Dynamic Response of
Nonlinear Systems
(1988) / Griffin M J

Solute Transport in Heterogeneous Reactive
Aquifers Using Particle and Mixed Finite
Element Methods
(1988) / Sovich T J

CARLETON UNIVERSITY

A Mechanism Free Plane Quadrilateral Element
with Rotational Degrees of Freedom and the
Associated Facet Shell Element
(1988) / Aziz O

An Interpretation and Validation of Noise
Contours for Land Use Planning Around
Airports
(1988) / Filho A J

Footings on Sloped Fills: Influence of a
Soil Reinforcement Layer
(1988) / Gnanendran C T

Effect of Foundation Uplift on the Seismic
Response of Steel Buildings
(1988) / Humdan Al Siwat J M

CARNEGIE-MELLON UNIVERSITY

Performance Simulation of Construction
Robots
(1988) / Adachi Y

Leasing Decisions, and a Spreadsheet
Programming Application
(1988) / Basak S

Harmonic Steady-State Vibration of a
Flexible Strip Bearing on a Viscoelastic
Halfplane
(1988) / Begole E H

Foundations on Nonlinear Inelastic Soil
(1988) / Dietrick J C

Improving Negotiation Skills with Game
Simulation
(1988) / Dudziak W J

Linear Feedback Control of a Single Link
Flexible Manipulator
(1988) / Heller M S

CARNEGIE-MELLON UNIVERSITY
(continued)

Evaluator: An Expert System for Evaluation
of Structural Systems
(1988) / Jain D

Reduction of Transient Beam Vibrations
Using Tuned Mass Dampers
(1988) / Kahrs M P

Assessment of the Potential for Automation
in the Design, Fabrication, and Erection of
Steel-Frame Structures
(1988) / King M G

LOGS: A Prototype Expert System to
Determine Stratification of Subsurface
Profile from Multiple Boring Logs
(1988) / Lok M H

Robotic Drilling Implementation in
Construction
(1988) / Miras H D

Spreadsheet Application for Evaluation of
Investments in Commercial Buildings
(1988) / Nabulsi R A

SEES: An Expert System for the Strength
Evaluation of Existing Structural Members
(1988) / Peters J F

Destruction of Iron-Complexed Cyanide by
Alkaline Hydrolysis
(1988) / Robuck S J

Chemical Degradation of Substituted Aromatic
Hydrocarbon Compounds by Manganese Oxide
(1988) / Shanmuganantha S

EDESYN: A Knowledge-Based Expert System
Shell for Preliminary Engineering Design
(1988) / Shukla S

An Expert System for Financial Status
Analysis
(1988) / Streubel C A

Application of Optimization Techniques to
Construction Project Finance
(1988) / Terasawa I

CASE WESTERN RESERVE UNIVERSITY

Generalized Dynamic Substructuring Utilizing
Mixed Finite Element and Modal Synthesis
Techniques and Development of Computer
Program DYSTAN 1987
(1988) / Abdallah A A

Diffusion of Water Into Adhesive Bonds
(1988) / Brewer D N

CENTRAL FLORIDA, UNIVERSITY OF

Continuous Time Convolution Model for
Generating Synthetic Hydrographs
(1988) / Brooks C

Optimization of Storm Sewer Construction
Costs Using a Microcomputer Program
(1988) / Fetter D E

Comparison of Subsurface Cavity
Investigations Using Earth Resistivity,
Seismograph and Ground Penetrating Radar
(1988) / Filler D M

Axisymmetric Wave Propagation in a Finite
Linear Elastic Cylinder
(1988) / Johnson B D

CLARKSON COLLEGE OF TECHNOLOGY

A Numerical Model for Ohio River Ice
Conditions
(1988) / Bjedov G

Contact Stress Distributions Beneath a Rigid
Circular Plate Resting on a Cohesionless
Mass
(1988) / Cooke R

Interpretation of Piezocone and Piezoblade
Tests in Clay
(1988) / Kabir M G

Shear Strength Behavior and Pore Pressure
Response of Natural and Reconstituted Silts
(1988) / Khader W M

The Effects of Anisotropic Consolidation on
the Liquefaction Potential of Mine Tailings
(1988) / Lavigne T A

Sediment Enrichment in Coastal Ice Covers:
An Explorative Experimental Research
(1988) / Li S

Deformation Characteristics of Rectangular
Plates Resting on Unilateral Edge Supports
(1988) / Mithwani M S

Response of Layered Soils to Seismic Loading
(1988) / Thilakaratne V

CLEMSON UNIVERSITY

Hydraulic Analysis of Alternative South
Carolina Curb Inlet Structures
(1988) / Bowman K N

The Experimental Determination of Large
Deflection Equilibrium Positions of
Cantilever Beams
(1988) / Carlson W S

The Effect of Modeling Techniques on the
Prediction of the Response of a Medium-Rise
Concrete Building Subjected to Lateral Loads
(1988) / Dillon J M

The Strength of Roof Anchorage in
Un-Reinforced Concrete Masonry
(1988) / Leland K B

Refinement and Analysis of the Quality
Performance Tracking System
(1988) / Patterson A L

A Comparative Study of Two Elastic-Plastic
Plate Bending Finite Elements
(1988) / Ravinder D

Investigation of Stress Concentrations
Around Circular Holes in Plates Subjected
to Bending
(1988) / Sarker A K

A Dynamic Test Procedure for Analyzing
Low-Rise Metal Buildings
(1988) / Sockalingam B

A Knowledge Based Expert System for the
Quality Performance Tracking System
(1988) / Sule M A

Study of Ebb Tidal Shoal Formation in
Movable Bed Tidal Inlet Model
(1988) / Wells W J

CLEVELAND STATE UNIVERSITY

Investigation of Zeta Potential as a
Controlling Parameter for Charge
Neutralization with Metal Salts
(1988) / Greenland F P

Reliability-Based Analysis of Contact Stress
Problems
(1988) / Powers L M

COLORADO STATE UNIVERSITY

Numerical Modeling of the North Boundary
Barrier System at the Rocky Mountain
Arsenal in Denver, Colorado
(1988) / Dwight D M

Constitutive Relationships for Collapse
In a Remolded Soil
(1988) / Hatton C N

Blast Induced Liquefaction of an Alluvial
Sand Deposit
(1988) / Jacobs P J

Thermal Dilution as a Method of Discharge
Measurement in Trapezoidal Channels
(1988) / Kattnig I M

The Discrete Vortex Method for Numerical
Analysis of Bluff Body Aerodynamics
(1988) / Kutz R F

A Numerical Investigation on Dense Plume
Dispersion Over Complex Terrain
(1988) / Lee T J

Electrohydrodynamic Flow in a Barbed Plate
Electrostatic Precipitator
(1988) / McKinney P J

Plan Morphology and Channel Alignment
of Rio Apure, Venezuela
(1988) / Smith M

Computer Aided Design of Light-Frame Structures
(1988) / Stonebraker C P

Mean and Peak Wind Load Reduction for Solar
Collectors
(1088) / Tan Z

CONCORDIA UNIVERSITY

Ultimate Bearing Capacity of Triangular
Shell Strip Footings on Sand
(1988) / Abd El-Rahman M

Methode Numerique de Selection de Logiciels
Pour la Petite Entreprise de Construction
(1988) / Gervais P N

Energy Storage Composite with an Organic PCM
(1988) / Khan M A

The Impact of the Occupational Health and
Safety Act on the Construction Industry in
the Province of Quebec
(1988) / Martel H P

Analysis of Construction Projects Using a
System Dynamics Methodology
(1988) / Sylvestre Y

Fast-Tracking of Construction Projects:
Analysis and Assessment
(1988) / Theberge P

CONNECTICUT, UNIVERSITY OF

Method Development for the Gas Phase
Analysis of Gasoline and Solvent
Contaminated Soil
(1988) / Cliff B L

Decompaction of Bituminous Concrete
Pavement Due to Cyclic Temperatures
(1988) / Descoteaux T J

Analysis of the Effects of Stagger in
Structural Tension Members
(1988) / Gulia F S

CORNELL UNIVERSITY

Measurement of Diffusivity on Undisturbed
Soil Cores by Capillary Rise
(1988) / Hawiger M M

Single Site and Regional Flood Frequency
Analysis with Historical Flood Information
(1988) / Jin M

A Study of the Laboratory and Field
Performance of a Semiconductor Soil Stress
Cell
(1988) / Kohl K M

Case Studies of Cracking oc Concrete
Dams--A Linear Elastic Approach
(1988) / Lin S S

The Effect of Shear on the Dewaterability
of Flocculent Suspensions and Related
Particle Characteristics
(1988) / Nayak C

The Behavior and Design of Noncontact Lap
Splices Subjected to Repeated Inelastic
Tensile Loading
(1988) / Sagan V E

An Educational Computer Graphics Program for
Examining the Behavior of Single
Two-Dimensional Finite Elements
(1988) / Shen T G

Hydrologic Models and Their Representation
in Stochastic Dynamic Programming Algorithms
for Reservoir Operation Optimization
(1088) / Siqueire Born P H

Fracture Analysis Code: A Computer-Aided
Teaching Tool
(1988) / Srinivasan M

Fatigue Behavior of Thick Steel Plates
Bent to a Low R/t Ratio
(1988) / Vossoughi H

Derivation of Balancing Curves for Multiple
Reservoir Operation
(1988) / Wu R S

DELAWARE, UNIVERSITY OF

Analysis and Design of Geotextile Reinforced
Granular Embankments Over Firm Foundations
(1988) / Boedeker R H

Simulation of Colloidal Destabilization with
Metal Coagulants
(1988) / Chern J M

DELAWARE, UNIVERSITY OF
(continued)

Erosion of a Curved Gravel Causeway Due to
Breaking Waves
(1987) / Han K S

Evaluation of On-Line Instrumentation for
Continuous Chemical Dose Control
(1988) / Kingery K M

The Influence of Loading History on the
Fracture Toughness of Short-Fiber Composites
(1988) / Parks B G

Random Vibration of Gravity Dam-Reservoir
System During Earthquake Excitation
(1988) / Wells J T

Numerical Simulation of Wave Induced
Turbulent Boundary Layers
(1988) / Wu N T

DREXEL UNIVERSITY

An Evaluation of Two Models for Yielding Systems.
(1987) / Adnane M A

A Contribution to the Analysis of Elastometric
Bearings
(1987) / Kartoum A

In-situ, Vacuum Assisted Steam Stripping to Remove
Volatile Pollutants from Contaminated Soils.
(1988) / Murphy V P

Computer Model Used to Simulate the Construction of
a House and the Effects of Changes in Manpower
Availability
(1988) / Rodriguez R

Strength Improvement of Cement Pastes with Fiber
Reinforcement Along with Epoxy Modification
(1988) / Shih J

DUKE UNIVERSITY

Comparative Sulfur Dioxide Reactivity of
Calcium Oxide Derived from Calcium Carbonate
and Calcium(OH)2
(1988) / Bruce K R

Mechanics of Flexible, pneumatic Structural
Elements for Robotic Limbs
(1988) / Ghattas J

Compacted Clay Soil Failure Mechanisms
Resulting from Long Term Exposure to a
Phosphogypsum Waste
(1988) / Kincaid L

A Study on Parallel Computational Schemes
for the Inverse Dynamics of Mechanical
Manipulators with Rigid Linkages
(1988) / Lu L Y

Biofilm Formation Effects on Flow Through
Porous Media
(1988) / Mueller V J

Structural Analysis of the Great Pyramid
Using Finite Element Techniques
(1988) / Stubig C E

Experiments to Reduce Vortex-Induced Loads
on Circular Cylinders
(1988) / Tinsley J C

FLORIDA, UNIVERSITY OF

The Effects of Dynamically Inserting the
Marchetti Dilatometer
(1988) / Basnett C

An Evaluation of Design Methods for Drilled
Shafts in Cohesionless Soils
(1988) / Burch S

Design, Development, and Use of a Field
Permeability Test (FPT) Apparatus and
Nondestructive Method for the Determination
of In Situ Permeability of Structural
Concrete
(1988) / Meletiou C

GEORGIA INSTITUTE OF TECHNOLOGY

A Geomorphologic Instantaneous Unit
Hydrograph Streamflow Model
(1988) / Kabouris I

HAWAII, UNIVERSITY OF

The Disinfection of Seawater Using
Ultraviolet Irradiation in Order to Inhibit
or Reduce Marine Microbiofouling
(1988) / Ono P R

Induced Mixing Characteristics in Shallow
Open Channel Flow
(1988) / Saito B Y

HOUSTON, UNIVERSITY OF

Behavior of Lightweight Concrete Columns
(1988) / Hussain S

Two-Parameter Soil Model for the Analysis
of Three-Dimensional Foundations on
Layered Media
(1988) / Mebarkia S

Mechanical Behavior of Polyester and Epoxy
Polymers and Polymer Concrete Systems
(1988) / Paul C

HOWARD UNIVERSITY

Cement Stabilization and Its Effect on the
Permeability of Saudi Arabian Sandy Soils
(1988) / Aboghalia S S

The Application of Finite Element Method to
the Analysis of Dental Structures
(1988) / Ahmed E M

A Statistical Analysis of the Rational
Formula
(1988) / Curry R D

Investigation of Thick Pile Caps
(1988) / Dagher R E

High Strength Concrete Using Synthetic
Aggregate Made of Recycled Cement
(1988) / Negahdary F

Design of a Chemical Tracking System
(1989) / Stone D A

IDAHO, UNIVERSITY OF

Probabilistic Modeling of Sloped Roof Snow
Loads
(1988) / Arnholtz D

IDAHO, UNIVERSITY OF
(continued)

Snow Loads on Glass Gabled Structures
(1988) / Baker R L

Estimating Ungaged Streamflow as a Function
of Climate and Watershed Characteristics
(1988) / Bourque F T

Predicting Roof Snow Loads on Gabled
Structures
(1988) / Giever P M

Evaluation of Hydrogeologic Siting Criteria
for Siting Hazardous Waste Management
Facilities in Idaho
(1988) / Hill B M

Design and Reliability of Cracked
Prestressed Girders by Supercomputer
(1988) / Khafagi B K

Predicting Cracking Moment Capacity of
Prestressed Decked Bulb-Tee Girders at
Any Age
(1987) / Koduah S

Seismic Vulnerability Analysis of Idaho
Public School Buildings
(1988) / Lavin M

Sesimic Design of Precast Concrete Curtain
Wall Panels
(1987) / Neal V

Random Problem Generation for Courses in
Structural Analysis
(1987) / Schroeder R D

Measurement of Crowd-Induced Dynamic Loads
(1988) / VanKleek P D

ILLINOIS INSTITUTE OF TECHNOLOGY

Design Code Implications of High Strength
Concrete
(1988) / Iyer V S

Analyzing Bidding Behavior to Predict
Contractor Performance
(1988) / Sami N

IOWA STATE UNIVERSITY

Safety Rating of Steel Sheet Pile Structures
(1988) / Askar Y I

Nonlinear Analysis of a Conveyor System
Subjected to Tornado Wind Loads
(1988) / Eberline D K

Fatigue of Simple One-Way Steel-Deck
Reinforced Floor Slabs
(1988) / Faust B D

Pile Tests with Vertical, Lateral, and
Combined Loads
(1988) / Johnson D E

Some Geotechnical Properties and Dehydration
Characteristics of Five Volcanic-Ash-Derived
Soils of Hawaii
(1988) / Kaspar C N

Fatigue of Composite Metal Deck Slabs
(1988) / Spellerberg L J

Boundary Conditions and Orientation
Behavioral Characteristics of Hollow-Core
Diaphrahms
(1988). / Tremel P M

KANSAS, UNIVERSITY OF

P-Approximation Curved Shell Elements for
Heat Conduction
(1988) / Abusaleh G

The Effects of Single and Joint Toxicity of
Atrizine and Alachlor on Three Non-Target
Aquatic Organisms
(1988) / Blackburn R A

Development of PC-Based Data Extraction and
Analysis for Automated Cone Penetrometer
(1988) / Chen W W

Design of General Cold-Formed Steel Shapes
(1988) / Glauz R S

Utilization of the Clegg Impact Hammer to
Predict California Bearing Ratio and Tire
Rut Depths of Operating Aircraft
(1988) / Grob J D

Cyclic Behavior of High Strength Concrete
Beams
(1988) / Hanks D L

LEHIGH UNIVERSITY

Sediment Deposition Modeling of Channel
Sound Systems
(1987) / Young C L

LOUISIANA STATE UNIVERSITY

Ion Shifts and PH Management in High Density
Shedding Systems for Blue Crabs and Red
Swamp Crawfish
(1988) / Allain P A

Biodegradation of Petroleum and
Petrochemical Wastes
(1988) / Forbes L W

An Anisotropic Distortional Yield Surface
Model
(1988) / Foroozesh M

Changes in the Early Hydration Behavior of
Portland Cement Used in the Solidification
of Hazardous Chemicals
(1988) / Frey F P

Constitutive Modeling of Metals Subjected
to Finite Elasto-Plastic Strains
(1988) / Ghaddar G M

Evaluation of the Leaching Properties of
Solidified Organic and Inorganic Wastes
(1988) / Herrera E

Evaluation of Transverse Joint and Subbase
Efficiency in Rigid Pavements Using
Nondestructive Testing
(1988) / Meroney M P

Determination of PH Gradients in
Electro-Chemical Processing of Kaolinite
(1988) / Putnam G A

Computer Assisted Dilatometer Testing
(1988) / Sarieddine H J

LOWELL, UNIVERSITY OF

Effect of the Number of Rebars on the
Damping in Reinforced Concrete Frames
(1988) / Bassil K N

Moment Coefficient for Concentrated Load
from Beams Reactions on Girders in Uniformly
Loaded Floor Systems
(1988) / Law K P

Use of Cement, Flyash and EER Admixture in
Soil Stabilization
(1988) / Li K

Cost Analysis of the Output Portion of a
Graphics Package
(1988) / Lupien E D

Trends in the Variation of Structural
Damping with Respect to Stiffness, Mass, and
Amplitude in Small Scale Three-Dimensional
Model Shear Buildings
(1988) / Peirent R

Voids in Fiber Reinforced Concrete Cylinders
(1989) / Samuel G T

MAINE, UNIVERSITY OF

Radon Progeny Buildup on Granular Activated
Carbon
(1988) / Bezbarua B K

An Overall Approach to Data Modeling for
Applications in Geographical Information
Systems
(1988) / Pullar D V

Consideration of Implementation Issues in
Aerotriangulation with GPS Derived
Exposure Station Location
(1988) / Zhao M

MANITOBA, UNIVERSITY OF

Stress Corrosion Cracking of Lac du Bonnet
Granite in Tension and Compression
(1988) / Bielus L P

Remote Fracturing Around Underground
Openings
(1988) / Carter B J

Models for Fire Station Location: A Review
and Improved Distance Estimation Method
Tested for Winnipeg
(1988) / Kersey L S

Statistical Analysis of Bank Erosion of the
Brahmaputra River in Bangladesh
(1988) / Khan M A

Investigation of Hydrometric Network
Characteristics Through the Use of Simple
Statistics
(1988) / Kiely D A

Basic Geomechanical Properties of the Dawson
Bay Formation
(1988) / Kroll D W

Welded Wire Fabric as Shear Reinforcement in
Concrete T-Beams Subjected to Cyclic Loading
(1988) / Pincheira J A

Multiple Shear Key Connections for
Load-Bearing Shear Wall Panels
(1988) / Serrette R L

MARQUETTE UNIVERSITY

Effects of Delvo on the Strength of Concrete
(1988) / Abdulkarim K

Effects of Acid Rain on Fiber Reinforced
Concrete
(1988) / Hamdan A

A Spread Sheet Computer Program for the
Design of Precast Concrete Connections
(1988) / Hayek M

Point Source Regulation of Storm Water
Discharges: Impact Analysis
(1988) / Lepak S

Analyzing the Dynamics of the Bucket of a
Walking Drag Line During the Swing
Operation
(1988) / Yeh H C

MASSACHUSETTS INSTITUTE OF TECHNOLOGY

Concrete Beam Column Joint Interactive
Design
(1988) / Banki A D

Model Synthesis for 3-D Soil-Structure
Interaction Problem
(1988) / Basaran C

Issues and Strategies for Improving
Constructibility
(1988) / Berenato D A

Modeling the Progressive Crushing of Shell
Structure with Computer Graphics
(1988) / Brodkin D

Development in China: A Feasibility Study
for a Mixed-use Real Estate Development
Project in the People's Republic of China
(1988) / Burns G R

Experimental Analysis of a Structural Model
Under Dynamic Loads
(1988) / Chang W W

Ice-Structure Interaction Accounting for
1 Fracture of Ice: A Nonlinear Finite
Element Approach
(1988) / Chen C

Automatic Generation of Construction
Schedules from Architectural Drawings
(1988) / Cherneff J M

Bearing Capacity and Settlement of
Foundations with Upward Flow Conditions Due
to Liquefaction
(1988) / Choi S J

The Effects of Inadequate Component:
Inspection on Facility Repair Projects
(1988) / Cowell J W

A New Analytical Approach to Study the
Effects of Joints Nonlinearities in Space
Structures
(1988) / Dergham S

Islamic Financial Principles and Their
Applications in Project Financing
(1988) / El-Husseini I A

Networking and Communications in an
Engineering Company
(1988) / El-Khaouam M M

MASSACHUSETTS INSTITUTE OF TECHNOLOGY
(continued)

New Concepts and Preliminary Design
Procedures for Flat Deployable Structures
(1988) / Gantes C

Analysis of Cell Cycle Specific Physiology
in Marine Phytoplankton by Centrifugal
Elutriation
(1988) / Gerath M W

Extreme Wind-Induced Facade Suction
Pressures on a Tall Building
(1988) / Gilbert T J

Soil Water Pathways in a Forested Watershed
in Central Massachusetts
(1987) / Gordesky B T

Detection of Leakage from Large Storage
Tanks Using Seismic Boundary Waves
(1988) / Halabe U B

Liquefaction Probability Mapping in Greater
Boston
(1988) / Hashash Y M

Interface Studies in Composite Materials
(1988) / Jao S H

Microwave Detection and Characterization of
Deep-Water Wave Breaking
(1988) / Jessup A T

Extraboard Personnel Scheduling in the
Transit Industry
(1988) / Kaysi I A

Response of Deep Foundations to Lateral
Forces
(1988) / Kraemer T C

Parametric Studies in Rock Fracture Flow
(1988) / Low L S

Evaluation of Accuracy of Measurement
Technologies and Their Impacts on
Infrastructure Management
(1988) / Madanat S M

Construction: The Foundation of National
Defense
(1988) / Martin G F

Transient Elastic-Plastic Thermal Analysis
of Line Heated Plates
(1988) / Ndumbaro P C

Reinforced Concrete Corbel - A Case Study
(1988) / Nguyen N T

An Evaluation of a Contractor's Position in
Project Financing: A Case Study of a
Japanese Construction Firm in the U.S.
(1988) / Nishikawa E

Program and Expert System Development for
Tutorial Applications in Engineering Geology
(1988) / Noack C A

Effects of Nonlinear Stiffness on Truss
Vibration
(1988) / O'Sullivan J J

A Financial Evaluation of the Eurotunnel
Project
(1988) / Okano S

Nonlinear Finite Element Analysis of
Tridimensional Reinforced Concrete
Structures
(1988) / Pagnoni T

Strategies for Conceptual Design of
Structures
(1987) / Pouangare C C

Numerical Modelling of Crack Propagation in
Tension Softening Materials
(1988) / Reyes O M

A Field-Usable Membrane-Probe Mass
Spectrometer
(1988) / Richardson L B

Groundwater Colloids in Two Mid-Atlantic
Coastal Plain Aquifers
(1988) / Ryan J N

Modification and Implementation of a
Computer Controlled Triaxial Apparatus
(1988) / Sheahan T C

Analysis of Chlorophyll Fluorescence in
Marine Phytoplankton: Interpretation of
Flow Cytometric Signals
(1988) / Sosik H M

Productivity in Construction: An Issue of
Analysis and Measurement
(1988) / Stathakis N

A Framework for Facility Modification
(1987) / Sydelko T G

Analysis of Claims in U.S. Construction
Projects
(1988) / Tanaka T

Stochastic Response of Tall Buildings with
Auxiliary Dampers
(1988) / Tazir Z H

On the Stability of Long Nonlinear Kelvin
Waves
(1988) / Tomasson G G

Proocedure for Investigation of Sample
Disturbance Using the Direct Simple Shear
Apparatus
(1988) / Walbaum M

Alternative Criteria for Selection of Public
Construction Contractors
(1988) / Walker J P

Model Penetration Anchor Into Rock
(1988) / Young S T

A Numerical Investigation of Unsaturated
Flow
(1988) / Zarba R L

Wave Propagation in Anisotropic Laminated
Rods
(1988) / Zinn M R

Part I: Refinements of the Geomorphologic
IUH. Part II: The Howard Topological
Channel Network Simulation Model Revisited
(1988) / van der Tak L D

MICHIGAN TECHNOLOGICAL UNIVERSITY

Interactive Load and Resistance Factor
Design of Two Dimensional Steel Frames
(1988) / Ayob M S

MICHIGAN TECHNOLOGICAL UNIVERSITY
(continued)

Deep Draw Molding of Wood Flake Composites:
Geometric Effects
(1988) / Baas T J

Frozen Compaction, Gradation Shift, and Thaw
Settlement of a Coarse Grained Soil
(1988) / Barker A E

Evaluation of Clegg Impact Tester on Snow
Roads in Antarctica
(1988) / Bott M W

A Microcomputer Program for the Analysis of
Two-Way Slab System: A Singularity Solution
(1988) / Ghusayni H N

Groundwater Recharge During Winter
(1988) / Harvey D C

Analysis and Testing of Toothed Metal Plate
Wood Truss Connections
(1988) / Reynolds G M

Design of Reinforced Concrete Floor Systems
Using Interactive Computer Graphics
(1988) / Said W J

Simulation of the Lifetime Behavior of
Structural Wood Systems
(1988) / Vacca P J

MINNESOTA, UNIVERSITY OF

An Experimental Investigation Into the
Behavior of Composite Open Web Steel Joists
(1988) / Alsamsam I M

Calibration of a Double Hardening
Constitutive Model for Sand by Using the
Improved MN-DOT Triaxial Testing System
(1988) / Alsiny A O S

Full-Scale Tests on Two Long-Span Composite
Open-Web Steel Joists
(1988) / Curry J H

Noise Generation by Air Bubbles in Water:
An Experimental Study of Creation and
Splitting
(1988) / Frizell K W

Modeling of Reservoir Suspended Solids
Patterns
(1988) / Hendrickson J S

An Interior Moment Resistant Connection
Between Precast Elements Subjected to
Cycle Lateral Loads
(1988) / Jayashankar V

Soil Physical Properties as Influenced by
the Addition of Municipal Compost
(1988) / Kreft D R

Experimental M-Theta Curves for Semi-Rigid
Composite Connections
(1988) / Loughlin J P

Environmental Studies: Part A: Feasibility
Study for CMA Production from Waste
Resources in Minnesota: Part B: Analysis
of Four Analytical Solutions of the
Transport and Adsorption of Soluble
Materials in Porous Media
(1988) / Lundquist J R

Analytic Elements for Flow Through Aquifer
Inhomogeneities and Leakage Through
Confining Layers
(1988) / Macneal R W

Testing of a Semi-Rigid Composite Connection
(1988) / McCauley R D

A Laboratory Investigation of the
Relationship Between Groundwater Velocity
and Mass Transfer from a Simulated Aromatic
Hydrocarbon Spill at the Water Table
Interface
(1988) / O-Dell C H

Influence of Laterial Load Resisting
Elements
(1988) / Park N Y

Dynamic Response Analysis of Shallow-Buried
Reinforced Concrete Arches
(1988) / Puglisi R D

Designs of Prestressed Concrete Systems with
Torsion
(1988) / Rivero J C

Experimental Investigations of Tip Vortex
Cavitation and Associated Noise
(1988) / Rogers M F

A Computer Program for the Analysis of
Catfish Production in Thermal Effluent
Aquaculture Systems
(1988) / Sawdey S L

The Planning and Design of a Mini-Hydropower
Facility
(1988) / Woods J L

An Interactive Simulation Program for
Intersection Design and Operational Analysis
(1988) / Yuan B

Numerical Simulation of Flow and Bed
Topography Along Channel Bend
(1988) / Zheng Y

MISSISSIPPI STATE UNIVERSITY

Investigation and Establishment of Certain
Response-Properties of Selected Shock-
Absorbing Materials and Systems
(1986) / Denson R H

Modeling the Movement of Dissolved Chemical Species
Injected into Deep Aquifers
(1986) / Erustun K

Color Removal from Groundwater by Ion Exchange
and Sludge Recycle Techniques
(1986) / Ficquette B N

An Introduction to Matrix Reduction Techniques
in the Analysis of Framed Structures by the
Stiffness Method
(1986) / Guillen O E

Evaluation of Revised Manual Compaction Rammers
and Laboratory Compaction Procedures
(1986) / Horz R C

Effect of Specific Gravity and Particle Diameter
on Universal Compression Index Equation
(1986) / Jahadi M R

MISSISSIPPI STATE UNIVERSITY
(continued)

An Evaluation of Factors Affecting Critical Void
Ratio Determination Using Monotonic R Triaxial
Tests
(1986) / Johnson H V

Load-Transfer Criteria for Numerical Analysis of
Axially Loaded Piles in Sand
(1986) / Mosher R L

Theoretical and Experimental Investigation of the
End Block of the Posttensioned Concrete Beam
(1986) / Sioushansian F R

Dynamic Shear Failure of Shallow-Buried Flat-
Roofed Reinforced Concrete Structure Subjected
to Blast Loading
(1986) / Slawson T R

An Investigation of the 3- (log)P Relationships
of Four Selected Soil Specimens Under One-
Dimensional Loading
(1986) / Templeton A E

The Determination of Elastic Lateral Buckling
Stresses for Tapered I-Beams with Flange
Discontinuities by Newmark's Method
(1986) / Turcotte L H

MISSOURI, UNIVERSITY OF (COLUMBIA)

An Analytical Study of the Nonlinear Interface
in Fibrous Composites
(1988) / Jin C

Transient Control of Hydraulic Capsule
Pipeline Bypass
(1988) / Woodson R J

MONTANA STATE UNIVERSITY

Hydraulic Effects of Biofilm Accumulation
in Simulated Porous Media Flow Systems
(1987) / Crawford D J

MONTREAL, UNIVERSITY OF

Analyse des Pressions Interstitielles Dans
les Excavations d'Argiles Sensibles
(1988) / Asselin R

Sensibilite aux Precipitations Acides des
Sediments d'Origine Glaciaire de la Region
du Lac Kempt, Quebec
(1988) / Aubertin A

Gestion des Structures: Surveillance et
Remise en Etat des Ouvrages en Beton
(1988) / Balkaloul Q E

Evaluation de la Resistance au Cisaillement
Argile-Acier a l'Aide de l'Essai de
Cisaillement Direct
(1988) / Beaudoin P

Developpement d'un Algorithme d'Evaluation
et de Prediction des Degradations Pour des
Chaussees Composites
(1988) / Beaulieu J

Etude de l'Action de Composes Antimoussants
sur l'Ozonation d'une Eau Usee Synthetique
(1988) / Beyne T

Recherche de l'Aluminium Dans l'Eau Potable
de Laval et de Montreal (Quebec) a la
Lumiere de ses Effets Potentiels sur la
Sante
(1988) / Bosisio M

Alea Seismique et Liquefaction Dans un Site
Donne
(1988) / Bouaou A

Modelisation de la Stratigraphie des Depots
Sedimentaires
(1988) / Boudali M

Le Comportement Dynamique d'une Argile
Sensible
(1988) / Boudjerada A

Analyse Elasto-Plastique des Colonnes a
Inertie Variable
(1988) / Coles Y

PRE2D: Logiciel Graphique, Interactif de
Preparation des Donnees Pour un Programme
d'Analyse par Elements Finis
(1988) / Dinu C

Evaluation de Deux Types de Beton de
Reparation
(1988) / Gamache D

Elimination de l'Interference des Chlorures
sur la Demande Chimique en Oxygene a l'Aide
de Nitrate d'Argent
(1988) / Leclair G

Developpement d'un Gravillonneur de
Laborratoire Pour l'Etude des Traitements
de Surface sur Chaussees
(1988) / Leclerc A M

Modelisation d'un Service Urbain de
Messagerie: Recherche d'un Decoupage Zonal
Optimal
(1987) / Ledoux R

Analyse des Systemes Zonaux Agreges
Assujettis a Des Contraintes de Compacite
de Forme
(1987) / Lefrancois A

Analyse de l'Ecoulement a Travers des
Barrages en Terre Lors du Remplissage du
Reservoir
(1987) / Leguy C

Effets Combines de la Preoxydation de la
Coagulation et de la Decantation sur les
Caracteristiques d'une Eau
(1988) / N'Dionque S

Etude de la Survie de Legionella Pneumophila
en Culture Pure et Mixte Dans l'Eau Potable
(1988) / Novello A M

Etude de l'Adsorption Rapide Dans un Filtre
au Charbon Actif Biologique
(1988) / Piotte D

Analyse Fonctionnelle d'un Logiciel Pour
la Localisation de Sites Potentiels
d'Enfouissement Sanitaire
(1988) / Roy M L

Sur la Relaxation d'une Argile Sensible
(1988) / Touchan Z

Analyse Sismique Pseudo-Bidimensionnalle des
Ourvages en Terre
(1988) / Touileb B N

NEBRASKA, UNIVERSITY OF

The Use of Vertical Electric Soundings to
Determine Groundwater Recharge Potential
(1988) / Curtis B A

Strength and Deflection of Reinforced
Masonry Walls
(1988) / Horton R T

Resilient Modulus Characterization of
Representative Nebraska Soils
(1988) / Page D M

An Expert System Based Risk Assessment for
Ground Water Protection
(1988) / Parsons J R

A Hydrologic Evaluation of Twenty-Four Small
Watersheds in Nebraska
(1988) / Riley T E

NEW BRUNSWICK, UNIVERSITY OF

Buckling of Steel Girders with Flange
Restraint
(1988) / Albert C

Factors Influencing Shopper Destination
Choice
(1988) / Badoe D A

Implementing a Land Information Network in
New Brunswick
(1988) / Coleman D J

Utilization of Thematic Mapper and
Multispectral Scanner Data for Crop
Classification and Monitoring
(1988) / Hashim M

A Unait Cost Analysis for Planimetric
Topographic Mapping
(1988) / Laroche S

A Study of Connections Between Hollow
Structural Section Columns and W Shape Beams
(1988) / Liew C K

Concrete Deterioration in Sea Water
(1988) / Lin J

Fiber Reinforced Concrete in Marine
Environment
(1988) / Ong C K

Aggregate Saturation in Lightweight Concrete
(1988) / Ooi O S

An Assessment of Cadastral Developments in
Sub-Saharan Africa
(1988) / Opadeyi J A

A System for the Matching of Digital Images
and Terrain Models
(1988) / Papanikolaou K

Low NOx Burner Coal/Sorbent Injection
Demonstration Project
(1988) / Pierrynowski R J

Structural Aspects of the Proposed Coal
Conversion for the Coleson Cove Generating
Station
(1988) / Ross T H

The Effects of Resampling on Digital Imagery
(1988) / Saleh R A

Out-of-Plane Behaviour of Concrete Masonry
Infilled Panels
(1988) / Seah C K

Dynamic Response of Masonry Infilled Steel
Frames
(1988) / Sofocleous C

Reinforced Glued-Laminated Wood Beams
(1988) / Tingley D A

The Application of Expert Systems in
Geographical Information Systems
(1988) / Uhlenbruck H M

Reducing Disagreement Between Stoke's and
Truncation Integration
(1988) / Zhang C

NEW JERSEY INSTITUTE OF TECHNOLOGY

Study of the Structure and Capacity of
Construction Industry in Pakistan
(1988) / Ansari T A

Thermal Desorption of Hazardous and Toxic
Organic Compounds from Soil Matrices:
Dichloromethane, Chloroform, Benzene,
Toluene, 1-Chloronaphthalene, 1, 2,
4-Trichlorobenzene
(1988) / Chemburkar A S

Softening Response of Concrete in Direct
Tension
(1988) / Cintora T

Development of a Rock Model Testing
Capability at the New Jersey Institute of
Technology and Testing of Plane Stress
Models Possessing a Central Circular
Opening with a Nearby Horizontal
Discontinuity
(1988) / Good D R

Determination of Project Duration for
Resource Constrained Projects Without Using
Resource Scheduling Procedure
(1988) / Karmakar R N

Recycled Plastics as Fillers in Polymer
Cement Concrete Composites
(1988) / Liu S J

Use of Latex Modified Concrete for Repairs
and Overlays
(1988) / Lo C H

Comparative Adsorption Studies on Clay Soils
(1988) / Mysore P R

Unsaturated Column Studies of Silty Sand
Using a Conservative Tracer
(1988) / Nallianathan C

Estimation of In-Situ Compression Index of
Contaminated Clays Using the Electrical
Method
(1988) / Ratnaweera P

Sorption Characteristics and Their Effects
on Geotechnical Properties of Clay Soils
(1988) / Rodrigo A M

Sorptive and Geotechnical Characteristics
of Kaolinite
(1988) / Shih C Y

NEW MEXICO INSTITUTE OF MINING & TECH.

Effect of Firing on Petrophysical Properties
of Berea Sandstone
(1988) / Ma S

NEW MEXICO STATE UNIVERSITY

Synthesis of a Biological Test Medium from
Primary Sludge
(1988) / Baraza A

NEW MEXICO, UNIVERSITY OF

A Finite Element for Arbitrarily Precise
Determination of Stress Intensity Factors
(1988) / Abdalla-Filho J

NEW YORK, STATE UNIVERSITY OF (BUFFALO)

A Constitutive Model for Anisotropically
Consolidated Clays
(1988) / Chopra M

Analysis of Waterbody Surface Heat Exchange
(1988) / Huang N C

Damage Assessment of Reinforced Concrete
Structures in Eastern United States
(1988) / Seidel M J

Liquefaction Potential Mapping for Upper
Manhattan in New York State
(1988) / Vijayakumar V

Experimental Study on Structural Behavior
of Plastic Liquid Storage Tank Under
Strong Ground Motions
(1988) / Wang C T

NORTH CAROLINA STATE UNIVERSITY

Evaluation of Stabilization Mechanisms of
Lime/Fly Ash Injected Slopes
(1988) / Baez-Satizabal J I

A Laboratory Model of a Multi-Layer System
for Non-Destructive Testing
(1988) / Eddy J L

NORTH DAKOTA STATE UNIVERSITY

Finite Element Analysis of Micropolar Elastic
Materials
(1988) / Tan S A

NOTRE DAME, UNIVERSITY OF

The Application of the Projected Membrane
Strain Method to Geometrically Nonlinear
Structural Analysis
(1988) / Couch B P

An Analysis of the Microscopic Hydraulic
Characteristics of a Discrete Rock Fracture
Using Finite Elements
(1988) / Deuell R J

Reliability of Multi Degree of Freedom
Structures Under Random Excitation
(1988) / Goswami S

Stochastic Analysis of High Permeability
Paths in the Subsurface: Two-Dimensional
Simulations
(1988) / Kovacs K J

An Evaluation of Borehole Temperatures at
Red Gate Woods, Cook County, Illinois
(1988) / Robinson R D

NOVA SCOTIA, TECH UNIVERSITY OF

Computerized Approach to the design of
Reinforced Concrete Footings
(1988) / Yalcin C M

OKLAHOMA, UNIVERSITY OF

Energy Method for the Analysis of Flexural
Behavior of a Rectangular Plate Resting on
an Elastic Halfspace
(1988) / Ko M

Experimental Determination of Shear Wall
Capacities Using Small-Scale Models
(1988) / Turley J B

OREGON STATE UNIVERSITY

Modeling Solute Transport by Centrifugation
(1988) / Celorie J A

The Fate of Hydrophobic Organic Compounds in Soil
Systems: A Kinetic Model and Sensitivity Analysis.
(1988) / Cork T D

Development of a Method to Elucidate Biodegradation
Pathways of Chlorinated One and Two Carbon Compounds
Using A Gas-Permeable-Membrane-Supported
Methylotrophic Biofilm
(1988) / Ryan K M

Laboratory Investigation of the Mechanics of
Raveling Soils
(1987) / Saunders G P

Long Cables Under Steady Ocean Loads with Coupled
Tension/Torsion
(1987) / Simpson R C

PENNSYLVANIA STATE UNIVERSITY

Shear and Flexural Strength of Reinforced
Concrete Beams Made with Crete Core Forms
(1987) / Aminmansour A

A Conceptual Design of Automated Real-Time
Data Acquisition Systems for Construction
Operations
(1987) / Bandyopadhyay A

The Application of Optimum Value Engineering
Design and Construction Techniques to the
U. S. Navy's Housing Construction Contract
Request for Proposal
(1988) / Giorgione M A

PRODAN: The Development of a Computer
Construction Productivity and Performance
System
(1987) / Griest D M

An Analysis of a Partial Versus Annual Flood
Peak Series for Pennsylvania Streams
(1987) / Johnson R C

A Comparative Study of American and Japanese
Construction Company Organization
(1987) / Kamegawa H

A Model Air Force Construction Quality
Management System
(1988) / Pocock J B

PENNSYLVANIA STATE UNIVERSITY
(continued)

The Specified Boundary Displacement Method
Applied to Fillets
(1987) / Proctor L

Determination of Energy Losses by Numerical
Solution of the Navier-Stokes Equations
for Incompressible Pipe Flow
(1988) / Reese R A

Use of Weather Radar to Investigate the
Effects of Temporal and Spatial Variations
of Rainfall on Hydrologic Forecasting
(1987) / Shedd R C

PITTSBURGH, UNIVERSITY OF

Velocity and Sediment Distribution Analysis
in Channels and Rivers
(1989) / Al-Yousfi A B

Bolted Timber Connections with Steel Side
Plates
(1989) / Call R D

Second Order Elastic Analysis of Plane
Frames
(1989) / Chen N

Three Dimensional Finite Element Analysis of
a TPS-Coated ISIS Implant
(1989) / Heyn C G

An Expert System in "C" for Evaluating the
Soil Liquefaction Potential of a Given Site
(1989) / Onyemelukwe O U

Commercialization Criteria for "Burnout
Control"
(1989) / Perry M T

PRINCETON UNIVERSITY

Fourier Transform Infrared Photoacoustic
Spectroscopy and Its Application to the
Study of Supported Metal Oxide Catalysts
(1988) / Alexander J

PURDUE UNIVERSITY

Wetlands and Drainage Inventory of Indiana
Interstate Highway Segments Using Aerial
Photographs
(1988) / Barbour R

The Reuse of High-Strength Bolts
(1988) / Betancourt M

Cyanide, Zinc, pH and Contact Time Effects
on Nitrifying Bacteria
(1988) / Doster A M

The Application of Routing Technologies to
the Problem of Snow Removal
(1988) / Haslam E P

Development of the IDOH Classification
System for Geotextiles
(1988) / Karcz D A

Processing of Digital Line Graph Data for
Decision Support Applications
(1988) / Kurmas P

A Study of Chemical Reaction Between Cement
Paste and Sea Water
(1988) / Matsukawa K

R-H Improvement Algorithm in Finite Element
Method
(1988) / Ning Y

Analysis and Modification of a Wind Wave
Prediction Model for the Great Lakes
(1988) / Zhang Y

QUEENS UNIVERSITY

Deformation Controlled Nonlinear Analysis of
Pre-Stressed Concrete Continuous Beams
(1988) / Kodur V K

The Rate of Consolidation and the Increase
of Undrained Shear Strength of Cohesive
Soils Under Vacuum Suctions
(1988) / Li A Y

The Shear Capacity of End Plate Connections
(1988) / Van Dalen M

Hydrologic Response of an Agricultural Field
Under Tile Drainage
(1988) / Whyte R J

RENSSELAER POLYTECHNIC INSTITUTE

Effect of Wind on Roof Snow Drifts
(1988) / Galanakis I F

Vibrations of Thin-Walled Composite Beams
(1988) / Kao C H

Finite Element Modeling of the Behavior of
Composite Materials
(1988) / Lambropoulos N D

Shear Deflection in Thin-Walled Multicelled
Orthotropic Composite Beams
(1988) / Melehan T P

The Effect of the Frequency of Cyclic
Loading on San Fernando Sandy Silt
(1988) / Normandeau D E

Transportation in Space
(1988) / Raess R

Dual Coulomb Friction System for Seismic
Isolation
(1988) / Zajda C H

SAN DIEGO STATE UNIVERSITY

Analytical Verification of Steel
Confinements on Failure Capacity of
Concrete Cylinders
(1987) / Ghazi M

SASKATCHEWAN, UNIVERSITY OF (SASKATOON)

Hydraulic Performance of the Framed Timber
Culvert
(1988) / Barber D B

A Model Study of Scour in a Short
Contraction
(1988) / Breland A B

Configurative Modification of a Cold
Temperature Rotating Biological Contractor
Process
(1988) / Dahlman R A

Engineering Properties of Potash Tailings
(1988) / Loi J L

SOUTH DAKOTA SCHOOL OF MINES AND TECHNOL.
(continued)

Fatigue Properties and Performance
Characteristics of Hooked End Steel and
Polypropylene Fiber Reinforced Concrete
(1988) / Venkatasamy V

Steel Fiber Reinforced Concrete
(1988) / Wang Z J

SOUTH FLORIDA, UNIVERSITY OF

Corrosion of Reinforcing Steel in Partially
Submerged Concrete
(1988) / Aguilar A

Development of Design Guidelines for
Pinellas County Road Underdrains
(1988) / Deschamps R J

Ultimate Load Capacity of Micro-Piles
(1988) / Jackson N M

The Effect of Ion Species on Concentration
Polarization
(1988) / Krinke S L

Qualitative and Quantitative Analysis of
Fly Ash Content for Portland Pozzolan
Cements
(1988) / Papadopoulos G

An Analysis of Seafood Processing Wastewater
and Potential Pretreatment Methodology for
Small Seafood Processors, Wakulla County,
Florida
(1988) / Verscharen J A

Potential Hydrologic Effects of Mining Near
an Area of Artesian Flow
(1988) / Zarbock H W

SOUTHWESTERN LOUISIANA, UNIVERSITY OF

Selective Calcium Hardness Removal in
Lime-Softening Process
(1988) / Oylowo L A

STEVENS INSTITUTE OF TECHNOLOGY

Adsorption of Organics on Flyash: A Study
of AdsorptionDesorption Behavior
(1987) / Kaouris M

Admixtures in Concrete
(1987) / Meyer S M

Computer Aided Stability Analysis of a
Deep Submersible
(1987) / Son K H

TENNESSEE TECHNOLOGICAL UNIVERSITY

Finite Element Analysis of Finite
Axisymmetric Piezoelectric Cylinders
(1988) / Cheng M F

Hydrologic Analysis of Suspended Sediment in
the Clinch and Powell Rivers
(1988) / Choate K D

Design and Operational Factors for
Biological Fixed-Film Reactors for Manganese
Fixation
(1988) / Chuang N R

Large Deflection Analysis of a Flex Nozzle
Using Finite Element Method
(1988) / Forsyth J A

A Feasibility Study: Modified Direct
Filtration of Cookeville's Drinking Water
(1988) / Gifford J S

Fracture Parameters of an Anisotropic Plate
from Whole-Field Displacements
(1988) / Harper D W

Efficient Finite Element Analysis Using
Microcomputers
(1988) / Johnson J L

Poisson's Ratio and Coefficient of Friction:
The Effect They Have Upon Frictional Forces
at the Interface of Concrete Caissons and
Limestone
(1988) / Jones B R

TENNESSEE, UNIVERSITY OF (KNOXVILLE)

Analysis and Plotting Capabilities of
GTSTRUDL Using the CDC Cyber System
(1987) / Henderson R C

Slip Surface in Sinkholes: An Analysis of
the Slipline of an Idealized Axisymmetric
Open Karst Pipe
(1987) / Yoon C J

TEXAS A AND M UNIVERSITY

Comparison of Two Terrain Analysis Models
(1988) / Arndt F J

Optimization of Composite Mode I Delamination
Fracture Specimens
(1988) / Brown J W

Alvenus Oil Spill Debris Disposal and the
Potential of Land Treatment
(1988) / Clark K G

A Comparative Evaluation of Laboratory Compaction
Devices Based on Their Ability to Produce Mixtures
with Engineering Properties Similar to Those
Produced in the Field
(1988) / Consuegra A E

The Relation of Coastal Water Level and
Postlarval Abundance of Estuarine-Recruiting
Species of the Northwest Gulf of Mexico
(1988) / Craddock P P

Improved Design Procedure for Embedded Plates in
Gravity Anchors for Precast Concrete Panels
(1988) / Fragomeli L F

Development of an Orthopedic Load Cell for
Stress Analysis of a Canine Tibia
(1988) / Green B W

The Impact of Crew Mix on Productivity and Labor
Cost of a Construction Work Activity
(1988) / Grosskopf M W

Time-Dependent Thermo-Mechanical Properties of
Aircraft Tire Material
(1988) / Hanson R R

A Knowledge-Based Object-Oriented Simulation
Framework for Construction Process Planning
(1988) / Hogue S

TEXAS A AND M UNIVERSITY
(continued)

Use of Explosives to Demolish Multistory
Steel Frame Buildings
(1988) / Landry C V

Evaluation of a Microplane Model for Progressive
Fracture in Concrete
(1988) / Loper J H

Findings in Seal Coat Design
(1988) / Palmer M A

Reservoir Outflow (Resout) Model
(1988) / Purvis S T

Flow Characteristics in an Irregular Spillway Model
(1988) / Scott M C

Reservoir Analysis Model for Battlefield Operations
(1988) / Sullivan G J

Two-Dimensional River Modeling
(1988) / Thompson J C

Energy Dissipation Characteristics of Rubber
Crash Cushion Elements
(1988) / Thompson M F

Application of the Green-Ampt Infiltration Equation
to Watershed Modeling with Estimated Parameters
(1988) / Warinner J S

Transverse Wave Loading on Partially Buried
Marine Pipelines
(1988) / Webb R E

A Knowledge Based Constructability Analysis Prototype
(1988) / Winter L J

The Role of Redundancy in Jacket-Type Offshort
Platforms
(1988) / Womble J E

TEXAS, UNIVERSITY OF (ARLINGTON)

A Comparative Study in Flexural Tensile
Bond Strength Normal to Bed Joints Between
Masonry Cement Mortars and Portland Cement
Line Mortars
(1988) / Ngowthannatas W

Evaluation of Masonry Mortar Properties and
Relationship Between Compressive Strength
of Two Inch Cube and Three Inch by Six Inch
Cylinder
(1988) / Singh R

TEXAS, UNIVERSITY OF (AUSTIN)

Tensile Strength of Polypropylene
Cementitious Panels
(1988) / Al-Ayyubi A S

Behavior of CTT-Nodes in Reinforced Concrete
Strut-and-Tie Models
(1988) / Anderson R B

The Effects of Withholding Mixing Water and
Retempering on Properties of Concrete
(1988) / Anderson S M

The Effect of Fly Ash on the Temperature
Rise in Concrete
(1988) / Barrow R S

Detailing of Structural Concrete Dapped-End
Beams
(1988) / Barton D L

Foundation Effects on the Inelastic Response
of Frames to Seismic Excitation
(1988) / Boubenider R

Characterization of Excavation Induced
Changes in Rock by Seismic Methods
(1988) / Camp W M

MACGRID: A Grid Structure Analysis Program
Utilizing the Macintosh Graphical Interface
(1988) / Canjura N E

Load-Deflection Behavior of Cast-in-Place
and Retrofit Concrete Anchors Subjected to
Static, Fatigue, and Impact Tensile Loads
(1988) / Collins D M

Proper Use of Superplasticizers in Readymix
Concrete Under Hot Weather Conditions
(1988) / Eckert W C

Finite Element Analysis of Externally
Prestressed Segmental Construction
(1988) / El-Habr K C

Stability of Amended Sand Permeated with Fly
Ash Leachate
(1988) / Eykholt G R

Investigation of the Flexural Properties of
Slurry-Infiltrated Fiber Reinforced Polymer
Concrete
(1988) / Farra Y

The Effect of Openings on the Cyclic
Behavior of Reinforced Concrete Infilled
Shear Walls
(1988) / Gaynor P J

MacBASP: The Development of a Graphical
Interface for the Program "BASP" on the
Macintosh Computer
(1988) / Gou Abi-jaoude A

Practices in Constructability
(1988) / Gronberg E M

Abrasion Resistance and Scaling Resistance
of Concrete Containing Fly Ash
(1988) / Hadchiti K M

Air Stripping of Volatile Organics in
Packed-Columns: Experiments and
Mathematical Modeling
(1988) / Handler N

Shear Capacity of High Strength Prestressed
Concrete Girders
(1988) / Hartmann D L

Computer Program for the Buckling Analysis
of Stiffened Plates Including a User
Interface
(1988) / Hindi A

Joint Behavior of Metal-Plate-Connected
Parallel-Chord Wood Trusses
(1988) / Inlow D N

Applications of Computer-Aided Design
Methods for Site Characterization in
Civil Engineering
(1988) / Jones N L

TEXAS, UNIVERSITY OF (AUSTIN)
(continued)

A Study of the Influence of Temperature of
the Substrate on the Construction of Thin
Bonded Portland Cement Concrete Overlays
(1988) / Koesno S

Analysis of Local Buckling of Thin-Walled
Sections
(1988) / Kulkarni J A

Failure Investigation of the El Faro
Building Vina Del Mar, Chile: March 3, 1985
(1988) / Labbe A R

Repair of a Reinforced Concrete Beam-Column
Subassemblage Subjected to Bidirectional
Cyclic Loading
(1988) / Martinez J E

A Numerical Model of Solute Transport in
Unconfined Aquifers
(1988) / McNew E R

Graphical Interface for Structured Analysis
on Desktop Computers: A Multi-Lingual
Approach
(1988) / Murthy M S

Characteristics of Steel Fiber Reinforced
Polymer Concrete and Portland Cement
Concrete Panels Subjected to Projectile
Impacts
(1988) / Ong C K

Examination of a One-Dimensional Linear
Theory of Consolidation for Soils Exhibiting
Secondary Compression
(1988) / Palmer M W

Behavior of Steel to Concrete Connections
Incorporating Adhesive Fillers
(1988) / Peeler D D

Distribution of Post-Tensioned Forces Prior
to Grouting Tendons
(1988) / Quade C E

An Evaluation of Models for Predicting
Infiltration Rates in Unsaturated Compacted
Clay Soils
(1988) / Reimbold M W

An Evaluation of High Molecular Weight
Methacrylate as a Repair Material for
Cracked Concrete
(1988) / Rodler D J

Alkali-Silica Reaction in Concrete
Containing Mineral Admixtures
(1988) / Schuman D C

Behavior of Cold-Formed Z-Section Members
in Compression
(1988) / Sudharmapal A R

Review of Repair Techniques for Earthquake
Damaged Reinforced Concrete Buildings
(1988) / Teran A

An Object-Oriented Data Representation for
Design of Reinforced Concrete Beams
(1988) / Varghese K

A Finite Difference Computer Program for
Vertical Flow of Water in Unsaturated Soil
(1988) / Wang L

Side Shear Testing for Drilled Piers
Socketed in Weak Rock
(1988) / Wendland S A

Methods of Analyzing and Factors Influencing
Frictional Effects of Subbases
(1988) / Wimsatt A J

Fretting Fatigue of Multiple Strand Tendons
in Post-Tensioned Concrete Beams
(1988) / Wollmann G P

TEXAS, UNIVERSITY OF (EL PASO)

Detection of Stiffness Losses in Two-Dimensional
Trusses
(1988) / Chen T C

Design of Clay Liners to Minimize the Effect of
Shrinkage Cracks on the Permeability of the
Liner
(1988) / Idris M Z

Computer Simulation of the Ultimate Pull-Out
Capacity of Anchor Plates Buried in Sand
(1988) / Limasalle P

Solution of the One-Dimensional Groundwater-
Contaminate Transport Governing Equation by the
Method of Lines with Cubic Hermite Polynomials
(1988) / Liu J C

Effects of Vertical Reinforcement on the Bearing
Capacity and Settlement of a Medium Dense Sand
(1988) / Lovelady P L

TOLEDO, UNIVERSITY OF

Solution of Theoretical Equations for
Acceleration Sensitive Piping System
Seismic Restraint Devices
(1988) / Abu-yasein O A

Finite Element Analysis of Driven Piles in
Cohesionless Soil
(1988) / Al-Khatib H A

Application of Strategic Planning Through
ORS II: A Case Study of the Toledo
Metropolitan Area
(1988) / Bdeiri I J

Theoretical and Experimental Investigation
of Moment and Shear Transfer at the Junction
of a Flat Plate and an Interior Column
(1988) / Liew K T

Synthetic Approaches to Pyrrolylethylenes
and Pyrrolylbutadienes
(1988) / Singh S P

U.S. NAVAL POSTGRADUATE SCHOOL

Laboratory Evaluation of Geotextile
Performance in Silt Fence Applications
Using a Subsoil of Glacial Origin
(1988) / Crebbin C

Volume Reverberation Measurements of
Sediments in the Laboratory
(1987) / Facada J F

Experimental Investigation of Damping
Characteristics of Bolted Structural
Connections for Plates and Shells
(1987) / Iverson T C

U.S. NAVAL POSTGRADUATE SCHOOL
(continued)

An Analysis of a Numerical Tidal Model
Applied to the Columbia River
(1988) / Koehler R B

Management Control of Facility Warranties in
NAVFAC (Naval Facilities Engineering
Command) Construction Contracts
(1987) / Nielsen T W

Groundwater Model Calibration for a
Hydrocarbon Plume in a Sandy, Surficial
Aquifer
(1988) / Scanlan S R

Development of Design Parameters for
H-Piles in Sand Using Static Analysis
(1988) / Ungard K

U.S.A.F. INSTITUTE OF TECHNOLOGY

A Comparison Study of Above Ground
Reinforced Concrete Cylindical Arches
Designed to Resist Blast Loads
(1988) / Boone T L

Construction Contract Claims
(1988) / Bowholtz D W

A Simplified Characterization of HVAC System
Design on Building Energy Use
(1988) / Reardon M K

Forecasting Building Maintenance Using the
Weibul Process
(1988) / Yeoman A K

UTAH STATE UNIVERSITY

Highway Drain Depth and Soil Stability
(1987) / Al-Himdani M K

Improving the Applicability and the
Versitality of the Chadwick Model for
Providing Short-Term Forecasts of the Level
of the Great Salt Lake
(1987) / Atkin W H

Instrumentation of the Steed Canyon
Landslide
(1987) / Brooks R K

Studies of the Persistence of Polycyclic
Aromatic Hydrocarbons in Two Unacclimated
Agricultural Soils
(1987) / Coover M P

Efficient Analysis and Design of Medium-Rise
Steel Buildings-Rigid Frame
(1987) / Krichi H

Clinoptilolite Amended Slow Rate Sand
Filtration for Improved Filter Performance
and Economics
(1987) / McNair D R

A Laboratory Study of Volatilization of
Hazardous Organic Constituents from Land
Application of Hazardous Wastes
(1987) / Reineman J A

UTAH, UNIVERSITY OF

The Confinement of High Explosives in a
Small Caliber Cartridge: Experiments
Leading to a Successful Design
(1988) / Andersson N H

Optimal Management of a Stratified Aquifer
System
(1988) / Santini M D

VERMONT, UNIVERSITY OF

Use of the One Dimensional Dispersion Model
in a Horizontal Flow Inhalation Chamber
(1988) / Nichols R B

VIRGINIA, UNIVERSITY OF

Phosphorus Dynamics Between the James River
and the Chesapeake Bay
(1988) / Fen C S

Evaluating Eutrophication Control
Alternatives for the Lower Neuse River,
North Carolina
(1988) / Huffman L G

WASHINGTON STATE UNIVERSITY

Control of Vibrations in Medium High Rise
Buildings Using Combined Friction
Viscoelastic Dampers
(1988) / Agarwal P K

Numerical Simulation of Two-Dimensional
Unsteady Flow in a Curved Channel
(1988) / Dammuller D C

Effects of Span-to-Thickness Ratio on the
Punching Shear Behavior of Reinforced
Concrete Slabs
(1988) / Lovrovich J S

Projectile Penetration in Dry Sand
(1988) / Meyer M R

Investigation of Dam-Break Flows
(1988) / Miller S

A Study of Basement Walls as Plate Elements
(1988) / Torabi M R

Characterizing the Effect of Moisture
Content on the Flexural Properties of Sugar
Pine and Black Spruce Dimension Lumber
(1988) / Vogt J J

WASHINGTON UNIVERSITY

Soil-Pile-Structure Interaction Effects on
the Seismic Response of a Shell of
Revolution
(1987) / Ahn K

Influence of Deformation and Embedment on
Cyclic and Monotonic Anchorage Strength for
Reinforced Concrete
(1987) / Belaloui K

Effects of Reversed Cyclic Loading on Local
Bond in Reinforced Concrete
(1987) / Bourouz A

General Loading of Semi-Rigid Connections
(1987) / Earson D L

Nonlinear Analysis Using a Simple Beam
Element
(1987) / Kasagi A

Development of an Hedonic Pricing Model of
Residential Property
(1987) / Kaufman C S

WASHINGTON, UNIVERSITY OF

Transfer of Force Between Reinforcing Bars
and Pretensioned Strand
(1987) / Abdie J L

Continuously Heat-Curved Mild Steel Plates
(1988) / Afeiche N E

Application of Analytical Close-Range
Photogrammetry in Refraction Study
(1987) / Alhomeida M I

An Integration of System Engineering Tools:
Expert Systems and Multi-Objective Analysis
(1987) / Berkshire P A

Kinetics of Methanotrophic Biodegradation
of Chlorinated Aliphatic Hydrocarbons
(1988) / Bjelland M

Estimating Aerial Photogrammetry Costs Using
a Computer Spread Sheet
(1988) / Bodnar A N

An Investigation Into the Effects of
Chhromate and Other Selected Factors on the
Colloidal Properties of Wyoming Bentonite
(1987) / Brisbine J M

Investigation of Oxygen Transfer Enhancement
in Biological Processes
(1987) / Deem S S

Seismic Performance of an Icelandic Precast
Concrete Structure
(1987) / Halldorsson A

Analysis of Laterally Loaded Piles with
Nonlinear Bending Behavior
(1987) / Heavey E J

An Expert System for Drought Management:
Development and Demonstration
(1987) / Holmes K J

Combination of Satellite Coordinates with
Terrestrial Coordinates
(1986) / Hsu P H

The Feasibility of Using Local Access Cable
TV as a Driver Information Tool
(1987) / Kirkemo G

Structural Analysis with Large Rotations
(1987) / Lund B A

Actual Tolerances Used in Cast-in-Place
Concrete Construction
(1987) / Marciniec T M

Forecast Uncertainty in Reservoir Operation
(1987) / Mishalani N R

The Effect of Lateral Pressure on the
Strength of Closed-Loop Anchorages
(1987) / Morikawa S M

A Preliminary Analysis of a Piecewise-Linear
Multiequilibrium Oscillator
(1987) / Nishikawa J A

An Evaluation of the Inverse Method of
Producing Error-Free Photocoordinates
(1987) / Rajehi A N

Risk-Based Selection of Monitoring Wells
for Assessment of Agricultural Chemical
Contamination
(1987) / Scheibe T D

Methane Oxidation in a Sparged Packed Bed
Reactor
(1988) / Seamons R M

A Theoretical Analysis of a Floating
Breakwater-Pendulum System
(1987) / Shaul L B

Bayes and Fuzzy Identification
(1987) / Shih Y

Aragonite Formation in Polar Soils
(1987) / Sletten R S

The Analysis of the Failure of a Six Story
Steel Building with Brace to Beam
Connections
(1988) / Tang M

The Effect of Alkalinity and pH on the
Corrosion of Mild Steel Pipe Exposed to
Potable Water
(1988) / Vanderwerff E

The Effect of Eccentricity of Lateral Forces
on the Strength of Loop Anchorages
(1988) / Weaver D L

Mean Stress Effects on Fatigue Near
Non-Structural Weldments
(1987) / Young R P

Local Bond Stress-Slip Relationships for
Deformed Bars for Monotonic and Cyclic
Loading
(1987) / Zouyed M A

Effect of Bonded Length and Concrete Cover
on Local Bond Stress-Slip Relationships
(1987) / Zouyed N B

WEST VIRGINIA UNIVERSITY

Design and Performance of a Vacuum Sewer
Flow Experiment
(1988) / Bowers R B

Application of Micro-Computer Based
Pre- and Post-Processors for Finite Element
Analysis of Fracture Zone Above Underground
Cavities
(1988) / Kang S K

Hybrid BEM-FEM Dynamic Analysis of Flexible
Strip-Foundations
(1988) / Kokkinos F T

Lateral Stability of Steel Stringers
Stiffened by Timber Decks
(1988) / Kumarjiguda S

WESTERN ONTARIO, UNIVERSITY OF

Full Scale Wind Response of Tall Structures
(1988) / Masciantonio A

WINDSOR, UNIVERSITY OF

Resistance to Uplift of Interior Footings
of Low-Rise Buildings
(1988) / Loong C W

Starred Angles Supporting Secondary Trusses
(1988) / Mok K H

Hydraulic Model and Prototype Studies of
Precast Cellular Concrete Shoreline
Protection Systems
(1988) / Patterson H R

WORCHESTER POLYTECHNIC INSTITUTE

Low Strain Level Shear Modulus of Slowly
Draining Soils
(1987) / Cerow G D

Structural Analysis with Answers Using
Micro-Computers
(1986) / LeBlanc S G

A Finite Element Analysis of Axisymmetric
Shell Structures
(1987) / Wood W M

WYOMING, UNIVERSITY OF

An Evaluation of Advanced Recreational
Signing
(1989) / Jones C P

Modified Finite Strip Method for the
Stability Analysis of Columns and Beams
(1989) / Kladianos J R

Initial Geotechnical Properties of a
Retorted and Combusted Oil Shale Solid Waste
(1989) / Rothwell R A

ALABAMA A - M UNIVERSITY

Inexact Reasoning Under Incomplete
Conditions in Rule-Based Expert Systems
(1988) / Lin M T

An Instructional Model Local Area Network
(1988) / Noble B J

ALABAMA, UNIVERSITY OF (HUNTSVILLE)

Mathematical Learning Theories and Genetic
Learning: A Framework for Analysis with
an Application to Concept Learning
(1988) / Goodloe J M

Development of an Object Oriented System in
a LISP Environment
(1988) / Winn H G

ALBERTA, UNIVERSITY OF

An Interactive Classifier Programming
Language
(1988) / Fenske K

Object Management, Protection and Scheduling
in FLEX
(1988) / Lau C

A Heuristic Approach to Query Optimization
(1988) / Meechan D J

Time Revisited
(1988) / Miller S A

A Computer Model of Knowledge Organization
and Strategy Shifts in Novice-Expert
Problem Solving
(1988) / Scharf P B

ARIZONA STATE UNIVERSITY

Time Efficient Implementation of Conway's
Sorting Algorithm
(1988) / Ahmed K U

User Interface for a Software Maintenance
Environment
(1988) / Barrios T E

Graphical Abstract Machine Generator (GAMG)
for ADA Software Environment
(1988) / Cai Y

Matching Points with Subpixel Accuracy in
Stero and Motion Analysis
(1988) / Chang A

Simulation of Fault-Tolerant Multiprocess
System
(1988) / Chang J

An Integrated Expert Database
Design -- Coupling Relational Database
and Prolog Systems
(1988) / Chen H M

Recognition of Partially Occluded Objects
(1988) / Choi I

Object-Oriented SERAP DBMS
(1988) / Chou Y C

Techniques to Maximize Transportability
Via a Set of CAIS Terminal Packages
(1988) / Clark D A

The Dempster-Shafer Theory of Evidence in
Reasoning: An Application in Marketing
(1988) / Cortes-Rello E

How People Use Large Vocabulary Systems
(1988) / Davids N S

An Automated Inspection System for Circuit
Card Assemblies
(1988) / Dayanand S

An Approach to Maintenance Training
(1988) / DiPrima E R

Detection of Surreptitious Insertion of
Trojan Horse or Viral Code in Computer
Application Programs
(1988) / Fastiggi M V

Curvature for Surface Interrogation
(1988) / Fayard L J

Parallel Algorithms and Their Analyses on
the Sequent Balance 21000
(1988) / Foster R

Maintenance Management Control System
(1988) / Fox D G

A Functional Programming Language for
Real-Time Embedded Systems
(1988) / Golovin P

Geometric Continuity Via Differential
Geometry and Gamma-Spline Interpolation
(1988) / Hansford D C

Applicability of the Simulated Annealing
Method for Tomographic Image Reconstruction
(1988) / Hill S E

ODDS: The Concurrent Fault Simulator
(1988) / Kapur S

Terrain Simulation Using a Model of Stream
Erosion
(1988) / Kelley A D

Using Three System Perspectives to Transform
Structured Analysis Into Object-Oriented
Design
(1988) / Khalsa G S

A Visual Navigation System Using Optic Flow
for Determining Surface Orientation and
Distance
(1988) / Kimberly H R

PetriNAS: A Requirements Specification Tool
Based on New Paradigms
(1988) / Kow C H

Visualized Software Engineering Environment
(1988) / Lee H M

Toward the Production and Application of an
Environment Type Definition Processor
(1988) / Levine D P

A Parallel-Processor Navigation System for
an Autonomous Vehicle
(1988) / Lovell R E

Parallel Sorting Algorithms on a
Mesh-Connected Computer
(1988) / Lung C

ARIZONA STATE UNIVERSITY
(continued)

Some Efficient Algorithms for Constructing
Perfect Hash Function from a Set of Mapping
Functions
(1988) / Mahbut F S

Information Retrieval Using Heuristic Search
(Version II): The Clustering Module and the
Query Handler
(1988) / Maini S

Intensional Horn Clause Logic as a
Programming Language, It's Use and
Implementation
(1988) / Mitchell W H

PARSL: A Prolog Analyzer for Restricted
Sublanguages
(1988) / Mittelmann R M

Curve and Surface Interpolation Using
Quintic Weighted Tau-Splines
(1988) / Neuser D A

IOGen: Toward an Automated Tool for
Production of Reliable and Valid Test Suites
(1988) / Norman M A

Simulation of a Multisensory Integration
System and Its Application in Modeling
Higher Order Cognitive Functions
(1988) / Novik E

Near Real-Time Shaded Display of Dynamic
Objects
(1988) / Ong L

Software Safety Assessment During Program
Maintenance
(1988) / Sheppard S S

Parallel Sorting Algorithms on a Hypercube
Multiprocessor
(1988) / Shyong D Y

Analyzing the Consistency of Semantic
Information Within a Maintenance Data Base
(1988) / Steckmesser K L

Efficient Hidden Surface Algorithms
(1988) / Vasconcellos J P

An Expert Database System Implemented in
C Prolog for the Engineering Design of
Electronic Modules
(1988) / White R C

Configurable Software Configuration
Management
(1988) / Wright D G

Low-Reynolds Number Modeling of Complex Flow
with and Without Density Variation
(1988) / Yoo G J

A New Software Design for DSL System 99
(1988) / Yuan M

Information Retrieval Using Heuristic
Search, Version 2: The Input/Output
Processor and the Knowledge Base Generator
(1988) / Yuen A F

Multimedia Document Management for RAP GDBMS
(1988) / Zainal Z

ARKANSAS, UNIVERSITY OF

Automated Visual Inspection to Improve
Quality Control in Manufacturing
(1988) / Allen D K

A Diagrammatic Environment for Simnet
Simulation Model Composition
(1988) / Hendricks S S

An Environment for Fortran Software
Development on the Personal Computer
(1988) / Justice E D

AUBURN UNIVERSITY

Dynamic Resoure Allocation in a Real-Time
Distributed Computing System
(1987) / Bate G T

Research and Development of a Microcomputer-
Based Revs Workstation Concept
(1987) / Cox R C

A Test Methodology for Query Utility Environment
for Software Testing
(1987) / Haga K D

The Solid User-Interface
(1987) / Head D M

Metamorphism and Deformational History of a Portion
of the Campbell Mountain and Suches 7-1/2 Minute
Quadrangles, Northern Georgia
(1987) / Heinrich N D

Conversion of Computer Software with the
Conversion Rule Description Language
(1987) / Jodis S M

A Menu Driven Computer Aided Design System for
Simulating Woven Fabric Structure and Coloring
(1987) / Lee M C

Grasp: An Executable Specification Language
and Support Environment for ADA
(1987) / Morrison K I

A Prototype Implementation of Retrieval-Oriented
Text Compression
(1987) / Tackett B D

Source Code Analysis for Error Detection and Metric
Derivation
(1987) / Welch C A

BALL STATE UNIVERSITY

"Hands-On" Computer Workshops for Improving
Microcomputer Literacy: Feasibility Studies,
Design, Layout, Workbooks
(1988) / Grady P N

An Analysis of Cost Influential Factors for the
Development of a Software Product and a
Hypothetical Method for Estimating Cost
Using the Cocomo and Putnam Models
(1988) / Missopoulos F S

Moessbauer Spectroscopy on the Apple Computer
(1988) / Moody K L

A Study on Relational Databases Through
Mathematical Theories of Relations and Logic
(1988) / Yu C

BOWLING GREEN STATE UNIVERSITY

An Analysis of the Use of Computer-Aided
Software Engineering Tools
(1989) / Huenemann K E

BRIGHAM YOUNG UNIVERSITY

PROMCOR: The Automated Selection and
Refinement of Instructional Modules Based
on Characteristics of a Learner Profile
(1988) / Ashton G L

A Multilayered Data Acquisition Subsystem
for Monitoring and Improving Performance
in Data Networks
(1988) / Blatter K L

Combining 3-D Input Tools with 3-D Output
Tools to Create an Interactive Environment
for Editing Primitive Volumetric Objects
(1988) / Broekhuijsen J A

Geographic Informations Systems: Are They
Worth It? A Cost/Benefit Analysis
(1988) / Brush S E

The Integration of Voice Into a General
Purpose Operating System
(1988) / Chan K M

Window Interactive Control Environment (An
Architecture for a Modular UIM.S)
(1988) / Chang G

A Framework for Product Verification Testing
(1988) / Choy S H

A Characterization and Categorization of
Higher Dimensional Presentation Techniques
(1988) / Cluff E

Controlling Computer Based Instruction
Curriculum and Student Knowledge
(1988) / Decker M L

Abstract Data Type Library User Interface
(1988) / Dittmer D

Preserving Behavioral Integrity in a Dynamic
Object-Oriented Environment
(1988) / Frank G A

Design of an Expert System for Generation of
Adaptive Recommendations
(1988) / Harker P L

Type Consistency Checking for Communicating
Independent Processes in a Transputer
Environment
(1988) / Hatch R D

Configuration Management for a Distributed
Operating System in a CIP/Transputer
Environment
(1988) / Hutcheson J E

OSA: An Object-Directed Methodology for
Systems Analysis and Specification
(1988) / Kurtz B D

Efficient Implementation of a Compiled
Object-Oriented Language
(1988) / McNeill T G

ORO: The Object-Relationship-Operation
Model
(1988) / Quass D W

Minimal Perfect Hashing Functions
(1988) / Radford D R

A Data Storage and Retrieval Method for
Large Databases: Based on Signature Files
and Superimposed Coding Methods
(1988) / Rayner P T

Sculptbox: A Volumetric Environment for
Interactive Design of Three-Dimensional
Objects
(1988) / Richardson A E

Dynamic Process Instantiation and
Configuration Verification in the
Distributed Programming Environment
(1988) / Sanders D S

A Communication System for the CIP Parallel
Language on a Network of Transputers
(1988) / Valimaki M T

Equivalence of an Abstract Data Type View of
Information Processing and a Knowledge and
Database View of Information Processing
(1988) / Weston J M

The Analysis and Selection of a Software
Methodology in the Application Development
Process
(1988) / Whale R C

Representing N-Dimensional Hyperplanes Using
Parallel Axes Computer Graphics: A
Demonstration Using Linear Programming
(1988) / Whiteley G R

BROWN UNIVERSITY

SAT: A Subsumption Architecture Tool for
Simulated Robot Control
(1988) / Arnold J E

CALGARY, UNIVERSITY OF

Polygon Mesh Modelling for Computer Graphics
(1988) / Allan J B

A Microcomputer-Based Knee Controller
(1988) / Campbell M M

Adaptive Predictive Text Generation and
the Reactive Keyboard
(1988) / Darragh J J

A Fast-Time Simulation of Air Traffic
Control
(1988) / Inkster J A

A Programming Environment for Well Log
Analysis
(1988) / Lowden B A

An Implementation of Ray Tracing Using
Multiprocessing and Spatial Subdivision
(1988) / Pearce A P

The Delta Chip
(1988) / Stone G D

Optimising Time Warp: Lazy Rollback and
Lazy Reevaluation
(1988) / West D R

CALIFORNIA STATE UNIV. (FRESNO)

A Comparative Study of Third-Generation
Three-and-a-Half Generation, and
Fourth-Generation Computer
(1987) / Yao J

CALIFORNIA STATE UNIV. (FULLERTON)

Neural Network Technology: The
Neurophysiological Approach to Artificial
Intelligence Challenges Von Neumann
Architecture and Expert Systems
(1988) / Kumagai L

CALIFORNIA STATE UNIV. (LONG BEACH)

Vertical/Horizontal Transmission Mechanism
for Non-Dichotomized Traits
(1988) / Cheng K S

Database Design for Social Science Research
(1988) / Elbanna I S

CALIFORNIA STATE UNIV. (SACRAMENTO)

A System Design for the ASCSUS Child
Development Center Using the Entity
Relationship Approach
(1988) / Bridges H E

Computer-Aided Instruction in Relational
Database Normalization
(1988) / Butler E C

A Syntax-Directed Editor for KAREL the Robot
(1988) / Chiou Y

The Cocomo Cost Estimation Model: A Case
Study
(1988) / Clark R

Knowledge Acquisition for Expert Systems
(1988) / Crothers R F

Performance Evaluation of the X.25 Packet
Switching Protocol Using Network II.5
Simulation Package
(1988) / Emdadi B

Requirements for a Computer Software System
for Use in Analysis of Large, Complex
Problems by Groups of Experts
(1988) / Goodwin S G

Evaluation and Enhancement of Distributed
Robot Control Environment Under UNIX
(1988) / Hung S H

Parallel Programming in Turbo PASCAL
(1988) / Lee S L

Application of Logic Programming Techniques
for the Specification and Simulation of
Transport Protocols
(1988) / Liu W

Development of an Intelligent Tutoring
System Designed to Facilitate Problem
Solving Capabilities for Computer Science
Students
(1988) / Ouellette D A

A Standardized Mouse Interface for Tect
Processing: Effects on Learning and
Performance
(1988) / Scanlan K

PASCORE, a PASCAL Language Implementation of
the Core Graphics System
(1988) / Scott D F

Ring LAN for PC Communication
(1988) / Sunaryo P H

An Implementation of a Projected Hessian
Updating Algorithm for Nonlinearly
Constrained Optimization
(1988) / Thomas R J

CALIFORNIA, UNIVERSITY OF (DAVIS)

Information Dissemination in Computer
Networks
(1988) / Agarwal R

Multiple Inheritance and Rule-Based
Delegation
(1988) / Almarode J E

Vector Conversion and Processing of
Document Images
(1988) / Brandt J W

Bidirectional Reflection Distribution
Functions from Surface Bump Maps
(1988) / Cabral B K

Image Segmentation in Computer Vision
(1988) / Chen S

Gaussian Techniques on Shared-Memory
Multiprocessors
(1988) / Darmohray G A

An Attribute-Based Filtering Tool
(1988) / Huang G G

Control of Discrete Event Systems
(1988) / Ma F Y

Problems in System Level Diagnosis
(1988) / Otsuka M

An Expert System Approach to Operating
System Design
(1988) / Rager N L

Computer Aided Design Via Quantitative
Feedback Theory
(1988) / Renn G S

Disk Scheduling Algorithms for a Multiple
Disk System
(1988) / Stabile J J

A High-Level Protocol for Distributed
Multiuser Sessions
(1988) / Sun Y Y

Classification of Seismic Signals Using
Artificial Neural Networks
(1988) / Talbot E B

An Operating System on the iAPX 286-Based
Real-Time Multiprocessor System
(1988) / Touranachun V

The Integration of MUMPS Into the DAISY
Environment
(1988) / Wolber D W

Evaluating the Performance of the Education
Machine by Simulation
(1988) / Wu J D

Protecting Digital Filters with Embedded
Convolutional Codes
(1988) / Zagar B G

CALIFORNIA, UNIVERSITY OF (LA JOLLA)

The Exchange of Control Information in a
UHF Radio Telephone System: A Case Study
(1988) / Kammer Y M

Software for the Development of Sliding
Block Encoders for Modulation Codes
(1987) / Tang J

CARLETON UNIVERSITY

Rationale: A Tool for Developing Knowledge
Based Systems that Reason by Explaining
(1988) / Abu-Hakima S

Some Learning Applications to Robot Motion
Problems
(1988) / Abularach M N

A Programmable Digital Signal Processor
Based Speech Feature-Extraction System for
a Modular Cochlear Implant Prosthesis
(1988) / Barszczewski P A

Graphically Defining Simulation Models of
Concurrent Systems
(1988) / Brauen H G

A Neural Controller for Collision-Free
Movement of Robot Manipulators
(1988) / Graf D H

Visibility for Binary Images on a
Mesh-Connected Computer
(1988) / Hassenklover A L

Direct Sequence SSMA Communication with
Some IC Realization
(1988) / Kemdirim D C

An Adaptive Mesh-Management Algorithm for
Three-Dimensional Finite Element Analysis
(1988) / McDill J M

A Study of Common Logic Design Errors and
Methods for Their Detection
(1988) / Mein G F

Directional Separability in Two and Three
Dimensional Space
(1988) / Nussbaum D

The Design of a Medical Knowledge Base
System for Ophthalmologists
(1988) / Ram G R

Adaptive Control for Robotic Manipulators
(1988) / Ramesh S

Propositional Dynamic Logic of Regular
Programs with Tests, Converse and Infinite
Computations
(1988) / Smith F J

Model Reference Adaptive Control of a
Flexible Manipulator
(1988) / Sutherland J L

A Rule and Case Based System for Learning
by Analogy
(1988) / Wang Z

A Functional-Test Specification Language
(1988) / Williams D L

CARNEGIE-MELLON UNIVERSITY

Pre-Deployment Validation of Software
Implemented Fault-Tolerant Systems
(1988) / Barton J H

Automatic Synthesis for Reliability
(1988) / Brennan A A

Predictive Subset Testing for IC
Performance
(1988) / Brockman J B

Analysis of Computational Complexity
Using Various Representations
(1988) / Bunting D R

Classification of Stop Constants in
Natural Continuous Speech
(1988) / Chigier B

A Hierarchial Problem Solving Architecture
for Design Synthesis of Single Board
Computers
(1988) / Gupta A P

Trend Analysis and Modeling of
Uni/Multi-Processor Event Logs
(1988) / Hansen J P

Feedback Control Synthesis for a Class of
Discrete-Event Systems Using Distributed
State Models
(1988) / Holloway L E

COAL: A New Framework for Performance
Evaluation
(1988) / Hughes K J

Automated Development of Control Software
for Discrete Manufacturing Systems
(1988) / Pathak D K

Computer-Aided Synthesis of Routine Designs
(1988) / Rehg J M

CASE WESTERN RESERVE UNIVERSITY

Type Checking and Inference in CaseDL
(1988) / Abad A

Load Sharing in a PC Local Area Network
Environment
(1988) / Badami A

Breakpoint Expression Evaluation
(1988) / Boughambouz I

An Ada Based Approach to MAP Network
Specification and Simulation
(1988) / Chandna A

A Unified Formulation of Circuit Reduction
for Testing
(1988) / Chiu S K

Akrti--A System to Draw Data Structures
(1988) / Desai T

An Expert System for Finite Element Analysis
of Electrochemical Cells
(1988) / DiCaprio P N

Towards Developing a Reflexive Partial
Evaluator for an Interesting Subset of LISP
(1988) / Guzowski M A

CASE WESTERN RESERVE UNIVERSITY
(continued)

An Extensible Interpreter for a Subset of
CaseDL Based on Prolog
(1988) / Harous S

An Explanation Shell for Expert Systems
(1988) / Lalee M

CaseDDB--A Data Base to Support the Case
Design Environment
(1988) / Robinson H L

Flavors Programming Tools
(1988) / Sheldon R A

Execution of Incomplete Specification
(1988) / Sudar M A

A Rule-Based Expert System for the
Diagnostic Interpretation of Evoked
Potentials in Clinical Neurology
(1988) / Xiao H

CENTRAL FLORIDA, UNIVERSITY OF

Development of Techniques to Perform
Simulation - Adaptation in a Simulation
Training Environment Using Expert System
Methods
(1988) / Bagshaw C E

GAPP Instruction Simulator
(1988) / Bauman R P

LANSIM: A Simulation Package for Estimating
Performance Characteristics of a Class of
Local Area Networks
(1988) / Buchner G C

A Cellular Algorithm for Data Reduction for
Polygon Based Images
(1988) / Caesar R J

A Display Peripheral for a Flight Simulator
Using a Three Dimensional Computer
Controlled Motor Drive Model
(1988) / Flowers K R

Research and Applications of Expert Systems
in Military Training
(1988) / Green P I

CHAR: The Fire Investigator's Aide
(1988) / Johnson P M

ADA Real-Time Performance Benchmarks for
Personal Computer Environments
(1988) / LaRoche S L

Direct Digital Control of a Liquid Level
Control Unit
(1988) / Lytle B V

The Design of a Systolic Architecture
to Implement Graphic Transformations
(1988) / Murray A A

The Aggregate Costs of Microcomputing
(1988) / Newkirk J R

An ADA Simulator for the Structured
Functional Simulation of a Digital
Signal Processing System
(1988) / Roney C P

Object-Oriented Development Technique for
Real-Time Software Systems: A Case Study
(1988) / Seldenright T S

Structured Development Techniques for
Real-Time Software Systems: A Case Study
(1988) / Seldenright V G

Development of a Bi-Directional Inference
Engine Using an Object-Oriented Approach
(1988) / Sidani T A

Survey of Local Area Networks for Low
End Systems
(1988) / Songprasit W P

Utilizing Object-Oriented Design to Reduce
the Effort Expanded on the Maintenance
Phase of a Software Project
(1988) / Storma S M

Comparison of Graphical Design Methodologies
(1988) / Woodard P S

CITY COLLEGE OF NEW YORK

Efficiency of Linear Programming Algorithms
(1988) / Dobric V

CLARK ATLANTA UNIVERSITY

An Automatic Learning of Grammar for Syntactic
Pattern Recognition
(1987) / Ofori P

CLARKSON COLLEGE OF TECHNOLOGY

A Computer Aided Engineering Package for
Control Systems
(1988) / Grace T B

SIMULATR: A Graphics Software Tool for
Robotics Education
(1988) / White R B

CLEMSON UNIVERSITY

Toward a Generic Description of Software
Design Methodologies and Environments:
Motivation, Issues and Form
(1988) / Badami S H

Microcode Based Reduced Instruction Set
Computer Architecture
(1988) / Bandyopadhyay A

Real-Time Control and Navigational Path
Planning for a Biped Robot
(1988) / Barry J D

Performance Analysis of Packet Fragmentation
in Interconnected Computer Networks
(1988) / Bijjahalli S S

A Distributed Computer System for Embedded
Realtime Control
(1988) / Camlin S L

Visual Classification of 2-D Polygonal
Objects Using a Neural Network Architecture
(1988) / Jamison T A

A Decomposition and Parallel Architecture
for the Geometric Transformation of Digital
Images
(1988) / Nay H

CLEMSON UNIVERSITY
(continued)

An Efficient Method for Simulating Token
Ring or Token Bus Access Protocols
(1988) / Panchmatia D M

A Rise Architecture Designed to Support
Functional Programming Environments
(1988) / Rodgers S D

A Prototype Testbed for Transient Computer
Communication Network
(1988) / Sinha B B

COLORADO STATE UNIVERSITY

Integration of a Trace Scheduling Optimizer in a
Retargetable Microcode Compiler
(1988) / Howland M A

Learning the Phonological Structure of a
Language--An Application for Boltzmann Machines
(1988) / Morris,C E

Integrated Approach to Instruction Scheduling
in High Performance Architectures
(1988) / Vin H M

CONCORDIA UNIVERSITY

Speeding up the Skeletonization of Binary
Patterns Using the Homogeneous
Multiprocessor
(1988) / Beffert H

Graph Algorithms for Database Schema Design
(1988) / Chang H

Classification of Digitized Curves
Represented by Signatures and Fourier
Descriptors
(1988) / El Busaeshi H

A Model and a Methodology for Distributed
Debugging
(1988) / Hamamtzoglou I

Development of a User-Friendly Menu Driven
Intelligent Front-End System for Large Frame
Based Knowledge Engineering Tools
(1988) / Kowalski A G

Database Access Using Voice Input and
Menu-Based Natural Language Understanding
(1988) / Menzies I M

A Proposal for Kernel Implementation of a
Window Facility for ASCII Terminals in the
Unix System V Operating System
(1988) / Mihordea M

Design of Distributed and Centralized
Rational Databases
(1988) / Nguyen H S

Polynomial Scaling
(1988) / Panah S

Experimental Evaluation of Distributed
Rollback and Recovery Algorithms
(1988) / Passier C

A Study of System Design Techniques in VLSI
(1988) / Prasad R N

Parallel Search of Game Trees
(1988) / Shved S

The Design and Implementation of a Systolic
Array Silicon Compiler
(1988) / Sun X

Issues in the Design and Implementation of a
Portable Natural Language Interface System
(1988) / Vincent P J

Mesh Generation for Finite Element Analysis
(1988) / Vo T N

CONNECTICUT, UNIVERSITY OF

A Syntactic Approach for Pattern Recognition
in Ultrasonic-Based Nondestructive Test
Applications
(1988) / Bansal V

Team Information Processing
(1988) / Bushnell L G

An Interactive Troubleshooting Unit for
Logic Circuits Using an Expert System
(1988) / Chan S W

Automatic Functional Design of Physical
Mechanisms
(1988) / Cuthill B B

Team Distrubuted Resource Allocation and
Management
(1988) / Kohn C

A Computational Cognitive Model of Expert
Knowledge and Reasoning in the Domain of
the Auditor's Going-Concern Judgement
(1988) / Krupka G R

Dynamic Analysis and Evaluation of Software
Designs: An Experimentation Analysis
Approach
(1988) / Law S P H

Tugboat: A Tool for Generating Software
Engineering Database Systems
(1988) / McDonald L

Routing in a Dynamically Reconfigurable
Distributed Database System
(1988) / Nagy S C

Design and Implementation of Session
Establishment in Mobile Networks
(1988) / Tran T T

A Graphic Performance Analysis Tool for
Software Design
(1988) / Wang J

Network Partitioning in a Dynamically
Reconfigurable Distributed Database System
(1988) / Wong D W H

DALHOUSIE UNIVERSITY

The Investigation of the Effect of Indexing
in Prolog for Bibliographic Retrieval
(1988) / Chen L S

DELAWARE, UNIVERSITY OF

Design, Implementation and Application of an
Extended Ground Reducibility Test
(1988) / Bundgen R

Common Sense Reasoning for Dynamic
Processes: A Case Study Using QP Theory
and Envisionment
(1987) / Chittajallu S S

DELAWARE, UNIVERSITY OF
(continued)

ALGER a System to Assist Beginning
Programming Students in Algorithm Design
(1987) / Isaacs E N

The Principle of Optimality in Dynamic
Programming
(1987) / Krishnamurti M

HPMDF: A Study of the Elements of Good
Message Relaying Style and a Practical
Implementation
(1987) / Lanbach M E

Subsumption Related Strategies for Reducing
the Search Space Generated by Resolution
Theorem Provers
(1988) / Nagarajan R

The Class Concept in Conceptual Modelling
(1987) / Tighe G C

Dynamic Window and Adaptive Time Out
Algorithms for Congestion Control in
Packet-Switched Computer Networks
(1988) / Tran T T

DREXEL UNIVERSITY

Impact of Expert Systems on Computer Networks
(1987) / Chaganty S

An Innovative Technique for Backbone Network Design:
Key to a Tool for Evaluating Decentralized and
Centralized Networks
(1988) / Chattopadhyay N G

Performance Estimates of Computer Networks Using
QNA: Simulation Validation
(1988) / Ghanta J

A Manchester Encoder/Decoder Made in Scalable CMOS
for Use in Local Area Networks at Ten Million Bits
Per Second
(1987) / Haynes C F

An Isolated Word Recognition System Using Condensed
Nearest Neighbor Approach
(1988) / Mansur A

An Integrated Services Digital Networks Terminal
Adapter Model Using Networks 11.5
(1987) / Pehlert W K

Qualitative and Analytical Studies of Implementing
User Help Expert System in Computer Networks
(1988) / Pitchai A S

Performance of Job Allocation Policies in
Multiprocessor Networks
(1987) / Ray A K

DUKE UNIVERSITY

Zyglotron: The Behavioral Simulator for
Blitzen
(1988) / Becher J D

Threshold Incomplete Factorization as a
Preconditioner of an Iterative Solution for
Nonsymmetric Systems of Linear Equations
(1988) / Corley C J

Pulsescript: A Flexible Language for
Waveform Specification
(1988) / Gunn D M

An Integrated Software Performance
Engineering Environment
(1988) / Martin C R

Near-Horn Prolog: A First-Order Extension
to Prolog
(1988) / Reed D W

An Extension of the Gypsy Information Flow
Semantics for Dynamic and Indexed Types
(1988) / Singer C D

Term-Rewriting Techniques for Logic
Programming I: Completion
(1988) / Smith M P

Closed-Form Transient Analysis of Markov
Chains
(1988) / Tardif H F

EAST TENNESSEE STATE UNIVERSITY

A Software Monitor for the Ingres Database
Management System
(1988) / Li X

Inlink: A Natural Language Interface for
Multiple Databases
(1988) / Martin D E

A Course in the Organization and Design of
Computer Programming Languages
(1988) / Tidwell R L

EAST TEXAS STATE UNIVERSITY

User Interfaces and User Interface
Management Systems: A Survey
(1988) / Burks M W

IBM Series/1 Minicomputer and IBM 5985 Color
Display Control Unit Emulation Software
(1988) / Chen S H

EMORY UNIVERSITY

On the Theory and Applications of the
Network Simplex Method
(1987) / Bennacer D

A Further Investigation Into the Causes
of Pathology in Game Trees
(1987) / Jerkunica B M

A System for Tool Based Solutions to
Application Problems
(1987) / Kim H R

Pairwise Orthogonal Orthomophisms of
Groups: A Complete Classification for
Groups of Order 12
(1987) / Lee J H

Estimation of a Constant in the Theory
of 3-Dimensional Random Orders
(1988) / Silverstein R S

FLORIDA ATLANTIC UNIVERSITY

Performance Analysis of Signaling System
Number 7 for ISDN Applications
(1988) / Eskamaei M

An Efficient Test Strategy for
Microprogrammable Minicomputers
(1988) / Franklin W A

FLORIDA ATLANTIC UNIVERSITY
(continued)

Delay Characteristics of File Transfer in
Token Ring Local Area Networks
(1988) / Horvath E I

HORN: An ECAD Database Structure
(1988) / Moster J A

Discrete Signal Representation Using
Triangular Functions
(1988) / Nallur P

Performance Analysis of Token-Passing Local
Area Networks
(1988) / Sadiku M N O

Flow Control in Interconnected Token Ring
Local Area Networks
(1988) / Varughese S B

FLORIDA STATE UNIVERSITY

An Entity-Relationship Approach to Grammar
Inference
(1988) / Biswas P K

Multiprocessor Implementation of Multidomain
Spectral Methods
(1988) / Druding D A

Arsbt, An Approximate Reasoning System
Building Tool
(1988) / Feng J

An Advisory Expert System for Supercomputer
Job Control Language
(1988) / Johnson K L

Applicative Frame-Based Languages
(1988) / Liao H S

A Perspective on Designing Knowledge-Based
Acceptance Tests for the Recovery Block Method
(1988) / Park S

Adaptation of the Isodata Clustering Algorithm
for Vector Supercomputer Execution
(1988) / Schow P H

A Design for a Diagnostic Expert System in the
Field of Education
(1988) / Szabo S

Semantic Networks for Solving Arithmetic Word
Problems: A Prototype Problem Solver
(1988) / Traynham K R

FLORIDA, UNIVERSITY OF

Comparative Evaluation of Expert System
Shells Currently on the Market with a
Correlation of Relationships Between Expert
System Shell Characteristics and Application
Topic Requirements
(1988) / Alexander J

A Task Planner for Discrete Assembly
(1988) / Ayoub R

PCxpert: An Expert System for Book
Preservation and Conservation
(1988) / Baati J

An Automated Control Structure for
Preliminary Mechanical Design
(1988) / Clinton J

A Methodology for Scenario-Based
Requirements Exploration
(1988) / Holbrook H

A Language Flow and Resonance Model of
Cognitive Processing
(1988) / Hutches D J

An Interactive Curve Generation Package
Utilizing the Programmer's Hierarchical
Interactive Graphics System
(1988) / Krell L C

Conceptual Data Management in the Integrated
Manufacturing Data Administration System
(1988) / Krishnamurthy V

Object Manipulation in the Object-Oriented
Semantic Association Model (OSAM)
(1988) / Kumar A

Natural Language Structure
(1988) / Miller V

Extending Software Project Information
Management Across the Development Phase
Boundary
(1988) / Mirkis A

Computer Simulations Used as Courtroom
Evidence
(1988) / Morie J

Molecular Dynamics Simulation of Glass
Structure
(1988) / O'Rear G

A Deterministic Conceptual Analyzer with
Look Behind
(1988) / Reed J H

Assigning Abstract Data Types to Real-World
Objects in RML
(1988) / Sarver T

The Design and Implementation of a Semantic
Query Processor in OSAM
(1988) / Soo C

GEORGIA, UNIVERSITY OF

A Parallel Implementation of Defeasible
Reasoning
(1988) / Gonsher R

An Implementation of an Extension to
Discourse Representation Theory:
Translating Natural Language to Discourse
Representation Structures to Prolog Clauses
(1988) / Goodman D M

Parsing German with GPSG: The Problem of
Seperable-Prefix Verbs
(1988) / Volk M

GUELPH, UNIVERSITY OF

An Analysis of the Economic Efficiency of
an Anaerobic Digestion System for a Hog
Operation
(1988) / Kelland D L

An Intelligent Interface for Research in
Dolphin Cognition
(1988) / Langton K B

A Dynamic Bandwidth Allocation Technique
for a Wide Area Using Fiber Optics
(1988) / Pell H L

GUELPH, UNIVERSITY OF
(continued)

Automated Synthesis of Digital Systems from
a Behavioral Description
(1988) / Shailesh S

Spatial Reasoning: A New Approach to
Global Routing
(1988) / Shun C B

Integrated Data Base Management Tools for
VLSI Design
(1988) / Yip L C

IDAHO, UNIVERSITY OF

A Study of Business Data Processing Form
1950 Through 1987 Encompassing Both
Mainframe and pc Environments with Special
Section Evaluating Computer-Aided Software
Engineering (Case) Tools Using the
Questionnaire Method
(1988) / Adams J P

Variations on Cooperative Computation and
Learning in Boltzmann Machines
(1987) / Barga R S

ILLINOIS INSTITUTE OF TECHNOLOGY

Integrity: From Data Dictionary to
Blackboard
(1988) / Lu J C

Noun Entries for a Relational Lexicon
(1988) / Van Berkum C

IOWA STATE UNIVERSITY

An Experimental Evaluation of Expression
Complexity Measures
(1988) / Dean P L

Program Parallelization and Architecture
Derivation for Systolic Implementations
(1988) / Gannett E W

Design of a High Level Tree Structured
Memory
(1988) / Lockard J R

Constructing the Minimization Diagram of a
Two-Parameter Problem
(1988) / Srinivasan S

Minimizing Overhead in the Dynamic Processor
Self-Scheduling Approach
(1988) / Wikstrom M C

JOHNS HOPKINS UNIVERSITY

Optical Constants of Infrared Active Phonons
as a Function of Frequency and Temperature
(1989) / Hoffmann C E

KANSAS, UNIVERSITY OF

Representation of Synchronization Primitives
and Concurrency Algorithms/Systems in a
Concurrency Method
(1988) / Brown T L

Random Number Generation on
Microcomputers: An Examination of
Various Algorithms
(1988) / Burger K R

On the Performance Tradeoffs of Locking
for Database Concurrency Control
(1988) / Chui P T

An Implementation of an Algorithm to Learn
from Examples, as an Aid to Design Expert
Systems
(1988) / Dean J S

Portable Asynchronous X.25 Protocol and
Bridge Design
(1988) / Lee K L

A New Paradigm for the Support of Updates to
View Extensions in a Relational Database
Management System
(1988) / Scherer E J

The G Graphics Programming Language
(1988) / Shelton J A

The Implementation of the Novice's Algorithm
Teacher (NAT): An Algorithm Development
Tool
(1988) / Smith S W

An Evidential Reasoning System
(1988) / Zarley D K

KENTUCKY, UNIVERSITY OF

Parallel Evaluation of Logical Expressions
(1988) / Dawson S E

A Comparison Between Sequential and Parallel
Implementations of a Deadlock Avoidance
Algorithm
(1988) / Druggan T

A Computer Solution to the Jigsaw Puzzle
Problem
(1988) / Easwar S

A Fast Architecture for Rule-Based Systems
(1988) / Griffen N

H(1): An Approach to Object-Oriented
Operating Systems
(1988) / Herrin E H

A Comparison of Some Basic Vertex Coloring
Algorithms as Adapted for Timetable
Construction and a Marketed Timetable
Producing Package
(1988) / Jones G D

A Biomechanical Simulation of Springboard
Diving
(1988) / Keller R A

A Fast PC Coprocessor for Compute Problems
(1988) / Miller W M

Remux: Remote Execution on Idle
Workstations
(1988) / Seshadri K

A Stream Based Terminal Server
(1988) / Shelton T S

LAMAR UNIVERSITY

Optimized String Storage and Retrieval
(1988) / Chen P M

The Design and Analysis of an Inference
Engine for the Expert System Shells
(1988) / Chen T A

LAMAR UNIVERSITY
(continued)

A Burn Schedule Generator for Hazardous
Waste Incineration
(1988) / Chiang M C

An Articulate Expert for Knowledge-Based
Tutoring of Complex Distance Problems
(1988) / Escamilla T D

A Performance Profile of IBM VSAM
(1988) / Greensfelder C A

A Design for the Automation of Production
Management in the Lamar University Computer
Center
(1988) / Harwood C W

The Complete Tokenizing of Storage
Records: The Production of a Concordance
(1988) / Hwang T

An Athletic Recruiting Information System
(1988) / Lu H R

Code Generators: A Comparison for Quality
Using Software Metrics
(1988) / Miller W

An Experiment in PC Based Optical Character
Recognition
(1988) / Sappington S B

Hazardous Waste Inventory Management System
(1988) / Stoudemayer L S

Joint Time-Frequency Distributions and Their
Application to Speech Signals
(1988) / Weng C Y

Logical Design of the Relational Structures
of the Database
(1988) / Zigiaris S

LEHIGH UNIVERSITY

Practical Issues for Expert Systems
(1987) / Koperna C A

Error-Correcting Codes in Byte-Organized
Memory Systems
(1987) / Park C S

LOUISVILLE, UNIVERSITY OF

Quantification of Microcirculatory Leakage Using
Computer Image Processing Techniques
(1988) / Banerji D

The Quantity and Quality of Information Flow in
Decision Systems
(1988) / Chang S K

Medical Expert System for Diagnosing Low Back Pain
(1988) / Espinosa A

Generalized Inverses of Large Matrices Using the
Schur Complement
(1988) / Frank N L

A Study of Software Psychology Metrics for Data
Management in User Interface Design
(1988) / Heineman, M H

A PC-Based Image Processing Workstation
(1988) / Liu H Y

A "Harmonium" Neural Network Implementation for
Interfacing With Expert Systems
(1988) / Magruder J K

Variables Associated with Change in Glasgow
Coma Scale
(1988) / Shroyer A H

Development of a Hospital Based Computer Support
Center
(1988) / Stearman J F

Design of a User Interface Management System
(1988) / Tang C

A Structured Methodology for Database Design
(1988) / Wang D T

Design and Implementation of a Cross Assembler
for the Phalanx System
(1988) / Zhang Y B

LOWELL, UNIVERSITY OF

Human Factors Implications of Text Display
Rates for Computer-Assisted Instruction
(1988) / Smaldone D J

MANITOBA, UNIVERSITY OF

The Cryptocomplexity of Four Difficult
Problems
(1988) / O'Connor L J

Multiprocessor System Design Validation
Using a Temporal Deductive System
(1988) / Yurkowski P M

MARQUETTE UNIVERSITY

Text Processing
(1988) / Chang Y

Static Analysis of a Vent Tube System with
the ANSYS Computer Program
(1988) / Chuang Y H

RFS Remote File Sharing
(1988) / Debruin R

G-Stop Facility in ATOMFT: User Manual
(1988) / Deshpande P

The Three Phase Commit Protocal and Its
Termination Protocals
(1988) / Ho Y S

Time Series and Dynamic Data System
Identification for the Computer-Aided
Aids Modeling and Analysis
(1988) / Sun G

Personal Computer Based Scheduling Software
and Its Application in a Multiphased
Construction Project
(1988) / Wade M

MASSACHUSETTS INSTITUTE OF TECHNOLOGY

A Network Element Based Fault-Tolerant
Processor
(1988) / Abler T A

Factory Development of Personal Computer
Software
(1988) / BenDaniel M M

MASSACHUSETTS INSTITUTE OF TECHNOLOGY
(continued)

Developing a Conceptual Design System Using
a Network-Oriented Knowledge Base
(1988) / Bichara A R

Development of Software Vertical Markets
(1988) / Binday M L

An Analysis of Factors that Influence
Management of Information Technology in
Multi-Divisional Companies
(1988) / Buzzard S H

Information Technology: Attitudes and
Implementation
(1988) / Carey D R

A Resampling Method for Reconstructing
Signals from Non-Uniformly Spaced Samples
(1988) / Cheng L

Upgrading a Retrieval Assistance System to
a Workstation/Windowing Environment
(1988) / Chong H F

Rapid Change in the Personal Computer
Market: A Quality-Adjusted Hedonic Price
Index, 1976-1987
(1988) / Cohen J M

An Analysis of Strategic Practices in a
Computer Services Company
(1988) / Corvi C

Graphical Editing of Composite Bezier
Curves
(1988) / Eisnman J A

An Analysis of Residential Demand for
Broadband Telecommunication Service
(1988) / Elkington H S

Technical Competitive Advantage in the
Software Industry
(1988) / Feld B A

Uncertain Reasoning for Improved Situation
Assessment
(1988) / Gans S S

Implementing Artificial Neural Networks in
Integrated Circuitry: A Design Proposal
for Back-Propagation
(1988) / Gilbert S L

Canadian Energy Demand after the Oil Shocks
(1988) / Greenberg P E

A 4.8 Kbps Multi-Band Excitation Speech
Coder
(1988) / Hardwick J C

Software Productivity
(1988) / Hennessy E J

A High-Level Signal Processing Programming
Language
(1988) / Hicks J E

Speech Recognition with a Perceptually
Based Front End
(1988) / Hopkins G A

Extracting Parallelism from Sequential
Programs
(1988) / Hsieh W C

Signal Representation and Reconstruction
from Sign Data
(1988) / Huang T T

Composing Process and Data Descriptions in
the Design of Software Systems
(1988) / Jackson D N

KOLA: Knowledge Organization Language
(1988) / Jang Y

Competitive Advantage in Personal Computer
Software Architecture
(1988) / Kishimoto M

Computational Origami: A Geometric Approach
to Regular Multiprogramming
(1988) / Lu H

Studies of Correlation Functions in Random
Medium Theory
(1988) / Nghiem S V

The Impact of Electronic Mail and Computer
Conferencing: A Case Study
(1988) / Phillips M W

Pseudo-Random Quantization for the
Minimization of Mean Square Bias and
Variance
(1988) / Preisig J C

MULTILOGO: A Study of Children and
Concurrent Programming
(1988) / Resnick M J

Performance Learning in Model-Based
Diagnosis
(1988) / Resnick P J

A Graphical Interface for Cooperative
Problem Solving
(1988) / Rohall S L

Distribution Strategies for PC Software
(1988) / Schuyler C

Using Goals and Strategies to Organize
Control Knowledge for Algebraic Manipulation
(1988) / Shukla R

Maestro - An RTLET Compatible Music Editor
and Musical Idea Processor
(1988) / Shyu R Y

Codec Characterization Using Digital Signal
Processing Techniques
(1988) / Smith S P

Design of a Network for Concurrent Message
Passing Systems
(1988) / Song P Y

The Potential Pitfalls in the
Conceptualization, Development, and
Implementation of Strategic Information
Systems
(1988) / Sosa G L

Transient Analysis of Some Open Queueing
Systems
(1988) / Stamoulis G D

Temporal Order Discrimination Using a
Multidimensional Tactile Display
(1988) / Stephan D A

A Relational Join Procedure as an Addition to SAS,
The Statistical Analysis System
(1988) / Wyatt S A

MISSOURI, UNIVERSITY OF (ROLLA)

Simulation of the IEEE 802.4 Token Bus
Access Protocol with Network II.5
(1988) / McIntosh D S

Intensity Blending of Computer Image
Generation-Based Displays
(1988) / Reidelberger E S

Deduction of a Functional Dependency from
a Set of Functional Dependencies
(1988) / Richardson J M

MONTANA STATE UNIVERSITY

ALCAD: A Computer Aided Design System for
Log House Construction
(1987) / Babcock R S

Universal Coding with Different Modelers in
Data Compression
(1987) / Kwan R C

Efficiency Measurements Between the Third
and Fourth Normal Forms of Database Schemes
(1987) / Shimura H

MONTANA, UNIVERSITY OF

Descriptive Study of Computer Spreadsheet
Applications for Grain Merchandising
(1988) / Orgas M C

A Comparison of the Petri Nets Model and
the Hoare Processes Model
(1988) / Shahpar B

MONTREAL, UNIVERSITY OF

L'Optimisation du Choix des Equipements de
Commutation Dans un Reseau Telephonique
Local
(1988) / Idrissi Kaitouni K

MURRAY STATE UNIVERSITY

The Characteristics of a Successful
Keyboarding Program
(1988) / Davie J K

Modeling and Analysis of the Distribution of
the Disease and Crime in the Commonwealth of
Kentucky Using a Georeferenced Information
System
(1988) / Li M

NEBRASKA, UNIVERSITY OF

A New Graph-Based Fault Simulation Algorithm
for Combinational Circuits
(1988) / Ke W

A Simulation of Parallel Processing
(1988) / Kess B L

Bi-Pushdown Automata
(1988) / Mei J

A Parallel Algorithm for Single Row Routing
(1988) / Roy A

The Data-Driven Execution Model for Parlog
(1988) / Tang C

Image Compression in the Color Domain
(1988) / Unverferth M J

NEW BRUNSWICK, UNIVERSITY OF

An Integrated Cross Software System
(1988) / Allain S E

Urban Forest Management Support System
Design
(1988) / Cline R M

Numerical Methods for the Solution of a
Second Order Fredholm Integro-Differential
Equation
(1988) / Gilmore K A

Fractals and Their Application
(1988) / Kalra P K

A Prototype PC-Based Client Information
System for the Alcoholism and Drug
Dependency Commission of New Brunswick
(1988) / Keilty R H

On the Complexity of Edge Domination
(1988) / Koilakos K

A File Transfer Protocol for VSPC
(1988) / Mitchell S V

Performance Modeling of Vectorized Monte
Carlo Codes
(1988) / Sarno R

Development of a Laboratory Base for a
Computer Operating Systems Course
(1988) / Velensky L S

Expert System for Diagnosing Resistance
Temperature Detectors
(1988) / Ward K

Real Time Process Control in a
Multiprocessor Distributed Environment
(1988) / Weeks J K

Interactive Generation of Interpolated
Objects
(1988) / Wu L

NEW JERSEY INSTITUTE OF TECHNOLOGY

Multiple-Screen Panning System
(1988) / Youssef A F

NEW MEXICO INSTITUTE OF MINING & TECH.

Recursive Query Processing Strategy
(1987) / Bhave N S

Deferred Rebalancing in AVL Trees
(1987) / Edwards D

Random Pattern Testability of FET Stuck-Open
Faults in CMOS Combinatorial Logic Circuits
(1988) / Perez R T

NEW MEXICO STATE UNIVERSITY

Information Retrieval Systems Using
Pathfinder Networks
(1988) / Fowler R H

Artificial Communicators: An Operating
System Consultant
(1988) / Mc Kevitt P

NEW MEXICO STATE UNIVERSITY
(continued)

Evaluation of a Two-Dimensional Finite
Element Program for Microcomputers
(1988) / Nungaray-Perez G E

Hycon: A Parallel Connectionist Simulator
(1988) / Plate T A

Developing a Marketing Guide for El Paso
Area Software Developers
(1988) / Royal K S

A Rapid Interface Prototyping Tool for the
IBM PC (RIPen)
(1988) / Vandenberg P D

Innovation and Continuity in On-Going
Software Documentation
(1988) / Warren K L

NEW YORK, STATE UNIVERSITY OF (BUFFALO)

Neuron Modeling Using Computer Architectural
Concepts
(1988) / Bauer S

NORTH CAROLINA STATE UNIVERSITY

An Intelligent Deductive Relational Database
System
(1988) / Cline S W

Graphics Algorithms for Parallel
Architectures
(1988) / Nagaraj V

Bounds on the Improvement of Restoration
Using Spatial a Priori Information
(1988) / Vora P L

NORTH DAKOTA STATE UNIVERSITY

The Benefit of Composite Semijoin
(1987) / Chen C S

"Cross Link" a Multilingual Work Processor
(1987) / Hakimzadeh H

A Local Area Network Query Processing Algorithm
Using Bit Vectors
(1988) / Reeves R L

A Request Ordering Concurrency Control Protocol
for Databases
(1988) / Richter R

KARDS, A Knowledge Representation System
(1988) / Thangiah S R

Vehicle Routing Algorithms Based on Spacefilling
Curves
(1988) / Walker R S

No Token Ring Protocol for Local Area Networks
(1988) / Yang C H

NOVA SCOTIA, TECH UNIVERSITY OF

Microcomputer Control of Metal Fatigue
Testing Systems
(1988) / Campbell D

OKLAHOMA, UNIVERSITY OF

Implementation of an Ungerboeck-Coded
Modulation Scheme on the TMS32010 Signal
Processor
(1988) / Ahmed S A

A Study and Comparison of the Fault
Detection Procedures for Semiconductor
Random Access Memories
(1988) / Ali I

Performance of Two Channel Reservation
Random Access in Local Networks
(1988) / Barghi H

Numerical Grid Generation and the Conjugate
Gradient Algorithm
(1988) / Bradley J A

The Recursive Linear Technique to the Load
Flow Solution
(1988) / Chang K K

The Impact and Control of Phase Shifters on
the Transmission Grid
(1988) / Craddock C E

A Sample Implementation of a Digital Logic
Design in a Large Scale Integration CMOS
Standard Cell Application Specific
Integrated Circuit
(1988) / Cramer K J F

Step and Repeat Control System Using
Microprocessor
(1988) / Haneef M

Short-Term Thermal Unit Commitment, a
Sequential Decision Approach
(1988) / Huang C

Analysis of Token Ring Protocols for
Integrated Voice/Data Transmission
(1988) / Huang S

Using Address Space Emulation to Enhance an
In-Circuit Test
(1988) / Ikelberry K D

A Microcontroller Based Technique for
Controlling the Speed and Phase of Small
Disk Drive Spindle Motors
(1988) / Irwin M J

Analysis of Fault Isolated and Repair Using
an Expert System
(1988) / Lambert J R

Development of a Computer Algorithm to
Recommend Overcurrent Device Selection and
Achieve Coordination on Distribution
Circuits
(1988) / Lankford C B

Comparison of Probabilistic Production Cost
Simulation Methods
(1988) / Lin M

Algorithms for Two-Layer River Routing
(1988) / Lin S

Algorithms for Shape Detection and
Recognition
(1988) / Nam R

Simulation of an Ethernet-Like Network
Interface Unit
(1988) / Nguyen H T

RENSSELAER POLYTECHNIC INSTITUTE
(continued)

Image Sequence Coding
(1988) / Bhargava S

A Methodology for the Use of Ada Software
in Embedded Real Time Applications
(1988) / Brennan T E

WINGRES: An INGRES-Based Database
Management System for the WARP Machine
(1988) / Burke D P

A C-Based Discrete Event Simulator for
Modeling Selected Local Area Network
Protocols
(1988) / Cannon M

Navagational Aids and Learning Styles:
Structuring Optimal Training for Computer
Users
(1988) / Cohan L A

Pattern Matching in Graphs and Hypergraphs
(1988) / Comly L T

Boundary Evaluation of Solid Models:
Algorithm Study and Implementation
(1988) / Crocker G A

An Automated Halstead Complexity Metrics
Analyzer for Quantitative Software
Assessment
(1988) / Doll J S

VISAGE: A Graphical Silicon Compilation
Environment
(1988) / Dragomirecky M

Hobbes's Social Contract Theory as
Implemented by a Computer
(1988) / Dunne J M

Implementation of Speaker Verification
Algorithms on the TMS32020
(1988) / Elser M G

The Effects of Menu Structure,
Categorization Scheme, and Practice on
Performance with a Menu-Driven Information
System
(1988) / Forster T A

The Interconnection of Local Area Networks
Using Integrated Services Digital Networks
(1988) / Gallagher M D

Tools for Particle Based Geometric Modeling
(1988) / Hersh J S

Digital Signal Processor for a High
Resolution Ultrasonic Imaging System
(1988) / Hijazi N A

Modeling of Mosis 3-Micron CMOS Cell Family
for the FRISC Design Environment
(1988) / Hua H K

Study of Continuous Phase Modulated Schemes
(1988) / Hulyalkar S N

A Model-Driven Approach to 3D Industrial
Visual Inspection
(1988) / Jaenicke R A

Network Delay Measurements on Ethernet
(1988) / Jolly C H

A PHIGS-Based Interactive Tutorial for
Advanced 3D Graphics
(1988) / Kader S E

The Uniform Grid Technique for Fast Line
Intersection on Parallel Machines
(1988) / Kankanhalli M

Real Time Control of Multilevel
Manufacturing Systems Using Colored
Petri Nets
(1988) / Kasturia E

Introduction of Variable Scoping and
Auxiliary Tree Functionality to a State
Tree-Based User Interface Management System
(1988) / Larson D J

An Approach to Automatic Coarse
Three-Dimensional Finite Element Mesh
Generation
(1988) / Lo J A

A Research Effort to Solve the Problem of
Software Loading of Hardware
(1988) / Majid N

Acceleration and Torque Feedback for Robotic
Control: Experimental and Theoretical
Results
(1988) / McInroy J E

Self-Test in Random Access Memories
(1988) / Merrill M C

Effects of Programming Expertise on the
Organization of Task Information
(1988) / Mikaelian D M

A Rapid Prototype for Computer-Aided
Control Engineering for Integrated
Aircraft Control Systems
(1988) / Mroz P A

A High-Level Description of Digital Terrain
Models Using Visibility Information
(1988) / Nair H

EPL - Equational Programming Language Code
Generation
(1988) / Nolan M A

Implementation Issues of Direct
Multivariable Model Reference Adaptive
Control
(1988) / Perkins B J

Design and Verification of a Butterfly
Processor for a Two Dimensional Fast Fourier
Transform Processor Suitable for Wafer Scale
Integration
(1988) / Philhower R A

Data Acquisition and Image Processing
System for the Focused Ion Beam
(1988) / Ramelson B M

Object-Oriented Software Interfacing: A
Generalized Diagnostic Platform
(1988) / Renze M B

A Path Recovery Scheme for the Truncated
Viterbi Algorithm
(1988) / Roethlisberger U

Issues in Distributed Heterogeneous
Databases
(1988) / Rudrakshi C

RENSSELAER POLYTECHNIC INSTITUTE
(continued)

Implementation of a Client for the Simple
Gateway Monitoring Protocol in a Unix
Environment
(1988) / Shikarpur U

Design and Analysis of Two New
Interconnection Networks
(1988) / Singh J

Extraction of a Single Voice from a Channel
Containing Multiple Simultaneous Speakers
(1988) / Sorensen J

Non-Uniform Rational B-Spline Curves in the
Programmer's Hierarchical Interactive
Graphics System
(1988) / Stevens B M

Algorithm Animation and Its Applications to
Computer-Aided Instruction
(1988) / Swaminathan R

A Study of Communication Overhead in an
Asynchronous Parallel Processing System
(1988) / Tan E

Parallel Processor Architectures for Image
Processing
(1988) / Tannenbaum D C

Applications of Expert Systems Technology on
the Development of DEX Hardware Diagnostics
Tools
(1988) / Thomas C A

Characteristics of Visibility Models
(1988) / Tserkezou P

Image Generation with an Ultrasonic Digital
Holographic System
(1988) / Van Etten J S

An Expert Systems Approach to the Legally
Oriented Forensic Tracking System
(1988) / Vijayaraghavan R V

RHODE ISLAND, UNIVERSITY OF

A Sketching Algorithm for Planar Mechanism Using
An IBM Personal Computer and Prime Medusa Concepts
(1988) / Cheng J

Analysis and Implementation of Karr's Algorithms
(1988) / Desai R

Design of a Domain Data Type Subsystem for a
Relational Data Base Management System
(1988) / Lee E M

An Interactive Computer Model for Multistage
Decision Making Under Uncertainty
(1988) / Liang K K

Decision Support Data Base Selection Expert
System for the Selection of Microcomputer
Data Base Software
(1988) / Nigam P

Ekress: An Expert System Shell for Learning
(1988) / Reynolds B

A Frame-Based Conceptual Analyzer for Natural
Language Processing
(1988) / Shen S J

Cela: An Interactive Graphic Editor and Analyzer
for the Design of Digital Combinational Circuits
(1988) / Sun V

Development of an Efficient Browsing Capability
for 2-D Image Data
(1988) / Tsai Y E

SAN DIEGO STATE UNIVERSITY

Influences in the Adoption of Home Computers
(1987) / Jacobs J H

Integrated Serviced Digital Network: A
Forecast of the Requirements of U.S.
Organizations
(1988) / Phillips A M

Temporal Information Processing
(1988) / Priebe C E

SAN FRANCISCO STATE UNIVERSITY

An Analysis of Instruction Set Usage of the
Intel 8086/88 Microprocessor
(1988) / Adams T

A Dynamic Authorization Subsystem Upon a
Relational DBMS
(1988) / Cheng E C

Hierarchial Bit Transposed Files
(1988) / Chengrian P K

HB Engine, A Microcomputer-Based Database
Engine
(1988) / Chou C

DBMS Security Through Encryption
(1988) / Falk A

Concurrency Control Subsystem for a
Relational Database System
(1988) / Lau F

Query Optimizer for Relational Database
System
(1988) / Lin L Y

Comparison of Schema-Based and Connection
Based Knowledge Representation Systems
(1988) / Lorenat M

Use of a Distributed Movie-Making System for
Presentation of Fluid Flow Data
(1988) / Robertoon D

An Investigation of the Application of the
Prolog-SQL Interface: PROSQL
(1988) / Silverman G

Principles of Converting dBase Applications
to dBase/SQL
(1988) / Sun Y J

QBF: Query by Forms
(1988) / Wang C C

SASKATCHEWAN, UNIVERSITY OF (SASKATOON)

Computational Colour Stereo Vision
(1988) / Brockelbank D C

Semantically Extending the LR Parser
(1988) / Dean T R

Problem Solving by Analogy in Novice
Programming
(1988) / Escott J A

SASKATCHEWAN, UNIVERSITY OF (SASKATOON)
(continued)

Array-Based Parallelism for Fine-Grain SIMD
Architectures
(1988) / Gammo L D

Finding Language Errors and Program
Equivalence in an Automated Programming
Advisor
(1988) / Huang X

Log-Tracker: An Integrated System for Human
Body Motion Analysis
(1988) / Long W

Disk Cache Management for Distributed
Systems
(1988) / Makaroff D J

Dynamically Restructuring Disk Space for
Improved File System Performance
(1988) / McDonald M S

Diagnosing Strategy Errors in SCENT
(1988) / Pospisil P R

An RLALR(k) Parser Generator
(1988) / Schmeiser J P

Characterising the Workload of a Distributed
File Saver
(1988) / Tourigny S R

Short-Term Scheduling in Shared Memory
Multiprocessing Systems
(1988) / Wasson B D

SOUTH CAROLINA, UNIVERSITY OF

Live Writer: An Interactive Editor
(1988) / Adams B D

The Career Advisor Expert System for the
Business School at USC
(1988) / Adams M H

Classification of Mangrove Forest Near
Laguna De Terminos, Mexico, Using Landsat
TM Data
(1988) / Boribalburibhand C

Stochastic Neural Networks: Multinomial
Conjunctoid Design and Development Based
on CMOS Technology
(1988) / Chen T

An Investigation of Cellular Automata
(1988) / Dube S

Investigation of Discrete Logarithmic
Encryption Algorithms
(1988) / Gamble R O

The CODASYL Network Database
(1988) / Hsiao L L

Morphological Classification of Color
Texture
(1988) / Huntsberger B A

Controlled Degradation of Resolution of High
Quality Flight Simulator Images for Training
Effectiveness Evaluation
(1988) / Kaip D D

Pascal Teaching Assistant (PTA)
(1988) / Lester C C

COBAS: Computer-Based Authoring System
(1988) / Leu J L

A Background Style-Checker for the Apple
Macintosh
(1988) / Lomas E M

An Intelligent Assistant for "vi"
(1988) / Matthews C T

A 3-D Solid Reconstruction from Serial
Sections
(1988) / Peroni M R

VIEWS: A 3-D Line Drawing Software Package
(1988) / Thakkar T N

A New Classification Method for Literary
Analysis: The Case of Jane Austen
(1988) / Upchurch C M

Parallel Graph Algorithms
(1988) / Vishywanathan S

The Preference Evaluator: A Model for the
Use of Feature Markers and Selectional
Restrictions in Natural Language
Understanding Systems
(1988) / Williamson D R

SOUTH DAKOTA SCHOOL OF MINES AND TECHNOL.

Texture-Based Cloud Classification
(1988) / Chen D W

Implementing an Electronic Mail System Based
on a High Performance Utility Library for
the TI-PC
(1988) / Wang S M

The Tordial Mesh Unit
(1988) / Wyman B

Natural Language Processing
(1988) / Zhang R

SOUTH DAKOTA, UNIVERSITY OF

Modifiable Accounting Application Programs
(1988) / Morrow S R

SOUTH FLORIDA, UNIVERSITY OF

A Simulator for Minimal Processor
Interconnect Strategies
(1988) / Bauer J T

A Comparison of Point-Valued and
Interval-Valued Reasoning Under Uncertainty
(1988) / Cheng P

Exhaust Nozzle Expert System
(1988) / Curtiss N B

An Environment for Object Recognition Using
Fourier Descriptors
(1988) / Eggert D W

Concurrency Control Based on Two-Phase
Locking in Distributed Database Systems
(1988) / Joo Y

Exploring Unknown Environments Using a
Hierarchical Workspace Representation
(1988) / Krause D M

Model-Based Two-Dimensional Industrial
Part Recognition
(1988) / Kung F K

TENNESSEE, UNIVERSITY OF (TULLAHOMA)
(continued)

Design and Implementation of a Postscript
Compatible System of graphics Primitive
Procedures
(1988) / McCormac J

Data Processing for a Fast Scanning
Spectroradiometer
(1988) / Sherrell F G

TEXAS A AND M UNIVERSITY

Parallel Distributed Processing Models Using the
Back-Propagation Rule for Studying Analytic and
Holistic Modes of Processing in Category Learning
(1988) / Bauer N K

Electronic Documentation Issues and Methods for the
Integration of Text and Graphics
(1988) / Berryman D R

An Object-Oriented Construction Kit Architecture
for Designing Man/Machine Interfaces
(1988) / Desoi J F

A Model for the Representation and Extraction of
Visual Knowledge from Illustrated Texts
(1988) / Koons D B

GSS: A User Interface Specification Framework
(1988) / Morris J G

Software Function Allocation Methodology
(1988) / O'Neal M R

Robot Task Planning in a Non-Ideal Blocks World
(1988) / Rangadass V

Encapsulating the Meta-Level Knowledge in
Distributed Artificial Intelligence
(1988) / Underbrink A J

Identity Verification by Keystroke Timing
(1988) / Usnick M C

A Hybrid of Connectionist and Production
System Models for Planning
(1988) / Veezhinathan J

TEXAS WOMEN'S UNIVERSITY

Rotations and Transformations of Conic
Sections and Quadric Surfaces
(1988) / Alves J R

Computer Graphics for Business Presentation
(1988) / Bangpipob S

Computer-Assisted Authoring System for the
Development of Federal Aviation
Administration Directives
(1988) / Hall B L

Designing a Personnel Data Base System with
DBASE III
(1988) / Han J Y

Automated Event Detection in
Electrocardiograms
(1988) / Kao Y G

Designing a Statistical Software System
(1988) / Ko A

An Online Bibliographic Search Package
(1988) / Najmabadi M

The Use of Telecommunications Devices
in the Adult Deaf
(1988) / Pflanx M J

Surrogate Data Input Devices for Seismology
(1988) / Thomas T S

Computer Animation Applications for
Microcomputers
(1988) / Wiser C C

TEXAS, UNIVERSITY OF (ARLINGTON)

A Multi-Planner Approach for Generating
Suspense Stories
(1988) / Atwood J R

The Design and Analysis of a Distributed
Database Management System
(1988) / Bhombal F W

A Fractal-Based Terrain Model Generator
(1988) / Earnest M M

Techniques of Mapping the Algorithms Into
the Higher Radix Hypercube Multiprocessors
(1988) / Gupta P

Query Processing in a Distributed Network
Database System
(1988) / Hoss A M

Toward a Learning Theory for Use in Computer
Tutorials
(1988) / Kern J E

A Systems Perspective to the Development of
a Rule-Based System for Project Management
Consultation
(1988) / Kramme K W

Instruction Scheduling of Pipelined
Processors
(1988) / Lindberg A H

Loop Coalescing and Data Dependency Matrix
Reduction Techniques for Systolic Array
Design
(1988) / Liou T

Problems in Temporal Database Transaction
Management
(1988) / Mak W K

Algorithm Transformation Based Techniques
for the Input/Output Control of Systolic
Array
(1988) / Mei H

A Method for Adaptive Boundary Approximation
of Octree Encoded Volumes by Cubic Surface
Patches
(1988) / Novobilski A J

Design of Flexible Manufacturing Systems
(1988) / Ortiz M C

Software Reliability Modeling: Software
Life Cycle Model
(1988) / Sheldon F T

A Neutral Network Controller for Robotic
Manipulators
(1988) / Smith J O

The ESCRO Realtime Specification Environment
(1988) / Talley T M

An Approach to a Theoretical Design of an
Assumption-Based Truth Maintenance System
for Defeasible Reasoning
(1988) / Wang S B

TEXAS, UNIVERSITY OF (AUSTIN)

Design of Memory Allocators
(1988) / Arni N V

Specification Techniques for Database Data
Models and Database Implementations
(1988) / Baugher M

Some Experiments with Confidence Interval
Generation in Steady-State Simulations
(1988) / Bean M C

Three Dimensional Object Recognition from
Range Data Using Curvature Guided
Extraction of Superquadrics
(1988) / Brady J P

Omnirec: A Character Recognition System
(1988) / Buchowski T J

Threads and Remote System Calls: A Case
Study of Their Utility and Performance
on Unix's RCP
(1988) / Bueche E C

On the Performance of Multiple Bus
Structures Under Non-Uniform Communication
Requirements
(1988) / Chatterjee A

Evaluation and Extension of Modechart: A
Tool for Real-Time System Specification
(1988) / Chellappa M

A Parallel Execution Model of Logic
Programming on IPSC
(1988) / Chen L G

Three-Dimensional Object Creation on the
Macintosh
(1988) / Conyne R B

A Code Generator and Optimizer for the
GRASPL Compiler
(1988) / Davies I R

Node Mapper, an Indexing Structure for the
Kyklos Database Machine
(1988) / Desai A Y

A Multiprocessor for the Solution of
Semiconductor Device Quation Problems
(1988) / Donohoe J L

A Retargetable Real Time Performance Monitor
(1988) / Dubey S R

Vincent: A Split Cell VLSI Editor
(1988) / Geraci B J

On the Applicability of Piecewise Affine
Models in Nonlinear Adaptive Equalization
(1988) / Hardwicke K

Performance Evaluation of Numerous Garbage
Collection Algorithms by Real-Time Simulation
(1988) / Heng S L

Analyze - An Expert Diagnostic System for
High Resolution Diode Laser Spectroscopy
(1988) / Ho T D

Development of a Reduced Time Interval
Partitioned Simulation Algorithm
(1988) / Kang S

Design of the Memory System for METRIC
(1988) / Kapur R

SMARTGEN: An Expert Test Generation
System
(1988) / Karim A

Parallel Heuristic Search on Multiprocessors
(1988) / Kunchitham R

Performance Comparison of Communication
Protocols in Multistage Interconnection
Networks
(1988) / Liano K

High Level Language Timing Analysis of TC2
Processor for Real Time System
(1988) / Lin C L

A Parallel Implementation of the Alpha-Beta
Search Algorithm
(1988) / Martinich L D

P-Graph: A Performance Evaluation and
Software Analysis Tool
(1988) / Matin A

The Theory of $1[1]/L[1]$-Optimal Feedback
Synthesis for Linear Systems
(1988) / Meendlovitz M A

Receiver Equalization
(1988) / Messer D D

A Microprogrammed Processor for a Graphical
State Programming Language
(1988) / Murphy G R

The Graph Display and User Interface for
P-Graph: A Program Analysis and Performance
Prediction Tool
(1988) / Petree M A

The TEGAS Compiler for the Switch Node
Amalgamation Program
(1988) / Poer L G

An Extended Graph Language for Parallel
Computation
(1988) / Ramgopal T T

Design of Reduced Order H(oo) Optimal
Controllers
(1988) / Romig J C

The Efficient Computation of Isovists
(1988) / Rumsey A A

Efficient Area Optimization for Hierarchical
Floorplans
(1988) / Sakhamuri P S

Nonlinear Iterative Techniques for Image
Restoration from Incomplete Spectral Data
(1988) / Sanders K D

Improvement of the Implicit Finite
Difference Code
(1988) / Santiago M

Determining Order Dependencies Among
Highly-Interconnected Persistent Objects
(1988) / Shadowens M B

TEXAS, UNIVERSITY OF (AUSTIN)
(continued)

Evaluation of Fault Simulation Algorithms
(1988) / Shivashankar V

Data Transfer Scheduling
(1988) / Somalwar K K

Exploratory Robot Learning of Sensory-Motor
Integration
(1988) / Spangler W S

GDB-Tool
(1988) / Sun S W Y

Model Building Using Qualitative Process
Theory
(1988) / Vinson T C

An Intelligent Disk-Cache for the Kyklos
Database Machine
(1988) / Virmani V

Quantifying Communication Cost and Failure
Probability for Distributed Systems
Comparison
(1988) / Waldecker B E

Implementing the Travelling Salesman
Problem on INTEL IPSC Parallel Computer
(1988) / Zard Abou Jaoudeh C E

TEXAS, UNIVERSITY OF (EL PASO)

Enchancements to 4.3 BSD Unix Serial Line Interface
(1988) / Almada A

System Editor for a Learning Expert System
(1988) / Chew T K

ANA: An Analogical Reasoning Expert System
(1988) / Chu S C

ALES: A Learning Expert System
(1988) / Goh T H

An Expandable Implementation of a Prolog Interpreter
(1988) / Jilg J M

Expansion and Application of Cubic Prolog
(1988) / Lew W

Updating Extensional Theories
(1988) / Patterson E L

Theorem Prover for Autoepistemic Logic
(1988) / Welborn C R

U.S. NAVAL POSTGRADUATE SCHOOL

Ship-Shore Packet Switched Communications
System and an Application in Hellenic Navy
(1987) / Agapiou E S

Document Generator Software Design that
Supports Turkish Alphabet
(1988) / Akinci M

An Extension to the Multilevel Logic
Simulator for Microcomputers
(1987) / Albuquerque J C

NALCOMIS/OMA (Naval Aviation Logistics
Command Management Information System for
Organizational Maintenance Activities)
Functional Considerations for Automating
Organizational Maintenance Activities
(1988) / Allen R T

Implementation of a Language Translator for
the Computer Aided Prototyping System
(1988) / Altizer C E

CD-ROM (Compact Disc Read Only Memory)
Library of the Future
(1988) / Avedissian H A

Distributed Computer Communications in
Support of Real-Time Visual Simulations
(1988) / Barrow T H

Development of an Instrument for Measuring
and Analyzing Client Satisfaction for Navy
Regional Data Automation Centers
(1988) / Birdwell R J

Performance Analysis of Aloha Networks
Utilizing Mualtiple Signal Power Levels
(1988) / Bochardt R L

Pseudo-Bayesian Stability of CSMA (Carrier
Sense Multiple Access) and CSMA/CD
(Collision Detection) Local Area Networks
(1988) / Boyana M A

A Demonstration of Interfaces Between
Automated Deployment Systems
(1988) / Breidert J E

ADA (Trade Name) as a Paedeuticc Tool for
Abstract Data Types
(1988) / Britnell R N

CD-ROM (Compact Disc Read Only Memory)
Library of the Future
(1988) / Butrym K P

Investigation and Implementation of an
Algorithm for Computing Optimal Search Paths
(1987) / Caldwell J F

Fortran Programs for Aerodynamic Analysis
on the Microvax/2000 CAD CAE Workstation
(1988) / Campbell J A

Accessing a Functional Database Via
CODASYL-DML Transactions
(1987) / Coker H

Implementation of a Compiler for the
Functional Programming Language PHI
(1987) / Cole E J

Naval Computer-Based Instruction: Cost,
Implementation and Effectiveness Issues
(1988) / Coleman D W

Deployment Planning: A Linear Programming
Model with Variable Reduction
(1987) / Collier K S

Implementation of a Compiler for the
Functional Programming Language PHI
(1987) / Connell J E

The Stylist: A Pascal Program for Analyzing
Prose Style
(1987) / Cool T C

Computer Aided Instructional Course
Authoring: An Examination of the
Pre-Release Version of Maestro in Developing
a Course in Nautical Rules of the Road
(1987) / Crabble C W

Microcomputer Software Support for Classes
in Aircraft Conceptual Design
(1987) / Cramer M C

U.S. NAVAL POSTGRADUATE SCHOOL
(continued)

Administrative (ZYB) Message Processing: A
Simulation and Analysis of Implementation
Strategies
(1988) / Cruz P A

Fault Diagnosis in Distributed Computer
Networks
(1987) / Dincer I

The Development of the Administrative
Sciences Personal Computer Network Tutorial
(1987) / Dyer G D

Long Haul Communications in the HF Spectrum
Utilizing High Speed Modems
(1988) / Ellis R H

Analysis of Existing Advanced Data Models
and Their Applicability as a Model for a
Multimedia Database Management System
(1988) / Enbody D M

Dynamic Positioning at Sea Using the Global
Positioning System
(1987) / Ezequiel A M

ADAMEASURE: An ADA (Trade Name) Software
Metric
(1987) / Fairbanks K S

UNIX Based Programming Tools for Locally
Distributed Network Applications
(1987) / Frank W C

A Conceptual Design of a Software Base
Management System for the Computer Aided
Prototyping System
(1988) / Galik D

Using Computer-Aided Software Engineering
(CASE) Tools to Document the Current Logical
Model of a System for DOD Requirements
Specifications
(1987) / Ganzer D A

Towards a Solution to the Proper Integration
of a Logic Programming System and a Large
Knowledge Based Management System
(1987) / Gorman J P

A Survey of Object Oriented Languages in
Programming Environments
(1987) / Haakonsen H

An Analysis of Acquisition Strategy
Planning for Major Navy Information Systems
(1987) / Haima J D

Design of an Addressable Memory Controller
(1987) / Ham B W

Automated Design of a Microprogrammed
Controller for a Fiinite State Machine
(1988) / Harmon J E

Design, Implementation, and Evaluation of a
Virtual Shared Memory System in a
Multi-Transputer Network
(1987) / Hart S J

Design and Development of a User Interface
for the Dynamic Model of Software Project
Management
(1988) / Haury C E

A Practical Application of Petri Nets in the
Software Safety Analysis of a Real-Time
Military System
(1987) / Hayward D L

A Multibody Dynamic Analysis of the N-ROSS
(Navy Remote Ocean Sensing System) Satellite
Rotating Flexible Reflector Using Kane's
Method
(1987) / Heffernan N F

Metor-Burst Communications: Is This What
the Navy Needs?
(1987) / Helweg G A

Application of Speech Recognition to the
Integrated Tactical Decision Aid (ITDA)
(1988) / Hill J K

Facility Design Validation of the
Information Systems Laboratory
(1987) / Holland T A

Temporal Data, Temporal Data Models,
Temporal Data Languages and Temporal
Database Systems
(1988) / Hom D D

An Examination of the Advanced
Communications Technology Satellite (ACTS)
and Its Application to the Defense Data
Network (DDN)
(1987) / Horner S C

Computer Assisted Instruction: Two Decades
in Perspective
(1987) / Hoskins T J

A Computer Program Package for Introductory
One-Dimensional Digital Signal Processing
Applications
(1988) / Hudik F E

SNAP/DDN (Shipboard Non-Tactical ADP
Program/Defense Data Network) Interface for
Information Exchange
(1988) / Hunt R W

A Static Scheduler for the Computer Aided
Prototyping System: An Implementation Guide
(1988) / Janson D M

An Evaluation of Automating Carrier Air
Traffic Control Center (CATCC) Status Boards
Utilizing Voice Recognition Input
(1988) / Jensen R D

A Computer Study of Air Defense Gun
Effectiveness
(1987) / Jung H D

Frequency Hopping with Analog Messages
(1988) / Kardisan S

The Formal Specification of Computer Systems
(1987) / Karrasch K

PC Software for the Teaching of Digital
Signal Processing
(1988) / Katzir Y

Benchmarking Preparation for and Aggregate
and Sorting Retrievals in the Multi-Backend
Database System
(1987) / Kelbe F E

U.S. NAVAL POSTGRADUATE SCHOOL
(continued)

A Simulation Study of Estimates of a First
Passage Time Distribution for a Semi-Markov
Process
(1987) / Kim S W

A Computer-Aided Instruction Program for
Teaching the TOPS20-MM Facility on the
DDN (Defense Data Network)
(1988) / Kim T W

An Abstract Interactive Graphics Interface
for the IBM/PC and Macintosh
(1988) / Ko-Hsin L

Task-Oriented, Naturally Elicited Speech
(TONE) Database for the Force Requirements
Expert System, Hawaii (FRESH)
(1988) / Larson V M

The Proposed Naval Postgraduate School
Campus Network: Computer Communications
for the 1990's
(1988) / Leahy K M

Optical Laser Technology, Specifically
CD-ROM (Compact Disc - Read Only Memory) and
Its Application to the Storage and Retrieval
of Information
(1987) / Lind D J

A Feasibility Study Using Chinese Speech as
a Command/Control Tool for Computer Systems
(1987) / Liu I K

Image Database Management in a Multimedia
System
(1988) / Lum V Y

Local Geoid Determination Using the Global
Positioning System
(1988) / Ma W M

The Near Real Time Information System
(1988) / Mahon F G

Benchmarking Preparation for and Aggregate
and Sorting Retrievals in the Multi-Backend
Database System
(1987) / Majors D S

Interconnecting Different Types of Local
Area Computer Networks
(1988) / Malkie D C

Image Database Management in a Multimedia
System
(1988) / Meyer-Wegener K

A Language Translator for a Computer Aided
Rapid Prototyping System
(1988) / Moefitt C R

Microcomputer Program Design Considerations
for the Novice User
(1987) / Moore D C

Development of an Instrument for Measuring
and Analyzing Client Satisfaction for Navy
Regional Data Automation Centers
(1988) / Morris P A

ADAMEASURE: An ADA (Trade Name) Software
Metric
(1987) / Nieder J L

A Survey of Automatic Code Generating
Software
(1988) / O'Brien S L

Editfont: An Interactive Font Editing
System
(1987) / Olivera H J

Computer Assisted Instruction: Two Decades
in Perspective
(1987) / Orrell J D

A Proposal for a Microcomputer Based System
to Automate the Marine Corps Crime
Statistics Reporting Program
(1988) / Paquette P E

A Relational Database Management System for
a ROK (Republic of Korea) Army Infantry
Division with Probabilistic Inventory
Control Model
(1987) / Park T

Implementation of a Parallel Multilevel
Secure Process
(1988) / Pratt D R

An Evaluation of Interactive Laboratory
System Software, ILS-PC/DOS, a Digital
Signal Processing Software Package
(1987) / Quintero T A

Software Reusability: A Decision Three
Model
(1988) / Randall W D

A Pad Router for the Monterey Silicon
Compiler
(1988) / Rexach C F

The Suitability of an Object-Oriented
Language for Prototyping and Abstract
Data Types
(1988) / Rowell M D

A Testure Analysis Approach to Computer
Vision for Identification of Roads in Aerial
Photographs
(1987) / Sando J M

Microprocessor Control of a Fast Analog to
Digital Converter for an Underwater Fiber
Optic Data Link
(1988) / Schlechte G L

Prototyping Visual Database Interface by
Object-Oriented Language
(1988) / Schuett R J

An Electromagnetic Compatibility Analysis of
the AN/URC-409 HE Wideband Communication
System as Installed on the LHD-1 Amphibious
Assault Ship
(1988) / Seaver G A

Software Tool Selection for a U. S. Navy
Software Maintenance Organization
(1987) / Sexton J

The Potential Benefits of Using
Teleconference Technology in the Classroom
Environment for U. S. Navy Training Courses
(1988) / Shannon J J

The Design and Testing of an Analog Optical
Communication Link Capable of the
Simultaneous Transmission of Four Frequency
Division Multiplexed Audio Signals
(1987) / Silvers M S

U.S. NAVAL POSTGRADUATE SCHOOL
(continued)

A Demonstration of Interfaces Between
Automated Deployment Systems
(1988) / Smart M J

An Evaluation of Automating Carrier Air
Traffic Control Center (CATCC) Status Boards
Utilizing Voice Recognition Input
(1988) / Spegele J J

A Database Design for a Unit Status
Reporting System
(1987) / Stebbins A J

The Impact of the Defense Data Network on
Naval Communications During the 1980's
(1988) / Stuckey V B

The Use of Coaxial Transmission Line
Elements in Log-Periodic Dipole Arrays
(1987) / Tarleton R E

Design and Implementation of a Debugger for
an Abstract Machine
(1987) / Victrum S

Semantic Shortcomings of Database Management
Systems Based on Relational Model
(1988) / Wall J S

The Near Real Time Information System
(1988) / Wise M R

Image Database Management in a Multimedia
System
(1988) / Wu C T

Implementation of Electronic Mail for
Informal Naval Communications
(1988) / Yelton H M

A Virtual Statistical Mechanical Neural
Computer
(1987) / Yost C P

Distributed Computer Communications in
Support of Real-Time Visual Simulations
(1988) / Yurchak J M

Accessing Hierarchical Databases Via SQL
Transactions in a Multi-Model Database
System
(1987) / Zawis J A

Distributed Computer Communications in
Support of Real-Time Visual Simulations
(1988) / Zyda M J

U.S.A.F. INSTITUTE OF TECHNOLOGY

Software Management Indicators: Managing
the Risk of Project Management
(1988) / Allen C C

Influence of Diagrams: Automated Analysis
with Dynamic Programming
(1988) / Baron C T

Quick Response Airborne Command Post
Communications
(1988) / Blaisdell R L

A Vectorized Hidden-Surface Algorithm
Implemented on the Cray-2 Supercomputer
(1988) / Donehower H R

Telephone Access to Wilford Hall Medical
Center and Its Busiest Appointment Centers
(1988) / Ellingswort M E

Reducing Software Costs Through
Standardization of Design and Configuration
Management
(1988) / Fetch W J

Optimal Server Scheduling to Maintain
Constant Customer Waiting Times
(1988) / Frey T J

A Source Code Analyzer to Predict
Compilation Time for Avionics Software Using
Science Measures
(1988) / Goepper E R

Design and Performance Analysis of a
Relational Replicated Database System
(1988) / Hanson J G

Effects of Voice Coding and Speech Rate on a
Synthetic Speech Display in a Telephone
Information System
(1988) / Herlong D W

Extending AFOTEC Software Evaluation
Guidelines
(1988) / Johnson S K

PHRASE: Instrumentation Software for the
IPSc Hypercube
(1988) / Kahl M A

A TCP/IP Gateway Innterconnecting AX.25
Packet Radio Networks to the Defense Data
Network
(1988) / Lebano T N

A Meta-Protocol Archetecture for Connecting
Computer Networks
(1988) / Lindsey J B

Design and Implementation of a Controller
and a Host Simulator for a Relational
Replicated Database System
(1988) / Lyon L A

Natural Language Generation
(1988) / Maybury M T

Auditory Models for Speech Analysis
(1988) / Maybury M T

A Systems Approach to the Aeromedical
Aircraft Routing Problem Using a Computer
Based Model
(1988) / McLain D R

Extensions to the Multilevel Programming
Problem
(1988) / Moore J T

A Prototypical Implementation of Galahab: A
Conceptual Modeling Language Using the
Object Paradigm
(1988) / Nickson M M

An Improved Data Collection and Processing
System
(1988) / O'Hair J R

Effects of Noise and Task Loading on a
Communication Task
(1988) / Orrell D H

U.S.A.F. INSTITUTE OF TECHNOLOGY
(continued)

The Integration of Information Systems
Planning Into the Plexsys Environment
(1988) / Pettigrew I

Multiattribute Multicommodity Flows in
Transportation Networks
(1988) / Popken D A

A Software System to Create a Hierarchical,
Multiple Level of Detail Terrain Model
(1988) / Roberts L

Implementation of the Netos Operating System
in Ada with Modifications to Allow Variable
Length Messages
(1988) / Rodriguez R

A Classification Methodology and Retrieval
Model to Support Software Reuse
(1988) / Ruble D L

Lo - Co Graf: Generating Maps to Support
Command and Control Crisis Management Using
Small Computers
(1988) / Saboi R

Computer Aided Design Parameters for Forward
Basing
(1988) / Salsano G M

Modeling the Merger of the Classified
Networks of the DDN: Blacker
(1988) / Swope R L

Design and Development of a Computer Based
Message Transfer System for the AF Logistics
Command Packet Radio Network
(1988) / Taris W J

Dynamic Analysis of Feedforward Neural
Networks Using Simulated and Measured Data
(1988) / Tarr G L

Circuit World: An Intelligent System for
Discrete Digital Logic
(1988) / Thatcher B J

Design and Implementation of a Diffusion
Link Laser Programmable Read Only Memory
and Automated Programming Station
(1988) / Tillie J J

Maintenance Metrics for Jovial (J73)
Software
(1988) / Tindell D R

A Comparison of Control Variates for
Queueing Network Simulation
(1988) / Tomick J J

Database Design and Land Battle Interface
for the Fast Stick Exercise
(1988) / Walker S A

Use of a Portable Microcomputer as a Data
Collection Tool to Support Integrated
Simulation Support Environments: A Concept
(1988) / Wilkinson J E

Improving the Survivability of a Stochastic
Communication Network
(1988) / Yim E

UTAH STATE UNIVERSITY

Study and Development of Expert Systems
(1987) / Agrawal R K

Auto Data Flow Diagram System
(1987) / Chyan L

Direct Simulation from a Model Specification
Language
(1987) / Pimentel R

UTAH, UNIVERSITY OF

Critical Path Determination in Path
Programmable Logic
(1988) / Brown J L

Modeling Systems of Rigid Bodies in Motion
Using Dynamics
(1988) / Elvins T T

A Multiprocessor Implementation of CSP
(1988) / Feng H

Modeling Techniques for the Optimization of
Hemodynamic Monitoring
(1988) / Gailey R N

Knowledge Representation and Algorithms for
a Gene Mapping Expert System
(1988) / Galland J

CAD-Based Grasp Synthesis
(1988) / Gatrell L B

Computer-Aided Geometric Design Based
Three-Dimensional Object Representations
for Computer Vision
(1988) / Ho C C

A Logical Approach to Program Analysis
(1988) / Krohynfeldt J J

Implementation and Applications of an
Annotated Functional Language
(1988) / Kuo W Y

EFEM: Expert System for Finite Element
Methods in Solid Modeler
(1988) / Lee H J

A Shading Method for Computer Generated
Images
(1988) / Malley T J

An Animation System Based on Functional
Dependencies and Inverse Kinematics
(1988) / McMinn G W

A Parallel Algorithm for Surface
Intersection
(1988) / Pates R F

PRT--A High Quality Image Synthesis System
for B-Splice Surfaces
(1988) / Peterson J W

Moped (A Portable Debugger)
(1988) / Pourheidari M

The Duck System: A Physician/Computer
Interface for Clinical Practice
(1988) / Reynolds C J

Data Translation Issues in a Hospital
Information Network
(1988) / Sawdey R J

UTAH, UNIVERSITY OF
(continued)

Display and Analysis of Three-Dimensional
Experimental Electrocardiographic Data
(1988) / Seeger A R

Interface of a Commercial Electrocardiogram
System to Help
(1988) / Tarng W P

Knowledge Frame Translator
(1988) / Wang W

Miranda as an Executable Specification
Language
(1988) / Yang H K

Physicians' Computerized Charting in Newborn
Intensive Care Unit
(1988) / Zheng Z

VERMONT, UNIVERSITY OF

Conformance Testing the Packet Level
Protocol of X.75 Implementations
(1988) / Remington H A

VICTORIA, UNIVERSITY OF

A Distributed Layout Compactor
(1988) / Byrne R G

The Computational Complexity of Edge
Colouring Restricted Graphs
(1988) / Cai L

A Modification to Earley's Algorithm for
Syntax Error Recovery
(1988) / Hagglund R D

A Practical Implementation of an Environment
for Programming-in-the-Large
(1988) / Klashinsky K B

Recognition and Ordering Algorithms
Concerning Global Inheritance in Lu
Factorizations
(1988) / Slater T A

Prototypes of Trace Specifications
(1988) / Wang Y

VILLANOVA UNIVERSITY

The Feasibility of Using an Integrated
Computer Aided Design/Finite Element
Analysis System to Automate Solutions to
Problems in Engineering Design
(1988) / Ranieri J

Evaluation of Database Management Systems
(1988) / Sinha S N

VIRGINIA, UNIVERSITY OF

A Global Object Code Optimizer
(1988) / Benitez M E

Rule-Based Focus of Attention in a
Dynamic Computer Monitoring System
(1988) / Blank G E

Resolution of State-Ambiguity in a Class
of Control Models
(1988) / Deal A L

Efficacy of Parallel Genetic Algorithms
(1988) / Hedge S U

Performance Analysis of the SAE AS4074.2
High Speed Ring Bus
(1988) / Minnich D W

Fault Tolerance in a Distributed
Hierarchical Control
(1988) / Pancerella C M

A Software Development Environment for
a Hypercube
(1988) / Rajiv P V

A Prototyping Environment for Distributed
Database Systems
(1988) / Ratner J M

A Study of Information Flow in Software
Development Environments
(1988) / Roh H

Performance Analysis of the FDDI Token Ring
(1988) / Simonson R A

State Errors in Distributed Hierarchical
Control Systems
(1988) / Taylor G R

A Meta File System
(1988) / Whitlatch J L

Experimental Evaluation of a Parallel
Programming Language
(1988) / Wise R M

WASHINGTON STATE UNIVERSITY

Commutation Monoids and Logics of Knowledge
(1988) / Iyer R V

A Radiosity Solution for Modeling Complex
Environments with Extensions for Nonplanar
Surfaces
(1988) / Kibler M K

Realistic Image Rendering: Advances in Ray
Tracing
(1988) / Wurzer R D

WASHINGTON UNIVERSITY

Grammar-Based Techniques for Interface
Design
(1987) / Cousins S B

The Design and Implementation of a
Multimodality PACS Development System for
Diagnostic Radiology
(1987) / D'Lugin J J

Computer-Aided Site-Directed Mutagenesis
(1987) / Kimmel H R

A Battery-Powered, Wearable Digital Hearing
Aid System
(1987) / O'Connell M P

Robustness Analysis of Multi-Input
Multi-Output Dynamics Systems
(1987) / Samet J

Fast Memory Access Using Content Addressable
Memory Modules
(1987) / Shalon T

WASHINGTON, UNIVERSITY OF

System Performance Analysis of the Sun
Network File System
(1987) / Brooks S

WASHINGTON, UNIVERSITY OF
(continued)

The DAS Version of Sendmail: An
Organization-Wide Aliasing Scheme for
Distribution System Running Sendmail
(1987) / Brunner F M

Heterogeneous Remote Procedure Call for
Franz Lisp
(1987) / Gosney K

The Design and Construction of a Prototype
Pyramid Machine
(1987) / Ling R

Object Recognition Using Structured Light
(1987) / McGarvey K J

Co-Edit, an Online Editing Tool for
Collaborative Writing
(1987) / Mueller M

Mail Systems for Personal Computers
(1987) / Rose S M

An Interactive Font Design Tool that
Produces Device Independent Font
Descriptions
(1987) / Rullman B R

Allocation Strategies for APL on the CHIP
Computer
(1987) / Schaad J L

Phoenix: An Interactive Curve Design System
Based on the Automatic Fitting of Hand
Sketched Curves
(1988) / Schneider P J

A Study of Multilevel Cache Memories
(1987) / Short R T

GeoLab: A Constraint-Based Geometer's
Laboratory
(1987) / Woolf M

WEST FLORIDA, UNIVERSITY OF

An Application of Modern Composite Design
Techniques to the Development of Real-Time
Data Monitoring Software
(1988) / Moore L C

Design and Implementation of a Mix
Environment for VAX/VMS Computer Systems
(1988) / Moore M J

A Practical Approach to Software
Engineering Education
(1988) / Raley R S

Datalink Subsystem Firmware for the MQM107
Subscale Drone
(1988) / Walthall D E

WEST VIRGINIA UNIVERSITY

Typesetting of Music Notation Using T(E)X
(1988) / Conrad P T

The Comparative Effects of a Programed
Instruction Format Workbook and the Lecture
Recitation Method on Entry Level Cad Skills
(1988) / Shaver R W

A Realization of Multiprocessing Garbage
Collection Algorithm for Rule-Based Expert
Systems
(1988) / Singh R

An Expert-System-Building Tool Employing a
Backward Chaining Inference Mechanism
(1988) / Wang W L

WESTERN MICHIGAN UNIVERSITY

Implementing Computer Use in 5th Grade
Curriculum: A Study of the Process of
Implementation and Subsequent Impact
(1988) / Davison W

Programmer and End User Communication in
Computer System Development
(1988) / Hart T

Reconstruction of a Real Object Using Stereo
Vision
(1988) / Vakalis I

WESTERN ONTARIO, UNIVERSITY OF

Database Modelling Concepts in an Object
Oriented Environment
(1988) / Chan T O E

Linear Algorithms for Clipping Convex and
General Simple Polygons
(1988) / Fung K Y

The Automated Cryptanalysis of Product
Ciphers
(1988) / Robbins L E

Practical Reductions of First Order Theories
(1988) / Szwarc H M

Design of User Interface Management Systems
(1988) / Ting H K

Issues of Name Management in Internet
Systems
(1988) / Triantafillou H

Solid Codes and Quasi-Completely Right
Disjunctive Languages
(1988) / Yu S S

WICHITA STATE UNIVERSITY

A Multi-Microcomputer Implementation of a
Direct Two-Dimensional Discrete Fourier
Transformation
(1987) / Boyd M J

A Practical LR Parser Maintaining Immediate
Error Detection
(1987) / Findley W D

A Symbol Table Management Facility Based on
a Relational Model
(1987) / Vore R L

WORCHESTER POLYTECHNIC INSTITUTE

Designing a Practical Distributed Text
Editor
(1986) / Barstow J C

A CAD Tool for Timing Analysis
(1987) / Bruen J

Monitoring Tools for an IEEE 802 Network
(1987) / Chiang P

WORCHESTER POLYTECHNIC INSTITUTE
(continued)

DSPL Acquirer: A System for the Acquisition
of Routine Design Knowledge
(1987) / Chiang Y T

Active User Participation in Information
System Development
(1986) / Danielson R E

An Adaptive Manufacturing Process
Description Language and Execution
Environment
(1987) / Garvey M D

Qualitative Reasoning About Shape and Fit
(1986) / Green D S

Distributed Multiprocessor Performance
(1987) / Huang C

NORMFIT: An Intelligent Tutor for Normal
Form Determination
(1987) / Huang X

Implementation of a Parallel Parser Driver
(1986) / Insinga A K

Explanation for Routine Design Problem
Solving
(1986) / Kassatly A A

An Expert System for Relational Database
Design
(1987) / Leamus L R

A General Purpose Traversal Mechanism for
Concurrent Logic Simulation
(1987) / Machlin D J

Language Processing Environment: Front End
Generator
(1986) / Masuck C M

Recognition of Neural Impulse Trains
(1987) / Packard C R

The Compilation of Cost Ordered Decision
Trees Within a Diagnostic Hierarchy from
Terminal Mode Rules
(1987) / Rollo P A

Computer Aided Fine Motor Skills Development
System
(1987) / Samara T K

Realistic Image Generation of Object Light
Sources
(1987) / Smith C D

XSAFE: A Prototype Expert for Security
Inspection of a VAX/VMS System in a Network
Environment
(1986) / Teng H S

Implementation of I.E.E.E. Standard 802.2
on an Ethernet Local Network
(1986) / Walton C B

WRIGHT STATE UNIVERSITY

Optimization of a Structured Query Language
for a Relational Database System
(1987) / Corcoran T J

An Empirical Study in the Modelling of
Heuristic Error Behavior
(1987) / Cotterman A J

An AI Approach to Machine Translation
(1987) / Drozdek A

Semantic Networks: A Cognitive Appraisal
(1987) / Dunlop C E

The Student Advisor: An Expert System
(1987) / Farrell C R

A Domain Independent Method of Comparing
Search Algorithm Run-Times
(1987) / Golden D J

Unidirectional Heuristic Search:
Improvements and Extensions
(1987) / Gregor A B

Stampede: Dialog and Graphics
(1987) / Haire G

Snowflake: A Stack Computer in Hardware
and Software
(1987) / Irwin J C

Data Processing of a Crooked Seismic Line in
an Area of Thick Glacial Till
Michigan
(1987) / Jones B C

AOC as a Teaching Tool
(1987) / Kolvrat K

A Computer Simulation of the Human Immune
System's Response to Viral Infection
(1987) / Lin B Y

Prolog SPION
(1987) / Molla S R

Edison-68
(1987) / Neikirk W L

Quantification in Logical Form
(1987) / Ray K A

A Rise-Stack Computer Architecture
(1987) / Rocheleau T J

Semantic Networks and Belief Spaces
(1987) / Sartorelli J

An Automated Computer Vision and Robotics
System for an Intelligent Workstation
(1987) / Scarpelli A J

MIMIC: A Forth Simulator and Microcode
Generator for Microcoded Computer Designs
(1987) / Simpson G F

Design of a Programmable Digital Filter
Using CMOS VLSI Technology
(1987) / Sweet F W

A Multiprocessor Avionics System for an
Unmanned Research Vehicle
(1987) / Thompson D B

A Natural Language Database-Query System
in Prolog
(1987) / Wu S Y

WYOMING, UNIVERSITY OF

Take Home Computer Program and Native
American Student Progress
(1989) / Baltes A K

ALBERTA, UNIVERSITY OF

Ultrasonic Attenuation in Cerium Aluminum
Below 4K
(1988) / Miner W A

CALGARY, UNIVERSITY OF

The Experimental Study of the Effects of Air
Flux on Low Temperature Oxidation Reactions
(1988) / Goulet D

FLORIDA, UNIVERSITY OF

Rippling of Asphalt Pavements in Response to
Low Temperatures
(1988) / Hardee H K

GUELPH, UNIVERSITY OF

Deviations from the Pair Potential
Contribution to the Structure of Krypton
at 237 K
(1988) / Youden J P

MANITOBA, UNIVERSITY OF

Carbon, Nitrogen, and Phosphorus Removal in
a Sequencing Batch Reactor at Low
Temperatures
(1988) / McCartney D M

MASSACHUSETTS INSTITUTE OF TECHNOLOGY

An Analysis of Analog CMO Coupling Circuits
at Liquid Nitrogen Temperature
(1988) / Yanney P R

PENNSYLVANIA STATE UNIVERSITY

Low Temperature Phase Relationships in the
System Silica-Aluminum Orthophosphate-Water
(1988) / Takahashi T

TEXAS A AND M UNIVERSITY

Total Plastic Strain and Electrical Resistivity in
High Purity Aluminum Cyclically Strained at 4.2 K
(1988) / Gehan J T

U.S. NAVAL POSTGRADUATE SCHOOL

Interface Characterization of Copper-Copper
and Copper-Silver-Copper Low Temperature
Solid State Bonds
(1987) / Dalbey R Z

Development of Methods for Low Temperature
Diffusion Bonding
(1987) / Muffler P A

U.S.A.F. INSTITUTE OF TECHNOLOGY

Low Temperature Photo-Luminescence Study of
Ytterbium Planted Into III-V Semiconductors
and Aluminum Germanium
(1988) / Colon J E

WESTERN MICHIGAN UNIVERSITY

Theoretical Thermodynamic Propperties of
Low Temperature Fluids
(1988) / Zahid Jamal Z A

AKRON, UNIVERSITY OF

DSPlab: A Signal Processing Environment
for the PC
(1989) / Adams D J

Design and Development of a Shared Memory
Multiprocessor System
(1989) / Anand S

Experimental Study of an Electromagnetically
Coupled Microstrip Patch Antenna
(1988) / Bobinchak J J

A Floating Point Processor Peripheral
(1989) / Bozzelli M P

Graphical Stability Methods for Numerical
Integration of Ordinary Differential
Equations
(1989) / Chicatelli S P

Ultrasound Overlapping Echos Identification
by Using Maximum Entropy Estimation Method
(1989) / Huo K

Word Boundary Detection by Using PLP Model
and a DTW-Based Speech Recognition System
(1989) / Jian X

A Vectorized Incomplete Choleski Conjugate
Gradient Algorithm for Large Linear Systems
of Equations
(1988) / Lin J

Model Order Reduction for Linear and
Nonlinear Systems
(1989) / Ma X

Parallel Processing: A Butterfly
Architecture
(1989) / Marosi S B

Implementation of High Speed Parallel Adders
Using Full CMOS VLSI Standard Cells
(1989) / Sharma N K

A Board-Level Text-to-Speech Conversion
System for the IBM Personal Computer
(1989) / Stimler W E

A Parallel Algorithm for Finite Element
Computation
(1989) / Subramaniam P

Maximum Entropy Method and Complex Maximum
Entropy Method Application to Phase
Estimation
(1989) / Vota J R

Design and Simulation of the Dynamically
Segmented Bus Architecture
(1989) / Xu B Q

ALABAMA, UNIVERSITY OF (UNIVERSITY)

An Intelligent Allocation Algorithm for
Parallel Processors
(1988) / Ananthram K G

A Computer Simulation Study of the Stability
of a Loss-Modulated Mode-Locked Laser
(1988) / Arnous A Z

A Theoretical Study of Single Side Band Mode
Locking of Lasers
(1988) / Badreddine B M

Efficient Parallel Architecture for Highly
Coupled, Real Time Linear System
Applications
(1988) / Barua S

Investigation of Weighted Least Squares
State Estimators Having Bad Data Detection
and Identification Capabilities
(1988) / El-Sayed S L

Economic Dispatch with On-Line Calculation
of Penalty Factors in an Electric Power
System
(1988) / Fallon S A

The Development and Integration of a Digital
Signal Processor Into an Analog
Communications Laboratory
(1988) / Kann D L

Computer Architecture for Intelligent
Real-Time Numeric and Symbolic Processing
(1988) / Mahadevan R K

A Direct-Execution Parallel Architecture for
the Advanced Continuous Simulation Language
(ACSL)
(1988) / Owen J E

Computer Graphics Expansion Board Using the
82786
(1988) / Patrick M

Sectoring Algorithm for Fast CGI
(1988) / Peswani S R

Identification of Helicopter Simulator
Dynamics
(1988) / Rhyne R E

Computer Architecture for Efficient
Algorithmic Executions in Real-Time Systems
(1988) / Saha A

Parallel Hardwood Architecture for Real-Time
Computer Graphics Algorithms
(1988) / Shirali N S

Scheduling Algorithm for Multiprocessor
Robot Arm Control
(1988) / Sidhu G S

A Theoretical Study of Synchronously Mode
Locked Dye Lasers Using the Coupled Mode
Approach
(1988) / Thati D

An Intelligent Processing Environment for
Real-Time Simulation
(1988) / Wells B E

ALBERTA, UNIVERSITY OF

Source Current Density Mapping of the EEG
(1988) / Arthur C R

Characterization of CMOS Magnetic Field
Sensors
(1988) / Briglio D R

An Analysis of a Broadband ISDN Signalling
System's Physical, Data Link, and Network
Layer Protocols for VLSI Implementation
(1988) / Brown J J

Linewidth and FM Noise Reduction of a 1.3
Micron GRECC Semiconductor Laser Using
Electrical Feedback
(1988) / Chan Y F

ALBERTA, UNIVERSITY OF
(continued)

Finite Element Mesh Generation for
Semiconductor Device Simulation
(1988) / Chau K Y

Thermal Measurements on Clearwater Shale
Utilizing a Transient Heat Probe and Direct
Electrical Heating
(1988) / Goulbourne D

Fabrication and Characterization of Silicon
Doping Superlattices
(1988) / McKinnon G H

An Interactive User Interface for a
Semiconductor Device Simulation Package
(1988) / Saad R M

Algorithms for Global Routing
(1988) / Singh R

Linewidth Reduction and Frequency Modulation
of a 1.3 Micron Semiconductor Laser with
Strong Frequency Selective Optical Feedback
(1988) / Somani A

A Sub-Optimal Amplitude Control System for
a Harmonic Oscillator
(1988) / Taylor L

ARIZONA STATE UNIVERSITY

Analysis of Multi-Exponential Signals in
Noise and Their Application in Deep Level
Transient Spectroscopy
(1988) / Alameh R M

Time-Domain Analysis of Nonlinear Microwave
Circuits Using a Frequency-Domain
Description of the Linear Subnetworks
(1988) / Arnold E N

Optimization of Deposition and Thermal
Processing Parameters in a-Sil-xCx:H and
a-Si:H/c-Si Solar Cells
(1988) / Banwari S K

Evaluation of Microwave Chamber Absorbing
Materials
(1988) / Brumley S A

A Study of Transient Transport Properties in
Photoconductivity Experiments
(1988) / Chamoun S N

Matching Points with Subpixel Accuracy in
Stereo and Motion Analysis
(1988) / Chang A

The Three-Dimensional Device Stimulator with
Impact Ionization Effect
(1988) / Chen Y S

Minimum Growth of Stabilizing Laws for
Nonlinear Feedback Systems
(1988) / Chu T M

The Methodology of Error-Control for Content
Addressable Memory and Its Design
(1988) / Day H C

A Vowel Detection Module for Use Within
an Automatic Speech Understanding System
(1988) / Earhart T R

Motion Compensation of Inverse Synthetic
Aperature Radar Images
(1988) / Fraser M L

Preparation and Characterization of N+/P
Indium Phosphide Homojunction Solar Cell
by Liquid Phase Epitaxy
(1988) / Friedrich U H

Transport Service Protocol for a Distributed
Operating System on LAN-Based Engineering
Workstations
(1988) / Fu M

Hardware Implementation of a Digital Neural
Network
(1988) / Fu W M

Toward a Non-Diverging Fast Decoupled Power
Flow Algorithm
(1988) / Gopalakrishnan L K

Aluminum-Germanium-Nickel Ohmic Contacts to
N- and P-Type Gallium Arsenide
(1988) / Haddock T B

Design of a Tuning Magnetic Power Supply for
the Kaon Accelerator
(1988) / Han B M

Preparation and Characterization of
Double-Layer AR Coating for Indium Phosphide
Solar Cells
(1988) / Hasan M S

Electronic Transport in Microstructures with
Defects
(1988) / Haukness B S

Kalman Filter Performance Evaluation with
White and Correlated Noise
(1988) / Holley S R

Effect of Solid Diffusion Source Impurities
on Silicon Devices
(1988) / Hsiu H L

Polycrystalline Silicon in Bulk-Barrier
Phototransistor for Image Element
(1988) / Hsu N

Measurements of Thin Water Films Using
Microwave Techniques
(1988) / Hurley R B

Fabrication and Characterization of an
Aluminum Gallium Arsenide/Gallium Arsenide
Laser Light Source for Fiber Gyroscope
Application
(1988) / Khawaja Y N

Optimization of a Single Photon Emission
Computed Tomography System with a Unique
Detector Configuration
(1988) / Kluksdahl E M

Charge Trapping in Amorphous Silicon
Nitrate Thin Films
(1988) / Krishna S

High Temperature Fluoride Corrosion Analysis
of Inconel 617 Thermal Energy Storage
Capsules
(1988) / Krishnamurthy M

BiCMOS Operational Amplifier
(1988) / Lee H J

ARIZONA STATE UNIVERSITY
(continued)

In-Flight HF Multipath Predictions
(1988) / Lee J U

Elastic Tunneling Through Ultra-Thin Silicon
Dioxide Films
(1988) / Limbert E J

Design Guidelines for Monolithic Microwave
Integrated Circuit Distributed Power
Combiner/Dividers
(1988) / Longbrake B A

An Improved Hybrid Channel Assignment Scheme
in High-Capacity Mobile Communication
Systems
(1988) / Lu C

An Integrated Graphics Environment for
Control System Design and Analysis
(1988) / Lyle D S

Optimal Losses for Distribution Feeders
(1988) / Magbool A

Laminar Natural Convection Heat Transfer in
a High Aspect Ratio Duct
(1988) / Majumdar D

Charge Trapping in Amorphous Silicon Nitrate
Thin Films
(1988) / Manglore R B

Research of Technological Barriers
Preventing High Efficiency Low Cost
Silicon Solar Cells
(1988) / Mih R D

Cadmium Telluride Crystal Gamma-Detector
Array for a Miniaturized Single Photon
Emission Computed Tomograph
(1988) / Murray D

Design, Construction and Evaluation of a
Computerized Electro-Optic Detector for
Use in a Mini-Spect
(1988) / Murty K N

Rank Conditions and the Multivariate Output
Pole Placement Problem
(1988) / Norris D E

Power Transmission Line Thermal Uprating:
An Expert System Approach
(1988) / Peng T

Comparison of Photovoltaic Energy Systems
for the Solar Village
(1988) / Pierce-French E C

Depletion Capacitance of P-N Junctions in
the Moderrate Forward-Biased Region
(1988) / Poo T K

Pattern Definition Using Carbon Enhanced
Vapor Etching
(1988) / Ramesh V

Electromagnetic Moment Modelling of the
Rectangular Spiral Inductor
(1988) / Romero F A

Ultra-Submicron Gate Gallium Arsenide
Field Effect Transistors
(1988) / Ryan J M

Innovative Test Structures for Electrical
Measurement of Thin Film Step Coverage in
VLSI Circuits
(1988) / Scheid D R

A Computer Simulation of a Digital LEC
Gallium Arsenide Crystal Growth System
(1988) / Schwartz J E

Source/Drain Formation Using Spin on Dopants
and Subsequent Rapid Thermal Diffusion
(1988) / Sheets G W

An Evaluation of Three Electronic Design
Automation Programs
(1988) / Sherman T R

Preparation and Characterization of
Amorphous Carbon Films
(1988) / Shinseki B S

Gallium Arsenide Integrated Optoelectronic
Devices
(1988) / Shiralagi K T

Calibration of a High Fidelity Communication
System: Transform Difference Method
(1988) / Sprute S E

Single-Mode Fiber Directional Coupler
(1988) / Tachasukij C

An Expert System for Black Start System
Restoration
(1988) / Thiyagarajah A

BiCMOS Sense Amplifier
(1988) / Tseng C

Automation of a Double Crystal (DC)
Diffractometer and DC Measurements
(1988) / Venkatram S

Signal Enhancement of Wigner-Ville
Time-Frequency Signals
(1988) / Wang C

A State Space Analysis of Some Classical
Frequency Domain Stabilization Schemes
(1988) / Williams J B

Application of Computer Vision to Physical
Measurements of Gallium Arsenide Crystals
(1988) / Wu C C

A High Performance CMOS Folded Cascode
Amplifier and Its Application
(1988) / Wu J C

ARMA Parameter Estimation of Noisy Speech
(1988) / Wygonski J J

Modeling of Gallium Arsenide Crystal Growth
Via System Identification
(1988) / Xu G

An Analytical Model of Threshold Voltage
for Narrow Width Fully Recessed Oxide MOSFET
Structures
(1988) / Yang C H

Design, Fabrication and Characterization of
a Silicon Bipolar Transistor
(1988) / Yu J Y

ARKANSAS, UNIVERSITY OF

A Multiplexed Dynamicallyy Adjustable
Switched-Capacitor CMOS Filter Bank Chip
(1988) / Basavapatna P

General Results in Optimal Control of
Discrete-Time Nonlinear Stochastic Systems
(1988) / Ciancetta M S

The Geometric Rectification of Radar
Images: An Interferometric Approach
(1988) / Hug J F

X Band Specular Reflectance Measurements
with Application to Radar Camouflage of
Fixed Installation
(1988) / Jacobs E L

A Touch Actuated Talking Book
(1988) / Jebasingam H

Image Processing Using Photorefractive
Crystals
(1988) / Khadr N M

Millimeter Wave Reflectance Measurements
of RCS Reduction Materials
(1988) / Long K R

A Neural Network for Shape Recognition
(1988) / Moorehead L B

The Robust Design of Discrete Time Control
Systems
(1988) / Niu X

Multiple Target Detection Using Optical Flow
(1988) / Viswanath H C

Characterization and Modeling of the
Aluminum Gallium Arsenide-Gallium Arsenide
Heterojunction Bipolar Transistor
(1988) / Woods B O

OHMIC Contacts to N-Gallium Arsenide Using
Indium on Gold Via a Heterojunction
(1988) / Yee W F

AUBURN UNIVERSITY

Yet Another Logic Simulator
(1987) / Abney L O

A User-Interactive Power-Flow and Transient-
Stability Program for the IBM PC/AT
(1987) / Burleson J A

A Digital Power System Monitor Analyzer
(1987) / Cox R J

The Soft/Hard Wet Anodization of Aluminum Oxide
and Its Use in a Thin Film Multilayered Capacitor
(1987) / Dickey J R

The Transient Electric Fields Near the Center
of a Multiply Excited Elliptical Loop Antenna
(1987) / Driscoll K T

Static State Estimation in Power Systems-Theory
and Computer Implementation
(1987) / Emesih V A

Algorithms for Feedback Control System Design
and Analysis
(1987) / Fields S L

Design and Fabrication of Multilayer Metal-Insulator
Metal Capacitors (MIMC)
(1987) / Frazier A B

Power Pattern Analysis of Thinned Planar Arrays
(1987) / Fromhold T W

Electromagnetic Analysis Using the Transmission
Line Modelling (TLM) Method
(1987) / German F J

A Home Network Using Carrier Sense Multiple
Access with Collision Detection as the Medium
Access Control Method
(1987) / Glassell J T

A Study in Modern Pole-Placement Control Design
with Least Squares System Identification System
Models
(1987) / Glassell R L

Stability Margins as Functions of System Parameters
in Kalman Filter Control Systems
(1987) / Golson C T

A High Performance Pulse-Width-Modulation Amplifier
(1987) / Haas H P

Modern Control Using a Microprocessor Based
Controller
(1987) / Howell L C

Processor Allocation of Program Graphs for a
Dataflow Multiprocessor
(1987) / Hwang D H

Singularity Expansion Analysis of Planar
Parallel Thin Cylinders
(1987) / Lindsey J M

An Investigation of the Effects of A/D Converter
Errors on a Fundamental Computer-Controlled
Phased-Array Beam-Forming Network
(1987) / Miller R M

Identification of Systems Using the Fast
Fourier Transform
(1987) / Olfati M

Digital Ask Demodulation in the Power Line
Communication
(1987) / Qian S

Reliability Analysis of the Solar Total Energy
Project (STEP), Shenandoah, Georgia
(1987) / Schneider W H

VHSIC Hardware Description Language (VHDL) with a
Data Compressor/Decompressor Example Application
(1987) / Smith J L

Efficient Current Expansion Modes for Frequency
Selective Surfaces--A Singularity Expansion
Method Approach
(1987) / Smith R G

Low-Pressure Metalorganic Chemical Vapor Deposition
of Gallium Arsenide on Gallium Arsenide Substrate
Using Triethylgallium and Solid Arsenic as Precursors
(1987) / Tong C C

An Algorithm for Dynamic Resource Allocation for a
Data Flow Computer
(1987) / Wang S J

A Method for Evaluating Distributed Control of
Textile Processes with Transportation Lag
(1987) / Watson C P

Visible Light Radiation and Device Characteristics
of Reverse-Biased Silicon Diffused P-N Junctions
(1987) / Williams C B

BRIGHAM YOUNG UNIVERSITY

Loop-Grain Modulators: A New Class of
Low-Power AM Modulators
(1988) / Fox D K

Methods of Adaptive Echo Removal
(1988) / Lang D A

Systolic Array Simulator: Systole
(1988) / Martell R W

Symmetric Routing: A Method to Schedule
Matrix Transposition and Other Permutations
in Binary N-Cube Networks
(1988) / McDanaiel S F

A High Frequency Narrowband Active Filter
for IF Applications
(1988) / Millward M C

Mode Characteristics and Second Harmonic
Enhancement in an Optical Fiber with
Non-Linear Thin-Film Coating
(1988) / Moon T K

Levels of Ordering and Consistence in
Shared-Memory Multiprocessors
(1988) / Nicholas K E

Vector Quantization Error Using Single Frame
and Sequential Frame Imagery
(1988) / Ostler R S

Azimuth Compression of SAR Data Using Least
Squares Estimation Techniques
(1988) / Remund B L

Change Detection Between Digital Image Pairs
(1988) / Talbot L M

BROWN UNIVERSITY

Management of Graphical Displays in a
Networked Experiment Environment
(1988) / Alvarado V M

BUCKNELL UNIVERSITY

The Design and Development of a Motorola
68010 Microprocessor System Based on the
VMEbus Specification
(1988) / Singh K

CALGARY, UNIVERSITY OF

The Cyber 205 Based Three Dimensional
Image Processing Facility
(1988) / Chan B P

Control of Multi-Modes of Oscillation
Using a Multi-Input Multi-Channel Controller
(1988) / Chebib S

The Design and Application of a High Quality
Three Dimensional Linear Trajectory Filter
(1988) / Fowlow T J

A High-Quality Single Chip Programmable
Digital Filter
(1988) / Green B D

Restoration of Gamma Camera-Based Nuclear
Medicine Images
(1988) / Hon T C

OTA Design and Biquad Filter Implementation
(1988) / LaFrenz J L

Advances in Design of Filters and Controls
(1988) / Lau K H

Design Techniques for Bilinear LDI Digital
Allpass Networks
(1988) / Lee L S

An Electrical Transmission Line Model of
a Pipeline
(1988) / Pal Z M

A Precision Data Acquisition System
(1988) / Paslawski D J

Expert Systems and Applications
(1988) / Soukaria A M F

CALIFORNIA STATE UNIV. (FULLERTON)

A Computer Program for the Analysis of
Erlang Distributed Continuous-Time Markov
Models
(1988) / Aberegg R N

Derivative Analysis of the Speech Contour
for Voice Recognition Using a Hopping FFT
(1988) / Budrovic M

Microprocessor Implementation of Vector
Control System for AC Induction Motor
(1988) / Chertov L

A Graphical System in Operational Evaluation
Modeling
(1988) / Dabrian A

Microprocessor Control for Pneumatically
Actuated Acoustical Pipe Organs Using the
Musical Instrument Digital Interface (MIDI)
Protocol
(1988) / Edwards S

CALIFORNIA STATE UNIV. (LONG BEACH)

Development of a Spacecraft Solid State
Memory
(1988) / Anderson S R

SCADS Software Computer Aided Design Systems
(1988) / Brown D M

Communication Via Meteor Trails
(1988) / De Angelis K J

Microcomputer Control of a pH Neutralization
Titration Process
(1988) / Elliott D R

Development of a Microprocessor-Based
Multifunction Microcomputer Board
(1988) / Lindberg M G

Tomographic Pollution Diagnostics
(1988) / Sharp G D

PC Based Time Series Analysis of Continuous
Events
(1988) / Uttamchandani V

CALIFORNIA STATE UNIV. (SACRAMENTO)

Optimal Pole Assignment by State Feedback
(1988) / Ali Marandi Ghoddousi M

CALIFORNIA STATE UNIV. (SACRAMENTO)
(continued)

EHV Transmission Line Power Carrying
Capability and Computer Program for Line
Compensation
(1988) / Boparai J S

Digital FM Analysis as Applied to the Art of
Music Synthesis
(1988) / Dame S G

Feedback Design Methods for a Nonlinear
Position Control Problem: A Case Study
(1988) / Doan C D

Modeling of a Servomechanism
(1988) / Hsu H C

Digital Echo Cancellation by Adaptive
Filtering
(1988) / Kalami E

Reduction of Noises and Effects of Beam
Hardening on X-Ray Computed Tomography
Images
(1988) / Kropas C V

A Vmebus-Based, High Speed, Single Board
Microcomputer
(1988) / Larson G A

Weighted Recursive Least-Squares Adaptive
Filter
(1988) / Najjar S A

Design of an Ethernet/Starlan Medium Access
Control Bridge
(1988) / Nishide K

Design of a Broadband Low Noise Gallium
Arsenide FET Amplifer Using Microstrip
Techniques
(1988) / Oliver C T

Continuous Mathematical Models Representing
Nonlinear Physical Phenomena
(1988) / Paunon G T

Design of a DSP Development Board for the
IBM PC/C, XT, AT
(1988) / Stanonis J E

Remote Operated Microprocessor Control
System
(1988) / Tory K D

Transient Stability Analysis of the
Hyatt-Thermalito Electric Power System
(1988) / Villalta E J

A Multiphase Multipolar DC Generator Design
(1988) / Worrell W

General Purpose Digital Signal Processing
System
(1988) / Wyman B H

CALIFORNIA, UNIVERSITY OF (DAVIS)

Development of Computer-Aided Design and
Analysis Programs for Digital Signal
Processing
(1988) / Mameesh M N

Finite-Precision Implementation of the
Speech Analysis and Synthesis Technique
(1988) / Nguyen T Q

CALIFORNIA, UNIVERSITY OF (IRVINE)

DFT Algorithms for VLSI Implementations
(1988) / Evangelista G

Design Role Crituqie for Logic Synthesis
(1988) / Iqbal S

A Study of Organic Optical Waveguides
(1988) / Minot K L

On-Line Identification and Real-Time Control
of a Pneumatic Robot Arm
(1988) / Nagarajan V

Negative Indecx-Change Analog Fresnel Lens
for Integrated Optics
(1988) / Norris J A

A New Approach to FIR Filter Design Using
B-Spline Functions
(1988) / Pang D

The Use of Conformal Transformations in the
Finite Element Solution of Electromagnetic
Field Problems
(1988) / Stefan S M

A Mathematical Solution to the Two
Dimensional Robot Path Planning
Problem: Solving a Global Optimization
Problem Using Embedding Methods
(1988) / Tominaga H N

Negative Index-Changed Hybrid Lens of Chirp
Grating and Analog Fresnel Types on Lithium
Niobium Oxide and Gallium Arsenide Planar
Waveguides
(1988) / Vu T Q

Capturing Processing of Acoustic Images
(1988) / Yang Y B S

CARLETON UNIVERSITY

Circuit Factor Compensation for SAW Filters
Using Modal Analysis
(1988) / Cameron T P

ALE - A Layout Methodology for Custom
Integrated Circuits
(1988) / Diesing N A

Distribution of Protocol Functions in an
Interworking-Based ISDN Access
Implementation
(1988) / Doble J

Burst-Trapping Codes for the Land Mobile
Data Channel
(1988) / Hum E N

Two-Dimensional Analytical Modelling of the
Short-Channel MOSFET
(1988) / Kendall J D

Noise Analysis and Simulation of Switched
Capacitor Circuits Using a Continuous
Time Circuit Simulator
(1988) / Kwan J

Modeling of a Wave Generator and the Design
of an Optimal Estimator
(1988) / Laurich P H

An Advanced Speech Coder Based on a Rate
Distortion Theory Framework
(1988) / LeBlanc W P

CARLETON UNIVERSITY
(continued)

Multi-State Sigma Delta Modulators
(1988) / Lee M C G

Rapid Densification of Borophosphosilicate
Glass
(1988) / Madsen L D

Improving Interconnect and Register
Allocation for the Behavioral Synthesis of
Digital Circuits
(1988) / Midwinter J K

VLSI Triangulation Processing for Machine
Vision
(1988) / Morris T D

High-Level Synthesis of Digital Circuits
Using Global Scheduling and Binding
Algorithms
(1988) / Paulin P G

A True Polysilicon Emitter Transistor
(1988) / Rowlandson M B

Gallium Arsenide Monolithic Microwave
Frequency Halver and Phase Detector for Use
in a Phgase Locked Oscillator
(1988) / Stapleton S P

CARNEGIE-MELLON UNIVERSITY

An Application General Design Algorithm for
Electromagnetic Optimization
(1988) / Allison S B

On Using a Tactile Sensor for Real-Time
Feature Extraction
(1988) / Berger A D

The Effects of Distortions on the
Performance of Optical Correlators
(1988) / Connelly J M

Simulation, Fabrication, and Testing of
Lotic Gates Designed for Deflecting and
Non-Deflecting Bubbles
(1988) / Cowan A

Frequency Domain Conditions for Parameter
Convergence of an Adaptive Multivariable
Identifier
(1988) / DeMathelin M

Simulation of Trumpet Tones Via Physical
Modeling
(1988) / Dietz P H

Sputter Deposition of Bismuth Substituted
Iron Garnets for Magneto-Optic Recording
(1988) / Eppler W

Applying Inductive Fault Analysis and
Symbolic Boolean Analysis in the Extraction,
Analysis and Testing of Bridging Faults
(1988) / Jan S

Efficient Modeling of Bipolar Junction
Transistors
(1988) / Johnson M

Automatic Generation of Kinematics for a
Reconfigurable Manipulator System
(1988) / Kelmar L J

Design Methodology for N-th Order Noise
Shaping Coders
(1988) / Kenney J G

Parallel and Systolic Algorithms for
Stochastic Image Relaxation
(1988) / McPheters M J

Evaluation by Photoluminescence of Processes
to Improve Gallium Arsenide
(1988) / Milliman M L

Propagation in One Micron Ion-Implanted
Contiguous Disk Bubble Memory Devices
(1988) / Nitzberg K R

Evaluation of Path Tracking Errors in Mobile
Robots: A Parameter Search Technique
(1988) / Pangels H M

FORS: Flexible Organizations
(1988) / Papanikolopoulos N P

Parallelism in Manipulator Dynamics:
Analysis and Implementational Issues for
High-Speed Control
(1988) / Ramos S

AWE: Asymptotic Waveform Estimator
(1988) / Ravichandran K

Domain Mode Ferromagnetic Resonance in
Low Q Garnet Thin Films
(1988) / Ren E W

Study of Magnetostatic Wave Phase Shift in
Moving Thin Films and Applications to
Rotation Rate Sensing
(1988) / Saunders A E

Electrical and Magnetic Properties of High
Temperature Superconductors
(1988) / Stamper A K

An Investigation of Noninvasive Airway Area
Determination from Acoustic Reflections
(1988) / Strowbridge C M

Asymptotic Waveform Estimation of Linear and
Nonlinear Circuits
(1988) / Virgilio D V

A Control Separation Approach to the
Specification and Analysis of Discrete State
Systems and Controllers
(1988) / Wilson R G

Constructive Piecewise Constant Table
Modeling for SPECS2
(1988) / Zhang X

CASE WESTERN RESERVE UNIVERSITY

Simulation of the Ada Rendezvous and
Mini-Map Network
(1988) / Campbell S T

Chamber-Type Oxygen Sensor
(1988) / Chu W S

Three Dimensional Simulation of Electron
Beam Guns
(1988) / Darkazanli A

Hardware, Software, and Mechanical Design
and Implementation of a Computer
Programmable Respirator
(1988) / Diaz-Garza P L

CASE WESTERN RESERVE UNIVERSITY
(continued)

Feature Measurement Using Geometric
Dimensional Tolerancing on Range Imagery
(1988) / Epner M J

Microwave Characteristics of Interdigitated
Photoconductors on a HEMT Struture
(1988) / Hill S M

Silicon-to-Silicon Direct Wafer Bonding
(1988) / Kido T

A Constant Flow-Rate Microvalve and Fine
Movement Microactuator Study Based on
Silicon and Micromachining Technology
(1988) / Park S

Circuit Equivalents for Piezoelectric
Transducers with Nonuniform Electroelastic
Parameters and Distortionless Signal
Propagation
(1988) / Raya J R

Three Dimensional-Computer Simulation of the
Potential Field in an Electron Gun
(1988) / Slimani D

Automated Determination of Possible Static
Hazards in Combinational Logic Circuits and
Cross Coupled Gates
(1988) / Yoke N J

CENTRAL FLORIDA, UNIVERSITY OF

SAW Draw, an Interactive Graphical Layout
System for Surface Acoustic Wave Devices
(1988) / Abbott J B

The Time Slot Interchange in a Digital
Central Office
(1988) / Almaalouf K G

Method for Sizing MOS Transistors for VLSI
(1988) / Amatangelo M J

Design of a Bandpass Oculometer
(1988) / Brandauer K

Polar Spectrum Coding
(1988) / Chapman D H

Significant Measurements of a Multiple
Target Tracking System Utilizing Munkre's
Algorithm as a Correlation Scheme
(1988) / Day N A

Phase Noise Analysis and Basic Measurement
Techniques
(1988) / Ferrell J A

Rise Time to Frequency Correlation of
Crosstalk in Coupled Microstrip Lines
(1988) / Fletcher C M

A Piezoelectric Actuated Scanning Mirror
System Utilizing a Type One Control Loop
(1988) / Gibbs D M

Multiple Signal Relative Entropy Spectral
Estimation
(1988) / Gregory D M

Using Combined Integration Algorithms for
Real-Time Simulation of Continuous Systems
(1988) / Harbor L K

Uniformity Correction for a Scanned Long
Wavelength Infrared Focal Plane Array
(1988) / Heinold G C

A New Method for Broadband SAW Diffraction
Analysis
(1988) / Hines J H

Application of Lattice Filters to Quadrature
Mirror Filter Banks
(1988) / Jaspers G R

Barium Borate in Optical Parametric
Oscillation
(1988) / Kerrigan M T

Distributed Temperature Sensors Based on
Modified-Clad Optical Fibers
(1988) / Kim Y H

A Comparison of Two Microcomputer Continuous
Simulation Languages
(1988) / Kirkland A B

Uniform Energy Discharge for Pulsed Lasers
(1988) / Koschmann E C

Three-Level Block Truncation Coding
(1988) / Lee D A

Computer Aided Design of Linear Phase Finite
Impulse Response Digital Filters
(1988) / Lyou W

Laser Induced Darkening in Semiconductor
Doped Glasses
(1988) / Malhotra J

A System for Measuring the Phase Difference
Between Two Points of a Propagating Laser
Beam
(1988) / Odhner J E

The Analysis of Phase Noise in Phase Lock
Loops
(1988) / Pavik D M

Numerical Solutions of Poisson's Equation
in a Superlattice Device
(1988) / Rollins N T

Optical Interferometric Formation of Phased
Array Antenna Foot Print
(1988) / Sherwood J E

An Adaptive Multi-Scene Correlation
Algorithm
(1988) / Vogt P E

A Numerical Analysis of the Resonant States
of Quantum Well and Superlattice Devices
(1988) / Whittaker A G

High Resolution Monopulse Tracking
(1988) / Young L A

CLARKSON COLLEGE OF TECHNOLOGY

Modeling and Analysis of Inverter-Fed
Induction Motors Using the Natural ABC Fram
of Reference and Network Graph Techniques
(1988) / Alhamadi M

A Technique for Detecting and Correcting
Intermittent Faults in Sequential Devices
(1988) / Hansel B

CLARKSON COLLEGE OF TECHNOLOGY
(continued)

Polygonal Representation: A Maximum
Likelihood Perspective
(1988) / Hemminger T

Design, Fabrication and Characterization of
a Double Heterojunction MODFET
(1988) / Limoncelli L

Post-Reconstruction Correction in
Quantitative Single
(1988) / Saha A

Implementation of Parallel Motion Analysis
Algorithms on Array Processors
(1988) / Shandilya A

Analysis of ESD Failure Threshold
Dependances on Parameter Variations of an
NMOS Protection Network
(1988) / Steele K

Comparison Study of Non-Linear Filters in
Image Processing
(1988) / Wang X H

Electrical Overstress in NMOS Silidiced
Devices
(1988) / Wilson D J

A Macro Approach to VLSI Testing
(1988) / Youngs J L

CLEMSON UNIVERSITY

Small Computer System Interface (SCSI) Host
Adapter Design
(1988) / Cantu M H

Measurement Techniques on Distorted
Waveforms
(1988) / Clapp M C

Target Injection Unit for the Enhanced
Modular Signal Processor
(1988) / Cooil R G

High Mobility, Very Thin Silicon-on-Saphire
Films
(1988) / Dabral S

Performance Evaluation of ISO Class 4
Transport Protocol
(1988) / Damle S P

Force Analysis and Control of Biped Robot
Contact with Environments
(1988) / Fan Y

A Study of the Changes in the Electrical
Characteristics of p-Substrate MOS
Capacitors Due to Charge Injection During
Accumulation and Inversion Biasing
(1988) / Hall M D

Track-While-Scan Radar Tracking Filters
(1988) / Kassel R J

Optimal Estimation Techniques for Load
Forecasting
(1988) / Lee L

Hardware Design and Implementation of a
Microcomputer-Based Test System for Pulsed
Electromigration Research
(1988) / Lin Y F

The Standard Cell Design of an 8259A
Programmable Interrupt Controller
(1988) / Mandyam B

Power System Network Modeling Techniques
(1988) / McManis R B

A Prototype PC Based In-Circuit Test Set
(1988) / Miles F M

Winchester DMA Interface PCA
(1988) / Nicholson H M

Optimal Estimation of Power System Frequency
Deviation with an Application to Load
Shedding
(1988) / Peterson W L

Digital Conference and Switching System
Design Description
(1988) / Pollak S J

Electromagnetic Penetration Through Narrow
Slots in Screens
(1988) / Reed E K

The Integrated Circuit Design and
Implementation of a CMOS Clock Generator
and Driver
(1988) / Sreeraman V

Three Approaches for Multilimbed Robots
to Work in the Reachable-but-Unorientable
(1988) / Su C

Three Phase Modeling of Distribution Systems
in the Presence of Harmonic Distortion
(1988) / Thompson R L

Application of the Bus Impedance Matrix in
Unbalanced Distribution Systems
(1988) / Thornton K P

A Comparison of Oxide Wearout Resulting from
Pulsed or DC Charge Injection on p-Type
Substrates
(1988) / Vigrass W J

A Low Resistance Test Set for Printed
Circuit Board Copper Thickness
(1988) / Wallace M C

The Determination of Partial Degradation
Mechanisms in Commercial ESD Protection
Circuits Using SEM Techniques
(1988) / Williamson P A

Three-Phase for Transient Stability Analysis
of Unbalanced Distribution Systems
(1988) / Zambrano V O

CLEVELAND STATE UNIVERSITY

A Theoretical Model of Indium Phosphide
Space Solar Cells
(1988) / Geier J V

The Use of Expert Systems in Fault Isolation
of Electrical Systems
(1988) / Pesec J

COLORADO STATE UNIVERSITY

Transport in Inversion Layers of
Silicon (111) Mosfets
(1988) / Brusenback P R

COLORADO STATE UNIVERSITY
(continued)

Electrical Properties of Anodic and Deposited
Silicone Dioxide MIS Structures on n-Type
Indium Antomony.
(1988) / Chen C W

Development of Non Contact Field Effect Measurement
Techniques for Semiconductor Surface Evaluation
(1988) / Dubey A

The Development of a Custom Robotics
Test Bed for the Puma-560
(1988) / Hunt G L

The Design and Implementation of an Automated
Scanning Spectral Radiometer System
(1988) / Johnson-Pasqua C M

Analysis of the Optical Distributed Arithmetic Unit
(1988) / Knittle C D

Sensing and Control of Tip Position of a
Single-Link Very Flexible Manipulator
(1988) / Kurker C M

Implementations, Algorithms and Architectures
for 2-D Recursive Filtering
(1988) / Lu T

A Semi-Numerical Solution to Wave Propagation
in Hexagonal Media
(1988) / Lusk M T

Image Interpolation Via Moments
(1988) / McCreanor R L

Remote Field Eddy Current Phenomena
(1988) / Nath S

Analysis and Design of Periodically Time-Varying
IIR Digital Filters
(1988) / Prater J S

Digital PID and Exact Terminal Controllers
for a Five Degree of Freedom Manipulator
(1988) / Wiemer D J

CONCORDIA UNIVERSITY

Performance of Various Line Coding Formats
on Optical Fiber Link
(1988) / Aribindi K P

A Design Technique for Digitally
Programmable Analog Active Filters
(1988) / Datta B K

Performance of a Frequency Hopped MFSK
Multi-Access Mobile Cellular Radio Network
Using Error Correcting Techniques
(1988) / Ellement C

Torque Pulsations and Currents in Induction
Motors with Unbalanced Voltages
(1988) / Khan A

Bulk and Epitaxial Growth and
Characteristics of Gallium Arsenide and
Lead Gadolinium Telluride
(1988) / Lacroix J M

Performance of M-ARY QAM in a
Line-of-Sight Multipath Fading Channel
(1988) / Lieu G M

Power System Generation Expansion
Optimization Using a Probabilistic
Screening Curves - Branch and Bound Hybrid
(1988) / Loud L W

Analysis of Mutual Coupling Effects in a
Finite Planar Circular Waveguide Array
(1988) / Nguyen D N

Two-Dimensional Modelling of Logical River
Ice Cover Melting Due to a Side Thermal
Effluent
(1988) / Plouffe P R

An Environment for Robot Trajectory Planning
and Generation
(1988) / Thompson S E

A Pulse-Width Modulated Series Resonant
Converter Operated at Resonant Frequency
(1988) / Vandelac J P

An Analysis/Simulation Program for Power
Electronic Circuits
(1988) / Vincenti D

Performance Analysis and Optimized Design of
CMOS Buffers
(1988) / Zhu Y

CONNECTICUT, UNIVERSITY OF

Model and Control of Post Operative Pain
(1988) / Liu F U

CORNELL UNIVERSITY

The Use of Buried Beryllium-Doped Layers to
Reduce Buffer Current in Low Noise
Pseudomorphic Modulation-Doped Field-Effect
Transistors
(1988) / Anderson S F

Formation of Quarter-Micron Silicon on
Insulator Isolated Islands of Substrate
Silicon by Selective Lateral Oxidation
(1988) / Arney S C

Algorithms for Matrix Transposition on a
Mesh-Connected Array Processor
(1988) / Azari N G

Gallium Arsenide/Aluminum Gallium Arsenide
Based Laser with Chemically Assisted Ion
Beam Etched Facets
(1988) / Behfar-Rad A

Design and Analysis of Gallium Arsenide
Schottky Barrier Photodiodes
(1988) / Brown A J

Spacial Distribution of the Generation
Lifetime in MOS Capacitors
(1988) / Choi J R

Sign-Sign Adaptive LMS
(1988) / Elevitch C R

Efficient Parallel Algorithms for Matrix
Operations
(1988) / Elster A C

Algorithms and Performance Analysis for the
Massively Parallel Processor
(1988) / Gutierrez M C

CORNELL UNIVERSITY
(continued)

Comparisons and Stability Analysis of
Linearized Equations of Motion for a Basic
Bicycle Model
(1988) / Hand R S

90 Faceted Turns for Gallium
Arsenide/Aluminum Gallium Arsenide
Infrared Waveguides
(1988) / Kim A S

A Process for Low Thermal Resistance for
the CW Operation of Laser Diodes
(1988) / Lestina M J

Fabrication and Characterizationof the
Planar Doped Barrier Spectrometer
(1988) / Lu S S

Statistical Study of Short Range (Distance
800 km) Trimpi Effects in the Northern
Hemisphere
(1988) / Martin S

Use of Stability Domain Distortions to
Determine Effectiveness of Linear Feedback
Control
(1988) / Naqavi S A

Expanded Rectangles: A New VLSI Data
Structure
(1988) / Quayle M S

Electron Beam Generation Based on Negative
Electron Affinity Photocathodes
(1988) / Sanford C A

A Comparison of Epitaxial Layers and Undoped
Semi-Insulating Substrates for Use in
Silicon Implantation Into Gallium Arsenide
(1988) / Schary A

On the Effect of Output Quantization in
Control Systems
(1988) / Williamson G A

An 8 Bit x 8 Bit Pipelined Multiplier
Simulated at 500 MHz
(1988) / Yoo C D

DELAWARE, UNIVERSITY OF

Evaluation of a Supervised Nonparametric
Classifier
(1987) / Armstrong M L

Bipolar Junction Transistors for Optical
Interconnections
(1988) / Berryhill J B

Design of an Integrated Silicon Fet:
Gallium Arsenide Led for Optical
Interconnects
(1988) / Hobart K D

Robust Smoothers Applied to Image Filtering
and Enhancement
(1987) / Hohman D S

Study of a Novel Multimedia Optical Fiber
Communication System
(1987) / Izzo M

Digital Signal Processing with
Microprocessors
(1987) / Kramer W T C

Grating Couplers for Optical Waveguides
(1988) / Levy J J

The Design of a Highly Available Electronic
Mail Repository
(1988) / Miller D P

Design and Implementation of a Network
Terminal Server
(1987) / Minnich K M

DREXEL UNIVERSITY

Reassessment of Matrix Solvers in Repeated Finite
Element Solution
(1988) / Anandaraj A W

Thin Film Amorphous Silicon Transistors for Liquid
Crystal Displays
(1988) / Anderson S C

Deposition of Yttrium-Barium-Copper-Oxide Thin
Films by RF Magnetron Sputtering of Single
Composite Targets
(1988) / Begley T A

Pilot for an Autonomous Mobile Robot
(1988) / Bhatt R P

Pilot Blackout Detection as a Multiobjective
Optimization Problem
(1988) / Chaudhary A H

Fabrication of Superconducting Yttrium Barium Copper
Oxide Thin and Thick Films: A Study of Zirconium
Dioxide Diffusion Barrier
(1988) / Choi J

Optical Domain Amplification
(1988) / Dutta R

A VLSI Compatible, Multilayer Planarization
Technique: DC Magnetron Sputtering Using
Applied Bias and Direct Substrate Heating
(1988) / Fissel M G

Path Planning in the Multi-Resolutional
Traversability Space
(1988) / Graglia P

Correction of TV Camera Distortions Using
Digital Techniques
(1988) / Gregoriou G K

On Integrating Motor and Weight Sensors in
Loss-in-Weight Feeding Systems
(1988) / Herscher H T

Magnetic Shielding of Incomplete Configurations
(1987) / Hoole N R G

Development of a Thin Film Deposition Technique
for Coating Fibers
(1988) / Ihsan M B

Computer-Assisted Method for Measuring Arterial
Diameters
(1987) / Knecht L B

Study of High Speed Fiber Optic Links for
Short-Haul Microwave Applications
(1988) / Koffman I

On Minimal Time Maneuver Detection
(1985) / Konstanzer G C

An Adaptive Parallel Error Correcting Scheme for
Binary Symmetric Communication Channels
(1988) / Lim C K

DREXEL UNIVERSITY
(continued)

Orthogonal Analysis of Multipulse-Excited LPC
Speech Coders
(1988) / Magner C A

An Analysis of Clutter Suppression Algorithms in
Ultrasonic Nondestructive Testing
(1988) / Murthy R

Application of Kalman Filtering in Inertial
Navigation System (INS) Aided by Global
Positioning System (GPS)
(1988) / Nam K

Fractal and View Scale Concepts in Dynamical
Analysis
(1988) / Nicoletti D W

Mobility Modeling of Silicon MOSFET's, with
Applications to Silicon on Sapphire
(1988) / Padmanabhan S

A New de-Embedding Technique for Characterization
of Electro-Optic Components
(1988) / Perino J S

The Development of an Ultrasound Exposure System
for Ultrasound Bioeffects Studies
(1988) / Poczobutt M T

The Channel Routing Problem and Its Solutions
(1988) / Rajput W E

Development of Relaying System for Induction
Motor Protection
(1987) / Ray G

Quantitative Microautoradiography
(1988) / Subrahmonia J

Recognition of Regions in Brain Images
(1988) / Waks A

Economical Finite Element Methods for the
Calculation of Eddy Current Problems
(1987) / Weeber K R

Design for Steering Accuracy in Optically
Controlled Phased Array Antennas
(1988) / Wilcox J J

Time-Optimal Robot Navigation Using the
Slack Set Method
(1988) / Zaharakis S C

DUKE UNIVERSITY

CNR Manipulation with Linear Filters
(1988) / Brown D G

Arsenic and Phosphorus Cooperatives
Dissusion Phenomena
(1988) / Deaton R

The SAR Pattern of a Phased Array of
Asymmetric Dipoles
(1988) / Dubal N V

The Palladium Encapsulating Layer and Its
Application to the Fabrication Process
(1988) / Gyllenhammer C R

Corner Point Identification of a 3
Dimensional Object
(1988) / Hermawan A K

A Multifunctional Bus-Monitor Module for Use in
Redundant Digital Processing Systems. The
Design of the Programmable Controller and VLSI
Implementation
(1988) / Markas A

A Comparison of the Vivid and Magic VLSI
Design Systems
(1988) / McMillan E N

Instruction Decoding for a Sustained
Performance Architecture
(1988) / Milburn B D

Frequency and Loss Factor Limits on Complex
Permittivity Measurements Using a Class of
Non-Invasive Probes
(1988) / Palmer W D

Modeling Multiwinding Transformers for
High-Frequency Application
(1988) / Skutt G R

Design of a Microwave Ion Source for
Electron Cyclotron Resonance Generated
Plasma
(1988) / Welch K A

FLORIDA ATLANTIC UNIVERSITY

A Simplified Jacobian Representation for
Robot Manipulation
(1988) / Alewine N J

Homomorphic Estimation and Detection of
Convolved Signals
(1988) / Cox S W

Finite Element Analysis of Inhomogeneous
Waveguides
(1988) / Dervain S T

Electromagnetic Radiation Absorption in
Block Models of Human Bodies
(1988) / Jong A Kiem R E

Optimally Polarized Monopulse Signals for
Target Parameter Estimation in Clutter
(1988) / Kang E W

Robot Singularities Under Small
Perturbations in the Kinematic Parameters
(1988) / Mohile A A

A Data Acquisition and Signal Processing
System for Analysis of Electromyogram
Signals
(1988) / Moller H C

Using Adaptive Controllers in the
Realization of a Position Control Scheme
(1988) / Samples R H

Development of Handprinting Character
Recognition Using Two-Stage Shape and Stroke
Classification
(1988) / Tse H

Modeling Errors in Kalman Filters
(1988) / Xu H

Interactive Computer Aided Digital Control
Design
(1988) / Yakali H H

FLORIDA STATE UNIVERSITY

An Approach to Implementing Linear Time-Varying
Digital Dynamic Models
(1988) / Negahdaripour G

FLORIDA, UNIVERSITY OF

A Comparison of Two Ferrite Materials for
Power Electronics Applications
(1988) / Amoni S

The Feasibility Study of a Semiconductor
Laser Based Polarization Detector for Blood
Glucose Monitoring in the Diabetic
(1988) / Arrieta I

High Speed Transient Measurement of Coplanar
Transmission Lines Utilizing Picosecond
Photoconductors Fabricated on an SOS
Substrate
(1988) / Atwater R

Computer-Aided Intelligent System to Train
Soldiers on the Utilization of the Satellite
Communications Terminal AN/TSC-85A
(1988) / Beatty W D

Implementation of Linear Filters in Parallel
Architectures
(1988) / Behara P

A Comparison of Hardware and Software
Implementations of Two Different Decoder
Algorithms for Redd-Solomon Codes
(1988) / Bonitz K L

Defects and Electrical Characterization of
Silicon-on-Insulator Materials and Devices
(1988) / Chen H S

Comparison of Smoothing Algorithms for
Electroencephalographic Signal Processing
(1988) / Choi H

Electron Beam Induced Current Analysis in
Dielectrically Insolated Pi-Silicon Tubs
(1988) / Chung B

A Graphic Semantic Data Definition Language
and a Graphic Browser for the Object
Oriented Semantic Association Model
(1988) / D'Souza T G

Manufacturing Effects on Stress-Induced
Grain Boundary Fractures in the Silicon
Aluminum Interconnect
(1988) / Dion M J

The Implementation of General-Purpose
Systolic Arrays for Digital Signal
Processing and Linear Algebra
(1988) / Euliano N R

Voltages at Both Ends of a Test Power Line
Induced by Lightening at the Kennedy Space
Center in 1986
(1988) / Georgiadis N

Optical Signal Processing Using Nonlinear
Waveguides in Semiconductor-Doped Glasses
(1988) / Han S

Frequency-Domain Transient Analysis of
Silicon Solar Cells
(1988) / Haney R E

A Backend Machine for Processing Text, Based
on Finite State Automata
(1988) / Hariharasubramanian P

Minimum-Mode-Size Low-Loss Titanium Diffused
Lithium Niobate Waveguides Operating at a
Wavelength of 1.3 Micrometers: Process
Optimization
(1988) / Horton T U

The Design and Implementation of a
Man-Machine Interface, Interactive Computing
Using Speech
(1988) / Hyun K

A Frame Buffer for Interactive Computer
Graphics
(1988) / Incirlioglu H

Analysis of a Clipping Algorithm
(1988) / Kalman A

3-Dimensional Object Recognition by Axial
Feature Signatures
(1988) / Lamba G

Fabrication and Characterization of
Potassium-Sodium-Ion-Exchanged Waveguides
(1988) / Miliou A N

Interactive Goal Driven Interface for
Digital Filter Design
(1988) / Orangio S

Development and Implementation of an Expert
System for Remotely Accessing a Relational
Database
(1988) / Ortiz A

Knowledge Base Design and Task Planning for
Manufacturing Workcells
(1988) / Pal S

Phase-Locked Loops Tracking Input Signals
of the Type $1/s(n)$
(1988) / Paphitis A

A Data Definition Language for the Object
Oriented Semantic Association Model and
Algorithms for Intelligent Query Processing
(1988) / Puranik S

A Database Design and Application
Development Tool Using an Enhanced
Functional Data Model Implemented in PROLOG
(1988) / Satya P V

Illumination Models for Real-Time Computer
Graphics
(1988) / Sofge D A

An Expert System Approach to the Academic
Counseling Problem
(1988) / Taner M

Numerical and Analytical Modeling of a
Photoconductive Switch
(1988) / Thompson S

Performance Analysis of a Differential MSK
System
(1988) / Tseng H W

The Design and Implementation of a Graphics
Interface for an Object-Oriented Query
Language
(1988) / Ty F S

FLORIDA, UNIVERSITY OF
(continued)

Slice Simulations of Analog, Neuron-Like
Circuits in an Optimization Problem
(1988) / Urbschat R

Development of a Strap-Down Digital Compass
(1988) / Vickers V

Syntactic Pattern Recognition of Spikes in
the Electroencephalogram
(1988) / Walters R

Frequency Domain Analysis of Multivariable
Systems with Structured Uncertainties
(1988) / Young P

HAWAII, UNIVERSITY OF

Design, Implementation and Testing of Data
Spreader/Despreader for Spread Aloha System
Using CMOS VLSI Techniques
(1988) / Jeevanjee N H

The Design, Analysis, and Optical
Implementation of a Class of Topologically
Similar Dynamic Multistage Interconnection
Networks
(1988) / Kaneshiro H S

Learning Algorithms for Feedforward Networks
(1988) / Keung D

A Network Model for Computer Reasoning
(1988) / Kurisu L M

Linear Edge Detection
(1988) / Ochiai J H

Design and Fabrication of a Novel Rectenna
Array for Electromagnetic Energy Conversion
and Infrared Detection
(1988) / Omid-Zohoor F

Stability Analysis and Digital Simulation of
a Biped
(1988) / Ranaweera A

HOUSTON, UNIVERSITY OF

Effect of Dissimilar Scale Factors for
Length and Diameter on Shallow Laterolog
Scale Model Tool Response
(1988) / Ansari M

Signal Processing Advisor (SPA): An
Intelligent Interface to Signal Processing
Tools
(1988) / Cespedes I

A Study of the Vector Form of the On Surface
Radiation Condition Method Applied to
Electromagnetic Scattering Problems
(1988) / Chang L

High Level Matching for Primitive-Based
Three-Dimensional Object Recognition
(1988) / Chao C H

A Theoretical and Computational Study of the
On Surface Radiation Condition Method in
Scalar Scattering Problems
(1988) / Chen S H

A Study Aid Expert System for Geophysical
Seismic Data Processing
(1988) / Dange S

Antenna Pattern Measurement Techniques for
Infinite Ground Plane Simulation Through
Edge Diffraction Elimination
(1988) / Delgado H

Analysis and Design of an Array of
Electromagnetically Coupled Microstrip
Dipoles
(1988) / Dinbergs S

A Laboratory Model of the Focused Induction
Logging Sonde
(1988) / Eshun K

The Determination of Root Mean Square
Velocities from Phase Angle Differences
Using the Hilbert Transform
(1988) / Gibbins M

Dielectric Constant Measurement of Saline
Solution at 1.1 GHz
(1988) / Han K

Geometric Analysis for the Primitive-Based
Three-Dimensional Object Recognition
(1988) / Hmam H

Analysis and Design of an Instantaneous
Floating Point Amplifier
(1988) / Ho N

Radiation and Propagation of Cylindrical
Leaky Waves on a Planar Dielectric
Multi-Layer Structure
(1988) / Ip A

Planar Transmission Line Excitation of
Dielectric Resonator Antennas
(1988) / Kranenburg R

Analysis and Design of a Traveling-Wave
Array of Vertical Monopoles in a Substrate
(1988) / Kwok S

Modelling the Electrode-Type Well-Logging
Sonde by the Method of Finite Elements
(1988) / Lam C

Design of an Automatic Analyzer and Grading
System that Evaluates Electronic Circuit
Performance Using SPICE
(1988) / Lee Y H

Conductivity Measurement at Low Frequencies
Using a Digital Bridge
(1988) / Lin C H

Design and Analysis of Electromagnetically
Coupled Microstrip Patches
(1988) / Manghnani P

Analysis and Design of a Large-Scale
Discrete-Time System Via the Matrix Sign
Function
(1988) / Mao J Q

An Intelligent System to Assist Table
Composition Using TEX
(1988) / Nulsen M

Computation of the Matrix-Valued Functions
and Their Applications
(1988) / Shen H

Numerical Methods for NDE of Stress in
Metals
(1988) / Shih M H

IOWA STATE UNIVERSITY
(continued)

A Personal Computer Based Serial Interface
for Graphic Display
(1988) / Jain S

Correcting Millimeter-Wavelength
Interferometer Data by Time-Series Modeling
and Kalman Filtering
(1988) / Koh Y

An Intelligent Printer Buffer
(1988) / Li P

Implementation of Voice/Data Integration on
a Token Ring Network
(1988) / Lio H

Graphical Computer Network Performance
Simulation
(1988) / Prayugo M

Slew-Rate Prediction for Bipolar Junction
Transistor Feedback Amplifier
(1988) / Santo H

Avalanche Build-up Time at High Gain in
a SAM-APD Structure
(1988) / Sargeant W L

A Monitor Tool for a Local Area Network
Based on Token Ring
(1988) / Tzeng S R

Interfacing a Speed Control Processor to the
IBM PC
(1988) / Woen T W

Digital Accelerometry: A Possible Clinical
Technique for the Evaluation of
Osteoarthritis of the Knee Joint
(1988) / Zhang A

Modeling of a Shield Wire Downlead for
Lightning Performance
(1988) / Zylstra P V

KANSAS, UNIVERSITY OF

A Combined LAPD/CSMA-CD Analytical Model
for the Study of Local Area Networks
(1988) / Ananthanpillai R

A Numerical Solution for Aperture Coupling
and Field Response of a Fat Cylindrical
Geometry
(1988) / Bakhtiari S

A Class-Oriented Voice Communication System
for Packet-Switched Networks: Description
and Performance Study
(1988) / DaSilva L A

Investigation of the Near Neighbor
Approximation in the Integral Equation
Modeling
(1988) / Kalbasi K

Effects of Tall Structures on Microwave
Communication Systems
(1988) / Kieu D C

KENTUCKY, UNIVERSITY OF

An Optical Encoder for Helical Motion
(1988) / Gohmann P A

Design of a Fast Hartley Transform Processor
Using a New Parallel Pipelined Architecture
(1988) / Stewart J S

Digital Position Control of Brushless DC
Motor
(1988) / Vrettos A

LAKE HEAD UNIVERSITY

Dielectric and Infrared Studies of Radio
Frequency Processes in Some Alcohols
(1988) / Mandal H

LEHIGH UNIVERSITY

A UNIX-Like File System for VM/CMS
(1987) / Bidwell P L

Helix Dispersion Measurements by
Perturbation of Traveling Waves
(1987) / Hoffman M H

Built-in Self Test for Memory Systems
(1987) / Lerner S A

Nitrogen Implanted Polycrystalline Silicon
Resistors
(1987) / Peridier C A

An Integrated Relative Humidity Sensor
(1987) / Silverthorne S V

An Investigation Into the Observed Supply
Current Variation of a Microwave Integrated
Circuit Chip
(1987) / Zidik M J

LOUISIANA STATE UNIVERSITY

Systolic Algorithms for Stable Matrix
Inversion
(1988) / Dharmarajan K R

Packing of Convex Polygons in 2-Dimensional
Non-Homogeneous Space
(1988) / Kothari R

Effects of Electron Beam Exposure and
Subsequent Annealing on the Dielectric
Breakdown Properties of Thin Silicon
Dioxide Films
(1988) / McSwain M D

Computational Complexity in Systolic
Arrays
(1988) / Nadhamuni S S

HALT: A Hardwood Lumber Training Program
for Graders
(1988) / Schwehm C J

Methods for Improving the Simulation Time of
Different Types of Interconnection Networks
(1988) / Somasegar S

LOUISIANA TECH UNIVERSITY

Local Area Network Laboratory Design
(1988) / Abrishamchian M L

A Time-Multiplexed Infrared Communicator
(1988) / Amiri J M

Measurement of Reactive Energy and Power
Factor Under Nonsinusoidal Conditions
(1988) / Khandani M

LOUISIANA TECH UNIVERSITY
(continued)

Microprocessor-Based Temperature Controller
(1988) / Mehrsa M

Design of a Microprocessor-Based Three-Phase
Converter
(1988) / Wright A M

A Synthesized Two Meter FM Transmitter
(1988) / Zarei-Toudeshki M

LOUISVILLE, UNIVERSITY OF

A Simple Complete Microcomputer in CMOS Technology
(1988) / Das A K

Dynamic Performance of Mos VLSI Inverters
(1988) / Joo Y S

Neural Network Based Computation and Pattern
Association Using the Minimum Energy Concept
(1988) / Kang M J

Generic Satellite Computer Interface Card for
MS-DOS Computers
(1988) / Levi A

A Microprocessor Based Telephone Dialer
(1988) / Raja A G

Design and Analysis of Mosfet Resistors in VLSI
(1988) / Rao S V

LOWELL, UNIVERSITY OF

Modeling of Short and Narrow Channel MOS
Transistor
(1988) / Agrawal R

Two-State Two-Phase Double Forward Converter
for Pulsed-Load Applications
(1988) / Jacobson B

Collision Avoidance Scheme for Robots in a
Cluttered Workspace
(1988) / Lee S

Fast Mellin Tranfm[sic] Algorithm
(1988) / Liu I C

Propagation Characteristics of a Corrugated
Dielectric Lined Circular Waveguide Feed
(1988) / McCormack P

Effect of Electrical Array Reconfiguration
(EAR) on Photovoltaic (PV) Powered
Centrifugal Pump
(1988) / Mulpur A K

Event Driven Simulation Algorithm for
Performance of CSMA/CD Protocol
(1988) / Patel R K

On Multiprocessor Dataflow Parallel
Pipelined Processors in Image Processing
(1988) / Simoes D D

MAINE, UNIVERSITY OF

A Surface Acoustic Wave Hydrogen Sulfide Gas
Sensor
(1988) / Falconer R S

Stroke Feature Character Recognition
Comparing Classical and Neural Network
Classifier Performance
(1988) / Shirvaikar M V

MANITOBA, UNIVERSITY OF

Analysis of Corona-Affected Surge
Propagation Along Monopolar Transmission
Lines
(1988) / Brankovic D V

Novel Schemes for Sidelobe Reduction in
Aperature Antennas
(1987) / Chamma W A

Recent Placement Methods and a New
Clustering Algorithm for Circuit Layout
(1988) / Doan T B

Moment Method Solution for Multi-Step
Discontinuities in Waveguides
(1988) / Hamid A M

Mathematical Modelling of the Contribution
of Vascular Volumetric Pulsation to the
Intracranial Pulse Wave
(1988) / Klassen P A

A System for 3D Tracking of Limb Movement
(1988) / Klein R

A Computer-Aided Personalized System of
Instruction
(1988) / Leung Y

Image Processing Using the Hough Technique
(1988) / Liao X

Simulation of Ringing and Cross-Talk on
Printed Circuit Boards
(1988) / McNeilly M E

Fault Modelling and Simulation and Built-in
Self-Test Methods for CMOS Circuits
(1988) / Nesbitt R J

Optimal Decoding of Run-Length Limited
Codes in a Magnetic Channel
(1988) / Oh Y

The Design of a Phased Array Microstrip
Antenna for Use on a Remotely Piloted
Vehicle
(1988) / Ollevier T E

An Iterative Procedure for Planning
Manipulator Movements
(1988) / Procca M J

Image Processing Algorithms and Their
Implementation in Silicon
(1987) / Sawh D R

Complex Wave Digital Networks Including the
Implementation of Even-Order Filters
(1988) / Scarth G B

The Combined Influence of Early Reflections
and Background Noise on Speech
Intelligibility
(1988) / Soulodre G A

Effects of Substrate Bias on the Properties
of Alpha-Silicon/Hydrogen Fabricated by
ECR Microwave Plasma CVD
(1988) / Tse T Y

Design of Digital Wave-State-Variable
Lattice Filters
(1987) / Webb C W

MANITOBA, UNIVERSITY OF
(continued)

A Pattern Recognition Based Arm Movement
Classifier with Increased Feature Space
(1988) / Wieler J A

Grapahics-Driven Electric Power Systems
Programs for the Personal Computer
(1988) / Wong K F

Modelling and Simulation of a Physiological
System Producing Ocular Pulse
(1988) / Yan Y

MARQUETTE UNIVERSITY

Implementing Real-Time Digital Signal
Processing Algorithms in Hardware
(1988) / Frigo F

Computerized Data Base Management: A
Decision Making Tool for Managers
(1988) / Kaufmann C

AC Electrical Behavior of Carbon and
Ruthenium-Oxide Based Resistors
(1988) / Khanam S

Digital High Voltage Probe
(1988) / Lin J H

The Graphic Entry Mode for Spice
(1988) / Lin S C

PC Based Real Time Data Acquisition and
Motor Control for a Wrist-Testing Device
(1988) / Luczyk W

Design of a Distributed Data
Acquisition/Analysis System
(1988) / Manikundalam R

Craniofacial Cephalometry: A Comparison
Between Two-Dimensional Radiography and
Three-Dimensional Computerized Tomography
(1988) / Montoure C

Electromagnetic Field Behavior of a
Tapered Iris in Rectangular Waveguide
(1988) / Natzke J

Transdermal Radio Frequency Coil Analysis
Using Network Analyzer Measurements for
an Artifical Heart Powering System
(1988) / Nedwek P

Analysis of a Voltage Controlled Current
Regulated PWM Induction Motor Drive
(1988) / Seibel B

Real Time Data Acquisition and Control of
a DC Motor Driving a Biomechanical Device
(1988) / Shastri J

Electrochemical Analysis of Microleakage:
Evaluation of Lateral Condensation,
Obtura, Ultrafil and Endotec Obturation
(1988) / Squitieri M

MASSACHUSETTS INSTITUTE OF TECHNOLOGY

Task-Level Robot Learning
(1988) / Aboaf E W

Formulation of Scattered Fields from Thick
Slots Cut in the Common Broadwall of Crossed
Waveguides
(1988) / Adams C J

Correlation of Permeability, Hydration, and
Crosslink Density in Polyelectrolyte
Membranes
(1988) / Adler K P

On the Analysis of the Hopfield Network: A
Geometric Approach
(1988) / Akra M A

System Identification for Small Continuous
Time Systems
(1988) / Alkhairy A S

The Photoconductivity of Carbon Fibers
(1988) / Alryyes A M

Control of Multi-Switch, Multi-Output Power
Converters
(1988) / Alvarez L S

Satellite Precision Pointing Using
Robustness Recovery
(1988) / Amsbury B W

Pulsed Operation of Five Mutually Coherent
Semiconductor Diode Lasers in an External
Cavity
(1988) / Anderson K K

Error and Flow Control Over Satellite
Links in Packet-Switched Networks
(1988) / Andrews E J

Simulation and Performance Analysis of
Trellis-Coded Modulation for Microwave
Digital Radio
(1988) / Animalu C N

An Adaptive Thermal Module for Transformer
Monitoring
(1988) / Archer D S

Power Distribution System Design by Multiple
Attribute Tradeoff Analysis
(1988) / Ariura Y

Effect of Multiple Scattering on the Radar
Cross Section of Polygonal Plate Structure
(1988) / Atkins R G

A Low Capacitance Vertical Power MOSFET with
an Integral Turn-Off Driver
(1988) / Bahl S

Network Synchronization in a PBX
Environment: An Application of Discrete
Control Correction
(1988) / Banerjee S

A WYSIWYG Program Generator for Interactive
Displays
(1988) / Bartholomew J K

Evidence for Anisotropic Mechanical
Properties of Fibroblast Membranes
(1988) / Basch R M

A Signature Schyeme Based on Trapdoor
Permutations
(1988) / Bellare M

An Investigation of a Small Bolus
Thermodilution Technique
(1988) / Bettadapur M S

Digital Signal Processor Implementation of
the Base Site Controller Data Recovery
Circuit
(1988) / Bordoloi A D

MASSACHUSETTS INSTITUTE OF TECHNOLOGY
(continued)

Time Scale Analysis Techniques for Flexible
Manufacturing Systems
(1988) / Caromicoli C A

The Effect of Flexible Walls on an
Acoustical Waveguide Transducer
(1987) / Caron G F

Kerr Electro-Optic Field Mapping
Measurements in Sulfur Hexafluoride
(1988) / Carreras R F

Parallel Computation in Model-Based
Recognition
(1988) / Cass T A

Limitations of Interpolative Modulators for
Oversampling A/D Converters
(1988) / Chao K C

Characterization of Gate Oxide Breakdown in
Analog Devices Semiconductor's CMOS Process
(1987) / Choi W Y

Fabrication of Lateral-Surface-Superlattice
MODFET's Using X-ray Lithography
(1988) / Chu W

A Dual Processor Vision System for Real-Time
Stereo and Motion
(1988) / Ciholas M E

Modeling of Delay in Emitter Coupled Logic
(1988) / Cohan J L

Current Drive on the Versator-II Tokamak
with a Slotted-Waveguide Fast-Wave Coupler
(1988) / Colborn J A

Describing the Control Systems of Simple
Robot Creatures
(1988) / Cudhea P W

A Survey of Nonlinear Modeling Techniques
and Their Application to Dial Line Modern
Echo Cancellation
(1988) / Culbert J A

A Radio Frequency DC/DC Square-Wave
Converter
(1988) / Culpepper B J

Parallel Algorithms for Polyphonic Pitch
Tracking
(1988) / Cumming D P

Identification of Resonant Electromechanical
Plants from Frequency Response Measurements
(1988) / DeAngelis F E

Low Frequency Acoustic Methods for Leak
Location in Underground Oil-Filled Conduits
(1988) / Delin K A

Congestion Control in Large Frame-Relay
Networks
(1988) / Desnoyers P J

Design and Implementation of an
Omnidirectional Wheeled Robot Motion
Controller
(1988) / DiLorenzo D J

Nondrian: A Two-Dimensional Graphical
Specification and Design Programming
Environment
(1988) / Fern K L

Coupled-Wave Analysis of Crossed Gratings
(1987) / Gaither S A

A Full Digital Implementation of an ACSB
Modem
(1988) / Gandhi S K

An Integrated Team Approach to the Redesign
of an Isolated Amplifier at Analog Devices
(1988) / Gatto D F

Interfaces for Subnetworks of Single Mode
Fiber Optic Networks
(1988) / Georgopoulos V C

Analysis of Values in the Radio Industry
(1988) / Greene I R

Bandwidth Limitations of a CMOS Hall-Type
Magnetic Field Sensor
(1988) / Hakkarainen J M

Performance Analysis of Retrial Access
Schemes for Optical Communication Networks
with Slow and Fast Tuning
(1988) / Hamdy W M

Performance Analysis of Quasi-Optimal,
Multipixel Laser Radar System Processors
(1987) / Hannon S M

Dynamic Response of ALOHA Systems with
Adaptive Retransmission
(1988) / Harman J S

A Memory Design for a Message Driven
Processor
(1988) / Hassoun S M

Imbedding Data Into Television Signals
(1988) / Hoque T I

Reactive Navigation for Mobile Robots
(1988) / Horswill I D

Design and Evaluation of a Linearizer for
Use with an L-Band Solid-State Power
Amplifier for Satellite Applications
(1988) / Huettig F R

A Thermally Isolated Microstructure Suitable
for Gas Sensing Applications
(1988) / Huff M A

A Monolithic BICMOS Radio Frequency (RF) to
Baseband Translator and Signal Conditioner
(1988) / Iacoviello D P

Resource Allocation Management: An Approach
to Automatic Test Program Generation
(1988) / Indech R G

A 50 MHz Frequency Synthesizer with 1 Hz
Resolution
(1988) / Ing E K

A Digital Model for Level-Clocked Circuitry
(1988) / Ishii A T

Parallel DNA Sequence Analysis
(1988) / Iyengar A K

An Intelligent Process Flow Language Editor
(1988) / Jayavant R

MASSACHUSETTS INSTITUTE OF TECHNOLOGY
(continued)

A State Observer for the Permanent-Magnet
Synchronous Motor
(1988) / Jones L A

Run-Time Negotiation of Value Representation
in a Distributed System
(1988) / Joseph A D

A Multiprocessor Based Design of a Digital
Beam Forming System
(1988) / Karagozyan K

A Spectroscopic Technique for Detection of
Intracellular Calcium Transients Using
Indo-1: A Means for Determining the
Electric Field Effects on Intracellular
Free Calcium
(1988) / Kenny J R

PATH: A Simulation-Based Transistor Sizer
(1988) / Kim J K

A Semi-Linear Filter Bank Model of the
Peripheral Auditory System
(1988) / Kim R Y

Performance and Design Issues of a SCRAM
Based Virtual Cache
(1988) / Kipnis J D

Design and Characterization of Polysilicon
Emitter Bipolar Transistors
(1988) / Krisch K S

A Post-Processor Model for Estimating
Channel Hot-Electron Emission
(1988) / Kumar A

The Selectively Delta-Doped Effect
Transistor Grown by Gas Source Molecular
Beam Epitaxy
(1988) / Kuo T Y

Automated Techniques for the Localization
of Software Bugs
(1988) / Kuper R I

Verification of a Dynamic Grinding Model
(1988) / Kurfess T R

Task Allocation for Efficient Performance
of a Decentralized Organization
(1987) / Lee C

The Effects of Rapid Thermal Annealing on
Gallium Arsenide Grown by MOCVD on Silicon
Substrates
(1988) / Lehman L L

A Cost-Effective Design of a Digital Signal
Processing Subsystem
(1987) / Li J

An Experimental Approach to Module
Diagnostics Using an Associative Memory
(1988) / Li R

Modularity in LISP Systems
(1988) / Loaiza J R

A Microfabricated Electrostatic Motor Design
and Process
(1988) / Lober T A

Central Processing in Sentence Reception
(1988) / Machado M E

Characterization of the Latissimus Dorsi
Muscle as a Power Source for Use in Cardiac
Assist
(1988) / Malek A M

Multi-Sensor Interface for an Implantable
Microprocessor-Based Pacemaker
(1988) / Margolis D A

Electricity Demand-Side Management in
Developing Countries: The Case of Costa
Rica
(1988) / Marin-von-Koller S

Message Monitoring on Quick-Silver
(1988) / Martin S R

The Design of a Versatile Microprocessor
Software Development Station
(1988) / Medley R A

Real Time Control, Acquisition, and Image
Processing for the Scanning Tunneling
Microscope
(1988) / Modiano A M

Measuring the Effects of Temperature and
Radiation on Minority-Carrier Recombination
Lifetime in n-Doped Bulk Silicon Using
MOS Capacitors
(1988) / Morgan J M

Theoretical and Experimental Studies of the
Action of Proteoglycan-Degrading Enzymes
Hyaluronidase, Chondroitinase, and Trypsin
on Articular Cartilage
(1988) / Morgenthaler A W

Parity Simulation of Two-Phase Hydraulic
Flow with Application to Nuclear Power
Plants
(1988) / Mweene L H

Synthesis and Structure of Sputtered Thin
Films of Chevrel Phase Superconductors
(1988) / Neal M J

An Object-Oriented Approach to Heterogeneous
Distributed Databases
(1988) / Ng G F

The Accurate Measurement of Intrinsic MOS
Capacitances
(1988) / Ngai J

Performance Measurement of Orphan Detection
in the Argus System
(1988) / Nguyen T D

A Comparison of Adaptive Predictors in
Sub-Band Coding
(1987) / Ning P C

A Framework for Real-Time Graphics on a
Multiprocessor
(1988) / Nussbaum D S

Tunable Long Wavelength Intensity Modulation
by Means of Intersubband Electroabsorption
in Quantum Wells
(1988) / Pan J L

A Threshold Strategy for a First in
First out Heterogeneous Two Server
Queing System
(1988) / Pankaj R K

MASSACHUSETTS INSTITUTE OF TECHNOLOGY
(continued)

Distributed Commit Protocols for Nested
Atomic Actions
(1988) / Perl S E

The Design, Construction and Evaluation of
a Real Time Optical Wave Front Sensor
(1988) / Petivan J L

Distribution of Computation in a
Multiprocessor System
(1988) / Phillips R A

Measurement of Heat Exchange and
Bronchospastic Response in Exercise-Induced
Asthma
(1988) / Potash R J

Control of a Reluctance Motor Using Constant
Frequency Excitation
(1988) / Potter D L

Recognition-Processing Systems for Speech
Enhancement: Application to Frequency
Lowering
(1988) / Power M H

Speed Considerations in an Actively Loaded
16kb Memory Cell
(1988) / Raman E R

An Object-Oriented Proof Environment for
Formal Hardware Verification
(1988) / Rege O U

Typechecking is Undecidable When Type is a
Type
(1987) / Reinhold M B

A High-Speed Data Acquisition System for
Multi-Station Flow Cytometry
(1988) / Roman P J

The Effects of Ion Implantation and
Annealing on the Properties of Titanium
Silicide Films on Silicon Substrates
(1988) / Rubin L M

Hierarchical Initial Placement on Gate
Arrays
(1988) / Rubin W E

Improving Control in Rule-Based Systems by
Symbolic Analysis of Data Patterns
(1988) / Santos W H

Diversity-Based Inference of Finite Automata
(1988) / Schapire R E

Encoding of Acoustic Resonance on the
Auditory Nerve
(1988) / Secker-Walker H E

Compacting VLSI Layouts Containing Octagonal
Geometry
(1988) / Seda S J

Constructing a Computational Environment for
Axiomatic Reasoning in Euclidean Space
(1988) / Shaw A C

Kerr Electro-Optic Field and Charge
Measurements in Transformer Oil
(1988) / Sheen D M

Phase-Lock Clock Generator for VLSI
Applications
(1988) / Sherred J L

Optical Amplification to Increase SNR
(1988) / Simon S A

Orthogonal Sub-band Image Transforms
(1988) / Simoncelli E P

Multistep Methods for Integrating the Solar
System
(1988) / Skordos P A

Optimization of Noise Figure in Microwave
Circuits with Multiple-Stage Feedback
(1988) / Soares S

A Simulation Environment for SCHEMA
(1986) / St. Pierre M A

High Temperature Superconductivity: Federal
Policies Towards the Commercialization of a
New Technology
(1988) / Stokes S J

The Optimal Polarimetric Matched Filter for
Achieving Maximum Contrast in Radar Imagery
(1988) / Swartz A A

An Adaptive Periodic Synthronizer Circuit
(1988) / Swiston R C

Thermal Annealing Effects on Boron-Implanted
HgCdTe Diodes
(1988) / Syz J K

Integrated Optic Laser Doppler Anemometer
(1988) / Szwarc B J

The Real-Time Digital Control of a
Regenerative Above-Knee Prosthesis
(1988) / Tabor K A

Analysis of Synthetic Tadoma System as a
Multidimensional Tactile Display
(1988) / Tan H Z

A Phase-Locked Loop Based High Speed Data
Communication Circuit for Parallel
Processing Computers
(1988) / Thompson B L

A Comparative Study of the Semiconductor
Industry of Taiwan and Korea
(1988) / Tsai W J

Speed and Parameter Estimation in Induction
Machines
(1988) / Velez-Reyes M

A Simplified Analysis of Variable Reluctance
Motors
(1988) / Walrath K E

Sensitivity of Multimicrophone Adaptive
Beamforming to Variations in Target Angle
and Microphone Gain
(1988) / Wei S M

Simulation of Inverter-Driven Six-Phase
Synchronous Machines
(1988) / Weil S

NuBus to PC/AT I/O Bus Converter
(1988) / Wolfe J V

MASSACHUSETTS INSTITUTE OF TECHNOLOGY
(continued)

Automated Semiconductor Wire Bond
Inspection Using Machine Vision Techniques
(1988) / Wu D S

Test Generation Guided Design for
Testability
(1988) / Wu P

An Analysis of Analog CMO Coupling Circuits
at Liquid Nitrogen Temperature
(1988) / Yanney P R

Grating Gate Si MOSFET for Study of Quantum
Transport Effects
(1987) / Yen A

Edge-Directed Image Sharpening
(1988) / Yun A J

Digital Control of an Electromechanically
Actuated Mirror for Laser-Based VLSI Repair
(1988) / Zahavi J S

Diffraction Loss in Urban Mobile Radio
Propagation
(1988) / Zancewicz G J

Partial Discharge Pulse Height Distribution
Analysis of Mixed Media Dielectric Systems
(1987) / von Guggenberg P A

MASSACHUSETTS, UNIVERSITY OF

A Self Re-Configuring Defect and Fault
Tolerant Random Access Memory for Wafer
Scale Integration
(1988) / Daruwala A M

Test Set Embedding in a Built-In Self Test
Environment
(1988) / Jansz W N

Low Sidelobe Microstrip Phased Arrays
(1988) / Kaufman B D

Segmentation of Multi-Look Speckled Images
(1988) / Labitt S G

On-Line Sensitivity Estimation for Load
Balancing in Real-Time Distributed Systems
(1988) / Lee J I

Lyapunov-Based Approach for Obstacle
Avoidance and Target Attainment
(1988) / Lee M

In Situ Study of the Reaction Between
Trimethylgallium and Ammonia
(1988) / Mazzarese D

A Graphic Network Management System
(1988) / Meyer A S

Ocean Surface Wind Speed Determination from
SMMR Brightness Temperature Data and
Applications to the SSMR I
(1988) / Morris J B

A Survey of Selected Data Flow Systems
Leading to the Proposal of a Novel Data
Flow Architecture
(1988) / Mottola J E

The Implementation of a Robust
Force/Position Controller on a Robotic
Manipulator
(1988) / Murah B

Fault Simulation and Test Generation for
Delay Faults
(1988) / Oomman B G

Computer Aided Design of a C-Band
Scatterometer
(1988) / Pazmany A L

Recursive Linear Smoothing of Images
(1988) / Punyagupta P

Synthesis of Series-Fed Rectangular
Microstrip Patch Arrays
(1988) / Sanchez V C

Adsorption Melchanism and Surface Reaction
of Ammonia on Gallium Arsenide
(1988) / Tripathi A

Design Issues for Higher-Level Protocols in
High Speed Communication Networks
(1988) / Van Leeemput G

Measurements of Atmospheric Liquid Water at
215 GHz
(1988) / Vandermark D

Segmentation of Spekled Images Using Complex
Data
(1988) / Vezina G

MIAMI, UNIVERSITY OF

Design of Broadband Matching Networks for
Microwave Amplifiers
(1988) / Dugue E I

Accuracy in Three-Dimensional Position
Verification of a Mobile Robot Using
Real-Time Moment Generation
(1988) / Hussain B

Protection and Coordination of Power Systems
with Co-Generation Units
(1988) / Mansour H A

An Expert System for Recovering 3D Shape and
Orientation from a Single Perspective View
(1988) / Shomar W J

MICHIGAN TECHNOLOGICAL UNIVERSITY

Electrical and Structural Characterization
of KNO(3) Thin Film Ferroelectric Memory
Devices
(1988) / Adams S E

A Simulation Technique for Nonlinear Control
Systems with Application to the Combined
Effects of Nonlinear Friction and Backlash
(1988) / Angeli P

A New Algorithm for Time Delay Estimation
Using Generalized Maximum Likelihood and
Kalman Filtering Techniques
(1988) / Bainbridge P A

Acceleration of Simulation for Mixed
Gate-Level and Behavioral-Level Modeled
Digital Electronic Designs
(1988) / Buchs K J

MICHIGAN TECHNOLOGICAL UNIVERSITY
(continued)

High Voltage Impulse Testing of
High-Pressure Oil-Filled (HPOF) Cable
Insulating Oil
(1988) / Foster J D

Functional Level Symbolic Simulator for
Digital Circuits
(1988) / Haveliwala A A

A Neural Network Based Multiplier
(1988) / Hu C

Characterization of Multi-Level
Interconnections on GaAs-Based VLSI
(1988) / Huang Y

Automatic Minimization and Design of Small
Digital Logic Circuits
(1988) / King D J

A Software-Based Phase-Locked Loop Signal
Tracking System
(1988) / Lund V

Implementation of an Automated Manipulator
Using an IBM 7540 Robot with Tactile Sensing
(1988) / Marts B S

An Investigation of Compensation Methods for
Systems Containing Time-Delay
(1988) / Murphy T J

Analysis and Synthesis of an Optimal
Ball-Balancing Control System
(1988) / Nadolsky D S

Electrical and Structural Characterization
of Mo/Si Contacts to Gallium Arsenide
(1988) / Patkar M P

A Tensile-Testing Apparatus for Use Under a
Microscope
(1988) / Peek D N

Improved Spectral Estimates of
Non-Stationary Processes Using a Modified
Phase Locked Loop
(1988) / Rau M

Test Procedures for Determining the
Dielectric Strength of Insulation Systems
Utilizing Silicone Fluids
(1988) / Simpson D E

Parametric Model Based Edge Detection for
IR Imagery
(1988) / Sullivan S P

A Silicon Compiler for VHDL
(1988) / Vasireddy V R

A VHDL Simulator
(1988) / Vyas M

MINNESOTA, UNIVERSITY OF

Numerical Approximation of 'Si(f)' for
Classical Diffusion '1/f' Noise in an
n + p Diode of Arbitrary Length
(1988) / Anderson J B

Hydraulic Servovalve Linearization Using
Dynamic Feedback
(1988) / Axelson S D

Current and Flux Distribution Within Thin
Film Inductor Geometries
(1988) / Benser E T

Simulation of Gallium Arsenide MESFET DCFL
Logic
(1988) / Bernhardt B A

Characterization of Rare Gas Radio Frequency
Discharges
(1988) / Bianchi T E

Robust Stability Bounds for State Space
Models
(1988) / Brauch J C

Dual Pressure Transducer Parameter and
Sensitivity Analysis
(1988) / Broderick J B

Pulse Detection in a Hard Disk Magnetic
Recording Channel
(1988) / Caddy R E

A Comparison of Four Magnetic Imaging
Methods
(1988) / Cameron G P

Temperature Dependence of Gallium Arsenide
Mesfet Direct-Coupled Field-Effect
Transistor Logic
(1988) / Conger J S

An Operating System Coprocessor Interface
(1988) / Dahl P J

Workstation-Based System-Level Simulation
Software Program Description
(1988) / Helfinstine A R

Intracardiac Electrical Impedance During
Sinus Rhythm and Ventricular Fibrillation
in Canines
(1988) / Huberty K P

Voluntary Cardiorespiratory Synchronization
and Health Enhancement
(1988) / Marble T N

System Design and Layout Studies for a High
Throughout Signal Processor Fabricated with
Gallium Arsenide LSI Configurable Cell
Arrays
(1988) / Naused B A

Flexible Media Single Disk Tester Manual
(1988) / Pentck C J

A Method of Assessing the Effect of
Parasitic Resistance on CMOS Digital Circuit
Performance
(1988) / Rabe R L

Knowledge-Based Spectral Estimation
(1988) / Recio A

A Loran-C Receiver
(1988) / Reilly T J

Field Emission Tips as Low Energy, High
Current Density, High Resolution Electron
Sources
(1988) / Schlotterback J W

An Experimental Study of Ion Bombardment
Energy Distributions in RF Discharges
(1988) / Toups M F

MINNESOTA, UNIVERSITY OF
(continued)

Feasibility Study of AC- and DC-Side Active
Filters for HVDC Converter Terminals
(1988) / Wong C

Intensity Analysis of Electron Diffraction
from Gallium Arsenide (110) Prepared by
Molecular Beam Epitaxy
(1988) / Zhou D N

MISSISSIPPI STATE UNIVERSITY

Two Dimensional MOS (n-Channel) Transistor
Incorporated Velocity Saturated Effects
(1986) / Ahmad A

Design and Analysis of High Voltage Impulse
Dividers
(1986) / Ahmad F H

A Process Development and Quality Evaluation
Test Chip for Double-Level Metal CMOS
(1986) / Ansel G M

Analytical Study and Design of Multi-Color
Pyrometer Systems
(1986) / Arabi A O

High Voltage DC Transmission Line Characteristics
and Potential for Interconnecting Libyan Systems
(1986) / Baiou A A

A Comparison of a Classical Guidance Law and
Three Modern Guidance Laws
(1986) / Baker R A

The Design of a Digital Acceleration Autopilot
for FAMMS
(1986) / Butler M L

Bubble Memory System for Motorola 6802
Microcomputer
(1986) / Chen L W

A Microwave Path Loss Test of a Defense
Metropolitan Area Telephone System
(1986) / Clark R F

The Design of a Digital Timing Control Unit
for Impulse Generator Control
(1986) / Crisler R W

Calculation of the Per-Unit-Length Impedance and
Admittance Matrices for Multiconductor
Transmission Lines
(1986) / Donohoe J P

Pulse Position Modulation and Demodulation
(1986) / El Murr N F

Analysis of the Space Telescope MA Low Data
Rate Communication Link
(1986) / El-Murr S C

Electrostatic Problems in Microelectronics
(1986) / Elkhouri G N

Emulation and Characterization of Digital
Controllers Realized with Fixed-Point Arithmetic
(1986) / Follett R F

Implementation of an Augmented Proportional
Navigation and Guidance Law using Kalman Filters
(1986) / Francis R C

Digitization of a Data Acquisition System for
Partial Discharge Pattern Recognition
(1986) / Gary S A

Test Structure Mask Realization for
CMOS/Refractory Metal Applications
(1986) / Hart H T

An Operating System for a Small Bubble Memory
(1986) / Hemler L G

A Tunable Notch Filter Using Adaptive Sampling
(1986) / Irwin R D

A Truth-Table-Based Approach for Testing
Digital Logic Circuits
(1986) / Jeng M J

The Capacitive Shaded Pole Induction Motor
(1986) / Khonsary S M

A Self-Contained Microprocessor-Based Data
Acquisition System
(1986) / Kuo J C

Power Grid Demand Control - An Alternative
to New Power Plant Construction
(1986) / Lane J W

A Microcomputer Controlled Cartridge Tape
Recorder System
(1986) / Lee C L

Distribution of High Bandwidth Signals Using
FDM (Frequency Division Multiplexing) Techniques
(1986) / Lupinetti F

System for Testing the Field Control Circuitry
of Brushless Synchronous Motors
(1986) / McAnelly M L

A General Purpose Microprocessor Controlled
Data Acquisition System
(1986) / McCleave B W

Experimental Measurement of Failure Rates
for Synchronizers
(1986) / Morgan K P

The Design of a STAR Gate Array for Space
Applications
(1986) / Newman W H

Design of A Hardware and Software Interface
Between A Microprocessor-Based System and A
20-ma-Current Loop TTY
(1986) / Pan J S

Frequency Response Simulation of Sampled-Data
Control Systems
(1986) / Pinion W I

The Design and Analysis of a Test Chip for
Evaluating Mask-to-Mask Alignment and Process
Variations for a Silicon Gate CMOS Technology
(1987) / Ramesh T K

An Observer/Controller Compensator Design for the
Advanced Gimbal System Using Modern Control
Techniques
(1986) / Robertson J A

Interconnection Studies for VLSI Fabrication
Technology
(1986) / Safdar S M

MISSISSIPPI STATE UNIVERSITY
(continued)

The Digital Computer Application in Electric Power
Systems with Analysis of the Existing Power System
in Eastern Libya, As A Case of Study
(1986) / Salah A S

The Comparison of Conventional (Voltage-Divider) and
Elector-Optic (Kerr-Effect) System in High Voltage
Measurements
(1986) / Sheissi G

Design of a VLSI Complexity Silicon-Gate CMOS Gate
Array
(1987) / Simpson J S

A Microcomputer-Based System for Testing and
Optimizing Distribution Transformer Cores
(1987) / Smith J C

An Energy Independent Dwelling
(1986) / Streeter K C

A Study of the Limitations and Applicability of
Direct Drive Techniques for Nuclear EMP Tests
(1986) / Tarbutton G L

A New Approach to the Analytical Design of
Compensators
(1986) / Tollison D K

A Survey of the Fast Fourier Transformation: A
Modern Digital Perspective
(1987) / Vassiliadis C A

A Computer Program for Analysis of a Rural Electric
System in Terms of System Performance and Energy Loss
(1987) / Whitman S D

Computational Methods for Determining z-Transforms
(1986) / Williamson M J

The Design of a Biotelemetry System to Monitor
Cervical Motility in Baboons
(1986) / Wiseman A P

On Common Mode Versus Differential Mode Drives of
Multiconductor Transmission Lines
(1986) / Younan N H

MISSISSIPPI, UNIVERSITY OF

Automatic Precision Microwave Measurements
in Waveguides
(1988) / Gundavahala A

Electromagnetic Scattering from Multiple
Strips in the Presence of a Circular
Conducting Cylinder
(1988) / Yu M C

MISSOURI, UNIVERSITY OF (COLUMBIA)

Selective-Pole Switching of Double-Circuit
EHV Line: Secondary Arc Current Calculation
(1988) / Al-Rawi A M

Obscured Square Aperture Effects on Depth
of Focus
(1988) / Bobo L B

Switched-Capacitor Networks in Operational
Amplifier Circuits
(1988) / Coleman M W

Study of Production Rate of A-Center in Float-Zone
Silicon
(1988) / Husain A

DCTUTOR: The Application of an Expert System
Shell as an Intelligent Tutoring System for
DC Circuit Analysis
(1988) / Linneman D A

Neural Network Systems for Recognizing Lines
in a Grey-Scale Image
(1988) / Mohandes M N

Image Analysis of Utility Maps Using Geometric
Properties
(1988) / Qiu H

Programming and Comparisons of the Algorithms
for the Discrete Fourier Transform
(1988) / Sung J J

MISSOURI, UNIVERSITY OF (ROLLA)

Optimizing Transformer Impedance for Short
Circuit Analysis
(1988) / Becker L S

Microcomputer Application for Power
Distribution System Overcurrent Protection
and Coordination
(1988) / Bernhard F V

Toward Interdisciplinary Education: The
Development of a Course in Automated
Manufacturing for Electrical Engineering
Students
(1988) / Dalton J S

Nonrectangular Windows for the Median Filter
(1988) / Dehner L G

Development of a Microcomputer-Based
Muscle Tonometer
(1988) / Gerling P G

Ultrasonic Film Thickness Gauge
(1988) / Haake J M

Investigations of Thermal Annealing Effect
on Interface States of SOS Devices Using
DLTS Method
(1988) / Horng D S

The Ethernet and the Validation of the
Network II.5 Simulator Program
(1988) / Jingle J G

A Multi-Channel Biotelemetry System for use
in Kinesiological Research
(1988) / Lickenbrock J A

Computer Aided Design of Microstrip
Antennas
(1988) / Martinosky J W

A Hardware Compiler for Erasable
Programmable Logic Devices
(1988) / Sarma R

Testing of Integrated Circuit Chips with
an IBM PC/XT Interface
(1988) / Vishwanathan N

The Center-of-Inertia Transform, with
Application to Electric Power Systems
(1988) / Waggoner D R

The Implementation of the Forth Language
on a RISC-Type Machine: The IMS T212
Transputer
(1988) / Yang F G

MISSOURI, UNIVERSITY OF (ROLLA)
(continued)

Maximum Log-Likelihood Estimation of
Discrete-Time Delay Function
(1987) / Yang J T

Bit-Slice Based Microprogramming Development
System Design
(1988) / Yeh W H

MONTANA STATE UNIVERSITY

Motional Electric Fields Associated with
Relative Moving Charge
(1987) / Klicker K A

MONTREAL, UNIVERSITY OF

Entrainement Quatre Quadrants par
Micropresseur et Fibres Optiques de
Machines a Courant Continu
(1988) / Arab A

Simulateur Fonctionnel a l'Aide d'une
Interface de Haut Niveau au Behavioral
Language Model
(1988) / Arel N

Etude des Inductances de Fuite d'un
Transformateur de Convertisseur
Quadri-Hexaphase
(1988) / Ba A D

Modelisation et Optimisation de
l'Approvisionnement en Cables Telephoniques
d'un Reseau d'Acces
(1988) / Beauregard A

Etude Comparative d'Algorithmes de Commande
de Manipulateur Robotique
(1988) / Benali R

Modelisation de l'Effecteur d'un Robot
Hydraulique en Vue de la Commande par
Micro-Ordinateur
(1988) / Bergeron J

Modele Quantitatif de l'Apport des Outils
Informatiques a la Formation en Genie
(1988) / Birikundavyi S

Etude et Realisation de Generateurs de
Structures Logiques Regulieres et
Semi-Regulieres
(1988) / Boubguira A

Analyse de Certains Problemes des Reseaux
Alternatifs/Continus Interconnectes
(1987) / Boutebel M

Systeme de Communication Seriel
Programmable a Multiples Points d'Acces
(1987) / Coderre P

Micro-Ordinateur Avec Convertion A/D D/A
Rapide
(1987) / Dada Z

Methode d'Etalonnage d'un Reflectometre
Six-Port Utilisant Quatre Positions d'un
Court-Circuit Mobile
(1988) / Demloj B

Design d'un Dispositif Actif
Hyperfrequentiel Multi-Ports
(1987) / Doucet M

Etude Experimentale de la Tension Disruptive
des Gaz en Champ non Uniforme
(1988) / Drapeau J F

Decodage Sequentiel Pour Canaux a Memoire
(1988) / El Hage S

Sur l'Analyse de la Performance d'Erreur
des Codes Convolutionnels
(1988) / El-Am A

Un Routeur Global Automatique de Circuits
Integres
(1988) / Gadiri A

Mesure de la Permittivite Complexe en
Hyperfrequence Utilisant un Systeme Six-Port
(1988) / Guay R

Utilisation de Modeles Probabilistes en
Analyse de Textures
(1987) / Lacasse V

Conception d'une Source de Tension de
Securite
(1988) / Larivee J F

Methode de Moments Dans la Solution d'une
Jonction en T
(1988) / Ly C D

Appareil de Mesure du Facteur de Deplacement
Pour Alimentation a Frequence Variable
(1988) / Mayaki F

Procedure d'Essais Statiques et Dynamiques
de Moteurs Pas-a-Pas
(1987) / Mongeau P

Conception d'un Transformateur de Puissance
Triphase a l'Aide d'un Ordinateur Personnel
(1988) / Oligny M

Evaluation des Tensions de Contact sur les
Reseaux de Distribution Souterrains 25kV
(1987) / Rajotte Y

Developpement d'un Analyseur de Reseau
Six-Oorts et Caracterisation de Transistors
Micro-Ondes
(1988) / Rivard J F

Reponse Frequentielle d'une Ligne de
Transmission Non-Uniforme Terminee par un
Systeme Resonnant
(1988) / Robichaud R

Etude sur la Pollution Harmonique: Normes,
Techniques de Mesure et de Compensation
(1988) / Siam A H

Simulation de Reseaux a Commutation de
Paquest par Routage Dynamique
(1988) / Sisto E

Analyse des Performances d'un Decodeur
Convolutionnel Bas sur l'Algorithme de
Viterebi Tronque
(1987) / Smith R

Simulation de Configurations d'Ordinateurs
Repartis de Systemes de Procede en Temps
Reel
(1988) / Tazi A

Contraintes Dielectriques dans les
Enroulements de Transformateurs THT
Dues a des Surtensions de Manoeuvre
(1988) / Tschudi D

MONTREAL, UNIVERSITY OF
(continued)

Etude Comparative des Methodes de
Compensation de l'Energie Reactive dans
lse Lignes Radiales Longues
(1988) / Wade C

Application d'Energie Pour le Sechage par
Micro-Ondes d'Equipement a Haute Tension
(1988) / Wakil M

Application du Compensateur Autosyntonisant
dans l'Industrie des Pates et Papiers
(1988) / Yak B L

NEBRASKA, UNIVERSITY OF

Susceptibility of Circuit Switched Networks
(1988) / Chang H

A Master Tape Processing System Designed for
the Requirements of High Speed Tape
Duplication
(1988) / Cheenne D J

Effects of Measurement Processing Order on
Nonlinear Applications of Sequential Kalman
Filters
(1988) / Dawson B D

Self-Tuning Control Simulation Study Based
on the Generalized Predictive Control Method
(1988) / Han S

Laser Etchihng of Metal Film Magnetic Disks
to Develop an Error Standard
(1988) / Leonhardt R W

The Characterization of Gallium Arsenide/
Aluminum Gallium Arsenide Heterostructures
by Variable Angle Spectroscopic Ellipsometry
(1988) / Merkel K G

Growth of Diamond by R. F. Plasma Assisted
Chemical Vapor Deposition
(1988) / Meyer D E

A Study of Empirical On-Line Tuning
Techniques for the Kalman Filter
(1988) / Olson R L

The Physical Mechanism of Triggering in
Trigatron Spark Gaps
(1988) / Peterkin F E

A Novel Technique for Designing Continuous
Time Filters in MOS Technology
(1988) / Prigeon J A

NEW BRUNSWICK, UNIVERSITY OF

Spectral and Correlation Techniques for
Identification and Estimation of GPS
Pseudo-Range Multipath Errors
(1988) / Bassett R T

Comparison of Operator Performance in
Alternative Myoelectric Control Strategies
for Multifunction Arms
(1988) / Daley T L

Sensor Array Processing of Evoked Potential
Signals
(1988) / Godin D T

Microprocessor Based Protection of Variable
Speed Drives
(1988) / Gokul P

Micropower Continuous-Time CMOS Circuits
(1988) / Jaafar M A

A Multiprocessor Fault Tolerant Digital
Controller
(1988) / Mullin R K

Adaptive Filters for Evoked Potentials
(1988) / Westwick D T

A Multiprocessor System for Loaded Telephone
Cable Fault Detection
(1988) / Zhou Z

NEW JERSEY INSTITUTE OF TECHNOLOGY

Learning System
(1988) / Adani B T

A Methodology for Designing Spacecraft
Onboard Processing Systems
(1988) / Ahmad M N

Temperature Dependent Reliability of Plastic
Packaged Microcircuits
(1988) / Ahmed W

RF Characterization of a Directly Modulated
Indium Gallium Arsenic Phosphide Buried
Heterostructure Laser Diode
(1988) / Alvarez C

Design of Bridge Inter-Connecting Token Ring
and Token Bus Local Area Networks
(1988) / Anupindi K

Multaid, a Consultant for
Multi-Microprocessor Network Selection
(1988) / Arruza B J

The Fast Fourier Transform in Assembly
Languages of Microprocessor 68000 and
Microcontroller 8052
(1988) / Baldawa H K

Multichannel Priority Interrupt System
(1988) / Bernal C

Data Flow Dependency Generator
(1988) / Chan C P

Waveform Analysis and Application to a
Model of the Egg Waveform
(1988) / Chang E R

The Switch Capacitor Biquad Filter and Its
Application on Active Filter Design
(1988) / Chang S C

Analysis and Design of a Surface
Microengineered Silicon Humidity Sensor
(1988) / Chen C Y

Investigation of Different CMOS DRAM Sense
Amplifier Configurations in VLSI
(1988) / Chiu B

A 3-D Tactile Sensor with Digital Output
(1988) / Cho S S

Optimal Cache Size in Microprocessor-Based
Multiprogramming System
(1988) / Christou C S

The Design of an Instrument Which Measures
Isotonic and Isometric Levator Muscle
Contractions
(1988) / Conigliari M F

NEW JERSEY INSTITUTE OF TECHNOLOGY
(continued)

Implementation of the Voice Processing
System
(1988) / Dan Y Y

Modeling and Simulation of Robot Arms with
Flexible Links
(1988) / Danis J G

C/KU Band Satellite Television Reception
System
(1988) / Economides C A

Effects of Peripheral and Microwave-Core
Heating on Survival in Hemorrhagic Shock
in Rats
(1988) / El-Assuooty A S

Charge Transfer Inefficiency of Buried
Channel Charged Coupled Device Registers
for Low Temperature Operation of Platinum
Silicon Infrared Image Sensors
(1988) / Esposito B J

Stepping Motor Control and Position
Monitoring
(1988) / Feng P K

An Investigation of Interrupt Handling
Schemes in Pipeline Processors
(1988) / Gami S R

The Tapered Dielectric Rod-Conical Antenna
(1988) / Georghiades N I

IBM/XT-Based Automatic Monitoring
Instrumentation System for Eating Behavior
Studies in Rats
(1988) / Gui B

Microcomputer-Controlled Respiratory Rate
Monitor
(1988) / Hartmann D J

A Computer-Based Microwave Interferometer
Instrumentation System as a Non-Invasive
Cardiopulmonary Monitor
(1988) / Ho C E

Performance of Single Dwell Serial
Synchronization Systems
(1988) / Hsiung C Y

A Silicon Photodiode Array for Soft X-Ray
Sensing
(1988) / Hu C

The Application of Zero-Sum Games to Low
Sensitivity Optimal Control
(1988) / Huang J

Reliability of Packaging in Microcircuits
(1988) / Iqbal Z

Failure Mechanisms of Hermetically Sealed
Solid Tantalum Capacitors
(1988) / Kazmi S R

An Investigation of Electrical and Optical
Properties of Reactively Sputtered Silicon
Nitride and Amorphous Hydrogenated Silicon
Thin Films
(1988) / Kim T H

System Performance Criteria in CDMA Networks
Using Gold Codes
(1988) / Kim Y H

Clinical Base Station for Portable Patient
Monitoring Equipment
(1988) / Kuang S

8085 In-Circuit Emulator for IBM Personal
Computer
(1988) / Lee J B

APAP: The Arithmetic Pipeline Analysis
Package
(1988) / Lin C C

An Analysis of the Second-Order, Type-One
Hibert Phase-Locked Loop
(1988) / Lin I S

Simulation Study of the Multi-h Continuous
Phase Modulation
(1988) / Lin M L

Microprocessor Controller for a Multi-Target
System
(1988) / Lin S

Independent Equations Method for Power
System State-Estimation
(1988) / Lin S G

Gallium Arsenide Standard Cell CAD Database
for Digital Design
(1988) / Liu N C

Development of 8052AH-BASIC Based Single
Board Computer II
(1988) / Lou S

Contrast Sensitivity Measurement Via Evoked
Potential
(1988) / Lu K G

ASICS Application on Voice Processing
Interface Design: GEGG Multiplier
(1988) / Luar G J

Design of a Fiber Optic Oxygen Sensor Based
on Immobilized Hemoglobin
(1988) / Mendu G R

Interface to 8052AH-BASIC Microcontroller
for a Multichannel Priority Interrupt System
(1988) / Mo C T

High Frequency Gaussian Beams Solution of a
Plane Wave Coupling Into Cylindrical Cavity
(1988) / Mourad C I

A Dynamic Set-Based Multiaccess Protocol for
Local Area Networks
(1988) / Nanavati P S

Arithmetic Pipeline Scheduler
(1988) / Pan C T

ISTEMCR: Integrated System for Truck
Engine Monitoring Via Cellular Radio
(1988) / Pan T S

Reliability Problems in Semiconductor Lasers
(1988) / Patel A R

Design Criteria of Coherent Optical Fiber
Communication System
(1988) / Patel R N

Interfacing of the Digital Spyghnomanometer
[sic] DS-165 to the Intel 8052AH BASIC Based
Single Board Computer
(1988) / Patel R P

NEW JERSEY INSTITUTE OF TECHNOLOGY
(continued)

NETEXPERT: An Expert System for Computer
Network Management
(1988) / Patel S

Image Enhancement and Edge Detection Using
Fuzzy Algorithm
(1988) / Patel S

Microprocessor Testing and Computer Aided
Diagnosability of Digital Circuits
(1988) / Patel S

Design and Analysis of an Optical-Powered
On-Chip Power Supply
(1988) / Patnaikuni R

Word Based Data Compression Using Arithmetic
Encoding
(1988) / Peckham C D

Testing of Ultrasonic Bonding in
Microcircuits
(1988) / Phiniotis I A

On Conventional and Harmonic Load Flow
Analysis
(1988) / Pradhan R B

Data Extraction and Analysis in Glucose
Monitor
(1988) / Qian C

Kalman Filtering Applied to the Analysis of
Brainstem Auditory Evoked Potentials
(1988) / Qian C A

Reliability Evaluation of Liquid Crystal
Displays Under Accelerated Test Conditions
(1988) / Rathore I S

Design and Development of an Infrared CCD
Camera
(1988) / Romanowich J F

Implementation of a Cross Coupled Phase
Locked Loop System with Closed Loop
Amplitude Control
(1988) / Sadeghi B

Study of Dynamic CMOS Families and
Development of Certain Guidelines for
Logic Synthesis
(1988) / Santurkar V

A Modified Extended Kalman Filter as a
Parameter Estimator for Linear Discrete-Time
Systems
(1988) / Schnekenburger B J

Microcomputer Control of an AC Induction
Motor Using a PWM Computational Intensive
Method
(1988) / Schung L W

An Investigation of Electrical and Physical
Properties of Reactively, Diode Sputtered
Silicon Nitride Thin Films
(1988) / Shah D M

Fault Location and Control Optimization
Planner (Expert System Approach)
(1988) / Shah F

Design of an Experimental Apparatus to
Measure the Charge and Mass of Ionized
Species in a Partially Ionized Cluster
Beam Deposition System
(1988) / Shah N K

Computer Analysis of the Electrocardiogram
Using Hardware Digital Signal Processing
(1988) / Shaikh M A

The Path Method in Power System State
Estimation
(1988) / Shirazi F A

LF Pulse Magnetic Field Generation and
Uniformity Studies
(1988) / Tan S

An Application of Bit Level Systolic Arrays
on Digitized Adaptive Array Processes
(1988) / Tsai S H

Analysis and Simulation of Chirp Radar
System
(1988) / Tsai T T

CMOS SRAM Design for Initializing
Programmable Logic Devices
(1988) / Vaidyanathan R K

Analysis of Spread Spectrum System Using
Direct Sequence Technique in Jamming
Environment
(1988) / Vyas K A

Attenuation Properties of an LP 11 Mode in
a W-Type Fiber Operating Below Cutoff
(1988) / Wang C C

Liquid Crystal Display and Thermal Printer
for a Single Board Computer
(1988) / Wang S J

Research, Development, and Evaluation of a
Solid State Transformer
(1988) / Wozniak R M

The Evaluation of the TI-Speech System on
PC Computer
(1988) / Wu Y

Schema Based Wiring Tool: A Computer
Network and CAD Design System
(1988) / Zhang J J

On-Line Blood Glucose Control System
(1988) / Zheng H

NEW MEXICO STATE UNIVERSITY

Zone Logic: A Nonprocedural Approach to
Control of Large Production Machines with
Distributed Microprocessors
(1988) / Ali M M

Optical Correlation Using Computer
Controlled Magneto-Optic Devices
(1988) / Baker T M

Calculation of Atmospheric Water Vapor
Profiles
(1988) / Finkner L G

A Feasibility Study on the Use of Streak
Cameras in Long-Range Depth-Imaging Systems
(1988) / Gentry S M

NEW MEXICO STATE UNIVERSITY
(continued)

Protection of Switched Capacitor Banks from
Overvoltages Due to Circuit Breaker Restrike
(1988) / Groenewold K J

The Image of Modern Digital Design Methods
on Microcomputer Performance, Reliability,
and Cost
(1988) / Ishak M

Design of a Fast Convolutional Masking Chip
in Three Micron CMOS Technology
(1988) / Moya J A

Analysis of a Slip-Extended Induction
Generator for Use in a Wind Energy
Conversion System
(1988) / Noteboom R R

Analytical Approaches to the Evaluation of
Islanding in Residential Photovoltaic
Systems
(1988) / Prasad N R

Use of the TI Deformable Mirror Device in
Optical Processing Systems
(1988) / Smith J Z

NEW MEXICO, UNIVERSITY OF

A Comparison Between Conventional and Neural Network
Clustering Algorithms
(1988) / Boujaoude E

Quantization Analysis of a Real-Time SAR Digital
Image Formation Processor
(1988) / Casebolt J K

On Fractals and Application of Newton's Algorithm
to Digital Filters
(1988) / Casey E

Image Restoration with Emphasis on Performance
Measurement
(1988) / Chang H L

Design of an Automated Software Driver for DSP
Development Systems
(1988) / Chang H Y

An Expert for Signal Classifications
(1988) / Chia B S

A New Method of Context Postprocessing in Optical
Character Recognition
(1988) / Chou C C

Predicting Seismic Signals from Borehole
Measurements Using Adaptive Filtering
(1988) / Clifford B

Design of a Screen Editor Which Supports Variable
Font Widths
(1988) / Cyrus W T

The Random Vector Performance Measure Comparison
Method for Edge Detectors
(1988) / Dominguez M A

Three Hardware Designs for Real-Time Change Detection
(1988) / Fong S Y

Image Transform Coding
(1988) / Frank E

Real-Time Seismic Event Detection and Source
Location Using a TMS32020 Digital Signal Processor
(1988) / Frederking J P

Evaluation of an Unstable Ring Resonator with
90 Degree Bean Rotation
(1988) / Holswade S C

A Connection Network for Robotic Gripper Control
(1988) / Horne W G

Design and Implementation of a Software System for
Pole-Zero Filter Design and Analysis
(1988) / Hu C A

Evaluation of a Lisp-Based Language for Implementing
Signal Processing Algorithms
(1988) / Kral D J

Investigation of Heterodyne Mixing Efficiency
Variations Due to Optimal Misalignments and
Aberrations
(1988) / Paranto J

A Radar Signal Processor: Analysis and Real-Time
Implementation
(1988) / Puri M

Optimal Design of Electronic Circuits
(1988) / Shyu C J

Assignment of Points to Three-Dimensional
Trajectories
(1988) / Sorsby C R

Data Retention Study of a 4K Eprom Embedded in a
Microcontroller
(1988) / Trodden T J

Anomalous Resistivity Measurements on Freshly
Cleaned (110) Silicon
(1988) / Wang Q

Effects of Bankwidth and F-Number on the Reflectivity
of Stimulated Brillouin Scattering
(1988) / Woolston T L

Chromium-Titanium Contact for SOI Structures
(1988) / Yu I

NEW YORK, STATE UNIVERSITY OF (BUFFALO)

Handwritten Script Recognition: A Hidden
Markov Model Based Approach
(1988) / Bahl P

Electromagnetic Acceleration of a Metallic
Carrier Plate: Experiment and Computer
Simulation
(1988) / Bernard M J

Microcomputer Based Data Acquisition and
Analysis in the Laboratory
(1988) / Cecere M

Single Clock Cycle Procedure and Process
Switching Through Hardware
(1988) / Czamara A

The Enhancement of Drilling Efficiency in
Carbon Steel Using Modified Pulse Shapes of
a Neodymium/YAG Laser
(1988) / Domankevitz Y

On Parallel Computation Using One-Sided
Jacobi
(1988) / Eggers W

NEW YORK, STATE UNIVERSITY OF (BUFFALO)
(continued)

An Investigation of the Effects of Ozone
on the Physical and Electrical Properties
of Polymer Insulation
(1988) / Gilmore J

Breakdown Studies of Sulfur Hexafluoride
Argon Gas Mixtures
(1988) / Khechen W

Characterization Methods of High Energy
Density Capacitors
(1988) / McDonald D

Shaded Surface Display (2 1/2D) of 3-D
Objects
(1988) / Rajagopal G

Study of Indium Phosphide
Metal-Semiconductor and Thin Insulator
Meta-Insulator-Semiconductor Schhottky
Diodes
(1988) / Reinhardt K

Picosecond Double Resonance Spectroscopy of
Excess Carriers in Aluminum Gallium
Arsenic-Gallium Arsenic Multiple Quantum
Well Structures
(1988) / Rossi T M

Low Frequency Linear Motion Inducer
(1988) / Sieracki D L

Modeling of Linear and Nonlinear Pulse
Forming Networks
(1988) / Stanton S

An Investigation of Dielectric and Metallic
Surface Plate Variations with Respect to
Tolerance Limits of Capacitors
(1988) / Stopher J

Model Relation of Barium Titanate for
Application Towards Fault Current Limiter
Investigations
(1988) / Wallman B

General Analysis of Saturating Iron Cores as
Fault Current Limiters on 145 kV Systems
(1988) / Zieman K

NORTH CAROLINA STATE UNIVERSITY

Recombination Dynamics in Indium Gallium
Arsenide/Gallium Arsenide Quantum Well
Heterostructures
(1988) / Benjamin S D

Optimal Problem Decomposition in the
Parallel Execution of Divide-and-Conquer
Algorithms
(1988) / Dholakia A

A Neural Network Processing Strategy for the
Detection of High Impedance Faults
(1988) / Ebron S D

A Heuristic Approach to Electric Power
Distribution System Feeder Reconfiguration
(1988) / Taylor T M

Phase Frame Representation of Multiple
Phase Transformers
(1988) / Tyndall B W

NORTH DAKOTA STATE UNIVERSITY

Real Time Acquisition and Analysis of
Respiratory Data
(1988) / Cain P W

An Experimental Speech Synthesis System Using a
Selective Waveform repetition and Folding Scheme
(1987) / Chu W S

Several Computer Methods of Determining the Network
Bus Impedance Matrix of Radial Distribution Network
(1988) / Lake J E

Optimal Control of Human Respiratory System
(1988) / Lin S L

Mass Storage of Analog and Digital Data Using the
Compact Laser Disc
(1988) / Nourbakhsh F

Operations Manual for the Applied Dynamics AD/4
(1987) / Ward S A

Use of Companding to Reduce Harmonic Distortion in
Sampled Analog Signal Systems
(1987) / Wehbe A

NORTHERN ILLINOIS UNIVERSITY

Design of Multiprocessor Systems and
Implementation of Adaptive Filters
(1988) / Gupta P K

Automatic Code Generator for Adaptive
Filters and Terminal Emulator for TMS32010
(1988) / Ranganathan G V

Writing Special Procedure and Subroutines
on TK Solver to Solve for Linear/Nonlinear
Electric Circuits
(1988) / Vohra P

NOTRE DAME, UNIVERSITY OF

Development of Data Acquisition Capabilities
for the Macintosh Plus Computer
(1988) / Carpenter P S

Power Spectral Density of Error Correcting
Line Codes
(1988) / Ehrman G J

Decentralized Detection: A Totally
Neyman-Pearson Optimal Design Scheme
(1988) / Francis J C

The Model Matching and Related Control
Problems: Conditions and Algorithms
(1988) / Gao Z

A Multiprocessor Data Acquisition System
for High-Energy Cosmic Ray Detection
(1988) / Markiewicz T J

Transport Service Over Computer Networks:
Design, Implementation and Specification
Methodology
(1988) / Saiya C N

Load Sharing in Networked UNIX Systems
(1988) / Subramanya K M

Distributing Information in Computer
Networks: A Basis for Distributed
Processing
(1988) / Turner D C

NOTRE DAME, UNIVERSITY OF
(continued)

The Analysis and Design of a Very Large
Scale Integration Very High Speed Adder
Circuit
(1988) / Vargas C O

Multiple Processors for Personal Computers
(1988) / Vaz T J

NOVA SCOTIA, TECH UNIVERSITY OF

A Two-Axis Control System for a
Photovoltaic Array
(1988) / Horton E P

OKLAHOMA, UNIVERSITY OF

Digital Signal Processor Based
Implementation of an MLSE Detector for
Generalized Tamed Frequency Modulation
(1988) / Abdallah R B

Design of a System Bus Monitor for Debugging
Hardware Failures on a 32-Bit Minicomputer
(1988) / Caro A M

An Algorithm for the Signal Flow Graph
Representation of Digital Filters
(1988) / Jalali A

A Comparison of Modern Frequency Estimation
Techniques
(1988) / Khelghati H

A Coherent Digital Demodulation Technique
for Continuous Phase Modulation
(1988) / Satarasinghe P J

Electric Power Production Simulation on a
Personal Computer
(1988) / Su G

Area Efficient Layouts for Routing Around
Rectangular Modules
(1988) / Teo K

OREGON STATE UNIVERSITY

COLAN III, A Control-Oriented LAN Using
CSMA/CD Protocol
(1987) / Eum D H

Acceleration of a Heuristic PLA Product Term
Reduction Program Through Complementation of
the PLA Specification
(1988) / Gilbert J D

Weather Influence on Loran-C
(1988) / Graulich M G

Instabilities in Indium Phosphide MIS Capacitors
by Min Tzuan Juana
(1988) / Juang M T

Design of COLAN II, a Control Oriented Local
Area Network
(1987) / Kao S K

Analytical Model of Gallium Arsenide MESFET Output
Conductance with Frequency and Temperature
Dependent Parameters
(1988) / Lam S C

Applications of Mathematical Programming
Techniques in Optimal Power Flow Problems
(1988) / Li R

CMOS Analog Design Using a Digital Gate Array
(1988) / Liu T C

Modeling of Subthreshold Current on Gallium
Arsenide MESFET and the Design of Voltage
Reference Circuit
(1987) / Or P K

An Expert System for Optimal Tuning of Adaptive
PID Regulators
(1988) / Saugen J D

COLAN IV, a Local Area Network for Communications
and Control
(1988) / Thye Y

A High-Swing CMOS Operational Amplifier Topology
(1988) / Yang J J

A High-Swing Class-AB CMOS Operational Amplifier
(1988) / Yu C

PENNSYLVANIA STATE UNIVERSITY

Finite Element Simulation of the
Atmosphere's Electromagnetic Response to
Charge Perturbations Associated with
Lightning
(1987) / Baginski M E

Processor Management and Task Scheduling in
a Partitionable Multi-Processor
(1987) / Choi K

Tracking Characteristics of Least Squares
Lattice Filters
(1988) / Clark W J

Modeling of Electromagnetic Response of
Metal Oxide Based Composites
(1987) / Datta S

An Experimental Determination of the
Intelligibility of Two Different Speech
Synthesizers in Noise
(1987) / De Paolis R A

Application of Ultraviolet Radiation in
Preoxidation Treatments of Silicon Wafers
(1987) / Duranko G T

Infrared Laser Amplification and Limiting in
Nematic Liquid Crystals
(1988) / Finn G M

Kalman Filter Detection of Abnormal
Biological Responses in Poultry Management
Systems
(1987) / Garnaoui K H

An Improved Algorithm for Text Segmentation
in Mixed Text/Graphic Documents
(1988) / Gattiker J R

Rocket Probe Instrumentation and Atmospheric
Electrical Parameter Measurements for
Thunderstorm Research
(1987) / Goodnow K J

Control Systems Analysis and Design Package
TOTAL
(1988) / Hossain S D

Design of a 17.3 to 17.7 GHZ Dualgate Mesfet
Mixer
(1988) / Kasody R E

PENNSYLVANIA STATE UNIVERSITY
(continued)

Magnetic Susceptibility of Nondegenerate
Semiconductors
(1988) / Klemkosky M A

An Active Detector Gallium Arsenide MMIC
Design
(1988) / Kline G L

The Design and Development of a Vehicle
Suitable for Use in a Flexible
Manufacturing Environment
(1988) / Kotzer D J

The Use of Charge Measurement for
Non-Equilibrium Metal Oxide Semiconductor
Characterization
(1987) / Lacroix D P

On the Reliability Evaluation of Multistage
Interconnection Network Based Multiprocessor
Systems
(1988) / Macaluso J J

Nonlinear Feedback Control of Robot Arm: A
Geometric Approach
(1987) / Marrekchi H

A Database VLSI Associative Join Module
(1987) / Petrie C R

An Analysis of Multipath Power Delay
Profiles
(1988) / Reyner G A

Performance Analysis of Slotted Ring Network
by Computer Simulation
(1988) / Salvi S B

A Uniform Solution of Double Knife-Edge
Diffraction
(1987) / Schenider M

Automated Encoding Method for Converting a
Raster-Scanned Image Into Vector
Representation in Graphics Applications
(1987) / Shih C

A Quantitative Method for Determining
Dynamic Response Characteristics of
Pyroelectric Materials
(1987) / Sopko J A

Quantization Error in Adaptive Arrays
(1987) / Umari M H

A Fault-Tolerant Task Mapping Method for
Multistage Network-Based Multiprocessor
Systems
(1987) / Yousif M S

PENNSYLVANIA, UNIVERSITY OF

The Conservation Law in Packet Switching
Networks and Optimal Policies for Circuit
Switching Networks with Two User Classes
(1988) / Chan W L

Analysis and Design of Current Injection
Devices
(1988) / Chao S

A Lagrangian Relaxation Algorithm for the
Design of a Broadcast Network
(1988) / Chung S P

Implementation of Neuts' Technique for
Solving Queueing Problems
(1988) / Forrest J

Field Induced Currents on Cables
(1988) / Garg A

A Tissue Model for Investigating Photon
Migration in Trans-Cranial Infrared Imaging
(1988) / Greenfield R L

The Systolic/Cellular System Assembler:
User's Guide
(1988) / Hartholz M A

A Compact Waya to Perform Interferometric
Sensing
(1988) / Hartley D N

Barium Titanate Ferroelectric Thin Film
from Organometallics and Rapid Thermal
Processing
(1988) / Hong X

Susceptibility of Spread Spectrum Signals
to Detection and Exploitation
(1988) / Horzempa J P

Noise Analysis of a Common Base and Common
Emitter Input Configuration for a High Speed
Bipolar Shaping Amplifier
(1988) / Hottenrott P A

Synchronization Performance of a Spread
Spectrum Local Area Computer Network
(1988) / Hughes I M

Development of a Single Link Manipulator
Test Rig
(1988) / Ibanez-Guzman J

Mean Frequency Estimation for Use with
Doppler Ultrasound Blood Flow Meters
(1988) / Lawee I P

Study of Localized Radar Cross Section of
Aircraft
(1988) / Lee W

High Resolution Microwave Imaging Radar
Facility with Extended Bandwidth
(1988) / Lee Y

Attraction Basins in Neural Associative
Memory
(1988) / Pancha G

Guided-Wave Structures Filled with Chiral
Materials: Chirowaveguides
(1988) / Pelet P F

Inverse Scattering for Surface Impedance
Objects
(1988) / Shiomos C M

A Cepstrum-Based Consonant Phenome Speech
Recognition System
(1988) / Stanfel L E

Design and Implementation of a
Time-to-Voltage Convertger/Analog Memory
for Colliding Beam Detectors
(1988) / Stevens A E

Clean Performance in Near-Field Inverse
Synthetic Aperture Radar Imagery
(1988) / Stockburger E F

PENNSYLVANIA, UNIVERSITY OF
(continued)

Integrated Knowledge-Based System for
Communication Network Design
(1988) / Vernekar A T

Generating Higher Order Harmonics with
Non-Linear Materials Deposited in
Micro-Machined Silicon Wafers
(1988) / Winslow J W

Comparative Performance Analysis of Fast
Packet Switches
(1988) / Yazdani H

A Portable Large Storage Transient Recorder
(1988) / Yu H

Near Field Imaging with FFT Processing
(1988) / Zhou C

Broadband Integrated Services Digital
Network Subscriber Access
(1988) / Zielke S

PITTSBURGH, UNIVERSITY OF

An Experimental Investigation of Surface
Effects on Lateral Magnetotransistors
(1989) / Chu H S

An Algorithm for Two Source Image Coding
(1989) / Kayhan A S

Nondestructive Testing of Ferromagnetic
Tubes Using the Remote Field Eddy Current
Effect
(1989) / Kilgore R J

Two-Source Image Coding by the Template
Method
(1989) / Kinsler M R

LQ Control of Rapidly-Varying Systems with
Application to Robotics
(1989) / Lodi S E

Application Requirements of Gate Turn-Off
Thyristors: A Critical Literature Survey
(1989) / Low K M

Tickle: An Integrated Design Capture and
Simulation Tool
(1989) / Mears L A

Inkas, An Inverse Kinematics Arm Solver
(1989) / Mu E

Bubble Noise Frequency Estimation of the
Anode Current Signal of an Aluminum
Electrolytic Cell
(1989) / Petrelli V A

Torque Disturbance Rejection in High
Accuracy Tracking Systems
(1989) / Profeta J

A Single Board Development System for the
6800 Microprocessor
(1989) / Richert G W

A Basic Assembler for the Texas Instruments
TMS32010 Digital Signal Processor
(1989) / Rumpf J A

Multifunction Analog I/O Subsystem for the
IBM PC and Compatibles
(1989) / Zhang Y

PRINCETON UNIVERSITY

Distributed Processing and Fiber Optic
Communications in Air Data Measurement
(1988) / Farry K A

PURDUE UNIVERSITY

Experiments with Non Centralized
Neuromorphic Control Architectures for
Autonomous Vehicles
(1988) / Azmak O

Brushless DC Synchro Drive System
(1988) / Bollenbach K H

Impedance Measurements of Underground Cable
Systems
(1988) / Bruce R N

Autocovatiance Functions of Ventricular
Fibrillation Waveforms and Outcome of
Defribrillation
(1988) / Cheng Y

An Investigation of Optical Fiber
Polarimetric Force Sensors
(1988) / Chua T H

The Inverse Nyquist Array Based Design
Method for Strongly Diagonal Dominant
Multivariable Systems
(1988) / Couturier C M

A Quasi-Dielectrically Isolated Bipolar
Transistor Using Epitaxial Lateral
Overgrowth
(1988) / Duey S J

File Input/Output Design Considerations for
a Distributed Operating System Implemented
on a Partitionable SIMD/MIMD Computer System
(1988) / Everett S F

A Microprocessor Controlled Valve Positioner
(1988) / Feldman M W

The Use of Image Processing Techniques for
the Analysis of Echocardiographic Images
(1988) / Garcia-Melendo E

Indium Antimony/Cadmium Telluride
Heterostructures Grown by Molecular Beam
Epitaxy
(1988) / Glenn J L

Creating a Cooperative Procedural
Environment for Feature Extraction Using
the Blackboard Architecture
(1988) / Hennessy D N

A Study of Plausible Reasoning in Expert
Systems for Troubleshooting
(1988) / Hori S

Delay Based Artificial Neural Systems
for VLSI
(1988) / Hughes C S

Parallel Analysis of Expert Systems
(1988) / Hung F F

A Fiber Optic Microbend Tactile Sensor Array
(1988) / Jenstrom D T

Alternative Modes of Communication Between
Man and Machine
(1988) / Keirn Z A

PURDUE UNIVERSITY
(continued)

SPARO: Spatial Relationship Query Based on
Spatial Decomposition
(1988) / Kerr R L

A Useful Generalization of the Theory of
Structural Stereopsis
(1988) / Kosaka A

One Dimensional Quantum Transport:
Characterization and Device Applications
(1988) / Lake R K

Dual Robot Control Hardware, Software System
and Experiments
(1988) / Lam Y

A Hybrid Method for Three Dimensional
Recogniation of Object Shape Features
(1988) / Marefat-Djahromi M

Computer Modeling of Electromagnetic Fields
for Cardiac Stimulation
(1988) / Mathine L A

Analysis and Modeling of a Single-Phase
Bifilar-Wound Brushless DC Motor
(1988) / Mayer J S

Routing Algorithms for Task Migration in
Hypercube Multiprocessors
(1988) / McSherry M A

Comprehending User Design Intentions by
Means of Dialog Interaction with Engineering
Design Knowledge
(1988) / Mohan L

On-Off Control of a Pneumatic Actuator
(1988) / Mouline A

Image Compression Using Real Transforms with
Reduced Complexity
(1988) / Nouira A

Pulse-Width Modulation Control of a
Three-Phase Inverter
(1988) / O'Connor S M

The Harmonic Impact of Rectifiers Served by
Unbalanced Three-Phase Sources
(1988) / Olejniczak K J

Simulation Studies of the Generalized Cube
Interconnection Network
(1985) / Ott J M

A Study of Polysilicon Contacted Emitter
Bipolar Transistors and Their Fabrication
Development
(1988) / Pak J H

A New Method of Determining Cardiac Output
Using Thoracic Electrodes and
Saline-Dilution Theory
(1988) / Patel U H

Capacitance-Voltage Characteristics of
MBE-Grown Aluminum/Zinc Selenide/Gallium
Arsenide MIS Capacitors
(1988) / Qiu J

An Artificial Intelligence Approach to
String Instrument Fingening
(1988) / Sayegh S

Development of an Epitaxial Lateral
Overgrowth Silicon Bipolar Transistor
Technology
(1988) / Siekkinen J W

Molecular Beam Epitaxy of Zinc Selenide on
Gallium Arsenide Epilayers for Use in MIS
Devices
(1988) / Studtmann G D

Constructive Desitn of Mechanical Systems
Through Functional Relationships
(1988) / Stutz J A

Large-Step Large Area Planarization of
Silicon Epitaxial Lateral Overgrowths
(1988) / Vedala S

Investigation of MBE-Grown Doped Zinc
Selenide Films Using Deep Level Transient
Spectroscopy
(1988) / Venkatesan S

Electronarcosis
(1988) / Walther T L

An Optical Fiber Rotary Displacement Sensor
with Intensity Modulation
(1988) / West S T

Development of a Baseline MODFET Fabrication
Process
(1988) / Whiteside M C

Measures of Controllability and an Energy
Approach to Controllability
(1988) / Wicks M A

A Study of Target Enhancement Algorithms to
Counter the Hostile Nuclear Environment
(1988) / Wright C H

Knowledge-Based Expert System for Casting
Design
(1988) / You I C

Nonlinear Matched Filtering
(1988) / Zeng M

QUEENS UNIVERSITY

Adaptive Decoupling of M1MO Systems with
Constrained Inputs
(1988) / Bennamoun M

On the Use of Input and Output Mappings to
Calculate Balanced Realizations
(1988) / Burns P D

Time Domain Analysis of Small Power Systems
Under Nonlinear Loading
(1988) / Burns S

Design and Development of a Computer
Controlled Rotary Test Facility
(1988) / Chang T W

QENS - An Enhance Version of the Electric
Network Simulator Program
(1988) / Cornel H C J

Analysis of Amplitude, Phase and Error
Distributions for Shadowed Mobile Satellite
Communication Channels
(1988) / Lee A C M

A Robotic Application for Determining Object
Location and Orientation in a Work Space
(1988) / Michaud J

QUEENS UNIVERSITY
(continued)

Three-Valued Logic for Self-Checking CMOS
Binary Circuit
(1988) / Pan S C Y

Modelling and Control of the Switched
Reluctance Machine
(1988) / Pattison L

Design of Linear Time Varying Recursive
Digital Filters
(1988) / Peters S D

Coloured Structured Petri Nets
(1988) / Smith J M

Microcontroller Control of Linear Induction
Motor Drive Systems
(1988) / Tong L C

Design of 1-D and 2-D Variable Recursive
Digital Filters
(1988) / Zarour R

REGINA, UNIVERSITY OF

Optimal Planning and Operation of
Multi-Purpose Reservoir Systems
(1988) / Abed A M

Computer Generation of Efficient Farm
Field Courses
(1988) / Liu G

RENSSELAER POLYTECHNIC INSTITUTE

Automated Inspection of Solder Joints Using
Thermal Signatures
(1988) / Akstens W S

Advances to an Expert System for
Computer-Aided Control Engineering
(1988) / Antoniotti A J

The System and Circuit Design of a Real Time
Display Interface for Multiprocessor
Computer Image Generation Systems
(1988) / Bins M H

A Real-Time Speaker Verification System
(1988) / Borkowski D G

The Effect of Ion Implantation Processing on
Semi-Insulating Gallium Arsenide Wafers
Using Photo-Enhanced Microwave Reflectance
(1988) / Campbell C S

Hybrid Control Applied to a Turret Drive
Positioning System
(1988) / Chisholm R V

An Automated Inspection System for the
Detection of C-Spring Defects
(1988) / Chow S S

A TMS32020 Based Digital Signal Processor
Design for Detecting the Sound of Breaking
Glass
(1988) / Chung G C

Parallel Implementation of the Kalman Filter
for Noise and Blur Removal from Digital
Images
(1988) / Damour K T

A Build-In Self-Test Floating Point
Co-Processor for Reduced Instruction
Set Computer
(1988) / Davis B R

A Theory of Acousto-Electro-Optic
Interaction in Anisotropic Media, Including
Polarization Effects
(1988) / DeCusatis C

Transform Trellis Coding of Images
(1988) / Divakaran A

Structural and Optical Characterization of
Gallium Arsenide Grown on Silicon and
Calcium Fluoride/Silicon
(1988) / Divakaruni S

Host - Coprocessor Performance Evaluation
for Typical Matrix Operations
(1988) / Dondiego M J

Evaluation of Motion Estimation Techniques
and an Image Stabilization System
(1988) / Dowling R C

MacVIDA - The Macintosh VLSI Interactive
Design Assistant
(1988) / Edgar C B

Determining the Location and Orientation of
a Robot Arm in Three Dimensions Using a
Single View
(1988) / Falsafi A

A Speaker-Personality Transformation System
Using Vector Quantized Vocoding
(1988) / Flemming K J

Sensitivity Optimization
(1988) / Goldberg R G

Laser Photoionization of Air for
Implementation of Noncontact Testing
(1988) / Gordon M T

An Improved Root-Locus Capability for
Pro-Matlab
(1988) / Graf R M

A Two-Time-Scale Design of Systems with
Lightly Damped Oscillatory Modes
(1988) / Hanley C J

Simultaneous Stabilization of Multiple
Systems by Using Pole Placement Techniques
(1988) / Hasekioglu A

Performance Evaluation of Pipeline
Architecture Machines for a Kalman
Filter Applications
(1988) / Helgerman E R

Practical Considerations in the Use of
Distributed Parameter Systems in
Modelling: A Control Point of View
(1988) / Helmicki A J

Colorizing Black-and-White Images
(1988) / Huebbner G

The Design and Implementation of a
Programmable Four Channel Keyboard for
Central Control of MIDI Performance Networks
(1988) / Ikeda K G

RENSSELAER POLYTECHNIC INSTITUTE
(continued)

Pilot Processing for Correction of Doppler
Shift and Multipath Fading Using
Transparent-Tone-In-Band Modulation
(1988) / Kolb S R

Integration of the Sky Computers VORTEX/AT
Array Processor Into the Workstation
Communications Simulator
(1988) / Kupiak D R

Bit-Slice Controller for Ultrasonic
Digital Holographic Image Reconstruction
(1988) / Liguori D J

An Arithmetic Logic Unit for the Fast
Reduced Instruction Set Computer (FRISC)
and CAD Tools for the FRISC Design
Environment
(1988) / Lopata D

An Automated Optical Inspection System for
Printed Wiring Board Defect Detection and
Component Placement Verification
(1988) / Muth W G

The Quality of Epitaxial Mercury(0.73)
Cadmium(0.27) Telluride Grown on (100)
Gallium Arsenide
(1988) / Natarajan V

The Application of Machine Vision Feedback
to Position Control in a Laser Soldering
Workstation for Electronics Manufacturing
(1988) / North R C

An Analysis of Modeling Clip
(1988) / O'Bara R M

Hardware Implementation of a Permutation
Network
(1988) / Pathak Y S

Design of a Radiation Hardened Pulse
Width Modulator
(1988) / Perreault G J

Design of Current Insertion Testing System
for the MK6 Guidance System
(1988) / Pouliopoulos J F

The Vehicle Intercomputer Communications
Chip (VICC)
(1988) / Preston R P

An All Digital Modulator/Demodulator for
Land Mobile Radio
(1988) / Rafter P G

MANNER (Multiple Access Using Node Number
and Reservation): A Protocol Suitable for
Satellite and High Speed Local Area Networks
(1988) / Ranjan K S

Hardware Development for a Digital
Ultrasonic Imaging System Using CAD Tools
(1988) / Ratnayake K B

Reconstruction of Images Blurred by Random
and Non Random PSF's
(1988) / Ravichandran B

C Implemented Vision Library (Civil)
(1988) / Repko M C

Cylindrical Curved Plate Energy Analyzer
for Use on the Heavy Ion Beam Probe
(1988) / Resnick J

Simulation of an Automated Cannon
Positioning System for a Large Bore
Self-Propelled Howitzer
(1988) / Rubin J F

Performance Evaluation of Estimation
Techniques for Target Tracking
(1988) / Savelloni M A

CMOS Standard Cell Implementation of
Processing Elements Within a Video Rate
Fully Recursive Two-Dimensional Filter
(1988) / Schack M W

Evaluation of Hidden Markov Models for Use
in a Speaker Verification System
(1988) / Scharer T J

Deterministic Resolution of Collisions in
CSMA/CD
(1988) / Schenker S J

Adaptive Jammer Suppression Using Decision
Feedback in a Direct-Sequence
Spread-Spectrum System
(1988) / Shah B

A Transform Domain Technique for Carrier
Synchronization and PN Code Acquisition in
a Direct Sequence Spread Spectrum Receiver
(1988) / Stern R A

The Hardware Implementation of a Digital
Demodulator for Analog FM Signals
(1988) / Thiel T E

Knowledge Representation Schemes for a
Document Analysis System
(1988) / Thomas M

ASIC Designs at CIE - Design and
Implementation of a Standard Cell Testchip
and a CAE Database Manipulator Chip
(1988) / Tiwary G

Growth and Properties of Zinc Selenide on
Gallium Arsenide by Organometallic Epitaxy
(1988) / Tyagi S

A Simulation of the Token Passing Bus
Priority Structure
(1988) / Valley S R

A Transparent Tone-In-Band Modulator for
Mobile Radio Communications
(1988) / Varghese A A

Navigation of the PUMA-600 Manipulator Using
a Vision System
(1988) / Whitson K J

RHODE ISLAND, UNIVERSITY OF

Design of an All-Optical Long-Distance
Soliton-Based Optical Fiber Communication
System Using Fluoride Fibers
(1988) / Abdelkader H A

Magneto-Optical Sensing Utilizing the
Faraday Effect in Bulk Cadmium Manganese
Telluride
(1988) / Adams J C

RHODE ISLAND, UNIVERSITY OF
(continued)

Development of a Simulation Environment for
Signal Processing Algorithms
(1988) / Ahuja S

Design of Single-Mode Heavy Metal Fluoride
Optical Fibers with Arbitrary Refractive
Index Profiles
(1988) / Bastien S P

Portable Muscle Stimulator Design for Transcutaneous
Functional Electrical Stimulation
(1988) / Berger N

Three Dimensional Reconstruction of Coronart
TreeLike Structures From Two Orthongonal Views
(1988) / Bergerson E

A Structured VLSI Implementation of a Hub Switch
R A Burst Switching Network
(1988) / Capozza P T

Edge Detection Algorithm for SST Images
(1988) / Cayula J F

Design of Distributed Feedback Semiconductor
Based External Cavity Lasers for Coherent
Communication Systems
(1988) / Engert C P

A Wideband Medium Power Microwave Amplifier
(1988) / Fascia A

A Monolithic LVDT Signal Processor
(1988) / Gamache R E

Performance Optimization of Abrasive Waterjet
Processes
(1987) / Hunt D

Trajectory and Parameter Estimation with
Measurements of Uncertain Origin
(1988) / Irza J W

Parallel Distributed Processing of Coronary
Angiograms
(1988) / Kottke D P

Mathematical Modeling of the Coronary Venous
Circulation
(1988) / Mo A P

On the Quantization of the Correlated Caussian
Source
(1988) / Narasimhan A

AR Spectral Estimation by Correlation Fitting
(1988) / Richards K P

Loss Measurement of Waveguide Modulators
(1988) / Tsai K F

ROCHESTER, UNIVERSITY OF

Detailed Routing for a VLSI Layout Based on
Net Ordering
(1988) / Coene J P

Signature Analysis of Robotic Part Placement
(1987) / Fullmer D M

Ultrasonic Absorption and Attenuation in
Mammalian Tissues
(1987) / Lyons M E

Finite Amplitude Propagation on Lossless and
Absorptive Media
(1988) / Reilly C R

Activity and Inhibition of Synthetic Lung
Surfactants
(1988) / Venkitaraman A R

ROSE-HULMAN INSTITUTE OF TECHNOLOGY

Suppressing Engine Knock in an Automotive
Spark-Ignition Engine Using a Retrofit
Circuit to Adaptively Retard Ignition
Timing
(1989) / Harding G L

The Theory and Applications of the Field
Network Analogue Technique
(1989) / Pfeifer C

Design of Image Processing Tools with an
Image Tracking Application
(1989) / Stetter M

SAN DIEGO STATE UNIVERSITY

Use of Digital Signal Processing Algorithms
to Realize Enhanced Resolution
Analog-to-Digital Conversion
(1988) / Ashkenasi M

Low Rate Image Coding
(1987) / Bhatt S N

Interference Rejection in Spread Spectrum
Communications
(1988) / Dayot S Y

An Efficient Resynchronization Algorithm
for a Sequential Decoder
(1988) / Dorr B L

Hardware Implementation of Vector DPCM
(1988) / Hang T

Variance Propagation Technique for
Theoretical Analysis of Error Propagation
(1988) / Krzeminski R H

Linear FC Chrip Pulse Compression for
Coherent Carbon Dioxide Laser Radars
(1987) / Lam N T

Programming Polynomial Discrete Fourier
Transforms and Convolutions: Algorithms
for Two Dimensional Signal Processing on
the IBM PC
(1987) / Liu Z

Design, Analysis, and Synthesis of an Eight
Channel Switchable Frequency Multiplexer
(1988) / Mayer K A

Image Sequence Coding
(1988) / Sanchez H W

CAD Tools for Hierarchical Design
(1988) / Shale R A

Improved VQ/MQ Algorithms with Medium-Band
Speech Coding Applications
(1987) / Tsai M C

Design and Verification of Microinstruction
Set of a Word-Slice Machine
(1988) / Wang H

A Study of Effects of Voltage Dips on Power
Supply Transients Stability
(1987) / Yung A C

SASKATCHEWAN, UNIVERSITY OF (SASKATOON)

A Microprocessor Based Power System
Stabilizer
(1988) / Bjornson A A

Data Compression by Using Symmetrical
Component Transform
(1988) / Chen C

Development of a Simulator for NMR Imaging
(1988) / Chu K C

Turbine-Generator Shaft Torsional Torques
Due to Network Disturbances
(1988) / Faried S O

Impacts of Generating Unit Unreliability on
Power System Adequacy and Costs
(1988) / Goel L K

Phase-Locked Loop Clock Extraction for
FDDI Systems
(1988) / Harris G K

A Study of the Standstill Frequency Response
Tests for Synchronous Machines
(1988) / Jin Y

Fast Adequacy Assessment of Composite Power
Systems
(1988) / Khan M E

Application of Digital Kalman Filters in
Power System Protection
(1988) / MacCormack J R

Design of an Interface Between a Vision
System and a Robot Controller
(1988) / Majumdar S

A K-Band Microwave Polarimeter
(1988) / Poettcker M P

Charge Transport Measurements on Amorphous
Selenium Photoreceptor Films
(1988) / Thakur R P S

Tactile Sensing System for an Industrial
Robot
(1988) / Vaidyanathan C S

An Adaptive and Self-Cognitive ACE Filter
(1988) / Ye X

Feasibility Study of Using a Digital
Computer as a Variable-Amplitude Variable
Frequency Oscillator
(1988) / Yonah Z O

SHERBROOKE, UNIVERSITY OF

Stereoscopie Appliquee a la Creation de
Representation 3D Pour la Vision
Artificielle
(1988) / Benakki A

Etude, Modelisation et Simulation du
Systeme Respiratoire
(1988) / Blondi M

Etude d'un Systeme de Communication par
Etalage Spectral a Synchronisation
Automatique
(1988) / El Mouine J

Etude et Developpement d'un Lien de
Transmission Transcutane Pour Prothese
Auditive Cochleaire
(1988) / Gauthier P

Contribution a la Reconstruction d'Images
Pour un Tomographe a Positrons a Haute
Resolution Spatiale
(1988) / Karuta B

Quantification Vectorielle Algebrique
Spherique par le Reseau de
Barnes-Wall: Application au Codage de
la Parole
(1988) / Lamblin C

Etude d'un Modele d'Audition et Application
d'un Critere de Distance Perceptuel au
Developpement d'un Codeur en Sous-Bandes
(1988) / Paillard B

SOUTH CAROLINA, UNIVERSITY OF

Laser Induced Fluorescence Imaging of Post
Arc Metal Vapor in a Triggered Vacuum Gap
(1988) / Abdalla M D

PEP: An Object-Oriented Architecture for a
Programming Environment Platform
(1988) / Anderson S R

Computing Elements for Silicon Neural
Networks
(1988) / Cho Y B

Self-Breakdown and Laser Initiated Breakdown
Studies of Plain and Solid Insulator-Bridged
Compressed Nitrogen Gas Gaps
(1988) / Foo K S

A Program for Analysis of Queueing Network
Models with Load Dependent Servers
(1988) / Hewitt I B

A Distributed Artificial Intelligence System
for Scheduling Automatic Guided Vehicles
(1988) / McElroy J F

Finite State Machine Application for VLSI
System Design
(1988) / Pratt M P

Optimal Planning of Phase-Shifting
Transformers for Line-Load Alleviation
During Contingencies
(1988) / Rylatt P E

High di/dt Pulse Switching with Involute
Structure Thyristors
(1988) / Sandaran V A

Secondary Electron Yield from Ceramic
Insulators as a Function of Primary Electron
Energy and Angle of Incidence
(1988) / Skipper M C

Partial Shape Recognition Using Simulated
Annealing
(1988) / Ulmer R M

SOUTH DAKOTA SCHOOL OF MINES AND TECHNOL.

Minimization of Exclusive-or Switching
Function Using Reed-Muller Polynomials
(1988) / Baker M J

Microprocessor Controlled Sonar Position
Measurement System for Industrial Robots
(1088) / Berg-Johansen R

SOUTH DAKOTA SCHOOL OF MINES AND TECHNOL.
(continued)

Method of Computerized Pattern Recognition
Using Compiler Techniques
(1988) / Bracht R

Compensation for Two Dimensional Position
Errors in Industrial Robots
(1988) / Ethiraj S

SOUTH DAKOTA STATE UNIVERSITY

The Fast Hartley Transform on a Parallel
Processor
(1988) / Lim B P

Microprocessor Control of a TTL Median
Filter Circuit
(1988) / Pladsen S W

Parallel Processing of the Fast Fourier
Transform
(1988) / Zhou K

SOUTH FLORIDA, UNIVERSITY OF

Maximum Likelihood Parameter Estimation for
Acoustic Transducer Calibration
(1988) / Ainsleigh P L

The Design of a 24 x 24 Bit Parallel
Multiplier Using Signed-Digit Representation
and Carry-Propagation-Free Addition
(1988) / Alyea S E

Modeling a Microstrip Discontinuity Using
the Method of Moments
(1988) / Austin-Lazarus P C

Electron Beam Probing of Integrated Circuits
with the Jeol JSM-840 Scanning Electron
Microscope
(1988) / Green R F

Architecture and Behavioral Simulation for a
Custom 32KBPS ADCPM Chip
(1988) / Guardiano G A

The Design of a CMOS Bit Slice Processor
(1988) / Hammel E C

Tandemable Microprocessor-Based-ADPCM
Test Bed
(1988) / Heathcock R B

Bidirectional Infrared Communications Link
for the Apple II Computer System
(1988) / Kelly N

Automatic Detection and Classification of
Epileptiform Transients in EEG
(1988) / Natour J D

Effect of Void Location on the
Electromigration Rate
(1988) / Nguyen P C P

Poiseuille Flow of a Mixture of Neutrally
Buoyant Particles in a Fluid
(1988) / Reinersman P N

Development and Construction of an Ultra
Sensitive Dual-Channel Noise Measurement
System
(1988) / Scott G J

Fault Tolerant Network Architectures for
Parallel Processing Systems
(1988) / Wills J M

SOUTHERN ILLINOIS UNIVERSITY

A Real-Time, Vectorized, Large Vocabulary
Speech Recognition Algorithm Using Parallel
Processing
(1988) / Blase G

STEVENS INSTITUTE OF TECHNOLOGY

Image Analysis from an Artificial
Intelligence Perspective
(1987) / Smith R

TENNESSEE TECHNOLOGICAL UNIVERSITY

Evaluation of Faults in High Phase Order
Transmission Systems
(1988) / Ahuja G S

Evaluation of Corner Detection Algorithms
(1988) / Davidson J M

Cascade Synthesis of Analog and Digital
Bounded Real Functions
(1988) / Fountain D W

The Effects of Load Characteristics on
Radial System Voltages
(1988) / Khurshid A

Estimation of Striking Distance of Lightning
Stroke to Overhead Lines
(1988) / Kotapalli A K

A 68000 Microprocessor Based Vision System
for Position Determination Via Moment
Invariants
(1988) / Pitts R M

A Two-Input Proportional-Integral-Derivative
Power System Stabilizer
(1988) / Smaili Y A

Numerical Determination of the Spatial
Electron Distribution in a Tokamak Plasma
(1988) / Wilkerson C E

TENNESSEE, UNIVERSITY OF (KNOXVILLE)

A 2-D and 3-D Robot Path Planning Algorithm
Based on Quadtree and Octree Representation
of Workspace
(1987) / Ali A K

The Network Interface Adapter: A Method for
Achieving Device Independent Extensions on
an Industrial Network
(1987) / Babin P D

Design and Implementation of a Parallel
Processing Machine for Artificial
Intelligence Applications
(1987) / Butler P L

A Comparison of 2D Fourier and Spatial
Techniques for Removing Noise in Real Time
X-Ray Radiography
(1987) / Carson D L

Optimum Transmitter Pulse Shapes for Maximum
Acoustical Excitation of Mechanical
Vibrations
(1987) / Castillo D R

TENNESSEE, UNIVERSITY OF (KNOXVILLE)
(continued)

Interactive Computer Graphics Systems
(1987) / Chandra P

Conversion of the HP3577A Network Analyzer
Mode of Operation with Low Noise and
Cross-Talk
(1987) / Crowley J E

A Digital Filter Design Procedure with
Application Program in AFL
(1987) / Eaton W J

Methods for Performing Discrete Fourier
Transform on the TMS32020 Microprocessor
(1987) / Frazier J W

Energy to Produce Electron-Ion Pair in
Sulfur Hexafluoride Mixtures
(1987) / Hilal Y H

Modeling and Control of a Helium Closed
Cycle Calibrating Cryostat
(1987) / Holcomb G D

A Computer Aided Design Software Package
for Logic Diagram Generation
(1987) / Hussain S K

Hybrid Phase-Locked Loop Synchronous Video
Detector for Television Receiver
(1987) / Iga H

Design of the Velocity Measurement System
Using a Heat Source and IR Detectors
(1987) / Nawathe D P

Automated Classification of Objects Through
Ultrasound Signature Analysis
(1987) / Senn D L

Design of a Captive Trajectory System
(1987) / Shuttleworth J G

A Low-Level Ion Beam Profile Monitor
(1987) / Simpson M L

A Firmware Programmable Measurement System
Controller
(1987) / Stultz P S

Design for a Thermal Imaging System Using a
Single Element Pyroelectric Detector as a
Sensing Element
(1987) / Vinson R K

Computer Programs for Designing
Two-Dimensional Recursive Digital Filters
Via Digital Spectral Transformations
(1987) / Vongtanaanek T

Digital Windowing of Transient Signals for
Fourier Spectrum Enhancement
(1987) / Young J W

TENNESSEE, UNIVERSITY OF (TULLAHOMA)

Feasibility of a Reflective Pressure
Transducer
(1988) / DeShetler W

The Development of a Digital Signal
Processor for FPA Data Acquisition and
Reduction
(1988) / Fugerer R H

TEXAS A AND M UNIVERSITY

Imaging and Evaluation of Latch-Up Sites in CMOS
Integrated Circuits
(1988) / Antoniou N

Disparity Coding: A Technique for Stero
Reconstruction
(1988) / Bell W B

An Algorithm for Faulted Phase and Feeder
Selection Under High Impedance Fault Conditions
(1988) / Benner C L

New Approaches to Interfacing Thermoelectric
Generators to the Load Bus in a Nuclear Space
Vehicle
(1988) / Brohlin P L

A Dual Tone Multiple Frequency Receiver Using a
Multiplexed OTA-C Filter Frequency Detection
Architecture
(1988) / Carcia G W

The Detection of High Impedance Faults Using
Random Fault Behavior
(1988) / Carswell P W

Standard Cell Implementation of a Micro-Control
Unit for a Prolog Unification Coprocessor
(1988) / Golnabi H

Fairness in Optimal Routing Algorithms
(1988) / Goos J A

Single Crystal Growth and Characterization of
Ilmenite, Iron Titanium Oxide for Electronic
Applications
(1988) / Gries B L

Simulation of SMES Connected to a Large Scale System
(1988) / Gulde J E

Feedback Effects on Laser Diodes
(1988) / Huda M B

Experimental Investigation of Active Microstrip
Patch Antennas for Power Combiner Applications
(1988) / Hummer K A

Investigation of Dielectric Overlay Microstrip
Circuits
(1988) / Klein J L

Designs for High Power, Single Mode Operation
in Broad Stripe Semiconductor Lasers
(1988) / Lai C P

Universal Signal Processing Method for
Multimode Reflective Sensors
(1988) / Larson R E

Parallel Algorithm and Computer Architecture
for Solving Burgers' Equation
(1988) / Lie H W

Symbolic Control of a Multi-Tasking Signal
Processor
(1988) / Lin S K

Range Image Analysis: Motion Estimation and
Object Recognition
(1988) / Mohan S

A Multiprocessor System for Real-Time Simulation
of Power Electronic Circuits
(1988) / Qawasmi N H

TEXAS A AND M UNIVERSITY
(continued)

A Knowledge-Based Approach for Power-Plant
Availability Analysis
(1988) / Reddy C S

Delay Modeling and Glitch Estimation for CMOS Circuits
(1988) / Shiau Y G

A Parallel Processing Algorithm for Large Scale
State Estimation
(1988) / Tapadiya P K

Low Frequency Noise Measurements of Resonant
Tunnel Diodes
(1988) / Villareal S S

Solid State Power Bus Controllers for
Aerospace Applications
(1988) / Villarreal T J

Bimodal Codebooks for Celp Speech Coding
(1988) / Woo H C

Fabrication of Porous Silicon Membranes
(1988) / Yue W K

On the Empirical Statistics of Parameter Estimates
in Parametric Modeling
(1988) / Zhu Y

TEXAS, UNIVERSITY OF (ARLINGTON)

Sliding Window Taylor Series Approximation
of the Energy Spectral Density
(1988) / Anderson L A

Rapid Method for Computer Recognition of
Handwritten Numerals
(1988) / Bhadsavle S Y

Modeling and Analysis of a Gallium Arsenide
Mesfet Considering Wave Propagation Effect
on the Electrodes
(1988) / Chang R J

A Comparative Study of Adaptive Control and
Sliding Mode Control for Robotic
Manipulators
(1988) / Cheung K W

Development of DCT Algorithms with DSP
Architectures in CAE Environments
(1988) / Ford S S

The Effects of an M of N Detection Criteria
on the Required Signal-to-Noise Ratio Based
on Various Target Fluctuations
(1988) / Frey T L

Ranging Performance of a Pulse Doppler Radar
(1988) / Glazner J K

The Usage of Alternative Green's Function
Formulations in the Analysis of Microstrip
Transmission Lines
(1988) / Grimm J M

W-Band Ferrite-Dielectric Image-Line Field
Displacement Isolators
(1988) / Guo Y

Fast Computation of the DFT Using the
Discrete Sine Transform
(1988) / Gupta A J

A New Method to Solve a System of Nonlinear
Equations with Applications to Engineering
Problems
(1988) / Halloran D J

Electron Density Profile in a Railgun
Simulator
(1988) / Headley C E

Determination of Natural Frequency Response
of a Power System for Automatic Generation
Control
(1988) / Jativa-Ibarra J A

Time Domain Analysis of Measured Frequency
Domain Radar-Cross-Section Data
(1988) / Jersak B D

A Novel Dielectric Multimeter Waveguide
(1988) / KcCroskie A I

Parameter Space Approach to Linear System
Stability Analysis Under Parameter
Variations
(1988) / Lai H M

Real Time Advisor for System Voltage Control
(1988) / Lai K C

Automation of Varian Gen II BME Growth
System Using IBM PC/XT
(1988) / Lin C

A 2-18 GHZ Power Divider
(1988) / Lou I J

Investigation of Non-Invasive Video
Digitization Techniques for Measurement
of Mandibular Performance
(1988) / Lu W C

High Resolution Motion Estimation in Images
Using Wigner Distribution
(1988) / Mandumula K

De-Embedding the Input Impedance of the
Flared Bilateral Slotline Antenna
(1988) / McCleese C

Fast Progressive Reconstruction of Images
Using the DCT
(1988) / Miran M

Gallium Arsenide RF Wafer Qualification
Using RF Probes
(1988) / Mitchell B H

Sampling and Reconstruction of NTSC Color
Signal at Sub-Nyquist Rate
(1988) / Shankar G

Investigation of Human Gait Performance with
a Robotic Vision System
(1988) / Tsaih J

Computer Design of an Optical Intensity
Homogenizer for Uniform Illumination of
a Surface
(1988) / Valentine C E

Dual Gate Gallium Arsenide Traveling
Wave Transistor
(1988) / Venpati S J

An Analysis of a Nonlinear Neural Network
(1988) / Ware C B

A Flexible Microprocessor Based Controller
for a Robotic Vision System
(1988) / Weathers J C

TEXAS, UNIVERSITY OF (AUSTIN)

Bispectral Inversion: The Construction of a
Time Series from Its Bispectrum
(1988) / Allison P S

Implementation of an Adaptive Nonlinear
Algorithm for Short-Term Load Forecasting
(1988) / Anderson G M

Remote Plasma Enhanced Chemical Vapor
Deposition of Silicon on Silicon
(1988) / Anthony B G

Pyramid Based Image Segmentation Using Multi
Sensor Data
(1988) / Asar H V

Implementation and Comparison of Automatic
Test Generation Algorithms
(1988) / Baeg S H

Input-Output Representations for Linear and
Nonlinear Systems
(1988) / Ball D J

An Analysis of Multiple Number of Faults per
Pass (MNFP) for Parallel Fault Simulation
(1988) / Baqai I

Feasibility of Very Large Time-Bandwidth
Signals in Active Sonar Systems
(1988) / Bartels K

A Library of Modular Routines for Generating
Test Patterns for Digital Circuits
(1988) / Belvin T H

Statistical Processing of Magnetotelluric
Data
(1988) / Booker J A

A Study of the Influence of Drift Energy on
Non-Stationary Transport in Submicron
Silicon Devices
(1988) / Bordelon T J

Transmission Properties of a Right Angle
Microstrip Bend with and Without a Miter
(1988) / Broumas A D

Polynomial Digital Filters and Applications
(1988) / Bui H Q

System Model and Performance Analysis of the
Distributed Instruction Set Computer
(1988) / Chai J M

OPS5C: A Treat Based OPS5 Compiler
(1988) / Chandra A

Effects of Rapid Thermal Oxidation on the
Nitride/Oxide Stacked Layer
(1988) / Chang W T

Layered Films of Aluminum/Titanium Nitride/
Tungsten Silicon/Silicon for Multilevel
Interconnects
(1988) / Chen J H

The Automatic Element Routine Generator: An
Automatic Programming Tool for Functional
Simulator Design
(1988) / Chuang C I

Investigation of Solid State Reaction of
Aluminum/Titanium/Silicon Ternary System
Under Controllable Deposition Conditions
(1988) / Chung I

An Analysis Program for the Study of
Particle Beam Probe Data from Thermonuclear
Fusion Experiments
(1988) / Duran-Gonzalez J S

Relational Database Structure for Storage
and Manipulation of Dependency Graphs
(1988) / Easwar S R

A Comparison of Methods for Estimating
Lyapunov Exponents from Experimental Data
(1988) / Feaster B B

Implementation of Task Level Data Flow
Language on the Hypercube
(1988) / Gulati R

A Hardware Performance Evaluation Monitor
(1988) / Gunja F Y

An Automated Standard Cell Layout Package
in a Multiprocessor Environment
(1988) / Guruswamy M

An Efficient Implementation of the
Two-Dimensional Semiconductor Device
Simulator, BAMBI, on the Cray X-MP
(1988) / Horne S C

Development and Verification of a Coaxial
Accelerator Simulation Code
(1988) / Ingram S K

Post-Compiling Method for Fine-Grained
Parallel Processing
(1988) / Ju D

CAD Algorithm for the Evanescent Mode
Waveguide Bandpass Filter with Non-Touching
E-Plane Fins
(1988) / Kong K S

Magneto-Optic Kerr Rotation Measurements of
Magnetic Multilayer Thin Films
(1988) / Kuhn P J

Wideband Monopulse Sonar Performance:
Cylindrical Target Simulation Using an
Acoustic Scattering Center Model
(1988) / Lacker S G

Machine Vision Inspection of Microwave
Circuit Board Foil Attachment
(1988) / Lawrence M R

Motorola 6811 Interface for Macintosh and
LPKF
(1988) / Lee A D

Qualitative Thermal Image Enhancement by
Histogram Modification
(1988) / Lee D H

An Image Processing Technique for the
Quality Assessment of TAB Outer Lead Bonds
(1988) / Manning A T

Architectural Requirements of an Engineering
Design Automation Compute Engine
(1988) / McDermott M W

TEXAS, UNIVERSITY OF (AUSTIN)
(continued)

Microcomputer Implementation of a
Comprehensive Process Characterization Test
Station and Program
(1988) / Miller D R

Nonlinear Analysis of a Monolithic QWITT
Oscillator
(1988) / Mortazawi Meybodi A S E A

A Comparative Analysis of Timing Models and
Scheduling Algorithms for Digital Simulation
(1988) / Pabari D M

System Design of Homopolar Generator-Based
Power Conditioning for Electric Accelerators
(1988) / Pappas J A

Applicability of Ferroelectrics as the
Capacitor Dielectric for ULSI One-Transistor
DRAM Cells
(1988) / Parker L K H

A Micro-Computer Plug-in Board for
Controlling a Laser Beam Shutter
(1988) / Prado F A

A Facility for the Display and Manipulation
of Dependency Graphs
(1988) / Ramachandran S

Design of a Real-Time Signal Acquisition and
Analysis System for Processing Low Angle
Propagation Statistics for the INTELSAT V-A
F-10 KU-Band Beacon
(1988) / Ranganathan M

Vertical Amplifier and Conversion Subsystem
for Digital Oscilloscope
(1988) / Raskin G D

Monte-Carlo Studies of Electron-Hole
Scattering and Minority-Electron in Gallium
Arsenide
(1988) / Sadra K

An Interpolating Parallel Ray Tracer
(1988) / Sankaran S

Solder Joint Inspection Using Thermal
Transient Characteristics
(1988) / Schellhase J C

Microprocessor and Analog Circuit Based
Gating Signal Generation for Harmonic
Frequency Inverters
(1988) / Sharma R

Modeling of Laser Microwelding of Multiple
Layer Assemblages of Thermal Conductors and
Insulators
(1988) / Shieh S

Effect of Third Harmonic Flux on Losses in
a Two-Pole Induction Machine
(1988) / Sidhartha R

A General-Purpose Object-Oriented
Simulation Environment
(1988) / Smith S P

The Human Conscious Operating System
(1988) / Starmack J R

The Automation of Crossbar Switch Design
Synthesis
(1988) / Taraporevale F P

Evaluating Testability Measures
(1988) / Underwood W C

A Broadband TEM Horn Antenna
(1988) / Vasanthakumara P K

CPLACE: A Standard Cell Placement Program
(1988) / Villarrubia P G

Parallel Standard Cell Placer Based on
Min-Cut
(1988) / Widjaja S T

A Single-Chip Microcomputer Based Small
Computer System Interfacea for the Atari
520 ST
(1988) / Williams S P

A Diagnostic Test System for the S-100 Based
Instrumentation System
(1988) / Wu Y J

Design, Implementation, and Evaluation of
STOIC on a 6811
(1988) / Yao L

Reaction Kinetics of Sputter-Deposited
Titanium on Silicon and Silicon Dioxide
Substrates During Rapid Thermal Annealing
(1988) / Yun E J

TEXAS, UNIVERSITY OF (EL PASO)

Loop Capturing Code Buffer for Prefetching
Instructions
(1988) / Bhagat A F

Realization and Performance Evaluation of a Digital
Filter that Used a Look-Up Table
(1988) / Cruz I

Numerical Analysis of the Completed Eigenvalue
Solution of Ridged Waveguide
(1988) / Granados G T

A Microcomputer-Based Multichannel Underwater
Acoustic Telemtry System 91988
(1988) / He Y Q

A Model for Luminescence in SRS: CEF3 ACTFEL
Display Devices
(1988) / Hung W

Electronic Characterization of Thin Film CDS/CDTE
Photovoltaic Cells
(1988) / Lyons L D

A Fast Polygonal Approximating Method for
Computer Vision
(1988) / Manara A H

Optimum Pulse Transmission Through Single-Mode
Optical Fibers and Low-Pass Filter Filtering
Effects
(1988) / Norte A D

An S D L C Based Data Acquisition System
(1988) / Parikh S R

Sound Localization Using Phase Audiometry
(1988) / Rhen L C

Levelized Distribution Transformer Load Management
by Synthetic Load Curve Generation
(1988) / Schulzke F P

Derivation of Photopic Transmittance and Range Using
Imaging Technology
(1988) / Stahoviak J

TEXAS, UNIVERSITY OF (EL PASO)
(continued)

Atmospheric Transmittance Modeling Using Cray
Supercomputer
(1988) / Tehrani-Movahed R

Interface of Local Area Network Controller
6800 Microprocessor
(1988) / Weng C C

TOLEDO, UNIVERSITY OF

Analysis and Design of a Burst Mode Digital
Demodulator Implemented Using a Digital
Signal Processor
(1988) / Bexten R L

A Phase-Controlled Parallel-Loaded
High-Frequency
(1988) / Blackburn S E

Representing Uncertainty in an Expert System
for Coronary Arterial Stenosis Diagnosis
(1988) / Freasier R E

A Steady-State Analysis of the Current-Fed
Parallel Resonant Converter
(1988) / Gerber S S

Applying Residue Number Theory, Fractal
Geometry and Quad-Tree Structures to Digital
Image Data Compression
(1988) / Goel B D

Design of an Adaptive Controller for
Generating Unaits Excitation and Governor
Systems
(1988) / Issa Y I

A Computer Simulation of a Concatenated
Coding Scheme Using Reed-Solomon and
Convolutional Codes for a Land Mobile
Satellite Direct Broadcast System
(1988) / Nissen C A

An Algorithm for Extracting the Left
Ventricular Boundaries from Thallium 201
Scintigrams
(1988) / Sarieh A

Color Identification and Separation Using
Multispectral Gray-Scale Images
(1988) / Shayestah S

A Study of Multiphase High Frequency Power
Distribution Systems Driven by Schwarz
Converters
(1988) / Shetler R E

Gain/Shape Transform Vector Quantization
(1988) / Tin N

TULSA, UNIVERSITY OF

E-Polarization Scattering Analysis of
Arbitrarily Shaped Two Dimensional
Dielectric, Magnetic and Perfectly
Conducting Bodies by Moment Method
(1988) / Baucke R C

Data Compression of Holter
Electrocardiograms
(1987) / Jalaleddine S M S

A Personal Computer Based Digital Facsimile
Coding Device with Inherent Data Compression
(1987) / Tromp H G

A Swept Frequency Method to Characterize the
Relative Permittivity and Permeability of
Homogeneous Materials
(1988) / Williams A R

A Background Method for Monitoring System
Performance Based on Cross Correlation Using
the MC68HC11 Single-Chip Microcomputer
(1987) / Yap C

TUSKEGEE INSTITUTE

Artificial Intelligence Applied to Speech
Recognition
(1987) / Bazile D R

Parametric Study of an Electric Arc for
Space Propulsion Applications
(1987) / Ghosh A K

Isolated Word Recognition Using
Maxima/Minima Probability Distribution
(1987) / Kadambe S L

High Speed Modern Design (Using Vector
Quantization)
(1987) / Nunn P D

Helium Speech Unscrambling
(1987) / Renuka S R

Adaptive Modelling Using Recursive Filter
Design
(1987) / Terutung H

U.S. NAVAL POSTGRADUATE SCHOOL

Calibration and Initialization of the NPS
Modified Infrared Search and Target
Designation (IRSTD) System
(1987) / Ayers G R

The Design of an Intelligent Multidisk
Control Module for VME Bus Based Systems
(1987) / Brooks S L

Numerical Analysis of Double Antennas:
Volume 1 and Volume 2
(1988) / Chafid A

Design of Survivable Shipboard HF Mast
Antenna Models Using the Numerical
Electromagnetics Code
(1987) / Choi I Y

Code Division Multiple Access Applied to
Fiber Optic Data Transmission
(1987) / Fischer T A

2 FSK/QPSK (Frequency Shift
Keying/Quadrature Phase Shift Keying)
Transmitter and Receiver: Design and
Performance
(1987) / Frostenson N A

Path Following Robot
(1987) / Goodway S G

Target Voltage Response in Reaction to
Laser Radiation
(1988) / Harkins R M

Multichannel data Transmission Through a
Fiber Optic Cable
(1987) / Hatzidakis F

Anti-Skywave AM Broadcast Antenna Design
(1987) / Hussain S

U.S. NAVAL POSTGRADUATE SCHOOL
(continued)

A Transition Radiation Experiment to Measure
the Electron Beam Modulation Induced by the
Free Electron Laser: A Design Study
(1987) / Joynson J E

Effects of Charge Distribution Within a
Particle Beam on the Sub-Cerenkov Radiation
(1987) / Jung Y S

Automatic Control of Robot Motion
(1987) / Kalogiros G P

Superconducting Technology for Electric
Propulsion
(1988) / Keamy E R

Performance Study of a Unipole Antenna with
Conventional and Elevated Radial Wire Ground
Screens
(1987) / Koutsouras D A

AR (Autoregressive) Modeling of Coherence
in Time Delay and Doppler Estimation
(1988) / Lee J

Diffraction Transition Radiation from
Periodic Electron Bunches
(1987) / Lee Y M

Representation of Nonstationary Narrowband
Random Processes and Their Application and
Effectiveness as Jamming Signals in Spread
Spectrum Communication Systems
(1987) / Low K M

A Study of LF (Low Frequency) Top-Loaded
Monopole Antennas Using Numerical Modeling
Techniques, Comparison to Scaled Test Model
Measurements
(1987) / Mahmud R

Use of a Coherent Square Wave Reference to
Demodulate BPSK (Binary Phase Shift Keying)
Carriers and a Visual Indicator of the
Quality of Received QPSK (Quadrature Phase
Shift Keying) Carriers
(1987) / Mehmet K

Investigation Into the Use of Texturing for
Real-Time Computer Animation
(1987) / Meier T W

Three-Dimensional Perspective Image
Generation from Sonar Bathymetry and
Imagery Data
(1988) / Myers R J

Natural Convection Immersion Cooling of an
Array of Simulated Chips in an Enclosure
Filled with Dielectric Liquid
(1987) / Pamuk T

Evaluation of Device Figure of Merit in
Microwave Switching Circuit Design
(1987) / Pibultip P

A Survey of Shipboard Combat Survivable
VHF/UHF (Very High Frequency/Ultra High
Frequency) Antennas
(1987) / Purvis H K

Photocurrent Generation from Basic Metals,
Utilizing a Short Pulsed ARF Excimer Laser
(1987) / Ringler T J

Design and Implementation of a Fiber Optic
RS232 Link
(1987) / Ryan J W

Direct Bit Detection Receiver Performance
Analysis for 8-DPSK (Differential Phase
Shift Keyed) Modulated Signals Operating
with Improper Carrier Phase Synchronization
(1987) / Sekerefelt M S

2 FSK/QPSK (Frequency Shift
Keying/Quadrature Phase Shift Keying)
Transmitter and Receiver: Design and
Performance
(1987) / Sonnefeld M D

Steady Flow Field Measurements Using Laser
Doppler Velocimetry
(1987) / Wilson R E

Digitally Programmable Active Switched
Capacitor Filters
(1987) / Yalkin C

U.S.A.F. INSTITUTE OF TECHNOLOGY

Speech Recognition Using Neural Nets and
Dynamic Time Wraping
(1988) / Barmore G D

The Impact of IEEE-1076 on VHDL
(1988) / Berk K J

Total Dose Response of Silicon-on-Insulator
Metal-Oxide-Semiconductor Field-Effect
Transistors
(1988) / Biwer M C

Back-Contact Vertical-Junction Solar Cell
(1988) / Carver M W

Digital Signal Processing for Detection of
Direct Sequence Signals Using the MODAC
Receiver
(1988) / Conner J R

Characterization of Metallic Coatings and
Thin Films Produced by Railgun Deposition
(1988) / DeLuca R J

Analysis and Design of Array Interconnection
Networks for Phased Antennas
(1988) / Downs J D

A Database Management System for Computer
Aided Digital Circuit Design
(1988) / Ehrhart S A

A Computer Simulation Analysis of
Conventional and Trunked Land Mobile Radio
Systems at Wright-Patterson Air Force Base
(1988) / Farrell T C

Optimization of Microwave Magnetoelastic
Delay Lines
(1988) / Feldman B F

A Digital Singel-Sideband Modulator for a
Digital Radio Frequency Memory
(1988) / Foltz T M

Performance Analysis of the Fiber
Distributed Data Interface in the Super
Cockpit Audio World
(1988) / Forlanda J E

Comparison of Photovoltic Energy Systems
for the Solar Village
(1988) / French P

U.S.A.F. INSTITUTE OF TECHNOLOGY
(continued)

Reverse Engineering VLSI Using Pattern
Recognition Techniques
(1988) / Fretheim E J

Object-Oriented Allocation of Resources in a
Tactical Communications Network
(1988) / Gier G R

Analytical Techniques for Tracking Filter
Implementation
(1988) / Gleason D

Voice Recognition and Artificial
Intelligence in an Air Traffic Control
Environment
(1988) / Hall R F

A Proototype Knowledge-Based System for
Developing Acquisition Strategies
(1988) / Hammell R J

Concentration Gradient Detection by
Near-Infrared Diode Laser
(1988) / Hancock D D

Digital Signal Design for Meteorscatter
Communications
(1988) / Jacobsmeyer J

Calculation of Carriers in Depletion Region
of Semiconductors with Capacitance-Voltage
Measurements
(1988) / Kim J H

F-16 Speaker-Independent Speech Recognition
System Using Cockpit Commands
(1988) / Kim P

Damage Effects on Thin Acrylic Targets
Caused by Infrared Radiation from a Pulsed
Carbon Dioxide Laser
(1988) / Kreifels T L

Investigation of a Hybrid Wafer Scale
Integration Technique that Mounts Descrete
Integrated Circuit Die in a Silicon
Substrate
(1988) / Mainger R W

Characterization of Semiconductor Device
Processing Etchout Solutions for Germanium
Arsenide and Platinum Germanium Arsenide
Layers Employed in Germanium Arsenide
Aluminum Germanium
(1988) / Mayhew B F

An Improved Method for Calculating Power
Density in the Fresnel Region of Circular
Parabolic Reflector Antennas
(1988) / Mize J E

Bit-To-Bit Error Dependence in Direct
Sequence Spread-Spectrum Multiple-Access
Packet Radio Systems
(1988) / Morrow R

Asynchronous Digital Regulators
(1988) / Ritchey V S

Enhanced Autonomous Face Recognition Machine
Volume I and II
(1988) / Sander D D

LMS Adaptive Filtering Applied to a
Microwave Arterial Pulse Monitor
(1988) / Simes B J

An Examination of a Simulated Microburst
Flow as Sensed by a Single Doppler Radar
(1988) / Smith E L

Determination of the Unstable States of the
Plasma in Semiconductor Devices
(1988) / Snyder M E

Robust Recognition of Loud and Lombard
Speech in the Fighter Cockpit Environment
(1988) / Stanton B J

Modeling and Performance of HF/OTH Radar
Target Identification Systems
(1988) / Strausberger O

On the Impulse Response of Monopulse Radars
(1988) / Tackett D L

The Ion-Assisted Deposition of Optimal Thin
Films
(1988) / Targove J D

Techniques in Thin Film Fabrication
(1988) / Thompson S D

A Microwave Measurement Technique for Complex
Constitutive Parameters
(1988) / Walker R N

Investigation of the Impedence Modulation of
Thin Films with a Chemically-Sensitive Field
Effect Transistor
(1988) / Wiseman J M

Electro-Encephalogram Based Adaptive
Estimation of Magneto-Encephalogram Signals
(1988) / Wood R A

8755 Emulator Design
(1988) / Woods J L

UTAH STATE UNIVERSITY

Signal Processing of Hydroxyl Airglow
Interferometric Data
(1987) / Chao S C

Infra-Red Guidance Aid for the Blind
(1987) / Juandy A

Stochastic Network Optimization
(1987) / Julien-Laferriere P

Development of Sliding Window and LDP of
Kermit for PDN
(1987) / Tan J K

UTAH, UNIVERSITY OF

Effects of Low-Level Static and Extremely
Low Frequency Time Varying Magnetic Fields
on the Electrical Impedance of Bilayer
Phospholipid Membranes
(1988) / Anderson A A

Variable Adaptive Narrowband Jammer
Suppressor
(1988) / Bullock S R

Three-Dimensional Modeling of Interstitial
Antennas for Microwave Hyperthermia
(1988) / Furse C M

A Study of Gallium Arsenide Logic Families
and Their Attributes
(1988) / Hatch M E

UTAH, UNIVERSITY OF
(continued)

Development of a Multipactor Pressure Gauge
for Use as a Noninvasive Vacuum Tube Sensor
(1988) / Kenny M B

A Conducting Elastomer Sensor for
Transducing Three-Dimensional Forces Into
Direct Digital Output
(1988) / Milne K B

Resonance Raman Spectroscopy of Porphyrin
Photosensitizers
(1988) / Mitchell J R

Calculated Response of Lossy Dispersive
Dielectric Models to Pulsed Plane Wave
Electromagnetic Radiation
(1988) / Moten K

Design and Simulation of a High Speed
Inductor Alternator
(1988) / Ruan M

Electromagnetic Analysis of a Radio
Frequency Liquid Level Measurement System
(1988) / Sorensen K W

A Parallel CMOS Architecture for an
Efficient Trellis Coding System
(1988) / Wang W

Characterization of an Evanescent Fiber
Optic Immunosensor Using Tetramethylrodamine
(1988) / Yoshida D E

Ultrasound Reflection Tomography: Compound
B-Scan Imaging with Synthetic Focus
Algorithm
(1988) / Yuan W

VERMONT, UNIVERSITY OF

Laser Diode Profiling Using a Charge Coupled
Photodiode Array
(1988) / Binder M S

MOSFET's Electrical Characteristics:
Comparison of Theories with Experiment
(1988) / Farah J S

Experimental Remote Data Acquisition and
Control System
(1988) / Goodheart M

Deep Level Transient Spectroscopy: Test
System Hardware Design
(1988) / Hosking T A

Two Graph Theoretic Reformulations of the
PLA Folding Problem and Algorithms for
Their Solution
(1988) / Lecky J E

Fast Signal Processing Using Multiple
Programmable Signal Processors: Algorithm
Decomposition and Multiprocessor
Configurations
(1988) / McGuire G A

VICTORIA, UNIVERSITY OF

VLSI Design of a Testable Processor Array
for Feature Extraction
(1988) / Aubry P P

VILLANOVA UNIVERSITY

A Method for Varying the Step Size in the
LMS Adaptive Algorithm
(1988) / Arnao M A

Electrical Transport Properties of Copper
Chloride Intercalated Graphite - Polymer
Composite Films
(1988) / Chen J Q

A Software Implementation of a Real-Time
Video System for Use with Digital Packet
Switched Communication Networks
(1988) / Giuffrida D G

Tone Reconstruction Using the Short-Time
Fourier Transport
(1988) / Piscitelli J L

VIRGINIA, UNIVERSITY OF

Design of a Computer Aided Manufacturing
System for Custom Contoured Wheelchair
Cushions
(1988) / Brienza D M

Distributed Sensor System Decision Analysis
Using Team Strategies
(1988) / Choe H C

Dynamic Response of Polymer Dispersed
Liquid Crystals
(1988) / Colvin S C

Quantization in Pyramid Progressive
Transmission of Images
(1988) / Grabb M L

A Decentralized Control Strategy for
Enhanced Army Field Power System
(1988) / Haberkorn W E

A Methodology for the Uninterpreted
Modeling of Digital Systems Using VHDL
(1988) / Hady F T

Segmentation of Textured Images by a
Maximum A-Posteriori Criterion Using Markov
Mesh Joint Density Models
(1988) / Ioannidis A

Automated Composition System for a
Hardware Description Language
(1988) / Jordan P R

Digital Control System Design for Active
Magnetic Bearings
(1988) / Keith F J

A Graphic Representation of VHDL Models
(1988) / Klenke R H

The Reliable Design of CMOS Synchronizers
(1988) / Ko T T

Power Spectra of Block Coded Modulation
(1988) / Lakshman M

Random Access Algorithms for Environments
with Capture
(1988) / Lyons D F

Symbol Synchronization for a Continuous
Phase Modulated Mobile-Satellite Modem
(1988) / Michelson S M

VIRGINIA, UNIVERSITY OF
(continued)

A Formalism of the Rete Algorithm in Terms
of Actor Theory
(1988) / Myers D C

The Gallium Arsenide Schottky Barrier
Membrane Diode
(1988) / Seidel L K

Dual Camera Representation of Images
(1988) / Sigda J P

An Improved DC-DC Converter and Adaptive
Controller for Wheelchairs
(1988) / Tsoi D

A Unified Model of Hardware/Software
Reliability
(1988) / Welke S R

Fault Tolerant Multiprocessor Digital
Control System Design for Active Magnetic
Bearings
(1988) / Yates S W

WASHINGTON STATE UNIVERSITY

Evaluation of Measurement Techniques for
Space Charge Density Near High Voltage DC
Power Lines
(1988) / Acord G C

HVDC Transmission Line Generated Corona
Behavior and Characteristics
(1988) / Collins H L

Innovations in Deep Level Transient
Spectroscopy
(1988) / Devries P D

Analysis of the Spatial, Temporal, and
Electrical Characteristics of Lightning in
Portions of the Northwestern United States
for 1985 and 1986
(1988) / Elkin P R

Analysis and Design of Gallium Arsenide
Integrated Series Impact Structures
(1988) / Lam N

Implementation of a Full-Wave/Quasi-Static
Hybrid Method for Analysis of Axially
Symmetric Thin-Wire Antennas with Capactive
Loads
(1988) / Mannikko P D

Gallium Arsenide Traveling-Wave Diodes for
Millimeter-Wave Frequency Applications
(1988) / Memaran-Dadgar A

Design of a Laboratory Bit-Slice
Microprogrammable Processor
(1988) / Ngo A H

Higher Radix Multiplication Algorithms for
VLSI Implementation
(1988) / Primlani K K

Techniques for the Development of a Language
for an Expert System
(1988) / Somanchi S V

A Cordic Hardware for High Speed Computer
Graphics
(1988) / Sy L Y

Modeling and Control of the Capek Robotic Arm
(1988) / Xu Z

WASHINGTON UNIVERSITY

Numerical Methods in Array Processing
(1987) / Cheng Y A

A New Method for Analyzing Auditory-Nerve
Discharge Patterns
(1987) / Karamanos N A

Simple Analytical Modeling of Gallium
Arsenide MESFET Nonlinear Behavior
(1987) / Kawai T

Inverse Synthetic-Aperture Radar Imaging
(1987) / Lewis R C

Efficient Computation of the FM Algorithm
in Time-of-Flight Positron-Emission
Tomography
(1987) / Liu S K

A New Method for EM Autoradiography:
Validation and Comparison to Crossfire
Analysis
(1987) / Roysam B

Resource Depletion and Substitution
Efforts in Tungsten
(1987) / Scherer A

Aluminum Gallium Arsenide Heterostructure
Waveguide Detector/Modulator
(1987) / Sherman G R

Classical Fault Analysis for MOS VLSI
Circuits
(1987) / Shing B L

WASHINGTON, UNIVERSITY OF

Impedance Characteristics of DH
Semiconductor Laser Diodes: Implication of
High-Level Injection Theory
(1987) / Al-Alusi M R

Composite Matched Filter Performance in the
Presence of Input and Processor Noise
(1987) / Amindavar H

Distributed Sorting on a Network of
IBM PC AT's
(1987) / Ananthaswamy A

An Interactive Speech Enhancement
Workstation
(1987) / Andrew R K

A Study of the Temperature Characteristics
of ISFET Sensors and Their Associated
Measuring System
(1987) / Aw C Y

A Neural Network Model Based on the Least
Squares Solution
(1987) / Baker J B

Experimental and Numerical Studies of
Internal Defibrillation Electrodes
(1987) / Barnett D W

Automated Detection of Senile Plaques in
Alzheimer's Disease Using Image Analysis
(1988) / Bartoo G T

WASHINGTON, UNIVERSITY OF
(continued)

Implementation of a Biomedical Virtual Image
Processor and Its Application to
Quantitative Microdensitometry
(1987) / Bauer G L

LANA: A Local Area Network Advisor Expert
Database System for Local Area Network
Design
(1987) / Benson D

Applications of a TMS32020 Digital Signal
Processor in Implementing Image Processing
Algorithms
(1987) / Blattenbauer J A

A Neural Network Signal Processor
(1987) / Bloor G J

Determination of Voltage Accuracy
Requirements for Parallel DC Bus Operation
of Power Converters
(1988) / Bocek J M

The Design and Implementation of a General
Purpose Low-Cost Bus Oriented Image
Processing System
(1987) / Budak P V

Model Reduction by Aggregation for Multiple
Input Multiple Output Systems
(1987) / Camps O I

Dynamic Simulation of Power Converters
(1988) / Chang H K

Software Development for the University of
Washington Graphics System Processor (UWGSP)
Image Processing System
(1987) / Chauvin J W

Adaptive Power Factor Controller
(1987) / Chen M

A Performance Analysis of Associative
Memories with Nonlinearities in the
Correlation Domain
(1987) / Choi J J

Diagnosis and Consensus in Distributed
Systems
(1988) / D'Souza G P

Simulation of the Open Systems
Interconnection Distributed Transaction
Processing Protocol Using OPNET
(1988) / Elkana A

Design and Implementation of a Distributed
Problem Solving Environment
(1987) / Fishkow M

Graphical Data Acquisition Using a Computer
Drawing Program
(1988) / Fu Y H

Efficient Calculation of Certain Receiver
Operating Curves
(1987) / Funk D E

Design of an Optical, High Frequency Voltage
Probe for Integrated Circuits
(1987) / Haueisen D C

An Architecture for Real Time
Back-Propagation of Fields
(1987) / Hill J M

A Study of Frequency-Wavenumber Power
Spectral Density Estimation Techniques
(1987) / Hippe R

A General Method of Noise Analysis for
Electroacoustic Transducers
(1987) / Hoard P S

A Case Study for Network Design Using
Industry Standards, Cost and Longevity
as Criteria
(1987) / Houglum S J

Built-in-Test-Equipment for Microprocessor
Based Systems
(1987) / Hua T

A Virtual Sorting Method
(1987) / Hunter D

A Waveform Verification System for PROM
Programmers
(1988) / Hutchinson J E

Miniature Silicon Based Liquid Junction
Reference Electrode
(1987) / Jin H

Implementation and Evaluation of Recoverable
Image Compression Techniques
(1987) / Jurgens P A

Virtual Image Processor: A Prototype
Implementation
(1987) / Kaucic R A

Gaseous Discharge Lamps as High Efficiency
Focused Light Sources
(1987) / Keh G C

An Adaptation of the Inertial Upper State
(IUS) Guidance Software
(1987) / Knoben D G

Model for a Velocity Sensitive Piezoelectric
Hydrophone
(1987) / Larson M D

Use of a Liquid Crystal Television as a
Programmable Spatial Light Modulator
(1987) / Leung W W

Electronically Commutated DC Motor Control
Using an Optical Encoder
(1987) / Li Z

Performance Analysis of a Robust MFSK
Communication System in Partial Band Noise
(1987) / Lintelman S

An Expert System Supplemented by a Database
Management System Serving as an Electrical
Distribution System Engineering Aid
(1988) / Long R

A Feedback Control Algorithm for Ballast
Control Systems of Offshore Production
Vessels
(1987) / Luk T H

Microwave Propagation and Scattering in a
Dense Distribution of Spherical Particles
(1987) / Mandt C E

An Administrative Database Using Microrim's
RBASE-5000
(1987) / Mar M

Development of an Adaptive Memory Structure
Based on Computer Models of Neural
Mechanisms
(1987) / Parkhurst W R

WINDSOR, UNIVERSITY OF

3-D Robot Vision Metrology and Camera
Calibration from Focus Information
(1988) / Cardillo R

CMOS VLSI Design Concepts and Practice
(1988) / Carr J T

Design of 1-D and 2-D Recursive Digital
Filters Using Hurwitz Polynomials
(1988) / Lee H J

Focusing of Images Using Spatial Operators
(1988) / Mameri M M

Digital Character Scaling by Contour Method
(1988) / Namane A

Custom Design of CMOS Read Only for VLSI
Residue Number System Hardware
(1988) / Raja P V R

Skeletonizations, Thinning and Thickening
Algorithms and Their Application to Arabic
Characters
(1988) / Tellache M

A Hierarchical Analysis and Verification
Methodology for Complex VLSI Systems
(1988) / Thomsen G A

A UNIX VLSI Design Workstation
(1988) / Yeung A

WORCHESTER POLYTECHNIC INSTITUTE

Modeling and Analysis of a High-Voltage
Power Converter
(1987) / Cummings M R

Multiple Accessing in Local Area ALOHA
Networks in the Presence of Capture
(1987) / Ganesh R

Linear Feedback Shift Register Testing of
VLSI Processors
(1987) / Reynolds S L

Digital Image Filtering Based on Local
Statistics
(1986) / Sherlock P H

Pressure Dependence of the Photoacoustic
Effect
(1986) / Uddin N

WRIGHT STATE UNIVERSITY

Design of Digital Autopilot Using Intel's
8097 Microcontroller
(1987) / Abbasi A A

A Computer System for an Automated
Workstation
(1987) / Johnson R E

Design Implementation and Analysis of a
Modular Digital Signal Processing System
(1987) / Schweitzer J R

Stability Robustness with Reduced Order
Observers
(1987) / Woo C H

WYOMING, UNIVERSITY OF

Optimal Dyadic Feedback Control of Multiple
Input Linear Systems
(1989) / Alam M S

Real-Time Multi-Channel Digital Filtering
Utilizing the WE DSP32 Digital Signal
Processor
(1989) / Birks R M

Optimal Scheduling of Power Generation with
Transmission Line Constraints
(1989) / Desai H D

Fast-Programmable Time Interval Generator
for Periodic and Aperiodic Waveform
Generation
(1989) / Howard R S

A Digital Estimator for the Speed Control
of an Induction Motor
(1989) / Overy E J

The Design of a Hearing Aid Radio Frequency
Link
(1989) / Smith M T

YOUNGSTOWN STATE UNIVERSITY

Implementation of a Digital Control
Algorithm with Multiplexed State Variable
Feedback
(1988) / Bassil I

Comparison of Parameter Estimation
Algorithms Implemented on a Personal
Computer
(1988) / Guo N C

Implementation of a Modern Control Algorithm
on a General Purpose Microcomputer
(1988) / Sarantopoulos A D

ALABAMA, UNIVERSITY OF (UNIVERSITY)

An Investigation of Diploid Genetic
Algorithms for Adaptive Search of
Nonstationary Functions
(1988) / Smith R E

ARKANSAS, UNIVERSITY OF

Filtration of Charged Aerosol Particles by
Electrified Fabric Filters
(1988) / Siag A M

Analysis of Magnetic Tap Flutter Effects on
the Spectrum of Pulse-Width Modulated Infant
Respiratory Signals
(1988) / Trulock D W

CALIFORNIA STATE UNIV. (FULLERTON)

The Nonlinear Oscillations of Disk Springs
Under Vibratory Loads
(1988) / Nordyke K

CALIFORNIA, UNIVERSITY OF (LA JOLLA)

Plane Stress Finite Element Implementation
in Instructional Program CALSD
Micro-Computer Version
(1987) / Abu-Ghazaleh N S

Computational Turbomachinery Flows Using
the Dawes Code
(1987) / Carroll M A

Liquid Film Tunnel
(1988) / Derango P J

Biomechanical Properties of Cultured
Osteochondral Allografts Using Linear
Biphasic Theory
(1988) / Field F P

Analysis of Geometrically Nonlinear Frames
by the Displacement Method
(1987) / Fuendeling R J

CLARK ATLANTA UNIVERSITY

Optimum problem Involving Integral Equations
(1987) / Twagiramungu I H

CLEMSON UNIVERSITY

Fracture of Discontinuous Metal Matrix
Composite Materials
(1988) / Albritton J R

Development of a Nonlinear Laminated Finite
Element
(1988) / Anandan V

An Experimental and Analytical Evaluation of
a Biaxial Test for Determining Shear
Properties of Composite Materials
(1988) / Barnett T R

On the Existence of Singular Stresses in
Radially Orthotropic Bodies
(1988) / Lamble P J

P-Version of the Finite Element Method
(1988) / Turimella J

A Study of Finite Element Remeshing
Procedures for Large Strain Problems with
Specific Application to the CFORM-FEM
Computer Code
(1988) / York A R

CORNELL UNIVERSITY

Computer Algebra Implementation of Lie
Transforms for Hamiltonian
Systems: Application to Nonlinear Stability
(1988) / Coppola V T

A Pseudoexponent for the Characterization
of Periodic and Chaotic Data Sets from
Forced Systems
(1988) / O'Reilly O M

FLORIDA, UNIVERSITY OF

Optimization and Sensitivity Analysis of
Structures Including Path Dependent
Processes
(1988) / Gunger M

The Effect of Acoustic Noise on Turbulence
Measurements
(1988) / Stone J G

IOWA STATE UNIVERSITY

Grid Optimization for the Boundary Element
Method
(1988) / Chakravarty R R

Boundary Variable Condensation in the
Solution of Multiple Domain Problems by
Boundary Elements
(1988) / Guha S

Vibrations of a Clamped Plate with Spring
and Mass Attachments
(1988) / Salazar-Vior J M

KENTUCKY, UNIVERSITY OF

Acoustic Scattering by Elastic Solids Using
the Boundary Element Method
(1988) / Goswami P P

The Boundary Integral Equation Method for
Half-Spaced Formulations in Two- and
Three-Dimensional Elastostatics
(1988) / Kondapalli P S

Structural Optimization of Laminated
Composite Plates and Cylindrical Panels
(1988) / Kumar N

Application of Two Surface Plasticity Theory
to Proportional and Non-Proportional
Cyclical Loading
(1988) / Mohamed Z II

LEHIGH UNIVERSITY

Acoustic Waves in a Cylindrical Duct Filled
with an Inhomogeneous Liquid
(1987) / Ozturk M

MARQUETTE UNIVERSITY

Mandibular Autorotation Following Maxillary
Impaction Surgery: A Careful Look at
Mandibular Movement
(1988) / Arnold J

The Double-Sealant Technique: An In Vitro
Evaluation of an Indirect Bonding Technique
(1988) / Bellon M

In Vitro Bond Strength Comparison of Metal
and Ceramic Brackets Using Autopolymerizing
and Light-Cured Resins
(1988) / Ostertag P

MARQUETTE UNIVERSITY
(continued)

Effect of Processing Parameters on Void
Content of Filament Wound Cylinders, Using
Taguchi Method of Experimental Design
(1988) / Ryan W

Analysis of Constrained Deformation of a
Circular Brush System
(1988) / Shia C Y

MASSACHUSETTS INSTITUTE OF TECHNOLOGY

Design and Construction of a Six Degree of
Freedom Parallel Link Platform Type
Manipulator
(1988) / Ismail A N

Design and Evaluation of Time Delay Control
for Active Magnetic Bearings
(1988) / Kondo F

An Integrated CAD/CAM System for Turning
Cylindrical Parts
(1988) / Kwok Y W

Dynamic Modelling and Control of a Magnetic
Bearing - Suspended Rotor System
(1988) / McCallum D C

Development and Verification of a Rotary
File Cutting Model
(1988) / Smith R P

Wave Impact Loads on Vertical Cylinders
(1988) / Zhou D

MICHIGAN TECHNOLOGICAL UNIVERSITY

Photoelastic Study of the
Trunnion-Roller-Ball Interface of a Tripot
Universal Joint
(1988) / Galarno M J

Elastic and Plastic Analysis of Stress and
Stability of Multiple Mine Openings in
Orthotropic Layered Rock
(1988) / Houghtaling T K

Surrogate Shoulder for Acromiclavicular
Joint Injury Due to Impacts
(1988) / Kenyon R R

MINNESOTA, UNIVERSITY OF

Thermoelastic Stability of Multiple Growth
Twins in Quartz and General
Geobarothermometric Implications
(1988) / Zanzotto G

MISSISSIPPI STATE UNIVERSITY

Dropped-Weight Simulation of Nuclear Surface Burst
Induced Ground Shock-A Feasibility Study
(1986) / Welch C R

NORTHEASTERN LOUISIANA UNIVERSITY

Numerical Analysis of Some Electro-Magnetic
Field Problems
(1988) / Ge F

NORTHERN ILLINOIS UNIVERSITY

A Study to Examine University Graphic Arts
Faculty in Terms of Supply, Demand, and
Their Importance to the Printing Industry
(1988) / Wilson D G

Design of Fundamental Mode Suppressed AT-cut
Quartz Resonators for Nontuned Third
Overtone Oscillators
(1988) / Yen C H

PENNSYLVANIA STATE UNIVERSITY

Study of Rheological Behavior of Bituminous
Materials by Indirect Tension Test
(1988) / Jia N Y

The Dynamic Response of a Fuzzy Dynamic
System
(1987) / Kim G

Processing and Mechanical Behavior of Tape
Cast and Laminated Silicon Carbide
Whisker/Alumina Composites
(1988) / Kragness E D

Calibration of a Method for the Measurement
of Residual Stress in Ceramics Using Knoop
Indentation Techniques
(1987) / Matsumura J M

Fatigue Crack Initiation and Propagation of
Induced Surface Flaw Under Biaxial Stress
Field
(1988) / Seibi A

A Hydrodynamic Analysis of Fluid Flow
Between Meshing Spur Gear Teeth
(1987) / Wittbrodt M J

RENSSELAER POLYTECHNIC INSTITUTE

Use of Auditory Cues to Reduce Visual
Workload
(1988) / Brown M L

SASKATCHEWAN, UNIVERSITY OF (SASKATOON)

Readable Maps for the Blind: A Study of
Graduated Tactual Area Symbols for
Ordinally Classed Quantitative Information
on Maps for the Visually Handicapped
(1988) / Hipkin S A

SOUTH CAROLINA, UNIVERSITY OF

Microcomputer Procedures for Management and
Plotting of Structural Geologic Data on
Stereographic Projection
(1988) / Michelin D

TEXAS, UNIVERSITY OF (ARLINGTON)

Implementation of the Strain Limit Method
of Determining the Yield Envelope for
Metals
(1988) / Belanus K

Composite Plate Buclking Investigation
Using REDUCE
(1988) / Forehand S M

Mode Localization Phenomena in Multispan
Beam System
(1988) / Kuo H D

Free Vibration Analysis of Point-Supported
Beam and Rectangular Plate Problem by
Computer-Aided Galerkin's Method
(1988) / Wang Y M

Buckling Analysis of Anisotropic Plates by
Using a Method of Weighted Residuals and
Symbolic Algebra Software
(1988) / Weber J A

TEXAS, UNIVERSITY OF (AUSTIN)

Implementation of Line Search Algorithms in
the Solution of Nonlinear Finite Element
Equations
(1988) / Bhagat R

Calculation of Energy Release Rate for 2-D
Rubbery Material Problems with Finite
Element Methods
(1988) / Chang J H

Simulation of Mass Transport with Reaction
in Laminar and Turbulent Boundary-Layer
Flow
(1988) / Geld J N

Determining the Parameters of Elastic
Strain Energy Functions
(1988) / Marusak R E

An Experimental Study of the Response of
Inelastic Thin Walled Tubes Under Combined
Bending and Tension
(1988) / Tischler J C

U.S. NAVAL POSTGRADUATE SCHOOL

A Three Dimensional Non-Singular Modelling
of Rigid Manipulators
(1987) / Altinok S

Designing an Automatic Control System for a
Submarine
(1988) / Babaoglu O K

Vibration Problems of Rotating Machinery
(1988) / Barber F L

SCMOS (Scalable Complementary Metal Oxide
Silicon) Silicon Compiler Organelle Design
and Insertion
(1987) / Baumstarck J E

An Expert System Interfaced with a Database
System to Perform Troubleshooting of
Aircraft Carrier Piping Systems
(1988) / Boozer P R

Preliminary Design of the Orion Attitude
Control System
(1987) / Chappell D C

An Expert System Interfaced with a Database
System to Perform Troubleshooting of
Aircraft Carrier Piping Systems
(1988) / Clayton I B

Theoretical Model of the Cathode Spot on a
Unipolar Arc
(1987) / Curtiss D H

A Computer Simulation Study of an Expert
System for Walking Machine Motion Planning
(1987) / Goodpasture R P

Microcomputer Control of a Hydraulically
Actuated Piston
(1987) / Grunther I

A Model of the Effect of Sparing and Repair
Turnaround Time of the Inertial Navigation
System on Aircraft Readiness for the F/A-18
(1988) / Hase C A

Modeling and Control of a Novel Robotic
Actuator
(1988) / Ingram J D

On Analysis of Viscoelastic Structures
(1987) / Kim J E

A Comparison of Six Repair Scheduling
Policies for the P3 Aircraft
(1988) / Latta P J

Vibration Response of Constrained
Viscoelastically Damped Plates: Analyses
and Experiments
(1987) / Maurer G J

Interactions Between Synoptic and Planatery
Scales of Motion
(1987) / Mcatee M D

A Microprocessor-Based, Solar Cell
Parameter Measurement System
(1988) / Oxborrow R R

Control System Simulation for a Single-Link
Flexible Arm
(1987) / Park K S

Machinery Diagnostics Via Mechanical
Vibration Analysis Using Spectral Analysis
Techniques
(1988) / Stamm J A

Development of a Boundary Layer Control
Device for Tip Clearance Experiments in an
Axial Compressor
(1988) / Tarigan M

Design of a Velocity and Position Control
Laboratory Servo System
(1987) / Ziegler M A

Control of a Flexible One-Link Manipulator
(1987) / Zouzias I E

U.S.A.F. INSTITUTE OF TECHNOLOGY

Development and Testing of a Device Capable
of Placing Model Piles by Driving and
Pushing in the Centrifuge
(1988) / Gill J J

Effects of Expansion Devices on the
Transient Response Characteristics of the
Air-Source Heat Pump During the Reverse
Cycle Defrost
(1988) / Peterson K T

Size Effects in Linear Elastic Fracture
Mechanics
(1988) / Pieri R V

UTAH STATE UNIVERSITY

A Comparison Study of the Attitudes of
Idaho's Industrial Arts/Technology Education
Instructors, Administrators, and Industrial
Managers Regarding Technology Education
(1987) / Thode B R

EAST TEXAS STATE UNIVERSITY

FFT for Arbitrary Length and Its Application
to Low Bit Rate Coding and Two Speaker
Separation
(1988) / Chien D T

SAN DIEGO STATE UNIVERSITY

Digital Methods for Intake-Retention
Functions
(1988) / Hope E P

TEXAS A AND M UNIVERSITY

Further Evaluation of Dose Estimation Using the
FBX Dosimeter
(1988) / Helfinstine S Y

A Study of the Response of a Gas Ionization
Chamber to Different Sources of Ionizing Radiation
(1988) / Zamble-Dieguez F E

U.S. NAVAL POSTGRADUATE SCHOOL

Comparison of P-3C Acoustic Processing
Capability with Acoustic Operator Capability
(1987) / Arnold R R

Root Placement with Transfer Function
Methods (Full State Feedback)
(1988) / Ates M

Measured Noise Performance of a Direct
Sequence Spread-Spectrum System and a
Comparison of Single-Vice Dual-Channel
Delay-Lock Loops
(1987) / Bartone C G

The Development and Structural
Characteristics of Dean Vortices in a
Curved Rectangular Channel
(1988) / Baun L R

An Advanced Study of Natural Convection
Immersion Cooling of a 3 x 3 Array of
Simulated Components in an Enclosure Filled
with Dielectric Liquid
(1988) / Benedict T J

Comparison of EMP (Electromagnetic Pulse)
and HERO (Hazardous Electromagnetic Effects
on Ordnance) Programs
(1988) / Bogan W R

Holographic Investigation of Solid
Propellant Combustion
(1988) / Butler A G

Separation in Time-Dependent Flow
(1988) / Butterworth W H

An Evaluation of the Impact of Variable
Temporal and Spatial Data Resolution Upon
IREPS (Integrated Refractive Effects
Prediction System)
(1987) / Dotson M E

Application of Voice Recognition Input to
Decision Support Systems
(1988) / Drake R G

Study of Vortices Embedded in Boundary
Layers with Film Cooling
(1987) / Evans D L

A Molecular Dynamics Simulation Study of
Liquid Metal Targets Using the Embedded
Atom Method
(1988) / Fisher M L

Feasibility Study of a Microprocessor
Controlled Actuator Test Mechanism
(1988) / Goode G L

The Characteristics of Reduced-Density
Channels in NH3-N2 Gas Mixtures
(1988) / Goodwin W A

Development of a Compact Apparatus for
Determining Compplex Parameters of Fluid
Filled Porous Solids by Impedance Techniques
(1988) / Grant S D

Possible Contributions of Lid Conditions
During Explosive Cyclogenesis
(1988) / Green C W

Unsteady Flow Field Measurements Using LDV
(Laser Doppler Velocimetry)
(1987) / Hedrick S D

The Application of Brian's Method to the
Solution of Transient Heat Conduction
Problems in Cylindrical Geometries
(1988) / Heinz K R

The Modeling of Viscoelastic Circular Plates
for Use as Waveguide Absorbers
(1988) / Hettema C D

Refractive Turbulence Profiling Via Binary
Source Intensity Scintillation Correlation
(1988) / Holland R R

Passive Vibration Control Using Viscoelastic
and Constrained Layer Beam Waveguide
Absorbers
(1988) / Horne C T

Numerical Field Model Simulation of
Full-Scale Fire Tests in a Closed
Spherical/Cylindrical Vessel with
Internal Ventilation
(1988) / Houck R R

Monte Carlo Calculation of Electron
Multiple Scattering in Thin Foils
(1988) / Jensen D C

Flow Visualization of the Airwake Around a
Model of a DD-963 Class Destroyer in a
Simulated Atmospheric Boundary Layer
(1988) / Johns M K

A Computerized Investigation Using the
Method of Images to Predict the Sound Field
in a Fluid Wedge Overlying a Slow Fluid
Half-Space
(1987) / Kaswandi C

Prototype Particle Size Analyzer
Incorporating Variable Focal Length Optics
(1987) / Kern L A

The Effects of Freestream Turbulence on
Airfoil Boundary Layer Behavior at Low
Reynolds Numbers
(1988) / Kindelspire D W

Natural Convection Liquid Immersion Cooling
of a Column of Discrete Heat Sources in a
Vertical Channel
(1988) / Knight D L

U.S. NAVAL POSTGRADUATE SCHOOL
(continued)

Tracking Procedure for Non-Normally
Distributed Measurement Errors
(1987) / Kukliansky A

Comparison of Model-Based Segmentation
Algorithms for Color Images
(1987) / Kupeli T

Technology Upgrade of a Silicon Compiler
(1987) / Malagon E G

Adaptive Control in Positioning a Rigid
Flexible Robot Arm
(1988) / Mardas C

Effects of Irregular Sea Surface and
evaporation Duct on Radar Detection
Performance
(1988) / Marom M

A Real-Time, Three-Dimensional Moving
Platform Visualization Tool
(1988) / McConkle C M

A Real-Time, Three-Dimensional Moving
Platform Visualization Tool
(1988) / McGhee R B

Thermal Imaging with AGA Thermovision 780
(1987) / McKaig T R

CMOS (Complementary Metal Oxide Silicon)
Cell Library for a Silicon Compiler
(1987) / Mullarky A J

Investigation of Thermoacoustic Heat
Transport Using a Thermoacoustic Couple
(1987) / Muzzerall M L

A Real-Time, Three-Dimensional Moving
Platform Visualization Tool
(1988) / Nelson A H

A Computer Code (USPOTF2) for Unsteady
Incompressible Flow Past Two Airfoils
(1988) / Pang C K

The Interaction of a Fluid Interface with a
Vortex Pair
(1988) / Petersen-Overton M D

Correction of Intertial Navigation System
Drift Errors for an Autonomous Land Vehicle
Using Optical Radar Terrain Data
(1987) / Rickenbach M D

Mechanical-Chemical Energy Transfer
Observations of Vaporific Explosions
(1988) / Rodriguez G

A Real-Time, Three-Dimensional Moving
Platform Visualization Tool
(1988) / Ross R S

Development of a New Method of Measuring the
Characteristic Impedance and Complex Wave
Number of a Porous Acoustic Material
(1987) / Schulz F F

Design Methodology Using the Genesil Silicon
Compiler
(1988) / Settle R H

Data Acquisition, Reduction and Analysis of
Shock and Vibration Testing Using a Super
Microcomputer
(1987) / Shafer R A

Forward-Biased Current Annealing of
Radiation Damaged Gallium Arsenide and
Silicon Solar Cells
(1987) / Staats R L

The Temporal and Spatial Acoustical Response
of a Point-Driven Fluid-Loaded Plate
(1987) / Stevens J G

Transient Three-Dimensional Heat Conduction
Computations Using Brian's Technique
(1988) / Watson J A

A Study of Natural Convection Cooling of
Multiple Discrete Heat Sources in a Vertical
Channel
(1988) / Willson T D

A Real-Time, Three-Dimensional Moving
Platform Visualization Tool
(1988) / Zyda M J

U.S.A.F. INSTITUTE OF TECHNOLOGY

Electrical Properties of Rare-Earth Ions of
Gallium Arsenide and Aluminum Gallium
Arsenide
(1988) / Chan Y K

Electrical Properties of Rare-Earth Ions of
Gallium Arsenide and Aluminum Gallium
Arsenide
(1988) / Hengehold R L

Space Nuclear Power Studies
(1988) / Tuttle R F

VIRGINIA, UNIVERSITY OF

Temperature and Density Measurements in a
Variable Pressure Hydrogen-Aid Flat Flame
Using Laser Raman Spectroscopy
(1988) / Haas F M

An Interrupt Drive Digital Controller for a
Magnetic Suspension System
(1988) / Lawson M A

Calibrated Tuner for Millimeter Wave
Device Evaluation
(1988) / Rothman M A

The Degradation of Electric Cable Insulation
Due to Gamma Radiation
(1988) / Tebay W M

ARIZONA STATE UNIVERSITY

Thin Layer Activation Method for Corrosion
Studies: Calibration Procedure
(1988) / Courtois M F

Effect of Hydrogen on Tensile Properties
of 2090 Aluminum-Lithium Alloy
(1988) / Kim S S

Pseudo Boundary Model and Feature Interface
for Design by Features
(1988) / Liou B B

Human Head Rotations: Biomechanics and
Neurocontrol
(1988) / Radhakrishnan S

BALL STATE UNIVERSITY

Developing Transparency Masters for Product and
Manufacturing System Design to Support Indiana's
Industrial Technology Education Curriculum
(1988) / Apple S D

Further Characterization of the Direct Injection
Nebulizer for Flow Injection Analysis and Liquid
Chromatography with Inductively Coupled Plasma
Spectrometric Detection.
(1988) / Avery T W

Construction of a Hybrid Vector which Allows for
Temperature Regulation of Expression of Cloned
Genes in Cyanobacterium, Synechocystis 6803
(1988) / Bae, I

A study of Deadlocks and Traps in Petri Nets
(1988) / Bagga K S

Cytochemical Localization of Adenosine Triphosphatase
in the Nuclear Envelope of Necturus Maculosus Oocytes
(1988) / Bergdall K M

A Comparative Analysis of the Performance of
Floating Point and Integer Based Line Drawing
Algorithms for Raster Displays
(1988) / Duerksen J L

Conformational Analysis of Phosphine Ligands,
Using Molecular Mechanics and Cone Angle Calculations
(1988) / Durst G L

Developing Transparency Masters for Introduction
to Construction Technology for Indiana's Industrial
Technology Education Curriculum
(1988) / Hobson D A

Creating an Ada Module Description Tool
(1988) / Rice R M

The Designing and Developing of Transparency
Masters for Introduction to Manufacturing
(1988) / Sexton R A

Analytical Model of an N+-P-P+ Concentrator Solar Cell
(1988) / Shaheen M

Using Ada Tasks (Concurrent Processing) to
Simulate a Business System
(1988) / Zahidin A Z

BRIGHAM YOUNG UNIVERSITY

Bias Characteristics of Saturated Second
Order Designs Used in Quality Engineering
Optimization
(1988) / Batchelor C L

Missing Cells in Designed Experiments: A
Statistical Strategy and Expert Assistant
(1988) / Higbee K T

An Automated and Expert Design System Based
on the General Linear Model
(1988) / Johnston R S

Comparison of ML and MPS Estimators for the
Weibull Distribution
(1988) / Lough R L

The AReas of Change Questionnaire: A
Re-Examination of Psychometric and Normative
Qualities
(1988) / Rodabough C B

Outlier-Robust Stratified Random Sample
Estimation
(1988) / Snow J L

CASE WESTERN RESERVE UNIVERSITY

FTIR Characterization of Polyimide on Copper
Substrates
(1988) / Kelley K A

Cure Behavior of the Epxoy Resin/Spiro
Orthocarbonate Slurry and Very Thin Epoxy
Resin Coatings on Steel
(1988) / Nigro J J

Investigation of Ethylene-Vinyl Alcohol
Copolymers for Contact Lenses
(1988) / Ou S H

Epitaxial Crystallization and Polymorphism
of Polyphenylene Sulfide
(1988) / Qian X

Development of the RIM Pultrusion Process
(1988) / Rotter G E

Synthesis and Characterization of Solid
State Polymerizable Alkyldiyne Dimers
(1988) / Walsh S P

Mechanical and Biological Testing of CFRP
Composites
(1988) / Wenz L M

DALHOUSIE UNIVERSITY

Errors-in-Variables Models
(1988) / Lee W N

A Comparison of Ordinal Categorical Data
Analyses of Paediatric Data
(1988) / Raad M A

Solution of Differential Equations by the
Method of Lie Symmetries
(1988) / Rai S K

Comparing the Performance of Semantic-Based
Concurrency Control Algorithms with
Two-Phase Locking
(1988) / Siddique S A

Multivariate Prediction Region in the
Presence of Missing Values
(1988) / Yung W T

FLORIDA, UNIVERSITY OF

Pattern Directed Preliminary Design with
Plates
(1988) / Boardman P

FLORIDA, UNIVERSITY OF
(continued)

Automated Abstract Design for Multiple Goals
(1988) / Bohren J

Managing Interacting Design Subproblems in
Incremental Preliminary Design
(1988) / Brown J P

ILLINOIS INSTITUTE OF TECHNOLOGY

Numerical Investigation of the Finite Length
Taylor Problem
(1988) / Zhao B

LAMAR UNIVERSITY

Selected Topics in Thermodynamics
(1988) / Chiang P P

Ethernet in a Non-Hierarchical Distributed
Control System
(1988) / Dunn W

The Effects of Mechanical Mixing on the
Activated Sludge Process
(1988) / Falgout M D

Detoxification of Ground Water by Stripping
Processes
(1988) / Hsiao K J

Selected Topics in Optimization
(1988) / Lin T H

Development of a Fluidized Bed Computer
Aided Design System
(1988) / Park S C

Multidimensional Digital Filters
(1988) / Tsai C H

MARQUETTE UNIVERSITY

A Proposed Algorithm for the Alignment of
a Strapdown Inertial Navigation System
(1988) / Ast O

High Voltage Measurement Using Surface
Acoustic Waves
(1988) / Bunni N

The Accuracy of Photogrammetric Measurements
Employed in the Analysis of Facial
Morphology
(1988) / Chu M

Investigation of Steady and Unsteady Flow
of an Incompressible Viscous Fluid Between
Two Concentric Rotating Spheres
(1988) / Gulwadi S

Facade Preservation Schemes and Factors
Affecting Cost
(1988) / Kennedy T

Measurement of Gas Flow Using Surface
Acoustic Waves
(1988) / Lee J M

Effects of J-Aggregating Dyes on Ionic
Conductivity of Silver Bromide Films with
Corresponding Optical Absorption
(1988) / Mahesh M

A Velocity Profile Generator
(1988) / Niemic R

Solutions of the Hamilton-Jacobi Equation
and Their Connection to Quantum Mechanics
(1988) / Pareek A

Object Oriented Design Techniques
Complement the Creation of AS4GLS
(1988) / Schmidt J

Burmester Theory and Linkage in Four-Bar
Mechanism Synthesis and Analysis
(1988) / Shon C Y

Nuclear Magnetic Relaxation in a Suspension
of Spheres
(1988) / Teng H

Quantifying Antimycotic Activity of Nystatin
Incorporated in Selected Denture Liners
(1988) / Truhlar M

Study of Interbilayer Forces In
Phospholipid Solvent Systems
(1988) / Varma V

Flow of a Power-Law Fluid in a Rayleigh
Step Bearing
(1988) / Yang D F

Statistical Thermodynamic Analysis of
Group IV Tellurides
(1988) / Yu T C

Spherical Bessel and Neumann Function
(1988) / Yu Y

MASSACHUSETTS INSTITUTE OF TECHNOLOGY

An Ultrasonic Scanning System for Fluid-Film
Detection in Cylindrical Journals
(1988) / Grube K W

MISSISSIPPI, UNIVERSITY OF

Hydraulic Geometry of Channel
Networks: Tests of Scaling Invariance
(1988) / Cadavid E E

Models for pH-Dependent Enzyme Deactivation
(1988) / Chitnis A

Optimization by Statistical Experimental
Methods of the Pultrusion Process for
Graphite-Epoxy Composites
(1988) / Donti R P

Effects of Simulated Dust Accumulation on
Summer Performance of Foil Radiant Barriers
(1988) / Jayanthi R P

Scattering by a Dielectric Cylinder
Partially Embedded in a Ground Plane
(1988) / Kolbehdari M A

Approximations to Creating a Distributed
Minix Operating System
(1988) / Naniwadekar D V

MTEEP, an Intelligent Expert System to
Design Mechanical Transmission Elements
(1988) / Ranjan S

Design and Implementation of a Retrieval
Component for an Intelligent Information
Retrieval System
(1988) / Sarji M A

SOUTHWEST TEXAS STATE UNIVERSITY

Projections in Normed Linear Spaces
(1988) / Barnhill W W

TENNESSEE, UNIVERSITY OF (KNOXVILLE)

An Optical Real-Time, Noncontact Vibration
Measurement Technique
(1987) / Belay T

Application of the Hilbert Transform to
Dopper Radar Signals from a Hypervelocity
Gun
(1987) / Cayse R W

Laminar Natural Convection Heat Transfer
Along a Needle with Thermal Radiation
(1987) / Lee G H

Automated Industrial Site Selection Using
Elevation Models and Digital Image
Processing
(1987) / McKamey M C

Analysis of Adhesion Test Methods and the
Evaluation of Their Use for Ion Beam Mixed
Metal/Ceramic System
(1987) / Pawel J E

Use of Extremely Sensed Data for Analysis of
Hard Rock North Seepage Area, Fabius Mines,
Alabama
(1987) / Polk-Conley A

Altternating Direction Implicit Iteration
for Iteration Matrices with Some Complex
Spectra
(1987) / Saltzman N G

Artin's Reciprocity Law
(1987) / Thakur U S

Development and Application of a Personal
Computer Based Finite Program
(1987) / Treat N L

TENNESSEE, UNIVERSITY OF (TULLAHOMA)

Constants of Motion in Fire Component Models
of Two Dimensional Turbulence
(1988) / Benjnood S M

Microprocessor Based Turbine Engine Image
Controller
(1988) / Date S

Aerodynamic Analysis of a Distortion Screen
wlth a Three-Dimensional Viscous Fluid
Dynamics Code
(1988) / Taylor W

Generation of Negative Entropy Spots in the
Wake of a Cylinder Embedded in a Sheared
Velocity Flow
(1988) / Wohlman R A

TEXAS SOUTHERN UNIVERSITY

Simplified Calculation of Calcium Carbonate
Saturation Index for Geopressured Energy
Wells
(1988) / Gholam V T

TEXAS, UNIVERSITY OF (AUSTIN)

Orientation of Objects in Euclidean Vector
Space Using Non-Euclidean Object Distances
(1988) / Kimbel J C

TOLEDO, UNIVERSITY OF

JCL Techniques to Automate an IBM Operating
System Based Data Center
(1988) / Fisher D

U.S. NAVAL POSTGRADUATE SCHOOL

Factored-Matrix Representation of
Distributed Fast Transforms
(1987) / Bainbridge R L

The Effects of a Pitched Field Orientation
on Hand/Eye Coordination
(1988) / Ballinger C J

Computer-Controlled Measurements of
Nonlinear Standing Waves
(1987) / Basaran H

An Improved User Interface for an
Interactive Graphics Figure Illustrator
(1987) / Beda T J

Computer Simulation Studies of Multiple
Broadband Target Localization Via Frequency
Domain Adaptive Beamforming for Planar
Arrays
(1988) / Behrle C D

Examinating the Reliability of a Hand
Geometry Identify Verification Device for
Use in Access Control
(1987) / Bright D C

Numerical Investigation of Orographically
Enhanced Instability
(1987) / Byrne G T

Estimating Reliability with Discrete Growth
Models
(1988) / Chandler J D

Flow Visualization by Laser Sheet
(1988) / Chlebanowski J S

A Computer Model Investigation of a Quad
Log-Periodic Array
(1988) / Christidis C

A Helicopter Submarine Search Game
(1988) / Chuan E C

Enhancements to the Command and Control
Workstation of the Future
(1988) / Cobb R

A Model for Evaluating the Effectiveness of
a Systematic Search for a Moving Target
(1987) / Colak U

Three-Dimensional Fractal Mountains
(1988) / Collins P J

A Weighted Covariance Approach
(1988) / Coughlin D R

Split-Levinson Approach to Autoregressive
Modeling
(1988) / Dicken W A

Flow Field Measurements Using Hotwire
Anemometry
(1987) / Doremus G J

Discrete Reliability Growth Models Using
Failure Discounting
(1987) / Drake J E

U.S. NAVAL POSTGRADUATE SCHOOL
(continued)

Horizontal Estimation and Information Fusion
in Multitarget and Multisensor Environments
(1987) / Fahmy A E

Test and Evaluation of the Transputer in a
Multi-Transputer System
(1987) / Filho J V

A Bayesian Method to Improve Sampling in
Weapons Testing
(1988) / Floropoulos T C

Design, Implementation, Building and
Evaluation of a Torus Double Transitive
Closure Network of Transputers
(1988) / Frazad Sasa J I

Experimental Studies of Joint Flexibility
for Puma 560 Robot
(1987) / Gonyier D K

Data Structures and Algorithms for
Supporting Glad Interfaces
(1988) / Grenseman P D

Microcomputer Applications with PC LAN
(Local Area Network) in Battleships
(1988) / Gulesen N

Segmentation of Noisy Images Using
Nonstationary Markov Fields
(1987) / Hactpasaoglu K

Single Phase Liquid Immersion Cooling of
Discrete Heat Sources in a Vertical Channel
(1987) / Hazard S J

ADAMEASURE: An Implementation of the
Halstead and Henry Metrics
(1987) / Herzig P M

Flow Visualization on a Small Scale
(1988) / Hixson R L

Infinite Impulse Response Notch Filter
(1988) / Jangsri V

A Data Analysis System for Unsteady
Turbulence Measurements
(1988) / Johnson D K

A Parallel Structure for On-Line
Identification and Adaptive Control
(1987) / Kim Y H

Implementation of Dynamic Control of a
Single-Link Flexible Arm Using a General
Micro-Computer
(1988) / Kirkland M

2-Dimensional Axisymmetric and 3-Dimensional
Finite Element Stress Analysis of the LHA-1
Class Superheater Header
(1988) / Kitchin D R

Methodologies for Resolving Anomalous
Position Information in Torpedo Range
Tracking Using Simulation
(1988) / Kroshl W M

Development of a Model Which Provides a
Total System Approach to Integrating Voice
Recognition and Speech Synthesis Into the
Cockpit of U. S. Navy Aircraft
(1988) / Lee M A

Survey of the Air Force P78-2 (Scatha)
Satellite Plasma Wave Data During Electron
Gun Operations
(1987) / Lowery D R

A Computer Simulation Study of Tripod
Follow-the-Leader Gait Coordination for a
Hexapod Walking Machine
(1987) / Lyman R L

Computer Simulation Model for Studying
Aircraft Take-Off Schedules at a Training
Air Force Base
(1988) / Macropoulos D G

Number of Samples Needed to Obtain Desired
Bayesion Confidence Intervals for a
Proportion
(1988) / Manion R B

Discrete Reliability Growth
(1988) / Markiewicz P A

A Prototype Simulation System for Combat
Vehicle Coordination and Motion
Visualization
(1988) / McConkle C M

Computer Model of an Aviation Gas Turbine
Test Facility
(1988) / Meaker M J

A Transfer Function Approach to Scalar Wave
Propagation in Lossy and Lossless Media
(1987) / Merrill T D

The Curve Fitting of Portable Wear Metal
Analysis
(1987) / Min B H

Unsteady Flow About Cambered Plates
(1987) / Munz P D

Pool Boiling of R-114/Oil Mixtures from
Single Tubes and Tube Bundles
(1987) / Murphy T J

An Analysis of Horizontal Temperature
Gradients and Heat Content in the Mixed
Layer and the Surface Forcing During
Patchex
(1987) / Murray J J

A Prototype Simulation System for Combat
Vehicle Coordination and Motion
Visualization
(1988) / Nelson A H

Structural Characteristics of Dean Votrices
in a Curved Channel
(1987) / Niver R D

An Interrogative Model of Computer-Aided
Adaptive Testing: Some Experimental
Evidence
(1988) / O'Donnell P A

Determination of Tafel Constants in
Nonlinear Polarization Curves
(1987) / O'Loughlin T E

Analysis of the Effect of Faulty Spares on
the Performance of Diagnostic Algorithms in
Reliable Systems
(1987) / Paktuna M

An Adaptive Lattice Algorithm for Spectral
Line Estimation
(1987) / Park I K

U.S. NAVAL POSTGRADUATE SCHOOL
(continued)

Image Texture Generation Using
Autoregressive Integrated Moving Average
(ARIMA) Models
(1987) / Rathmanner S C

Numerical Field Model Simulation of Full
Scale Fire Tests in a Closed Spherical
Cylindrical Vessel
(1987) / Raycraft J K

Cross-Correlation of Uniform Cyclic
Difference Set Sequences
(1987) / Rogers D L

Frequency-Sampling Design of Two-Dimensional
Fir Digital Filters with Nonuniform Samples
(1987) / Rozwod W J

Calibration of a Triple Wire Probe for
Turbulence Measurements
(1987) / Selman G J

A Study of Finite Difference and Finite
Element Vertical Discretization Schemes for
Baroclinic Prediction Equations
(1987) / Shapiro B G

An Investigation of Finite Difference and
Finite Element Vertical Schemes for the
Baroclinic Prediction Equations
(1988) / Sloniker D E

Measured Probability Density Function of a
Phase-Locked Loop (PLL) Output
(1987) / Topcu M

Computer Aided Design for linear Control
State Variable System (SVS)
(1987) / Unlu I

An APL Workspace for Conducting
Nonparametric Statistical Inference
(1987) / Vagts W F

A Three-Dimensional Nonsingular Simulation
of Rigid Manipulators
(1988) / Verbos R M

A Comparative Analysis of a Generalized
Lanchester Equation Model and a Stochastic
Computer Simulation Model
(1987) / West T A

Adaptive Identification by Systolic Arrays
(1987) / Willsi P A

An Improved User Interface for an
Interactive Graphics Figure Illustrator
(1987) / Wu C T

U.S.A.F. INSTITUTE OF TECHNOLOGY

Finite Element Analysis of Shell Structures
with an Eighteen-Mode Three Dimensional
Solid Element Based on a New Mixed
Formulation
(1988) / Ausserer M F

A Decision-Based Methodology for Object
Oriented Design
(1988) / Barnes P D

Prototype design of a Space War Game for the
Air War College
(1988) / Bishop B S

Numerical Simulation of Leading-Edge Vortex
Rollup and Bursting
(1988) / Brandt S A

A Multiple Attribute Decision Analysis of
High-to-Medium-Altitude Air Defense Command
and Control Systems
(1988) / Broadnax G A

Sample Introduction Using the Hildebrand
Grid Nebulizer for Plasma Spectrometry
(1988) / Brotherton T

A Broad-Band Array of Aperture Coupled
Cavity-Backed Slot Elements
(1988) / Burns J M

Use of Commercial Satellite Imagery for
Surveillance of the Canadian North by the
Canadian Armed Forces
(1988) / Chekan R J

Position, Scale and Rotation Optical Pattern
Recognition for Target Extraction and
Identification
(1988) / Childress T G

General Results in Optimal Control of
Discrete-Time Nonlinear Stochastic Systems
(1988) / Ciancetta M S

Dealing with Uncertainty in Chemical Risk
Analysis
(1988) / Clement D S

An Analysis of Modeling Satellite Data in
Air Land Combat Models
(1988) / Cordner T G

Eignevalues and Eigenvectors of Equations of
Motion Involving Fractional Derivatives
(1988) / Devereaux M

A Multiple-Valued Logic System for Circuit
Extraction to VHDL 1076-1987
(1988) / Dukes M A

Design Methods for Optimizing Efficiency of
the Series Resonant Converter Operating
Above Resonance
(1988) / Fahy F A

Coupled Integral Equation Solution for Two
Dimensional Bistatic TE Scatter from a
Conducting Cavity-Backed Infinite Plane
(1988) / Fairbanks R R

A Position, Scale, and Rotation Invariant
Holographic Associative Memory
(1988) / Fielding K H

Design Requirements for a Decision Support
System for the Dynamic Retasking of
Electronic Combat Assets
(1988) / Fletcher C D

Deploy: An Airlift Deployment Scheduler for
Real-Time Crisis Action Planning
(1988) / Foster R M

Measuring Atmospheric Free Radicals Using
Chemical Amplification
(1988) / Ghim B T

Effects of Battle Damage Repair on the
Natural Frequencies and Mode Shapes of
Curved Rectangular Composite Panels
(1988) / Goodwin W P

U.S.A.F. INSTITUTE OF TECHNOLOGY
(continued)

A Comparison of Variable Selection Criteria
for Multiple Linear Regression: A
Simulation Study
(1988) / Hansen R J

Synthesis of Resistive Taper to Control
Scattering Patterns of Strips
(1988) / Haupt R L

A Modification of the Bluemax II Terrain
Following Model to Incorporate Radar
Detection
(1988) / Hearrell J D

Nonlinear Optimization Involving Polynomial
Matrices and Their Generalized Inverses
(1988) / Hill R R

Adaptive Design of the Non-Combatant
Evacuation Operation Decision Support System
for U. S. Central Command
(1988) / Hippenmeyer W F

The Aspect Ratio Dependence of the Attractor
Dimension in Taylor-Couette Flow
(1988) / Hirst D A

A Decision Analytic and Expert System
Approach for Implementing Rules of
Engagement of a Strategic Defense System
(1988) / Hollenga D

Analysis of C-17 Avionics Automatic Test
Equipment Basing Alternatives
(1988) / Humlie M S

Integrated Tactical Warning and Attack
Assessment Communications Media
Survivability Definition Study and Decision
Support Model
(1988) / Jacobs M G

Modeling the Rated Force Using Network Flow
and Goal Programming Techniques
(1988) / Jameson R S

A UTD Scattering Analysis of Pyramidal
Absorber for Design of Compact Range
Chambers
(1988) / Joseph P J

A Diagnostic System Using Boolean Reasoning
(1988) / Kainec J J

Redesign and Rehost of the Big Stick Nuclear
Wargame Simulation
(1988) / Kalili R M

Graphical Man-Machine Interface for an
Integrated Evaluation Environment
(1988) / Kanski J J

Super-Resolved Imaging Using the Hartley
Transform and a Hemispherical Object Space
(1988) / Kelly S L

Temporal Clustering in the Multi-Target
Tracking Environment
(1988) / Kelso T S

Hardware Implementation of a BCH
Encoder/Decoder and Interface
(1988) / Leclair N R

Real-Time Display of Time Dependent Data
Using a Head-Mounted Display
(1988) / Lorimor G K

Problem Specific Applications for Neural
Networks
(1988) / Lutey M K

Simulating Rule-Based Systems
(1988) / Mahaba N M

Distributed Discrete-Event Simulation Using
Variants of the Chandy-Misra Algorithm on
the Intel Hypercube
(1988) / Mannix D L

Discrete Event Simulation Model
Decomposition
(1988) / Matthes S R

Sensor and Actuator Selection for Fault
Tolerant Control of Flexible Structures
(1988) / McCasland W N

Bistatic Resonant Scattering Measurements
of Special Shapes
(1988) / McCool S W

Modeling the Pave Low Helicopter Pilot
Career Field
(1988) / McKoy J C

The Characterization of Gallium
Arsenide/Aluminum Gallium Arsenide
Heterostructures by Variable Angle
Spectroscopic Ellipsometry
(1988) / Merkel K G

A Fault Tolerant Self-Routing Computer
Network Topology
(1988) / Mitchell T L

A Vortex Panel Method for Potential Flows
with Applications to Dynamics and Controls
(1988) / Mracek C P

Estimating Regional Fluxes of
Evapotranspiration and Sensible Heat from
Measurements of the Planetary Boundary Layer
(1988) / Munley W G

Method for Obtaining Empirical Correlations
for Predicting Crack Propagation in a
Burning Solid Propellant Grain
(1988) / Nimis J A

The Radar Cross Section of a Two Dimensional
Dielectric Coated Cylinder Using On-Surface
Radiation Conditions
(1988) / Oetting W D

Design of an Officer Assignment and Career
Planning Decision Aid for Electronic
Security Command
(1988) / Paulsen R A

A Computational Study of Coherent Structures
in the Wakes of Two-Dimensional Bluff Bodies
(1988) / Pearce J A

Search - An Interactive Computer Program for
Optimizing Two-Variable Unconstrained
Experiments or Simulation
(1988) / Ploetner B G

Adding Map-Based Graphics to the Theater
War Exercise
(1988) / Quick D A

BALL STATE UNIVERSITY

Determination of Household Radon Air and Water
Concentrations for Selected Homes in East-Central
Indiana, Utilizing Activated Charcoal Canister
and Liquid Scintillation Techniques
(1988) / Dewus M A

The Impact of Residential Wood Combustion on Indoor
Particulate Matter Levels by Randall P. Kirk
(1988) / Kirk R P

COLORADO STATE UNIVERSITY

Comparison of Downward and Horizontal Facing
Filter Cassettes for Aerosol Sampling Bias
(1988) / Nickelson D

Controlling Airborne Lead in Indoor Firing Ranges
(1988) / Robbins S K

IDAHO, UNIVERSITY OF

Local Air Quality Impacts of the University
of Idaho's Wood Fired Boiler
(1988) / Edward S G

INDIANA UNIVERSITY

Indoor Air Pollution: Volatile and
Semi-Volatile Organic Compounds in
Residences
(1988) / Anderson D J

KENTUCKY, UNIVERSITY OF

Adjustment of Wind Fields for Application
in Air Pollution Modeling
(1988) / Mathur R

LOUISIANA STATE UNIVERSITY

Modeling Short Range Dispersion from Area
Sources of Air Pollution
(1988) / Chitgopekar N P

MASSACHUSETTS, UNIVERSITY OF

Use of Controlled Low Velocity Air Patterns
to Improve Operator Environment at
Industrial Work Stations
(1988) / Chamberlin L A

A Comparison of Three Methods for Measuring
Respirable Suspended Particles in the
Indoor Environment
(1988) / McDonald S M

Development of a Methodology Using Several
Computer Geocoded Databases in the
Calculation of a Measure of Potential
Exposure
(1988) / Saboda K

Local Exhaust Ventilation for Atomic
Absorption Spectrophotometers
(1988) / Tiernan T A

NEW JERSEY INSTITUTE OF TECHNOLOGY

Determination of Polycyclic Aromatic
Hydrocarbons from Incineration Emission
Samples
(1988) / Cebula R
(1987) / Zheng X

Analysis of C2 and C4 Hydrocarbons in
Ambient Air. Section I
Quantification of Volatile Organics Emitted
from a Landfill. Section II.
(1988) / Chang T S

The Quantitative Determination of
Formaldehyde in Ambient Air (Analytical
Method)
(1988) / Lin C

An Improved Method of Gas Chromatographic
Determination of Trace Organic Vapors in
Ambient Air Collected in Canisters
(1988) / Shen Y

Separation and Characterization of
Biologically Significant Chemical Classes
in Airborne Particulate Organic Matter:
Measurement of Selected Chemical Classes
in Freshly Collected Air Samples
(1988) / Sun X

Formation of Light Hydrocarbons and Soot in
the Pyrolysis Reactions of Chloro,
Dichlorobenzene, Acetylene and Benzene in
an Atmosphere of Hydrogen
(1988) / Zhu L J

PENNSYLVANIA, UNIVERSITY OF

A Continuous Flow Dissolved Oxygen Packed
Bed Microbial Toxicity Monitoring System
(1988) / Yee C J

TEXAS SOUTHERN UNIVERSITY

A Study of Radionuclides as Found in the
Aqueous Environment Near a Coal Fired
Electric Power Plant
(1988) / Wilkerson D F

TEXAS, UNIVERSITY OF (AUSTIN)

Precipitation Scavenging of Sulfates
(1988) / Lawlis R S

U.S. NAVAL POSTGRADUATE SCHOOL

Study of Electrostatic Modulation of Fuel
Sprays to Enhance Combustion Performance in
an Aviation Gas Turbine
(1987) / Manning W W

An Experimental Investigation of the
Ignition and Flammability Limits of Various
Hydrocarbon Fuels in a Two-Dimensional Solid
State Fuel Ramjet
(1987) / Wooldridge R C

Optical Sizing of Soot in Gas Turbine
Combustors and Exhaust Augmentor Tubes
(1987) / Young M F

U.S.A.F. INSTITUTE OF TECHNOLOGY

Impact of an Increase in Atmospheric Methane
on the Temperature Field and Subsequent
Climatic Effects
(1988) / Morrison J S

WASHINGTON STATE UNIVERSITY

Developing and Testing of Passive Ozone
Monitor
(1988) / Huang Y

WASHINGTON, UNIVERSITY OF

Chlorinated Emissions in Smoke from Buring
Forest Fuels
(1987) / Reinhardt T E

WESTERN ONTARIO, UNIVERSITY OF

Air Infiltration in High-Rise Buildings
(1988) / Grabau P

BROWN UNIVERSITY

The Effect of Induced Piezoelectric Charge
on the Subthreshold Characteristics of
Gallium Arsenide Mesfets
(1988) / Cooper L S

Design and Implementation of the LEMS
Microphone Array for the Aquisition of
High Quality Speech Signals
(1988) / Jones T L

Void Growth and Coalescence in Porous
Plastic Solids
(1988) / Koplik J

Oxidation Behavior of Silicon Carbide
Whisker-Reinforced Alumina
(1988) / Lin F

Investigation of Stress Effects on the DC
Characteristics of Gallium Arsenide Mesfets
Through the Use of Externally Applied Loads
(1988) / McNally P J

Mechanics And Fracture Mechanisms of Large
Aspect Ratio Solder Joints Subject to
Plastic Constraint
(1988) / Ranieri J P

Crack Growth in Transformation Toughened
Ceramics: Compression-Compression and
Tension-Tension Cyclic Loads
(1988) / Sylva L A

A Pressure-Shear Plate Impact Experiment for
Studying the Effect of Changing Strain Rate
on the Plastic Response of Metals
(1988) / Tong W

Order Determination in Linear Differential
Systems Identification
(1988) / Yoo D H

CALIFORNIA STATE UNIV. (LONG BEACH)

Estimation of the Cohesion Parameters of
Parylene C by Swell Technique
(1988) / Leighton E M

Dynamic Analysis of Delamination Growth by
Finite Element Method
(1988) / Ngo H C

CALIFORNIA, UNIVERSITY OF (DAVIS)

The Prediction of the Centerline Density of
an Atomic Beam of Mercury Issuing from a
Finite-Slit-Shaped Converging Nozzle
(1988) / Bahowick S M

The Removal of Suspended Solids Using
Overland Flow Systems
(1988) / Bauser A M

Instrumentation and Methodology for Analysis
of Mechanical Energy in Cycling
(1988) / Beard A M

A New Algorithm for System Identification
from Frequency Response Information
(1988) / Berchin G J

High Directivity Coupled-Line Directional
Couplers
(1988) / Bickford J D

Determination of the Mechanical Properties
of a Column of Polymeric Foam
(1988) / Biesiada T A

A 5 x 5 Tactile Sensor Array and Analog
Interface Electronics
(1988) / Blumenthal J M

Investigation of Chain Dynamics with
Application to Noise
(1988) / Bothwell S L

Performance Characteristics of a High
Resolution Liquid Crystal Lens
(1988) / Brinkley P F

Quadratic Incentive Coordination and
Parallel Decomposition for Non-Convex
Optimal Control Problems
(1988) / Bromberg M C

A 16-Bit D/A Converter Based on Delta
Sigma Modulation
(1988) / Brown J E

An Experimental Investigation of the Droplet
Combustion of HAN-Based Propellants
(1988) / Call C J

The Effect of Compression Ratio on
Performance of a Single Cylinder Spark
Ignition Engine Fueled with Producer Gas
Generated from Rice Hulls
(1988) / Camacho I R

Iron and Manganese in Stratified Reservoirs
(1988) / Cartwright E B

A Stochastic Model for Rainfield Arrivals as
Observed on a Radar Scope
(1988) / Chen Z

Finite Element Analysis and Evaluation of a
Three-Dimensional Plate Theory
(1988) / Cho H

An Expert System for 31-Phosphorus Nuclear
Magnetic Resonance Spectral Analysis and
Interpretation
(1988) / Chow J L

An Analytical and Experimental Comparison
Between Eccentrically Braced Frames and
Disposable Knee Bracing Frames
(1988) / Corte A

A Model of Eddy Conductivity for
Non-Equilibrium Thermal Boundary Layers
(1988) / Cotter C M

Coupling of the Finite and Boundary Element
Methods in Elastostatics
(1988) / Cox J V

Design Automation of Easily Testable PLA
(1988) / Daud N A

Modulating Stiffness for Suspension Control
in Ground Vehicles
(1988) / Dehpanah K

The Computer Emulation of an Engineer's
Calculation and Sketch Sheet
(1988) / Drisko K

Applying the Quantitative Feedback Theory
to the Design of Digital Controllers for
Manipulators
(1988) / Duncan G A

CALIFORNIA, UNIVERSITY OF (DAVIS)
(continued)

Microbial Transformations of Selenium in
the Presence of Nitrate
(1988) / Elkholy N M

Development of a Lighting System for Human
Powered Vehicles
(1988) / Empey D M

Flow Visualization of Superposed
Flows: Study of Interfacial Shape and
Stability
(1988) / Falsafi S

The Incentive Coordination Method for a
Class of Constrained Long Horizon Optimal
Control Problems
(1988) / Fu D

Control and Regulation of Open Channel Flow
(1988) / Garcia A

Diffusion, Reaction and Adsorption on
Heterogeneous Catalytic Surfaces
(1988) / Garza V J

Implementation of an Active-R O-Pamp Circuit
Using Semi-Custom Technology
(1988) / Gee K B

A New Technique to Investigate
Wear-Corrosion Processes in Dental Material
(1988) / Gibbs C A

Prefractal Properties of Vortex Filaments
(1988) / Gleckler P J

Multivariable Optimization of Bicycling
Biomechanics
(1988) / Gonzalez H K

An Optimal Receiver Using a Time-Dependent
Adaptive Filter
(1988) / Greene C D

An Evaluation of Mechanical Expansion
Anchors
(1988) / Harrington C N

An Experimental Study of Entrainment Effects
of Heat Transfer from a Surface with a Fully
Developed Impinging Jet
(1988) / Hechanova T E

Emissions of Volatile Organic Compounds
from Municipal Solid Waste Landfills
(1988) / Herrera T A

A Multi-Link Model of a Ski Jumper
(1988) / Hibbard R L

Criteria for Better Interpolation of Cross
Sections for Improved Accuracy of Water
Surface Profile Computations
(1988) / Hilts D

An Expert System for Composite Material
Based Pressure Vessel Design
(1988) / Huang D D

Droplet Combustion of Mixtures of
Chlorinated Hydrocarbons
(1988) / Hwang R J

Coprocessor Implementation of a Digital
Spectrum Analyzer
(1988) / Japait S

The Behavior of Silt Under Triaxial Loading
(1988) / Jeong S

The Sequestration and Biodegradation of
Toluene by Activated Sludge in a Sequencing
Batch Reactor
(1988) / Jepsen D T

Visual Display of a Robotic Arm Using the
Hewlett-Packard Advanced Graphics System
(1988) / Kemp S G

Optimization of Fixed and Fluidized Bed
Freezing Process
(1988) / Khairullah A M

Electrical Dispersion in Relation to Double
Porosity, Consolidation, Compression Index,
Swell Index, and In-Situ Permeability of
Soils
(1988) / Knapp A Q

Evaluation of a Traveling Bridge Granular
Filter for Wastewater Reclamation
(1988) / Kolivodiakos N

The Rise and Fall of Diesel Cars: A
Consumer Choice Analysis
(1988) / Kurani K S

A Study of Two Efficient Access Protocols
for Unidirectional Broadcast Bus Networks
(1988) / Lantz A C

Robust Independent Robot Joint Control:
Design and Experimentation
(1988) / Lasky T A

Identification of Analog Systems from
Discrete Data
(1988) / Lawrendra J

Analog Delta Sigma Modulators with Reduced
Sensitivity to Circuit Non-Idealities
(1988) / Levinson R A

A Modified Horizontal Bridgman Furnace for
Reducing Dislocations During Single Crystal
Growth
(1988) / Lucas M L

Interactive Remote Applications in an Image
Processing Environment
(1988) / Maeder H P

Role of Sand Filters in Achieving Adequate
Disinfection of Secondary Effluent by Using
Chlorine
(1988) / Manglik P K

The Low Cycle Fatigue Response of
Mechanically Alloyed Aluminum
(1988) / Mercer M E

Determination of Vertical Temperature
Profiles at Point Loma San Diego
(1988) / Meyer G K

Flexibility of the Human Knee Due to Axial
and Vargas/Valgus Moments in Vivo
(1988) / Mills O S

Intake-Manifold Dynamics for Automobile
Internal Combustion Engine
(1988) / Miyano T

A Study of Fresh Produce Spatial
Distribution Patterns in the United States
(1988) / Monzon J S

CALIFORNIA, UNIVERSITY OF (DAVIS)
(continued)

Quantitative Physiology of Hybridoma Cell
Culture for Monoclonal Antibody Production
(1988) / Mosher C S

Modelling of Headloss at River Intersections
(1988) / Nam I G

Measuring Adsorption Equilibrium Constants
(K) and Intraparticle Diffusion Coefficients
(D[i]) for Adsorbing Species and Internal
Tortuosity Factors in Small-Pore Particles
in a Packed-Bed Using Arrested Flow
Chromatographic Method with Adsorbing Gases
(1988) / Oh M

Vehicle Directional Response and Control
Enhancement Through Coordinated Steering and
Braking
(1988) / Oliver D B

Force Calculations for a Continuum Model of
Lattice Systems
(1988) / Osterkamp T A

Residual Stress Measurements of a
Yttrium-Barium-Copper-Oxide Ceramic
Superconductor and a Basalt Glass-Ceramic
by X-Ray Diffraction
(1988) / Park J S

A DCT/IDCT Unnified Architecture Chip
(1988) / Parkhurst J R

The Design of a Practical Control System
for a Double Disc Grinder
(1988) / Petersen F C

The Effects of Arching and Consolidation
Around Transducers on Pore Pressure
Measurements in Clay
(1988) / Sathialingam N

A Study on Lattice Adaptive Filters
(1988) / Savoye P

Transportation/Land Use Policy Simulation
(1988) / Setiawan W

Performance Evaluation of Detectors for
Cyclostationary Signals
(1988) / Spooner C M

A Study of the Sorption of Selected Trace
Organic Compounds Found in Landfill Gas
(1988) / Stiegler L C

A New Strategy to Approach the Nyquist Limit
for Data Transmission Using Practically
Realizable Filters
(1988) / Subasinghe D

An Experimental Study of Ultrasonic Ranging
Elements for Factory Automation and Mobile
Robot Applications
(1988) / Tsui H R

The Effects of Finite Sampling and Additive
Noise on Image Reconstruction from the
Radon Transform
(1988) / Tsujimoto E M

An Adaptive, Nonlinear, Time Dependent,
Interference Rejection Technique
(1988) / Vetter F J

Investigation of the Saturation
Magnetization of Ferromagnetic Thin Films
(1988) / Wallace D J

Design of a Simple Digital Video Mixer
(1988) / Walsh J W

Nonlinear Computer Model of an Active Air
Suspension
(1988) / Wanner M C

Modeling and Simulation of a General Wankel
Compressor
(1988) / Welch M S

A Computer Code for Calculating Subcooled
Boiling Pressure Drop in Forced Convective
Tube Flows
(1988) / Wong C F

Prediction of Bruising on the Golden
Delicious Apple Due to Impact
(1988) / Yazdani R

Dielectric Method for Laboratory and
In-Situ Soil Characterization
(1988) / Yogachyandran C

CARNEGIE-MELLON UNIVERSITY

Secondary Sources of Titanium: A Materials
System Approach
(1988) / Eveille Raj G

Fiber Optic Residential Subscriber
Network - Policy Issues Analysis
(1988) / Ferrante F E

Identification of the Direct Distribution
Model from Regionalized Mechanistic Models
of Aquatic Acidification
(1988) / Labieniec P A

Explanation and Understanding of
Multi-Attribute Decision Problems
(1988) / Simons M

The Role of Applied Knowledge Industrial
Competitiveness: A Comparison of Engineers
in the U.S. and Japan
(1988) / Zahray W P

CASE WESTERN RESERVE UNIVERSITY

Transfer Function Estimation from Sampled
Data
(1988) / Boenawan S

A Successive Linear Programming Approach to
Solve the Correction Problem of Groundwater
Contamination
(1988) / Luo Y

CENTRAL FLORIDA, UNIVERSITY OF

Heavy Metal Removal by Sedimentation of
Street Sweepings in Stormwater Runoff
(1988) / Brabham M E

A Study of the Reaction Kinetics for the
Thermal Oxidation of Dilute n-Heptane in
a Plug Flow Reactor
(1988) / Ramsey S H

CLEMSON UNIVERSITY

Managing Soil Loss at a Major Construction
Site: Alternatives to the Conventional
Approach
(1988) / Braswell S G

Modeling Crop Uptake of Sludge-Borne Cadmium
(1988) / Brown S M

Determination of Biodegradation Kinetic
Parameters for 4-Nitrophenol and
2-Chlorophenol Using Electrolytic
Respirometry
(1988) / Dang J S

Dynamics of Lysine and 2-Chlorophenol
Removal by Two-Membered Continuous Cultures
of Bacteria
(1988) / Davis K L

The Distribution and Mass Loading of
Polychlorinated Biphenyls in Lake Hartwell
Sediments
(1988) / Germann G G

Microbially Mediated Reductive
Dehalogenation of Small Chain Chlorinated
Organics
(1988) / Johnson B R

A Survey of Indoor Radon in South Carolina
by the Passive Charcoal Adsorption Technique
(1988) / Jones M D

Using HEC-6 to Model Sediment Transport in
an Eastern Mountain Stream
(1988) / Kosik D R

Evaluation of the Error Propagation
Technique for Uncertainty Analysis of Select
Solutions of the Advective-Dispersive
Transport Equation
(1988) / Pollog T E

Simulation Studies of the Transient Response
of Activated Sludge Systems to Biodegradable
Toxic Shock Loads
(1988) / Santiago I

An Economic Evaluation of Sewage Sludge
Drying and Incineration Process
(1988) / Schwarz S

The Development of a Nonlinear Parameter
Estimation Technique for Use with
Biodegradation Data
(1988) / Wang X

An Examination of Institutional Impediments
to Rational Nonpoint Source Pollution
Control
(1988) / Warren C B

Design of Single-Sludge Systems: The
Modified Ludzack-Ettinger and Bardenpho
Processes
(1988) / Youker B A

COLORADO SCHOOL OF MINES

Soil and Sediment Chemistries in Relation
to Water Quality of Colorado Mountain Lakes
(1988) / Margulies T D

COLORADO STATE UNIVERSITY

Control of Pneumatic Noise Using
Ducting and Sound Absorbing Enclosures
(1988) / Casciano J R

Characterization of Lead Fume Exposure During Gas
Metal Arc Welding on Carbon Steel
(1988) / Larson J K

FLORIDA, UNIVERSITY OF

The Relative Influence of Current Reduction
by Seagrasses on Sediment Nutrients and
Seagrass Growth in High and Low Nutrient
Waters: A Simulation Model and Field
Observation
(1988) / Bartleson R

Measurement of Vertical Ozone Concentration
Profile in a Slash Pine Forest
(1988) / Cavender K

Effects of Waste Phosphatic Clay and
Organic Amendments on Plant Uptake and
Extractability of Selected Heavy Metals
(1988) / Gonzalez R

Microcomputer-Based Spreadsheet and CAD
System for Inventory and Analysis of Small
Quantity Generators of Hazardous Wastes
(1988) / Knowles L

Groundwater Model Calibration for a
Hydrocarbon Plume in a Sandy, Surficial
Aquifer
(1988) / Scanlan S

Enumeration and Control of Nitrifying
Bacteria in Automatic Sampling Equipment
(1988) / Stevens C

Network Energy Expenditures for Subsystem
Production
(1988) / Tennenbaum S

The Role of Mycorrhizae in Reclamation of
Phosphate Mined Lands by Ecological
Successional Processes
(1988) / Wallace P

GUELPH, UNIVERSITY OF

The Enhanced Production of Volatile Fatty
Acids from Cellulose
(1988) / Eszes V R

A Mathematical and Experimental Approach to
Modelling Pollutant Movement Through
Snowpacks
(1988) / Forward R J

The Effect of Freezing Parameters on Ground
Beef Patty Quality
(1988) / Hanenian R A

Aerodynamic Separation of Alfalfa Haylage
Into High and Low Protein Fractions
(1988) / Mkomwa S S

A Finite Element Model for Drainage in a
Brookston Clay
(1988) / Peters E J

HOUSTON, UNIVERSITY OF

Radium Removal from Water by Manganese
Dioxide Adsorption and Diatomaceous Earth
Filtration
(1988) / Patel R

Biodegradation of Chloroform and
Trichloroethylene by Methanotrophic Bacteria
(1988) / Patterson A

ILLINOIS INSTITUTE OF TECHNOLOGY

Solution and Precipitate Characteristics for
Hydrolysis - Precipitation of Copper(II) in
the Presence of Nitrate
(1988) / Boice R E

Characterization of Respirable Particulate
Matter in Athens Greece
(1988) / Constantine V

INDIANA STATE UNIVERSITY

Postfire Vegetation Change Detection Using
Landsat MSS and TM Data
(1988) / Jakubauskas M

JOHNS HOPKINS UNIVERSITY

Reductive Dissolution of CoOOH: Microbial
Exudates and Natural Organic Matter
(1989) / Boswell K L

Reductive Dissolution of Manganese Oxide
Particles by Dimethylphenols
(1989) / Kolodny N

Hydrolysis of Carbaryl: The Effect of
Mineral Surfaces
(1989) / Torrents A

LEHIGH UNIVERSITY

A Tool to Study Dynamic Phenomenon of a Four
Bar Linkage and Its Application in Balance
Program Verification
(1987) / Brooks D R

Automatic Grip Selection
(1987) / Connolly M L

Etching Radiation Treated Polyimide Films
in Oxygen/Nitrogen Fluoride and
Oxygen/Carbon Tetrafluoride Plasmas
(1987) / Dibble E

Improving Productivity Via a Real-Time
Database Management System and Local
Area Network
(1987) / Donofrio K J

A Feasibility Study of the Finite Element
Analysis Method Applied to the Redesign for
Manufacturability of Mechanical Products
(1987) / Ginzburg M

A Kinetic and Mass Transfer Model to
Describe Nonuniformities in a Radial Flow
Plasma Reactor
(1987) / Russo W M

Product Design Principles that Support
Manufacturing with Expert System
Considerations
(1987) / Sears K H

LOUISIANA STATE UNIVERSITY

The Relationship of High Ozone Levels from
Anthropogenic Sources to Synoptic Weather
Types in Baton Rouge, Louisiana: 1981-1985
(1988) / Goldberg A S

Chronic Ethyanol Ingestion: Its Role in the
Mutagenicity of Arylamines and Aromatic
Nitro Compounds
(1988) / Traynor C A

LOWELL, UNIVERSITY OF

Review of PCB Destruction Methods
(1988) / Freedman S B

Effects of Ozone on Total Coliform as a
Function of pH and Temperature
(1988) / Karotseris D

Optical Analysis of Secondary Solar
Concentrators
(1988) / Muriel M B

Conversion from High Enriched Uranium Fuel
to Low Enriched Uranium Fuel for the
University of Lowell Research Reactor
(1989) / Stoddard J E

Development, Testing, and Implementation of
a Secondary Order Correction to Generalized
Perturbation Theory
(1989) / Swanbon G A

Cost Benefit Analysis on None New England
Passive Solar Homes
(1988) / Walsh E J

Effects of Sulfur Dioxide and Radiation on
Plants
(1988) / Yemoh-ndi J

MARQUETTE UNIVERSITY

An Electrochemical Evaluation of
Microleakage of Retrofilling Materials
(1988) / Buratti M

The Electrical Conductivity and Transference
Number of Yttria Stabilized Zirconia
(1988) / Yin M Y

MASSACHUSETTS INSTITUTE OF TECHNOLOGY

Experimental Investigation of Nitrogen-
Containing Compounds in Turbulent Diffusion
Flames
(1988) / Bahadori T

Promoting the Development of Alternative
Waste Disposal Technologies Under the
Massachusetts Superfund Law
(1988) / Baskir J N

Logic Based Approach to Factory Design
(1988) / Brereton M F

Evaluation of a House Simulator Model as a
Tool for the Study of Customer Response to
Demand Side Management
(1988) / Byrd J G

Integrated Energy Systems: An Analysis of
Natural Gas Consumption and Electricity
Generation Interactions in Oklahoma
(1988) / Dembinski M M

MASSACHUSETTS INSTITUTE OF TECHNOLOGY
(continued)

The Principles of Design Applied to
Engineering and Policy
(1988) / Filippone S F

The Engineering Research Center: An
Experiment in R & D Policy
(1988) / Johnston J E

Modelling Complex Coordination Processes: A
Case Study of Engineering Change Management
(1988) / Lin F

Laboratory Hazardous Waste Management: A
Case Study at the Massachusetts Institute
of Technology
(1988) / Skumanich M

The Design and Fabrication of an Automated
Shear Stud Welding System
(1988) / Zieglerr A G

Is There an Optimal Power Purchase for
Electric Utility Companies?
(1988) / de Rycke T

MICHIGAN TECHNOLOGICAL UNIVERSITY

Probablistic and Deterministic Keyblock
Analyses for Excavation Design
(1988) / Hoerger S F

Contaminant Transport Models in Layered
Porous Media: Application of the Orthogonal
Collocation Method
(1988) / Im J S

Optimal Control on Underground Mine Fire
(1988) / Zhou X

MONTANA COLLEGE OF MINERAL SCI. & TECH.

Revegetation of Silver Bow Creek Tailings:
A Greenhouse Study
(1988) / Nemgar T

MONTANA STATE UNIVERSITY

Filter Ripening Sequence Reduction by
Physical and Chemical Variations of
Backwashing
(1987) / Cranston K O

Methods to Improve Environmental Monitoring
Network Design
(1987) / Cunnane M

Improving the Removal of Phenols in the
Biological Portion of the Conoco Oil
Refinery Wastewater Treatment Plant,
Billings, Montana
(1987) / Gilman G

MONTANA, UNIVERSITY OF

Information Resource Requirements for
Facility Managers: A Case Study of Montana
Deaconess Medical Center Engineering and
Maintenance Department
(1988) / Jacobson N G

MONTREAL, UNIVERSITY OF

Utilisation de la Spectroscopie
Photoacoustique Pour la Detection d'Eau
Absorbee Dans les Polymeres
(1988) / Bordeleau A

Application des Fibres Optiques Unimodales
Effilees
(1988) / Gonthier F

Determination et Quantification de Liens
Chimiques Dans le Alpha-Silicon/Hydrogen
Par XPS
(1988) / Izquierdo R

Simulation Numerique du Transistor a
Induction Statique a Base de Silicium
Amorphe
(1988) / Kemp M

Fabrication, Caracterisation et Optimisation
de Transistors MESFET sur l'Arseniure de
Gallium
(1988) / Lagarde C

Application de l'Optique Integree sur Verre
a le Detection des Contraintes
(1988) / Sequin F

Caracterisation d'Echantillons de Silicium
Amorphe Hydrogene par Mesures Seebeck
(Puissance Thermoelectrique)
(1988) / Tremblay M

NEW JERSEY INSTITUTE OF TECHNOLOGY

Degradation of 1, 2-Dichlorobenzene and
1, 2, 4-Trichlorobenzene in Water Using
Ozone, Hydrogen Peroxide and Ultraviolet
Radiation
(1988) / Chen W J

Estimation of Modified Proctor Densities
Based on Visual Identiification
(1988) / Damodaram R

Expert Systems Approach for Means of
Egress Analysis
(1988) / Dontas D N

Adsorption and Kinetics of Adsorption by
Flyash in Solution
(1988) / Jogimahanti M K

Analysis of Selected Pollutants in Complex
Media
(1988) / Luo S

Optimal Design Aspects of Unit Treating
Hazardous Wastes Via Biodegradation: A
Theoretical Approach
(1988) / Saghafi F

Use of Recirculation Reactor to Study
Biodegradation of Aroclor 1242, a
Polychlorinated Biphenyl
(1988) / Sanji B S

Column Base Plate Under Eccentric Loading
(1988) / Sheu J P

Pentachlorophenol and Oxygen Consumption of
Immobilized Mixed Bacterial Culture
(1988) / Tsadwa T S

Friedman's Model of Calculating the Optimum
Bid Price Using dBASE III Plus
(1988) / Uzel S

Decontamination of Contaminated Clay Soils
by Thermal Treatment
(1988) / Weerasinghe C S

NEW JERSEY INSTITUTE OF TECHNOLOGY
(continued)

A New Hazardous Waste Treatment Technology
Utilizing Low Power Density Microwave Energy
(1988) / Windgasse G E

Preparation and Formulation of Novel Dental
Adhesives Based on Polyphosphonate and
Silane Compound
(1988) / Wu Y

NEW MEXICO INSTITUTE OF MINING & TECH.

Shock Consolidation of Cubic Boron Nitride
(1988) / Hao J

OKLAHOMA, UNIVERSITY OF

Interactions and Effects of Oxygen, pH, and
Mercury Concentration on Partitioning of
Mercury Into Solution and Lahontan Reservoir
Sediments
(1988) / Trent M A

PACIFIC, UNIVERSITY OF THE

Evaluation of FFT and LPC Methods in the
Analysis of Time Series: A Comparative
Study
(1988) / Fakhral-Deen A A

A 68000-Based Produce Sorting Microcomputer
(1988) / Haidamus R A

PENNSYLVANIA STATE UNIVERSITY

Estimating Auqifer Properties and Original
Oil in Place in Water-Drive Reservoirs: A
Comprehensive Study
(1987) / Bouchard A J

Experimental Investigation of Crossflow
Effect on Oil Recovery in Waterflooding
(1987) / Boukadi F

Foam Generation and Propagation in Porous
Media
(1987) / Mfonfu G B

The Effect of Electric Field on Mechanical
Fracture in Electroceramic Devices
(1988) / Taylor D J

Use of the Proximity Effect in Hearing Aid
Microphones to Increase Telephone
Intelligibility in Noise
(1987) / Wilson K R

PRINCETON UNIVERSITY

An Investigation of the Effects of a
Longitudinal Heating Element on a Laminar
Boundary Layer Over a Flat Plate
(1988) / Moore C D

A Study of the Spatial Variability of
Microscopic Solute Transport in Dispersive
Flows Using Fiber-Optic Sensors
(1988) / Nielsen J M

Unsteadiness of Shock Wave/Turbulent
Boundary Layer Interactions with Dynamic
Control
(1988) / Selig M S

RHODE ISLAND, UNIVERSITY OF

The Development of a Programmable Assembly
System
(1987) / Donovan C

Versatile, Three-Fingered Grippers for Assembly
and Insertion Processes
(1987) / Field G L

SHERBROOKE, UNIVERSITY OF

Simulation de la Charge Hydraulique a
Ville Mercier
(1988) / Bachand G

Etude sur la Valorisation du Biogaz
d'Enfouissement Sanitaire
(1988) / Bajeat P

Syntheses des Impacts Relatifs aux Projets
de Dragage et Utilisation d'Ecrans Flottants
Comme Mesure de Protection
(1988) / Baril J

L'Environnement et la Sante sur le
Territoire du Departement de Sante
Communautaire de l'Hopital de l'Enfant-Jesus
(1988) / Bolduc D

Etude sur le Compostage des Boues de
Stations d'Epuration Municipales
(1988) / Couture F

Guide de l'Intervenant Municipal ou
Industriel en Matiere de Protection des
Eaxu Souterraines
(1988) / Deschamps M

La Surveillance Appliquee au Processus
Federal d'Evaluation et d'Examen en
Matiere d'Environnement: Objectifs,
Definitions et Procedures
(1988) / Dudley L

Traitement des Eaux de Lixiviation de Site
d'Enfouissement Sanitaire par la Tourbe de
Saphigne
(1988) / Gillespie A

Effet d'une Grande Energie de Pluie sur
l'Erosion Hydrique de Sols Agricoles dans
les Cantons de l'Est
(1988) / Lalime M A

Caracterisation par Microscopie des Fibres
Residuelles Resultant des Reacteurs Lors du
Processus de la Valorisation de la Biomasse
(1988) / Perreault N

L'Ozonation et Son Application au Traitement
des Boues de Station d'Epuration des Eaux
Usees Domestiques en Vue d'une Disposition
par Epandage Agricole
(1988) / Robic S

Etude des Corridors d'Ecoulement du Lac
Saint-Louis (Quebec)
(1988) / Robitaille M

Les Procedes de Traitement de Dechets
Dangereux par Solidification/Stabilisation:
Mise en Situation par le Procede Allemand GFS
(1988) / Ross A

SHERBROOKE, UNIVERSITY OF
(continued)

Caracterisation, Dynamique Trophique et
Impacts Environnementaux des Lacs
Artificiels de la Region de l'Estrie,
Province de Quebec
(1988) / Soucy C

Complement a l'Etude Limnologique 1986 du
Comite Charmes
(1988) / Tremblay L

SOUTHERN ILLINOIS UNIVERSITY

An Evaluation of Regulatory Requirements for
Illinois Domestic Wastewater Treatment
Plants Under the Superfund Amendments and
Reauthorization Act of 1986, Title III, the
Emergency Planning and Community
Right-to-Know Act
(1988) / Accardo M H

Sustained Middle Income Housing Development:
An Important Link to the Economic
Revitalization of East St. Louis, Illinois
(1988) / Brooks C J

The Effects of Agricultural Policy on
Resource-Use Efficiency
(1988) / Little C A

Citizen Participation in Governmental
Hazardous Waste Regulatory Actions:
Practical Adaptations for Use in University
Programs
(1988) / Luly C B

Nitrate Levels in Groundwater in Boone and
Winnebago Counties, Illinois
(1988) / Meyer J

Paleolimnology of Phantom Lake, King
County, Washington
(1988) / Vallarino J

An Option for Managing Household Hazardous
Waste
(1988) / Wallace M A

TEXAS A AND M UNIVERSITY

Electrostatic Charge Generation during Impeller
Mixing in Two-Phase Systems
(1988) / Hernandez A

Development of a Safety Analysis System for the
Offshort Personnel and Equipment Transfer Process
(1988) / McKenna M G

EMI Shield Enhancement Through the Addition of
Copper Coated Glass Fibers
(1988) / Montanye J R

Toxic Species Evolution from Guayule Fireplace Logs
(1988) / Soderman K L

TEXAS, UNIVERSITY OF (AUSTIN)

Immiscible-Phase Flow in the Subsurface and
Implications for Facilitated Transport
(1988) / Shultz S A

Multisolute Adsorption of Toxic Organic
Compounds onto Soil
(1988) / Yoon C G

U.S. NAVAL POSTGRADUATE SCHOOL

An Investigation of Acoustic Cavitation
Produced by Pulsed Ultrasound
(1987) / Bruce R L

Computer Aided Software Engineering (CASE)
Environment Issues
(1987) / Frey W K

Measurements of Gas Turbine Combustor and
Engine Augmentor Tube Sooting
Characteristics
(1987) / Grafton T A

Further Development of a One-Dimensional
Unsteady Euler Code for Wave Rotor
Applications
(1987) / Johnston D T

Stress Analysis of the LHA-1 Class
Superheater Header by Finite Element Method
(1987) / Kaufmann J W

Effect of Fiber Diameter on the Reliability
of Composites - Automated Laser Diffraction
Implementation
(1987) / Kunkel J S

Analytical and Experimental Studies of Beam
Waveguide Absorbers for Structural Damping
(1988) / Lee G G

Composite Failure Criterion - Probabilistic
Formulation and Geometric Interpretation
(1988) / Lim J C

An Investigation of Acoustic Cavitation
Produced by Pulsed Ultrasound
(1987) / Middleton R D

Numerical Electromagnetic Models of Cube
Shaped Boxces - An Initial Investigation
for Near-Field Prediction of HF Shipboard
Environments
(1987) / Molina C R

The Thermal Behavior of Film Cooled
Turbulent Boundary Layers as Affected by
Longitudinal Vortices
(1987) / Ortiz A

Passive Multipath Target Tracking in an
Inhomogeneous Acoustic Medium
(1987) / Shefi A

Background Gas Pressure Dependence of
Unipolar Arcing on Soda Lime Glass and
Plastic Induced by a Carbon Dioxide
Pulsed Laser
(1988) / Wojtowich A R

U.S.A.F. INSTITUTE OF TECHNOLOGY

Effect of Freestream Turbulence on a
Two-Dimensional Cascade, with and Without
Surface Roughness, at High Reynold Number
(1988) / Absar S

A Method of Determining Neutron Dose to a
Human Phantom
(1988) / Archuleta M G

Determination of Deflections of the Vertical
Using the Global Positioning System
(1988) / Aristov A

U.S.A.F. INSTITUTE OF TECHNOLOGY
(continued)

Response Surface Analysis of Stochastic
Network Performance
(1988) / Bailey T G

Joint Probability Data Association on
Tracking Multiple Munitions Fragments
(1988) / Beyers R J

Development of Techniques to Relate Radon
Levels in Homes in the Dayton Area to Local
Geology and Fill Material
(1988) / Bouchard J P

Automatic Liquid Agent Detection:
Determining the Distribution of Chemical
Agent Droplet Mass
(1988) / Bowman B A

The Manufacture of Non-Linear Optical
Polydiacetylene Thin Films Using
Poly-Benzyl-L-Glutamate as an Alignment
Matrix
(1988) / Budelier J A

Time Resolved Photo-Luminescence of
Ytterbium in Indium Phosphide
(1988) / Bumgarner T F

An Educational Expert System Shell
Integrating Object-Attribute-Value Triples
and Frames
(1988) / Clark E G

An Analysis of Noise Reduction Using Back
Propagation Neural Networks
(1988) / Cox K S

Capacity of Human Operator Using Smart Stick
Controller
(1988) / Cozzone A

An Approximation Technique for Solving a
System of Fredholm Integral Equations for
Asymmetric Detector Response Functions
(1988) / Daniel R B

Development of a Shock Capturing Code for
Use as a Tool in Designing High-Work Low
Aspect Ratio Turbines
(1988) / Driver M A

Nonlinear Finite Element Analysis of
Gebneral Composite Shell
(1988) / Egan G S

Comparison of Split-Film and X-Film
Measurements in 2-D Flow
(1988) / Fisk T E

Structural Optimization Including
Centrifugal and Coriolis Effects
(1988) / Gans H D

Evaluation of the Spatial and Temporal
Characteristics of the Conductive PRIZ
(1988) / Gardner P J

Neutron Multiplication in Beryllium
(1988) / Hartley R S

Investigation of Boundary Layer Disturbances
Caused by Periodic Heating of a Thin Ribbon
(1988) / Haven B

Geomagnetic Field Energy as a Source of
Thrust
(1988) / Hildenbrandt S R

Design of a Syntax Validation Tool for
Requirements Analysis Using Structured
Analysis and Design Technique (SADT)
(1988) / Jung D H

Modeling the Effects of Heavy Charged
Particles on MOSFETS
(1988) / Kattner K M

Design of Water Tunnel to Measure Wall
Pressure Tunnel Blockage and Wake Effects
(1988) / Lautenbach K A

Collision Energy Transfer Mechanisms Between
Singlet Oxygen and Monofluoride
(1988) / Lee W M

Frequency Self-Locking of a Laser Diode
(1988) / Lukacs S F

A Detection Theory Analysis of Visual
Display Performance
(1988) / Mabry T R

Verification and Limitations of the
Monostatic-Bistatic Radar Cross Section
Relationship Derived by Kell
(1988) / MacLennan J

Impact of Riblets Upon Flow Separation in a
Diffuser Section
(1988) / Martens N W

The Design and Implementation of a Graphical
VHDL User Interface
(1988) / Matechik S M

A Survey of Digital Beamforming Techniques
and Current Technology
(1988) / McCord J E

Aircraft Tracking with Duel TACAN
(1988) / McCormak C J

Concept and Design of an Auditory
Localization Cue Synthesizer
(1988) / McKinley R

A Computer Model for Analysis of
Dielectrically Enhanced Single-Zone Tunnel
Dryers
(1988) / Melendez W

The Response of Single Elements Mercury
Cadmium Telluride and Indium Antimony Solid
State Infrared Detectors to Millimeter Wave
Electromagnetic Energy
(1988) / Moore T D

ISA: A Prototype Intelligent Scheduling
Assistant for Defense Acquisition Management
(1988) / Moran J L

Analysis and Simulation of a Pseudonoise
Synchronization System
(1988) / Morgan F A

Analysis of Synthetic Aperature Radar
Imagery Using Synthetic Discriminant
Functions
(1988) / Munipalli S

U.S.A.F. INSTITUTE OF TECHNOLOGY
(continued)

A Digital Rate Controller for the Control
Reconfigurable Combat Aircraft Designed
Using Quantitative Feedback
(1988) / Neumann K N

Multiple Model Adaptive Tracking of Airborne
Targets
(1988) / Norton J E

Optical Activity and Its Influence on
Photo-Refractive Materials
(1988) / Padden R J

Performance of an Adaptive Matched Filter
Using the Griffiths Algorithm
(1988) / Pasko P D

Selection of a Frequency Sensitive QFT
Weighting Matrix Using the Method of
Specified Outputs
(1988) / Philips W D

Modified Backword Error Propagation for
Tactical Target Recognition
(1988) / Piazza C C

Hardware Modeling with VHDL Simulation
(1988) / Pompilio D G

The Effects of Incidence Angle and Free
Stream Turbulence on the Performance of a
Variable Geometry Two-Dimensional Compressor
Cascade
(1988) / Poniatowski E M

Design and Implementation of a PC-Based
Network Simulator for Engineering Education
(1988) / Puckett R D

Investigation of a Self-Aligned, Strained
Channel, Semiconductor Insulator
Semiconductor Field
(1988) / Rapp D W

Transverse Shear Considerations in
Anisotropic Plates
(1988) / Reams R H

Robotic Tacticle Sensor Fabricated from
Piezoelectric Polyvinylidene Fluoride Films
(1988) / Reston R R

An Analysis of Platinum Silicide and Indium
Antimonide for Remote Sensors in the
Three-to-Five Micrometer Wavelength Band
(1988) / Schoon N F

An Investigation of Constitutive Models for
Predicting Viscoplastic Response During
Cyclic Loading
(1988) / Shaffer D A

Performance of Preconditioned Iterative
Methods in Computational Electromagnetics
(1988) / Smith C F

Implementation of Phase Interpolartion
Estimator (PIE) to Accurately Estimate a
Single Tone Frequency in Noise
(1988) / Spaulding E M

Effect of Two Body Motion on Radar Beam
Quality for Various Distributed Sparse Array
Configurations
(1988) / Spencer M G

Analysis and Simulation of an Audio Adaptive
Equalizer
(1988) / Strasburger J R

Solution of Potential Flow Past an Elastic
Body Using the Boundary Element Technique
(1988) / Taylor N

Multiple Model Based Robot Control:
Development and Initial Evaluation
(1988) / Tellman L D

Effects of Different Surface Treatments. In
the Tensile Bond Strength of Polymethyl Meth
Acrylate Processed Against Chemically Etched
Ticonium 100
(1988) / Tiffany R L

Quadratic Optimal Control Theory for
Viscoelastically Damped Structures Using a
Fractional Derivative Viscoelasticity Model
(1988) / Walker R N

Angle of Attack and Sideslip Estimation
Using an Inertial Reference Platform
(1988) / Zeis J E

UTAH STATE UNIVERSITY

An Adjunctive Indicator in Nonpoint
Pollution
(1987) / Eberl S G

Alternate Stilling Basin Designs for a
USBR Type-II Basin
(1987) / Meacham J F

Design of Low Frounde Number Stilling
Basins
(1987) / Nielsen K D

A Proposed Methodology to Prepare a
Relative Seismic Slope Stability Map of
Davis and Salt Lake Counties, Utah
(1987) / Rathbun D J

Sediment Yield from Landslips in Steep
Mountain Watersheds in Utah
(1987) / Weston A

WASHINGTON STATE UNIVERSITY

Measurements of Organic Acids and Their
Precursors in the Remote Marine Troposphere
(1988) / Arlander D W

An Evaluation of the Effectiveness of the
Washington Main Street Program
(1988) / Dickinson P W

Numerical Simulation of Hydraulic Jump
(1988) / Gharangik A M

Toxicity of Toluene and o-Xylene to
Pseudomonas Testosteroni in the
Starvation/Survival Mode
(1988) / Kong S

Pacific Northwest Recycling Initiatives
Economics Versus Energy Savings
(1988) / Nicholls M H

Design, Installation, and Evaluation of a
Free Hydrocarbon Recovery System Following
Subsurface Gasoline Contamination
(1988) / Petrich C R

WASHINGTON STATE UNIVERSITY
(continued)

Inorganic Chemistry of Precipitation Near
Palmer Station, Antarctica
(1988) / Rauen C D

Concentrations, Trends, and Seasonal Cycles
of Atmospheric Methane, Carbon Monoxide, and
Carbon Dioxide at Palmer Station, Antarctica
(1988) / Schilling K J

Partitioning of Number 1 Fuel Oil Components
Between Solid and Aqueous Phases
(1988) / Wulf D

WASHINGTON, UNIVERSITY OF

Definition of Developable Hull Surfaces with
High Level Computer Graphics
(1988) / Aguilar G D

Contributions of Windows and Isovists to
the Judged Spaciousness of Simulated Crew
Cabins
(1987) / Al-Sahhaf N A

The Use of Computer Simulation for
Predicting Erection-State Rework
Probabilities for Ships' Hull Blocks
(1987) / Giesy P J

Engineering Aspects of Pen Farming Finfish
in Marine Waters
(1987) / Haddon P

Computer Aided Process Planning (CAPP) for a
Circuit Board Assembly Line
(1988) / Piyarali A

WEST VIRGINIA UNIVERSITY

Comparison of Functional Strength in Various
Protective Clothing Ensembles
(1988) / Seremetis E

WORCHESTER POLYTECHNIC INSTITUTE

Relationship Between Ignition Source
Strength and Design Inerting Concentration
of Halon 1301
(1986) / Das A

Time Dependent Calculation of Visibility in
Compartment Fires
(1987) / Davoodi H

Risk Assessment in Telephone Exchanges
(1986) / Johnson P F

Physical Modeling of Horizontal Vent Flows
Using Helim and Air
(1987) / Shanley J H

Predicting Thermal Radiation from Large
Open Pool Fires
(1987) / Shokri M Y

ALBERTA, UNIVERSITY OF

A Comparison of Two Soils from the Milk
River Ridge, Southwest Alberta
(1988) / Brierley J A

Allocation and Dynamics of Carbon in Two
Soils Cropped to Barley
(1988) / Dinwoodie G D

Prediction of Saturated Hydraulic
Conductivity in Peat
(1988) / Sherstabetoff R D

ARKANSAS, UNIVERSITY OF

Growth and Soil Water Use of Double-Cropped
Wheat and Soybeans
(1988) / Daniels M B

Spatial and Temporal Variability of a
Sharkey Clay Cropped to Soybeans
(1988) / De Angulo J

Chemical Measurement of Nitrogen
Mineralization in Arkansas Soils
(1988) / Hattey J A

Spatial Variability of Physical Properties
of Captina Soils
(1988) / Wood L S

AUBURN UNIVERSITY

Effects of Subsurface Applied Lime and Phosphorus
on the Growth of Sorghum on an Acid Mine Overburden
(1987) / Boozer R T

Landscape Positions, Soil Properties, and Erosion
in the Coastal Plain and Limestone Valley Regions
of Alabama
(1987) / Branch R T

Response of Sericea Lespedeza to Herbicides and
Weed-Control Systems
(1987) / Jones J D

Aging of Pearlmillet Leaves as Affected by
N Fertilization
(1987) / Rose P A

BOSTON COLLEGE

The Geochemistry and Petrogenesis of the
Sharpners Pond Plutonic Suite, Northeastern
Massachusetts
(1988) / Loftenius C J

BROOKLYN COLLEGE

The Application of the Pixe Method to
Geochemical Analysis
(1988) / El-Tabakh M

CALIFORNIA, UNIVERSITY OF (DAVIS)

Selenium Absorption and Translocation by
Plants
(1988) / Atkins C E

Isotopic Composition of Soil Carbon Dioxide
and Groundwater in an Irrigation Cotton
Field, San Joaquin Valley, California
(1988) / Bell R B

Duripan Genesis on Granitic Pediments of the
Mojave Desert
(1988) / Boettinger J L

Nutrient Constraints in a Traditional Rain
Fed Cropping System of the Andean High
Sierra in Ecuador
(1988) / Bossio D A

Fitting of Mixture Model for Evaluation of
Water and Salinity Stress on Sorghum Yield
(1988) / Chang W L

Genesis of Arid, Alpine Soils on Glacial
Moraines in the White Mountains of
California
(1988) / Morris S A

COLORADO SCHOOL OF MINES

The Aerobic Biodegradation of Quinoline and
Analogs from a Creosote-Contaminated Site in
Pensacola, Florida
(1988) / Bennett J L

The Application of an Ion-Interaction Model
to Solubilities of Some Alkaline-Earth
Carbonates, Thorium Oxide and Uranium Oxide
in Brines
(1988) / Ranville J F

Clay-Organic Interactions
(1988) / Sadowski R M

Relationship of Trace Elements in Soils and
Plants to Petroleum Production in Eagle
Springs Oil Field, Railroad Valley, Nevada
(1988) / Saeed M A

Geochemical Evolution and Depositional
History of Sediment in Modern and Ancient
Saline Lakes: Evidence from Sulfur
Geochemistry
(1988) / Tuttle M L

COLORADO STATE UNIVERSITY

Improved Method for the Control and Monitoring
of Redox in Soils
(1988) / Brennan E W

CONNECTICUT, UNIVERSITY OF

Thiocynate: A Potential Inhibitor of Plant
Growth from Crucifer Residue
(1988) / Lee R F

Effect of Nitrogen Source on the Yield and
Nitrogen Recovery of Corn and the Use of
June Soil Nitrate Level as a Nitrogen
Availability Index
(1988) / Morris T F

CORNELL UNIVERSITY

An Analysis of Soils and Land Use
Interaction in Tompkins County, New York
(1988) / Asfaw N

The Issue of Lime Requirement Revisited:
The Shoemaker, Maclean and Pratt Single
Buffer Method and Cornell Soil Test in
Comparison to the Standard Incubation Method
(1988) / Chowoe J S

An Alternate Method for the Determination
of Cations in Solution in Small Soil Samples
(1988) / Martinez G A

FLORIDA, UNIVERSITY OF

Cloning and Characterization of a Toxin Gene
from Bacillus Sphaericus
(1988) / Ali A N

Assessment of Lithologic Versus Pedologic
Origin of Sandy-Over-Loamy Boundary in Some
Associated Florida Soils
(1988) / Cabrera Martinez F

Phosphorus and Nitrogen Effects on Root
Colonization of Onions by Three Vesicular
Arbuscular Mycorrhizal Fungi
(1988) / Neal L

Characterization and Evaluation of Enriched
Fly Ash as Fertilizer
(1988) / Valle Melo J

GEORGIA, UNIVERSITY OF

Analysis of Organic and Inorganic Sulfur
in Forest Soils
(1988) / Caldwell P R

Geochemical Aspects of Eocene-Oligocene
Volcanism and Alteration in Central Utah
(1988) / Kim C S

Analysis of the Maganese Content of the Fort
Payne Formation of Northwestern Georgia and
Northeastern Alabama
(1988) / White R L

GUELPH, UNIVERSITY OF

A Theoretical and Wind Tunnel Investigation
of the Effects of Ridge Roughness on Aeolian
Thresholds and Sediment Flux Rates for
Three Soils
(1988) / Alberico J J

Quaternary Paleoenvironments and Paleosols
of the Kathmandu Basin, Nepal
(1988) / Dongol G M

Fermentation Products of Manure and Plant
Residues as Carbon Sources for Denitrifying
Bacteria in Soil
(1988) / Paul J W

Adsorption of Boron by Clays and Limed Soils
and Boron Availability to Alfalfa in
Relation to Soil Solution Chemistry
(1988) / Su C

The Effect of Tillage Systems on Dynamics of
Soil Surface Water Content, Carbon Dioxide
Flux and Microbial Biomass
(1988) / Zhai R

HAWAII, UNIVERSITY OF

Effect of Shifting Frequency on Liveweight
Gain of Grazing Steers
(1988) / Alvarez A

Erosional Relations in a Short-Rotation
Eucalyptus Plantation on a Typic Hydrandept
(1988) / Schultz J M

IDAHO, UNIVERSITY OF

Influence of Drill/Press Wheel Configuration
on Spring Wheat Growth and Development
(1987) / El-Nagar G

Phosphorus Decrease in a Southern Idaho
Soil
(1988) / Freeborn L L

Evaluation of Lime Requirement Tests and
the Effects of Soil pH Manipulation on
Wheat and Lentils in a Northern Idaho Soil
(1988) / Mohebbi S

Irrigation and Nitrogen Management Effects
on Spring Pea Seed Yield and Quality
(1987) / Raymond M A

ILLINOIS STATE UNIVERSITY

Effects of Inorganic Nutrient Treatment,
Soil Micor-Organisms, and Vesicular
Arbuscular Mycorrhizae on the Growth of
Big Bluestem (Andropogon Gerardii)
(1988) / Bentivenga S P

Biomass Production and Microclimate of
Burned and Unburned Hill Prairie as
Influenced by Slope
(1988) / Teesdale D

MAINE, UNIVERSITY OF

Modeling Nitrogen Dynamics Using Object
Oriented Programming
(1988) / Crosby C J

Evaluation of the Effects of Wood-Ash
Amendment on Forest Soils
(1988) / Unger Y L

MANITOBA, UNIVERSITY OF

The Effect of Various Amendments on C and N
Dynamics of Active Organic Matter and
Aggregate Stability in Different Soils
(1988) / Janzen R A

Effects of Simulated Erosion on Canola
Productivity
(1987) / Kenyon B E

MASSACHUSETTS INSTITUTE OF TECHNOLOGY

Geochemistry of Hawaiian Dredged Lavas
(1988) / Gurriet P C

MASSACHUSETTS, UNIVERSITY OF

A User-Build System for Automated
Monitoring and Controlling of Controlled
Atmosphere Apple Storages
(1988) / Kaminsky K S

MINNESOTA, UNIVERSITY OF

Root Biomass Dynamics for Two Salix Clones
Grown on a Histosol
(1988) / Chakoian L M J

Moisture Regimes, Morphological
Characteristics, and Hydraulic Properties
of Five Toposequences in the Wadena
Drumlin Field
(1988) / Harris J K

The Effect of Selected Soil Variables on
Corn Grain Yields
(1988) / Miller M B

Sulfur Fertilization to Improve Yield and
Quality of Corn and Alfalfa in Minnesota
(1988) / O'Leary M J

MINNESOTA, UNIVERSITY OF
(continued)

A Model to Predict Water Availability for
Crop Growth in East Africa
(1988) / Robinson M R

Nitrogen Availability for Corn as Affected
by Urea-Ammonium Nitrate Placement,
Ammonium Thiosulfate, and Simulated Rainfall
(1988) / Zadak C G

MISSISSIPPI STATE UNIVERSITY

Characterization and Productivity Potentials of
Selected Prime Farmland Soils in Mississippi
(1986) / Cairns J

MONTANA STATE UNIVERSITY

Influence of Nuclear and Cytoplasmic Factors
on Cellus Initiation in Anther Culture of
Wheat
(1987) / Becraft P W

Improvement of Chickpea Stand Establishment
in Cool Soils
(1987) / Bollinger S A

Water Retention and Flow Characteristics of
Six Calcarious Soils of Montana
(1987) / Browne M F

Determining Verticillum Wilt Resistance in
Alfalfa
(1987) / Bruce M R

Band Applications of Elemental Sulfur
Inoculated with Thiobacillus Thioparus to
Enhance Nutrient Availability
(1987) / DeLuca T H

The Soil Adsorption, Mobility, Degradation,
and Residual Properties of AC 222,293
(1987) / Fellows G M

Relationship Between Seed Vigor Testing and
Field Performance of 'Regar' Meadow
Bromegrass
(1987) / Hall R D

Analysis of Genetic Diversity in Barley
Using Restriction Fragment Length
Polymorphism Analysis
(1987) / Kramer M G

Effects of Harvest and Insecticide Treatment
on Predominant Species of Ground Beetles in
Alfalfa and Sainfoin
(1987) / Lester D G

Evaluation of Seed Vigor Tests for Safflower
(1987) / Luth D

Weed Control in Alfalfa Grown for Seed
(1987) / Stannard M E

An Evaluation of Plant Drought Stress
Parameters in Spring Wheat
(1987) / Touray K S

NEW MEXICO INSTITUTE OF MINING & TECH.

Geology and Geochemistry of the Hembrillo
Canyon Succession, San Andres Mountains,
Sierra and Dona Ana Counties, New Mexico
(1987) / Alford D E

Sulfur Dioxide and Particle Emission from
Mount Etna, Italy
(1988) / Andres R J

Geochemical Study of the Cochiti Mining
District, Sandoval County, New Mexico
(1987) / Apodaca L E

Geochemical Analyses of Ore Fluids from
the St. Cloud-U.S. Treasury Vein System,
Chloride Mining District, New Mexico
(1988) / Behr C

Geology and Geochemistry of a Proterozoic
Superacrustal and Intrusive Sequence in the
Central Wet Mountains, Colorado
(1988) / Lanzirotti A

The Emission of Gases and Aerosols from
Mount Brebus Volcano, Antarctica
(1988) / Meeker K

Geology and Geochemistry of the Proterozoic
Age Alder Group, Central Mazatzal
Mountains, Arizona
(1988) / Reed J R

NEW MEXICO STATE UNIVERSITY

Effects of Polyacrylamide on Soil Crusting
and Seedling Emergence
(1988) / Caminade B M

Sulfur Oxidation in Soils and Mine Spoils as
Influenced by S Source, Nutrient Amendment,
and Thiobacillus Inoculation
(1988) / Haffner W M

Effects of Sulfuric Acid on P Availability
and P-Fractionation in Two Calcareous Soils
(1988) / Khorsandi F

Effect of Sulfur, Thiobacillus, and
Nutrients on Soil pH and Plant Growth
(1988) / Ramolemana G M

NORTH CAROLINA STATE UNIVERSITY

Comparisons of Magnesium Sources and
Relationships Between Soil Acidity Estimates
Among Lime Trials in North Carolina
(1988) / Ngachie V

NORTH DAKOTA STATE UNIVERSITY

Soil Water Climatology for north Dakota
(1988) / Brenk P C

Influence of Waterlogging on Extractable Soil
Phosphorus
(1988) / Cudworth D K

Effects of Revegetation and Erosional Dynamics on
Panspots Genesis and Distribution in Western
North Dakota
(1988) / Hopkins D G

Nitrogen Source and Zinc Nutrition of Two Legumes
(1988) / Mavi H

Gypsum Occurence in Soils on the Margin of
Semipermanent Prairie Potholes Wetlands
(1987) / Steinwant A L

NORTHEASTERN LOUISIANA UNIVERSITY

Soil-Borne Bacteriophage Tolerance to
Selected Biochemical Soil Amendments
(1988) / Weng W

OKLAHOMA, UNIVERSITY OF

Geochemistry of Mafic Rock Units of the
Southern Oklahoma Aulacogen, Southwestern
Oklahoma
(1988) / Aquilar J

Transient Flow Through Porous Media
(1988) / Blatz S K H

A Geochemical and Isotopic Study of the
Garber-Wellington Aquifer, Cleveland
County, Oklahoma
(1988) / Scott T D

OLD DOMINION UNIVERSITY

Provenance of Sands Within the Lower
Chesapeake Bay Based on Ilmenite Composition
(1987) / Council E

Linear Sand Deposits in Southeastern
Virginia Coastal Plain: A Fourier Shape
Analysis and Trace Element Content Study
of Ilmenites
(1987) / Evans A

Diagenesis Within the Lucayan Limestone
Great Abaco Island, Little Bahama Bank
(1988) / Goeke M

Trace Elements in Middle Ordovician
Carbonates: Implications for Diagenesis
(1988) / Strauss R

Depositional and Diagenetic History of the
middle Ordovician Carbonates of the
Shenandoah Valley, Northern Virginia
(1987) / Weyenberg L

OREGON STATE UNIVERSITY

Calibration of Crop Water Use with the Penman
Equation Under United Arab Emirates Conditions
(1988) / Mohammed F H

Relationship Between Pore Size, Particle Size,
Aggregate Size and Water Characteristics
(1987) / Wu L

PENNSYLVANIA STATE UNIVERSITY

Metal Concentrations in Soil, Vegetation,
and Microtus Pennsylvanicus on Strip Mines
Reclaimed Conventionally and with Municipal
Sludge
(1987) / Alberici T M

Soil Amendments to Reduce Cadmium and Zinc
Uptake by Plants Grown on Zinc-Smelter
Contaminated Soils
(1988) / Bowers M E

Metal Extraction from Contaminated Soils
Using Chelating Agents
(1988) / Linn J H

The Effects of Acid Precipitation on Two
Soils of Pennsylvania
(1987) / Schuyler S S

RHODE ISLAND, UNIVERSITY OF

Geochemistry of Tributyltin in Coastal Waters:
An Experiment in a MERL Mesocosm
(1988) / Adelman D

SASKATCHEWAN, UNIVERSITY OF (SASKATOON)

The Soil Residual Properties, Mobility and
Canola Tolerance of DPX-7881: A
Sulfonylurea Herbicide
(1988) / Beckie H J

Variable Rates of Nitrogen Fertilizer
Application with Slope Position
(1988) / Belyk M B

Soil Residual Herbicides: Moisture
Conservation, Wheat Production, Persistence
(1988) / Gerwing P D

The Effect of Erosion on Soil Productivity
and Related Soil Properties
(1988) / Verity G E

STEPHAN F. AUSTIN STATE UNIVERSITY

Geochemical and Hydrogeologic Investigation
of Low-pH Conditions in South-Central Rusk
County, Texas
(1988) / Hayes J J

Some Effects of Water Discharge Through the
Mount Enterprise Fault Zone on Surface Water
Composition in Southern Rusk
(1988) / Reiner S

Cation Affinity for Diagenetic Materials in
the Hensel Sandstone (Cretaceous) Gillespie
County, Texas
(1988) / Smith C D

A Heavy Metals Analysis of Lone Star
Reservoir, Lone Star, Texas
(1988) / Witt W T

TENNESSEE, UNIVERSITY OF (KNOXVILLE)

Soil Solution Composition and Orchardgrass
Germination as Influenced by Simulated Acid
Rain in a Field Environment
(1987) / Haun G W

TEXAS A AND M UNIVERSITY

Mutagenic Potential of Plants Grown on a Soil
Amended with Mutagenic Municipal Sewage Sludge
(1988) / Fiedler D A

Response of Pearl Millet to Soil Moisture in One
Agroclimatological Zone of Niger, West Africa
(1988) / Gandah M

Water Infiltration Studies of the Major Rice
Producing Soil Series of the Texas Gulf Coast
(1988) / Nesmith D M

Arsenic Species in Soil Solution and Plant
Uptake of Arsenic Under Flooded Conditions
(1988) / Onken B M

Evaluation of Vegetative Cover on Reclaimed Land by
Color Infrared Videography Relative to Soil
Properties
(1988) / Pfordresher A A

Pesticide Fate in an Above Ground Disposal System
(1988) / Vanderglas B R

Contour Strip Rainfall Harvesting for Cereals
Production on Sandy Soils in Niger (West Africa)
(1988) / Zaongo C G

TEXAS, UNIVERSITY OF (AUSTIN)

Sorption and Removal of Chlorinated Volatile
Organic Compouunds from Soil
(1988) / Atere-Roberts S O

Detoxification and Immobilization of
Chlorophenols in Soil
(1988) / Dasappa S M

TULSA, UNIVERSITY OF

Geochemical Biomarker Study of the Woodford
Shale in the Witcher Field, Oklahoma County,
Oklahoma
(1987) / Smith T N

TUSKEGEE INSTITUTE

Relationships Between Sweet Potato
Phenotypic Characteristics and Root Yield
(1987) / Barnes J A

UTAH STATE UNIVERSITY

A Global Solar Irradiance Climatology of
an Intermountain Region
(1987) / Dobry E W

Effects of Vesicular-Arbuscular Mycorrhizae
Grown Under Drought Stress and Their
Influence on Organic Phosphorus
Mineralization
(1987) / Duce D H

An Analysis of Apparent-Motion Vectors in,
and the Structure of a Mid-Lattitude
Sporadic E Layer Using a 2.66 MHZ Radar
(1987) / Halderman T D

Evaluation of Foliar Versus Soil Applied
to Alfalfa
(1987) / Hasanein B M

The Effect of Fertilizer Placement
Geometry on Eroded and Noneroded Complex
Soils of the Clarkston Watershed
(1987) / Hirpa H

Effects of Phosphorus on No-Till, Minimum
Till and Conventional Till Irrigated Field
Corn
(1987) / McKay J A

Boron Composition of Alfalfa in Utah as
Related to Soils and Irrigation Waters
(1987) / Radtke R N

An Analysis of Upper Atmospheric Parameters
Derived from the Obsevation of Meteor Echoes
by a 2.66 MHz Radar
(1987) / Turek R S

VICTORIA, UNIVERSITY OF

The Growth Response of Secondary Vegetation
to Silvicultural Treatments, Soil and Site
Conditions in the Carnation Creek Wateshed,
Vancouver Island
(1988) / Hays W J

WASHINGTON STATE UNIVERSITY

General Soil Map of Washington State
(1988) / Boling M S

The Effect of Water Deficit and Panicle
Anatomy on Wild Oat Seed Dormancy and
Abscisic Acid Content
(1988) / Heimbigner B L

Embryonic ABA Sensitivity During Grain
Development and Heritability of Grain
Germinability and Embryonic ABA Sensitivity
in One Winter Wheat Cross
(1988) / Ho Y

The Effect of Slope Position, Aspect, and
Cultivation on Organic Carbon Distribution
in the Palouse
(1988) / Rodman A W

Genetic Variation for Nitrogen Accumulation,
Agronomic Performance, and Melting Quality
of Spring Barley Production with Direct
Drilling
(1988) / Tillman B A

Electrophoretic Mobility and Monovalent
Cation Selectivity of Three Reference
Clay Minerals
(1988) / Xu S

Nitrogen Fertilization and Nitrification
Inhibition Effects on Growth and Nodular
Activity in Soybean
(1988) / Zahid M A

WEST TEXAS STATE UNIVERSITY

A Stream Sediment Geochemical Survey of the
Eastern Half of the Capitan Mountains:
Lincoln County, New Mexico
(1988) / Ellinger S T

WEST VIRGINIA UNIVERSITY

Biodegradation of a Dilute Waste-Oil
Emulsiton Applied to Soil
(1988) / Elsavage R E

ARIZONA STATE UNIVERSITY

Scanning Electron Microscope Techniques for
Discrimination of Magmatic and Hydromagmatic
Pyroclasts: 79 Deposits
(1988) / Komorowski J C

BROCK UNIVERSITY

Geochemistry and Petrogenesis of the
Proterozoic Pater Metavolcanic Suite,
Spragge, Ontario
(1987) / Lorek E G

CALIFORNIA STATE UNIV. (LONG BEACH)

Relation Between Clay Extraction Techniques
and Experimentally Determined Chemistry and
Mineralogy of Clays--Implications for
Interpretation of Clay Chemistry
(1988) / Anders N E

Gravity Survey Across the Simi Fault and
Engineering Geology Implications Oversized
Plates
(1988) / Collender J M

The Stratigraphy and Structure of a Portion
of the Las Vegas Range, North-Central Clark
County, Nevada
(1988) / Kraft R P

Environmental Stratigraphy of
Post-Dunderberg Carbonate Strate, Nopah
Formation, (Upper Cambrian), Southern
Great Basin
(1988) / McCutcheon K F

Gravimetric Study of Lone Pine Canyon, San
Bernardino County, California
(1988) / Phibbs J K

A Gravity Interpretation of Structural
Features in the Cajon Pass Area, San
Bernardino County, California Plates
(1988) / Spindler R H

Field Investigation of the Aftershocks from
the October 1, 1987 Whittier Narrows
Earthquake
(1988) / Yi D W

CALIFORNIA, UNIVERSITY OF (DAVIS)

Geophysical Investigation of the Ward Valley
Aquifer, Lake Tahoe, California
(1988) / Niblack R L

COLORADO SCHOOL OF MINES

Statistical and Deterministic Deconvolution
Applied to Vertical Seismic Profiling
(1988) / Al-Qahtani A

Computerized Geometric Correction of Ground
Penetrating Radar Images
(1988) / Bochicchio R

An Expert System to Assist in Processing
Vertical Seismic Profiles
(1988) / Crisi P A

In Situ Soil Air Stripping: Analysis of
Data from a Project Near Benson, Arizona
(1988) / Johnson J J

CORNELL UNIVERSITY

New Estimates of Crustal Motion in the
Western United States
(1988) / Coyne J M

Physical Properties and Mechanical State of
an Artificial Silty Clay as a Simulation of
Deformation of Deep Sea Sediments at
Convergent Plate Boundaries
(1988) / Hou G

Regional Investigations of Recent Vertical
Crustal Movements in the Eastern United
States, Using Precise Releveling Data
(1988) / Jurkowski G A

Deep Crust of the Colorado Plateau: The
COCORP Reflection Surveys
(1988) / Lundy J

Structural Geometry, Sequence, and
Kinematics of the Black Pine Mountains,
Southern Idaho: Implications from the
Cover Rocks to Metamorphic Core Complexes
(1988) / Wells M L

The Cordilleran Foreland Thrust Belt and
Transition to the Hinterland in Northwestern
Montana from COCORP and Industry Seismic
Reflection Data
(1988) / Yoos T R

GEORGIA INSTITUTE OF TECHNOLOGY

Modeling of Seismic Coda, with Application
to Attenuation and Scattering in
Southeastern Tennessee
(1988) / Ogilvie J S

The Role of Biogenic Hydrocarbons in the
Photochemical Production of Tropospheric
Ozone in the Atlanta, Georgia Region
(1988) / Richardson J L

IDAHO, UNIVERSITY OF

Analysis of Landslide Occurrences in Idaho
(1988) / Adams W C

Geological Engineering Assessment of the
Bosie Foothills, ADA County, Idaho
(1988) / Beck C C

Hydrologic Summarization and Management
Classification of Tributary Basins in the
Upper Snake River Basin
(1988) / Broadhead R

Rock Discontinuity Properties and Ground
Water Flow in an Underground Lead-Zinc
Mine, Coeur d'Alene Mining District, Idaho
(1987) / Haskell K

Geotechnical Investigation of the Grouse
Creek Slope Failure
(1988) / Hultman W A

Spatial and Temporal Analysis for Idaho
Earthquake Hazard Assessments
(1988) / Jones A L

Induced Polarization: A Geophysical Method
for Detecting Metal Contaminated Ground
Water
(1988) / Krom T D

IDAHO, UNIVERSITY OF
(continued)

Evaluation and Verification of the
Rockpack/Backpack Rock Slope Stability
Computerized Analysis Packages
(1988) / Lamb K J

Methods of Analysis in Geological
Engineering
(1987) / Wittreich C D

INDIANA STATE UNIVERSITY

The Use of Thermal Infrared Multispectral
Scanner Data for Geochronologic Mapping of
the Cima Volcanic Field, San Bernadino,
California
(1988) / Levine N

IOWA STATE UNIVERSITY

Aspects of Pre-Illinoian Glaciation in
Southwest Iowa
(1988) / Schilling K E

LAURENTIEN UNIVERSITY

Gold Mineralization at the Kremzar Property
(1988) / Kwok K M

LEHIGH UNIVERSITY

Ground-Water Hydrology of the Raritan Bay
Area, New Jersey: Hydrogeologic Framework
and Simulation of the Freshwater/Saltwater
Interface Under Predevelopment Conditions
(1987) / Declercq E P

MAINE, UNIVERSITY OF

A Study of Morphologic Variation from
Bulimina Marginata to B Aculeata, 7,000
Years Before Present to Recent, in the
Gulf of Maine
(1988) / Burgess M V

Varying Rates in Local Relative Se-Level
Rise Controlling the Sediment Budget and
Sedimentologic Evolution of Northern
Casco Bay, Maine
(1988) / Hay B W B

The Mass Balance and Climatic Sensitivity of
North Cascade, Washington and Coast Range,
Southeast Alaska Glaciers
(1988) / Pelto M S

Flow Characteristics of Byrd Glacier,
Antarctica
(1988) / Scofield J P

Basal Thermal Regime of the Antarctic
Ice Sheet
(1988) / Shields E L

The Geochemistry of Chlorites as a Function
of Mineral Assemblage, Bulk Composition and
Sulfur Fugacity in the Rangeley Area,
Western Maine
(1988) / Teichmann F

The Occurrence of Staurolite and Its
Implications for Polymeta Morphism in the
Mount Washington Area, New Hampshire
(1988) / Wall E R

The Late Holocene Evolution of the Lubec
Embayment
(1988) / Walsh J A

Argon/Argon Mineral Ages from Southwestern
Maine: Evidence for Late Paleozoic
Metamorphism and Mesozoic Faulting
(1988) / West D P

MANITOBA, UNIVERSITY OF

Copper Oxysalts: The Jahn-Teller Effect
and Its Structuural Implications
(1988) / Eby R K

Structural Geology of the Vein System in the
San Antonio Gold Mine, Bissett, Manitoba,
Canada
(1988) / Lau M H

Alteration Zones in the Flin Flon Area,
Manitoba
(1988) / Mare M P H

Palaeomagnetic Analyses of the Leaf
Rapids Area in Manitoba
(1988) / Strobel G

MICHIGAN TECHNOLOGICAL UNIVERSITY

A Shallow Seismic Reflection Study Along the
Bear Lake Road, Hancock, Michigan
(1988) / Fan C M

Hydraulic Field Testing in a Shallow
Fractured Basalt: Applications to In-Situ
Solution Mining
(1988) / Mentel G J

Evaluation of Bioextraction Techniques for
In-Situ Mining of Copper Sulfide Ores in
Michigan
(1988) / Miller T J

MINNESOTA, UNIVERSITY OF

Application of a Full-Scale Dump Point
Stability Experiment to Engineeering
Analysis
(1988) / Crum D A

Numerical Analysis of Buckling Phenomena in
Layered Rock Media
(1988) / Papamichos E

Adjustment of Ventilation Network Parameters
Using Weighted Least Squares Optimization
(1088) / Rellier F

MONTANA COLLEGE OF MINERAL SCI. & TECH.

Design and Application of a Multi-Frequency
Data Collection System for the Study of
Rock Properties
(1988) / Miller B L

Three Dimensional Gravity Modeling
Techniques with Application to the Ennis
Geothermal Area
(1988) / Semmens D

MURRAY STATE UNIVERSITY

The Use of Computer Graphic Representations
of Statistical Training Class Distributions
as an Aid for Analyzing and Understanding
Statistical Distance
(1988) / Graham D L

NEW MEXICO STATE UNIVERSITY

A Hydrogeological Investigation of the Las
Cruces Geothermal Field
(1988) / Gross J T

NEW YORK, STATE UNIVERSITY OF (BUFFALO)

Liquefaction Potential Evaluation of
Surficial Deposits in Buffalo, New York
(1988) / Baumgras L

NEW YORK, STATE UNIVERSITY OF (FREDONIA)

Spatial Analysis of Precipitation in
Western New York and Northwestern
Pennsylvania
(1988) / Kamakaris D

Site Evaluation of the Town of Harmony
Landfill
(1988) / Schumacher M

OKLAHOMA, UNIVERSITY OF

Biomarker Characterization of Evaporite
Carbonate Source Rocks and Crude Oils, and
Oil-Source Correlations, in the Black Creek
Basin, Alberta, Canada
(1988) / Clark J P

A Surface to Subsurface Study of the
Sycamore Limestone (Mississippian) Along
the North Flank of the Arbuckle Anticline
(1988) / Cole T

Geochemistry of the Sandy Creek Gabbro,
Wichita Mountains, Oklahoma
(1988) / Diez de Medina D M

Brachipod Biostratigraphy and Biofacies
Analysis of the Marble Falls Formation
(Pennsylvanian) of Central Texas
(1988) / Dihrberg E E

Three Dimensional Photoelastic Study of
Abrupt Changes in Strike Associated with
Force Folding
(1988) / Lin C

Lineament Study from Satellite Imagery of
the Promontory Mountains in Utah
(1988) / Prucha C P

Two-Dimensional Front Tracking Model for
Microcomputers
(1988) / Rezigh A A

A Theoretical Study of the Flow of Slightly
Compressible Non-Newtonian Fluids in
Eccentric Annuli
(1988) / Smith S D

OLD DOMINION UNIVERSITY

Recent Benthic Foraminifera of Breton and
Stake Islands, Northern Gulf of Mexico
(1988) / Collins E

The Hydrogeology of Nags Head Woods, Dare
County, North Carolina
(1987) / Emry J S

Foraminiferal Biostratigraphy and
Paleoenvironmental Analysis of a Deep Well
in Vermillion Parish, Louisiana
(1988) / Orndorff A

QUEENS UNIVERSITY

Seismic Detection of Collapse Structures:
Case Study of Rocanville Mine, Saskatchewan
(1988) / Bostock M G

SAN DIEGO STATE UNIVERSITY

The Geology, Petrography, Geochemistry, and
Geochronology of the Tres Harmanos-Santa
Clara Region, Baja California, Mexico
(1987) / Chadwick B

The Effects of Acid Rain on Southern
California Lake Water Sediment Systems
(1988) / Frankel M G

Geochemistry and Petrography of the Santiago
Peak Volcanics, Santa Margarita and Santa
Ana Mountains, Southern California
(1988) / Gorzolla Y R

Geology of the Southern Sierra Calamajue
Area: Structural and Stratigraphic Evidence
for Latest Albian Compression Along a
Terrane Boundary, Baja California, Mexico
(1987) / Griffith R C

Geology of the Granite Mountain Area:
Implications of the Extent and Style of
Deformation Along the Southeast Portion of
the Elsinore Fault
(1988) / Lampe C M

Geohydrologic Investigation of Vallecito
Valley, San Diego County, California
(1987) / Martin D C

Neotectonics of the Northern Elsinore
Fault, Temescale Valley, Southern California
(1988) / Millman D E

SASKATCHEWAN, UNIVERSITY OF (SASKATOON)

A Two-Dimensional Field Study of the 1985
Ice Island Reflection Experiment
(1988) / Burianyk M J A

The Sedimentology and Stratigraphy of a
Gravelly Meander Lobe in the Saskatchewan
River, Near Nipawin, Saskatchewan
(1988) / Campbell J E

Albian Foraminiferal Bisotratigraphy of a
Key Borehole in Northeastern British
Columbia
(1988) / Harrison S M

Stratigraphy of the Viking Formation and
Newcastle Member of the Ashville Formation
in Saskatchewan
(1988) / Koziol B L

Quaternary Geology and Drift Prospecting,
Dawn Lake Area Northern Saskatchewan .
(1988) / Millard M J

A Seismic Study of the Haughton Crater,
Devon Island, Canadian Artic
(1988) / Scott D L

The Origin of Water Leaks in Saskatchewan
Potash Mines
(1988) / Wittrup M B

SOUTH CAROLINA, UNIVERSITY OF

Derivation of an Empirical Pore Model
Consistent with a Variety of Physical
Properties
(1988) / Brumfield D S

Column Experiments Examining Reductions in
Permeability of Clayey Sand from the
Application of a Neutral Sodium-Nitrate
Solution: Importance to Solute Transport
Modeling
(1988) / Dennehy K F

Estimation of Seamount Isostatic
Compensation in the Western Pacific
(1988) / Freymueller J T

The Thermal Maturation and Hydrocarbon
Potential of the Oligocene of the Northern
Sinai, Egypt
(1988) / Gaither D W

Radon Emanation from Synthetic and Natural
Solids: Implications for the Existence of
Nanopores in Rocks and Minerals
(1988) / Glicksberg D H

The Stratigraphy and Structural Setting of
Early Jurassic to Mid-Cretaceous Sedimentary
Rocks in the Southern Neuquen Basin,
Argentina
(1988) / Harder M W

The Nature of Continental Sedimentation
Seaward of the Ganges River System
(1988) / Hariu T M

An Evaluation of the Hydro-Geochemical
Parameters Affecting the Quality and Volume
of Leachate from Acidic Sandstone
(1988) / Johns K B

The Effect of Anomalous Delta-18-O Values on
Stratigraphic Interpretations: Delaware
Basin, West Texas
(1988) / Mucciarone D A

An Integrated Geological and Geophysical
Study of the Central South Carolina Coastal
Plain
(1988) / Muthanna A

Isovolumetric Geochemical Weathering of Some
Granitic Rocks in the Carolina-Piedmont
(1988) / Nelson G K

Groundwater Flow Patterns in a Forested
Beach Ridge-Swale System Adjacent to a
South Carolina Salt Marsh
(1988) / Nittrouer P L

Modern Sedimentary Processes in the
Wilmington Canyon Area, U. S. East Coast
(1988) / Sanford M W

Computer Simulation of Isolated Carbonate
Platforms: Judy Creek, a Case Study
(1988) / Scaturo D M

Geochemical Characterization of
Metavolcanics from the Carolina Slate
Belt, South-Central South Carolina
(1988) / Shelley S A

Quantitative Dynamical Geology of Pinedale
Anticline, Green River Basin, Wyoming and of
Sleipner Area, Norwegian North Sea: Two
Dimensional Computer Simulations
(1988) / Wei Z

SOUTH DAKOTA SCHOOL OF MINES AND TECHNOL.

Petrographic and Petrophysical
Characteristics of the Upper Cretaceous
Turner Sandy Member of the Carlile Shale,
Todd Oil Field Weston County, Wyoming
(1988) / Charoen-Pakdi D

The Use of the Pitzer to Determine
Saturation of Trona and Other Associated
Minerals
(1988) / Mutlu H

Hydrogeology of the Spearhead Lake Area,
Hubbard County, Minnesota
(1988) / Schubbe D L

TEXAS, UNIVERSITY OF (AUSTIN)

Hydrogeology of the Hickory Sandstone
Aquifer, Upper Cambrian Riley Formation,
Mason and McCulloch Counties, Texas
(1988) / Black C W

Paleoecology of the Eocene Wheelock Member
of the Cook Mountain Formation, in Western
Houston County, Texas
(1988) / Caskell B A

Foraminiferal Biofacies of Middle to Late
Paleogene Rocks in the Western San Emigdio
Mountains, California
(1988) / Cervantes M A

Geochemical Flow Modeling of the Supergene
Alteration of Porphyry Copper Deposits
(1988) / Chieruzzi G O

Structural and Metamorphic Constraints on
Fault Displacement Between Coherent
Blueschist Terranes Near Ball Mountain,
Eastern Belt, Franciscan Complex, Norther
California
(1988) / Copeland W B

Meteoric Water Penetration in the Frio
Formation, Texas Gulf Coast
(1988) / Donnelly A C

The Mapping of Tectonic Features in the
Ocean Basins from Satellite Altimetry Data
(1988) / Gahagan L M

An Assessment of Future Coastal Land Loss in
Galveston, Chambers, and Jefferson Counties,
Texas
(1988) / Germiat S

Oblique Slip Faults in the Northwestern
Picuris Mountains of New Mexico: An
Expansion of the Embudo Transform Zone
(1988) / Hall M S

Geology of the Southeastern Termination of
the Cordillera Central, South-Central
Hispaniola, Greater Antilles
(1988) / Heubeck C E

The Role of Bacteria in the Deposition and
Early Diagenesis of the Posidonienschiefer,
a Jurassic Oil Shale in Southern Germany
(1988) / Hiebert F K

TEXAS, UNIVERSITY OF (AUSTIN)
(continued)

Interpretation of Seismic Signal and Noise
Through Line Intersection Mis-ties Analysis
(1987) / Huston D C

Reservoir Characterization for Numerical
Simulation of Mesaverde Meanderbelt
Sandstone, Northwestern Colorado
(1988) / Jones J R

Hydrogeologic Controls on Underflow in
Alluvial Valleys - Implications for Texas
Water Law
(1988) / Larkin R G

Late-Stage Alleghanian Wrenching of the
Narragansett Basin, Rhode Island
(1988) / Mahler J P

Tectonic History and New Isochron Chart of
the South Pacific
(1988) / Mayes C L

Geologic Evolution of the Northern
Newfoundland Basin
(1988) / Meador K J

Multistage Dolomitization of the Portoro
Limestone, Liguria, Italy
(1988) / Miller J K

Facies Relations and Controls on Artesia
Group Deposition in the Matador Arch Area,
Texas
(1988) / Nance H S

The Morphology, Depositional Setting, and
Evolution of the Paleocene Lavaca Submarine
Canyon System, Northwest Gulf of Mexico
(1988) / Paige R E

Deposition and Diagenesis of a Late
Guadalupian Barrier-Island Complex from the
Middle and Upper Tansill Formation
(Permian), East Dark Canyon, Guadalupe
Mountains, New Mexico
(1988) / Parsley M J

Temperature Variations and Their Relation to
Groundwater Flow, South Texas, Gulf Coast
Basin
(1988) / Pfeiffer D S

Stratigraphy of the Tertiary of the Middle
Magdalena Basin (Columbiia), Central and
Northern Parts
(1988) / Ramirez-Serafinoff R E

Late Quaternary Sedimentology and Evolution
of the Continental Slope (Long. 94
Degrees-95 Degrees), Northwest Gulf of
Mexico
(1988) / Satterfield W M

Sandstone and Shale Diagenesis of the Frio
Formation (Oligocene), Texas Gulf Coast: A
Close Look at Sandstone/Shale Contacts
(1988) / Sullivan K B

Chronostratigraphy, Depositional Rates,
Continental Margin Progradation, and Growth
Fault Dynamics Within the Tertiary Wedge,
San Marcos Arch, Northwest Gulf of Mexico
(1988) / Travis D S

Seismic Stratigraphy and Tectonic Evolution
of the Stord Basin, Northern North Sea
(1988) / Weatherill P M

Development of a Structural Framework from
Seismic Reflection Data
(1988) / Wood B L

Deposition and Diagenesis of the Jurassic
Smackover Formation, Hatter's Pond Field,
Southwest Alabama
(1988) / Worrall J G

Source Process Study with the Inclusion of
the Effects of Near-Source Bathymetric
Structure for Submarine Events in the Gulf
of California
(1988) / Zemlicka G

Structure and Stratigraphy of the Central
Cordillera Septentrional, Dominican Republic
(1988) / de Zoeten R

TEXAS, UNIVERSITY OF (EL PASO)

Modeling the Feasibility of Strontium Remobilization
from Geothite Surfaces in a Saturated Tuff at Yucca
Mountain, NYE County, Nevada
(1988) / Dicke C A

Geology and Genesis of the Palm Park and Horseshoe
Barite Deposites Southern Caballo Mountains, dona
Ana County, New Mexico
(1988) / Filsinger B

Transport and Fate of Organic Chemicals in Bandelier
Tuff at Los Alamos National Laboratory Chemical
Waster Site
(1988) / Fisher M A

The Depositional Environments of the Late Permian
Rustler Formation, in the Vicinity of the Waste
Isolation Pilot Plant (WIPP) Site, Southeastern
New Mexico
(1988) / Holt R M

Experimental Determination of Reactive Tracer
Suitability for Groundwater Tracer Tests
(1988) / Newman B D

Genesis of the Vein Deposits Along the Great Master
Lode, Phillipsburg Area, Northern Black Range,
Sierra and Catron Counties, New Mexico
(1988) / Pearson J W

The Dewey Lake Formation: End Stage Deposit of a
Peripheral Foreland Basin
(1988) / Schiel K A

Geology of Northern Sierra De Palomas Chihuahua,
Mexico
(1988) / Sivils D J

TOLEDO, UNIVERSITY OF

Behavior of Introduced Solutes and
Evaluation of a Solute Transport Model for
a Low-Gradient Sand Bed Stream in Northwest
Ohio
(1988) / Ellison A B

Hydrogeology of the Bellevue-Castalia Area,
North-Central Ohio, with an Emphasis on
Seneca Caverns
(1988) / Kihn G E

WASHINGTON, UNIVERSITY OF

Subglacial Conditions Inferred from Deposits
of the Puget Lobe, Washington: Till
Characteristics, Magnetic Data and
Calculalted Basal Drag
(1987) / Brown N E

Tectonic Evolution of the Lake Van Area,
Relation of the North Anatolia and Zagros
Deformation Belts: Application of the Large
Format Camera (LFC) Imagery
(1987) / Bushara M N

Setting and Origin of Exhalative Bedded
Barite and Associated Rocks of the Roberts
Mountains Allochthon in North-Central Nevada
(1987) / Dube T E

Carbon and Oxygen Isotopic Composition of
Holocene Lake Sediments from Okanogan
County, Washington
(1987) / Forbes J

Glacial Stratigraphy and Landscape Evolution
of the North-Central Puget Lowland,
Washington
(1987) / Haase P C

Morphometric Analysis of Pleistocene Glacial
Deposits in the Kigluaik Mountains,
Northwestern Alaska
(1987) / Kaufman D S

Two-Phase Separation and Fracturing in
Mid-Ocean Ridge Gabbros at Temperatures
Greater than 700 Degrees C
(1987) / Kelley D S

Structure and Hydrothermal Alteration
Associated with Epithermal Gold-Silver
Mineralization, Wenatchee Heights,
Washington
(1987) / Margolis J

Deformation Near Lillooet, British Columbia:
Its Bearing on the Slip History of the
Yalakom Fault
(1987) / Miller M

WINDSOR, UNIVERSITY OF

Behaviour of Five Organic Pollutants in
Aqueous Environments
(1988) / Chodola G R

Deterioration of Carbonate Building Stones
Due to Acid Solutions
(1988) / Larbi E Y

WRIGHT STATE UNIVERSITY

Delineation of the Bedrock Topography and
Potentialo Aquifers by Gravity and Seismic
Refraction Techniques
(1987) / Acharya A

Analysis and Modeling of Gravity Anomalies
Related to the Knox Unconformity
(1987) / Armstrong W M

A Stream-Sediment Reconnaissance Survey of
the Northern Snowcrest Range, Madison
County, Montana
(1987) / Burkhart P A

Sedimentary Environment and Provenance of
the Snowbird Group, OCOEE Supergroup in
Tennessee and North Carolina
(1987) / Cooper B B

A Gravity Survey of Darke County, Ohio
(1987) / Daugherty C M

Chemical Characteristics of Ground Water in
a Portion of Greene County, Ohio
(1987) / Fenno D P

An Analysis of the Cedar Bog Hydrologic
System Through the Use of a
Three-Dimensional Groundwater Flow Model
(1987) / Hillman D L

A Hydrogeochemical Investigation of
Inorganic Constituents in Surface Water and
Ground Water in the Vicinity of a Disposal
Site in Clark County, Ohio
(1987) / Hoose L

Olistoliths in the Yellow Breeches Member of
the Wilhite Formation and Their Significance
in Determining the Tectonostratigraphic
Position of the OCOEE Supergroup
(1987) / LePain D L

A Shallow Seismic Reflection Study of the
Mad River Buried Valley at Huffman Dam, Ohio
(1987) / Middlebrooks P K

Geochemical Analysis of Cold Creek and
Blayne Spring, Ennis, Montana
(1987) / Monks K S

Depositional and Diagenetic History of a
Devonian Coral and Stromatoporoid Biostrome,
Falls of the Ohio River, Louisville,
Kentucky
(1987) / Paguette D E

Utilization of Strontium Ratios to Trace
the Groundwater Flowpath Into Cedar Bog
Memorial Swamp, Champaign County, Ohio
(1987) / Price S G

An Analysis of the Randomness of Soil
Compactors in the Mini-Sosie Seismic System
(1987) / Ratliff T W

A Hyudrogeochemical Profile Analysis of
Groundwater Discharge Zones Within Cedar
Bog, Champaign County, Ohio
(1987) / Ricketts B M

A Relationship Between Gravity and the
Erosional Highs on the Know Unconformity in
Morrow County, Ohio
(1987) / Sweazy C L

Data Processing of a Seismic Line in an Area
of Thick Glacial Till Using Surface
Consistent Automatic Statics: Osceola
County, Michigan
(1987) / Wittoesch D F

Subsurface Structure of the Know, Chazy, and
Trenton Formations in Morrow County, Ohio
as Determined from Well Log Data
(1987) / Zuberi S A

ACADIA UNIVERSITY

Volcanism and Geochemistry of Parts of the
Endeavour Segment of the Juan de Fuca Ridge
System and Associated Seamounts
(1988) / Leybourne M

Geology of the Sporting Mountain Area,
Southeastern Cape Breton Island, Nova Scotia
(1988) / Sexton A J

The Depositional Environment of the Late
Carboniferous, Coal-Bearing Upper Thorburn
Member of the Stellarton Group, Pictou
Coalfield, New Glasgow, Nova Scotia
(1988) / Snow R

AKRON, UNIVERSITY OF

Dolomitization, Diagenesis and
Paleoenvirohnments of a Middle to Upper
Silurian Shallowing Upward Sequence Exposed
Near Peebles, Ohio
(1988) / Blauch M E

The Sedimentology and Stratigraphy of a
Transgressive Barrier at Sheldon's Marsh
State Nature Preserve, Erie County, Ohio
(1988) / Bray T F

The Depositional Environments and Early
Diagenetic Controls on the Mineralogy of
the Kittanning Members, Allegheny Group,
Eastern Ohio
(1988) / Brocculeri T

The Hydrology of Eastern Franklin and
Western Townships, Summit County, Ohio
(1988) / Garvey J T

Computer-Aided Mapping and Statistical
Analysis of the East Canton Oilfield, Rose
Township, Carroll County Northeastern Ohio
(1988) / Green S R

An Analysis of 222Rn Soil Gas Concentrations
in the Serpent Mound Area, Southwestern Ohio
(1988) / Heirendt K M

A Gravity Survey of a Portion of the
Cuyahoga River Valley Near Brecksville, Ohio
(1989) / Hennessy T L

Dolomitization and Textural Analysis of the
Lower Middle Devonian Carbonates of
North-Central Ohio at Sandusky Crushed Stone
Parkertown Quarry, Erie County, Ohio
(1989) / Kendall R L

Terminoglacial/Proglacial Depositional
Environments in Late Pleistocene Sediments
of Erie County, Pennsylvania
(1989) / Rokovich R A

ALABAMA, UNIVERSITY OF (UNIVERSITY)

A Kinematic Model for Interpreting the
Structural Evolution of the Citronelle Dome
(1988) / Cottingham J P

Petrology and Geochemistry of a Portion of
the Rockford Granite, Coosa County, Alabama
(1988) / Gomolka J J

Petroleum Geology and Geochemistry of
South-Central Escambia County, Alabama
(1988) / Hall D R

Stratigraphy and Depositional Environments
of the Mississippian-Pennsylvanian Parkwood
Formation on a Portion of the Northwest Limb
of the Cahaba Synclinorium, Jefferson and
Bibb Counties, Alabama
(1988) / Leverett D E

Depositional and Diagenetic Characteristics
of the Smackover Formation in the
Mississippi Interior Salt Basin of Southwest
Alabama
(1988) / Moss N E

Provenance and Depositional Environment of
the Straven Conglomerate, Pottsville
Formation, Cahaba Synclinorium, Central
Alabama
(1988) / Osborne T E

The Post-Knox Unconformity and Its
Relationship to Bounding Stratigraphy,
Alabama Appalachians
(1988) / Roberson K E

Factors Influencing the External Morphology
of Stromatoporoids in the Upper Silurian
(Pridoli) of Central New York
(1988) / Williams M S

ALBERTA, UNIVERSITY OF

Sedimentation and Diagenesis of the Upper
Devonian Kakisa Formatiohn, Trout River
Area, North West Territory
(1988) / Fyvie D J

Late Glacial Geology of the Finlay River
Valley, British Columbia
(1988) / Leslie L E

Expert Systems Applications in Hydrogeology
(1988) / McClymont G L

The Late Quaternary Lacustrine Record from
the Upper Cataract Brook Valley, Yoho
National Park, British Columbia
(1988) / Reasoner M A

Fluid Inclusion and 8[18] 0/Study of the
Precious Metal-Bearing Veins of the Wheaton
River District, Yukon
(1988) / Rucker P D

An Electrochemical Study at 25 Degrees C of
the Leaching Behaviour of Gold and Silver
in Inorganic and Organic Solutions
(1988) / Sayer C J

Surface Karst on Grand Cayman Island,
British West Indies
(1988) / Squair C A

ARIZONA STATE UNIVERSITY

The Development and Distribution of Surface
Lava Textures at the Mount St. Helens Dome
(1988) / Anderson S W

Diagenetic Changes in Texas Gulf Coast
Sediments
(1988) / Brady S C

Theoretical Energetics of the Formation of
Iron Nickel Alloy and Magnetite in Olivine
(1988) / Doorn S S

ARIZONA STATE UNIVERSITY
(continued)

The Silicate Inclusions of Group IAB Iron
Meteorites: Implications for Metal-Silicate
Segregation and Core Formation
(1988) / Herpfer M A

The Effect of Fine-Grained Minerals on the
Coulomb-Parameter Values of the Debris-Flow
Fluid-Phase
(1988) / Kafura C J

Influence of Lithology on Alluvial Fan
Morphometry, White and Inyo Mountains,
California and Nevada
(1988) / Lecce S A

A Stable Isotopic Investigation of the
Mazatzal Peak Quartzite: Implications for
Fource Terranes
(1988) / Neet K E

Control of Random, Two-Dimensional
Disturbances in a Boundary Layer
(1988) / Pupator P T

Modeling Volcano Morphology
(1988) / Schuver H J

ARKANSAS, UNIVERSITY OF

The Effect of Agriculture on the Quality of
Ground Water in a Karstified Carbonate
Terrain, Northwest Arkansas
(1988) / Adamski J C

Structural Geology of the Enola, Arkansas
Earthquake Swarm
(1988) / Burroughs R K

Geology of the Cass and Yale 7.5 Minute
Quadrangles of Franklin and Johnson
Counties, Arkansas
(1988) / Duran W K

Petrography and Environmental Facies Within
Dolomitized Ordovician Strate, Northwest
Arkansas
(1988) / Holland K D

Data Preparation and Analysis Techniques
Used in Modeling the Sparta Aquifer in
Eastern Arkansas
(1988) / Kilpatrick J M

A Hydrogeological and Geochemical Site
Characterization of a Proposed Landfill
Site, Searcy County, Arkansas
(1988) / Kresse T M

Evaluating the Significance of
Stream-Aquifer Interaction in the
Mississippi Alluvial Plain of Eastern
Arkansas Using Hydrograph Analysis and a
Numerical Ground-Water Flow Method
(1988) / Plafcan M

The Effects of Rainstorm Events on the Water
Chemistry of a River with Emphasis on the
Heavy Metal Content
(1988) / Wickliff D S

Nonfan Systems: An Alternative to Submarine
Fan Systems for Marginal Sedimentation,
Upper Jackfork Group (Pennsylvania) of
Arkansas
(1988) / Zimmerman K A

AUBURN UNIVERSITY

The Geology of Base and Precious Metal-Bearing Quartz
Veins in Hall and Gwinnett Counties, Georgia
(1987) / Allen N E

Stratigraphic Analysis and Depositional Environments
of the Smackover Formation in Concecuh Basin,
Escambia County, Southwestern Alabama
(1987) / Esposito R A

Carbonate Lithofacies and Diagenesis of the Upper
Part of the Knox Group (Lower Ordovician) and
Overlying Basal Lenoir Limestone (Middle Ordovician),
Appalachian Fold and Thrust Belt, Central Alabama
(1987) / Pruneau J C

Sedimentary Facies, Depositional Environments, and
Sea-Level Cycles of the Upper Cretaceous Mooreville
Chalk (Campanian), West-Central Alabama
(1987) / Wylie J A

BALL STATE UNIVERSITY

The Use of Soil Characterization Information in
the Correlation of Wisconsinan-Age Glacial Drift
in Randolph County, Indiana
(1988) / Anderson N P

BAYLOR UNIVERSITY

The Effects of Urbanization on Stream Flow,
Sediment Transport, and Sediment Deposition
in the White Rock Prairie, Central Texas
(1988) / Allen E A

A Hydrogeologic Assessment of the Ozan
Formation, Central Texas
(1988) / Barrett D P

Relationship Between Basin Parameters and
Landform Configuration, Lampasas Cut
Plain, Central Texas
(1988) / Brown T E

Slope Failure Along the Urbanizing White
Rock Escarpment, Central Texas
(1988) / Campbell K S

Hydrogeology of a Portion of the Washita
Prarie Edwards Aquifer: Central Texas
(1988) / Cannata S L

A Geological and Geophysical Assessment of
the Hammond Field Area, Zavala County, Texas
(1988) / Carrillo V G

Hydrogeology of a Taylor Marl Flow System
Backland Prairies, Central Texas
(1988) / Guillette B

Structural Analysis of the South Flank of
the Sweetwater Uplift Carbon County, Wyoming
(1988) / Homan K S

Regional Stratigraphy of the Paleocene
Midway Group, East Texas Basin
(1988) / Oldlani M

Geology and Flow Systems of the Hickory
Aquifer in the San Saba County Area, Texas
(1988) / Pettigrew R J

Aspects of Leon River Drainage History with
Implications to Other Central Texas Streams
(1988) / Tharp T L

BOSTON COLLEGE

Crustal Modeling in Maine Through the
Simultaneous Inversion of Gravity and
Magnetic Data
(1988) / Coblentz D D

A Three-Component Seismic Refraction Survey
in Northwestern Arizona
(1988) / Hussey V K

Stratigraphic, Structural and Magnetic
Investigations of the Mare Hill Area,
Aroostook County, Maine
(1988) / Lacombe P J

Crustal Structure in Maine as Determined
from Nodeling Teleseismic P Waveforms
(1988) / Lory R E

BOWLING GREEN STATE UNIVERSITY

Geologic Study of the Volcanic Units in the
Bald Mountain-Castle Rock Region, Chaffee
County, Colorado
(1989) / Bade C

Automatic Identification of Digitally
Recorded Seismic Events at the Bowling Green
State University Seismic Observatory (BGO)
(1989) / Benko T S

An Isoseismal Study of Northwest Ohio and
Southeast Michigan Based on Data from the
January 31, 1986 Geauga County, Ohio
Earthquake
(1988) / Bogner K A

Petrography and Paragenesis of the Siderite
Orebody, Macleod Mine, Wawa, Ontario
(1989) / Cohen M H

Illite Crystallinity of the Middle and Upper
Ordovician Allochthonous and Autochthonous
Rocks in the Great Valley of Central
Pennsylvania
(1989) / Epstein R L

Cation Chemistry of Detrital Garnets as a
Provenance Indicator in Beach Sands from
Southern California
(1989) / Graves A E

Precious Metal Geochemistry of the Helen
Siderite, Macleod Mine, Wawa, Ontario
(1989) / Harding T A

Origin and Significance of Slickensides in
the Upper Devonian Hampshire Formation,
Central Maryland
(1989) / Hodkiewicz P

Petrography and Chemistry of Several
Igneous Bodies Along the Northeastern
Boundary of the Wolf River Batholith,
Wisconsin
(1989) / Holzel F R

Biochronology, Based Upon Gastropoda, of
the Pinecrest Beds of the Tamiami Formation
at Sarasota, Florida
(1989) / Larsen J B

An Integrated Investigation of the Bowling
Green Fault Using Multispectral Reflectance,
Potential Field, Seismic, and Well Log Data
Sets
(1988) / VanWagner E

Diagenetic Changes and Trends in an Upper
Ordovician Shallowing-Upward Sequence
(1989) / Zuiderhoek T E

BRIGHAM YOUNG UNIVERSITY

The Distribution of Iron in Staurolite at
Room and Liquid Nitrogen Temperatures
(1988) / Alexander V D

Silicified Wood of Thuja, Robinia and
Quercus from Miocene Rocks Near Wall
Canyon Creek, Northwestern Nevada
(1988) / Call V B

Analysis of the Grand Wash-Reef
Reservoir-Gunlock Fault Zone, Washington
County, Utah, and Mohave County, Arizona
(1988) / Hammond B J

Geology of the Jump Creek 7 1/2 Inch
Quadrangle, Carbon County, Utah
(1988) / Hansen C D

Geology and Tectonics of the Jungapeo,
Michoacan, Mexico Area and Its Relationship
to the Mexican Volcanic Belt
(1988) / Petersen M D

BROOKLYN COLLEGE

Paleo-Depth of Burial of Surface-Exposed
Paleozoic Carbonates in Arbuckle
Mountains, Oklahoma
(1988) / Glash S J

BROWN UNIVERSITY

Volcanic and Tectonic Evolution of Large
Impact Basins on Mars
(1988) / Wichman R W

CALGARY, UNIVERSITY OF

Reflection Seismic Study of a Shallow Coal
Field in Central Alberta
(1988) / Lyatsky H

Depositional Systems of the Mississippian
MC-3 (Alida Beds), Pierson Field, Manitoba
(1988) / Olsen-Heise K E

CALIFORNIA STATE UNIV. (FRESNO)

Carbonate Petrology and Paleoenvironment of
the Riepe Spring Limestone, East Central
Nevada
(1987) / Poole T C

CALIFORNIA, UNIVERSITY OF (DAVIS)

Hydrothermal Alteration at the Pacific Mine,
Placerville, El Dorado County, California
(1988) / Considine K A

Secular Variation from Lacustrine Records
from Western North America: 0-40,000
Years BP
(1988) / Hanna R L

Thermobarometry Across the Aluminosilicate
Triple-Point Bathograd of Southwestern
New Hampshire
(1988) / Hartl P D

Geology of the Mountain Pass Area, San
Bernardino County, California
(1988) / Nance M A

Paleoenvironmental Analysis of the Upper
Cretaceous (Coniacian-Campanian) Chico
Formation, Northeastern Sacramento Valley,
California
(1988) / Russell J S

CARLETON UNIVERSITY

The Permafrost Regime in the MacKenzie
Delta-Beaufort Sea Region, North West
Territory and Its Paleoclimatic Implications
(1988) / Allen D M

Stratigraphy and Alteration of Gabbroic
Rocks Near the San Antonio Gold Mine in the
Rice Lake Area, Southeastern Manitoba
(1988) / Ames D E

Sedimentology and Stratigraphy of the
Neohelikian Elwin Formation, Uppermost
Bylot Supergroup, Borden Rift Basin,
Northern Baffin Island
(1988) / Knight R D

Stratigraphy, Structure and Volcanogenic
Sulphides in Sicker Group Volcanic Rocks
on Mt. Sicker, Vancouver Island, British
Columbia
(1988) / MacRobbie P A

Geology of the Ellington Volcanic
Sedimentary Complex, Great Bear Magmatic
Zone, Northwest Territories
(1988) / Pelletier K S

CITY COLLEGE OF NEW YORK

Geohistory Analysis of the Colville Basin:
North Slope, Alaska
(1988) / Gullett C D

COLORADO SCHOOL OF MINES

An Assessment of the Long-Term Hydrologic
Effects of Artificial Recharge on the
Denver Groundwater Basin Using Computer
Simulation Methods
(1988) / Aikin A R

A Sensitivity Analysis of Factors that
Determine Rock Slope Stability at the
Lariat Clay Pit South-Central Jefferson
County, Colorado
(1988) / Anthony J W

Investigation of the Corrosion
Characteristics of Aluminum-Lithium and
Aluminum-Lithium-Copper Weldments in a 3.5
Percent Sodium Chloride Solution
(1988) / Beverini G

An Integrated Geological and Geophysical
Study of the Structural Boundary Between
the San Luis Basin and the San Juan Sag,
South-Central Colorado
(1988) / Covarrubias O

Trace Element and Rare Earth Element
Variation in Fluorites Collected from Skarn
and Epithermal Mineral Deposits in the
Sierra Cuchillo Area, South-Central New
Mexico
(1988) / Eppinger R G

Geology, Neotectonics and Geological Hazards
of the Mount Rose 7.5 Minute Quadrangle,
Northern Tahoe Basin, Nevada
(1988) / Lewis R L

Origin of the Dolomite-Anhydrite Intervals
of the Upper Minnelusa Formation, Powder
River Basin, Northeastern Wyoming
(1988) / Miller K A

Laramide and Post-Laramide Deformation of
the Northern Mosquito Range Between Fremont
and Hoosier Passes, Eagle, Lake, Park, and
Summit Counties, Colorado
(1988) / Oppenheimer W L

Geology and the Hydrocarbon Fairway of the
Codell Sandstone Member of the Late
Cretaceous Carlile Formation, Denver Basin,
Colorado
(1988) / Panigoro H

Stratigraphy and Palynology of the Tullock
Formation of the Fort Union Group
(Paleocene), McCone County, Montana
(1988) / Serenko T J

Field Studies and Modeling Analysis of the
Roan Creek Landslide, Garfield County,
Colorado
(1988) / Umstot D

A Modeling Methodology for Performance
Assessment and Design Optimization for
Engineered Covers at Tailings Disposal
Sites
(1988) / Wright W E

COLORADO STATE UNIVERSITY

Landslide Evaluation, Green Mountain
Reservoir, Colorado
(1988) / Anderson S A

Llama Use on Public Lands
(1988) / Arndt C A

Effects of Aspen Cutting on Soil Moisture
(1988) / Clark G M

Evaluation of a 3-D Diagnostic Wind Model: Nuatmos
(1988) / Connell B H

Surficial Geology and Quaternary History of the
Central Plains Experimental Range, Colorado
(1988) / Davidson J M

Alteration and Paragenesis of the Paradise
Peak Gold/Silver Deposit
(1988) / Dobak P J

Geology of the Taylor Silver Deposit,
White Pine County, Nevada
(1988) / Edwards J M

Decreasing Sediment and Salt Loads in the Colorado
River Basin: A Response to Arroyo Evolution
(1988) / Gellis A C

Microwave Measurement of Liquid Water in Snow
(1988) / George D M

Geology and Gold Potential East of the Vulcan
Mine, Gunnison and Saguache Counties, Colorado
(1988) / Hunsaker E L

Inorganic Nitrates in the South Platte River Basin,
Colorado: Adams, Larimer and Weld Counties
(1988) / Hussain N U

Hydrogeology of Fractured Basement Complexes - A
Case Study from the Kurunegala District of Sri Lanka
(1988) / Jayasena H A

COLORADO STATE UNIVERSITY
(continued)

An Engineering Geology Site Investigation of the
Proposed Poudre Dam Site Larimer County, Colorado
(1988) / Kaplin J L

Geology of the Telegraph Mine Tectono-Hydrothermal
Breccias, San Bernardino Co., California
(1988) / Lange P C

The Effects of Storm Trajectory on Precipitation
Chemistry in Rocky Mountain National Park
(1988) / McLaughlin P

Discrimination of Rock Units in North-Central
Colorado by Cluster Analysis of Airborne Gamma
Spectrometry
(1988) / Moll S H

Economic Geology of Part of the Gold Brick
District, Gunnison County, Colorado
(1988) / Neff L M

The Planimetric Patterns of Experimental Tension
Fractures and Their Geomorphic Significance to Mars
(1988) / Pranger H S

Migmatites and Mafic Boudins of the Gannett Peak
Area, Sublette and Fremont Counties, Wyoming
(1988) / Seanor C E

The Effect of Climate on Drainage Basin
Characteristics in Central Texas
(1988) / Sortman V L

Geology and Geochemistry of the Smuggler Mine
Silver-Lead-Zinc-Copper-Barium Manto Deposits
Aspen, Colorado
(1988) / Stegen R J

Structural Geology and Tectonic Setting of the
Cherry Creek Metamorphic Suite, Southern
Madison Range
(1988) / Sumner W R

Sedimentology and Architecture of Gilbert- and
Mouth Bar-Type Fan Deltas, Paradox Basin, Colorado
(1988) / Wood M L

DALHOUSIE UNIVERSITY

Clay Mineral Distribution in Mesozoic and
Cenozoic Strata of the Labrador Shelf
(1988) / Douma M

The Study of the Glacial Stratigraphy and
Sedimentation of the Sheldonn Point Moraine,
Saint John, New Brunswick
(1988) / Nicks L P

The Geology and Geochemistry of the Sangster
Lake and Larrys River Plutons, Guysborough
County, Nova Scotia
(1988) / O'Reilly G A

The Antigonish Basin of Maritime Canada: A
Sedimentary Tectonic History of a Late
Paleozoic Fault-Wedge Basin
(1988) / Prime G A

Geochemical Discrimination of the
Peraluminous Devonian-Carboriferous
Granitoids of Nova Scotia and Morocco
(1988) / Richard L R

DELAWARE, UNIVERSITY OF

Change in the Diatom Assemblage of Rehoboth
Bay, Delaware and Environmental Implications
(1987) / Beasley E L

The Taxonomy and Paleoecology of the Fresh
Water Ostracoda of Pickerel Lake, South
Dakota
(1987) / Smith A J

DUKE UNIVERSITY

Relationship Between Physical Condition of
the Carbonate Fraction and Sediment
Environments: Northern Shelf of Puerto Rico
(1988) / Benes P S

Clastic Dispositional Processes in Response
to Rift Tectonics in the Malawi Rift, Africa
(1988) / Bishop M G

The Acoustic Stratigraphy of Lake Malawi,
East Africa
(1988) / Flannery J W

An Analysis of Replenished Beach Design on
the U.S. East Coast
(1988) / Leonard L A

Storm Sediment Transport as Indicated by
Benthic Foraminifera, North Insular Shelf,
Puerto Rico
(1988) / Lightner J T

A Petrographic and Geochemical Study of the
Metavolcanics from the Proterozoic Bou Azzer
Ophiolite Complex, Morocco
(1988) / Naidoo D D

Sedimentation Off Deltas and Border Faults
in Northern Lake Malawi: Evidence from
High-Resolution Acoustic Remote Sensing and
Gravity Cores
(1988) / Ng'ang'a P

Tectonics, Structure, and Sedimentation of
the Lake Victoria Basin, East Africa
(1988) / Rach N M

Modern Processes in a Contentinal Rift
Lake: An Interpretation of 28 kHz Seismic
Profiles from Lake Malawi, East Africa
(1988) / Scott D L

Basement Controls on the Architecture of the
Lake Malawi Rift Zone, East Africa
(1988) / Versfelt J W

Extensional Faulting in the Mark Area
(1988) / Winters A T

EAST CAROLINA UNIVERSITY

Paleoecological Interpretation of a Late
Pleistocene Marine Fossil Bed in Eastern
North Carolina
(1987) / Auch T W

Trace Element and Stable Isotopic Trends
Associated with Diagenesis of Selcted
Benthic Foraminifers from Miocene Sediments
(1988) / Jones W E

Diagenesis of Benthic Foraminifera in the
Miocene Pungo River Formation of Onslow
Bay, North Carolina Continental Shelf
(1987) / Moretz L C

EASTERN KENTUCKY UNIVERSITY

An Investigation Into the Movement of an
Agricultural Pesticide Within the
Groundwater System of a Karst Swallet
(1988) / Devilbiss T S

Depositional Environments of a Chesterian
(Upper Mississippian) Sandstone Body in
Breckinridge County, Kentucky
(1988) / Ross B C

Borehole Television Logging: A New Tool for
Downhole Evaluation in the Devonian Shale
(1988) / Wilson R W

EMORY UNIVERSITY

A Field Study of Flow Through Fractured
Shale and Its Comparison to Two
Mathematical Models
(1988) / Diprima L J

Paleoecology of a Floodplain Lake in the
Durham Sub-Basin of the Deep River Basin
(Late Triassic), North Carolina
(1988) / Renwick P L

EMPORIA STATE UNIVERSITY

Small-Pebble and Heavy-Mineral Composition
of Glacial Deposits in Northeastern Kansas
(1988) / Nutter B L

FLORIDA, UNIVERSITY OF

The Distribution of Radioactivity in the
Surficial Deposits of Levy, Marion, and
Citrus Counties, Florida
(1988) / Abbott T

The Hydrogeologic System of Trail Ridge
Above the Basal Clays Near Folkston, Georgia
(1988) / Burklew R

Distribution of Magnesium Oxide,
Polygorskite and Other Minerals in a North
Florida Phosphorite
(1988) / Dufresne D

The Distribution of Gamma Radiation in the
Surficial Deposits of the Florida Panhandle
(1988) / Hansen J

The Carbon Cycling in an Anoxic Lake Basin,
Johnson Pond, North Central Florida, Using
Carbon Isotopes as a Tracer
(1988) / Japy K

The Differentiation and Classification of a
Sand Underlying Typical Trail Ridge Sands
Near Screven, Wayne County, Georgia
(1988) / Mallard E

A Bouguer Gravity Survey and Computer Model
of North Florida and South Georgia
(1988) / McNeely P

Heat Flow and Heat Production in the Ozark
Plateau Region, Missouri
(1988) / Meert J G

Temperature Variation in the Upper Part of
the Floridan Aquifer, Alachua County,
Florida
(1988) / Mickle A

Characterization of Carbonate Rocks from the
Saline Zone of the Floridan Aquifer,
Pinellas County, Florida
(1988) / Quinn H

Correlation Between Uranium Levels,
Phosphate, and Biostratigraphy in the
Central Florida Phosphate Mining District
(1988) / Schult M F

Geochemical Investigation of Some Archean
Banded Iron Formations from Southwest
Montana
(1988) / Stanley D R

Contamination Potentials of Three Aquifer
Systems in a Microfractured Karst Terrain,
Alachua County, Florida
(1988) / Stransky R

Postranial Skelton of Trachytherus
(Mammalia, Notoungulate) with an Evaluation
of Dentition
(1988) / Sydow H

Late Pliocene-Early Pleistocene Water Mass
and Dissolution Patterns in the High
Latitudes of the Southeast Indian Ocean
(1988) / Turner J

FORT HAYS KANSAS STATE UNIVERSITY

Taphonomic Evidence for Predation and
Scavenging of Teleoceras, with a Description
of the Camelidae from the Minimum Quarry
Local Biota of North-Central Kansas
(1988) / LaGarry H E

Stratigraphy, Paleontology, and
Paleobiogeography of Lower Vertebrates
from the Cedar Mountain Formation (Lower
Cretaceous), Emery County, Utah
(1988) / Pomes M

Geology of the Stockade Mountain 15'
Quadrangle, Malheur and Harvey Counties,
Oregon
(1988) / Stimac J P

Geology of the Fairport Quadrangle, Ellis
and Russell Counties, Kansas
(1988) / Thompson N W

Stratigraphy and Depositional Environments
of the Middle Jurassic (Callovian) Ralston
Creek Formation, Beulah-Wetmore Area,
South-Central Colorado
(1988) / de Albuquerque J S

GEORGIA, UNIVERSITY OF

Ground Truth Evaluation of the Continuous
Surficial Sediment Sampling System for
Marine Pollution Assessment
(1988) / Culp R A

Gold in Placer and Saprolite Deposits
(1988) / Durrett J F

Paleomagnetic Analysis of Recent Abyssal
Sediments from the Northeast Argentine Basin
(1988) / Elrod M M

Timing and Extent of Caledonian Orogeny
Within a Portion of the Western Gneiss
Terrane, Senja, Northern Norway
(1988) / Hames W E

GEORGIA, UNIVERSITY OF
(continued)

Sematic Impurities in Some Commercial
Georgia Kaolins
(1988) / Jones T

Petrographic and Geochemical Relations
Between the Rocks on the North and South
Limbs of the Chibougamau Anticline:
Assimilation of Roof Rocks, Crystallization
and Residual Liquid Compositions in the
Dore Lake Complex, Quebec, Canada
(1988) / Lapallo C M

Foraminiferal Biostratigraphy of a Portion
of the Chesterian Series (Upper
Mississippian) in Cores from Blount and
Chandler Mountains, Alabama
(1988) / Millar W W

Flint Kaolins in the Georgia Coastal Plain
(1988) / Moskow M G

Ichnology, Depositional Environment, and
Paleoecology of the Upper Pennington
Formation (Upper Mississippian), Dougherty
Gap, Walker County Georgia
(1988) / Sheehan M A

Electrical Resistivity Characteristics of
Water-Bearing Fractures in the North Georgia
Piedmont
(1988) / Steflik M

Ilmenite Alteration and Heavy-Mineral Suites
of Lynches River, South Carolina Piedmont
and Coastal Plain
(1988) / Trein E A

Electromagnetic Investigation of Groundwater
in the Georgia Piedmont: Applications to
Water Well Siting and Waste Site Leakage
(1988) / Tuck F C

Surface Geophysics and Contaminant Transport
at a Hazardous Waste Landfill, Clarke
County, Georgia
(1988) / Zungailia E J

IDAHO STATE UNIVERSITY

Investigations Into Fractured Granitic Rock
by Detailed Geologic and Geophysical Methods
(1988) / Beem L I

Stratigraphy of the Wood River Formation in
the Eastern Boulder Mountains, Blaine and
Custer Counties, South-Central Idaho
(1988) / Burton B R

Kinematic History of a Ductile Shear Zone in
the Wind River Mountains, Sublette County,
Wyoming
(1988) / Dvoracek D K

Geology and Petrology of Archean Migmatites
and Gneisses, Mount Baldy Area, Horseshoe
Lake Quadrangle, Wind River Mountains,
Wyoming
(1988) / Reubelmann K L

Neogene--Quaternary Basin--Fill History of
the Pahsimeroi Valley, Idaho
(1988) / Ungate C A

An Ecological and Sedimentological
Evaluation of a Relocated Cold Desert Stream
(1988) / Vinson M R

Stratigraphy and Depositional Environments
of the Brigham Group, Cub River Area,
Franklin County, Idaho
(1987) / Zahn P D

IDAHO, UNIVERSITY OF

The Pre-Tertiary Geology of the Lower Salmon
River Canyon Between Packers Creek and Billy
Creek, West-Central Idaho
(1988) / Borovicka T G

Reconnaissance Glacila Geology of the Hoboe
Valley, Atlin Provincial Wilderness Park,
British Columbia, Canada
(1988) / Campbell R B

Coastal Geomorphology of Puget Bay, Alaska
(1987) / Chaney G

Geochemical Exploration in a Geothermal
Area Using Sulfur Gases in Near-Surface
Soils
(1988) / Dolenc M R

Paleontology, Stratigraphy, and Depositional
Environment of the Zorritas Formation,
Sierra de Almeida, Northern Chile
(1988) / Fisher L L

The Quaternary System of the Eastern Spokane
Valley and the Lower, Northern Slopes of the
Mica Peak Uplands, Eastern Washington and
Northern Idaho
(1988) / McKiness J P

Stratigraphy and Carbonate Petrology of the
Lower Ordovician Garden City Formation,
Southeast Idaho and Northern Utah
(1988) / Parsley G P

Geology, Petrology, and Alteration of the
Sorrel Spring Syenitic Complex, Custer
County, Idaho
(1987) / Schalck D

Late Pleistocene and Holocene Geologic
History of the Taylor-Hilgard Portion of the
Southern Madison Range, Southwestern Montana
(1988) / Shaw C W

Reconnaissance Geology of the Cayuse Point
7.5 Minute Quadrangle, Elmore County, Idaho
(1987) / Slavik H

Quaternary Terrace Formation in the Little
Salmon River Basin
(1988) / Welford M R

Geology and Orientation Geochemistry of a
Portion of the Carico Lake Quadrangle,
Lander County, North-Central Nevada
(1987) / Westfall J

INDIANA UNIVERSITY

Metasomatism Associated with Ductile
Deformation in a Granodioritic Gneiss,
Central Connecticut
(1988) / Ambers C P

Analysis of Muddy Siliciclastic Rocks and
Provenance Determination
(1988) / Bangs C L

Depositional and Diagenetic History of the
Lower Mississippian Stobo Carbonate Mound,
Monroe County, Indiana
(1988) / Becker M J

INDIANA UNIVERSITY
(continued)

The Effect of Aqueous Chemical Environments
on the Shock Failure and Rubblization of
Green River Oil Shale
(1988) / Bogardus J W

Three-Dimensional Velocity Structure and
Precise Earthquake Locations on the
Calaveras Fault in the Morgan Hill Area,
California
(1988) / Chang C

Study of Selected Sandstone Reservoirs in
the Indiana Portion of the Griffin
Consolidated Field that Lies Above New
Harmony
(1988) / Fituri H S

Seismic Migration by Phase Shift Plus
Interpolation
(1988) / Hettenhausen R L

Occurrence and Biochronology of Middle
Mississippian Branchiopods of the Ramp
Creek Formation and Harrodsburg Limestone,
Indiana and Kentucky
(1988) / Hirt D S

A Subsurface Study of the Ste. Genevieve
Limestone (Mississippian) in the
Northeastern Warrick County, Indiana
(1988) / Johnson J J

Interpretation of the Depositional
Environment of the Hardinsburg Formation,
Gibson County, Indiana
(1988) / Laurin P R

Generation, Testing, and Filtering of
P-Tau Transforms Using the Fourier Transform
(1988) / Madrid G A

Refraction Residual Analysis of a
Crytoexplosion Structure Near Kentland,
Indiana
(1988) / Markisohn D B

Enhanced Geotomography and Its Application
to the Kentland, Indiana Impact Site
(1988) / Meyerholtz K A

Stratigraphy and Environmental Analysis of
the Siberia Limestone Member (Tobinsport
Formation) and the Leopold Limestone Member
(Branchville Formation) of Perry and Dubois
Counties, Indiana
(1988) / Perucca M

The Origin of Kaolin Contained in the
Whitemud Formation, Southern Saskatchewan,
Canada
(1988) / Pruett R J

A Sulfur Isotope Study of the Pyrite to
Pyrrhotite Conversion
(1988) / Snyder K

Seismicity and Crustal Structure of the
Pamir-Alai Region in Soviet Central Asia
(1988) / Swanson W A

IOWA STATE UNIVERSITY

A Regional Analysis of Calcite Twinning
Strain in the Bighorn Mountains, Northern
Wyoming
(1988) / Carson D W

Equatorial Waves Simulated the NCAR
Community Climate Model
(1988) / Cheng X

Chemistry of the Near-Surface Groundwater,
Great Salt Plains, Alfalfa County, Oklahoma
(1988) / Slaughter C B

Implications of an Oelian Sandstone Unit of
the Basal Morrison Formation, Central
Wyoming
(1988) / Weed D D

KANSAS, UNIVERSITY OF

Sandstone Diagenesis in an Interbedded
Carbonate-Siliciclastic Sequence, Virgilian
Holder Formation, New Mexico
(1988) / Bowman M W

Shallow Structure in Southeast Kansas Using
Short-Period Surface Wave Dispersion
(1988) / Hildebrand G M

The Fracture Pattern of North-Central Kansas
and Its Relation to Hydrogen Soil Gas
Anomalies Over the Midcontinent Rift System
(1988) / Johnsagard S K

The Effects of Near-Surface Geology on
Shallow High-Resolution Seismic Data Quality
in Northeastern Kansas
(1988) / Kalik A J

Geochronology and Geochemistry of Plutonic
and Volcanic Rocks in the Early Proterozoic
La Ronge Domain, Saskatchewan
(1988) / Livesey C L

A geophysical Study of the Hill's Pond
Lamproite, Woodson and Wilson Counties,
Kansas
(1988) / Markezich M A

Uranium-Lead Geochronology of Precambrian
Rocks in the Beaverlodge Area, Northwestern
Saskatchewan
(1988) / Persons S S

The Stratigraphy of the Quaternary Alluvium
in the Great Bend Prairie, Kansas
(1988) / Rosner M L

Origin of Dolomite in the Tamabra Formation
(Mid-Cretaceous), East-Central Mexico
(1988) / Stephens B P

KENTUCKY, UNIVERSITY OF

Geology of the Upper Elkhorn No. 3 Coal in
the Eastern Kentucky Coal Field
(1988) / Assad J

A Hydrogeologic Investigation of the
Stamping Ground, Kentucky Area
(1988) / Dimmock P

Detailed Subsurface Mapping in Central
Kentucky with Electrical Resistivity Methods
(1988) / Galceran C M

An Evaluation of the Drastic System in
Assessing Groundwater Pollution Potential
for Parts of Marshall, Fayette, and Madison
Counties, Kentucky
(1988) / Lovins E E

LAKE HEAD UNIVERSITY

Reactions of Ni(II) with NaBh4 and NaBh3 CN
in the Presence of Mono-and Bidentate
Phosphines Under CO Atmosphere
(1988) / Mirza H A

The Study of the Third Dimension in the
Thunder Bay Silver Veins
and Stable Isotape Results
(1988) / Sherlock R L

A Theoretical and Experimental Investigation
of Pressure Solution
(1988) / Steele O J

LAURENTIEN UNIVERSITY

Silurian (Llandovery-Wenlock) Patch Reef
Complexes of the Chicotte Formation,
Anticosti Island, Quebec
(1988) / Brunton F R

Upper Ordovician and Lower Silurian
Stratigraphy and Paleontology of Southampton
Island, Northwest Territories
(1988) / Dewing K

Depositional Framework and
Paleoenvironmental Interpretation of the
Heiberg Formation: Sverdrup Basin, Canadian
Arctic Archipelago
(1988) / Foley S L

LOUISIANA STATE UNIVERSITY

A Description of the Seasonal and Spatial
Variability of Thermohaline Properties in
the Lombok Strait and the West Flores Sea
Regions, Indonesia: 1985
(1988) / Arief D

Clay Mineral Diagenesis of the Pumpkin
Valley Shale at Oak Ridge, Tennessee
(1988) / Baxter P M

Lithofacies Architecture of the Tocito
Sandstone, Northwest New Mexico
(1988) / Bergsohn I

An Investigation of the Mid-Continent
Gravity High
(1988) / Coakley B J

Controls on the Fluxes of Chloride and
Sulfate Within the Louisiana Reach of the
Mississippi River
(1988) / Coleman J B

Morphology of Two Late Pleistocene Submarine
Canyons, Northwest Gulf of Mexico
(1988) / Eumont J C

Sedimentology and Stratigraphy of the
Semilla Sandstone Member of the Mancos
Shale, North-Central, New Mexico
(1988) / Fleming T F

Early Diagenetic Ferroan Carbonates in a
Subsiding Marsh Sequence, Terrebonne Parish,
Louisiana
(1988) / Moore S E

Permeant-Induced Changes in the Permeability
and Pore, Structure of Unconsolidated
Sediments
(1988) / Raffensperger J P

Pore Water Chemistry in Sediments of an
Abandoned Mississippi River Delta Complex
(1988) / Snow D D

Clay Mineral Diagenesis in the Lower Eocene
Wilcox, Lockhart Crossing Field, Livingston
Parish, Louisiana
(1988) / Strickler M E
Fluid Inclusion
A Seismic Stratigraphic Investigation of the
Late Wisconsin Mississippi Fanlobes
(1988) / Wagner J B

LOUISIANA TECH UNIVERSITY

Paleoenvironmental Reconstruction of the
Upper Claiborne, Jackson, and Lower
Vicksburg Groups (Upper Eocene and Lower
Oligocene) at Little Stave Creek, Clarke
County, Alabama
(1988) / Jordan K B

A Regional Analysis of Sedimentation and
Diagenesis of Porosity Trends in the Lower
Cretaceous Rodessa Formation, Northern
De Soto Parish, Louisiana
(1988) / Patrick M S

MASSACHUSETTS, UNIVERSITY OF

Systematic Trends in the Ancient Fracture
Features of the Tharsis Region, Mars
(1988) / Francis R A

Paleosecular Variations in Lake Sediment
Cores from Northern Ellesmere Island, Canada
(1988) / Leung J C

Modelling and Interpretation of a Major
Gravity High in South-Central Massachusetts
(1988) / Maalouf G Y

Sedimentology and Diagenesis of Jurassic
Lacustrine Sandstones in the Hartford and
Deerfield Basins, Massachusetts and
Connecticut
(1988) / Meriney P E

Geochemistry of the Granite-Gabbro Complex
of Vinalhaven Island, Maine: An
Investigation of the Commingling of Mafic
and Fesic Magmas
(1988) / Mitchell C B

Evidence of Transpressional Tectonics,
Southeastern Terminus of East Pryor
Mountain, Pryor Mountains, Montana and
Wyoming
(1988) / Orrell S A

The Relationship Between Strain and Magnetic
Susceptibility Anisotropy in Mylonitized
Henderson Gneiss
(1988) / Shannon M C

Hydrogeologic Investigation of a Municipal
Landfill in Broome County, New York
(1988) / Soukup J J

Geometry and Deformation History of
Mylonitic Rocks and Silicified Zones Along
the Mesozoic Connecticut Valley Border
Fault, Western Massachusetts
(1988) / Stopen L E

Chemical Stratigraphy of Lavas Exposed in
the Walls of Two Pit Craters on Kilauea
Volcano, Hawaii
(1988) / Wenz K P

MICHIGAN TECHNOLOGICAL UNIVERSITY

Geology and Geochemistry of Hydrothermal
Alteration Associated with Precious Metal
Mineralization in the Clark Creek Region,
Marquette County, Michigan
(1988) / Baxter D A

Verticle Petrologic Changes of the
Jacobsville Sandstone at Rice Lake Hole
Number 1
(1988) / Bowers M C

A Comparison of Model Approaches for
Evaluating Groundwater Flow and Transport
(1988) / Hagley M T

A Preliminary Study of the Relationship
Between Pore Geometry and Mass Transport in
Porous Media Using Fractal Geometry and
Digital Image Analysis
(1988) / Nordeng S H

Magmatic Processes, Evolution and Mantle
Source Characteristics Contributing to the
Petrogenesis of Midcontinent Rift Basalts:
Portage Lake Volcanics, Keweenaw Peninsula,
Michigan
(1988) / Paces J B

MINNESOTA, UNIVERSITY OF

Some Impacts of Drainage Ditch and Road
Construction on Red Lake Peatland,
Northern Minnesota
(1988) / Bradof K L

Drift Prospecting in Eskers in Northeastern
Minnesota
(1988) / Brown T R

The Relationship Between the Basal Zone and
Cloud Zone Copper-Nickel Sulfides, and the
Significance of Mafic Pegmatites, Minnamax
Property Duluth Complex, Minnesota
(1988) / Ervin S J M

The Relationship Between the Basal Zone and
Cloud Zone Copper-Nickel Sulfides, and the
Significance of Mafic Pegmatites, Minnamax
Property Duluth Complex, Minnesota
(1988) / Ervin S M

Large Felsic Flows in the Keweenawan North
Shore Volcanic Group in Cook County,
Minnesota
(1988) / Fitz T J

Petroology and Sedimentatiion of the Archean
Seine Group Conglomerate and Sandstone,
Western Wabigoon Belt, Northern Minnesota
and Western Ontario
(1988) / Frantes J R

A Late Holocene Vegatational Sequences from
the Southeast Missouri Ozarks
(1988) / Huber J K

Sand Dunes on the Anoka Sand Plain
(1988) / Keen K L

Sedimentary Features, Lithology, Structure,
and Heavy Mineral Suites of the Quetico
Metasediments, Superior Province, Ontario
(1988) / Pezzutto F

Paleoenvironments of the Bay of Carthage
Based on Ostraode Fauna
(1988) / Rausch D E

Glacial History and Climatic Change in the
Central Peruvian Andes
(1988) / Seltzer G O

Contact Metamorphism and Partial Melting of
the Bluegrass Creek Suite, Central Laramie
Mountains, Wyoming
(1988) / Spicuzza M J

Analysis of Small Scale Structures Developed
During Monoclinal Folding: Biebel
Monocline, Gunnison, Colorado
(1988) / Wright S F

MISSISSIPPI STATE UNIVERSITY

Application of Photo-Optical and Landsat MSS Data to
Monitor Surface Aggregate Mining: Lowndes and Monroe
Counties, Mississippi
(1986) / Barcellona B

The Elusa Oikoumene: A Geographical Analysis of an
Ancient Desert Ecosystem Based on Archaeological,
Environmental, Ethnographic, and Historical Data
(1986) / Elliott J D

The Relationship Between Physical Setting and
Sense of Place
(1986) / Nitz K M

Calcareous Nannoplankton of the Tupelo Tongue
(Upper Cretaceous), Lee County, Mississippi
(1986) / Salomon R A

Trace Fossils of the Tallahatta Formation
(Claiborne Group) in East-Central Mississippi
(1986) / Yip F Y

MISSOURI, UNIVERSITY OF (COLUMBIA)

Structural Analysis of Shear Zones in the
Central Vermilion District, NE Minnesota
(1988) / Bidwell M E

Plagioclase Compositions in the Dufek Massif
Anorthosites, Dufek Intrusion, Antartica
(1988) / Haensel J M

Experimental Thermal Maturation of Type III
Organic Matter: Kinetics and Catalysis
(1988) / Wheatley T L

MISSOURI, UNIVERSITY OF (ROLLA)

Cathodoluminescence Petrography of
Phosphate Grains in Jurassic (Aalenian)
Sedimentary Iron Ores, Alsace-Lorraine,
France
(1988) / Karakus M

Optimum Offset Distance for Seismic Velocity
Analyses and CMP Stacking
(1988) / Ku J W

The Application of Cathodoluminescence
Microscopy to the Study of Sparry Dolomite
from the Viburnum Trend, Southeast Missouri
(1988) / Voss R L

MONTANA COLLEGE OF MINERAL SCI. & TECH.

Geology of the Bone Basin Area, Madison and
Jefferson Counties, Montana
(1988) / Ahmed S

Detailed Water-Level Data and an Evaluation
of Mathematical Approaches for Near-River
Monitoring
(1988) / Newcomer D R

Geology of the Calvert Hill Area, Beaverhead
County, Montana
(1988) / Truckle D M

MONTANA STATE UNIVERSITY

Timing and Mechanism of Formation of
Selected Talc Deposits in the Ruby Range,
Southwestern Montana
(1987) / Anderson D L

Soil Development, Morphometry, and Scarp
Morphology of Fluvial Terraces at Jack Creek
Southwestern Montana
(1987) / Bearzi J P

Protolith and Tectonic Setting of an
Archean Quartzofeldspathic Gneiss Sequence
in the Blacktail Mountains, Beaverhead
County, Montana
(1987) / Clark M L

The Occurence and Timing of Gold
Mineralization at the Red Pine Mine,
Western Tobacco Root Mountains, Southwestern
Montana
(1987) / Kinley T M

Laramide Basement Deformation in the
Northern Gallatin Range and Southern
Bridger Range, Southwest Montana
(1987) / Miller E W B

Precambrian Geology of Lake Plateau,
Beartooth Mountains, Montana
(1987) / Richmond D P

Archean Geology of the Spanish Peaks Area,
Southwestern Montana
(1987) / Salt K J

Evolution of the Eocene Avon Volcanic
Complex, Powell County, Montana
(1987) / Trombetta M J

Sedimentology, Provenance, and Tectonic
Implications of the Cretaceous Newark
Canyon Formation, East-Central Nevada
(1987) / Vandervoort D S

MONTANA, UNIVERSITY OF

The Stratigraphy and Sedimentology of the
Middle Proterozoic Snowslip Formation in
the Lewis, Whitefish and Flathead Ranges,
Northwest Montana
(1988) / Ackman B C

The Role of Water in the Formation of
Granulite and Amphibolite Facies Rocks,
Tobacco Root Mountains, Montana
(1988) / Angeloni L M

Particle-Size and Chemical Control of Metals
in Clark Fork River Bed Sediment
(1988) / Brook E J

Distribution and Concentration of Metals in
Sediments and Water of the Clark Fork River
Floodplain, Montana
(1988) / Brooks R

The Geology and Mineralization at the Omar
Copper Prospect, Baird Mountains
Quadrangle, Alaska
(1988) / Folger P F

Late Proterozoic and Early Cambrian
Structural Setting of Western Montana and
the Stratigraphy and Depositional
Environment of the Flathead Sandstone in
West-Central Montana
(1988) / Kruger J M

The Relationship Between Electrical
Resistivity and Hydraulic Conductivity in
Two Fractured Bedrock Aquifers in Western
Montana
(1988) / Lazuk R

Hydrogeology of the Jocko Valley,
West-Central Montana
(1988) / Thompson W R

Hydrogeology of the Hamilton North and
Corvallis Quadrangles, Bitterroot Valley,
Southwestern Montana
(1988) / Uthman W

MONTREAL, UNIVERSITY OF

Transport de Sediments en Ecoulement Mince
et Morphologie du Lit Autour d'un Obstacle
(1988) / Boyer C

Interpretation Sedimentologique de Depots
Pyroclastiques des Monts Vulsini, Italie
Centrale
(1988) / Brissette F

Dynamique des Ecoulements et du Transport a
une Confluence de Cours d'Eau Naturels a Lit
Sablonneux
(1988) / De Serres B

Geologie et Geomorphologie de la Colline
Blanche, Reggion de Temiscamie, Quebec
(1988) / Gagnon L

Vers une Approche Perceptuelle Automatisee
de la Generalisation Cartographique des
Lignes
(1988) / Gauthier C

l'Utilisation des Fractales Dans la
Description et la Representation de Surfaces
Topographiques
(1988) / Gravel G

Etablissement d'un Modele Geostatistique a
la Mine Doyon, Canton Bousquet, Abitibi,
Quebec
(1988) / Jutras M

Desertification du Milieu Sahelien et
Teledetection Spatiale
(1988) / Kane R

l'Approche Tridimensionnelle des Surfaces
de Drainage
(1988) / Lemieux C

l'Erosion du Cratere du Nouveau-Quebec
(1988) / Marsan B

l'Evolution de la Plate-Forme Greseuse de
l'Ile du Cap aux Meules, Iles de la
Madeleine
(1988) / Paquet G

Etude Structurale et Petrographique du
Gisement S-50 a la Mine Kiena, Val d'Or,
Quebec
(1988) / Quirion D

NEBRASKA, UNIVERSITY OF

Nitrate-Nitrogen Profiles Documenting Land
Use Practice Effects on Ground Water in
and Around Sidney, Nebraska
(1988) / Bryda A P

Investigation of the Thermal Regime in a
River-Aquifer System Near Ashland, Nebraska
(1988) / Chu T

Paleobiology and Paleoecology of Crinoids
from the Lower Stanton Formation (Late
Pennsylvanian, Missourian) of the
Mid-Continent United States
(1988) / Holterhoff P F

Functional Morphology and Paleoecology of
North America Entelodonts (Artiodactyla,
Entelodontidae)
(1988) / Joeckel R M

Chert Within the Upper Member of the
Pennsylvanian Hermose Formation:
Southeastern Utah
(1988) / Kuntz G B

Taphonomy of the Valentine Railway Quarry
"B" Bone Bed (Late Barstovian),
North-Central Nebraska
(1988) / McCool K E

Oligocene Calcareous Nannofossil
Biostratigraphy, Sedimentation Rates and
Unconformities from Site 540, DSDP Leg 77
(1988) / Moran M J

Shallow Ground Water Denitrification
Associated with an Oxidation-Reduction Zone
in Hall County, Nebraska
(1988) / Parrott J D

Thermal Maturation in Microfossils from the
Oread Formation (Pennsylvanian),
Midcontinent, U.S.A.
(1988) / Rankis-Ikstrums L V

A Preliminary Investigation of Hydrocarbon
Contamination: Alliance, Nebraska
(1988) / Van Noort P J

Kettle Hole and Sandur Development at the
Margin of the Casement Glacier, Glacier Bay
National Park, Alaska
(1988) / Watts L J

NEW BRUNSWICK, UNIVERSITY OF

Lower Devonain Volcanic and Sedimentary
Rocks of the Eastport Formation, Southwest
New Brunswick
(1988) / Fay V K

NEW MEXICO INSTITUTE OF MINING & TECH.

The Relationship Between Mineralogy,
Groundwater Chemistry and Groundwater
Flow in the Moreno Hill Formation
(1987) / Ardito C P

Engineering Geology of the Northern
Estancia Valley, North-Central New Mexico
(1987) / Barrie D

Geology and Geochemistry of Early
Proterozoic Supracrustal Rocks from the
Western Dos Cabezas Mountains, Cochise
County, Arizona
(1987) / Bowling G P

Geology of the Southern Caoncito de la Uva
Area, Socorro County, New Mexico
(1987) / Brown K B

Precambrian Geology of the Upper Brazos Box
Area, Rio Arriba County, New Mexico
(1988) / Gabelman J L

Whiterock (Lower Middle Ordovician)
Cephalopod Fauna from the Ibex Area,
Millard County, Western Utah
(1987) / Gil A V

Carbonate Sedimentology of the Virgilian
Part of the Horquilla Formation, Big
Hatchet Peak Area, Hidalgo County, New
Mexico
(1987) / Gramont B

Geology of the Navajo Gap Area Between the
Ladron Moauntains and Mesa Sarca, Socorro
County, New Mexico: A Structural Analysis
(1987) / Hammond C M

The Estimation of Recharge Through
Unsaturated Porous Media Using the Flux
of Deuterium
(1988) / Hobbs T C

Geological, Paleomagnetic and Geophysical
Studies at Jones Camp Dike, Socorro County,
New Mexico
(1987) / Jochems T P

Strain Energy and Jointing in Coal and
Adjacent Sediments
(1987) / Linden R M

Subsurface Analysis of the (Permian Abo
Formation in the Lucero Region, West-Central
New Mexico
(1987) / Little G E

Geochemistry and Tectonic Setting of the
1700 Ma Alder and Red Rock Groups from
Tonto Basin, Arizona
(1988) / Noll P D

An Interpretation of the Depositional
Setting for the Sugarite Coal Zone of the
Raton Formation, Located Near the City
of Raton, New Mexico
(1987) / Perry P L

Geochemistry of Valle Grande Member Ring
Fracture Rhyolites, Valles Caldera, New
Mexico
(1987) / Spell T L

Seismic Refraction Investigation of the
Shallow Subsurface of the Lower Rio Salado,
Northwest of San Acacia, New Mexico
(1988) / Zody S P

NEW MEXICO STATE UNIVERSITY

Flow Structures and Zonation of Ring
Fracture Domes and Their Relation to Ore
Control in the Mogollon Mining District,
Catron County, New Mexico
(1988) / Avery D C

Lithofacies, Stratigraphy and Cyclic
Sedimentation in a Mixed Carbonate and
Siliciclastic System, Red House Formation
(Atokan), Sierra County, New Mexico
(1988) / Kalesky J F

Stratigraphy and Structures of the Escalante
Mine Formation Iron County, Utah
(1988) / Williamson A L

NEW MEXICO, UNIVERSITY OF

Geology and Geochemistry of Mid-Tertiary Volcanic
Rocks in the Eastern Chirichua Mountains
(1988) / Bryan C R

Characterization and Genisis of Palogonite and
Authigenic Minerals, Hanauma Bay and
Coco Crater, Oahu, Hawaii
(1988) / Cowan R E

Geochemistry and Petrogenesis of Rocks with
Shoshonitic Affinities, Crandall-Sunlight Region,
Absaroka Volcanic Field,Wyoming
(1988) / Erskine D W

Depositional Environments in an Upper Triassic Lake,
East-Central New Mexico
(1988) / Hester P M

Depositional History of Lower Cretaceous Strata in
Northeastern New Mexico: Implications for Regional
Tectonics and Sequence Stratigraphy
(1988) / Holbrook J M

Quaternary Alluvial Sequence of the Upper Pecos
River and a Tributary, Glorieta Creek,
North-Central New Mexico
(1988) / Karas P A

Petrology and Petroginesis of Alkali Basalts and
Their Associated Inclusions from the Elephant
Butte Area, Sierra County, New Mexico
(1988) / Kelly J C

Sources of Clasts in Terrestrial and Lunar
Impact Melts
(1988) / McCormick K A

Behavior, Speciation and Environmental Impact of
Selenium at Bosque Del Apache National Wildlife
Refuge and Poison Canyon, New Mexico
(1988) / Persico J L

Fine-Grained, Millimeter-Sized Objects in Ordinary
Chondrites and Their Relation to Chondrules and
Matrix
(1988) / Recca S I

Structural and Metamorphic Evolution of Proterozoic
Rocks in the Northern Taos Range, Taos County,
New Mexico
(1988) / Smith R F

Glacial Chronology and Soil Development in Winsor
Creek Drainage Basis, Southernmost Sangre de Cristo
Mountains, New Mexico
(1988) / Wesling J R

NEW YORK, STATE UNIVERSITY OF (ALBANY)

The Providence Island Formation in the
Northern Appalachia Region - A Lower-Lower
Middle Ordovician Analogue to Recent
Arid-Semiiarid Tidal-Slat Carbonates of the
Persian Gulf Trucial Coast
(1987) / Hernandez M R

Structural Analysis Across the Northeast
Boundary of the Taconic-Allochthon,
West-Central Vermont
(1987) / Hoak T E

Dolomitization of the Hatch Hill Arenites
and the Burden Iran One
(1987) / Hofman P M

Granitic Pegmatities in the Southeastern
Adirondacks: Their Use as Indicators of
Temperature Pressure, and Fluid Conditions
During a Late Stage of the Grenville Orogeny
(1987) / Mihalich J P

Geology, Geochemistry, and the Geochronology
of the Lems Ridge Olistostrome, Klamath
Mountains, California
(1987) / Ohr M

Geology of the Northern Baie Verte
Peninsula, New Foundland, Canada
(1987) / Stella P J

Structural Investigations on Experimentally
and Naturally Produced Slickenslides
(1987) / Will T M

Identification for Endmembers for Magma
Mixing in Little Sitkin Volcano, Alaska
(1987) / Wolf D A

NEW YORK, STATE UNIVERSITY OF (BUFFALO)

An Investigation of DSDP Leg 86 Ash Layers
Bearing on Stratigraphy and Volcanic History
of the Northwest Pacific
(1988) / Erb D

Seismic Stratigraphy of the Long Island
Platform on the U.S. Atlantic Continental
Margin
(1988) / Jowett R

The Geology of Pine and Crater Buttes: Two
Basaltic Plains Constructs on the Far
Eastern Snake River Plain, Idaho
(1988) / Mazierski P

A Geochemical Mineralogic and Petrographic
Study of the Middle Ordovician Trenton
Group in New York State
(1988) / Panus E

The Regimen of Grand Union Glacier and the
Glacial Geology of the Kigluaik Mountains,
Seward Peninsula, Alaska
(1988) / Przybyl B J

NORTH CAROLINA STATE UNIVERSITY

A Climatology of Air Parcel Trajectories
Into Mount Mitchell, North Carolina
(1988) / Seabaugh S S

NORTH CAROLINA, UNIV. OF (CHAPEL HILL)

A Geochronological Study of the Grenville
Front Near Coniston, Ontario, Canada
(1987) / Bullock W R

The Geology and Stratigraphy of the
Northeast Quarter of White Cross Quadrangle,
Orange County, North Carolina
(1987) / Chiulli A T

Volcanic Stratigraphy in the Carolina Slate
Belt Near Chapel Hill, North Carolina
(1987) / Eligman D

Petrology of the Mid-Cretaceous Aguas Buenas
and Rio Maton Limestones in Southeast
Central Puerto Rico
(1987) / Kaczor L

NORTH CAROLINA, UNIV. OF (CHAPEL HILL)
(continued)

Interpretation of the Radioactive Fallout
Record from the Annual Bands of Montastrlea
Annularis, Broward County, Florida
(1987) / Madzsar E M

The Structural Geology of the Worthington
Mountains, Lincoln County, Nevada
(1987) / Martin M W

Origin of the Keefer Sandstone (Middle
Silurian) of Northeastern West Virginia
and Western Maryland
(1987) / Meyer S C

An Investigation of the Proterozoic
Basement-Ashe Formation Boundary West of
the Grandfather Mountain Window,
Northwestern North Carolina
(1987) / Mies J W

Lithostratigraphy and Molluscan
Biostratigraphy of the River Bend Formation
at the Martin Marietta Quarry, New Bern,
North Carolina
(1987) / Rossbach T J

Foraminiferal Biostratigraphy of the
Pollocksville and Haywood Landing Members
of the Belgrade Formation, Oligocene/
Miocene, North Carolina
(1987) / Russell M D

An X-Ray Diffraction Study of Displacive
Phase Transitions in Terrestrial Tridymite
(1987) / Smelik E A

A Numerical Study of Caustics and
Polarizations Produced by Seismic Waves
Interacting with Three-Dimensional
Geological Structures
(1987) / Sorauf C M

A Structural, Stratigraphic and
Petrochemical Study of Willow Creek Canyon,
Range, Nye County, Nevada
(1987) / Wright S D

Geochemistry of Volcanic Rocks from the
Grandfather Mountain Window, Northwestern
North Carolina
(1987) / Zmoda A J

NORTHEASTERN LOUISIANA UNIVERSITY

A Study of the Depositional Environment and
Diagenesis of the Hill Member of the Rodessa
Formation in Caddo Parish, Louisiana
(1988) / Alexander N J

Depositional and Diagenetic Interpretation
of the Sligo (Pettet) Formation, Pine Island
Field, Caddo Parish, Louisiana
(1988) / Amerson M D

Clay Mineral Analysis of the Dolet Hills
Formation (Paleocene) in Desoto Parish,
Louisiana
(1988) / Fohn K M

Depositional Environment of the Mitchell
Tongue of the Gloyd Member, Rodessa
Formation (Lower Cretaceous), Duty,
St. Mary, and Walnut Hill Fields, Lafayette
County, Arkansas
(1988) / Fontana R V

Diagenesis, Facies Development, and
Depositional Environment of the Lower
Tuscaloosa Formation, Thompson Field,
Amite County, Mississippi
(1988) / Hippensteel D L

Diagenesis and Porosity Development of Lower
Tuscaloosa Sandstones in Wilkinson and Amite
Counties, Mississippi, and St. Helena
Parish, Louisiana
(1988) / Hopper J T

Depositional Environment and Diagenesis of
the Gloyd Member of the Rodessa Formation,
Naconiche Creek Field, Nacogdoches County,
Texas
(1988) / Kelly G E

Tectonic Boundaries of the Eastern Gulf
Coast of North America
(1988) / Leonard C

A Paleoenvironmental Analysis of a New
Species of Permian Rostroconch Recovered
from the Upper Hueco Formation, Robledo
Mountains, Dona Ana County, New Mexico
(1988) / Loggins S K

Depositional Environment and Diagenesis of
the Benbrook Member of the Goodland
Formation (Lower Cretaceous), Sabine Parish,
Louisiana
(1988) / McPhearson R D

Geology and Geochemistry of the Turkey Hill
Gold Mine Area, Lumpkin County, Georgia
(1987) / Mittler R W

Heavy Mineral Analysis of Selected Loess
Samples from Franklin and Catahoula
Parishes, Louisiana
(1987) / Slapp K P

NORTHERN ARIZONA UNIVERSITY

Sedimentology of an Eocene Volcanic Mudflow
Deposit: Beaver Rim, Fremont County,
Wyoming
(1988) / Baughman R L

Sedimentologic and Morphologic Analysis of
Modern Macrotidal Deposits, Puerto Penasco,
Sonora, Mexico
(1988) / Brown K

Facies and Depositional Environments of the
Coal-Bearing Upper Carbonaceous Member of
the Wepo Formation (Upper Cretaceous)
Northeastern Black Mesa, Arizona
(1988) / Carr D A

A Kinesmatic and Geometric Structural
Analysis on an Early Proterozoic
Crustal-Scale Shear Zone: The Evolution
of the Shylock Fault Zone, Central Arizona
(1988) / Darrach M E

Sedimentology and Provenance of the Upper
Cretaceous Iron Springs Formation,
Southwest Utah
(1988) / Fillmore R P

Late Cretaceous Ductile Deformation,
Metamorphism and Plutonism in the Piute
Mountains, Eastern Mohave Desert
(1988) / Fletcher J M

NORTHERN ARIZONA UNIVERSITY
(continued)

A Geological Investigation of the Early
Proterozoic Irving Formation, Southeastern
Needle Mountains, Colorado
(1988) / Gonzales D A

Stratigraphy and depositional Environment of
the Surprise Canyon Formation, and Upper
Mississippian Carbonate - Classic Estuarine
Deposit, Grand Canyon, Arizona
(1988) / Grover P

Cinder Cone Breaching Events at Strawberry
and O'Neill Craters, San Francisco Volcanic
Field, Arizona
(1988) / Harwood R

Architectural Element Analysis and
Depositional History of the Upper Petrified
Forest Member of Chinle Foramtion,
Petrified Forest National Park, Arizona
(1988) / Johns M E

The Structure and General Stratigraphy of
the Western Half of the Payson Basin, Gila
County, Arizona
(1988) / Muehlberger E W

Geologic History of the House Mountain Area,
Yavapai County, Arizona
(1988) / Ranney W

Geochemical Facies Analysis of the Surprise
Canyon Formation in Fern Glen Channelway,
Central Grand Canyon, Arizona
(1988) / Shirley D H

A Geophysical and Geochemical Investigation
of Selected Collapse Features on the
Coconino Plateau in Northern Arizona
(1988) / Waters J P

NORTHERN ILLINOIS UNIVERSITY

Nitrate Occurrence in Ground and Surface
Waters, DeKalb County, Illinois
(1988) / Andrews W J

The Geology of the Nephi 7.5' Quadrangle,
Central Utah
(1988) / Biek R F

Sediment Transport and Deposition in a
Temperate Glacial Fjord, Glacier Bay, Alaska
(1988) / Cowan E A

Development of a Kilowatt-Plus Helium
Microwave-Induced Plasma System as an
Analytical Atomic Emission Source
(1988) / Cull K B

Spectroscopic Studies of Solutes Embedded
in Porous Vycor Glass
(1988) / Hsu W

Modeling Point Source Dispersion Inside a
Stagnation Zone Created During Stable
Onshore Flow Into Coastal Complex Terrain
(1988) / Keating J W

Pleistocene Paleoenvironmental Record of the
Planktonic Foraminifera of the North-Central
Gulf of Mexico
(1988) / Liu S T

Hydrogeology and Hydraulic Characteristics
of a Fractured Dolomite Aquifer at an EPA
Superfund Site, Ogle County, Illinois
(1988) / Olson D N

Infiltration at a Northeastern Illinois
Landfill
(1988) / Price B C

OKLAHOMA, UNIVERSITY OF

Seismic Stratigraphy of the Upper
Pennsylvanian Swope Limestone, Comanche
County, Kansas and Woods County, Oklahoma
(1988) / Austin M N

Structural Implications of Thermal Maturity
Data: Sawtooth Mountains, Montana
(1988) / Becker D G

The Depositional History of the Lower Deese
Group (Middle Pennsylvanian), Ardmore Basin,
Oklahoma
(1988) / Liesch A R

Seismic Stratigraphic Analysis of the
Cretaceous Carbonates of Paso Caballos
Basin, Guatemala
(1988) / Mazariegos Alfaro R A

Paleomagnetic and Rock Magnetic
Investigation of the Relationship Between
Hydrocarbons and Authigenic Magnetic
Minerals in the Permian Lyons Sandstone,
Northern Front Range and Denver Basin,
Colorado
(1988) / McCollum R A

Geological History and Sedimentary
Environment of the Muskogee Oilfield,
Oklahoma
(1988) / Perez J M

Seismic Analysis of a Complex Structure in
Stephens County, Oklahoma
(1988) / Shinol J H

OREGON STATE UNIVERSITY

Stratigraphy, Depositional Environments, and
Diagenesis of the Devonian Simonson and Guilmette
Formations in the White Pine, Egan, and Schell
Creek Ranges, White Pine County, Nevada
(1987) / DeDen F M

The Geology of the East Central Desolation Butte
Quadrangle, Grant County, Oregon
(1987) / Elfrink N

The Geology of the Elk Mountain-Porter Ridge
Area, Clatsop County, Northwest Oregon
(1988) / Goalen J S

The Geology, Igneous Petrology, and Mineral
Deposits of the Ataspaca Mining District,
Department of Tacna, Peru
(1988) / Horning T S

Geologic Structure of the Upper Ojai Valley
and Chaffee Canyon Areas, Ventura County,
California
(1988) / Huftile G J

Relations Between Geology and Mass Movement
Features in a Part of the East Fork Coquille
River Watershed, Southern Coast Range, Oregon
(1987) / Lane J W

OREGON STATE UNIVERSITY
(continued)

Balanced Structural Cross section of the Western
Salt Range and Potwar Plateau, Pakistan:
Deformation Near the Strike-Slip Terminus of an
Overthrust Sheet
(1987) / Leathers M R

The Effects of the Mazama Tephra-Falls: A
Geoarchaeological Approach
(1987) / Matz S E

Structural Interpretation of Seismic Reflection
Data From the Eastern Salt Range and Potwar
Plateau, Pakistan
(1988) / Pennock E S

Geology and Mineralization of the Seaman Gulch
Area, East Shasta Mining District, Shasta County,
California
(1988) / Prihar D W

Stratigraphy and Depositional Environments of the
Mississippian Rocks, Garnet Range-Bearmouth Area,
Granite County, Western Montana
(1988) / Schneider R C

Effects of Earthflows on Stream Channel and
Valley Floor Morphology Western Cascade Range,
Oregon
(1988) / Vest S B

OREGON, UNIVERSITY OF

Spectroscopic and Calorimetric Studies of
Olivine Glass and Crystal and the Origin
of Ultramafic Liquid Immiscibility
(1988) / Cooney T F

Boiling and Acidification in Epethermal
Systems: Numerical Modeling of Transport
and Deposition of Base, Precious and
Volatile Metals
(1987) / Spycher N F

PENNSYLVANIA STATE UNIVERSITY

Engineering Geologic Parameters and Their
Relationship to Roof Falls in a Coal Mine
on the Appalachian Plateau, Pennsylvania
(1987) / Blackmer G C

A Geochemical and Structural Study of the
South Mountain Batholith of Southern Nova
Scotia for Tin and Uranium Mineralization
(1987) / Brickey D W

Relationship of Remotely Sensed Spot Data to
Infiltration Capacity of Surface Mined Land
in Central Pennsylvania
(1988) / Guebert M D

The Mid- to Late-Nineteenth-Century United
States Census as a Focus for Positivist
Geographic Thought
(1988) / Hannah M G

Aspects of Petrography, Palynology and
Inorganic Constituents of Subbituminous
Coals from Canyon Creek, Alaska: A High
Latitude Tertiary Coal Field
(1987) / Lamberson M

Infiltration Characteristics and Hydrologic
Modeling of Disturbed Land, Moshannon,
Pennsylvania
(1987) / Lemieux C R

Shallow Marine Sedimentary Processes Along
the Late Devonian Catskill Shoreline in
Pennsylvania: Storm Versus Tidal Influence
(1987) / Loule J P

Periodicities in Color Banding in Sphalerite
of the Upper Mississippi Valley District
(1987) / Mason S E

Two-Variable Mapping: A Practical Case for
the Soil Map
(1987) / Mc Lay W J

Numerical Modelling of Chloride Migration in
a Fractured Bedrock Aquifer, Bear Creek,
Pennsylvania
(1987) / Mosconi D A

Geomorphic Significance of Longitudinal
Stream Profiles in Fluviokarsts
(1988) / Sasowsky I D

Agriculture, Culture, Technology: Amish
Agriculture in Pennsylvania's Kishacoquillas
Valley
(1987) / Siegel C I

Trough Cross-Stratification Geometry and
Methods for Determination of Paleoflow
Direction
(1988) / Tchinda F

The Variability of German Winter Temperature
in Relation to Human Performance and Its
Implications for Tactical Military Operation
(1987) / Yeshnik M A

PENNSYLVANIA, UNIVERSITY OF

The Structural and Metamorphic History of
the Cream Valley Fault Zone, Near
Conshohocken, Pennsylvania
(1988) / Howard C T

PRINCETON UNIVERSITY

Spatial Variation of SMART 1 Array
Earthquake Ground Motions
(1988) / Cunniff J M

PURDUE UNIVERSITY

Calculating Crack Density Values from
Compressional and Shear Wave Velocities
(1988) / Baud R D

Argentian Zinc-Iron Tetrahedrite-Tennantite
Thermochemistry
(1988) / Ebel D S

An Integrated Geophysical Study of the
Southeastern Extension of the Midcontinent
Rift System
(1988) / Fox A J

Hydrodynamic and Hydrologic Investigation
of a Coastal Landfill in Southern Lake
Michigan
(1988) / Hoover J A

Determining Aquifer Transmissivity and
Storativity with Recirculating Pumping
Tests
(1988) / Hunt P J

Seismic Study in the Ouachita System of
Southern Arkansas and the Adjacent Gulf
Coastal Plain
(1988) / Jardine W G

PURDUE UNIVERSITY
(continued)

A Sedimentologic and Paleontologic
Investigation of the Tres Montes Region,
Southern Chile
(1988) / Krishna P P

Further Geologic Investigation of a
Superfund Landfill Site, Montgomery County,
Pennsylvania
(1988) / Loughnana B K

Dust Devil Behavior at White Sands Missile
Range and Relationships to Local Surface
Boundary Layer Meteorology
(1988) / McClelland T M

The Biostratigraphy of Early Tertiary
Macroinvertebrates from the La Meseta
Formation, Seymour Island, Antarctic
Peninsula
(1988) / Stilwell J D

QUEENS COLLEGE

Stream Sediment Geochemistry in the Mine
Hill Area, Roxbury Quadrangle, Western
Connecticut
(1988) / Loomba L

REGINA, UNIVERSITY OF

Age of Rotational Landslides in the West
Block of the Cypress Hills, Saskatchewan
and Alberta
(1988) / Goulden M R

Water Erosion Risk-Assessment and Mapping
Based on the ARC/INFO Geographic
Information System
(1988) / Yong X

RENSSELAER POLYTECHNIC INSTITUTE

Porphyrin Absorption Bands Isolated from
Whole Rock Reflectance Spectra
(1988) / Holden P N

Surficial Geology and Glacial History of the
Little Falls Quadrangle, Central Mohawk
Valley, New York
(1988) / Maynard D

RHODE ISLAND, UNIVERSITY OF

A Seismic Refraction Investigation of the
Hydrogeology of Fractured Bedrock
(1988) / Hanson L G

SAN DIEGO STATE UNIVERSITY

An Evaluation of Potential Lateral Saltwater
Intrusion in the Ocotillo-Coyote Wells
Groundwater Basin, Imperial County,
California
(1987) / Mark D L

Evaluation of Hydraulic and Hydrogeochemical
Parameters of a Deep Fracture Groundwater
System in the Area of Cuyamaca - Julian, San
Diego County, California
(1987) / Ross C G

Distribution, Preservation and Paleoecology
of Benthic Foraminifera from McMurdo Sound,
Antarctica
(1988) / Vaughn N R

SOUTH CAROLINA, UNIVERSITY OF

Ground Water Mineralization in the Carolina
Slate Belt
(1988) / Clark S L

SOUTH DAKOTA SCHOOL OF MINES AND TECHNOL.

A Sedimentologic and Stratigraphic Study of
the Paleocene Fort Union Formation in the
South Cave Hills of Harding County, South
Dakota
(1988) / Best W A

Geochemistry of the Epigenetic Uranium
Bearing Cretaceous Lakota Formation,
Southern Black Hills, South Dakota
(1988) / Blake B J

Taphonomic Analysis of the Miocene Flint
Hill North Locality Bennett County, South
Dakota
(1988) / DeTample C A

A Paleochannel of the Tertiary White River
Group Near Fairburn, South Dakota
(1988) / Kleiter K J

Determination of Longitudinal and Lateral
Sediment Distribution in Reservoirs
(1988) / Sabtan A A

Lunar Regolityh Breccias and Soils:
Comparative Petrology, Chemistry, and
Evolution
(1988) / Simon S B

SOUTHERN METHODIST UNIVERSITY

Experimental Determination of Seismic Source
Characteristics for Small Chemical
Explosions
(1988) / Grant L T

Carbonate Apron Deposition of the Bigfork
Chert (Middle to Upper Ordovician), Western
Ouachita Mountains, Oklahoma
(1988) / Kasulis P F

An Analytical Technique for Estimating the
Deflection Profile, Mechanical Thickness,
and Failure of the Oceanic Lithosphere
(1988) / Pattey P D

SOUTHWESTERN LOUISIANA, UNIVERSITY OF

Quaternary and Environmental Geology of
South-Central Ascension Parish, Louisiana
(1988) / Christians G L

Biostratigraphy, Paleoecology and Taxonomy
of the Marginulina Zone in Cameron and
Calcasieu Parishes
(1988) / Desselle B A

The Paleoecology and Biostratigraphy of the
Anahuac Formation in Parts of Vermilion and
Acadia Parishes, Louisiana
(1988) / Hall C

Quaternary and Environmental Geology of
Southwestern St. Martin Parish, Louisiana
(1988) / Rouly K

Depositional Environments and Distribution
of Updip Lower Tuscaloosa Stringer Member
Sandstones
(1988) / Rutherford M S

SOUTHWESTERN LOUISIANA, UNIVERSITY OF
(continued)

The Origin of Late Quaternary Non-Marine
Evaporite Sequence in the Gran
Desierto/Bahia Adair Region, Sonora, Mexico
(1988) / Sinitierre S M

A Reconnaisance Mineralogical and Chemical
Study of Interdistributary Sediments from
Two Borings Taken in the Atchafalaya Basin
(1988) / Zanghi E M

STEPHAN F. AUSTIN STATE UNIVERSITY

Depositional History and Diagenesis of the
Noodle Creek Limestone (Wolfcampian),
Wallace Ranch and North Rough Draw Fields,
Kent County, Texas
(1988) / Brown R B

Identification and Geological Significance
of Petrified Wood from the Oligocene
Catahoula Formation, Jasper County, Texas
(1988) / Chadwick M L

The Influence of Phylloid Algae on Carbonate
Sedimentation in the Winchell Limestone
(Canyon Group), North-Central Texas Ranger
Quarry, Texas
(1988) / Davis R A

Correlation of the Woodford Formation in
South-Central Oklahoma Using Gamma-Ray
Scintillation Measurements of the National
Background Radiation
(1988) / Jolly G D

Distribution of Facies, Depositional and
Diagenetic Environments of Knowles Limestone
(Cotton Valley Group-Upper Jurassic)
DeSoto-Sabine Parish, Louisiana
(1988) / Mahbubullah A K M

A Depositional Environmental Analysis of the
Moenave Formation, Zion National Park,
Springdale, Utah
(1988) / Queen E B

Dolomitzation in the Ellenburger Group, Val
Verde Basin, West Texas, and Its
Relationship to Brecciation
(1988) / Strange W D

TEMPLE UNIVERSITY

The Study of Two Archean Paleosols and
Implications for the Precambrian Atmosphere
and Environment
(1988) / Foster R W

Hydrothermal Reaction of Groundwater with
Basalt from the Entablature of the Cohassett
Flow, Grande Ronde Formation, Hanford,
Washington: The Effect of the Introduction
of Fresh Solution to the System at 200 and
300 Degrees Centigrade, and 30 MPA
(1988) / Gardiner M A

The Role of Ductile Deformation in the
Formation of Large Garnets on Gore Mountain,
Southeastern Adirondacks
(1988) / Goldblum D R

Anatomy of a Single Punctuated Aggradational
Cycle, Keyser Formation, Silurian-Devonian,
Central Pennsylvania
(1988) / Hamilton F K

Stratigraphic Analysis of the Lower New
Scotland Formation: An Episodic Perspective
(1988) / Side D M

The Origin and Implication of the Steep
Gravity Gradient in the Vicinity of the
Martic Zone, Southeastern Pennsylvania
(1988) / Song T S

The Geometry and Conditions of Deformation
of the Rosemont Shear Zone, Southeastern
Pennsylvania Piedmont
(1988) / Valentino D W

TENNESSEE, UNIVERSITY OF (KNOXVILLE)

The Paleoenvironmental Interpretation of the
Lower Silurian Rockwood Formation at Green
Gap, Whiteoak Mountain, Southeastern
Tennessee
(1987) / Easthouse K A

Sedimmentology of the Clinch Sandstone
(Lower Silurian) Along Clinch Mountain,
Northeastern Tennessee
(1987) / Fischer M W

The Structural Geology of a Part of the
Pulaski Thrust Sheet Near Boone Dam:
Sullivan and Washington Counties, Tennessee
(1987) / Harlow G E

Stratigraphy and Sedimentology of the Middle
Ordovician Sevier Shale Basin Near Avens
Bridge, Washington County, Virginia
(1987) / Howze B D

Comparative Diagenetic Histories of the
Holston and Rockdell Formations, Middle
Ordovician of Northeastern Tennessee
(1987) / Jernigan D G

Sedimentology of the Clinch Sandstone
(Lower Silurian) at Powell Mountain,
Southwestern Virginia
(1987) / Marks G T

Internal Sediments Within Ore Bodies of the
Mascot-Jefferson City Zinc District, East
Tennessee
(1987) / Matlock J F

Geometric and Mesofabric Analysis of the
Hossfeldt Anticline--Eustic Syncline and the
Lombard Thrust Zone Near Three Forks,
Southern Montana
(1987) / Mitchell M M

Structural Geology and Geometries of the
Denton Duplex Along the Frontal Blue Ridge,
Near Hartford, Tennessee
(1987) / Robert L C

Sedimentology of a Fluvial-to-Marine
Transition: The Chilhowee Group (Lower
Cambrian), Southeast Tennessee
(1987) / Skelly R L

Deformation of Ordinary Chondritic
Meteorites
(1987) / Sneyd D S

TEXAS A AND M UNIVERSITY

Structure and Distribution of Abnormal Pressures
in the Vicksburg Formation (Oligocene), Hinde
Field, Starr County, Texas
(1988) / Barrell K A

TEXAS A AND M UNIVERSITY
(continued)

Sedimentology and Holocene Stratigraphy of a
Carbonate Mangrove Buildup, Twin Cays, Belize,
Central America
(1988) / Bond G B

Development and Chemical Quality of a Ground-Water
System in Cast Overburden at the Gibbons Creek
Lignite Mine
(1988) / Borbely E S

Late Holocene Isotopic and Sedimentologic
Records Contained in Carbonate Lagoonal
Cores, Northern Little Bahama Bank
(1988) / Canova J L

Depositional Environment and Reservoir
Characteristics of the Upper Frio Sandstones,
Willamar Field, Willacy County, Texas
(1988) / Caram H L

Origin of Gaseous Hydrocarbons in East-Central
Texas Groundwaters
(1988) / Coffman B K

Engineering Geology and Geohydrology of the
Burkeville Confining System Northeast of
Conroe, Texas
(1988) / Daniel J H

Deposition of the Woodbine-Eagleford Sandstones,
Aggieland Field, Brazos County, Texas
(1988) / DeDominic J R

A Solute Transport Model Calibration Procedure
as Applied to a Tritium Plume in the Savannah
River Plant F-Area, South Carolina
(1988) / Edwards D E

The Effects of Composition and Bedding on Log
Response, Yowlumne Sandstone, Kern County, California
(1988) / Fortner D W

Depositional Environment of the Yates Formation
in Kermit Field, Winkler County, Texas
(1988) / Gormican S C

Use of Superposition and the Extended Pulse Model
to Evaluate the Contaminant Transport Parameters
of Variably Source-Loaded Plumes
(1988) / Hankins D W

The Lithology, Environment of Deposition, and
Diagenesis of the Queen Formation at McFarland,
McFarland North, and Magutex Queen Fields,
Andrews County, Texas
(1988) / Holley C E

A Mechanical Model of Early Salt Dome Growth
(1988) / Irwin F A

Analysis of Monoclinal Folds Associated with the
Brittmore Fault in Northwest Houston, Texas
(1988) / Jones J A

Influence of a River Valley Constriction of
Upstream Sedimentation
(1988) / Kinnebrew Q

Depositional and Diagenetic Characteristics of a
Phylloid Algal Mound, Upper Palo Pinto Formation,
Conley Field, Hardeman County, Texas
(1988) / Lovell S E

The Environment of Deposition and Diagenesis of the
Shattuck Member of the Queen formation (Gudalupian,
Permian) at Caprock Queen Field, Chavez County,
New Mexico
(1988) / Malicse J A

The Distribution of and Sources for Quartz Silt
Deposited on the Northern Gulf of Mexico
Continental Shelf
(1988) / Peterson M A

A Postmortem Assessment of Environmental Compliance
of a High-Level Radioactive Waste Respository,
Hanford Site, Washington
(1988) / Petrini R H

Facies Analyses and Environment of Deposition for
the Jurassic "A" Zone of the "Mulussa" (DOLAA) Group,
in the Homs Block, Syria
(1988) / Quintana M A

The Kinematics of Debris Flow Transport Down a Canyon
(1988) / Santi P M

Reservoir Rock-Property Calculations from Thin
Section Measurements
(1988) / Sneed D R

Depositional Environment of the Caballos Formation
San Francisco Field, Neiva Sub-Basin, Upper
Magdalena Valley, Colombia
(1988) / Sneider J S

Depositional Environment of Downdip Yegua
(Eocene) Sandstones, Jackson County, Texas
(1988) / Whitten C J

A Model of Stable Geomorphic Form for an
Urban Stream, College Station, Texas
(1988) / Woodhouse E G

Depositional Environment of Upper Devonian Gas
Producing Sandstones, Westmoreland County,
Southwestern Pennsylvania
(1988) / Work R D

TEXAS CHRISTIAN UNIVERSITY

Depositional and Diagenetic History of the
West Smyer Field, Hockley County, Texas
(1988) / Griffin A F

Depositional Systems in the Dockum Group
(Upper Triassic), Scurry and Mitchell
Counties, Texas
(1988) / Ness D L

Isolated Bioherms in the Tansill Formation
(Guadalupian), Dark Canyon, Guadalupe
Mountains, Southeastern New Mexico
(1988) / Ward R J

TEXAS, UNIVERSITY OF (ARLINGTON)

The Distribution of Brick Red Lucite in the
Western North Atlantic: Evidence from
Visible Light Spectra
(1988) / Barranco F T

Proximal Flank Facies of the May 18, 1990
Ignimbrite: Mount St. Helens Washington
(1988) / Beeson D L

A Remote Sensing Study of the Volcanoes of
the Central Andes Between 21 Degrees and
24 Degrees
(1988) / Cheek C A

TEXAS, UNIVERSITY OF (ARLINGTON)
(continued)

Depositional History of the Lower Permian
Abo Formation and Lower Meseta Blanca Member
of the Yeso Formation, North Central New
Mexico
(1988) / Crabaugh J P

Geoelectric Stratigraphy and Subsurface
Evaluation of Quaternary Deposits at
Cooper Basin, Northeast Texas
(1988) / Darwin R L

Trace Element Geochemistry and Stable
Isotope Constraints on the Petrogenesis of
Cerros del Rio Lavas, Jemez Mountains,
New Mexico
(1988) / Duncker K E

Fusillade Biostratigraphy of Late Paleozoic
Inliers in the Edwards Plateau, Central
Texas
(1988) / Ford J W

Volcanological Investigation of the
El Cajete Series, Jemez Mouantains, New
Mexico
(1988) / Kircher D E

Carbonatite Genesis: A Geochemical Link
Between Carbonatite and Silicate Magmas
from Brava, Cape Verde Islands
(1988) / Miller D J

Chemical and Mineralogical Nature of
Phosphate Nodules from the Upper Cretaceous
of North Central Texas
(1988) / Reynolds S K

Sedimentology and Depositional Environment
of the Basement Sand (Lower Cretaceous),
West Texas
(1988) / Romanak M

Grain Morphology and Origins of Fine
Ash Deposits from the 180 A. D. Taupo
Eruption, Central North Island, New Zealand
(1988) / Talbot J P

Facies and Depositional Environments of the
Boquillas Formation, Upper Cretaceous of
Southwest Texas
(1988) / Trevino R H

TEXAS, UNIVERSITY OF (AUSTIN)

Eocene and Younger Volcanism on the Eastern
Bank of the Sierra Madre Occidental: Nazas,
Durango, Mexico
(1988) / Aguirre-Diaz G J

Mid-Cretaceous Unconformities in the East
Texas Basin and the Sabine Uplift
(1988) / Badachhape A R

Depositional Development and Fluvial
Architecture of the Narrabeen Group,
Illawarra District, Sydney Basin, Australia
(1987) / Reynolds S A

TOLEDO, UNIVERSITY OF

Stratigraphy and Depositional Environment
of the Bayport Limestone of the Southern
Michigan Basin
(1988) / Ciner A T

An Investigation of the Compatibility of
Zirconium with the Structures of Zeolites
Na-A and Faujasite
(1988) / DeMasi A E

An Investigation of Slope Failure Along the
Maumee River from Grand Rapids, Ohio, to
Toledo, Ohio
(1988) / Dimit G

Application of a Finite-Difference
Ground-Water Flow Model in the Evaluation
of the Simi-Confined Carbonite Aquifer of
Hancock County, Ohio
(1988) / King M A

Ground-Water Resources of Putnam
County, Ohio
(1988) / Stuckey G H

A Structural Investigation of the Room
Temperature Phase Transformation in
Maximum Microcline
(1988) / Zhang H

TULSA, UNIVERSITY OF

The Seismic Stratigraphy and Neogene
Evolution of Lime and Yaquina Basins,
Central Peru Forearc
(1988) / Ballesteros M W

Depositional Evolution of Trench Slope
Basins Shumagin Margin, Alaska
(1987) / Edelhoff E G

The Tectonic Framework and the Cretaceous
to Cenozoic Evolution of the East-Central
Philippines
(1987) / Florendo F F

Resedimentation and Diagenesis of
Barremian-Aptian Shallow-Water Carbonates
from the Galicia Margin, Eastern North
Atlantic at Ocean Drilling Program Site 641
(1987) / Germann S H

The Paleodepositional Environments of
Seamount 853, East Mariana Basin, Western
Pacific Ocean
(1987) / Orban C R

Geometric Evolution of a Flood-Oriented
Carbonate Shoal, Spanish Cay, Little
Bahama Bank
(1987) / Passaretti M L

UTAH STATE UNIVERSITY

Structural Geology of the Northern Part of
Clarkston Mountain, Malad Range, Utah
(1987) / Green D A

Spatial and Temporal Landslide Distribution
and Hazard Evaluation Analyzed by
Photogeologic Mapping and Relative Dating
Techniques, Salt River Range, Wyoming
(1987) / Rice J B

Hydrogeology and Hydrochemistry of Springs
in Mantua Valley and Vicinity, North-Central
Utah
(1987) / Rice K C

The Surficial Geology and Neotectonics of
Hansel Valley, Box Elder County, Utah
(1987) / Robison R M

Geochemical diagenesis of the Cretaceous
Woodbine Formation, East Texas Basin, Texas
(1987) / Wuerch H V

UTAH, UNIVERSITY OF

Geology of the Lothidok Range, Northern
Kenya
(1988) / Boschetto H B

The Geology and Genesis of the Apex
Gallium-Germanium Deposit, Washington
County, Utah
(1988) / Bowling D L

Geomorphic and Tectonic Significance of
Quarternary Sediments in Southern White
River Valley, Nevada
(1988) / Diguiseppi W H

Characterization of Tephra from the
Shungura Formation Southwestern Ethiopia
(1988) / Haileab B

Depositional Environments, Petrology, and
Diagenesis of the Basal Limestone Facies,
Green River Formation (Eocene), Uinta Basin
Utah
(1988) / Little T M

Stratigraphy, Depositional Environments, and
Petrology of the Lowermost Moenkopi
Formation, Southeastern Utah
(1988) / Ochs S

Glaciation and Quaternary Geomorphology of
the Blacks Fork Drainage, High Uinta
Mountains, Utah and Wyoming
(1988) / Schlenker G C

Bioerosion by Intertidal Endolithic
Invertebrates: Puerto Penasco, Sonora,
Mexico
(1988) / Stearley R F

Fracturing in the Wasatch Fault Zone:
Implications for Fluid Flow, Rock Mass
Strength and Earthquake Processes
(1988) / Thompson T R

Sedimentology and Tectonic Setting of
Conglomerate in Paleocene North Horn
Formation, Spanish Fork Canyon and Mount
Nebo Areas, Central Utah
(1988) / Wachtell D L

Sedimentology and Depositional Environments
of the Permian Lower Cutler Group: Turk's
Head, Canyonlands National Park, Utah
(1988) / Weis M A

VERMONT, UNIVERSITY OF

The Geology and Petrology of the Eastern
Mesozoic Belt, Northern Sierra, California
(1988) / Christ G

Depositional Environment of the Lower Oak
Hill Group, Southern Quebec: Implications
for the Late Precambrian Breakup of North
America in the Quebec Reentrant
(1988) / Dowling W M

Commingling of Magmas in the Belknap
Mountains Complex, Central New Hampshire
(1988) / Whittemore A S

VIRGINIA STATE COLLEGE

The Downstream of Floodwalls on the James
River at Richmond, Virginia
(1989) / Proper M S

Geology of the Centerville Quarry, Fairfax
County, Virginia
(1989) / Sheffey R G

WASHINGTON STATE UNIVERSITY

Gold Mineralization in Low Angle Faults
American Girl Valley, Cargo Muchacho
Mountains, California
(1988) / Branham A D

Folding and Faulting in the Footwall of the
Diversion Thrust, North-Central Montana
(1988) / Johnson D M

Computer Analysis of Remote Sensing and
Geologic Datasets
(1988) / Johnson L K

WEST TEXAS STATE UNIVERSITY

Predicting Ages from Canopy Measurements for
Two Species of Juniperus in the Texas
Panhandle
(1988) / Allison P S

Tertiary Ash Flow Tuffs in the Southern
Quitman Mountains and Their Relationship to
Calderas in the Eagle and Northern Quitman
Mountains
(1988) / Babb J L

Cenozoic Geology of the Eastern Half of the
La Flor Quadrangle, Durango and Chihuahua,
Mexico
(1988) / Bartolino J R

Depositional History and Hydrocarbon
Potential of the Lower Douglas "A" Sandstone
(Upper Pennsylvanian) in a Portion of
Hemphill County, Texas
(1988) / Sparling K D

The Effects of Subsurface Dissolution of
Permian Salt on the Deposition, Stratigraphy
and Structure of the Ogallala Formation
(Late Miocene Age) Northeast Potter County,
Texas
(1988) / Wilson G A

WEST VIRGINIA UNIVERSITY

Stratigraphic Analysis and Syndepositional
Structural Influence on Cambrian Units of
the Rome Trough in West Virginia
(1988) / Allen J P

Paleontology of the Meadville Member of the
Cuyahoga Formation (Mississippi) in Medina
County, Ohio
(1988) / Bird D L

Stratigraphy of the Upper Monongahela Group
in Portions of Greene County, Pennsylvania
and Monongalia County, West Virginia
(1988) / Brazzle S A

A Study of Dewatering Effects at Three
Longwall Mines in the Northern Appalachian
Coal Field
(1988) / Dixon D Y

WEST VIRGINIA UNIVERSITY
(continued)

Low Pressure Metamorphism in the Orrs
Island-Harpswell Neck Area, Maine
(1988) / Dunn G R

Structural Analysis of the Silurian-Devonian
Cover on the Smoke Holes, West Virginia
(1988) / Gerritsen S S

Deformation and Metamorphism of the
Aluminous Schist Member of the Setters
Formation, Cockeysville, Maryland
(1988) / Hall P S

Petrology and Diagenesis of the Big Injun
Sandstone Clay County, West Virginia
(1988) / Hannah S B

An Evaluation of Point-Source Limestone
Intriduction as an Ameliorative Procedure
to Reduce Acidity in Two Low Buffer Capacity
Streams
(1988) / Ivahnenko T I

Stratigraphy and Depositional Environments
of Upper Devonian and Lower Mississippian
Sandstones of Southeastern West Virginia
(1988) / Jewell G A

Depositional Environments of the Upper
Chemung and Lower Hampshire Formations of
East-Central West Virginia
(1988) / McDowell R J

Ground Water Chemistry as an Exploration
Tool for Natural Gas in Southwestern West
Virginia
(1988) / Zurbuch J S

WESTERN KENTUCKY UNIVERSITY

The Use of Down-Hole Video to Evaluate
Parameters Affecting Removal of Storm Water
Runoff Into a Karst Aquifer in Bowling
Green, Kentucky
(1988) / Reeder P P

WESTERN MICHIGAN UNIVERSITY

Definition of the B Salt Edge of Southern
Limb of the Michigan Basin Using Seismic
Techniques
(1988) / Fici H

A Seismic Analysis of Reefs in the Traverse
Limestone of Allegan County, Michigan
(1988) / Henderson W

Electrical Resistivity as an Approach to
Evaluating Brine Contamination of
Groundwater in the Walker Oil Field,
Ottawa County, Michigan
(1988) / Koehler J

An Investigation of a Soil Gas Sampling
Technique and Its Applicability for
Detecting Gaseous PCE and TCA Over an
Unconfined Granular Aquifer
(1988) / Mayotte T

Geometry and Kinematics of the Central
Snowcrest Range: A Rocky Mountain Foreland
Uplift in Southwestern Montana
(1988) / McBride B

The Significance of Organic Complexing in
the Mobility of Iron in the KL Landfill
Leachate Plume, Kalamazoo, Michigan
(1988) / Rudder J

Fission-Track Geochronology and
Geothermometry of Late Cretaceous-Early
Tertiary Epizonal Plutons, in and Adjacent
to the Lombard Thrust Fault, Southwestern
Montana
(1988) / Talanda J

WESTERN ONTARIO, UNIVERSITY OF

Comparative Geology of Gold-Bearing Archean
Iron Formation, Slave Structural Province,
Northwest Territories
(1988) / Ford R C

Geology of the Archean Turbidite-Hosted
Gatlan Gold Occurrence, Wallace Lake
Greenstone Belt, Southeast Manitoba
(1988) / Gaba R G

Genesis of Intra-Till Sorted Sediment
Layers Catfish Creek Till Bradville Ontario
(1988) / Hart B R

Geology and Control of Sulphide Deposition
of the J and L Massive Sulphide Deposit,
Southeast British Columbia
(1988) / McKinlay F T

WESTERN WASHINGTON UNIVERSITY

The Depositional Environment, Petrography,
and Tectonic Implications of Informally
Named Middle to Late Eocene Marine Strata,
Western Olympic Peninsula, Washington
(1988) / Adams B N

The Geology and Petrology of the Fifes Peak
Formation in the Cliffdell Area, Central
Cascades, Washington
(1988) / Carkin B A

Hydrothermal Alteration of Serpentinite
Associated with the Devils Mountain Fault
Zone, Skagit County, Washington
(1988) / Graham D C

The Effects of Urbanization on the Water
Balance of the Fishtrap Creek Basin,
Northwest Washington and South Central
British Columbia
(1988) / Lindsay C S

Changes in Grain-Size Distribution in a
Gravel-Bed Stream Due to a Point-Source
Influx of Fine Sediment
(1988) / Maloy J A

Mid-Tertiary Volcanic Rocks of the
Timberwolf Mountain Area, South-Central
Cascades, Washington
(1988) / Shultz J M

The DeForest Creek Landslide and Sediment
Transport in Deer Creek, Skagit County,
Washington
(1988) / Thompson J N

The Effects of a Mid-Foreshore Groundwater
Effluent Zone on Tidal-Cycle Sediment
Distribution
(1988) / Turner R J

WICHITA STATE UNIVERSITY

A Quantitative Examination of Alluvial Fans
on the Piedmont of the San Juan Mountains,
South-Central Colorado
(1987) / Bamberger M J

Study of Hydrocarbon Production and
Potential Production of the Ireland
Sandstone (Douglas Group, Upper
Pennsylvanian) in Northern Barber County,
Kansas
(1987) / Bober D R

Porosity and Permeability Analysis in
Mississippian System, St. Louis Limestone
Formation, Damme Field, Finney County,
Kansas Utilizing Petrographic Image Analysis
(1987) / Dietterich R J

A Subsurface Study of an Alga-Mount Complex
(Toronto Limestone, Upper Pennsylvanian) in
Parts of Barber, Kingman, and Pratt
Counties, Kansas
(1987) / Grommesh M W

Study of the Ireland Sandstone (Douglas
Group, Upper Pennsylvanian) in East Central
Kansas
(1987) / Henning L G

Factor Analysis in the Interpretation of
Petrology of Rambler Area, Newfoundland,
Canada
(1987) / Kambampati M V

Relationship Between the Reservoir
Properties of the Cattleman Sandstone
(Cabaniss Formation, Cherokee Group,
Pennsylvanian) Montgomery County, Kansas
(1987) / Patterson J M

Geology and Regional Tectonic Significance
of Blanca Peak Area, Huerfono County,
Colorado
(1987) / Reynolds T B

Fluvial Sedimentology and Associated Bone
Distribution in the Northwestern Corner
(Steveville Region) Dinosaur Provincial
Park, Alberta, Canada
(1987) / Salisbury J P

Depositional Environment and Petrographic
Image Analysis of Porosity Development of
a Middle Pennsylvanian Series Hydrocarbon
Reservoir in the Midcontinent Region
(1987) / Scheffe G L

Fourier Grain-Shape Analysis of Sediments
from the Oregon Continental Shelf
(1987) / Swilley G K

Aspects of the Depositional and Diagenetic
History of the Pleistocene Miami Oolite in
the Southern Florida Keys (Big Pine Key to
Key West)
(1987) / Zimmerman P J

Local Facies Relations of the Dennis
Formation (Kansas City Group, Pennsylvanian
System), in Northwestern Kansas
(1987) / Zink L

WINDSOR, UNIVERSITY OF

Solution-Generated Collapse (SGC) Structures
and Formation-Fluid Hydrodynamics in
Southern Manitoba
(1988) / Bacopoulos I

An Investigation of Hydraulic Connection of
Sand Lenses in Silty Clay Till Deposits of
Essex County
(1988) / Brathwaite S L

A Sedimentologic and Stratigraphic Study of
Carboniferous Through Jurassic Strata of
Kenya
(1988) / Karanja S

Late Quaternary Paleoceanography in Kane
Basin, Canada and Greenland
(1988) / Marentette K

Isotopic and Geochemical Study of
Groundwater in the Eastern Province of Kenya
(1988) / Mwangi M P

A Stratigraphic and Sedimentologic Study of
the Cretaceous and Tertiary Strate of Kenya
(1988) / Nyagah K

WYOMING, UNIVERSITY OF

Callovian and Oxfordian Foraminifera of
Northern and Western Wyoming
(1989) / Caparco N J

Tectonothermal Controls of Hydrocarbon
Maturation and Siliciclastic Source/
Reservoir Diagenesis, Eocene Misoa
Formation, Bolivar Coastal Field, Maracaibo
Basin, Venezuela
(1989) / Manske M C

Geology of a Precambrian Igneous/Metamorphic
Sequence, Laramie Range, Wyoming
(1989) / Marlatt G G

Foreland Structure and Karstic Ground Water
Circulation in the Eastern Gros Ventre
Range, Wyoming
(1989) / Mills J P

Provenance, Dispersal, and Tectonic
Significance of the Evanston Formation and
Sublette Range Conglomerate, Idaho-Wyoming-
Utah Thrust Belt
(1989) / Salat T S

Structural Obstruction of Recharge to the
Paleozoic Aquifer, Denver-Julesburg Basin
Along the Laramie Range, Wyoming
(1989) / Wiersma U M

BOSTON COLLEGE

Geophysical Investigations of the Proposed
Micronesian Trench
(1988) / Bessette R P

BOWLING GREEN STATE UNIVERSITY

Neotectonic Study of the South-Central
Portion of the Mojave Desert Utilizing
Potential Field and Multispectral Remote
Sensing Geophysical Techniques
(1989) / Etter K H

CALGARY, UNIVERSITY OF

Code Improvement Through Flow Analysis
(1988) / Andrews K O M

The Complex Resistivity Response of
Mineralized Rocks at Sub-Freezing
Temperatures
(1988) / Calvert H T

Design and Applications of a Diversity
Stack Computer Program
(1988) / Dufresne D

The Effect of Wettability and Pore Geometry
on Three-Phase Fluid Displacement in Porous
Media
(1988) / Kong V W T

Origin of Columnar Jointing in Recent
Basaltic Flows, Garibaldi Area, Southwest
British Columbia
(1988) / Lee L J

Attenuation of the Effect of Harmonic
Distortion on Synthetic Vibroseis Data Using
an Exact Wave-Shaping Filtering Method
(1988) / Lortie J

Investigation of Inhomogeneous Body Waves in
an Elastic/Anelastic Medium
(1988) / Slawinski M A

A Systematic Wettability Study of Selected
Alberta Oil Reservoirs
(1988) / Steffes D A

Fluvial Geomorphology, Sedimentology and
Paleohydrology of the Piedmont East of
Glacier National Park, Montana
(1988) / du Toit C

COLORADO SCHOOL OF MINES

"STAFOID" An Expert System to Identify
Seismic Tape Formats
(1988) / Agena W F

An Expert System to Evaluate Geologic Risk
for Onshore Domestic Petroleum Exploration
(1988) / Lessenger M A

Seismic Prestack Analysis of a Mission
Canyon Reservoir, Wiley Field, Bottineua
County, North Dakota
(1988) / Marhoefer J

Seismic Stratigraphic Investigation of the
Minnelusa and Goose Egg Formations, Western
Crook County, Wyoming
(1988) / Shoemaker S W

Correlation and Prediction of the Earth's
Electromagnetic Field
(1988) / Worthington E W

CONNECTICUT, UNIVERSITY OF

A Seismic Retraction and Electrical
Resistivity Evaluation of a Fault Zone Near
the University of Connecticut Landfill,
Storrs, Connecticut
(1988) / Girioni M J

The Determination and Reduction of Ship
Noise
(1988) / Zhang Z

HAWAII, UNIVERSITY OF

Possible Effects of Sea Level Rise on
Kailua, Oahu and Implications for Pacific
Atolls
(1988) / Rodgers D L

Indo-Australia Plate Motion Determined from
Australia and Tasman Sea Hotspot Trails
(1988) / Yan C Y

MASSACHUSETTS INSTITUTE OF TECHNOLOGY

The Relationship Between Plate Curvature
and Elastic Plate Thickness: A Study of
the Pen-Chile Trench
(1988) / Judge A V

Topographic Reconstruction from Radar Images
(1988) / Matarese J R

Structural Style and Kinematic History of
the Active Panamint-Saline Extensional
System, Inyo County, California
(1988) / Sternlof K R

MIAMI, UNIVERSITY OF

The Formation and Preservation of
Submariliths on the South Florida Shelf
(1988) / Berler D H

The Reliability of Paleotemperatures from
Cretaceous DSDP Inoceramus
(1988) / Child C J

The Late Pleistocene Deep Water History of
the Venezuela Basin in the Eastern
Caribbean Sea
(1988) / Cofer-Shabica N B

Distribution and Petrography of Late
Cenozoic Dolomites Beneath San Salvador and
New Providence Islands, the Bahamas
(1988) / Dawans J M

Satellite Laser Ranging and Geologic
Constraints on Plate Tectonic Motion
(1988) / Douglas N B

The Control of Uranium Concentrations in
Pleistocene Planktonic Foraminifera
(1988) / Harrison S A

Experimental Compaction Studies of Lithic
Sands
(1988) / Kurkjy K A

The Evolution of Shallowing-Upwards Reef to
Oolite Sequences at the Leeward Margin of
Caicos Platform, B. W. I.
(1988) / Waltz M D

MICHIGAN TECHNOLOGICAL UNIVERSITY

A Magnetotelluric Transect of the Oregon
Coast Range
(1988) / Kitchen M R

MINNESOTA, UNIVERSITY OF

A Comparison of Several Land Seismic
Impulse Sources
(1988) / Nelson J C

MURRAY STATE UNIVERSITY

A Quantitative Comparison of an Isodata and
a Single-Pass Clustering Algorithm Applied
Over Simulated Multispectral Image Data Sets
(1988) / Cobb C M

OKLAHOMA, UNIVERSITY OF

Characterization of Possible Kerogen
Percursors by Pyrolysis-Gas Chromatography
(1988) / Galvez Sinibaldi A S

Experimental Investigations of Three
Dimensional Kinematics at Sharp Terminations
(Corners) in Forced Folds
(1988) / MacLeod N S

An Analysis of Geophysical Anomalies in
North-Central Oklahoma and Their
Relationship to the Midcontinent Geophysical
Anomaly
(1988) / Nixon G A

OREGON STATE UNIVERSITY

High Resolution Seismic Refraction Study of the
Uppermost Oceanic Crust Near the Juan De Fuca
Ridge
(1987) / Poujol M

PRINCETON UNIVERSITY

Pressure-Saturation Relations for Water,
TCE and Air in Sand
(1988) / Turrin R P

PURDUE UNIVERSITY

Permeability of Young Oceanic Crust
(1988) / Ballotti D M

The Seismic Properties of Mafic Volcanic
Rocks of the Reweenawan Supergroup and
Their Implications
(1988) / Lippos C S

Seismic Velocity Analysis of Shot Point 16
from the PASSCAL Ouachita Experiment
(1988) / Lyslo J A

An Integrated Geophysical Study of the
Grenville Front in Lake Huron
(1988) / Smith J G

STEPHAN F. AUSTIN STATE UNIVERSITY

Structure and Seismic Stratigraphy of
Post-Ouachita Paleozoic to Jurassic
Sediments of Southeastern Arkansas,
Northeastern Louisiana, and West-Central
Mississippi
(1988) / Sutley W C

TEXAS A AND M UNIVERSITY

Geology and Tectonics of Mendana Fracture Zone
Between Latitudes 10 Degrees S - 12 Degrees S and
Longitudes 80 Degrees W - 83 Degrees W
(1988) / Bullion D N

Time Series Analysis of Paleomagnetic Polarity
Transition Zones
(1988) / Celaya M A

Seismic Stratigraphy and Structure of the Barter
Island Sector of the Western Beaufort Sea
(1988) / Coffman J D

Late Tertiary Paleomagnetic Data from Leyte,
Philippines: Implications for Philippine
Fault Zone Motion
(1988) / Cole J T

Seismic Model Study of Fault and Fold Geometries
(1988) / Hand L M

Sonar Probing of Concrete
(1988) / Mims J H

Estimation of the Direction of Remanent
Magnetization: An Inverse Method Using
the Phase Spectrum of a Magnetic Anomaly
(1988) / Moriarty T D

Characterization of a Sandstone Reservoir Using
Seismic Methods: Yowlumne Field, Kern County,
California
(1988) / Proust R D

Paleomagnetism of Paisano Volcano, Texas
(1988) / Ryan D

High Resolution Seismic and Paleomagnetic Study
of the Structure and Sedimentation of Sweet and
Phleger Banks, Northern Gulf of Mexico
(1988) / Singleton S W

In Situ Orientation and Calibration of
Three-Component Downhole Seismometers
(1988) / Smith G A

Three-Dimensional Seismic Stratigraphic Study of
Minnelusa Formation, Powder River Basin,
Campbell County, Wyoming
(1988) / Walters D L

TEXAS, UNIVERSITY OF (AUSTIN)

Near-Surface Velocity Reconstruction and
Diving Wave Tomography: Erawan Field, Gulf
of Thailand
(1988) / Sauve J A

An EMAP Survey of the Southern Wind River
Overthrust, Wyoming
(1988) / Williams J B

TEXAS, UNIVERSITY OF (EL PASO)

Geophysical Anomalies in Southwestern New Mexico and
Adjacent Areas
(1988) / DeAngelo M V

The Seismicity of Northern Tanzania and Southern
Kenya and It's Tectonic and Implications
(1988) / Yarwood D R

TULSA, UNIVERSITY OF

Correlation of Ocean Drilling Program Site
679 Well Log and Seismic Reflection Data
(1987) / Burger J C

UTAH, UNIVERSITY OF

Controlled-Source Audio-Frequency
Magnetotelluric Responses of Three
Dimensional Bodies
(1988) / Butterworth N A

A Study of Seismicity and Spectral Source
Characteristics of Small Earthquakes:
Hansel Valley, Utah and Pocatello Valley,
Idaho Areas
(1988) / Chen G J

Electrical Properties of Clay-Bearing
Sandstones
(1988) / Groenewold J C

A Finite Difference Simulation of Seismic
Wave Propagation and Resonance in Salt
Lake Valley, Utah
(1988) / Hill J A

Precision Gravity Reobservations at
Yellowstone National Park, Wyoming,
1977-1987
(1988) / Hollis J R

The Feasibility of Imaging Near-Vertical
Incidence Lower Crustal and Upper Mantle
Reflectors from Seismic Refraction Data
(1988) / Kjos E J

Extremal Inversion of Seismic Reflection
Data from the 1986 Passcal Basin and Range
Seismic Experiment
(1988) / Matulevich M M

Use of Transient Electromagnetic Soundings
to Correct Static Shifts in Magnetotelluric
Data
(1988) / Pellerin L D

WASHINGTON, UNIVERSITY OF

Crustal Structure and Seismicity of the
Northern Gorda Ridge
(1987) / Johnson D A

The Subduction Geometry Beneath Western
Washington from Deconvolved Teleseismic
P-Waveforms
(1987) / Lapp D B

Coda-Q Analysis at Mt. St. Helens and Its
Implications for Volcanic Seismology
(1987) / McClurg D C

Compression of Gamma-Nickel Silicate
Spinel: A Silicate Spinel, and a
Geophysical Implication
(1987) / Nelson C J

WESTERN ONTARIO, UNIVERSITY OF

Evaluation of Various Processing Techniques
Applied to Seismic Data Recorded from an Ice
Island in the Arctic Ocean, Canada
(1988) / Cox T P

Mise-a-la-Masse Study of Prismatic
Bodies: Model Tank and Field Experiments
(1988) / King J G

Investigations in Shallow Reflection
Seismology
(1988) / Shortt T A L

Analysis of the Data of the 1984
Kapuskasing Seismic Experiment
(1988) / Wu J

WYOMING, UNIVERSITY OF

Converting Spinner Magnetometers for
Computer-Controlled Data Collection and
Reduction and Paleomagnetic Study of
Volcanic and Volcaniclastic Rocks from the
Absaroka Mountains, Northwestern Wyoming
(1989) / Isbell W B

P-SV Wave Processing of Sonic Logs
(1989) / Munoz P

ALABAMA, UNIVERSITY OF (HUNTSVILLE)

A Knowledge Based Decision Aid for Research
and Development Project Evaluation
(1988) / Wakefield G S

ALFRED UNIVERSITY

Interacting New Facilities and Location
Allocation Problem
(1988) / Brown G F

Robot Selection: A Fuzzy Set Multicriteria
Decision Making Approach
(1988) / Singh H P

ARIZONA STATE UNIVERSITY

Development of Process Control Procedure
Using Attributes Data and an Investigation
of Expert System Application
(1988) / Date S C

Voice Recognition and Artificial
Intelligence in an Air Traffic Control
Environment
(1988) / Hall R F

A Qualitative Simulation Based Heuristic
for Assisting Organizational Acceptance of
Computer Integrated Manufacturing
(1988) / Paul B K

Roads: A Real-Time Operator Advisor Design
System for Expert Systems in Process Control
(1988) / Sullivan E

Smart Help for Operator Performance
(Shop): Methodologies and Requirements
Definition for Shop Floor Help Systems
(1988) / Wall M L

ARKANSAS, UNIVERSITY OF

A Comparison of One and Two Person Lifts in
the Performance of Manual Material Handling
Tasks
(1988) / Lewis D M

An Object-Oriented Programming Approach to
Modeling Components of a Manufacturing
System Using Smalltalk-80
(1988) / Ogle M K

Analysis of Automated Extraction for a
Storage/Retrieval Carousel
(1988) / Stanley D A

Real-Time Microcomputer Control and
Integration of an Automated Guided Vehicle
System
(1988) / Whitt W A

AUBURN UNIVERSITY

Normal Quantile Approximations of Certain
Symmetrical Linear Combinations
(1987) / Bowman V A

A Study of Voice Control of Robots
(1987) / Cheng S H

The Algebraic Structure of a b[k] Factorial Design
(1987) / Chih W W

A Study of Off-Angle Viewing Effects on the
Legibility of Information Presented on a
Video Display Unit
(1987) / Chu R W

The Influence of Individual Moderators on the
Relationship Between Job Enrichment and
Organizational Commitment
(1987) / Hodges C R

Work-in-Process Inventory Control for Repetitive
Manufacturers in a Material Requirements Planning
Environment
(1987) / Ledbetter M E

A Method for Determining the Alias Structure of a
Taguchi-Based Statistical Experimental Design
(1987) / Schultz L A

An Experimental Investigation of the Effects of
Data Models and Task Complexity on the Performance
of the End User
(1987) / Thompson N G

The Third and Fourth Moment Limits Required to Achieve
an Adequate Normal Approximation for a Linear
Combination of Beta Variates
(1987) / Wu J C

BOSTON COLLEGE

Technology and Inertia
(1988) / Davis H F

BRIGHAM YOUNG UNIVERSITY

NC Tool Path Verification Utilizing
Stereographic Display
(1988) / Bell K R

A Rule-Based Approach Employing Feature
Recognition for Engineering Graphics
Characterization
(1988) / Belnap B K

Automatic Avoidance of Numerical Control
Tooling Fixtures
(1988) / Bodine B G

Multi-Point Flame Speed Determination in an
Internal Combustion Engine
(1988) / Bonin M P

Economic Modeling for Electronic Assembly
(1988) / Carter M L

Macromechanical Properties Characterization
of an Advanced Polymeric Composite Made by
Filament Winding
(1988) / Graff M C

Design for Testability
(1988) / Haas J D

Incremental Forming of Large Fiber
Reinforced Thermoplastic Parts
(1988) / Hauwiller P B

Electronic Component Selection Attributes
(1988) / Klem P E

A Literature Review of Dimensional Change
of Steel During Heat Treatment for Stock
Reservation
(1988) / Liu C C

Robot Path Following Accuracy and
Repeatability
(1988) / Mao G

Computer-Aided Design Parts Libraries:
Structure, Access, and Implementation
(1988) / Metcalf R P

BRIGHAM YOUNG UNIVERSITY
(continued)

Applications of Group Technology for
Assembly Intensive Manufacturing
(1988) / Midgley B W

Continuous Scallop Height Tool Paths for
Sculptured Surfaces
(1988) / Mortensen F L

Product Labeling
(1988) / Novak R S

Development of a Design Concept Storage
and Retrieval System
(1988) / Orr R L

Space Filling Curves in Tool Path
Applications
(1988) / Takezaki Y

A Comparison of Automated Versue Manual
Techniques in the Manufacture of Orthotics
(1988) / Wilkes K S

CALIFORNIA STATE UNIV. (LONG BEACH)

Location Decision-Making in Manufacturing:
A Three City Study
(1988) / Moshier S P

CASE WESTERN RESERVE UNIVERSITY

An Interactive Approach for Multiple
Objective Integer Programming with
Application to the Assembly Line Balancing
Problem
(1988) / Araki S

A Hierarchical Intelligent Controller for
Complex and Uncertain Processes
(1988) / Berger K M

A Method for Off-Line and On-Line
Multiobjective Optimization for Machining
(1988) / Deviprasad J

A Unified Approach to Predictive Control
(1988) / Dong Y

Process Parameter Estimation Based on
Integral Measures
(1988) / Kuchar A

Multiple Objective Control System Design
Using Constrained Optimal Output Feedback
(1988) / Kyr D E

State Tolerance Relations in Deterministic
and Stochastic Automata
(1988) / Radivoyevitch T

Finite State Algebraic Systems and Trellis
Codes
(1988) / Trott M D

CENTRAL FLORIDA, UNIVERSITY OF

MRP, JIT and OPT: A Comparison on the
Production Floor
(1988) / Denicole-Christopher D M

A Student Performance Evaluation Method for
an Intelligent Simulation-Based Tutoring
System
(1988) / Dixon C M

A Proposed System for Determination of
Percent Cloud Cover
(1988) / Emrich C L

Management Considerations of Productivity
Issues for Local Government
(1988) / Gilbertson S M

Geographic Information Systems: The
Developer's Perspective
(1988) / Henris J B

Interactive Video Based Training Systems
(1988) / Hightower R H

Some Reasons Why New Technologies Are Not
Always Successful - Specifically Directed
Toward CIM
(1988) / Lai S F

A Dynamic Performance-Based Student Model
for an Intelligent Simulation Training
System
(1988) / Sargeant J M

An Overview of Machine Vision
(1988) / Telchi J C

Economic Design of Lot-by-Lot Acceptance
Sampling by Atttributes
(1988) / Wall M S

A Learning Curve Model for Interrupted
Production
(1988) / Zarkoob B M

CLEMSON UNIVERSITY

An Investigation of Localized and Global
Approaches for Tiled Window Arrangement
(1988) / Chen J F

Analyzing the Effect of Installing
Diagnostic Equipment on a Manufacturing Cell
(1988) / Lahoti A K

A Simulation Model to Examine Equipment
Delays in Adaptive Control Machining
(1988) / Minton E T

A Dynamic Programming Approach for the
Design of Machining Centers
(1988) / Moody B R

A Cost Based Breakeven Analysis of Military
Standard 105D Acceptance Sampling Plan
Versus Statistical Process Control
(1988) / Motamedi A

An Investigation of Area Coding Techniques
for Monochromatic Visual Displays
(1988) / Mukherjee S

Development of an Algorithm to Generate a
Computer-Generated Time Formula Using
Multiple Regression Analysis
(1988) / Mukhopadhyay S

Integrated Workcell Control Methodologies
(1988) / Tissue J A

Design of a movement Control Training Aid
Using Visual Interactive Simulation
(1988) / Zargan C S

CORNELL UNIVERSITY

Earnings and Occupational Characteristics of
Men and Women: A Cross-Sectional and
Longitudinal Study
(1988) / Cheikh N E

A Cognitive Model of Successful and
Unsuccessful R&D Projects
(1988) / Knight D B

Industrial Decline and Readjustment: The
Experience of Minnesota's Iron Range
(1988) / Lund M A

The Utility of an Unstructured Setting for
the Assessment of Personality and
Interpersonal Behavior
(1988) / Seibert S E

FLORIDA, UNIVERSITY OF

Maximum Flow Labeling Algorithm for Pure
Processing Networks with Single Processing
Node
(1988) / Horak T

A Mathematical Programming Approach to the
Group Technology Part Family/Machine Group
Formation Problem
(1988) / Lee E K

Development of a Personal Computer Control
Program for Enhancement of Undergraduate
Instruction of Industrial and Systems
Engineering Students
(1988) / Weldon C

GEORGIA INSTITUTE OF TECHNOLOGY

Determination of Cutting-Tool Inventory
Levels in a Flexible Manufacturing System
(1988) / Graver T W

An Analysis of the State Space Graph for
Integer Permutation with Application to
Local Search
(1988) / Thompson B M

HOUSTON, UNIVERSITY OF

A Searching Procedure for Finding the
Underlying Probability Distribution Using
Visual Assessment and Statistical Techniques
(1988) / Chiu N

Resolution of Complex Reliability Networks
Using Symbolic Reasoning
(1988) / Patel R

An Analysis of Transition Management from an
Engineering Management Perspective
(1988) / Sitton R

Roboedit: A Case Study in Knowledge-Based
Welding Robot Programming Systems
(1988) / Sullivan E

IDAHO STATE UNIVERSITY

Boiler Design Considerations for the Freon
Loop Testing Facility
(1988) / Elatrash M S

Probabilistic Determination of Setpoints
(1989) / Sehgal S

IOWA STATE UNIVERSITY

Optimization for Deterministic Line
Balancing Problem with Unscheduled Downtimes
(1988) / Amin A V

Performance Analysis of a Generic Naval C2
System by Use of Timed Petri Net
(1988) / Choi B W

Analysis of an Adaptive Geometric Moving
Average Technique for Control of Metal
Cutting Operations
(1988) / Cyr P D

Investigation of Certain Properties of
Present Worth and Rate of Return Methods
of Analysis
(1988) / Deng J

A Study of Persistence Among Undergraduate
Women Engineering Students
(1988) / Evans M A

A Micro-Computer Based Vision System for
Quality Control of Finished Parts
(1988) / Heng K J

A Microcomputer Simulation Program to
Compare Mutually Exclusive Alternatives
Under Risk
(1988) / Kostka D B

The Development and Evaluation of a Graphics
Preprocessor for GPSS on Auto-CAD
(1988) / Labban J A H

Productivity Measurements of Information
Services
(1988) / Larson K W

Predictive Models of Hand Torque Strength
for Electrical Connectors
(1988) / Ma X

The Taguchi Methods of Quality Control
Examined: with Reference to
Sundstrand-Sauer
(1988) / Moller-Wong C L

LEHIGH UNIVERSITY

Smalltalk-80: The Use of Solving
Statistical Problems
(1987) / Chang C

A Model for Error Analysis of Robot
Manipulators
(1987) / Hammel G

Improvement of Software Reliability
Prediction: Piecewise Weibull Failure Rate
Model
(1987) / Park C

Computer System Development Productivity
Measurement and Improvement
(1987) / Pazak J M

A Career Development Program for a
Successful Manager of Information Services
(1987) / Regina J L

LOUISIANA STATE UNIVERSITY

Knowledge Based Selection of a Forecasting
Technique - A Prototype
(1988) / Chainani S R

MASSACHUSETTS INSTITUTE OF TECHNOLOGY
(continued)

Building a Better Workforce: The Corporate
Role in Education
(1988) / Gordon B S

Strategic Management and Human Resource
Practices and Experiences in Automotive
Joint Ventures Established in Mexico, as
a Strategy of Competitive Advantage
(1988) / Gutierrez M

Development of the Organization Structure
of a Global Business Unit
(1988) / Heiskanen T P

Service Quality Frameworks in the Service
Industry
(1988) / Hoech J

A Study of Information Technology Innovation
in High Technology Firms
(1988) / Homer P B

The Effects of Bilateral Aviation Agreements
on the Market for International Passenger
Service from the U.S. to Europe
(1988) / Hopkins B

An Object-Oriented Approach Towards
Enhancing Logical Connectivity in a
Distributed Database Environment
(1988) / Horton D C

Information Technology Planning for a
Rapidly Growing Computer Distributor
(1988) / Kershner S K

Sun Microsystems and a Strategic Analysis of
the Technical Workstation Industry
(1988) / Kessle-Hunter K L

Electricity Conservation at New England
Electric System: Economic and Strategic
Issues
(1988) / King C N

What Makes Technical Professionals Leave and
then Return: Career Anchors in a Federal
R & D Laboratory
(1988) / Kittredge R J

The Role of Defense Spending in the
Development of the U.S. Electronics Industry
(1988) / Koch L

Structural Design of Information Systems for
Computer-Integrated Manufacturing
(1988) / Koefoot E B

The Strategic Evolution of a Young Software
Company
(1988) / Lamp E D

The Management/Engineering Interface:
Redesign for Improved Manufacturing TPO8
(1988) / Laney A R

The Bounded Rationality Constraint:
Experimental Design and Analysis
(1988) / Louvet A A

The B-1B Bomber: A Program History
(1988) / Magiawala K R

Modeling and Evaluation of Expert Systems in
Decision Making Organizations
(1988) / Marie-Joseph Perdu D

CONSTRUCTION: The Foundation of National
Defense
(1988) / Martin G F

Comparative Case Studies of the Introduction
of New Manufacturing Systems
(1988) / McCan S A

An Analysis of the Korean Semiconductor
Industry with Comparisons to Japan
(1988) / Moran P B

An Investigation of the Implementation of
Taguchi Method Quality Engineering in Five
U.S. Corporations
(1988) / Morrow J A

Success Factors in Strategy of Suppliers of
Computer-Integrated Manufacturing Hardware
and Software
(1988) / Murphy B D

The Use and Management of Information
Technology in a Multinational Company
(1988) / Nash L S

The Structure of Industrial Marketing
Organizations: A Comparative Study
(1988) / Ofner A G

Towards a Theory for the Design and
Operation of Manufacturing Systems
(1988) / Parthasarathy S

Financial Contracting in Construction
Management
(1988) / Pastore R M

Information Technology Planning for a
Rapidly Growing Computer Distributor
(1988) / Pastoriza J J

Solution of Some Problems in Decentralized
Detection by a Large Number of Sensors
(1988) / Polychronopoulos G H

Innovation in the Development of
Just-in-Time Delivery Services in the U.S.
Automobile Parts Industry
(1988) / Proft M R

Acquiring Modern Technology: Make, Buy or
Borrow?
(1988) / Rangan S

The Economics of Product Design: A Model
and an Application
(1988) / Sabanoglu J

Study of Probabilistic Noise in
One-Dimensional Images
(1988) / Saias A I

Optoelectronics Technology Strategy: A Case
Study of Selected U.S. Firms
(1988) / Schexnayder M C

Teamwork and Participative Management:
Elements of Successful Implementation
(1988) / Schwartz R J

A Comparative Analysis of the U.S. and
Japan: Communications Competitiveness
(1988) / Seese L A

Product Development in the Automobile
Industry: Corporate Strategies and Project
Performance
(1988) / Sheriff A M

MASSACHUSETTS INSTITUTE OF TECHNOLOGY
(continued)

Effect of Consumer Categorization Behavior
on New Product Sales Forecasting
(1988) / Srinivasan K V

The Proposed Chemical Weapons Convention and
U.S. National Security: A Policy Assessment
(1988) / Stern J E

Management of Technical Design by
Non-Technical Managers
(1988) / Strauss L I

A Study of Production Loading in a Job Shop
(1988) / Sy-Quia G C

Information Technology and Human Resource
Management: A Case Study of Organizational
Learning
(1988) / Villiger D E

Evaluation of Increased Productivity to an
American Production Line: Application of
Japanese Total Quality Control Policies
(1987) / Vitry d'Avaucourt B

Spatial Logistics Management: Analysis of
Potential Applications
(1987) / Wong C

Chinese Motor Vehicle Industry: Technology
Strategy for the Future
(1988) / Xue Q

The U.S. and Japanese Machine Tool
Industries: Applications of Flexible
Manufacturing Systems
(1988) / Zaumseil D R

The Use of Simulation in Automated,
Just-in-Time Manufacturing Systems: A Case
Study of a Disk-Drive Assembly Line
(1988) / van Doorne A W

MASSACHUSETTS, UNIVERSITY OF

A Rule-Based Expert System for Fixturing on
a CAD System Using Flexible Fixtures E-CAFFS
(1988) / Alladin S

VDT Workstation Design: The Effect of
Adjustable Furniture and Glare on
Performance of a Classification Task
(1988) / Dawes S M

Routing in Open Finite Queueing Networks
(1988) / Gosavi H D

A Generic Simulator for Continuous Flow
Manufacturing
(1988) / Krishnaswamy C

Feature Reasoning for Automatic Robotic
Assembly
(1988) / Liu H C

Rules for an Expert Fixturing System on a
CAD Screen Using Flexible Fixtures
(1988) / Lyu P C

Demonstration of a Generation Effect Under
Text Editing Conditions
(1988) / Nnaji P H

Wind Turbine Layout for Large Scale
Windfarm Applications
(1988) / Schlange T

Selection of Graphic Symbols for Visual
Displays: The FI Model
(1988) / Tanner N S

Probing the Operator's Mental Model of
System State Categories Using
Multidimensional Scaling
(1988) / Zubritzky M C

MIAMI, UNIVERSITY OF

Policy Rules Affecting AS/RS Throughput: An
Evaluation Using Computer Simulation
(1988) / Alvarez E J

A Computer Interface for Direct Numerical
Control in CIM
(1988) / Beruvides M G

Physiological Models and Guidelines for the
Design of Prolonged Arm Lifting Tasks
(1988) / Tritar M

MICHIGAN TECHNOLOGICAL UNIVERSITY

Identifying the Human Factors Associated
with the Corporate Implementation of Desktop
Publishing
(1988) / Aller B M

Project Management for the Renovation of a
Unit Operations Laboratory
(1988) / Furtaw W J

Implementation of New Process and Management
Technology: A Case Study
(1988) / Gray V C

A Market Analysis and Applications of
Quality Function Deployment in the Banking
Industry: A Case Study
(1988) / Harris D G

Simulation of Material Design of Flat Rolled
Steel Manufacturing for Improved
Productivity
(1988) / Hoerger J D

An Evaluation of Statistical Process Control
in the Mineral Core Plant, Weyerhaeuser,
Marshfield, Wisconsin
(1988) / Kraai E D

Patient Transportation Quality and
Productivity Control with Automated Decision
Support
(1988) / McKay S V

Functional Integration for the Development
of a Vendor Certification Program at Mercury
Marine Division, Brunswick Corporation
(1988) / Riutta B L

MINNESOTA, UNIVERSITY OF

A Formulation of Diagnostic System on Fault
Localization
(1988) / Hu L

Dynamic Estimation of Job Lead Time
(1988) / Vig M M

A Simulation of a High Volume, Two Product
Machining Cell Under CFM
(1988) / Weaver G M

MISSISSIPPI STATE UNIVERSITY

Scheduling A One-Machine Job Shop with Job
Precedence Constraints
(1986) / Emplaincourt M C

Estimating Drafting Time for Pre-engineered Metal
Buildings
(1986) / Fort G A

Aspects of Engineering Economy Studies Under
Inflation
(1986) / Jordan J D

A Comparison of the New Lot-Sizing Methods With The
Least Total Cost Method and A Study of Inflation
Effects on an MRP System
(1986) / Kim J K

A Study of the Relationship Between Group Technology
and Material Requirements Planning
(1986) / Lakhmani G

Effective Heuristic Scheduling Algorithms to Minimize
Makespan on Parallel Non-Identical Processors
(1986) / Lee K D

An Investigation of the Comparative Rate of Return
(1986) / Park Y P

A Network Scheduling Procedure for a Manufacturing
Plant with Limited Resources
(1986) / Saha G G

A Study of the Japanese Kanban System and a
Comparison with Material Requirements Planning
(1986) / Shin M N

Design of a Mattress Facility
(1986) / Sisman Z B

JSPS: A Dynamic Job Shop Computer Simulator for
Use in Evaluating Priority Dispatching Rules
(1986) / Sun Z B

Heuristic Algorithm for School Bus Routing
(1986) / Tsai B Y

Systems Analysis of Sea Water Desalination
Plants in Libya
(1986) / Zahri A M

MISSOURI, UNIVERSITY OF (ROLLA)

Application of the Analytical Hierarchy
Process in Considering the Effects of Risk
and Time Pressure on Decision Making
(1988) / Daily G D

Packaging Engineering Education Qualitative
Need Assessment
(1988) / Sachs M D

The Extent of the Customer Activated
Industrial Product Idea Paradigm
(1987) / Strahorn D C

A Systematic Transition to Flexible
Manufacturing
(1988) / Webber D L

McDonnell Douglas Corporation's Five Keys
to Self-Renewal: An Evaluation of Their
Development and Implementation
(1988) / Winkler S L

MONTANA COLLEGE OF MINERAL SCI. & TECH.

Identification of Potential Gaseous Health
Hazards and Quantification of Benzene,
Toluene, and Xylene at a Pilot Coal
Beneficiation Plant
(1988) / Clark R

Identification and Quantification of Gaseous
Contaminants Evolving from Rosebud Coal from
Simulated Coal Beneficiation Process
(1988) / Faulkner D P

Evaluation of the Particle Size/Mass
Distribution of Quartz and the Particle
Size/Mass Distribution of Talc
(1988) / Vega C C

MONTREAL, UNIVERSITY OF

La Securite du Travail au Laboratoire de
Robotique de l'Ecole Polytechnique de
Montreal
(1988) / Alaurent R

Interprete Graphique Pour la Norme Videotex
sur une Station Bureautique Evoluee
(1988) / Asfazadour R

Implantation Exacte de Transactions a l'Aide
de Machines Sequentielles
(1988) / Babin G

Le Transfert de l'Apprentissage Dans le
Milieu de Travail: Etude de Cas en Milieu
Hospitalier
(1988) / Beaudin L

Systeme Integre de Mesure et d'Analyse de
la Performance Budgetaire: Thheorie et
Application
(1988) / Bellazzi C

Modex 11: Logiciel Servant a Resoudre une
Version du Probleme de Planification de la
Production d'un Parc a Predominance
Hydroelectrique
(1988) / Bergeron P

Conciliation et Mediations a l'Occasion de
la Greve a Marine Industrie en 1984 et 1985
(1988) / Bourque A

La Dynamique de l'Evolution du Transfert
Technologique au Japon et son Impact sur
les Investissements Japonais a l'Etranger
(1988) / Boutin F

Modelisation des Arrets Multiples d'Autobus
Pour les Reseaux de Transport en Commun
(1988) / Bouzaiene Ayari B

Prevision a Court Terme de la Demande
d'Electricite a L'Aide de Modeles de
Series Chronologiques
(1988) / Brousseau D

La Conception Interactive de Caracteres
Typographiques
(1988) / Bur J

Etude d'Implantation et Analyse de
Rentabilite d'un Centre de Conditionnement
de Verre et Aluminium
(1988) / Chaarani A

Adaptation des Techniques d'Affichage
Tridimensionnel aux Reseaux de Transport
(1988) / Claude Y

MONTREAL, UNIVERSITY OF
(continued)

Le Contrat Individuel de Travail a Duree
Determinee: Etat Actuel du Droit Quebecois
(1988) / Pharand D

Etude des Particularities du Droit du
Point de Vue de la Conception d'un Systeme
Expert et Realisation de Chomexpert
(1988) / Poulin D

Validation d'un Modele Strategique de
Transport des Marchandises Multi-Produit
et Multi-Modal
(1988) / Rukazamuheto M

Une Etude Comparative de l'Alignement de la
Formation sur la Strategie de Developpement
Dans l'Entreprise
(1988) / Scott Y

Developpement d'un Systeme Expert Pour
l'Aide a la Conception de Systemes
d'Information
(1988) / St-Jacques C

Deux Nouveaux Algorithmes Exacts Pour le
Probleme des M Voyageurs de Commerce
(1988) / Trinh K N

Probleme d'Affectatin des Ressources Pour
la Distribution des Boissons Gazeuses
(1988) / Trottier F

Impact des Technologies Informatiques
CAO/FAO sur le Contenu, les Exigences et la
Productivite de l'Emploi Individuel
(1988) / Trudel H

Analyse Statistique de l'Evolution
Temporelle des Precipitations Acides au
Quebec
(1988) / Vachon M

NEBRASKA, UNIVERSITY OF

The Effects of Display Complexity in
Information Processing
(1988) / Batra S

A Qualitative Comparison of Three
Micro-Computer Simulation Packages with
Animation: GPSS/PC, Simfactory,
Siman/Cinema
(1988) / Glaser C A

Improvement of Electro-Discharge Machining
Performance by Radio Frequency Control and
Orbital Motion Movement
(1988) / Royo G F

NEW JERSEY INSTITUTE OF TECHNOLOGY

The State and Future of Televised
Professional Training Systems
(1988) / Brady R M

Utilization of the Theory of Constraints
Approach for Improving a Manufacturing
Enterprise Profitability: A Case Study
(1988) / Chaut Y

A Model for Estimating the Cost of
Developing an Expert System
(1988) / De Troia L V

A Methodology for Evaluating Prospective
Investments in Flexible Automation
(1988) / Eikenbusch W A

Micro CAD/CAM System Selection and
Implementation
(1988) / Haas C G

The Strategic Planning Dilemna in
American Manufacturing
(1988) / Kiray J A

A Model for the Determination of Market
Based Pricing of Telecommunication Products
(1988) / Sullivan K J

Development of a Knowledge-Based System for
Human Error Analysis
(1988) / Sung B C

Evaluation of Word Processing Cost
Effectiveness
(1988) / Tedesco N E

The Changing Role of the Electric Utility
Rates Manager
(1988) / Tutela R J

Robotization for Assembling Small Parts
(1988) / Valese F J

Computerizing the Human Resource Information
System
(1988) / Wang D C

NEW MEXICO STATE UNIVERSITY

Building Blocks for Computer-Integrated
Manufacturing
(1988) / Cox B E

NEW YORK, STATE UNIVERSITY OF (BUFFALO)

Visual Search Strategies: An Interfixation
Distance Approach
(1988) / Baveja A

A Heuristic Appoach to Solving the
Generalized Orienteering Problem
(1988) / Brown K

Psychophysical Evaluation of Handle Design
(1988) / Chi C F

Evaluation of Algorithms for Combining
Independent Data Sets in a Human
Performance Expert System
(1988) / Gawron V

The Equity of Risk in the Transportation of
Hazardous Materials
(1988) / Gopolan R

File Allocation and Report Assignment in
a Large Distributed Database Network
(1988) / Hartman B

A Mathematical Programming Approach to
Reconstructing Missing Data for a Bill
of Materials
(1988) / Kolb G

Multiple Routing for Equity of Risk in the
Transportation of Hazardous Materials
(1988) / Kolluri K

Use of Convex Cones in Solving Multiple
Criteria Linear Programming Problems
(1988) / Prasad S

A Simulation of HMO Laboratory Care Suites
(1988) / Wessel M

NORTH CAROLINA STATE UNIVERSITY

Computer Aided Selection of Automatic
Identification Systems
(1988) / Kilgroe C D

A Finite Method of Feasible Directions for
Solving Linear Programming Problems: Theory
and Computational Results
(1988) / Luh H P

Intelligent Distributed Control System for
Multiple Manufacturing Cells
(1988) / Terribile M T

NORTH DAKOTA STATE UNIVERSITY

Chain-Code Algorithm for a Vision-Based-System
Using Micro D-Cam camera
(1987) / Abouzeid M G

Design of a Generalized Numeric Control Postprocessor
(1988) / Fuller P B

Off-Line Programming and Kinematics Simulation of
the Armdroid and NND robots
(1988) / Rong-Yuan W

OKLAHOMA, UNIVERSITY OF

Automated Guided Vehicle Systems: A Look at
Guide Path Configurations
(1988) / Basgall J R

Implicit Model Reference Control with Feed
Forward Structure for a Class of Plant in
Discrete-Time State-Space
(1988) / Chen T

A Universal Approach for Single-Row Routing
(1988) / Chen Y

An Expert Heuristic Selection System for
Project Scheduling
(1988) / Khuzema A

A Systematic Approach to Design Constraint
Covergence When Integrating Manufacturing
Facilities
(1988) / Shore S D

Economic Impact of Integrated Energy Systems
in Oklahoma
(1988) / Subbiah V

Effect of Display Type on Operator
Performance in a Process Control Situation
(1988) / Sundar K M

On-Line Tool Wear Sensing by Using Acoustic
Emission
(1988) / Tjiang N T

OREGON STATE UNIVERSITY

Simulation of Mechanized Log Harvesting System
(1987) / Wiese C

PENNSYLVANIA STATE UNIVERSITY

Model Based Automated Inspection of
Industrial Parts Using Machine Vision
(1987) / Anand S

A Proposed Approach to the Design of
Flexible Inspection Fixtures
(1987) / Buley D T

A One-Time Optimal Process Adjustment Using
Mathematical Programming
(1987) / Davachi A

Comparison of In-Transit Routing Policies
for Automated Guided Vehicle Systems
(1988) / Davis D A

Comparison of Dynamic Muscle Strength
Measurements on Two Isokinetic Dynamometers
(1987) / Fotouhi D M

Mental Workload and Accuracy in Visual
Inspection
(1987) / Gibson D C

Guide Path Design for Automated Guided
Vehicle Based Flexible Manufacturing System
(1987) / Goetz W G

An Investigation of Perturbation Analysis
(1987) / Johnson M E

Knowledge Representation Schemes in
Artificial Intelligence Systems: A
Comparative Study
(1987) / Joo S H

Gate Size for Sound Castings in Ceramic
Shell Mold
(1988) / Kajita T

Shortest Flow Time (SFT) and Least Penalty
Method (LPM): Two Heuristics for the
Scheduling of N-Nonpreemptive, Nondivisible
Jobs on M-Nonidentical Parallel Machines
(1988) / Lin C W

An Implementation Scheme for Distributed
Transaction Processing
(1987) / Mannai D N

Analysis of the Inertial Properties of Head
Protective Assemblies and Their Effect on
Head/Neck Kinematics in Support of
Integrated Night Vision System Use in
Ejection Seat Aircraft
(1987) / Mc Cauley D S

A Simulation Analysis of Robotic Assembly
Systems
(1987) / Mc Gill S D

Analysis of Training, Retention, Transfer,
and Retraining Performance Using Learning
Curve Parameters
(1988) / O'Rourke S A

Generation of Efficient Numerical Control
Tool Paths Using an Interactive Computer
Aided Design System
(1988) / Rose M E

Computational Testing and Analysis of a
Modification of the Karmarker Linear
Programming Algorithm
(1987) / Schall K C

Analytical Simulation Models for the Design
of Miniload AS/R Systems
(1987) / Smith A D

PENNSYLVANIA STATE UNIVERSITY
(continued)

A Methodology for the Validation of
Terminating Simulation Models Using
Profile Analysis
(1988) / Sticklin S W

Development of a Controller and Graphical
Emulator for a Mini-Automated Storage and
Retrieval System
(1988) / Strong D B

High Speed Electroslag Welding
(1987) / Su C T

The Development of a Computer-Aided
Inspection System
(1987) / Traband M T

Scheduling and Machining of Nonidentical
Jobs on Nonidentical Parallel Machine Cells
(1987) / Watson E F

PURDUE UNIVERSITY

Automatic Cutter Selection and Cutter Path
Generation for Prismatic Parts
(1988) / Bala M V

The Effect of Social and Cognitive Factors
on the Performance of Text-Editing: An
Investigation Into Problem-Solving
Strategies
(1988) / Bringelson L S

Surface Tracking/Model Creating Robot Using
Learning Control and Infrared Distance
Sensing
(1988) / Brown P K

A Knowledge-Based Design Procedure for
Intelligent Manufacturing Systems
(1988) / Floss P

AMPS - An Automatic Manufacturing Process
Planning System
(1988) / Kanumury M

On Horn and Related Clause Systems in
Propositional Logic
(1988) / Montanez-Martinez M A

A Unified Decision Environment for Product
Design and Economic Justification
(1988) / Noble J S

Random Generation of Traveling Salesman
Problems Using Clique Tree Inequalities
(1988) / Rais A

Cell Design and Configuration in Automated
Electronics Assembly
(1988) / Raman G G

The Effect of Residual stresses on the
Strength of Ground Ceramics
(1988) / Samuel R

CLAMPS - Automated Fixturing in a Flexible
Manufacturing Environment
(1988) / Shah J M

TREEMAKER: A Tool for Constructing
Probability Trees with Radom Parameters
(1988) / Stouppe M S

Flow Knowledge Representation and Layout
Evaluation for Manufacturing Systems
Layout Design
(1988) / Stump D J

Storage Location Characterization and
Analysis of Assignment Policies for
Warehousing Systems
(1988) / Thompson L L

RENSSELAER POLYTECHNIC INSTITUTE

Microcomputer Procedures for Simulation
(1988) / Echeandia E A

Simulation of an Heuristic Rule to Compute
the Production Quantity of Slow-Moving
Inventory Items
(1988) / Montero A S

A Sequence Based Similarity Coefficient for
Grouping Parts and Machines in Cellular
Manufacturing
(1988) / Snigar D A

Implementation of the Specialists and
Intelligent Device Interfaces
(1988) / Tarbox G H

SAN DIEGO STATE UNIVERSITY

Attitudes of San Diego Area Management
Information Systems and Data Processing
Managers Towards the Uses of Expert Systems
in Business Applications
(1987) / Briggs R O

The Comparison of Concurrency Control in
Centralized and Distributed Databases
(1987) / Dharachan A

Designing and Automated Shipment Information
System for a Large Multinational
Manufacturing Company
(1987) / Garbus B L

Improving Human-Computer Interaction:
Analysis of Issues and Techniques
(1988) / Hashem M M

The Implementation of Local Area Networks
(1987) / Kromminga G D

Computer Automation of Job Analysis and
Description
(1988) / Kulwiec M F

Marketing the Lease: A Computer-Based
Algorithm
(1987) / Smith M E

Techniques for Diagramming and Critiquing
User-Computer Dialogs
(1987) / Wang C

Integrated Services Digital Network
(1987) / Yusoff Y M

SOUTH FLORIDA, UNIVERSITY OF

Parameter Design for a Wave Solder Process
(1988) / Rao R L

Off-Line Quality Control Using Parameter
Design for Wave Soldering Machine
(1988) / Sharma N

SOUTH FLORIDA, UNIVERSITY OF
(continued)

Optimization of Cooling Water System for
Thermal Power Plants
(1988) / Som P

Modeling and Analysis of High-Level Data
Link Control Protocol
(1988) / Yoong Y K

TEMPLE UNIVERSITY

A Coupled Knowledge-Based System for Control
System Design
(1988) / Palumbo N F

TENNESSEE TECHNOLOGICAL UNIVERSITY

Economic Analysis of Fossil-Fuel Power
Plants
(1988) / Chen H M

Use of Statistical Tools in Quality Control
(1988) / Malmquist K W

TENNESSEE, UNIVERSITY OF (KNOXVILLE)

A Justification Methodology for the
Evaluation of Local Area Network
Alternatives
(1987) / Liggett H R

The Calculation of Reliability for Complex
Systems
(1987) / Lyle M K

An Analysis of the Critical Parameters that
Impact the Success of a Bar Code System
(1987) / Stumb P C

TENNESSEE, UNIVERSITY OF (TULLAHOMA)

Management of the Decatur, Alabama
Refrigeration Upgrade
(1988) / Bowling R F

Product Transfer to Puerto Rico: A Study
Case
(1988) / Bryant M

An Application of the Principles of "In
Search of Excellence" at the Project
Management Level
(1988) / Elrod P

Analysis of the Quality Action Documentation
System
(1988) / Felch M

Just-in-Time Implementation at Samsonite
Furniture of Murfreesboro, Tennessee
(1988) / Hollady A

Research Administration at UTSI
(1988) / Johnson R W

An Analysis of AEDC Environment for
Nurturing Creativity and Implementing
New Technologies
(1988) / Linkous P

Management Audit of Lannom Company
(1988) / Linneweber C

Implementation and Operation of a Quality
Circle in a Government Contractor Design
Engineering Office
(1988) / Mayberry B R

Just-in-Time Implementation at Samsonite
Furniture of Murfreesboro, Tennessee
(1988) / Schaller D P

An Analysis of AEDC Environment for
Nurturing Creativity and Implementing
New Technologies
(1988) / Stevens D

Technology Transfer in a Testing
Organization
(1988) / Truesdale R

A Management Analysis of Gresham, Smith and
Partners
(1988) / Utley D R

An Analysis of AEDC Environment for
Nurturing Creativity and Implementing
New Technologies
(1988) / York C

Management Audit of Lannom Company
(1988) / Zapata J

TEXAS A AND M UNIVERSITY

A Comparison of Nuclear Reactor Control
Room Display Panels
(1988) / Bowers F R

Design of a Multicriteria Performance Model for
Jet Systems
(1988) / Coulibaly S

An Analysis of Font Styles on the Screen of a
Computer Terminal
(1988) / Eason J L

Framework for Database and Modelbase in an
Information Based Simulation System
(1988) / Ghoshal D

Considerations of Driver Preferences and
Performance for Selection of Electronic
Automobile Instrumentation
(1988) / Harwood R S

A Network Approach for Identifying Minimum-Cost
Aircraft Routing and Fuel-Allocating Decisions
(1988) / Kabbani N M

An Evaluation of Back Support Devices for Females
Involved in Lifting Tasks
(1988) / Sherwood C A

A Comparison of Three Army Chemical-Biological
Protective Masks: Their Effects on Field of
View and Visual Acuity
(1988) / Takao M J

TEXAS, UNIVERSITY OF (AUSTIN)

Component Allocation and Partitioning for a
Multiple Head Sequential Surface Mount
Placement Machine
(1988) / Woodruff J B

TEXAS, UNIVERSITY OF (EL PASO)

Computer-Aided Operation and Management of the Lift
(1988) / Ahmed M M

An Expert System for Selecting Solution Methods for
Nonlinear Optimization
(1988) / Deshmukh A M

TEXAS, UNIVERSITY OF (EL PASO)
(continued)

An Interactive CAD/CAM Software Package for a CIM
Workcell
(1988) / Liou F S

Economic Design of an X-Bar Control Chart with
Non-Exponential Times Between Process Shifts
(1988) / Liu S F

TOLEDO, UNIVERSITY OF

Development and Material Selection for a
State of Charge Indicator
(1988) / Pudlo R

TORONTO, UNIVERSITY OF

An Investigation of the Effects of Operating
Experience and Participation in a Motorcycle
Operating Course on the Perception of Risk
Associated with the Operation of a
Motorcycle
(1988) / Mitchell D A

Scheduling N Jobs on M Machines in a
Deterministic Environment
(1988) / Woodbury N L

An Approximate Reasoning Approach for
Aggregate Production Planning
(1988) / Zhao Z

U.S. NAVAL POSTGRADUATE SCHOOL

Three Dimensional Visual Display for a
Prototype Command and Control Workstation
(1988) / Abner M D

Survey and Recommendations for the Use of
Microcomputers in the Naval Audit Service
(1987) / Augustine M

Analytical Evaluation of UNREP (Underway
Replenishment) Methods Using the Model BFORM
(Battle Force Operation Replenishment Model)
(1988) / Barnaby S L

The Viscoceilometer and Its Tactical
Applications
(1987) / Bridges M A

An Analysis of the Advanced Traceability
and Control System Goals
(1987) / Bruner C D

A Data Base Design for a Multimedia C2
Workstation in Support of RESA (Research,
Evaluation and Systems Analysis)
(1987) / Carroll M F

Toward an OPTAR (Operating Target)
Allocation Model for Surface Ships the
Pacific Fleet
(1988) / Cataland J A

Comparison of PASCAL and the DBASE III Plus
Language in Programming an Inventory
Management System
(1987) / Chang T

A Review of the Evolution of Naval Data
Automation and the Optical Media Mass
Storage Alternatives Related to Naval
Aviation Technical Documentation
(1987) / Clarey R J

Evaluation of Work Distribution Algorithms
and Hardware Topologies in a
Multi-Transputer Network
(1988) / Cloughley W R

Standardization of Hull, Mechanical, and
Electrical Equipment (H,M&E) Inventory
(1987) / Corbett J C

A Template for the Selection and Array of
Inventory as an Aid in the Development of
Evacuation Plans
(1987) / Dietz J L

The Impact of Independent Research and
Development Regulations on Companies not
Required to Negotiate Advanced Independent
Research and Development Agreements
(1987) / Drew C C

Network and Database Design in Support of
the Joint Theater Level Simulation
(1988) / Dunn C

Network and Database Design in Support of
the Joint Theater Level Simulation
(1988) / Evans S H

Measuring Losses of Learning Due to Breaks
in Production
(1988) / Everest J D

Data Administration for the Rapid
Acquisition of Manufactured Parts
(1988) / Fads C T

Contracting Principles: A Conceptual
Framework for Their Identification and
Validation
(1987) / Fawbush J A

Using Cost Realism to Improve the Source
Selection Process
(1987) / Hall W E

A Feasibility Study of Relating Surface Ship
OPTAR (Operating Target) Obligation Patterns
to Their Operating Schedules
(1988) / Hanson C D

Preliminary Work on the Command and Control
Workstation of the Future
(1988) / Harris F E

An Analysis of the Advanced Traceability
and Control System Goals
(1987) / Honcycutt T W

The Application of Cost-Benefit Analysis in
Raising the Noncompetitive Small Purchase
Threshold
(1987) / Howard R L

A Comparative Study of GDSS (Group Decision
Support System) Use: Empirical Evidence and
Model Design
(1987) / Hughes A L

A Feasibility Study of Relating Surface Ship
OPTAR (Operating Target) Obligation Patterns
to Their Operating Schedules
(1988) / Kiker K L

An Evaluation of a Joint Replenishment
Inventory Model with Radom Demands
(1987) / Kim W B

U.S. NAVAL POSTGRADUATE SCHOOL
(continued)

Solution of Large-Scale Multicommodity
Network Flow Problems Via a Logarithmic
Barrier Function Decomposition
(1988) / Lange H

Survey and Recommendations for the Use of
Microcomputers in the Naval Audit Service
(1987) / Lapoint T

A Simulation Study of Estimates of System
Availability
(1988) / Lee C H

Application of a Database System for
Inventory Management of Classified Material
(1987) / Lim R M

A Conceptual Design of a Multimedia DBMS
(Database Management System) for Advanced
Applications
(1988) / Lum V Y

MACCAD (Macintosh Computer Aided Design),
Computer Aided Design Tool for System
Analysis
(1987) / MacDonald K

Implementation of the Runge-Kuttta-Fehlberg
Method for Solution of Ordinary Differential
Equations on a Parallel Processor
(1987) / Mayo C F

Surface Polariton Resonances and Reflectance
on a Bigrating
(1987) / Melendez G J

A Conceptual Design of a Multimedia DBMS
(Database Management System) for Advanced
Applications
(1988) / Meyer-Wegener K

Strategic Planning, Polaris, and Tomahawk:
Technological Imperative Hypotheses
(1987) / Norris D T

A Conceptual Level Design for a Static
Scheduler for Hard Real-Time Systems
(1988) / O'Hern J T

Performance Work Statements: Significant
Problems in the Preparation Process
(1987) / Paddock C D

Contracting Issues Associated with Reduction
of Repair Turnaround Time Within the
Contract Depot Maintenance (CDM) Program
(1987) / Petty R E

Competitive Procurement of Electrical Power
(1987) / Ray J L

An Evaluation and Analysis of the United
States Navy
Air-Launched Missiles and Supply Support
Alternatives for the Naval Air Systems
Command Omnibus Program
(1987) / Ripperton J G

A Feasibility Study in Path Planning Using
Optimization Techniques
(1987) / Sanders D W

Data Administration for the Rapid
Acquisition of Manufactured Parts
(1988) / Smith P A

Solving the Multicommodity Trans-Shipment
Problem
(1987) / Stantec C J

Contracting: A Systematic Body of Knowledge
(1987) / Thornton C L

A Comparative Study of GDSS (Group Decision
Support System) Use: Empirical Evidence and
Model Design
(1987) / Webb D H

Design and Implementation of a Prototype
Graphical User Interface for a Model
Management System
(1988) / Wyant M A

Preliminary Work on the Command and Control
Workstation of the Future
(1988) / Yurchak J M

Preliminary Work on the Command and Control
Workstation of the Future
(1988) / Zyda M J

U.S.A.F. INSTITUTE OF TECHNOLOGY

Optimization of Inventory Levels for the
Air Force Commissary Service
(1988) / Britt R J

Analysis of Design-Related Problems from an
Industrial Project
(1988) / Glavan J R

Modeling the Scheduled Preventive
Maintenance of a Linear System
(1988) / Gunes E M

A Methodology for Scenario-Based
Requirements Exploration
(1988) / Holbrook H

The Application of Statistical Process
Control in Nonmanufacturing Activities
(1988) / Olson L W

Decreasing Nonconformance of Parts in the
Air Force Supply System
(1988) / Stone M A

UTAH STATE UNIVERSITY

Furrow Geometry Changes as a Function of
Flow Rate, Slope, Time of Flow and Distance
from the Inlet
(1987) / Gezehegn L K

UTAH, UNIVERSITY OF

The Effects of Stochastic Random Variables
on a Just-in-Time System with Kanbans
(1988) / Sadri C S

Optimization of a Unimodal, Discrete Event
Supply Support for
and Davies, Swann and Campey Search
Technique
(1988) / Sun R H

VIRGINIA, UNIVERSITY OF

An Empirical Investigation of Local
Neighborhood Search Based Algorithms for
Job-Shop Scheduling
(1988) / Shekhar C

WEST VIRGINIA UNIVERSITY

Evaluation and Selection of a Materials
Management and Control System
(1988) / Clark M K

The Use of Simulation in the Analysis of
Cart Flow for a Hospital Case Cart System
(1988) / Landers J J

A Simulation Study of Spare Ratio in
Management of Bus Rolling Stock
(1988) / Niaki S T

AUBURN UNIVERSITY

Effects of Irrigation Rate and Media Type on
Growth of Trees in Large Containers
(1987) / Martin C A

CALGARY, UNIVERSITY OF

Control of Irrigation Return Flows in
Southern Alberta
(1988) / Lipowska E

Evapotranspiration in an Irrigated Wheat
Field
(1988) / MacQuarrie P A

CALIFORNIA, UNIVERSITY OF (DAVIS)

The Response of Three Year-Old Thompson
Seedless Grapevines to Drip and Furrow
Irrigation in the San Joaquin Valley
(1988) / Araujo F J

Salinity Control and Reducing Ground Water
Quality Degradation by Trickle Irrigation
with Zero Percent Leaching
(1988) / Mansoubi A

Infiltration Under Surge Flow Irrigation
(1988) / Purkey D R

Field-Wide Infiltration Variability in
Furrow Irrigation
(1988) / Tarboton K C

The Effects of Saline Irrigation Water on
the Growth and Development of Safflower
(1988) / Weyrauch R L

COLORADO STATE UNIVERSITY

Uniformity of Trickle Irrigation
(1988) / Mizyed N R

IDAHO, UNIVERSITY OF

Computer Aided Design and Hydraulic Analysis
of Solid Set Turf Irrigation Systems
(1988) / Beck D B

An Economic Impact Analysis of the 1987
Drought in the Arrowrock Division of the
Boise Irrigation Project
(1988) / DeWitt C E

LOUISIANA STATE UNIVERSITY

Simulation of Surface Irrigation Systems
(1988) / Yu F X

MAINE, UNIVERSITY OF

A Study on the Water Use of Potato Under
Supplemental Irrigation
(1988) / Ramanan K

MANITOBA, UNIVERSITY OF

Rainfall Data Generation Pemali-Comal
Irrigation Area, Java-Indonesia
(1987) / Darsono S

Movement of Pesticides to Ground Water in an
Irrigated Soil
(1988) / Krawchuk B P

A Reconnaissance Study of the Potential for
Large Scale Irrigation of the Canadian
Prairies
(1988) / McPhail G D

MASSACHUSETTS INSTITUTE OF TECHNOLOGY

Evaluating Investments in Large Public
Irrigation Projects: A Case Study of the
San Francisco River Basin in Brazil
(1988) / Kettelhut J T

MONTANA STATE UNIVERSITY

Simulation of Irrigation and Reservoir Water
Use in the Canyon Ferry Drainage Basin
(1987) / DeLuca D K

NEW MEXICO STATE UNIVERSITY

Economic Analysis of Water Scheduling
Alternatives: The Case of the Elephant
Butte Irrigation District, New Mexico
(1988) / Bustillos L C

The Impact of Increasing Water Prices on
Agricultural Production: A Case Study of
the Elephant Butte Irrigation District,
Dona Ana County, New Mexico
(1988) / Sloan D K

NORTH DAKOTA STATE UNIVERSITY

Effects of Irrigation Water Quality on the Physical
and Chemical Properties of Four Slowly Permeable
North Dakota Soils
(1988) / Costa J L

OREGON STATE UNIVERSITY

Irrigation Water and Plant Density Effects on the
Epidemiology of Aerial Stem Rot of Potatoes
(1987) / Cappaert M R

The Demand for Electricity in Western U.S. Irrigated
Agriculture: A Dual Cost Function Analysis
(1988) / Connor J D

SASKATCHEWAN, UNIVERSITY OF (SASKATOON)

A Socio-Economic Analysis of Adoption of
Irrigation in Saskatchewan
(1988) / Manning S F

U.S.A.F. INSTITUTE OF TECHNOLOGY

Estimating Evapotranspiration of an
Irrigated Surface from Upwind and Downwind
Vertical Profiles of Temperature and
Humidity
(1988) / Zehr D F

UTAH STATE UNIVERSITY

A Design and Evaluation Method for
Supplementary Mechanized Irrigation
Development
(1987) / Araujo E M

Wetting Front Instability in Layered Soils
(1987) / Cheraghi S A

Field and Model Studies of Irrigation
Effects on Groundwater in Saline Alluvium
(1987) / Cole L E

An Alternative Management System for the
Tapacuma Irrigation Project, Guyana
(1987) / Forsythe V C

UTAH STATE UNIVERSITY
(continued)

Water Table and Salinity Movement in an
Irrigated Silty clay Soil Above an Artesian
Aquifer
(1987) / Javaid R

A Comparison of Calcium Carbonate Saturation
Indexes related to Occurance and Solution of
Water Well Encrustation and Corrosion
(1987) / Karahaliloglu A F

Effects of Precision Land Levelling on
Basin Irrigation
(1987) / McClung J A

An Irrigation Management Simulation Model
for Microcomputers
(1987) / Mulkay L M

Evaluation of Large Scale Irrigation on
Small Fields
(1987) / Nader H

Parameter Sensitivity Analysis for the
Hydrodynamic, Zero-Inertia and Kinematic
Wave Hydraulic Models
(1987) / Oliveira Filho J C

Modeling Dry Bean Yield as Influenced by
Irrigation, Planting Date, and Location
(1987) / Stewart F R

WASHINGTON STATE UNIVERSITY

Assessing Risk in Partial Irrigation of
Columbia Basin Winter Wheat
(1988) / Hibbs R A

Performance of Low Pressure Set-Move
Irrigation Sprinklers
(1988) / Hodges K J

WEST VIRGINIA UNIVERSITY

Agriculture and Nonrenewable Resources: The
Case of Irrigation on the Ogallala Aquifer
(1988) / Walker A R

WYOMING, UNIVERSITY OF

Field Evaluation of a Feedback Control
Applied to a Furrow Irrigation System
(1989) / Recio F A

FLORIDA ATLANTIC UNIVERSITY

Effects of External Disturbance on the Flow
Development Near the Boundary of a Work
Table
(1988) / Auche E B

Calcareous Deposits in Simulated Fatigue
Cracks of Cathodically Protected Steel in
Seawater
(1988) / Davidson K D

Structural Integrity of Calcareous
Deposit-Rehabilitated Marine Reinforced
Concrete
(1988) / Dohlen S

Transmission of Vibrational Power in Joined
Structures
(1988) / Gibert T M

Vibrational Power Flow in Thick Connected
Plates
(1988) / McCollum M D

The De-Dopplerization of Acoustic Signatures
(1988) / Mouches J M

Characterization of Environmental Cracking
of Prestressing Steel Under Cathodically
Polarized Conditions in a Simulated Concrete
Environment
(1988) / Narayanan P K

Power Flow Analysis of Simple Structures
(1988) / Rassineux J L

An Analytical and Experimental Investigation
of a Multicellular Box Beam Bridge System
Using an Acrylic Model
(1988) / Ravichandran T

Fatigue of High Strength Steels in Sea Water
(1988) / Sablok A K

The Role of Calcium Hydroxide in the
Maintenance of Passivity of Steel
(1988) / Sohanghpurwala A S

Stability and Response of Suspension Bridges
Under Turbulent Wind Excitation
(1988) / Sternberg A

HAWAII, UNIVERSITY OF

Wave- and Current-Induced Viscous Drift
Forces on Floating Platforms
(1988) / Chitrapu S A

Molecular Diffusion of Oxygen and Nitrogen
in Seawater
(1988) / Oney S K

Numerical Grid Generation and Nonlinear
Waves Generated by a Ship in a Shallow-Water
Channel
(1988) / Qian Z

LOUISIANA STATE UNIVERSITY

Degradation of Azinphosmethyl and
Pendimethalin in Modified Sediment Materials
and Effects of These Pesticides on the
Crayfish, Procam-Barus Clarkii, in
Laboratory Microcosms
(1988) / Boyd W

Effect of Hydrogen Sulfide on Growth,
Nutrient Uptake, and Metabolic Energy
Level in Two Marsh Macrophytes
(1988) / Koch M S

Subtidal Circulation at Calcasieu Pass
(1988) / Lee J M

MASSACHUSETTS INSTITUTE OF TECHNOLOGY

Vibration Problems of Rotating Machinery Due
to Coupling Misalignments
(1988) / Barber F L

Advanced Laser Fluorescence Measurements of
Lubricant Film Behavior in a Diesel Engine
(1988) / Billian S A

Rough Surface Ice Scattering Using the Exact
Integral Equation Method with the P.E.
(1988) / Bouxin H C

The Response of Delaminated Composite Panels
to Air Blast Loading
(1988) / Byrtus J E

On the Optimum Propeller Loading with
Inclusion of Duct and Hub
(1988) / Caja A S

Flow Induced Vibration of a Non-Constant
Tension Cable in a Sheared Flow
(1988) / Capozucca P

The Design of an Interactive Algebraic
Surface Editor
(1988) / Carlson R H

Robust Estimation of Ocean Currents from
Moving Vessels Using Loran C
(1988) / Davis R W

Infrared Signature of Ships at Sea
(1988) / Delaney M E

Computer-Assisted Teaching of Marine
Hydrodynamics
(1988) / Denson L A

Interaction Between a Two-Dimensional Wake
and the Free Surface at Low Froude Numbers
(1988) / Dimas A A

Time Series Analysis of Ocean Waves
(1988) / Dommermuth D G

An Application of Group Technology to Naval
Ship Repair
(1988) / Duca S C

An Analysis of Full-Scale Experimental Data
on the Dynamics of Very Long Tethers
Supporting Underwater Vehicles
(1988) / Engebretsen K B

Computational and Experimental
Investigations of the Flow Around Cavitating
Hydrofoils
(1988) / Fine N E

Computer Aided Routing of Combination
Carriers
(1988) / Garmilla M

Non-Destructive Techniques for Inspection of
Arc Spot Welds
(1988) / Hays K T

MASSACHUSETTS INSTITUTE OF TECHNOLOGY
(continued)

The Impact of Group Technology-Based
Shipbuilding Methods on Naval Ship Design
and Acquisition Practices
(1988) / Heffron J S

Non-Linear Material Three Degree of Freedom
Analysis of Submarine Drydock Blocking
Systems
(1988) / Hepburn R D

Maritime Fraud: The Neglected Burden
(1988) / Katsikis V I

Superconducting Technology for Electric
Propulsion
(1988) / Keamy E F

Earthquake Resistant Submarine Drydock Block
System Design
(1988) / Luchs J K

Distortion and Residual Stresses in a Welded
Stiffened Ring
(1988) / Machuca B B

Strategic Analysis of Free Port Investment
(1988) / Morova P

Automatic Generation of Flexible Fixturing
Layout for Sheet Metal Application
(1988) / Nam J H

Propeller Onset Flow Alterations Due to a
Non-Axisymmetric Duct
(1988) / Nicholson S J

In-Process Control of Distortion During
Aluminum Butt Welding
(1988) / Park S

Elastic Analysis of a Circular Toroidal
Shell Using Computer Modeling
(1988) / Powell R S

A Hydrodynamic Study of a Submerged Vehicle
(1988) / Shields L A

Penetration Effects of the Compound Vortex
in Gas Metal-Arc Welding
(1988) / Spencer J P

Comparative Naval Architecture of Modern
Foreign Submarines
(1988) / Stenard J K

Design Impact of Propulsion Plant Variations
on a Small Combatant Ship
(1988) / Stergiakis C L

Ship Acquisition Lessons Learned/Feedback
Loops for U.S. Government Agencies
(1988) / Stiglich J F

The Effects of Resin, Reinforcement, and
Interleafs on Impact Resistance of a
Fiberglass Reinforced Plastic Boad Hull
Laminate
(1988) / Stroh L D

An Approach to Ship Acquisition Strategy
Development for U.S. Government Agencies
(1987) / Strong S C

Computer Simulation of Icebreaker Dynamics
(1988) / Suchanekk B J

Residual Strength of a Ship After an
Internal Explosion
(1988) / Surko S W

Wave Propagation in Beams Having Double
Resonant Loading
(1987) / Taylor P D

Detection of Broken Rotor Bars in Induction
Motors Using Stator Current Measurements
(1988) / Welsh M S

MIAMI, UNIVERSITY OF

Evaluation of Models of the Time-Averaged
Vertical Distribution of Sand Suspended by
Waves
(1988) / Gonzalez E A

Moving Boundary Exact Solution to
Underwater Shock-Structure Interaction
(1988) / Khangaonkar T P

Implementation of an Expert System for
Design of Single-Point Subsurface Moorings
(1988) / Kumaran S

Interaction Between Ocean Waves and
Surface Currents
(1988) / Leipold Y N

NORTH CAROLINA STATE UNIVERSITY

Turbulent Structure and Wind Wave
Interaction in the Marine Surface Layer
During SUPERDUCK 1986
(1988) / Kang M J

A Stable Isotope Evaluation of the Blake
Outer Ridge (BOR) Deep Sea Deposit
(1988) / Palczuk N C

NOVA SCOTIA, TECH UNIVERSITY OF

Parametric Study of Ship-Hull Vibrations
(1988) / Bartran I M

Dynamical Analysis of a Crane Barge in
Regular Waves
(1988) / Kumar R

Transient Nonlinear Waves Generated by a
Moving Pressure Disturbance on a Free
Surface
(1988) / Lu Q

OREGON STATE UNIVERSITY

The Static and Cyclic Pullout Behavior of Plate
Anchors in Fine Saturated Sand
(1988) / Petereit R A

Kinematics and Return Flow in a Closed Wave Flume
(1987) / Ramsden J D

Wave Forces on Concrete Pipes and Plates Used as
Seabed Artificial Reef Units
(1987) / Shin O

RHODE ISLAND, UNIVERSITY OF

The Non-Gaussian Response of Single-Degree-of-
Freedom Systems to the Drag Force
(1988) / Dixit S

Enforcement of State Coastal Resources
Management Regulations: The Rhode Island
Coastal Resources Management Council
(1988) / Kaiser D W

RHODE ISLAND, UNIVERSITY OF
(continued)

Socioeconomic Aspects of Small-Scale
Commercial Fisheries Development in
the Republic of the Marshall Islands
(1988) / Mandich C A

Dredging Analysis of Mid-Sized U.S. North
Atlantic Ports
(1988) / Masters M H

The Statistical Properties of Structural
Response to Simulated Wave
(1988) / Muin M

Application of System Identification Techniques
for Calibration of Underwater Acoustic Transducers
(1988) / O'Neill D J

Motion Limitations of New England Fishing
Vessels in Various Local Sea States
(1988) / Thibodeaux J L

Hydrodynamic Modifications to Scallop
Dredges for Improved Efficiency
(1988) / Vaccaro M J

Fatigue Damage Estimation of the Offshort Structure
Under Long-Term Nonstationary Wave Loadings
(1988) / Zhao D

TEXAS A AND M UNIVERSITY

Heavy Lift Crane/Derrick Barge Stability Analysis
(1988) / Loesch R M

Random Wave Forces on a Free-to-Surge
Vertical Cylinder
(1988) / Sajonia C B

The Effects of Flow Rate on In vivo
Fluorescence Measurements
(1988) / Sweet S T

The Effect of Pipe Spacing on Marine Pipeline Scour
(1988) / Westerhorstmann J H

U.S. NAVAL POSTGRADUATE SCHOOL

An Analysis of Building a Submarine Base in
the Arctic
(1988) / Best T J

Flow Visualization of the Airwake of an
Oscillating Generic Ship Model
(1987) / Biskaduros J L

An Investigation Into the Feasibility of a
Specialized Allowance of Critical Spare
Parts for Gas-Turbine Class Ships
(1987) / Bogott K W

A Study of Model Based Maneuvering Controls
for Autonomous Underwater Vehicles
(1987) / Boncal R J

Anti-Submarine Warfare: A Strategy Primer
(1988) / Breemer J S

Application of the Analyysis Phase of the
Instructional System Development to the
MK-105 Magnetic Minesweeping Mission of the
MH-53E Helicopter
(1987) / Broughton D S

Experimental Verification of AUV (Autonomous
Underwater Vehicle) Performance
(1988) / Brunner G M

Visualization of the Flow Field Around an
Oscillating Model of the USS Enterprise
(CVN-65) in a Simulated Atmospheric
Boundary Layer
(1988) / Cahill T A

Flow Visualization of the Airwake Around a
Model of a Tarawa Class LHA in a Simulated
Atmospheric Boundary
(1988) / Daley W H

Preliminary Design and Cycle Verification
of a Digital Autopilot for Autonomous
Underwater Vehicles
(1988) / Delaplane S W

Experimental Investigation of the Effects
of Underwater Exposure on the Damping
Characteristics of Bolted Structural
Connections for Plates and Shells
(1988) / Durham R W

Estimation Forces Acting on an Underwater
Vehicle with G.P.S. (Global Positioning
System) and Kalman Filtering
(1988) / Easton J K

Some Experiments with Underwater Acoustic
Returns from Cylinders Relative to Object
Identification for AUV Operation
(1988) / Farren M A

Excess Outfitting Materials in Ship
Construction Navy (SCN) Shipbuilding
Programs: An Analysis of the Initial
Allowance Development Process
(1987) / Feierabend R H

An Analysis of an Auto-Alert Sonobuoy
Detection Model
(1988) / Filanowicz R W

An Acoustic Bubble Density Measurement
Technique for Surface Ship Waters
(1987) / Hampton S W

The Impact of Group Technology-Based
Shipbuilding Methods on Naval Ship Design
and Acquisition Practices
(1988) / Heffron J S

Coordinated Steering of a Surface Ship
(1987) / Lee S S

Earthquake Resistant Submarine Drydock Block
System Design
(1988) / Luchs J K

A Computer Simulation Study of Rule-Based
Control of an Autonomous Underwater Vehicle
(1988) / MacPherson D L

The Maritime Strategy and Soviet Submarine
Launched Cruise-Missiles: Implications for
the U. S. Navy
(1987) / Majewski E J

An Examination of the Outporting Ship
Program Implemented in Response to the
Increased Program Size of the Ready
Reserve Force
(1988) / McFarland J M

Decoy Effectiveness in a Multiple Ship
Environment
(1988) / Sengel C

U.S. NAVAL POSTGRADUATE SCHOOL
(continued)

NROSS (Navy Remote Ocean Sensing System): A
Force Multiplier
(1987) / Spohnholtz C L

A Conceptual Design Study of a Hovering
System Controller for an Autonomous
Underwater Vehicle
(1987) / Thompson C A

An Analysis of Selected Surface Ship OPTAR
(Operating Target) Obligation Patterns and
Their Dependency on Operating Schedules and
Other Factors
(1987) / Williams T D

Cost Estimating Relationship Associating
Engineering Drawing Quality with
Installation Cost Growth for USN Ship
Alterations
(1988) / Willstatter K

WASHINGTON, UNIVERSITY OF

Design of Breakwater for Small Boat Basin at
Neah Bay, Washington
(1987) / Chaudhry A R

Regulating Marine Safety in the Arctic
(1987) / Maguire D M

WESTERN ONTARIO, UNIVERSITY OF

En Experimental Study of Wind and Wave
Effects on a Semi-Submersible Offshore
Platform
(1988) / Helliwell S P

AKRON, UNIVERSITY OF

Effects of Fillers and Plasticizers on EPDM
and Neopren Gasket Compounds
(1988) / Alamolhoda A A

Elastic-Quasi-Dynamic Solution of the
Mallory Tube Test
(1989) / Assaad M C

The Extent of Correlation Between the Dipole
Moment and the Chain Dimensions for
Polymethylmethacrylate
(1989) / Ghosh K

Critical Comparison of Methods of
Measurement of Carbon Black Dispersion in
Rubber Compounds
(1989) / Lee S

The Thermodynamics and Kinetics of Phase
Transitions in Poly(Azomethine Ether) Liquid
Crystalline Polymers
(1989) / Lipinski T M

Mechanism of Phenolic Tackifier Molecules
(1988) / Magnus F L

Processing-Structure-Property Relationships
in Cold PEEK Poly(Ether Ketone) and PPS
Poly(P-Phenylene Sulfide)
(1989) / Niemeyer M F

Peroxide Vulcanization of Rubber at High
Temperatures
(1988) / Pham H Q

Synthesis and Characterization of Block
Copolyimides Containing Oxethylene Linkages
(1988) / Schriver A J

Thermoplastic Fiber-Reinforced Composites
Based on Liquid Crystalline Polymers
(1988) / Swaminathan S

Resorcinolic and Cobaltic Bonding Agents
for Brass - Natural Rubber Adhesion
(1989) / Tultz J P

Processing-Structure-Property Relationships
in Biaxially Stretched and Tubular Blown
Thermoplastic Elastomer Films
(1989) / Wang M

Strength and Adhesion to Glass of
Plasticized and Tackified Isoprene Rubber
(1989) / Yu S

Effect of Highly Structured Carbon Black on
the Viscosity of Ethylene Propylene Rubber
(1988) / Zaddi E H

ALABAMA, UNIVERSITY OF (UNIVERSITY)

An Experimental Investigation of the
Chemical Vapor Deposition of Silicon Nitride
Coating on Carbide Cutting Tools
(1988) / Chen C H

Computer Simulation of Solidification in
Investment Casting
(1988) / Huang H

Interaction Between Silicon-Carbide Particle
and the Solid-Liquid Interface in Aluminum
Based Metal Matrix Composites
(1988) / Kacar A S

Solidification of Monotectic Alloys in
Normal and Low Gravity Environments
(1988) / Singh A K

AUBURN UNIVERSITY

Effects of Thermal Aging on Precipitation
Development of Alloy HT-9
(1987) / Chen P S

Deformation Mechanisms and Fracture Processes
in Welded 2219-T87 Aluminum Plate
(1987) / Chen W F

Synthesis and Characterization of Modified
Barium-Ferrite
(1987) / Paig Y J

CALIFORNIA STATE UNIV. (LONG BEACH)

The Effects of a High Humidity Environment
on the Fatigue Life of Aluminum Alloy
20240-T351
(1988) / Voris H C

CARNEGIE-MELLON UNIVERSITY

Symposium on Manufacturing Systems-Design,
Integration and Control
(1988) / Hazen F B

A Model for the Mechanical Stresses Induced
by Head Disk Contact
(1988) / Keremes J J

Elastic-Plastic Compaction of a
Two-Dimensional Assemblage of Particles
(1988) / O'Donnell T P

CASE WESTERN RESERVE UNIVERSITY

The Effect of Calcium Treatment on
Mechanical Properties and Inclusion
Morphology of High Sulfur 8620 Steel
(1988) / Doty M L

An Investigation of the Ingot Cracking
Sensitivity of a Low Carbon Bearing Steel
(1988) / Miraglia S J

A Desulfurization Model Based on Top Slag
Control, for Ladle Refining Steel
(1988) / Rhoads C A

Stress Rupture Behavior of a Silicon Carbide
Coated, Low Modulus Carbon/Carbon Composite
(1988) / Rozak G A

Failure Mechanisms for Protective Coatings
on Polymer Surfaces Exposed to Atomic Oxygen
(1988) / Rutledge S K

Indentation Fracture and Plasticity in
Silicon
(1988) / Rybicki G E

Viscous and Viscoelastic Behavior of a Glass
Cylinder in Uniaxial Compression
(1988) / Sakoske G E

Ultrahigh Temperature Ceramic-Ceramic
Composites
(1988) / Wang D

The Effects of Reheating Alloy Steel
Ingots in an Oxygen Enriched Atmosphere
(1988) / Woszczynski S J

CORNELL UNIVERSITY

Synthesis and Characterization of Glassy
Main Chain, Thermotropic Liquid Crystalline
Polyesters Based Upon a Highly Substituted
Phenylene-Terephthalate Mesogenic Unit with
Flexible Spacers
(1988) / Delvin A M

Thermodynamic Aspects of Sintering in Glass
and Glass-Ceramic Powder Compacts
(1988) / Ducamp V C

DREXEL UNIVERSITY

Strip Casting of Plain Carbon Steel by
Spray Deposition
(1987) / Annavarapu S

Interactions of Lithium with Lithium Borates
(1988) / Bae B S

Adhesion of Composite Resin to Dentin: The role
of the Oral Environment and Bonding Agents
(1987) / Ellison S R

Statistical Analysis and Characterization of the
Effect of Casting Parameters and Heat Treatment
on the Tensile Properties and Microstructure of
A-356 Aluminum Silicon Castings
(1987) / Grecu D M

Filtration of Aluminum Deoxidized Steel
(1987) / Jones S C

Fatigue Studies of Underaged Precipitation
Hardened Aluminum
(1988) / Kudrolli A R

Characterization and Cold Compaction of PEEK Powders
and PEEK/Nickel Powder Blends
(1988) / Reilly J J

Physical/Thermal Relationships of Synthetic Mold
Powders for Continuous Casting of Steel
(1988) / Sankaranarayanan S

3-D Graphite Fiber Reinforced Polymer Composites
for Orthopedic Applications
(1988) / Sharda A N

Fracture Toughness of Polyester Casting Resins
(1988) / Souders S D

EAST TEXAS STATE UNIVERSITY

XPS Study of the CVD/W/Si and W/TiSi(2)
Interfaces: Interfacial Fluorine
(1988) / Carlisle J A

FLORIDA, UNIVERSITY OF

Influence of Sol Properties on Coating Dense
and Porous Substrates
(1988) / Chu P Y

Alumina Sol-Gel Coatings Formed by
Electrophoretic Decomposition
(1988) / Dalzell W

Hydrophilic Surface Modifications of
Bisphenol-A Polycarbonate and Polypropylene
by Gamma Radiation-Induced Graft
Polymerization
(1988) / Daniesl K

Room Temperature Oxidation of Nickel(110)
Exposed to High Oxygen Pressures
(1988) / Dykstal C

Evolution of Physical, Mechanical, and
Microstructural Properties of Liquid-Phase
Sintered Porous Bronze Bearings
(1988) / England S

Biosurface Properties of Ocular Implant
Polymers
(1988) / Hofmeister F

Suspension Processing of Aluminum
Oxide/Zirconium Oxide/Silicon Carbide
Whisker Composites
(1988) / Rojas O E

Wear Behavior of Dental Composite
Restorative Materials
(1988) / Sarrettt D C

Fracture Studies of Silica by Molecular
Dynamics
(1988) / Swiler T P

Factors Affecting Enhanced Drying of Silica
Gels
(1988) / Zutshi A

GEORGIA INSTITUTE OF TECHNOLOGY

Quantitative Fractographic Analysis of
Aluminum (20) 3/Aluminum-2.5 Percent
Lithium Metal Matrix Composite
(1988) / Feinberg-Ringel K S

Corrosion Behavior of Palladium-Cobalt and
Palladium-Copper Alloys in Artificial
Saliva
(1988) / Rasera V

Damage Evolution, Characteization and
Durability of B/Al Composites
(1988) / Webb G

HOUSTON, UNIVERSITY OF

Fretting Studies of Some FCC Metal Alloys
with the Aid of Fretting Maps
(1988) / Odfalk M

Nondestructive Determination of Materials
Textures by Ultrasonic Techniques
(1988) / Spies M

IDAHO, UNIVERSITY OF

Application of Acoustic Emission for Remote
Sensing Utilizing the Kalman Filter for
Noise Suppression and State Variable
Estimation
(1988) / Light M D

ILLINOIS INSTITUTE OF TECHNOLOGY

Systematics of the Rare Earth-Nickel (RE-NI)
Binary Alloy Systems
(1988) / Pan Y Y

Wheel and Rail Contact Analysis and Damage
Study in Laboratory Simulation
(1988) / Qiu W

Steel Grade Optimization of a Forged Load
Binder
(1988) / Sarkar R

IOWA STATE UNIVERSITY

Thermal Conductivity of La3-xRxS4
R = Samarium, Ytterbium, and Europium
(1988) / Kokos G B

IOWA STATE UNIVERSITY
(continued)

Primary Creep Analysis of a Commercial
High Alumina Low-Cement refractory Castable
(1988) / Lundeen B E

Preparation and Sintering of La-Doped Yttria
Powders Precipitated by Continuous Methods
(1988) / Nissen M J

Metastable bcc Mischmetal-Magnesium Alloys
(1988) / Sabariz A L R

JOHNS HOPKINS UNIVERSITY

Effect of Alloying Elements on the Corrosion
Resistance of Rapidly Solidified Magnesium
Alloys
(1989) / Makar G L

KENTUCKY, UNIVERSITY OF

Solid State Gas and Vapor Sensors
(1988) / Kim J S

The Reduction of Variably Loaded and Heat
Treated Nickel on Silica Impregnated
Catalysts
(1988) / Maginnis M A

Erosivity of Coal-Water Slurries
(1988) / Puentes E

LAURENTIEN UNIVERSITY

Fragmentation of Sudbury Hard Rock by High
Frequency Cyclic Loading
(1988) / Kordgharachorloo F

Fractal Description of Fineparticle Systems
(1988) / Trottier R A

LEHIGH UNIVERSITY

Examination of the Lamellar Morphology in
Group IA and IIID Iron Meteorites
(1987) / Kowalik J A

Effects of Multiple Doping (Magnesium
Oxide, Iron Oxide) and Pores on the
Sintering of Alumina
(1987) / Shao J

MARQUETTE UNIVERSITY

Effect of Boron on Discontinuous
Precipitation in JBK-75, A Modified
A-286 Type of Austenitic Stainless Steel
(1988) / Chiang H W

MASSACHUSETTS INSTITUTE OF TECHNOLOGY

SiC Whisker Reinforced MoSi(2)
(1988) / Carter D H

Gas-Metal Arc Welding in Pure Argon
(1988) / Eickhoff S T

Pore Pressure Induced Cracking (PPIC): A
Theoretical and Experimental Analysis
(1988) / Fitz-Patrick R P

Non-Destructive Techniques for Inspection
of Arc Spot Welds
(1988) / Hayes K T

Magnetic Properties of Rapidly-Solidified
Metallic and Ceramic Alloys
(1988) / Henderson L J

Fabrication of Cast Particulate Reinforced
Metals Via Pressure Infiltration
(1988) / Klier E M

Spectroscopy of Tm(3+): YLF as a Laser
Material for Diode Laser Pumping
(1988) / La Rosa G

Parameters Affecting Vapor-Liquid-Solid
Growth of SiC Whiskers at 1450 Centigrade
(1988) / McCarthy B E

The Succinonitrile-Water System:
Thermodynamic Analysis and Measurement of
Physical Properties
(1988) / Ng L H

Magneto-Optic Kerr Effect of TbFeCo Film
(1988) / Ushioda A

Low Cycle Fatigue of Alloy 718 in Cryogenic
Environment
(1988) / Vergara Aimone J A

Attrition Processing of Ti(3)Al-TiC
Particulate Reinforced Composites
(1988) / Warwick M J

Processing of Thick Film Dielectrics
Compactible with Thin Film Superconductors
for Microwave Signal Processing Devices
(1988) / Wong S C

Methods of Determining the Impurity and
Carrier Profiles in Implanted GaAs
(1988) / vanDuyne L S

MIAMI, UNIVERSITY OF

Aspects of Salt Gland Functioning in the
Green Turtle, Chelonia Mydas
(1988) / Nicolson S W

MINNESOTA, UNIVERSITY OF

XPS Study of Reaction and Stability at
Polyimide Interfaces with Metal and
Semiconductor Overlayers
(1988) / Anderson S G

Grinding Media Wear and Its Effects on the
Surface Properties of Coal
(1988) / Baker M D

Block Copolymer Diffusion in Selective
Solvents
(1988) / Birch A P

Oxidation and Reduction of Rhodium and
Iridium Particles on Silicon Dioxide and
Aluminum Dioxide Supports
(1988) / Burkhardt J J

Corrosion and Electrodissolution Studies of
Copper and Cobalt with Microelectrodes and
the Quartz Crystal Microbalance
(1988) / Chandler C J

Role of Sodium Silicate in Phosphate
Flotation
(1988) / Dho H

Thin Film Characterization by Electron
Microscopy
(1988) / Erickson D D

MINNESOTA, UNIVERSITY OF
(continued)

Oxidation of Silicon Carbide in an Alkali
Vapor Environment
(1988) / Pareek V K

Photoelectrochemical and Structural
Investigation of Anodic Oxide Films on
Titanium
(1988) / Tyler P S

Fracture Mechanics of Heavily Filled
Particulate Composites
(1988) / Ubel F A

NEW MEXICO INSTITUTE OF MINING & TECH.

Microstructures and Mechanical Properties of
B2 and L2 Compounds in the Nickel Aluminum
Tantalum System
(1987) / Chen C W

Studies on Effects of Corrosion Inhibitors
Upon Sulfide Stress Cracking Behavior of
Steel
(1988) / Drewien C A

Five-Spot Displacement at Unit Mobility
Ratio in a Reservoir with Heterogeneous
Permeability
(1988) / Heravi N E

Aluminizing of Low Alloy Steels for Sulfide
Stress Corrosion Resistance
(1988) / Isbasaran H

Behavior of Interface in Alumina/Glass
Composite
(1988) / Maheshwari A

Gasification kinetics and Utilization of
New Mexico Coals for Direct Reduction of
New Mexico Iron Ores
(1988) / Tosyali O

NORTH CAROLINA STATE UNIVERSITY

Mechanical Alloying of Immiscible Components
(1988) / Gross S S

Topographical Measurement of Integrated
Circuit Patterns Using SEM Digital Image
Processing
(1988) / Lee J H

Lateral Cracks and Erosion in Glass
(1988) / Pathak D K

NOTRE DAME, UNIVERSITY OF

Enhanced Removal at Edges of Brittle
Materials During Solid Particle Erosion
(1988) / Ahmed T

OKLAHOMA, UNIVERSITY OF

Frequency Estimation Using Simulated
Annealing
(1988) / Dinh H P

A Study of the Adsorption of Binary Anionic
Surfactant Mixtures on Alpha Aluminum Oxide
(1988) / Lopata J J

Tensile Properties of an AISI 1137 Steel in
the Longitudinal and Transverse Directions
(1988) / Ponnle Y O

Toxicity Screening and Risk Evaluation of a
New Sulfur-Containing Surfactant
(1988) / Smith M L

OLD DOMINION UNIVERSITY

Molting in the Mature Female Blue Crab,
Callinectes Sapidus Rathbun
(1988) / Haven K J

PENNSYLVANIA STATE UNIVERSITY

X-Ray Diffraction Residual Stress Analysis
of Inconel Alloy 600 U-Bend Tubes
(1987) / Ivkovich D P

Examination of the Slow Fracture
Characteristics of a Metal Particulate
Glass Composite
(1987) / Jessen T L

Diffusion of Cobalt-57 in the Surface Layers
of Alumina and Nickel Oxide
(1987) / Macey E P

Phase Relationships and Mechanical
Properties in the Zirconia-Hafnia-Magnesia
and Zirconia-Yttria-Magnesia Systems
(1987) / Manning S L

Ultra-Low Permittivity Porous Silica Thick
Films for Gallium Arsenide IC Packaging
(1987) / Mohideen U

Surface Reaction Layer Formation on Polished
Fluorozirconate Glass During Exposure to
Water and Aqueous Hydrogen Fluoride
Solutions
(1987) / Phelps A W

Aluminum Nitride Substrates Prepared by Arc
Plasma Spraying
(1987) / Pickrell D J

Carbon Black - Polyethylene Composites for
PTC Thermistor Applications
(1987) / Rohlfing L L

An Investigation Into the Roles of Internal
Stress and Crack-Twin Interactions on the
Fracture Behavior of 95 Molybdenum 1 Percent
Lead Zirconate/5 Molybdenum 1 Percent Lead
Titanate (95/5 PZT)
(1987) / Shellman D

Characterization and Utilization of Metal
Finishing Wastes from Ferrous and
Non-Ferrous Foundry Operations
(1987) / Urenovitch L A

PENNSYLVANIA, UNIVERSITY OF

DLTS as an Optimization Tool for Evaluation
of the Performance and Stability of
Alpha-Silicon/Hydrogen p-i-n Photovoltaic
Cells
(1988) / Kalina J E

Oxidation, Reduction, and Stability Studies
in the Barium Lanthanum/Copper/Oxide [Delta]
System
(1988) / Katzan C M

The Effects of Silicon-Nitride and
Silicon-Carbide Coatings on the Oxidation
Behavior of Carbon/Carbon Composites
(1988) / Sloan J W

Effect of Composition on High and Low
Temperature Embrittlement of Low Alloy
Steels
(1988) / Solaski T P

PITTSBURGH, UNIVERSITY OF

Orientation Studies on Doubly Oriented,
Crystallizable Poly(Ethylene Terephthalate)
(1989) / Berg E M

The Freeze Drying of Sulfate Salts to
Produce Undoped and Titanium-Doped High
Purity Aluminas
(1989) / Beyer L A

The Characterization of the Bonding
Mechanism of Electrolessly Deposited Copper
on Aluminum Oxide
(1989) / Cotte J M

Electrical Properties of Exfoliated-Graphite
Filled Polyester Based Composites
(1989) / Foy J V

Electron Microscopy of Yttria-Doped Alumina
(1989) / Kalnas C E

Hot Corrosion of Silica and Alumina
(1989) / Lawson M G

Experimental Test of the Molecular
Entanglement Theory of Shear Thinning in
Entangled Polymers
(1989) / Matty J T

The Correlation Between the Tear Energy of
Elastomers and Linear Viscoelasticity
(1989) / Rosner M J

PURDUE UNIVERSITY

Application of the Laser Flash Technique
to Measurement of the Thermal Conductivities
of Liquid Oxides
(1988) / Doolittle J S

X-Ray Scattering from Superlattices
(1988) / Rajagopalan S

QUEENS UNIVERSITY

A Case Study of High Frequency Microseismic
Monitoring Prior to a Rockburst
(1988) / Madsen D A

A Study of the Production of Bulk Amorphous
and Nono-Crystalline Nickel-Phosphorus
Alloys Through Electrodeposition
(1988) / McMahon G S

Ore Reserve Estimation of Tabular
Orebodies: A Geostatistical Approach
(1988) / Pareja L D

The Effect of Corrosion Products on the
Corrosion of Rolled 304L Stainless Steel
in Boiling 65 Percent Nitric Acid
(1988) / Perera L M

High Frequency Microseismic Monitoring
Applied to the Determination of Stress
Levels in Hard Rock Mines
(1988) / Semadeni T A J

RENSSELAER POLYTECHNIC INSTITUTE

A Simplified Buckling Analysis for
Open-Section Stiffened Composite
Compression Panels
(1988) / Barlow S R

Ceramic Matrix Composites Via Organometallic
Precursors
(1988) / Boisvert R P

Literature Survey: The Application of
Acoustic Emission to the In-Process
Inspection of Austenitic Stainless Steel
Welds
(1988) / Campbell D R

Melt Rheology of Iron Powder/Wax Mixtures
(1988) / Carpenter B J

The Gel-to-Glass Conversion in Silica:
Determination of Structural Changes Through
the Electron Paramagnetic Resonance of
Copper(II) Complexes
(1988) / Darab J G

Interfacial Elastic Strain Measurement in
Epitaxial Indium Antimony-Gallium Arsenide
Using Indirect Lattice Imaging
(1988) / Gong R

The Mechanism of Strength Loss by Thermal
Degradation of a Phenolic Resin Matrix
Composite
(1988) / Hays W D

Gravitational Effects on Liquid Phase
Sintering
(1988) / Kipphut C

Microstructure and Grain Growth of Mullite
Prepared by the Sol-Gel Method
(1988) / Lee C

Electron Paramagnetic Resonance Studies of
the Corrosion of Ion Implanted Glasses
(1988) / MacGregor A I

The Effect of Temperature and Environment
on the Tensile and Cyclic Behavior of
IC-221 and IC-218
(1988) / Matuszyk W

Continuous Thermal Processing of Silicon
(1988) / McRae J M

Evolution of Microstructures in Tin-Bismuth
Alloys
(1988) / Meloro L A

The Effect of Boron, Chromium, Oxygen Active
Elements, and Inert Fiber Reinforcements on
the High Temperature Oxidation Behavior of
Nickel Aluminum
(1988) / Montrym J P

Mechanical Properties and Microstructural
Features of Alumina Fiber Reinforced Nickel
Aluminum Matrix Composites
(1988) / Moore B V

Formation of Cobalt Silicides by Rapid
Thermal Annealing
(1988) / Sitaram A R

Structural and Chemical Analysis of Dental
Enamel from Magnesium and Sodium Fluoride
Supplemented Rats
(1988) / Spencer P

RENSSELAER POLYTECHNIC INSTITUTE
(continued)

Microstructural and Thermal Characteristics
of Mushy Zone
(1988) / Tsutsumi K

Weld Ductility Studies in a Tin-Modified
Copper-Nickel Alloy
(1988) / Tumuluru M

The Role of Powder Characteristics in
Binder Incorporation for Injection Molding
Feedstock
(1988) / Warren J

SOUTH DAKOTA SCHOOL OF MINES AND TECHNOL.

Analysis of Dynamic Properties and Modelling
for a Composite Laminate Robotic Arm
(1988) / Lin J C

Design of Dispersion Strengthened Multiphase
Stainless Steels
(1988) / Qadri J H

STEVENS INSTITUTE OF TECHNOLOGY

Passivation of Deep Level Defects and
Shallow Doping Levels on Gallium Arsenide
and Aluminum Gallium Arsenide by Hydrogen
Plasma Exposure
(1987) / Nabity J C

TEXAS, UNIVERSITY OF (ARLINGTON)

Investigation of the Effects of Orientation
and Physical Aging on the Creep
Characteristics of Polyarylene Sulfide 2
Using Thermomechanical Analysis
(1988) / Sullivan L L

TEXAS, UNIVERSITY OF (AUSTIN)

Interfacial Chemistry of Zinc and Indium
Thin Films on Mercury Cadmium Telluride
(1988) / Greene W T

Part One: The Synthesis and Properties
Characterization of Bi(2)O(2)S. Part
Two: Identification of the 90K High T(c)
Superconductor and Its Phase Relationship
with La(3)Ba(3)Cu(6)O(14+y)
(1987) / Hsu H M

High-Energy/High-Rate Consolidation of
Tungsten and Tungsten-Based Composite
Powders
(1988) / Raghunathan S K

U.S. NAVAL POSTGRADUATE SCHOOL

Corrosion Performance of High Damping Alloys
in 3.5 Percent Sodium Chloride Environment
(1987) / Akhtar S

Composite Reliability Enhancement Via
Preloading
(1987) / Bell D K

Composite Materials at High Temperatures
(1988) / Bess C R

The Effect of Heat Treatment and Cyclic
Strain Amplitude on the Damping Properties
of Iron-Chromium Based Alloys
(1988) / Childs J L

A Comparison of High Damping Shape Memory
Alloys with Copper-Manganese-Based and
Iron-Chromium-Based Alloys
(1987) / Cronauer J T

Data Acquisition and Control for Multiple
Composite Life Tests
(1988) / Emery J W

Characterization of High Damping
Iron-Chromium-Molybdenum and
Iron-Chromium-Aluminum Alloys for Naval
Ships Application
(1988) / Ferguson D B

The Age Hardening Response of
Thermomechanically Processed Aluminum
Magnesium Lithium Alloys
(1987) / Ferris W F

Characterization of the Corrosion Behavior
of High Damping Alloys in Seawater
(1987) / Fscue W D

Processing of Aluminum Alloy 2090 for
Superplasticity
(1988) / Groh G E

The Effect of Temperature on the Tensile
Properties of HSLA-100 Steel
(1987) / Hamilton J E

Composite Reliability Enhancement Via
Reloading
(1988) / Jones M C

Finite Element Analysis of Laminated
Composite Plates
(1988) / Lee M H

An Investigation of the Hot Corrosion
Protectivity Behavior of Platinum Modified
Aluminide Coatings on Nickel-Based
Superalloys
(1987) / Malush R E

An Electron Microscopy Study of Tweed
Microstructures and Premartensitic Effects
in High Damping 53-Copper 45 Manganese
2-Aluminum Alloy
(1988) / Mayes L L

Microstructural Characterization of HSLA-100
GMA-Weldments
(1987) / Mickelberry K D

A Comparison of Tripping Behavior of Wide
and Narrow Flanged "I" and 'Z' Stiffened
Panels
(1987) / Miller R B

Determination of the Origin of Self-Pumped
Phase Conjugation in Barium Titanate
(1987) / Moore T R

Optimizing Superplastic Response in Lithium
Containing Aluminum-Magnesium Alloys
(1987) / Munro I G

Microstructural Characterization of the
Heat Affected Zone of HSLA (High Strength
Low Alloy)-100 Steel GMA (Gas Metal Arc)
Weldment
(1987) / Potkay G P

U.S. NAVAL POSTGRADUATE SCHOOL
(continued)

The Effect of Processing and Superplastic
Deformation on Ambient Ductility of
Aluminum-10 Percent Magnesium-0 Percent
Zirconium
(1987) / Solomos D K

Thermomechanical Processing of Aluminum
Alloy 2090 for Grain Refinement and
Superplasticity
(1987) / Spiropoulos P T

Investigation by Differential Scanning
Calorimetry of Microstructure in a
Superplastic Aluminum-Magnesium-Zirconium
Alloy
(1987) / Stewart D L

Effects of 67.5 MeV Electron Irradiation on
Yttrium-Barium-Copper-Oxide and Gadolinium-
Barium-Copper-Oxide High-Temperature
Superconductors
(1987) / Sweigard E L

Fiber-Optic Implementation of MIL STD-1553:
A Serial Bus Protocol
(1987) / Wester R S

The Influence of Total Strain, Strain Rate
and Reheating Time During Warm Rolling on
the Superplastic Ductility of an
Aluminum-Magnesium-Zirconium Alloy
(1987) / Wise J E

An Investigation of Unipolar Arcing at
Atmospheric Pressure in Aluminum 2024 and
Aluminum Coated Glass Slides
(1987) / Woodson S W

U.S.A.F. INSTITUTE OF TECHNOLOGY

Crack Growth of a Titanium Aluminide Alloy
Under Thermal Mechanical Fatigue
(1988) / Burgess D G

The Effects of Ply Dropoffs on the Tensile
Behavior of Graphite/Epoxy Laminates
(1988) / Cannon R K

Damping Characteristics of Advanced Carbon
Carbon and Tehrmoplastic Materials
(1988) / Doederlein T A

A Study of the Failure Characteristics of a
Thermoplastic Composite Material at High
Temperature
(1988) / Fisher J M

Application of the Boundary Element Method
to Fatigue Crack Growth Analysis
(1988) / Kelly T C

A Study of Failure Characteristics in
Thermoplastic Composite Material
(1988) / Martin R J

Fracture Toughness Testing of a Ceramic
Matrix Composite at Elevated Temperature
(1988) / Mol J H

Investigation of Failure Modes in Fiber
Reinforced Ceramic Matrix Composites
Advisor
(1988) / Moschler J W

Surface Passivation of Gallium Arsenide
(1988) / Racicot R J

Dimunition and Longitudinal Splitting of
Carbon Fibers Due to Grinding
(1988) / Seibert J F

Macrocrack Multiple Defect Interaction
Considering Elastic, Plastic, and
Viscoplastic
(1988) / Smith L K

Investigation of Crack Growth in Titanium
Aluminide at Elevated Temperatures
(1988) / Staubs E A

Determination of Plutonium-Beryllium Source
Strength by Manganese Activation
(1988) / Whitworth P F

A Study of Damage Tolerance in Curved
Composite Panels
(1988) / Wilder B L

UTAH, UNIVERSITY OF

Steric Interactions of Monodisperse Silica
Dispersion
(1988) / Chang S

The Analysis and Characterization of
Gallium Arsenic Phosphorus: Zinc
Light-Emitting Diodes (LEDs)
(1988) / Chen M E

Order-Disorder Transformation in Lithium
Iron Oxide-Lithium Aluminum Oxide System
(1988) / Chen Z C

Phase Transformation in the Lithium Aluminum
Oxide - Lithium Iron Oxide System of
Aliovalent Dopants
(1988) / Kim S J

Metal-Ferroelectric-Semiconductor Interface
(1988) / Radhakrishnan G

Effect of Carbon Fiber Surface Properties on
the Adhesion to Organic Polymers
(1988) / Yon K Y

VERMONT, UNIVERSITY OF

Characterization of Low-Frequency Intensity
Noise and Spectral Characteristics of a
Laser Diode
(1988) / Anderson B

Degradation of n-MOSFET's at Liquid Nitrogen
Temperature
(1988) / Weber C C

VIRGINIA, UNIVERSITY OF

The Icosahedral to Crystalline
Transformation in Aluminum-15 Copper-12.5
Magnesium-20 Lithium-16.5
(1988) / Diehl P E

Synthesis of Aluminum-Lithium Alloys with
High Lithium Content
(1988) / Gaillard M C

Effect of Dispersoid Particles on Recovery
and Recrystallization of Aluminum Alloys
(1988) / Gudmundsson H

Creep of Cast 390 Aluminum Alloys
(1988) / Jones B R

VIRGINIA, UNIVERSITY OF
(continued)

Growth and Characterization of GEL-XSNXTE
for Thermoelectric Utilization
(1988) / Madey M A

The Physics of Mechanical Alloying: A
First Report
(1988) / Maurice D R

The Microstructure and Mechanical Properties
of High Temperature Aluminum Composites
(1988) / Pollock W D

Effects of Hot Rolling and High Temperature
on Aluminum-2124 Silicon Carbide Whisker
Composites
(1988) / Schueller R D

The Microstructure, Deformation, and
Fracture of (Aluminum, Nickel)3 Titanium
Lithium-2 Intermetallic Alloys
(1988) / Turner C D

Factors Affecting the Stability and
Performance of Lead Dioxide Anodes in a
Hexavalent Chromium Generator
(1988) / Vora R J

Electrochemical Sensors to Monitor the
Corrosion of Reinforcing Steel in Concrete
(1988) / White S A

WASHINGTON STATE UNIVERSITY

Factors Affecting Superplastic Stability in
an Aluminum-Lithium-Copper-Zirconium Alloy
(1988) / Ash B

Photoluminescence of Transition Metals in
Silicon
(1988) / Beauchaine D A

Extension of Metal Cutting Tool Life by
Electrospark Alloying
(1988) / Brown E A

Acoustic Emission - Fracture Strength
Relations of AL/GR Composites
(1988) / Easterday J D

Effect of pH and Maleic Anydride on
Inhibition of Corrosion by Sulfite Ion
(1988) / Feng X

An Automated Evaluation of Formability of
Aluminum-Lithium Alloys and Some Aluminum
Alloys in a Uni-Axial State of Stress
(1988) / Ha J B

Crack Growth Behavior During LME of High
Strength Aluminum Alloy by Mercury
(1988) / Liu Y

Finite Element Modeling for Interlaminar
Stresses in Composite Plates
(1988) / Newton J A

Equilibrium Hydrogen Pressures in
Vanadium-Aluminum Alloys
(1988) / Park J Y

The Recrystallization of an
Aluminum-Lithium-Zirconium Based Alloy
(1988) / Ren B

Coalescence of Aluminum Alloy During Salt
Melting Process
(1988) / Taghiei M M

Wear Resistant Surface Coatings Deposited on
Titanium-Aluminum-Vanadium by Electrospark
Alloying
(1988) / Topinka W A

A Study on Machined Subsurface and Tool Wear
(1988) / Xie Q

WASHINGTON, UNIVERSITY OF

The Strengths of Aluminosilicate Proppants
in Stress Corrosion Conditions
(1987) / Armstrong W D

Oxidation Behavior of Dense Beta-Sialon
(1987) / Chung Y H

Compaction and Sintering Phenomena in
Titanium-Copper Alloys
(1987) / Daimellah A

Chromia Impregnation Strengthening of
Alumina
(1987) / Hamano Y

Low Temperature Oxidation of Aluminum
Nitride
(1987) / Jones S

Microhardness Anisotropy of Single Crystal
Sapphire and Sapphire Solid Solutions
(1988) / Kaji M

Microstructural and Fractographic
Characterization of B(4)C-Al Cermet
(1987) / Kim G

Microstructural Analysis of CVD Aluminum
Oxide Film on Silicon
(1987) / Kim J

Preparation and Microstructure Evolution of
Hierarchically Clustered Powders
(1988) / Martin C B

Powder Metallurgy and Sintering Kinetics of
a Eutectoid Titanium-Copper Alloy
(1988) / Nash R A

Elevated Temperature Fracture Properties of
CVI Silicon Carbide Matrix Continuous
Ceramic Fiber Composites
(1987) / Peussa J T

Textural effects on the fracture Resistance
of Polycrystalline Aluminum Dioxide
(1987) / Salem J A

Characterization of Copper-3 Percent Cobalt
P/M Alloys
(1987) / Salhi S M

Pressureless Sintering of Short-Fiber
Reinforced Mullite
(1987) / Yamada R

WINDSOR, UNIVERSITY OF

Computer Calculation of Aluminum Corner of
Aluminum-Telluride-Beryllium Phase Diagram
(1988) / Fang W

c-Axis Oriented Barium Titanium Silicon
Oxide Thin Films
(1988) / Kawa P M

Beryllium-Enhanced Precipitation in
Aluminum-Magnesium-Silicon Alloy: Role of
Nucleation Entropy and Interfacial Energy
(1988) / Tian X

WORCHESTER POLYTECHNIC INSTITUTE

Application of Microcomputer Spreadsheets in
Undergraduate Materials Processing Courses
(1986) / Benda C M

An Experimental Investigation of the
Degradation of a Cobalt Chromium Aluminum
Yttrium Coating During High Temperature
Oxidation
(1987) / Chartier D M

Microstructure-Dielectric Property Relations
in Glass-Ceramics
(1986) / Fitzsimmons T J

Crevice Corrosion of Implant Alloys
(1987) / Hughes M Y

The Development and Characterization of a
Low Dielectric Constant Thick Film Material
(1986) / Kellerman D W

A Finite Difference Model for the
Degredation of High Temperature Oxidation
Resistant M-Chromium Aluminum Yttrium
Coatings
(1987) / Lee E Y

Microstructure and Properties of Hot
Isostatic Pressed Silicon Nitride for
Application in Ceramic Bearings
(1987) / Leo T P

The Deformation of P/M Rene' 95 Under
Isothermal Forging Conditions
(1987) / Morra J E

The Effects of Sputter Deposition
Parameters on 10-Titanium 90-Tungsten
Diffusion Barriers
(1987) / Oparowski J M

YOUNGSTOWN STATE UNIVERSITY

Correlation of the Fracture Toughness of
Titanium with Modified Instrumented Charpy
Impact Test Results
(1988) / Einfalt J

AKRON, UNIVERSITY OF

Life Prediction of Prestressed Forging
Dies - An Analytical Approach
(1989) / Boyer R M

Design of a Totally Implantable Artificial
Bladder and Sphincter
(1988) / Kellackey K M

Extreme Operating Conditions of a Power
Plant Boiler Feedwater Pump as Induced by
Component Excitations
(1989) / Manos M G

Biomechanical Evaluation of Microarterial
and Microvenous Anastomoses in the Rabbit
(1989) / Marcinek S A

Evaluation of Acrylic Cementation Techniques
in the Treatment of Tumors of the Bone
(1989) / Merholz W J

A PC Based On-Site Generation Economic
Screening Model
(1989) / Miller W C

A Model for Enzyme Enhanced Antibiotic
Diffusion
(1988) / Senary M K

Blood Protein and Cellular Deposition on
Deforming Surfaces
(1989) / Sutphin C M

Combined Finite Element - Experimental
Technique for Determining Tensile Properties
of Artificial Ice
(1988) / Xian X

ALABAMA, UNIVERSITY OF (HUNTSVILLE)

Application of Disturbance Accommodating
Control Theory to Structural Dynamics
Control
(1988) / Dalton E C

Predicting Compression Set in Elastomeric
Materials After Partial Stress Relaxation
(1988) / Ledbetter F E

Stress Intensity Factors for a Part-Through
Flaw in a Hollow Cylinder
(1988) / Meyers C A

ALBERTA, UNIVERSITY OF

A System for Evaluating Spinal Fixation
Devices
(1988) / Evenson R

The Design, Instrumentation, and Performance
of a Refrigerated Marine Icing Wind Tunnel
(1988) / Foy C E

ARIZONA STATE UNIVERSITY

Dynamic Analysis of Helicoidal Bars
(1988) / AlGhamdi S A

Introduction of Local Modes in Large Space
Structures for Vibration Suppression
(1988) / Balkema K J

Implementation of Feature Mapping and
Reasoning Shell with Application to Group
Technology Coding
(1988) / Bhatnagar A S

Measurement of Local Heat Transfer on
Shrouded Rotating Disks with Jet Impingement
(1988) / Bosch G

An Investigation of the Point Matching
Technique for the Solution of Two
Dimensional Stress Problems
(1988) / Cegalis J

Investigation of the Plane-Workspace of Two
Special 3-R Manipulators
(1988) / Charif M M

Mixed Convection Along a Wavy Surface
(1988) / Ghosh Moulic S

Rule-Based Feature Recognition: Concepts,
Primitives and Implementation
(1988) / Hwang J L

A Study of Activity-Induced Bone Remodelling
(1988) / Johnson E M

An Automated Data-Acquisition System for
Local Mass Transfer Measurement with
Naphthalene Sublimation
(1988) / Kapat J S

Triparametric Volumes for Flow Modeling
and Visualization
(1988) / Kersey S N

Simulation of an Optical Inspection System
(1988) / Koury E P

Flow-Quality Improvements in the Arizona
State University Unsteady Wind Tunnel
(1988) / Mousseux M C

The Use of Implicit Functions in Solid
Modeling: An Investigation
(1988) / Nagasuru D

Feedforward Control of Superheated Steam
Temperature in a Power Plant
(1988) / Pan Z

Prediction of Limit Cycles in Nonlinear
Control Systems with Multiple Nonlinearities
(1988) / Pillai V K

Feature Based Object Decomposition for
Finite Element Mesh Generation
(1988) / Razdan A

Rotation-Fixed End Conditions for Conical
Piloting as an Assembly Method
(1988) / Seguara F J

Turbulent Mass Flux Measurements in the
Developing Region of a Free Binary Gas Jet
(1988) / Zhu J Y

Dynamical Analytsis and Optimal Control of
a Flexible Robot Arm
(1988) / Zhu W

ARKANSAS, UNIVERSITY OF

A Study of Transient, Low Intensity Heat and
Mass Transfer in Soil Utilizing Gamma Ray
Densiometry
(1988) / Gartman B C

Measurement of Viscosity of Multiphase
Fluids Between Rotating Plates
(1988) / Person T E

ARKANSAS, UNIVERSITY OF
(continued)

The Effects of Fluid Levels in the Iverson
Wet Clutch Assembly
(1988) / Richardson D E

Detection and Characterization of Fatigue
Damage in 6061-T6 Aluminum Using the
Magnetic Bridge Sensor
(1988) / Woodward M R

A Cinematographic and Finite Element
Analysis of the Influence of Soft Tissue
Movement on Electrogoniometric Data
(1988) / Young M

AUBURN UNIVERSITY

Numerical and Experimental Analysis of a Stefan
Program with Contact Conductance
(1987) / Hamm J M

A Computer Program for the Kinematic Analysis of
Planar Four-Bar Mechanisms
(1987) / Hubbert T E

Oscillatory Flows of Viscoelastic Liquids
(1987) / Spain J D

BRIGHAM YOUNG UNIVERSITY

Cross Flow Jet Mixing Measurements in a
Simulated Entrained-Flow Coal Gasifier
(1988) / Braithwaite D J

Development of Two-Dimensional Tolerance
Modeling Methods for CAD Systems
(1988) / Chun C S

A Generalized Method of Finding Deflections
in Robot Linkages
(1988) / Connell J A

Robot Path Following Using Tool Axis Control
(1988) / Davison B D

Robot Path Following Using Tool Axis Control
(1988) / De Davison B

Strain Hardening in Torsionally Strained
OFHC Copper Tubing
(1988) / Field D P

Optimal Design of a Turbine Blade Cooling
Passages
(1988) / Glover G W

On the Correlation of Misorientation with
Grain Growth in Commercial Pyurity Aluminum
(1988) / He Y

Model and Design of a Mechanical Tendon
System
(1988) / Johnstun C R

An Efficient Method for Iterative Tolerance
Design Using Monte Carlo Simulation
(1988) / Larsen D V

Nonlinear Tolerance Analysis Using the
Direct Linearization Method
(1988) / Marler J D

An Interferometric Study of Heat Flow in a
Channel with Discrete Heat Sources
(1988) / Martin T J

A General Four Way Hydraulic Valve Model
and Valve Orifice Geometric Parameter
Identification
(1988) / Martinez M W

Convective Heat Transfer from
Two-Dimensional Discrete Heat Sources
Along a Vertical Channel
(1988) / McEntire A B

The Application of Convolution Methods to
the Statistical Analysis of Mating Hole
Patterns
(1988) / Nielsen D B

Characterizing Nonlinear Kelvin-Voigt and
Three-Element Anelastic Systems from
Hysteresis Data
(1988) / Palmer A J

The Structured Singular Value as a
Robustness Measure for Control System
Design
(1988) / Philbrick D O

CICATS: Cadam Interface to Computer-Aided
Tolerance Selection
(1988) / Rime C N

A Practical Method for Three-Dimensional
Tolerance Analysis Using a Solid Modeler
(1988) / Robison R H

Measurements of Local Heat Transfer
Coefficients: Results for an Axisymmetric
Single-Phase Water Jet Impinging Normally
on a Flat Plate with Uniform Heat Flux
(1988) / Stevens J W

Correlation Between Chemical Stimulation and
Flow Resistance of Platelet Aggregates
(1988) / White J V

Comparison of Methods for Determining a
Complete Orientation Distribution Function
(1988) / Wright S I

Fall Cutting Management of Alfalfa and Its
Relation to Carbohydrate Level, Yield and
Stand Persistence
(1988) / Zaifnejad M

BROWN UNIVERSITY

The Mechanical Properties of Coronary
Arteries and the Stress Distributions in
Atherosclerotic Vessels
(1988) / Kenny S M

BUCKNELL UNIVERSITY

General Formulation for the Solution of 2-D
Problems in Elasticity Via Complex Variables
and Collocation
(1988) / Fischer W C

A Fatigue Crack Initiation Study of Hot
Isostatically Pressed Cast Aluminum Alloys
(1988) / Peterec R H

A General Complex Variable Collocation Code
for 2-D Elastic Fracture Analysis
(1988) / Wu B

CALGARY, UNIVERSITY OF

Large Deformation Behaviour of an Inflatable
Arch Structure
(1988) / Balas L

CALIFORNIA STATE UNIV. (SACRAMENTO)

Electrical Brush Colling
(1988) / Alereza R

A Dosimetry Method for the
Radioimmunotherapy of B-Cell Chronic
Lymthocytic Leukemia
(1988) / Amorino G P

Correlations of the Spectrum Parameters of
the EEG with Anesthetic Concentration
(1988) / Anvari B

Preliminary Design of a 25 MW Geothermal
Power Plant
(1988) / Falaki F

Cooling of Electronic Equipment
(1988) / Khatib-Shahidi S

Heat Transfer by Convection and Radiation
in Turbulent Boundary Layer Flow
(1988) / Mesbah M

A Comparison of Features Extracted from
Fourier and Parametric EEG Power Spectra
(1988) / Ottah O

Dimensional Accuracy Generated from
Computerized Axial Tomography Data of Bone
Models
(1988) / Pang A K

Data Compression Techniques for a Diphone
Based Speech Synthesis System
(1988) / Stephens S L

Diphone Concatenation in an Unlimited
Vocabulary Speech Synthesis System
(1988) / Thornton L M

Computer-Aided Optimization of Dynamic
Systems Using Multi-Stage Monte Carlo
Technique and Bond Graph Modeling
(1988) / Tran M N

Computer Aided Approach 3-Dimensional
Vehicle Dynamics Using Bond Graph Modeling
Method
(1988) / Tse K B

A Spatial Representation of Vertical Foot
Forces Using a Microcomputer-Based Graphics
System
(1988) / William C L

CALIFORNIA, UNIVERSITY OF (DAVIS)

Development of Colonic Myoelectrical
Activity in the Nonhuman Infant Primate
(1988) / Hom D M

Analysis of Microradiographs from Bone
Biopsies by Means of Automatic Digital
Image Processing Techniques
(1988) / Papamichos T

Carrier Detection of Duchenne Dystrophy
Using Signal Processing Techniques
(1988) / Preising B M

CALIFORNIA, UNIVERSITY OF (IRVINE)

Design, Analysis and Testing of a High
Bandwidth, Force Controlled Hydraulic
Robot Arm
(1988) / Desai J O

An Investigation of Coupled Natural
Convection and Radiation Heat Transfer
Across Two Inclined Air Layers Separated
by a Semi-Transparent Nongray Film
(1988) / Fuehrer P L

Adaptive Control of Systems with Varying
Effective Order
(1988) / Gu Y

Physical and Mechanical Processes in Drying
of Calcium Phosphate Gels
(1988) / Lu B

Infrared Analysis of Water in Combustion
Products
(1988) / Montgomery T A

Three Degree-of-Freedom Planar Robot Path
Planning
(1988) / Naqvi S A

Science Experiments Plan for Pool
Fires: Preignition Processes
(1988) / Schiller D N

The Workspace of Two Adjacent Planar
Manipulators, One with One Link, Another
with Two Links, Holding an Object of One
Dimension
(1988) / Tsai Y M

CALIFORNIA, UNIVERSITY OF (LA JOLLA)

Quantitative Dynamic Analysis of
Post-Junctional Leukocyte Accumulation
(1988) / Breit G A

An Analysis of Optimization Techniques as
Applied to Vascular Networks
(1988) / Quintero L J

Dynamics of Magnetic Recording Sliders Using
Laser Doppler Interferometry
(1988) / Riener T A

CARLETON UNIVERSITY

Study of the Tire-Terrain Interaction Using
the Finite Element Method
(1988) / Chen S T

Evaluation of Stress Intensity Factors in
Multi-Cracked Thick-Walled Cylinders Using
Finite Element Methods
(1988) / Kirkhope K J

Flue Gas Waste Energy Recovery Optimization
with Spray Recuperators
(1988) / Lee D H

A Derivation of Gross Thrust for a
Sea-Level Jet Engine Test Cell
(1988) / MacLeod J D

Fatigue Crack Growth in Semi-Elliptical
Surface Cracks
(1988) / McFadyen N B

Design of a Research Scale Pressurized
Fluidized Bed Coal Gasifier
(1988) / McKinney D H

Axial Flow Compressor Modelling for Engine
Health Monitoring Studies
(1988) / Muir D E

CARLETON UNIVERSITY
(continued)

Three-Dimensional, Finite Deformation,
Thermal-Elasto-Plastic Finite Element
Analysis
(1988) / Oddy A S

Refuvenation of Gas Turbine Discs
(1988) / Pishva S M

Boundary Integral Equation Fracture
Mechanics Analysis Using the Sub Domain
Method
(1988) / Quesnel P

Damage Tolerance Based Life Extension of
Turbine Discs: A PFM Approach
(1988) / Roodnick G M

CARNEGIE-MELLON UNIVERSITY

Calculation of Wall Shear Rate Acting on a
Canine Carotid Artery Perfused In Vitro
(1988) / Johnson G A

Application of a Mathematical Model to
Experimental Data of Arterial Wall Transport
(1988) / Neumann S J

CASE WESTERN RESERVE UNIVERSITY

Performance Evaluation and Failure Analysis
of an Implantable Neuromuscular Stimulation
System
(1988) / Banks S A

A Refractive Index Matched Facility to
Study Solid-Liquid Multiphase Flows Using
Laser Velocimetry
(1988) / Bhunia S K

A Mathematical Model for the Respiratory
Central Pattern Generator
(1988) / Botros S M

Simulation and Modeling of Multidegree of
Freedom Mechanical System for Space
Applications
(1988) / Chang R Y

Control of Reactionless Robots
(1988) / Chen J L

Temperature and Albumin Effects on
Adsorption of Bilirubin from Standard
Solution Using Anion-Exchange Resin
(1988) / Davies C R

An Investigation of Roller Coaster Dynamics
and Track Bank Angle Optimization
(1988) / Dubovec J

An Experimental Study of Opposed Flow
Diffusion Flame Extinction Over a Thin
Fuel in Microgravity
(1988) / Ferkul P V

Local Failure Stress Analysis and Design of
Intramuscular Electrode Leads
(1988) / Fu S

Evaluation of the Selspot II
Three-Dimensional Tracking System as a Means
of Tracking Upper Extremity Kinematics
(1988) / Hart R L

Engine Cylinder Pressure Measurement and
Performance Analysis
(1988) / Hwang S C

Computer Aided Graphical Design of Four-Bar
Mechanisms
(1988) / Isble C J

Computer Aided Design of Stamping Die
(1988) / Jeong T

Development and Evaluation of a Finger Joint
Angle Sensor
(1988) / Jespersen E

A Noninvasive Mapping of the Electrical
Activity of the Heart
(1988) / Kavuru M S

The Effect of Cooling on the Calibration of
Gardon Heat Flux Sensors
(1988) / Krane M H

A Microcomputer-Based System for Real-Time
Beta-to-Beat Measurement of Intracardiac
Conduction Time Intervals
(1988) / Lu B

The Fatigue Damage Behavior of a Single
Crystal Superalloy
(1988) / McGaw M A

Implementation of Just-in-Time Production by
Part Orientation Through Vision-Aided
Robotics
(1988) / Miller J A

Control of Elbow Extension in the C5
Quadriplegic Using Functional Neuromuscular
Stimulation
(1988) / Miller L J

Experimental Determination of Aerodynamic
Damping in a Three-Stage Transonic Axial
Flow Compressor
(1988) / Newman F A

Computer Simulation of Functional Gages for
Geometrically Dimensioned Parts
(1988) / Orwig D B

The Effects of Static Field Inhomogeneities
on Magnetic Resonance Images
(1988) / Parrish T B

The Development of a Peroxidase Method for
Measuring Transmural Concentration Profiles
of Macromolecules Across the Rat Aorta
In-Vivo: Its Application in a
Lipopolysaccharide Endothelial Cell Injury
Model
(1988) / Penn M S

A Numerical and Experimental Study of the
Meniscus Shape on Combined Thermocapillary
and Natural Convection Flow
(1988) / Platt J A

Infrared Surface Temperature Measurements
for the Surface Tension Driven Convection
Experiment
(1988) / Pline A D

Imbalanced Biphasic Electrical Stimulation:
Muscle Tissue Damage
(1988) / Scheiner A

CASE WESTERN RESERVE UNIVERSITY
(continued)

A Multichannel Implantable Telemeter for
Acquisition of Physiological Data
(1988) / Schild J H

A Discrete-Event Model of Paraplegic FNS
Gait
(1988) / Simon E S

Experimental Study of Bypass Transition in
a Boundary Layer
(1988) / Suder K L

The Design and Mofification of Student
Laboratory for Electric and Hydraulic Motors
(1988) / Sutardja K

Analysis and Simulation of Electrohydraulic
Pilot Relief Valve
(1988) / Tompkin E J

The Manufacturing of an Intermuscular
Electrode
(1988) / Utter J C

The Use of Multigrid Techniques in the
Solution of the Elrod Algorithm for a
Dynamically Loaded Journal Bearing
(1988) / Woods C M

A Noise Study of a Cold-Heading Machine
(1988) / Yang W L

The Effect of a Trio-Ventricular Syschrony
on Stroke Volume During Ventricular
Tachycardia in Man
(1988) / Yeh E C

Spectrophotometric Studies of Intracellular
pH and Oxygen Tension in Hamster Diaphragm
Muscle
(1988) / Zhang R

CENTRAL FLORIDA, UNIVERSITY OF

Parametric Shape Generation of Aerodynamic
and Structural Objects for Aerodynamic,
Structural and Radar Cross-Section Analysis
(1988) / Basil F C

Investigative Study of the Reliability of a
Last Row Rotating Blade of a Low Pressure
Steam Turbine
(1988) / Beal W S

The Design of a Dual Method Metal-Organic
Chemical Vapor Deposition System
(1988) / Cox D B

Boundary Layer on a Continuous Flat Plate
in Parallel Flow with Similar and
Non-Similar Boundary Conditions
(1988) / Feber V H

The Application of Transfer Matrices in the
Dynamic Analysis of Planar Elastic
Mechanisms
(1988) / Lindenberg R K

Computerized Data Acquisition in
Experimental Applications
(1988) / Meyer J A

Investigation of Surface and Subsurface
Deformation in Sliding Wear
(1988) / Mohr P J

Finite Element Analysis of a Glass Fiber
Inclusion
(1988) / Southland J R

The Feasibility Analysis of Cooling a
Residential Space Using an Underground
Air Pipe Coupled with an Air Chimney
(1988) / Trinh N K

CLARKSON COLLEGE OF TECHNOLOGY

Three-Dimensional Numerical Modeling of
Fluid Flow and Particle Transport in Clean
Rooms
(1988) / AbuZeid S

Compliance Control with Stick-Slip Friction
for Robots
(1988) / Desikan R

A Fluid Dynamics Study of a Microcontaminant
Particle Removal Process
(1988) / Gale G W

Remote Temperature Mapping of Enclosures by
Infrared Radiometry
(1988) / Giannola P S

Measurements of the Temperature Distribution
Inside Warm Air Jet by Holographic
Interferometry and Thermocouple Probes
(1988) / Han T K

Fault Analysis and Reliability of Robotic
Systems
(1988) / Krishnaswami K

An Investigation of the Variation of the
Coefficient of Friction in the Strip
Tension-Friction Test
(1988) / Lin Y C

Optimal Path Planning for Robotic
Manipulators
(1988) / Muthuswamy S

Experiments on the Fracture of Freshwater Ice
(1988) / Nigam D

Comparison Testing of Solar Air Heating
Systems with Phase Change Energy Storage
(1988) / Penik M A

A Study of Friction Conditions During Strip
Drawing with Strip Surfaces Having Zero and
Complete Permeability
(1988) / Rajan G

Computational Modeling of Infiltration and
Groundwater Flow in Mesoscale Heterogeneous
Geohydrologic Region
(1988) / Rao S N

Investigation Into the Use of Automation to
Sample Ash as it Passes out of a Stoker-Fed
Combustion Chamber
(1988) / Robertson K J

Design of a Computer Controlled Air Motor
for Robotic Application
(1988) / Schwindt J R

A Second Order Skew Upwind Weighted
Differencing Scheme for Convection-Diffusion
Computations
(1988) / Zheng X

CLEMSON UNIVERSITY

The Osteoporotic Vertebral Centrum Under
Compressive Load: An Analytical Assessment
of Mechanical Integrity
(1988) / Adams D J

Superheated Steam Drying of Spherical
Particles in Packed Beds
(1988) / Atalas B

An Electron Microscope Method for Polymer
Surface Analysis
(1988) / Boatman S E

Computer Aided Manufacture of Forging Dies
(1988) / Cherukuri N M

An Expert System for Stress Analysis of Hip
Prostheses
(1988) / Dingankar A T

Correlation of Vertebral Body Strength with
QCT Mineral Density Measurements
(1988) / Donovan D M

Gait Analysis in Normal Children and in
Cerebral Palsy Children Receiving a
Selective Posterior Lumbar Rhizotomy
(1988) / Eldridgge N E

A Comparison Analysis of Independent
End-Effector Control Methods in Robot
Assisted Material Removal
(1988) / Guvent L

A Polyvinylidene Fluoride Transducer for the
Intra-Articular Determination of Joint
Contact Pressure
(1988) / Hall J E

A Partially Resorbable, Small Diameter
Vascular Graft
(1988) / Hare W D

The Design and Fabrication of a Composite
Bone Screw
(1988) / Hookano E

Improved Methodology for Production of
Uniform Thickness Polymer Film Using an
Extrusion Die
(1988) / Lewis A S

Intelligent Medical Record System for the
Cardiac Catheterization Laboratory
(1988) / Liang Y C

An Easy-to-Use Computer Algorithm for the
Synthesis of Plate Cams Using Spline
Functions
(1988) / Manly J B

Dynamic Modeling, Simulation, and Control of
a Single-Link Flexible Manipulator with
General Motions
(1988) / Neriya S V

A Determination of Fracture Gap Displacement
with Flexible Internal Fixation Plates
(1988) / Porter M R

Natural Convection from Heated Plates on
Large Enclosures
(1988) / Ramanathan S

Positional Error Analysis of Harmonic Drive
Gearing
(1988) / Ransom T R

Investigation on the Measurement and
Modeling Aspects of a Single Link Flexible
Manipulator
(1988) / Ravishankar B N

Dynamic Modeling and Analysis of an Air Lift
Sampler System
(1988) / Rogers M A

Characteristics of Two Phase Air: Liquid
Flow in Vertical Tubes
(1988) / Shevick D S

A Study of Natural Convection Cooling of
Printed Circuit Boards on a Vented Cabinet
(1988) / Sorbye S

Transient Characteristics of Heated
Turbulent Jets in Enclosures
(1988) / Tiwari R K

Analysis of a Titanium-Coated Dacron Velour
(1988) / Vaskelis P S

CLEVELAND STATE UNIVERSITY

Gripper-Based Position/Force Sensor Design
(1988) / Benabdellah M

Heat Transfer in the Thermal Entrance
Region of a Laminar Non-Newtonian Falling
Liquid Film
(1988) / Nee Y L

Analysis of Boundary Layer Development as a
Function of Chamber Pressure in the Lewis
1030:1 Area Ratio Rocket Nozzle
(1988) / Smith T A

Simulating Transitional Flow and Heat
Transfer Over Flat Plates and Circular
Cylinders Using A L-Epsilon Turbulence Model
(1988) / Sullivan T J

Heat Transfer in a Laminar Cylindrical
Wall Jet
(1988) / Yuko J R

COLORADO STATE UNIVERSITY

Generalized Three-Dimensional Variational Geometry
(1988) / Beaty P L

Influence of Aperture Height and Width on
Interzonal Natural Convection
(1988) / Boardman C R

Laminar Convective Heat Transfer in
Partitioned Enclosures
(1988) / Caroutas C A

Two Dimensional and Temperature Effects on Acoustic
Cavity Performance in Rocket Engines
(1988) / Dodd F E

Aperture and Prandtl Number Effect on Interzonal
Natural Convection
(1988) / Neymark J

Automated Digital Skull - Face Superimposition
in Human Identification
(1988) / Nickerson B A

Artificial Neural Systems
(1988) / Sjostrom A M

A Structured Approach to Process Representation
(1988) / Small D H

Implementation of Deterministic and Stochastic
Industrial Experimentation Methods
(1988) / Woods K L

CONCORDIA UNIVERSITY

Finite Volume Solutions of the Euler
Equations on Cartesian Grids
(1988) / Chandrasekhar R

Ride Dynamics of Heavy Vehicles Using Local
Equivalent Linearization Technique
(1988) / Dhir A

Application of Sensitivity Analysis to
Parameter Changes in Nonlinear Hydraulic
Control System
(1988) / Farahat S

A Submodel Component Reanalysis Technique
for Liquid Tanker Body Design Using the
Finite Element Method
(1988) / Gupta V

The Torque-Thrust Coupling Effect in Twist
Drills and Its Influence on Drill
Characteristics and Drilling Forces
(1988) / Kramadhati N

The Development and Analysis of an
Intelligent Robotic Work-Cell
(1988) / Montor T

Evaporator Modelling Under Conditions of
Dry, Wet and Frosted Finned Surfaces
(1988) / Oskarsson S P

On the Stream Function Solution of Inviscid
Transonic Flow Problems
(1988) / Tata V R

CONNECTICUT, UNIVERSITY OF

The Comparison of Two Mean Frequency
Estimates for Ultrasound Doppler Waveforms
Using a Microcomputer Based System
(1988) / Cote G L

Dynamic Analysis of a High Rise Structure
(1988) / Ebert K D

Optimal Path Synthesis of Robotic
Manipulators
(1988) / Elmali H

Study of Some Cases of Synthesis of Plane
Trusses with and Without Buckling
(1988) / Tseng P C

CORNELL UNIVERSITY

AI-Based Generative Process Planning for
Rotational Parts
(1988) / Jagdale S S

An Expert System Environment for Resin
Selection and Troubleshooting Problems
in Injection Molding
(1988) / Jong W R

Elastohydrodynamic Lubrication Analysis of
Dynamically Loaded Joaurnal Bearings: A
Model Approach
(1988) / Kumar A

ISCOP: An Expert System for Tool Selection
and Determination of Machining Parameters
(1988) / Melkote R N

Continuum Modeling of Deformation Processes
on Powders
(1988) / O'Donnell M C

Determine Object Location Using a Stereo
Disparity Vision System
(1988) / Pherwani R M

Graphic Simulation and Verification of
Fixe-Axis NC Machining
(1988) / Sambandan K

Inclusion of Particles in the Modeling of
Atmospheric Turbulence
(1988) / Sanghi S

Experiments on the Free Fall of an
Inductively Heated Sphere Through a Fluid
with a Strongly Temperature-Dependent
Viscosity
(1988) / Suresh N C

Development of the Integrated Model
Environment
(1988) / Warkentin R H

DELAWARE, UNIVERSITY OF

Vortex Sheet Modeling of Surface Waves
(1987) / Adopley J A K

Sudden Expansion in Gas-Solid Flow
(1988) / Arefmanesh A

Effects of Open Holes (Molded-in and
Drilled) and Low Velocity Impact Damage on
the Compressive Strength of Woven Carbon
Fiber Reinforced Polymer Composites
(1988) / Ghaseminejhad M N

Low Velocity Impact Damage of Composite
Materials
(1988) / Godil M A

Interlaminar Fracture Toughness of a
Three-Dimensional Composite
(1988) / Guenon V A F

A Post-Buckling Analysis Applied to Creping
Mechanics
(1987) / Hopkins D A

Fatigue Accumulation in Automobiles Due to
Highway Waviness
(1987) / Jvs P

Synthesis of an Elastic Mechanism
(1987) / Kishore A

An Investigation of the Strength Translation
Phenomenon in Kevlar 49/Epoxy Composite
Test Samples
(1988) / Maas D R

Fracture Toughness and Fatigue of an Alumina
Fiber Reinforced Magnesium Metal Matrix
Composite
(1988) / Magata A

Computation of a Supersonic Anisymmetric
Inlet Flow
(1988) / Marcozzi M D

Friction and Wear of Advance Thermoplastic
Composites: Unidirectional and Woven Carbon
Fiber/Polyetheretherketone (PEEK) System
(1987) / Mody P B

DELAWARE, UNIVERSITY OF
(continued)

The Effect of Stitching on In-Plane and
Interlaminar Properties of Carbon-Epoxy
Fabric Laminates
(1987) / Ogo Y

Thick-Walled Composite Material Pressure
Hulls: Three-Dimensional Laminate Analysis
Considerations
(1988) / Peros V

Friction and Wear Behavior of Unidirectional
and 3-Dimensionally Braided Fiber
Fb/Aluminum Composites
(1987) / Reese E D

A Unified Approach to Modeling the Dynamic
Characteristics of an Actively Controlled
Vertical Rotor
(1988) / Reimers A E D

Characterization and Modeling of 3-D Braided
Metal Matrix Composites
(1988) / Remond O G

Characterization of Mode I, Mode II, and
Mixed Mode Delamination of Thermoset and
Thermoplastic Composites
(1987) / Rothchilds R J

The Effect of Temperature on the Material
Damping of Graphite/Epoxy Composites in a
Simulated Space Environment
(1987) / Spirnak G T

Modeling and Correcting Scatter in
Industrial Applications of Computerized
Tomography
(1987) / Straining R J

Static and Fatigue Strength of Notched
Graphite-Peek Fabric Composites
(1988) / Tase A S

On the Vibrations of Helical Rods
(1987) / Tsay H S

Experimental Study of Mixed Convection in a
Horizontal Porous Annulus
(1987) / Vanover D E

DREXEL UNIVERSITY

A Generalized Finite Difference Method for Pipe
Flow Problems
(1988) / Balcarek R J

Optimal Controller Design for Manned Flight
Simulator
(1988) / Knaak E R

A Microcomputer Approach to Integrated CAD/CAM
Systems
(1988) / Kolluri S P

A Gas Turbine Engine Emissions Model as a Function
of Engine Operating Conditions, Fuel Properties
and Combustor Geometry
(1987) / Marx M M

Dynamic Stability of Elastic Linkage Systems
(1988) / Nagarajan S

Atomization in a Confined Turbulent Shear Layer
(1987) / Tallio K V

A Technique for Calculating the Contact Area of
the Human Subtalar Joint
(1988) / Tihansky C E

Thermal Behavior of a Rotating Roll Applied to
Metal Forming Processes
(1987) / Tong S

DUKE UNIVERSITY

An Experimental Study of Inertia and Wall
Effects for a Large Sphere Oscillating in
a Fluid-Filled Cylinder
(1988) / Coffey B E

A Constant Temperature Thermistor Anemometer
(1988) / Gottwald J A

Transient Shear Stress Distribution on a
Small Sphere in Turbulence
(1988) / Kwon K Y

A Lagrangian Description of 1-D Compress
Flow with Heat Addition
(1988) / Mendivil D F

Asymptotic Modal Analysis of a Rectangular
Acoustic Cavity
(1988) / Peretti L F

Kinetics of Biomass Growth and Xanthan
Production by Xanthomonas Campestris When
Limited by NH(3)-N and Glucose Respectively:
Experimental and Modeling Analysis at Three
Initial Glucose Concentrations
(1988) / Schweickart R W

Time Optimal Control of a Robotic
Manipulator
(1988) / Yang J

The Coordination of Multiple Robotic
Manipulators
(1988) / Young R F

FLORIDA ATLANTIC UNIVERSITY

Numerical Solution of Two-Dimensional
Incompressible Flow About an Airfoil
(1988) / Barnes R S

A New Approach to the Inverse Kinematic
Analysis of Redundant Robots
(1988) / Dutta P S

Response of Secondary Systems to Seismic
Excitation
(1988) / Hu Lung J

An Application of Perturbation Techniques
to the Vibration Analysis of Structures with
Viscoelastic Coating Modifications
(1988) / Kelly W J

Feasibility Study of High Performance
Submarine Craft
(1988) / Yao J

FLORIDA STATE UNIVERSITY

The Design and Control of a Micron-Accurate Robot Arm
(1988) / Ossib Z S

FLORIDA, UNIVERSITY OF

An Algebraic Grid Generation Technique for
Complex Two- and Three-Dimensional Spatial
Domains
(1988) / Bailey R T

FLORIDA, UNIVERSITY OF
(continued)

Motion Analysis of an Articulated
Transporter/Manipulator System
(1988) / Chiang S C

A Boundary Element Method for the Solution
of Inverse Heat Conduction Problems with
Partially Unknown Boundary Geometries
(1988) / Choi C

The Synthesis of Non-Circular Gears
Incorporating Involute Profiles
(1988) / Dooner D

Detection of Polycyclic Aromatic
Hydrocarbons in an Axisymmetric Laminar
Diffusion Methhane/Air Flame Visible Laser
Induced Fluorescence
(1988) / Dufflocq Flores M

Development of a Translating Double Pendulum
and a Computer Model to Determine the
Accuracy of an Analytical Photogrammetric
Motion Analyzer
(1988) / Fitzgerald D

Pilot Scale Boiler Study of Sulfur
Hexafluoride, and Emissions of Carbon
Monoxide, Carbon Dioxide, Oxygen and
Unburned Hydrocarbons as Surrogates for
Verification in Hazardous Waste Destruction
Removal Efficiency
(1988) / Fournier D J

Kinetic Considerations in the Hybrid Control
of Robotic Manipulators
(1988) / Griffis M W

Time-Optimal Trajectory Planning by
Variation of Path and Robot Location
(1988) / Groman B

Design and Simulation of a Tri-Mode Robotic
Vehicle
(1988) / Haukoos D

The Design, Construction, and Operation of a
100 HP Boiler/Incineration Unit for the
Simulation of Hazardous Material Destruction
Under Laboratory Conditions
(1988) / Hopmeier M

A Computer Simulation of Plastic Strain
Criterion to Predict the Cyclic Life of
Composite Materials
(1988) / Hurt T

Kinematics of Manipulators with Degrees of
Freedom Distributed in Open-Loop and
Closed-Loop Subchains
(1988) / Hwang R T

Factors Affecting the Performance of the
Large-Angle Scattered Intensity Ratioing
Technique Used by the Integrated Particle
Sizing System for Measurements of Particle
Size
(1988) / Iisager B

Numerical Simulation of the Turbulent Flow
Field in a Motored Two-Dimensional Wankel
Rotary Engine
(1988) / Li Z

Displacement Analysis of Two Special Cases
of the TTT Manipulators
(1988) / Lin W

Design and Construction of a Ejector System
with a Variable Molecular Weight Primary
Fluid
(1988) / Pfahler J N

Two-Dimensional Modelling of a Concrete
Pavement's Thermal Response to Varied
Weather Conditions
(1988) / Ryder E

Three-Dimensional Numerical Simulation of
the Flow Field Inside a Motored Wankel
Engine
(1988) / Steinthorsson E

Motion Planning for Planar Robots in an
Obstacle-Strewn Environment
(1988) / Wang B W

Feedforward and Feedback Strategies Applied
to Controlling a Servo-Driven Flexible Arm
(1988) / Wells R L

Sorbtion Behavior of Hygroscopic Materials
and Its Effect on Building Moisture and Air
Quality
(1988) / West M K

A Rule-Based Elimination and Refinement
Algorithm in Two-Dimensional Optimum
Structural Design
(1988) / Wu C Y

Self-Excited Vibrations in Oil Drilling
(1988) / Zamudio C

GEORGIA INSTITUTE OF TECHNOLOGY

An Experimental Study of the Automation of
Thermoplastic Composite Processing
(1988) / Andersen B J

An Investigation of the Drying Cycle of a
New Generation of Dishwashers
(1988) / Cowell S D

Automatic Correction of Robot Programs
Based on Sensor Calibration Data
(1988) / Duggan M S

Finite Element Analysis of Stress in
Prosthesis Implanted Canine Femur
(1988) / Groome I M

Experimental Verification of a Model of a
Two-Link Flexible, Lightweight Manipulator
(1988) / Huggins J D

Improvement of Harmonic Balance Solution
Routines for Nonlinear Systems
(1988) / Kirksey J F

Steady and Pulsatile Flow Visualization in
the Human Abdominal Aorta
(1988) / Moore J E

Specific Heats of Aqueous Lithium Bromide
Solutions
(1988) / Moran J P

Application of a Coronary Circulation
Computer Model to the Human
(1988) / Olsmats H M

Practical Applications of Complex Joint
Angle Solutions for Robotic Manipulators
(1988) / Park Y S

GEORGIA INSTITUTE OF TECHNOLOGY
(continued)

Analysis of Singular Configurations for
Robotic Manipulators
(1988) / Pohl E D

Damage Rate Approaches for Nickel-Base
Superalloys
(1988) / Reynolds G J

HAWAII, UNIVERSITY OF

Thermogravimetric Studies of Cellulose and
Lignin and the Determination of Chemical
Reaction Kinetics and Mechanisms
(1988) / Cooley S A

A Study of Catalytic Reaction Chemistry of
Five and Six Carbon Sugars in Near Critical
Water
(1988) / Leesomboon T

HOUSTON, UNIVERSITY OF

On the Mechanics and the Mechanisms of
Fracture in Polymeric Materials
(1988) / Balzano M

An Investigation of Necking and Effects of
Necking-Induced Molecular Orientation on
the Mechanical and Fracture Behavior of
Polycarbonate
(1988) / Buisson G

Laminar Boundary-Layer Momentum-Thickness
Effects on Two-Stream Mixing Layers
(1988) / Chachere A

An Automatic Image Processing Technique for
Simultaneous Multipoint Measurements in
Unsteady Fluid Flows
(1988) / Jain S

Scattered Light Investigation of
Three-Dimensional Stress Fields
(1988) / Murthy N

Liquid Phase Processing of Yttrium Barium(2)
Copper(307-X) Superconductors: A Parametric
Study of Microstructural, Mechanical and
Superconducting Properties
(1988) / Selvamanickam V

Application of Finite Element and Expert
System Methods to a Study of Shoulder
Dystocia
(1988) / Sharma V

Design and Development of Engineering Aids
to Evaluate the Birthing Process
(1988) / Sorab J

Partitioning and Guided Search: A
Generalized Approach to Problem Solving
in Preliminary Design
(1988) / Srinivasan S

Computer-Aided Multi-Objective Design of
Spacecraft Thermal Control Systems
(1988) / Struble C

Phase Dependence of Coherent Structures in a
Two-Stream Mixing Layer
(1988) / Tung C

Global Cubic Equation of State for
Thermodynamic Properties
(1988) / Vandermarliere P

Low-Cycle Corrosion-Fatigue of 2024-T4
Aluminum and 1045 Steel in Salt Water
(1988) / Zaini M

The Compromise Decision Support Problem: A
Fuzzy Formulation
(1988) / Zhou Q J

HOWARD UNIVERSITY

Effects of Torsion and Curvature on Viscous
Flow in Helical Tubes
(1988) / Damavandi M A

Higher Approxiimations to Viscous Flow in
Narrow Curved Gaps
(1988) / Farmanara F

Position Assessment of the Robot End
Effector Using Three Linear Cable
Transducers
(1988) / Goodwyn V L

Natural Convection in a Partially Blocked
Vertical Channel
(1988) / Oliver D

Design Sensitivity Analysis and Maximum
Frequency Design of Symmetric Laminated
Angle-Ply Plates
(1989) / Ramachandran S

IDAHO, UNIVERSITY OF

An Analysis of Shape Factors in the
Prediction of Thermodynamic Properties
Using the Principle of Corresponding States
(1988) / Ahmad F

A Linear Least-Squares Regression Algorithm
for the Optimization of Thermodynamic
Equations of State
(1988) / Bjornn K R

Characterization of the Effects of
Three-Dimensional States of Stress on
Damping of Laminated Composites
(1988) / Hwang S J

Determination of Temperature and Phase
Distributions in Irradiated
Uranium-Plutonium-Zirconium Fuel
(1988) / Lahm C E

Damping Properties of Thin Polymers in
Dynamic Shear
(1987) / Rao V

ILLINOIS INSTITUTE OF TECHNOLOGY

Experimental Investigation of the Wear
Behavior of Magnesia-Partially Stabilized
Zirconia
(1988) / Benetatos N V

IOWA STATE UNIVERSITY

Use of Multiple Nodes in the Implementation
of the Boundary Element Method for Problems
in Zoned Media
(1988) / Agarwal R

Heat Transfer in Water-Fluidized Beds
(1988) / Burken L W

Photoacoustic Detection of Unborned Carbon
in Airborne Fly-Ash
(1988) / Dona A R

IOWA STATE UNIVERSITY
(continued)

A Study of Tractor-Semitrailer Ride Using
Computer Simulation
(1988) / Dusenberry D L

Coal Water Mixture Combustion Mechanisms in
Fluidized Beds
(1988) / Gregory J W

Measurement of Turbulent Stresses in an
Oscillating Boundary Layer Flow
(1988) / Johnson W A

Capture of Fine Particles on Charged
Moving Spheres
(1988) / Liu X

Computer Simulation of a Spin-Stabilized
Space Simulator with Liquid Fuel Stores
(1988) / Obermaier L A

Development of Blade-Wake Total-Pressure
Loss in an Axial-Flow Compressor Stage
(1988) / Reding J D

Simple Computer Models for Hierarchial
Processing Applied to Sensory Receptive
Fields
(1988) / Solessio E C

A Nondestructive Method of Determining
Residual Stress in Cold Worked Nodular
Iron Crankshafts
(1988) / Thompson R D

A Mathematical Model of a Thermal Energy
Storage System Utilizing Spherically
Encapsulated Phase Change Material
(1988) / Timmer K J

Comparison of Several Active Suspension
Strategies Using Computer Simulation
(1988) / Warner P K

JOHNS HOPKINS UNIVERSITY

Ambulatory Monitoring of Vital Signs for
Anxiety Studies
(1989) / Amaresan M

Calcium Kinetics in the White Skeletal
Muscle Sarcoplasmic Reticulum of the New
Zealand White Rabbit
(1989) / Nemerson S F

A Study of the Electric Field Distribution
in the Heart in the Internal Defibrillation
(1989) / Shankar B

KANSAS, UNIVERSITY OF

Initial Graphics Exchange Specification
(IGES) Flavoring for Improved Computer
Aided Design (CAD) Data Exchange
(1988) / Bradford J R

Finite Element Analysis of a Metal Matrix
Tensile Coupon with a Central Slot
(1988) / Buck G

The Development of a Prototype Flexible
Fixturing System for Prismatic Parts
(1988) / Frechette S P

Optimal Air-to-Air Heat Exchangers for
Heat Recovery from Residential Ventilation
Air
(1988) / Haberlein R A

Flow Visualization Techniques for a
Convection Oven
(1988) / Hardin R A

The Microscopic Aspects of Fracture in Gray
and Compacted Graphite Cast Irons
(1988) / Holmgren S D

Least-Squares Finite Element Solution for
the Inverse Problem of Aquifer
Transmissivity
(1988) / Huels C R

Measurements of Natural Convection in a
Membrane Stratified, Water-Saturated
Porous Medium with Downward Convection
(1988) / Mouzouris Y A

P-Approximation Axisymmetric Shell Elements
for Heat Conduction in Generally
Orthotropic and Laminated Composite
Materials
(1988) / Orth N J

Microstructural Aspects of Fracture in
Malleable Iron
(1988) / Pourladian B

Geometrically Nonlinear Formulation for the
Three Dimensional Curved Beam Elements with
Large Rotations
(1988) / Sorem R M

A Symmetric Finite Element Formulation for
the Inverse Problem of Aquifer
Transmissivity
(1988) / Tidjani L

Analysis of Submersible Oil Well Cable Armor
Materials for Use in Sour Well Environments
(1988) / Zachariah D R

KENTUCKY, UNIVERSITY OF

Predicting Wear Distribution in Slurry
Pipelines
(1988) / Cader T

Experimental Developments in the Combustion
of Crude Oils
(1988) / Elam S K

Scaling Flashover Phenomenon in Enclosure
Fires
(1988) / Jolly S

A Frequency-Dependent Foundation
Representation for Rotor Response Analysis
Using the Finite Method
(1988) / McMains T H

Compensator Synthesis Using Factorization
Approach
(1988) / Patwardhan A

Analysis and Correction of Sound Intensity
Errors Caused by Phase Mismatch
(1988) / Sonday E J

Flame Spread Into a Quiescent Oxidizing
Medium Under Microgravity
(1988) / Srikantaiah N

Pyrolysis Studies of Charring and
Non-Charring Materials
(1988) / Venkatesh S

LEHIGH UNIVERSITY

Interactive Design and NC Machining of
Free Form Surfaces
(1987) / Tarhan M S

Analysis and Testing of a Switched
Inert-Hydraulic Servo-Transformer
(1987) / Tentarelli S C

LOUISIANA STATE UNIVERSITY

Natural Convection in Enclosures with Offset
Baffles
(1988) / Jetli R

Recogniition and Classification of
Straight-Line Elements in Free-Hand
Engineering Sketches
(1988) / Law C B

Dynamic Sculpting Using Instant Centers
(1988) / Naik S J

Meteoroid Shield Effectiveness Analysis
(1988) / Ravichandran N

Natural Convection from a Horizontal Disk
Situated at the Bottom of a Top-Vented
Enclosure
(1988) / Siddiqui S A

Characterization of Fly Ash Using Electron
Beam and X-Ray Beam Techniques
(1988) / Soroczak M M

Abrasive Wear Characteristics of
Pre-Corroded Plain, Nitrided and TiN Coated
SAE 4130 Steel
(1988) / Vaingankar M P

A 3D Tendon Path Planning Method for Surgery
of the Hand
(1988) / Yoon I

LOUISIANA TECH UNIVERSITY

Generalized Task-Based Quantitative
Assessment of Robotic Aids for Augmentative
Manipulation by Physically Disabled Persons
(1988) / Abraham P M

An Adaptive Keyboard Filter Routine for Use
by Individuals with Cerebral Palsy
(1988) / Wadsworth C T

LOWELL, UNIVERSITY OF

Analysis of Thermal Microcracking in
Graphite/Polyimide Composite Materials: A
Study of Fiber Volume Effects
(1988) / Albert K M

The Effect of Load Variability on the
Loss-of-Load probability of Stand-Alone
Photovoltaic/Storage Systems
(1988) / Bloom D R

Mechanics of Interleafed High-Temperature
Polymer Composites
(1989) / Dardzinski P V

Optimization of Mass and Stiffness Matrices
Using a Generalized Inverse Technique on
the Measured Modes
(1988) / Leung R K

Mathematical Model of the Chemical Vapor
Deposition of Silicon Nitride
(1989) / Lynch J M

Analysis and Design of an Active Multiple
Degree-of-Freedom Accelerometer
(1989) / Pinto G A

Orthonormalization of Rigid Body Modes and
Experimental Flexural Modes
(1988) / Wu C H

MAINE, UNIVERSITY OF

Structural Testing System Design for Steel
Alate Shear Wall Test
(1988) / Chen R

A Comparison of the Flow Fields in a Gas
Turbine Vane Passage and in a Bent Duct
(1988) / Simonds M E

MANITOBA, UNIVERSITY OF

KBGT: A Knowledge-Based System for Group
Technology
(1988) / Ibrahim W M

Development of a Multitasking Control
Environment for a Flexible Manufacturing
System
(1988) / Judt C P

Development of a Flexible, Microcomputer
Based, Three-Axis Machine Tool Controller
(1988) / Kostyniuk T M

Snow Clearing Vehicle Routing: The Postman
Problem in a Hierarchical Network
(1988) / Liu D Q

Heat Recovery from Horizontal Tubular
Condensers of Refrigeration Systems
(1988) / Nitheananda T

A Logical Model of Integrated Manufacturing
(1988) / Russell J S

Thermo-Mechanical Treatment of Aluminum
Copper Magnesium Alloys
(1988) / Singh R

Improvement of an Artificial Safety Knee
Joint
(1988) / Slagerman G L

The Investigation of Effect of Stratified
Liquid Storage on the Performance of
Thermosyphon Solar Domestic Hot Water
Systems
(1988) / Sujumnong M

The Use of Zeta Bankruptcy Analysis in a
Simulation Model to Study the Economic
Benefits of Computer Integrated
Manufacturing
(1988) / Yu E C

MARQUETTE UNIVERSITY

Dispersion of Signal Averaged Pulsatile
Blood Flow Waveforms Due to a Variable
Heart Rate
(1988) / Adamson N

MARQUETTE UNIVERSITY
(continued)

Joining of Ceramic/Metal to Bone
Interface - A Review
(1988) / Chiu C H

(31)P Nuclear Magnetic Resonance
Spectroscopy of Bone Material for
Evaluation of Osteoporosis
(1988) / Srinivasan R

MASSACHUSETTS INSTITUTE OF TECHNOLOGY

Sliding Behavior of Material Pairs at Room
Temperature and at Low Temperature
(1988) / Aized D

Determining the Load in the Canine
Prosthetic Hip In-Vivo Using an Instrumental
Prosthesis
(1988) / Allan C D

Fault Diagnosis Based on Causal Structure:
A Boolean Matrix Algebra Approach
(1988) / Baker W L

Design of Robot-Operated Modular and
Adaptable Fixturing Systems
(1988) / Barrientos J H

Analysis of Wilson's Method for Predicting
the Efficiency of Axial-Flow Turbines
(1988) / Bazan A

Notch Filter Control of Magnetic Bearings to
Improve Rotor Synchronous Response
(1988) / Beatty R

Steering Law Design for Redundant Single
Gimbal Control Moment Gyro Systems
(1987) / Bedrossian N S

Viscously Damped Joystick for Proportional
Control by Tremor Patients
(1988) / Beringhause S

An Acoustic Emission Study of the Wear
Mechanisms of Oxide Magnetic Rigid Disks
(1988) / Besen M M

Advanced Laser Fluorescence Measurements of
Lubricant Film Behavior in a Diesel Engine
(1988) / Billian S A

Automation of Modular Fixturing
(1988) / Blacker S J

Modelling the Transient Temperature and Heat
Transfer for the Piston Ring, Lubricant
Film, and Liner of a Reciprocating Engine
(1988) / Boisclair M E

The Optical Properties of Silicon Oxynitride
Dielectric Waveguides
(1986) / Bossi D E

Investigations on Group Delay in Structures
and Acoustical Spaces
(1988) / Boulahbal D

Transient Measurement of Gas Permeability in
Closed-Cell Foam Insulation
(1988) / Brehm T R

Automated Handling of Flexible Materials
(1988) / Briggs J C

A Study of the Bauscyhinger Effect and
Cyclic Plasticity in Spheroidized Steels
(1988) / Bronkhorst C A

Response of Delaminated Composite Panels to
Air Blast Loading
(1988) / Byrtus J E

An Expert System for Dynamic Multiple
Attribute Decision Making in Job Shop
Scheduling
(1988) / Cavalieri A J

Electrophysiological Recovery of Peripheral
Nerves Regenerated by Biodegradable Polymer
Matrix
(1988) / Chang A S

Computer-Aided Surgery: An Interactive
Simulation Systems for Intertrochanteric
Osteotomy
(1987) / Chang G H

Three-Strand Rope Behavior in Tension and
Torque
(1988) / Chen J

Human-Interactive Simulator for Underwater
Remotely Operated Vehicle
(1988) / Chin K P

Determination of Mechanical Properties of
LPCVD Silicon Nitride Thin Films
(1988) / Crowe K E

Design of a High Pressure Steam Chamber for
Heat Setting in False-Twist Texturing
(1988) / Curran E C

The Design and Analysis of an Actively
Augmented Vibration Isolation System Using
Classical and Linear Quadratic Feedback
Control Techniques
(1988) / Demeo M E

Interaction Between a Two-Dimensional Wake
and the Free Surface at Low Froude Numbers
(1988) / Dimas A A

Camera-Positioning System Coupled to
Manipulator End Effector: An Operator-Aid
for a Remotely Operated Submersible
(1988) / Doherty N

Application of Group Technology to Naval
Ship Repair
(1988) / Duca S C

An Analysis of Full-Scale Experimental Data
on the Dynamics of Very Long Tethers
Supporting Underwater Vehicles
(1988) / Engebretsen K B

A Method for 2-D Visualization of a
Propagating Spherical Turbulent Premixed
Flame
(1987) / Ferreira E A

Multivariable Sliding Mode Control Applied
to a Turbofan Aircraft Engine
(1988) / Fludder S M

An Expert System for Programming Robotic
Cloth Manipulation Tasks
(1988) / Foley M F

Hardware Design and Implementation of Dual
Direct-Drive Servo System
(1988) / Fridman M

MASSACHUSETTS INSTITUTE OF TECHNOLOGY
(continued)

A Computer Aided Interactive Model for the
Design of the Manufacturing Process for a
Submarine Hull
(1988) / Gallo M E

Inherent Limitations of State-of-the-Art
Inspection Equipment for Duplex Bearing
Pairings
(1988) / Gangitano A J

The Design and Analysis of a Large Angular
Range, Two-Axis Flexure Assembly
(1988) / Gin B A

Development of a New Wall Shear Stress
Gauge for Fluid Flow
(1988) / Gur Y

Inertia Friction Welding of Aluminum Alloys
for Space Repair Applications
(1988) / Guza D E

Gait Orthosis Combining Controllable Damping
and Muscle Stimulation
(1988) / Hausdorff J M

Forced Nonlinear Internal Waves with
Applications
(1988) / Hedayati Z

Vapor Condensation on a Turbulent Liquid
Interface, for Application in Low-Gravity
Environments
(1988) / Helmick M R

The Effect of Aging on the Mechanical
Behavior of Metallic Glass Pd(40)Ni(40)P(20)
(1988) / Hu K J

Force Control of the MIT Direct Drive Arm
(1988) / Klobucher M A

Underground Hydraulic Fracture: Secondary
Fracturing and the Application to Primary
Fracture Diagnostics
(1988) / Klohr J D

Evaluation of a Variable-Configuration-Die
Sheet Metal Forming Machine
(1988) / Knapke J A

The Design and Implementation of a Pick
Motion Sensor for Excavator Guidance
(1988) / Koselka H A

Road Damage Caused by Heavy Vehicles:
Busses Versus Trucks
(1988) / Laby K P

Analysis of the Time Delay Controller for
Linear, SISO System
(1988) / Leung Y F

Transformations and Stability of Electrical
Machines in Nearly Ideal Operation
(1988) / Liu X Z

The Formation of Fouling and Slagging
Deposits in Pulverized Coal Combustion
(1988) / Loehden D O

A Freeze-Drying Process for Fabrication of
Polymeric Bridges for Peripheral Nerve
Regeneration
(1988) / Loree H M

Generation and Evaluation of Mechanical
Assembly Sequences Using the Liaison-
Sequence Method
(1988) / Lui M M

Estimation of Isometric Recruitment Curves
of Electrically Stimulated Muscle
(1988) / MacLean K E

The Design of a Lightweight Elbow Prosthesis
Emulator
(1988) / Mansfield J M

Application of Nonlinear Control to Gas
Turbine Engine Fuel Flow Tracking
(1988) / Martin J F

Effects of Force and Visual Feedback on
Space Teleoperation
Implications
(1988) / Massimino M J

Hard Sludge Sampling from the Primary Side
of a PWR Steam Generator
(1988) / May D R

Distributed Parameter Active Vibration
Control of Smart Structures
(1988) / Miller S E

Automated Three Dimensional Inspection of
Single Curvature Surfaces Using Moire
Interferometry
(1987) / Moens L C

Measurement of Wall Shear and Wall Pressure
Downstream of a Honeycomb Boundary Layer
Manipulator
(1988) / Moller J C

Cooldown Simulation for Compact Ignition
Tokamak (C.I.T.), Poloidal Field (PF) Coils
(1988) / Mottahed B D

Factors Influencing Heterogeneous Lung
Emptying During Forced Expiration
(1988) / Nielan G J

Predicting the Acoustics of an Experimental
Machine Enclosure Using Finite Element and
Boundary Element Techniques
(1988) / Oppenheimer C H

Water Permeability of Mouse Ova and
Predicted Osmotic Behavior During Freezing
(1988) / Orrico M R

A Systems Approach to the Torque Control of
a Permanent Magnet Brushless Motor
(1987) / Paul B J

The Effect of Liquid Inventory on the
Performance of a Closed Two-Phase
Thermosyphon
(1988) / Petroff C

A Complete Helium Saturated-Vapor-
Compression Cycle with a Two-Phase Expander
and a Saturated Vapor Compressor
(1988) / Phamduy T

Active Control of Bending Waves at Acoustic
Frequencies
(1988) / Pines D J

Observer Design for a Two Link Planar Robot
Arm: Theory and Implementation
(1988) / Potamianos P

MASSACHUSETTS INSTITUTE OF TECHNOLOGY
(continued)

Equipment Model for the Low Pressure
Chemical Vapor Deposition of Polysilicon
(1988) / Prueger G H

Automated Visual Inspection of General
Curvature Surfaces Using Moire
Interferometry
(1988) / Reidemeister E P

The Effects of Crystallographic Orientation
on Stress in a Gas Turbine Nozzle's Single
Crystqal Vane
(1988) / Rock P J

Arm Motion in Crank Turning
(1988) / Russell D L

Microyield Behavior of Single Crystal
Silicon
(1988) / Russell R J

Hard Sludge Sampling from the Secondary Side
in PWR Steam Generators
(1988) / San Pedro R I

Stability/Performance Trade-Offs for
Computer Controlled Manipulators
(1988) / Sharon S

Tactile Interface for Three-Dimensional
Computer-Simulated Environments:
Experimentation and the Design of a
Brake-Motor Device
(1988) / Smith M J

Prediction of Distortion in Welding Ring
Stiffened Cylinders
(1988) / Song H

Coordinated Control of Machine Tool
Axes: Noncircular Turning
(1988) / Sparks A G

Performance and Design Aspects of
Radial-Inflow Turbines
(1988) / Spinney C

Heat Transfer and Performance Calculations
in a Rotary Engine
(1987) / Stanten R A

Design and Control of a Six-Degree-of-
Freedom Platform with Variable Admittance
(1988) / Stelman N M

Computer Simulation of Icebreaker Dynamics
(1988) / Suchanek B J

Residual Strength of a Ship after an
Internal Explosion
(1988) / Surko S W

Design and Development of a Thin-Film
Resistance Temperature Detector Based Heat
Flux Transducer for a Ceramic Surface in a
Diesel Combustion Chamber
(1988) / Szydloski D J

The Real-Time Digital Control of a
Regenerative Above-Knee Prosthesis
(1988) / Tabor K A

The Effects of Distributed and Centralized
Task Execution in Distributed Command,
Control, and Communication Systems
(1988) / Tani D M

The Dynamic Behavior of Nylon and Polyester
Rope Under Simulated Towing Conditions
(1988) / Toomey C J

A Strategy for Near Optimal Robotic
Manipulator Operations Subject to
Statistical Variations
(1988) / Urbank M G

Time Optimal Trajectory Planning for Mobile
Manipulators
(1988) / Vance E E

Analysis of Fracture Initiation in Surface
Cracked Plates
(1988) / Wang Y

Parameter Studies of the Pavement Damage
Caused by Tractor Semi-Trailer Trucks
(1988) / Warburton D J

Implementation Issues in Sliding Mode
Automotive Power Train Controllers
(1988) / Weeks R W

Design and Control of a High Speed
Mini-Robot for Submicron Assembly
Applications
(1988) / Yahiaoui M

A Network Model of the Human Aqueous
Outflow System
(1988) / Yan D B

Improving the Efficiency of Time-Optimal
Path-Following Algorithms for Robot
Manipulators
(1988) / Yang H S

Control Law Designs for Active Suspensions
in Automotive Vehicles
(1988) / Yue C

MASSACHUSETTS, UNIVERSITY OF

Designing with Features: The Acquisition
and Organization of Features
(1988) / Cunningham J J

Aging and Fatigue Behavior of Titanium
Carbon Coated Optical Glass Fibers
(1988) / Francisco N P

Failure Analysis of Silane-Originating
Reaction Bonded Silicon Nitride with and
Without a Pre-Oxidation Treatment
(1988) / Gennari P A

A Theoretical Model for Fluid Transport in
the Interstitial Space
(1988) / Hatzimanolakis E

Determination of the Critical J-Integral
Value Independent of Initial Crack Sizes
and Specimen Length
(1988) / Joe C R

Iterative Respecification: A Computational
Model for Automating Parametric Mechanical
System Design
(1988) / Meunier K L

Measurement of Crack Tip Inelastic
Deformation Using Digital Image Correlation
(1988) / Michail S K

MASSACHUSETTS, UNIVERSITY OF
(continued)

Fractal Analysis of Rubber Wear Surfaces and
Debris and Fractal Analysis of Model Pore
Structures and Its Application to Carbon
Black and Other Real Systems
(1988) / Stupak P R

An Application of MACSYMA in Fluid Mechanics
(1988) / Syu C

MIAMI, UNIVERSITY OF

The Effect of Cardiovascular Drugs on the
Polarization and Recovery of Platinized
Porous Platinum Iridium Pacemaker Electrodes
(1988) / Accorti P R

Photochemical Induction of Spinal Cord
Injury in the Rat by Intravenous Erythrosin
B and Argon Laser Irradiation
(1988) / Cameron T L

Computer-Based Control of Classical
Conditioning Sessions and an Expert System
for the Evaluation of Conditioned Heart
Rate Responses in the Rabbit
(1988) / Chang J T

Microcomputer Assisted Determination of
Regional Myocardial Function and Infarct
Size in a Coronary Occluseion-Reperfusion
Rabbit Model
(1988) / Cideciyan A V

The Effect of Left Ventricular Acute
Ischemia on the Endocardial Evoked Response
(1988) / Desai T H

Determination of Bioelectric Sources in a
Volume Conductor from External Measurements
of the Electric Potential
(1988) / Georgiou M F

Study of Solar-Hydrogen Production Methods
and Their Comparison
(1988) / Lutfi N

MICHIGAN TECHNOLOGICAL UNIVERSITY

Thermal Energy Distributions in Workpieces
During Abrasive Waterjet Cutting
(1988) / Ansari A

Analysis of Three-Dimensional Surface
Texture
(1988) / Boudreau B D

Multivariate System Analysis Using Data
Dependent Systems and Fourier Transform
Methods
(1988) / Dabrowski S

Development and Analysis of a Frequency
Response Function Estimator Based on the
Correlated Contaminant Model
(1988) / DeClerck J P

Droplet Formation and Breakup for a
Prototype Swirl Injector
(1988) / Ellcey E C

Low Pressure Preheated Fuel Injection for
Gas Turbine Use
(1988) / Furry D E

The Influence of Interference-Fits on the
Strength of Shafts in Bending
(1988) / Garnett M D

A Structural Analysis of a Femorotibial
Intramedullary Nail Using Experimental
Mechanics and Finite Element Analysis
(1988) / Gruenberg S B

The Development of an Age Hardenable
Copper-Tin-Nickel Gear Bronze
(1988) / Gugel M D

Elastic Stability and Nonlinear Behavior of
Imperfect Shells with Initial Stresses
(1988) / Ju X

Time Accurate Simulation of Diesel Spray
Structures
(1988) / King E W

Computations for the Flow of Pure Vapor
Undergoing Film Condensation Between
Parallel Plates
(1988) / Kiziyalli Y C

Analysis of the Combustion of Four
Alternative Fuels in a Two-Stroke, Direct
Injection Diesel Engine Using High Speed
Photography
(1988) / Masterson D T

A Simplified Building Energy Analysis Model
(1988) / Meyer J R

A Data Dependent Systems Approach to
Filtering Noisy Images
(1988) / Morihara R H

Design and Development of a New Prosthesis
for Above Knee Amputees: Computer
Simulation and Experimental Evaluation of
Sagittal Plane Kinematic and Kinetic
Responses of a Prosthetic Limb
(1988) / Nazre A A

CARS - Controlled Automobile Response
Simulation - Computer Simulation Program
(1988) / Parsons T J

Validation of an Engine Deposit Energy
Transport Model Utilizing Internal Energy
Generation
(1988) / Penfold J N

A Heat Transfer and Stress Analyses of the
Integrated Evacuated Stationary
Concentrating Solar Collector
(1988) / Reed M J

Analysis of Using Experimentally Determined
Complex Frequency Response Functions to
Obtain System Matrices of Vibratory Systems
(1988) / Sim E S

Turbulent Heat Transfer Coefficients and
Pressure Drop Characteristics in a Straight
Circular Tube Downstream of a 180 Degree
Bend
(1988) / Walavalkar A A

The Sampling of Polynuclear Aromatic
Hydrocarbons in the Particulate and Vapor
Phases of Dilute Diesel Exhaust
(1988) / Waldenmaier D A

A Study of the Compositional and
Morphological Changes in a Two-Stroke Engine
Combustion Chamber Deposit During
Accumulation and Abatement
(1988) / Welch B L

MINNESOTA, UNIVERSITY OF

An Experimental Study on Physical
Human-Machine Interaction
(1988) / Anderson B J

The Influence of a Ceramic Particle Trap on
the Size Distribution of Diesel Particles
(1988) / Baumgard K J

Development of a Thin Film Zinc Oxide
Transducer on a Tungsten Carbide Tool Insert
(1988) / Bischoff B J

Compliant Motion Control Without Force
Measurements
(1988) / Bouklas K G

Inverse Solutions for Winding Processes and
Their Applications
(1988) / Boutaghou Z E

Performance of the High Yield Technology
PM-100 Particle Flux Monitor
(1988) / Caldow R

Design and Implementation of a Computer
Controlled 3 DOF Robot Finger
(1988) / Cameron L C

Impedance Control of a Robot Finger for
Contact Stability
(1988) / Cameron L C

Measurements in Thermal Plasma Jets
(1988) / Capetti A

On the Role of Structural Non-Linearity in
the Behavior of Self-Excited Chatter
Vibration in Turning
(1988) / Carlson A L

Ful Thermal Development of a Radiatively
Participating Fluid in Poiseuille Flow
(1988) / Chikh S

On the Stability of Interacting Humans and
Robotic Systems
(1988) / Foslien W K

Sieving Caputre of Particles by Microporous
Membrane Filtration Media
(1988) / Grant D C

Robotic Deburring of Parts with Unknown
Geometry
(1988) / Her M G

A Theoretical and Experimental Study on
Physical Human-Machine Interaction with
Constrained Motion
(1988) / Hessburg T M

A Robotic Design Program for the MacIntosh
Computer
(1988) / Ibele J L

Development and Field Testing of a Sulfur
Aerosol and Gas Monitor Using a Pulsed
Precipitator
(1988) / Keady P B

Natural Convection Mass Transfer from
Inclined Cylinders with Active and Inert
Bases Using an Electrochemical Technique
(1988) / Khan V

Finite Element Method for the Analysis of a
Material Property Identification Algorithm
for Pressbrake Bending
(1988) / Kim S

Two-Dimensional Model for Plasma Sintering
of Alumina
(1988) / Konstadinidis I X

Some Numerical Methods for the Solution of
Inverse Deformation Problems
(1988) / Maniatty A M

A Robotic Design Program for the MacIntosh
Computer
(1988) / Mirth J A

Tool Condition Monitoring in Multi-Tooth
Face Milling Using Instrumented Inserts
(1988) / Moser T M

Joule-Thomson Coefficients in Air, Nitrogen,
and Argon: Analysis of Experimental,
Analytical and Theoretical Methods
(1988) / Muraoka M J

Design of Harmonic Controllers for Servo
Systems to Achieve Tracking and Disturbance
Rejection of Repetitive Signals
(1988) / Narayanan K

Internal Conduction Influences on Cylinder
in Crossflow
(1988) / Ng W Y

Internal Conduction Influences on Cylinder
in Crossflow
(1988) / Ng Y W

A Study of the Oxidation of Sulfide/
Hydrosulfide on Glassy Carbon Platinum,
Graphite, and Cobalt Electrodes
(1988) / Nygren K

The Use of Force Data to Counteract
Additional Loads in Robotics
(1988) / Osowski T J

Laminar Natural Convection Over a Vertical
Cylinder Attached to a Horizontal Plate at
the Bottom
(1988) / Phan N G

A Knowledge-Based System for Mechanical
Design
(1988) / Rosen D W

Investigation of the Effect of Flow Swirl on
Heat Transfer Inside a Cylindrical Cavity
(1988) / Salce A

A Second Law Simulation of the Kalina Cycle
with a Water-Ammonia Mixture Working Fluid
(1988) / Schue D R

Heat Transfer from a Flat Surface to a Row
of Impinging Circular Air Jets Including
the Effect of Entrainment
(1988) / Seol W

On the Parameters that Control the Quality
of a Plasma Arc Cutting Torch
(1988) / Shell D B

MISSOURI, UNIVERSITY OF (COLUMBIA)
(continued)

Analytical Treatment of Piston Motion in a
Diesel Engine
(1988) / Tjan K J

Aerosol Reactor for Production of Metal
Oxide Particles
(1988) / Todd M P

MISSOURI, UNIVERSITY OF (ROLLA)

An N-Space Approach to the Analysis of
Nonclassically Damped Dynamic Systems
(1988) / Gharpurey C M

Natural and Mixed Convection Along Slender
Vertical Cylinders with Nonuniform Heating
(1988) / Heckel J J

Effect of Temperature on the Acoustic
Bending Response of a Flat Panel
(1988) / Koai W

An Analytical and Experimental Investigation
of a Latent Thermal Management System
(1988) / O'Dell M P

Performance Characteristics of a Lean
Turbocharged Spark Ignition Engine Fueled
with Methanol
(1988) / Pannone G M

Mapping of Input Link Ground Pivot Locations
for GCRR and GCCC Four-Bar Mechanisms
(1988) / Rodgers W J

Coal Disintegration Using High Pressure
Water Jets: Mechanism study
(1988) / Wijeratne L H

Geneva Mechanism
(1988) / Zhang B

MONTREAL, UNIVERSITY OF

Etude de Stabilite d'une Plaque
Rectangulaire Soumise a une Excitation
Parametrique
(1988) / Afrite Bennani A

Realisation d'un Systeme d'Imagerie par
Resonance Magnetique (IRM) Pour Fins
d'Enseignement
(1988) / Allard L

Simulation Numerique d'une Boucle Thermique
(1988) / Aube F

Etude Thermique Dans la Lyophilisation de
Lait par Micro-Ondes
(1988) / Ben Souda K

Effect de la Reutilisation sur les Proprietes
Physiques dees Catheters d'Angiographie
Jetables
(1988) / Bentolila P

Composites Thermoplastiques: Polypropylene
Renforce de Paillettes de Verre
(1988) / Bouti A

Controle du Deplacement d'un Robot
Manipulateur le Long d'une Trajectoire
Pre-Determinee
(1988) / Fillion Y

A Low Cost Vision System for Airfoil
Inspection
(1988) / John F

Methode d'Analyse de la Vie en Fluage des
Aubes de Turbine Sous des Conditions
d'Operation Variables
(1988) / Julien A

Une Methode de CAO d'un Bloc Perfore Pour
Oleohydrauliques (Montage par Cartouches)
(1987) / Kamoso L M

Conception d'un Appareil Pour lEtude in
Vitro de la Reparation Ligamentaire Sous
Stimulation Mecanique
(1988) / Labelle S

Etude de la Propagation Oblique d'Ondes
Acoustiques Spheriques Dans un Guide d'Ondes
et de Leur Rayonnement a Son Extremite
(1988) / Lahlou R

Conception d'un Logiciel Pour la Stimulation
Electrophysiologique Cardiaque
(1988) / Li Z

Etude des Propietes Mecaniques d'un
Ligament Croise Anterieur Synthetique du
Genou Humain
(1988) / Masson M

Etude de la Mise en Forme des
Thermoplastiques Estampables
(1988) / Mathieu Y

Etude sur le Comportement en Fatigue des
Plaques Minces Avec et Sans Joints Soudes
(1988) / Merabtine A

Influence des Charges sur la Mise en Oeuvre
et les Proprietes d'un Polyethylene a Haute
Masse Moleculaire Chauffe Par
Radiofrequences
(1988) / Ouellet C

Etude de Contraintes a l'Interieur
d'Attaches
(1988) / Radacovici A

Systeme d'Acquisition et de Traitement de
Signaux Respiratoires
(1988) / Rivest J F

Considerations Techniques et Economiques
Dans la Prise de Decision en Matiere de
Gestion des Technologies Medicales: 2
Exemples
(1988) / Rochon A

Convection Naturelle Dans un Milieu Poreux
Sature Confine Entre Deux Cylindres
Concentriques Horizontaux
(1988) / Rousse D R

Analyse Dynamique des Plaques Circulaires
Non-Uniformes
(1988) / Selmane A

Commande par Micro-Ordinateur d'un Banc
d'Essai Pour Moteurs Hydrauliques
(1988) / Tremblay P

NEBRASKA, UNIVERSITY OF

Effect of Sensitization on Hydrogen
Transport in Austenitic Stainless Steels
(1988) / Davis D A

NEW JERSEY INSTITUTE OF TECHNOLOGY
(continued)

Investigation of the Characteristics of the
Kerf and the Surface Generated in the Course
of Cutting Titanium with Abrasive Water Jets
(1988) / Vora A

Closed Loop Machining in Precise Turning
(1988) / Wei C S

NEW MEXICO, UNIVERSITY OF

Development of a Coordinate Measuring System for
CNC Machining of Complex Surfaces
(1988) / Ahghar M

Prediction of Optical Phase Degradation Through
Deterministic Media
(1988) / Clark T T

An Experimental Study of the Burn Rate of
TiH1.65/KC104 Pyrotechnic Under Confinement
(1988) / Hingorani-Norenberg S L

Unsteady Aerodynamic Loading of a Model in a
Shock Tube
(1988) / Nelson I I

NEW YORK, STATE UNIVERSITY OF (BUFFALO)

Instructional Robot Calibration, Control,
and Programming
(1988) / Beiter P J

Digital Simulation of a Pneumatic Pressure
Regulator
(1988) / Benzoni A M

On the Synthesis of Multibar Linkages
(1988) / Dhingra A

Image Analysis of Fiber Orientation in Short
Fiber Composite Material
(1988) / Fritz G

System Identification in the Microcomputer
Environment
(1988) / Garcia E

The Analysis of a Nonlinear
Thermoviscoelastic Three-Bar Subject to
a Kinematic Disturbance
(1988) / Graesser E

Steady and Unsteady Friction Behavior at
Dry and Lubricated Contacts
(1988) / Hess D P

The Effects of Yaw Error and Instabilities
on the Aerodynamic Performance and
Structural Design of a Horizontal Axis
Type with Turbine
(1988) / Koutelas C

Interactive Robot Motion Planning in a
Two-Dimensional Workspace by Optimization
Methods
(1988) / Lai M J

Dynamic Modal Survey of V-5 Compressor
(1988) / Lee-Glauser G

Minimum Model Error Estimation of Modal
Truncation Errors
(1988) / Lin J C

Automatic Generation of Engineering Drawings
and Process Routings for Valve Spool in a
CAD Environment
(1988) / Pequignot S

On the Dynamics of Damped, Distributed
Composite Laminate Systems
(1988) / Popp G

Taguchi Methods and Analytical Optimization
Strategies
(1988) / Raju C

Symbolic Computation of Derivatives of User
Supplied Functions for Dynamics Analysis
and Optimization
(1988) / Ren A

Automatic Generation of Apt Code from a
Part Drawing Data Base
(1988) / Sanka R

System Identification of a Chaotic System
(1988) / Tong P H

Use of Green's Function in Heat Conduction
with Change of Phase
(1988) / Zhang X

NORTH CAROLINA STATE UNIVERSITY

Frost Growth Model for Laminar Airflow
Between Parallel Plates
(1988) / Hargrove P D

Thermohydraulic Modeling of a Monogroove
Heat Pipe Condenser
(1988) / Henson R A

Elastic Emission Polishing
(1988) / Loewenthal M D

An Experimental Performance Analysis of
Southern Exposure Frame Walls, Trombe Walls
and Sunspace Interface Walls at the North
Carolina State University Solar House
(1988) / Nielsen D H

Anisotropy in Plastic Torsion
(1988) / Sheikh-Ahmad J Y

Thermal Contact Conductance of Various
Metals at Elevated Temperatures
(1988) / Skidmore M M

NORTH DAKOTA STATE UNIVERSITY

Kinematic Analysis of a Power Take-Off Shaft with
Two Universal Joints and a Bent Center Shaft
(1988) / Kim S H

Finite Element Structural Model of the Human Finger
(1988) / Resh R A

Three Dimensional Structural Finite Element Code
for Machine Frame Analysis
(1988) / Willey S L

Force Sensor for Adaptive Control of a Robot
(1988) / Yeung-Bae R

NOTRE DAME, UNIVERSITY OF

Modeling and Parameter Identification of
Dynamic Systems with Application to Aircraft
Landing Gear
(1988) / Bacarro J M

NOTRE DAME, UNIVERSITY OF
(continued)

Trailing Vortex Interaction with a Thin
Airfoil
(1988) / Fang J

Natural Convection in a Tilted Trapezoidal
Enclosure
(1988) / Kumar V

NOVA SCOTIA, TECH UNIVERSITY OF

Stress Analysis of the Proximal Femur and
the Mechanical Performance of a Femoral
Total Hip Replacement Component
(1988) / Brennan D P

State Variable Formulation of Dynamic
Systems
(1988) / Callaghan R J

Thermal Performance of Heat Exchangers with
Tube/Shell Diameter Ratios Between 0.1 and
0.3 at Low Reynolds Numbers
(1988) / Garg R K

OKLAHOMA, UNIVERSITY OF

Moving Region: Recognition for Facial
Paralysis Diagnosis
(1988) / Park J

Finite Element Stress Analysis of Human
Tooth with and Without a Restoration
(1988) / Salari M

Radiative Heat Exchange Between Two
Adjacent Plates Using a Monte Carlo Analysis
(1988) / Vatsyayana C V

OREGON STATE UNIVERSITY

Natural Convection from a Horizontal Cylinder
Parallel to a Heated Vertical Wall
(1987) / McCoy T J

A Study of Laser Processed Intermetallics
(1988) / Roub E R

Microwave Sintering of Alumina-Titanium Cermets
(1987) / Wrzesinski W R

PENNSYLVANIA STATE UNIVERSITY

Wall Shear Stress Measurements in the Penn
State Ventricular Assist Device Using
Hot-Film Anemometry
(1987) / Baldwin J T

Development of an Aortic Pressure Observer
for the Penn State Electric Ventricular
Assist Devices
(1987) / Bennett B S

Caviation Inception in Rectangular
Two-Dimensional Slots
(1987) / Blair W G

A Finite Element Technique for Obtaining
Brainstem Surface Voltages from Acoustic
Brainstem Response Data
(1987) / Bliton A C

A System for Multineuron Data Acquisition
(1988) / Brim K W

A Comparison of Three Coal-Derived, Middle
Distillate, Synthetic Fuels in a Single
Cylinder DI Diesel Engine
(1987) / Buzza T G

Characterization of Histidase: A Study of
Product Inhibition, Covalent Modification,
and pH Dependence
(1988) / Celiker I

An In-Vitro Study of the Systemic and
Pulmonary Performance of the Penn State
Single-Port Assist Device
(1987) / Christiansen J E

Road Roughness Calculation for the Automated
Loading Facility
(1987) / Dari C R

A New Material for Ultrasound Tissue
Phantoms with Variable Acoustic Properties
(1987) / Della Corna L V

Multi-Electrode Determination of Resistivity
in Layered Media
(1987) / Erickson J R

An Experimental Investigation of
Particulates in Boron-Containing Diffusion
Flames
(1987) / Funari M J

Microprocessor Control of High-Speed Wind
Tunnel Stagnation Pressure
(1987) / Fung Y T

A Design Methodlolgy for the Reduction of
Induction Noise from Four-Cylinder Internal
Combustion Engines
(1987) / Heydrick D S

Kinematic Synthesis for Dynamic Response
(1988) / Idlani S B

Design of Cooling Towers Using the
Effectiveness-NTU Method
(1987) / Jaber M H

Application of Light-Scattering Techniques
to the Measurement of Particulates in
Boron-Containing Flames
(1987) / Khan A

Lumped Mass Modelling of Flexibility in
Planar Robot Manipulators
(1988) / Lee W Y

Numerical Analysis and Optimization of
Objects Subject to Unsteady Heat Conduction
(1988) / Madison J V

The Effect of Continuous Positive Airway
Pressure on Indices of Ventilatory Effort
in Humans
(1987) / Mc Donough J M

Modal Analysis of the Front Fork of an
Off-Road Motorcycle Using the Finite Element
Method
(1987) / Moujahed A

Double Electrode Through-the-Arc Seam
Tracking for Automated Electric Arc Welding
(1987) / Sochor T A

Application and Evaluation of the Intensity
Techniques for Sound Power Measurements of
a Complex Source
(1987) / Taschuk E G

PENNSYLVANIA STATE UNIVERSITY
(continued)

Implementation and Evaluation of a
Suboptimal Controller for the Penn State
Electric Motor Left Ventricular Assist
Device
(1987) / Tirinato J A

The Transcutaneous Energy Transmission
System: Electromagnetic Field Studies and
Shielding Considerations
(1987) / Trumble D R

Development of a Dual Beam Reflection Laser
Vibrometer for the Measurement of Dynamic
Structural Displacements and Rotations
(1987) / Walsh J A

Compliant Two-Link Robot Manipulator Model
for Testing Adaptive End-Point Control
(1987) / Wieand J S

On the Mathematical Modeling and
Optimization of Grinding Circuits
(1987) / Yildirim K

Iterative Methods for Design Sensitivity
Analysis
(1988) / Yoon B G

PENNSYLVANIA, UNIVERSITY OF

Study of Structural Force Coupling in
Specific Non-Linear Structures Subject
to Earthquakes
(1988) / Arasteh M D

The Effect of Joint Compliance on Static and
Dynamic Response of a Robot Manipulator
(1988) / Buschman N A

Simulation of a Two Link Robot Manipulator
with Elastic Members
(1988) / Hinds J C

Heat Clearance Properties of a Radiatively
Heated Biological Tissue
(1988) / Klemick S G

Symbolic Computation in Heat Transfer
Analysis
(1988) / Palis M A

Kinematic Design, Analysis, Simulation, and
Animation of Serial Manipulators Via
Paul/Rosa Notation
(1988) / Park B N

A Theoretical Analysis of the Effect of
Inertia Forces on the Behavior of
Tube-to-Tubesheed Expanded Joints
(1988) / Portonova S J

Iterative Solution to the Inverse Kinematic
Problem of Robots
(1988) / Siraj S

Modeling of Electric Arcs in Ball-Bonders
(1988) / Vacek D J

An Analytical Evaluation of FLUENT, a
General Purpose Computer Program Developed
by Creare, Inc. for Modeling Fluid Flow
(1988) / Wisnosky K W

PITTSBURGH, UNIVERSITY OF

A Numerical Study of Confined Turbulent
Flows
(1989) / Al-Arabi A M

Transient Thermal Stresses of a Slab Solid
Propellant
(1989) / Chen C Y

Study, by Finite Element Methods, of a Shear
Fracture Test Specimen, Applied in Bond
Strength Tests for a Brittle Adhesive
(1989) / Condrac R F

A Design Analysis of Multilayered Cylinders
of the Ninth Order
(1989) / Congelio C E

A Numerical Analysis of a Thermally
Expandable Fluid Subjected to Thermal
Hydraulic Transients
(1989) / Conroy C T

The Development of an Industrial
Applications Laser Laboratory Including a
Prototype Carbon Dioxide Laser Beam
Delivery System
(1989) / Gates G H

CAD/CAM Design of Small Double Pipe Heat
Exchangers Using Personal Computers
(1989) / Ghaben J K

Temperature Measurement in Solid-Gaseous
Two Phase Flow with Emphasis on Pulverized
Coal-Air Mixtures
(1989) / Grimberg M

An Evaluation of Ceramic Beam Specimens
with Crack Tips in a Shear Stress Field
(1989) / Heiple R W

Flow Modeling of Shell and Tube Heat
Exchanger with the Duval Computer Code
(1989) / Jones O C

Thermoviscoelastic Stress Analysis of the
Solid Propellant in a Cylindrical Form
(1989) / Lin T H

Creep Buckling Design Limits for Hollow
Circular Cylinders Under Axial Compression
and External Pressure
(1989) / Livingston J M

Velocity Measurements in Hydrocyclones by
Using Laser Doppler Anemometer
(1989) / Shen H Q

Numerical Study of Natural Convection Heat
Transfer Between Parallel Plates with
Non-Newtonian Fluids
(1989) / Sithambaram R

Computer-Aided Simulation of Plastic
Deformation Processes
(1989) / Suer M

An Investigation Into Utilizing Dynamic
Impact Testing as a Means to Detect
Interlaminar Flaws in Laminated Composite
Plates
(1989) / Testa M F

The Transient Thermal Stresses of a
Spherical Solid Propellant Cooled
Symmetrically by Forced Convection
(1989) / Yang J L

PURDUE UNIVERSITY

Enhancement of Pool Boiling from a Simulated
Microelectronic Heat Source
(1988) / Anderson T M

A Reliability Study of Cooling Systems
Serving Computer Rooms
(1986) / Banduric R D

Multivariable Control of a Biaxial Machine
Tool
(1988) / Burhoe J C

Analysis of the Control of a Kamyr Digester
Based on the Purdue Kamyr Digester Model
(1988) / Butler A C

Dynamic Modeling and Optimal Control Design
of Passive and Active Electro-Magnetic
Acoustic Transducers
(1986) / Carey D M

Solidification of Binary Substances in
Rectangular and Cylindrical Cavities
(1988) / Christenson M S

Aerodynamics of an Oscillating Cascade with
Flow Separation
(1988) / Eley J A

Numerical and Experimental Study of the
Effects of Thermo/Diffusocapillary
Convection on Solidification of a Binary
Substance
(1988) / Engel A H

Finite Element Analysis of the Patella with
and Without Prosthesis
(1988) / Faux D R

Multiobjective Probabilistic Design of Gear
Trains Utilizing Modified Game Theory
(1988) / Freiheit T I

Plasma Processes at a Cathode Spot in a
Magneto-Plasma-Dynamic Thruster
(1988) / Goodfellow K D

Refrigerant Mass and Mass Flow Rate for Each
Component in a Split Heat Pump System
(1988) / Grzymala T E

Comparison of Steady and Unsteady Secondary
Flows in a Turbine Stator Cascade
(1988) / Hebert G J

Determination of Local Convection
Coefficients for Shrouded Pin Fin Arrays by
Naphthalene Sublimination and Foil Heating
Techniques
(1988) / Heidmann J D

A Numerical Investigation of Hydrogen
Ignition
(1988) / Hitch B D

Evaluation of a New Spot-Electrode Array to
Noninvasively Monitor Changes in Cardiac
Output by Impedance Cardiography
(1988) / Kiesler T W

Investigation of an Optical Method for Rapid
Recalibration of an Automated Machining
Center
(1988) / Lopez J

Enhancement of Forced Convection Boiling
from a Simulated Microelectronic Heat Source
in a Rectangular Channel
(1988) / Maddox D E

A Comparative Parametric Stability Study for
Flexible Cam-Follower Systems Due to Varying
Cam Profiles
(1988) / Mahyuddin A I

Predicting the Onset of Nucleate Boiling in
Wavy Free-Falling Turbulent Liquid Films
(1988) / Marsh W J

The Development of Fracture Toughness Test
Computer Control Software and the Evaluation
of an Aluminum-2.6 Lithium-0.09 Zirconium
Alloy's Fracture Properties
(1988) / McKeighan P C

Optical Particle Sizing in a High
Temperature Environment
(1988) / Miles B H

Concentration Measurements in a Cold Flow
Model Annular Combustor Using Laser Induced
Fluorescence
(1988) / Morgan D C

Representation and Analysis of Size and
Geometric Tolerances for Part Assemblies
(1988) / Pandit V

An Object-Oriented CAD Framework
(1988) / Paul A P

A Compact Optical Hole Sensor for Robotic
Manipulations
(1988) / Rajadhyaksha M M

Nonlinear Nonplanar Dynamics of a
Parametrically Excited Inextensional
Elastic Beam
(1988) / Restuccio J M

On the Acoustics of Shell Enclosed
Compressors with Special Attention to Gas
Pulsations on the Suction Side
(1986) / Roys B J

Time Series Modeling of Nonlinear Systems
(1988) / Seidel D K

Real Time Collision Avoidance of a Planar
Manipulator with an Interfering Single Link
Arm
(1988) / Seitz B D

Kinematics of Two Cooperating Planar Robots
(1988) / Sissom B J

The Development of an Intra-Oral Stimulator
for Use in Speech Motor Control Research
(1988) / Sisson D E

Transient Thermal Fracture of Bonded
Dissimilar Materials
(1988) / Smith C S

Drop Size Investigation of an
Electrostatically-Assisted Fan-Spray
Atomizer
(1988) / Snyder H E

Aerodynamic Detuning and Aeroelasticity of
a Supersonic Turbomachine Rotor
(1988) / Spara K M

PURDUE UNIVERSITY
(continued)

Design of a Tracking Manipulator Using
Ultrasonic Sensors
(1988) / Stamper R E

Preliminary Experiments in an Investigation
of Flat Plate Film Cooling Effectiveness
(1988) / Sweeney P C

A Numerical Analysis of the Influence of
Internal Nozzle Geometry on Diesel Fuel
Injection
(1988) / Tennant P A

Application Oriented Robot Programming
System for a Workcell Simulator
(1988) / Till P K

The Representation of Mechanical Assemblies
in a Computer-Aided Design Environment
(1988) / Tilley C M

Analysis of the Thermal Performance of
Griffin: An Energy Efficient Residence
(1988) / Torcellini P A

Development of a Heat Flux Gauge for a
Partially Insulated Internal Combustion
Engine
(1988) / Tree D R

Finite Element Analysis of the
Three-Dimensional Euler Equations in a
Rotating Turbomachinery Blade Passage
(1988) / Turner D A

An Object - Oriented Approach to
Feature-Based Design with Tolerancing
(1988) / Turner G P

Heat Transfer to Water Sprays
(1988) / Valentine W S

Robot Path Planning and Obstacle
Avoidance: A Design Optimization Approach
(1988) / Venkataraman S G

The Inverse Kinematics of a Three Cylindric
Robot
(1988) / Vierstra B C

The Effect of Inlet Condition on Pulsatile
Flow Through a Curved Tube
(1988) / Walker J D

Computations of Gas-Particle in Axisymmetric
Nozzles
(1988) / Wang Y

Vibrational Power Flow Analysis of Rods and
Beams
(1988) / Wohlever J C

QUEENS UNIVERSITY

Computer Aided Tool Path Programming for a
Numerical Control Facility
(1988) / Fritz D W

An Environmental Simulator for Thermal
Performance Testing of Solar Collectors
(1988) / Johnson G P

Design and Construction of a Temperature
and Force Sensing Gripper
(1988) / Nshama W M E

Development of an Expert for Process Fault
Diagnosis
(1988) / Stewart D A

Design and Analysis of a Tibial Interface
Component for Knee Arthroplasty
(1988) / Willings T J

RENSSELAER POLYTECHNIC INSTITUTE

Development of Enhanced Ceramic Surfaces for
High Temperature Waste Heat Recovery
(1988) / Armstrong R D

Describing and Tracking Evolving Bodies in
Automated Two Dimensional Metalforming
Analyses
(1988) / Ashley R

Use of Physical Simulation in the
Instruction of Flexible Manufacturing
System
(1988) / Aslam S

An Interactive Graphics Application for
Computer Aided Development of Inspection
Programs for Coordinate Measurement Machines
(1988) / Atkins N W

Design for Assembly of Aerospace Structures:
A Qualitative, Interactive Approach
(1988) / Baum J P

Study of Transfer Films from Magnetic Tapes
(1988) / Blanc P

Nonivasive Measurement of Cardiac Output by
Expired Carbon Dioxide Monitoring
(1988) / Browne J A

An Investigation of Self-Upsetting Rivet
for Fatigue Life Improvement
(1988) / Chandra J B

Comparing Current Patterns Used in Electric
Current Computed Tomography
(1988) / Cheng K

Design and Development of a Supercritical
Shaft Test Rig
(1988) / Conley W P

Film Cooling of Turbine Blade Trailing
Edges
(1988) / DelPorte S H

Stress Analysis of a Hydromechanical
Transmission Input Housing
(1988) / DiLeonardo P

An Electronic Module Digitizing Computer
Program
(1988) / Dolan D

Combustion as a Source of Light
(1988) / DuPlessis L P

Friction and Film Thickness Measurement in
Hydrodynamic Journal Bearings
(1988) / Fina F L

An IMU Tolerance and Misalignment Study
Using VSA Software Analysis Package
(1988) / Frederick M S

Simulated Adaptive Force Control of
Scalloped Surface Grinding for Application
to a Robotic Die Finishing System
(1988) / Gatlin D S

RHODE ISLAND, UNIVERSITY OF
(continued)

An Analytical and Experimental Investigation of
Anisotropic Polymer Composites at High Temperatures
(1988) / Mendelsohn D L

Wave Propagation in Unsaturated and
Saturated Porous Media
(1988) / Prakash V

A Gripper-Based Tactile Sensor Based on
Plate Theory
(1988) / Teixeira M

Dynamic Analysis and Redesign of a Post Notching
Machine and Feeding System
(1988) / Tsai R F

ROCHESTER, UNIVERSITY OF

An Analytical Displacement Calculation for
Optically Generated Surface Acoustic Waves
(1988) / Leong K J

ROSE-HULMAN INSTITUTE OF TECHNOLOGY

Development of a Finite Element PC Program
for Dynamic Analysis of Structures
(1989) / Bao C

An Improved PC Finite Element Analysis of
Acoustic Cavities
(1988) / Cao C

The Adaptation of a Helium Bubble Generator
for Flow Visualization in a Wind Tunnel
(1988) / Danner D B

The Analysis of a Film Tower Die
Utiliziang the Ansys Finite Element Package
(1988) / Greer R D

Analysis of Wave Power Conversion for a
Two-Dimensional Device with Oscillating
Water Column
(1988) / Jiechi X

Application of the Component Element Method
to the Impact Damped Harmonic Oscillator
(1988) / Jones R E

Part I. Coupling and Characterization of
Thin Films. Part II. Integration of a
Laser Diode Into a Digital Speckle Pattern
Interferometer
(1988) / Krumpholz E

Development of a Force Analyzing Ransom Rest
for Use in the Evaluation of Handgun Recoil
Reduction Devices
(1988) / Lozier T

Time History Study of a Classical Cantilever
Ream Damped by Internal Mechanical Means
(1988) / Nale T A

Finite Element Analysis of Bus Frames Under
Simulated Crash Loadings
(1989) / Parakh Z K

Vibration Characteristics of Composite
Structural Systems
(1989) / Vedder J

The Biocompatible, Chemical and Mechanical
Evaluation of a Quartz Fiber Reinforced
Polybutadiene Composite Hip Femoral
Component
(1989) / Wack M A

A Comparison of the Euler, Timoshenko, and
Levinson Methods for the Prediction of
Vibration in Low Aspect Ratio Rectangular
Prismatic Beams
(1988) / Whiteman G L

SAN DIEGO STATE UNIVERSITY

Laser Assisted Electrochemical Hole
Machining
(1987) / Bagheri H

Inverse Kinematics for 9-Degree of Freedom
Robot Manipulators
(1988) / Evans R E

Parabolized Navier Stokes Calculations for
Solid Rocket Nozzle
(1987) / Gregg T A

A Visual Study of Turbulent Spots in a
Kerosene Medium
(1987) / Guillaume D W

Minimizing the Volume Measurement Errors
of Fluid Reservoirs in a Multi-Attitude
Environment
(1988) / Lukens M A

Angle of Attack Loss for Flat Plate
Cascade
(1988) / Mohssenzadeh M

A Study in Adsorption Air-Conditioning
Using Silica Gel
(1988) / Momtaz R W

Theoretical and Experimental Analysis of
Stress and Displacement Between
Non-Conforming Contacts
(1988) / Nourani A

Nonlinear Finite Element Modeling of
Superplastic Forming (2D-SPF)
(1988) / Sadeghi R S

Design and Development of a Robotic
Interaction System Keying on Non-Repetitive
Human Input
(1987) / Scheerer J J

A 3-D Kinematic Model for the Biomechanical
Analysis of Lower Limb While Cycling
(1988) / Shen H

Optimization Using the Finite Element
Method as Applied to a Planar Truss Design
(1987) / Traganza W T

SASKATCHEWAN, UNIVERSITY OF (SASKATOON)

A Study of Gas Mixing and the Single-Breath
Diffusing Capacity
(1988) / Berneche J A

Feasibility study of a Hydraulic Elbow
Loader
(1988) / Hayes S J

Operating Strategies for Sunspaces in Canada
(1988) / Lumbis A J

SASKATCHEWAN, UNIVERSITY OF (SASKATOON) (continued)

Power Spectral Analysis of Myoelectric
Signals
(1988) / Shankar S

Portable Arrhythmia Monitor
(1988) / Suresh H K

Energy Conservation by Adding Extra Layers
of Polyethylene and Thermal Curtains for
Commercial Greenhouses Located in Cold
Climates
(1988) / Tse P W

SHERBROOKE, UNIVERSITY OF

Simulation Numerique de la Mesuree de la
Puissance Acoustique Obtenue par
Intensimetrie
(1988) / Agnan J L

Contribution a l'Etude des Thermopompes en
Regime Transitoire et Permanent: Etude
Experimentale et Simulation
(1988) / Ajam B

Etude de la Propagation du Son sur les
Surfaces Courbees
(1988) / Berry A

Etude du Comportement Mecqnique d'un
Composite Stratifie Graphite/Epoxy Soumis
a des Sollicitations Monotones et Cycliques
(1988) / Dupont L

Une Nouvelle Methode Pour Predire la Rupture
en Fatigue des Materiaux Composites
(1988) / Fawuz Z

Etude Numerique des Phenomenes d'Instabilite
Thermohydraulique Induits Dans les
Ecoulements Diphasiques Liquide-Vapeur
(1988) / Kraitem M

Identification des Microendommagements Dans
les Composites Carbone-Epoxyde par l'Analyse
Temporelle et Spectrale des Sighnaux Detectes
par les Techniques d'Emission Acoustique
(1988) / Maslouhi A

Calcul des Forces Aerodynamiques
Instationnaires sur les Pales d'une Turbine
d-Aerogenerateur a Axe Vertical
(1988) / Masse B

Synhtese de Poudres Ultrafines Alliees
Aluminium-Cuivre Dans un Reacteur a
Plasma a Induction HF
(1988) / Morel C

Convection Mixte en Regime Laminaire dans
un Tuyau Incline Soumis a un Flux de
Chaleur Constant a la Paroi
(1988) / Nguyen C T

Caracterisation de l'Endommagement des
Materiaux Composites Carbone-Epoxyde
Sollicites en Compression Statique et
Cyclique
(1988) / Roberge J

Etude du Comportement Dynamique de la
Pompe Centrifuge en Regime Diphasique
(1988) / Tran C

SOUTH CAROLINA, UNIVERSITY OF

Determining Small Planar Rotations of
Cylinders Using Computer Vision
(1988) / Carman D B

Quantitative Non-Destructive Evaluation of
Composite Structures Through Image
Correlation Acquired Strain Fields
(1988) / Chen H S

Design and Performance Analysis of Nonlinear
Model-Based PID Feedback Control for Robot
Manipulator
(1988) / Chiou Y C

Computer Vision and Graphics Assisted
Off-Line Robot Programming Methodology
(1988) / Choi H S

The Development and Application of Machine
Vision for Inspection of Electronic
Assemblies
(1988) / Davis W G

Process Optimization Technique for the
Circuit Board Inspection Machine
(1988) / Menon U S

Comparison of Feedback Control Laws for
Robotic Arm Manipulators
(1988) / Mrad N

FRP/Plywood Panel Stiffness
(1988) / Mullininx R B

A Model of Transient Heat Transfer in
Muscle Tissue
(1988) / Sun W

Non-Destructive Evaluation of Strength of
G1-Ep (0 Degree/90 Degree)s Composite
Through Image Correlation Acquired Strains
(1988) / Waikar R J

SOUTH DAKOTA SCHOOL OF MINES AND TECHNOL.

An Ihntegrated Software Environment for
FMS Design Using Simulation and Expert
System Techniques
(1988) / Balakrishnan R

An Expert System for Computer Integrated
Design and Manufacture of Plate Cams
(1988) / Basha S S

Use of Artificial Intelligence in the
Control of Two Cooperating Robots
(1988) / Chen S M

An Investigation of Air Flow and Heat
Transfer Characteristics in a Fluidized
Bed Heat Exchanger
(1988) / Crandall W C

Application of State Space Analysis in
Servohydraulic Controls
(1988) / Hamilton G M

Temperature Control of a Salt Specimen in a
Materials Testing Machine
(1988) / Nieland J

Thermal Stress Effects of Hot Air Jet
Exhaust Blasts on Concrete Airfield Pavement
(1988) / Rajpal I S

SOUTH DAKOTA SCHOOL OF MINES AND TECHNOL.
(continued)

Statistical Model for Error Compensation of
a Robot Using Calibration Technique
(1988) / Singh A K

Evaluation of Penumatic Actuator for Use in
Robotic End Effectors
(1988) / Subbarao K C

The Effect of Bi-Axialitly on Failure in
Composites
(1988) / Wang H H

SOUTH DAKOTA STATE UNIVERSITY

Forced Vibrations of Hollow Cylindrical
Structures
(1988) / Nunez Briceno D J

SOUTH FLORIDA, UNIVERSITY OF

An Analysis of the Wind Flow Around the
Space Shuttle and Service Structure
(1988) / Lawless P B

Stress Analysis and Evaluation of Lifting
and shipment of a 130 Ton Steam Generator
(1988) / Misvel M C

Dynamic Response of a Thin Rectangular Plate
with Fixed and/or Simply Supported Edges
Determined by the Finite Difference Method
(1988) / Staples J F

The Development of a Slip Sensor Evaluation
Apparatus
(1988) / Stipe M E

The Design of a Free-Standing Hydrofoil
Loadbox
(1988) / Wuthrich J

SOUTHERN METHODIST UNIVERSITY

The Use of Parallel Processors for the
Optimization of the Geometric Control of a
Robot Manipulator
(1988) / Ghais H

STEVENS INSTITUTE OF TECHNOLOGY

Simulation of the Dynamic Response of a
High-Speed Slider - Crank Mechanism with an
Elastic Connecting Rod Using Abaqus
(1987) / Cha E

TENNESSEE TECHNOLOGICAL UNIVERSITY

Non-Destructive Technique of Determining
Plastic Deformations Using Pulse
Transmission and Acoustic Emission
Transducers in 304 Stainless Steel
(1988) / Abitorabi M

Residual Mechanical Properties of Tough
Matrix Composites After Low Velocity Impact
(1988) / An C C

A Computer Simulation for a
Thermodynamically-Controled
Char-Recirculation Biomass-Gasification
Power Plant
(1988) / Anderson D E

Gross Motion Synthesis of RSSR and RSSP
Space Mechanisms
(1988) / Bookout P S

Application of Instantaneous Invariants to
the Synthesis of Spherical Linkages
(1988) / Bunduwongse R

Optimum Synthesis of Multi-Loop Linkages Via
Coupler Curves of Four-Bar Linkages and
Error Minimization of the Loop Closure
Equations
(1988) / Burke D M

Flaw Detection Methods Based on Digital
Signature Analysis for Rotating Machinery
Quality Control
(1988) / Demirdogen A C

Computational Study of Unsteady Heat
Transfer Problems Related to High-Energy
Lasers
(1988) / Kurkal K R

A Flow Device for Tailored Velocity
Profiles
(1988) / Moore C C

Simulation Comparison of Control System for
Sequoyah Nuclear Plant
(1988) / Ollat X

Investigation of the Use of SUPERTAB to
Perform Finite Element Analysis of a Shell
Roof and a Pipe Tee
(1988) / Saxon J B

Performance Analysis of a Fossil Power Plant
Using On-Line Simulation Software
(1988) / Shih Y J

Performance Investigation of an Olympic
Competition Smallbore Rifle by Vibration
Signature Analysis
(1988) / Slonena R R

Mobility Criterion and Classification
Scheme of Geared Five-Bar Linkages
(1988) / Tay K C

Super Curve Generation and Its Applications
(1988) / Wang S C

An Investigation of Fracture and Fatigue
Properties of ARALL-2
(1988) / Wilson C D

Experimental Study of Velocity Profile
Variation Downstream of HVAC Duct Fittings
(1988) / Yoon H G

TENNESSEE, UNIVERSITY OF (KNOXVILLE)

A Study of Low Frequency Combustion
Instability in Rocket Engine Preburners
Using a Heterogeneous Stirred Tank Reactor
Model
(1987) / Bartrand T A

Optimal Torque and Energy Paths in the First
Four Joints of a Redundant Robotic Arm with
a Study on Singularities
(1987) / Bitar E M

Performance Evaluation of a Dual-Speed
Heat Pump
(1987) / Carr J R

Flow Induced Vibration Within Centrifugal
Gas Compressors
(1987) / DeSteffano T D

TENNESSEE, UNIVERSITY OF (KNOXVILLE)
(continued)

Development and Analysis of a Gas Cantilever
Coriolis Mass Flowmeter Concept
(1987) / Denny A G

Fuel-Nitrogen Conversion in a Supercharged,
Spark-Ignition Engine
(1987) / Denny J C

Economical Feasibility of a Ground-Coupled
Heat Pump for the Knoxville, Tennessee Area
(1987) / Griffith W A

Altitude Test Cell Rocket Diffuser Heat
Transfer and Erosive Environment Modeling
(1987) / Jordan J L

Matrix Method Application to Vibrations in
Beam Type Periodic Structures
(1987) / Kilani A W

Development of an Acoustic Levitation
Apparatus for a coal Pyrolysis Study
(1987) / Marquis J E

Heat Exchanger Characteristics Required for
a Fluid-Cooled Venturi
(1987) / McAmis R W

TENNESSEE, UNIVERSITY OF (TULLAHOMA)

Heat Transfer in the Aeropropulsion System
Test Facility Mixing Header
(1988) / Deibner G W

Feasibility Study for Methods of Determining
Crystal Phase of Aluminum Oxide Rocket
Exhaust Particles
(1988) / Dill K M

Design and Preliminary Testing of an
Experimental Probe for Measuring Sootblowing
Effectiveness and Controlling Ash Deposition
in CFFF Boiler
(1988) / Gopathi K

Finite Element Thermal and Stress Analyses
of Thin Cylindrical Rings of Variable
Geometry Subjected to High Heat Flux
Conditions
(1988) / Groff M B

Aircraft Turbine Engine Thermodynamic Power
Cycle Performance Comparisons
(1988) / Hastings J

Performance Study of a Continuously
Variable Stroke Internal Combustion Engine
(1988) / Jennings B R

Thermal Investigation of a Signal Chopping
Device Within a cryogenic Test Chamber
(1988) / Little J

Determination of Surface Heat Transfer
Coefficients of Hypersonic Wind Tunnel
Aerodynamic Models Using Emmissivity
Characteristics of Thermographic Phosphor
Paints
(1988) / Lowrey G A

Heat Transfer Analysis of CFFF Superheater
by the Effectiveness NTU Method
(1988) / McMinn W T

A Locally Implicit Scheme for Navier-Stokes
Equation
(1988) / Nayani S N

Linearity and Spectral Characteristics of
Integrating Sphere Sources
(1988) / Nicholson R A

Sensitivity Study of the Effects of PDF
Models on the Predicted Structure of
Turbulent Combustion Flow
(1988) / Peng X

Comparison of Two-Dimensional and Three
Dimensional Computational Fluid Dynamics
Solutions for a Vectoring Turbine Engine
Exhaust Nozzle
(1988) / Pruffert M

A Finite Difference Scheme for the Analysis
of High Frequency Sound Radiation
(1988) / Raviprakash G K

A Water Tunnel Simulation of the Internal
Flow of a Segmented ARC Heater
(1988) / Rigney S

TEXAS A AND M UNIVERSITY

Design and Construction of a Learning End-Point
Sensor
(1988) / Aguirre L A

Stiffness Analysis of an Intramedullary
Nail Loaded in Torsion
(1988) / Boyd J G

Recycling and Alloying of Polyethylene Terephthalate
(1988) / Bugg M S

Evaluation of a Direct Evaporative Roof Spray
Cooling System
(1988) / Carrasco A D

Determination of the Dynamic Young's Modulus,
Shear Modulus and Internal Friction as a Function
of Temperature and Microstructure in Uranium -
2.4 Wt Percent Niobium
(1988) / Chancellor W M

Experimental Measurements and Methods for Data
Analysis to Determine the Rotordynamic
Coefficients of a Labyrinth Seal
(1988) / Choi S K

Gear Coupling Effects on Rotordynamics
(1988) / Clark R W

Temperature Dependence of Stiffness and
Damping for Several Metal and Polymer
Matrix Composites
(1988) / Cribb V N

Experimental Studies on the Group Combustion
of Coal Char Particles
(1988) / Dahdah T F

Techniques to Quantify Vascular Ingrowth in
Soft Tissue Biomaterials
(1988) / Estridge T D

The Design and Modeling of a Compliant Manipulator
(1988) / Fancello A

Synthesis of Controllers for Prespecified
Performance in Linear Uncertain Systems
(1988) / Franchek M A

TEXAS A AND M UNIVERSITY
(continued)

A Shear Deformable, Doubly Curved Finite Element
for the Analysis of Laminated Composite Structures
(1988) / Fuehne J P

Discriminant Analysis of Heart Valve Sounds During
Respiratory Static Volume Conditions
(1988) / Hatzopoulos S D

An Analysis of the Progression of Damage in
Laminated Composites Under Transient Loading
(1988) / Havelka J J

A Comparison of Experimental and Theoretical Results
for Labyrinth Gas Seals with Honeycomb Stators
(1988) / Hawkins L A

A New, Efficient Computational Model for the
Prediction of Fluid Seal Flowfields
(1988) / Hibbs R I

Experimental Investigation of an Electroviscous
Damper for Rotordynamics Applications
(1988) / Hogue M S

The Effect of Respiratory Static Volumes on
Heart Valve Sounds
(1988) / Im J J

An Investigation Into the Cutting Forces Generated
for Sharp and Dull End Milling Tools Used in
Flexible Manufacturing Systems
(1988) / Jones A S

Rotordynamic Coefficients and Leakage
Flow of Parallel-Grooved Liquid Seals
(1988) / Kilgore J J

The Design, Construction, and Calibration of a
Heat Exchanger Calorimeter
(1988) / Lovelady J P

Design Fixation: An Inhibition of Creativity and
Quality in Engineering Design
(1988) / Mauldin R G

Enhanced Finite Element Analysis Using the
Patran-G Pre- and Post Processor Program
(1988) / Murry M L

A Robot Manipulator Calibration Procedure
with Experimental Verification
(1988) / Padavala S S

Development of a Room Air Conditioner Design Model
(1988) / Penson S B

The Effects of Expansion Devices on the Transient
Response Characteristics of the Air-Source Heat
Pump During the Reverse Cycle Defrost
(1988) / Peterson K T

Group Ignition and Combustion of a Cloud of
Char Particles Under Transient Conditions
(1988) / Ramalingam S C

The Assessment of Mixing/Solid Suspension in a Slab
Tank Due to Vibratory Agitation
(1988) / Ramsey C J

Identification of Kinematic Parameters Using
Several Models with Experimental Verification
(1988) / Rho J J

Cycle Simulation of Coal-Fueled Engines Utilizing
Low Heat Rejection Concepts
(1988) / Roth J M

An Analysis of Energy Use in Two Texas State Agencies
(1988) / Rothbauer K C

Group Pyrolysis, Ignition, and Combustion
of a Spherical Cloud of Coal Particles
(1988) / Ryan W R

Ultrasonic Rayleigh Wave Inspection
of Unit Train Rotary Couplers
(1988) / Salkowski C L

Heat Transfer and Flow Characteristics of Cooling
Channels in Turbine Blades
(1988) / Saxena A

The Effect of Alternate Defrost Strategies on the
Reverse-Cycle Defrost of an Air-Source Heat Pump
(1988) / Schliesing J S

Variable Structure Control of a Flexible Manipulator
(1988) / Shelburne K B

Geometry Dependence of Crack Growth Resistance
Curves in Thin Sheet Aluminum Alloys
(1988) / Stricklin L L

Study of the Simultaneous Heat and Mass Transfer in
Two-Dimensional Porous Media
(1988) / Suh Y B

Determination of Young's Modulus, Shear Modulus and
Mechanical Damping as a Function of Temperature and
Microstructure for Uranium-2-Wt Percent Molybdenum
Using the Pucot
(1988) / Varughese J V

Measurement of the Young's Modulus and Internal
Friction of Single Crystal and Polycrystalline
Copper, and Copper-Graphite Composites as a
Function of Temperature and Orientation
(1988) / Wickstrom S N

Experimental Investigation of Velocity Biasing in
Laser Doppler Anemometry
(1988) / Wiedner B G

Pump Impeller-Shroud Leakage Path Forces:
Their Effect on a Jeffcott Rotor
(1988) / Williams J P

Anger Camera Image Generation with Microcomputers
(1988) / Williams K M

An Experimental Validation of a Theoretical
Charpy-Fracture Toughness Correlation
(1988) / Zapata J E

Pressure Drop and Heat Transfer Distributions
in Three-Pass Rectangular Channels with Rib
Turbulators
(1988) / Zhang P

TEXAS, UNIVERSITY OF (ARLINGTON)

Implementation of Automatically Programmed
Tools System Processor (APT IV) on a
Microcomputer
(1988) / Abolfathi F

An Interactive Graphic Finite Element
Preprocessor Incorporating Automatic
Mesh Generation
(1988) / Bhatia R P

Design and Construction of a High Current
Electron Injector Utilizing Laser Back
Illuminated Photoemission
(1988) / Choe J H

TEXAS, UNIVERSITY OF (ARLINGTON)
(continued)

Effect of High Speed Machining on Surface
Integrity of AISI 4340 Steel
(1988) / Deshpande R S

Structural Optimization with ADS/NASOPT
(1988) / McConnell J C

Computer Integration of a Robotic Abrasive
Waterjet Cutting System
(1988) / Simmons D R

Equivalent Stabilization Controller for
Disturbance Rejection in the Presence of
Measurement Noise
(1988) / Sobhani M

The Constrained Lyapunov Problem Having a
Specific Output Feedback Control Structure
(1988) / Wilson T A

TEXAS, UNIVERSITY OF (AUSTIN)

Design of a Current Collection System for a
High Magnetic Field, Air Core Homopolar
Generator
(1987) / Aanstoos T A

Quantitative Characterization of Arterial
Tissue Thermal Damage
(1988) / Agah R

Empirical Models of Convection Drying Rate
Curves for Nonhygroscopic Industrial
Materials
(1988) / Armstrong M P

Finite Element Analysis of Heat and Mass
Transfer During Planar Freezing of a
Binary Solution
(1988) / Barta A B

Alterations to the Vasoactivity in the
Peripheral Cutaneous Microcirculation
Following Burn Insult
(1987) / Blake G K

Comparison of Spot, Segment and Band
Electrode Arrays in Impedance Cardiography
(1988) / Bue G C

Computer Image Processing Technique for
Analysis of Tire Contact Pressures
(1988) / Chan G P

Control of Air/Fuel Ratio During Rapid
Transients of a Computer-Controlled Engine
(1988) / Chiu J P

Development of a Scanner and Signal
Processing System for Laser Anemometry
(1988) / Ciancaeelli C R

In Vitro Evaluation of a Pulsatile Flow
Blood Pump for Pediatric Application
(1988) / Dullye K K

Dynamic Behavior of Surface Mounted Chip
Components During Solder Reflow
(1988) / Ellis J R

Thermal Analysis of Brush Gear for a High
Current, High Speed, Continuous Duty
Homopolar Generator
(1987) / Everett J E

A Hydrodynamic Study of Film Flow Over Rough
Surfaces with Application to Falling Film
Absorption
(1988) / Ewert M K

Frequency-Domain Digital Filtering
Techniques for the Removal of Power Line
Noise from the Electrocardiogram
(1988) / Ferdjallah M

Thermal Analysis of Surface Mounted Chip
Components During Reflow Soldering Using
Finite Element Analysis
(1988) / Goh R T

Dynamics of an Undersea Cable-Towed
Transducer Platform
(1988) / Gomez B J

Automatic Visual Inspection of Printed
Circuit Boards
(1988) / Gonzalez Olivares V M

The Effect of Inertial Coupling in the
Dynamics and Control of Flexible Robotic
Manipulators
(1988) / Graves P L

Dielectrically-Assisted Drying of a Porous
Nonhygroscopic Packed Bed
(1988) / Grolmes J L

Predicting the Wear of Specimens Having
Different Geometries for Constant Velocity
and Unlubricated Sliding
(1988) / Guerra R J

Design, Fabrication, and Testing of a
Railgun Bore Diameter Measuring Tool
(1988) / Harville M

Mechanical and Electrical Properties of
Glass, Carbon, and Glass-Carbon Fiber
Reinforced Epoxy
(1988) / Headifen G R

Elastic-Plastic Deformation of
Unidirectional Fiber Reinforced Composites
(1988) / Hsu F J

User-Interactive General Serial Manipulator
Analysis Program (MAP)
(1988) / Kang H

Commissioning the Rotor-Bearing Systems in
the Balcones 60-Megajoule Modular Homopolar
Power Supply at the Center for
Electromechanics
(1988) / Kitzmiller J

A Parametric Array for Use as an Ultrasonic
Proximity Sensor in Air
(1988) / Lee Y S

Real Time Eigen Vector Decomposition of
Single-Trial Visually Evoked Potentials
(1988) / Lichter P A

Modeling and Simulation of a Stewart
Platform Type Parallel Structure Robot
(1988) / Lim G K

Control of Flexible Robotic Manipulators
(1988) / Lin S H

Experimental Evaluation of Human Body Motion
Analysis Techniques
(1988) / Lombrozo P C

TEXAS, UNIVERSITY OF (AUSTIN)
(continued)

Development of an Empirical Model for the
Drying of Hygroscopic and Non-Hygroscopic
Materials
(1988) / Meiners A C

A Computer Model for Analysis of
Dielectrically Enhanced Single-Zone
Tunnel Dryers
(1988) / Melendez W

A Close Tolerance Positioning System Design
for the Manufacture of Dot Matrix Printheads
(1988) / Morgan S A

Study of Embedded Thermoelectric Coolers
for Cooling Integrated Circuits
(1988) / Petrakis G O

A Phenomenological Model for the Near Wall
Structure Due to a Turbulent Re-Attching Flow
Behind a Backward Facing Step
(1988) / Philips P K

Experimental Study of Heating Patterns and
Power Reflection of Microwave Hyperthermia
Antennas on Simulated Muscle Tissue
(1988) / Pressly M D

An LED Array, Microcomputer-Controlled
Visual Evoked Potential Stimulus Unit
(1988) / Samani D M

A Macintosh II Based Instrument for
Measuring Thermal Conductivity of Tissue
Using Self-Heated Thermistors
(1988) / Santos R R

An Optimum Orientation Investigation of
Grid-Interactive Single-Axis Tracking
Flat-Plate Photovoltaic Arrays for Austin,
Texas
(1988) / Sloan C M

Pharmacodynamic Differences Between Quabain,
Digoxin and Digitoxin in Unsedated Rabbits
(1988) / Sly M K

Digital Signal Processing Methods for
Time-Based Biomechanical Signals
(1988) / Smith T J

Numerical and Experimental Investigation of
the Influence of the Soret Effect on Natural
Convection in a Binary Liquid
(1988) / Srinivasan R

Analysis and Evaluation of the Performance
of a Robotic Assembly Line
(1988) / Stavros S P

ORS Synchronization of a Preload Responsive
Extracorporeal Heart Pump
(1988) / Stewart V M

Elastic-Plastic Deformation of Geopressured
Reservoir Rock
(1988) / Sun M M

Dynamic Modeling of Electromagnetic
Launchers
(1988) / Thakore A K

Experimental Study on the Effects of
Temperature on an Oxygen Generation System
(1988) / Walshak D B

EEG Correlates of Manual Control Tracking
(1988) / Wen T S

Feasibility of a Computerized Robotic Laser
Capable of Inducing Lesions of Predetermined
Size for Ophthalmic Treatment
(1988) / Yang Y

Three-Dimensional Computer Analysis and
Simulation of Human Body Joint Forces and
Torques with Application to the Upper
Extremity
(1988) / Ziegler J M

TEXAS, UNIVERSITY OF (EL PASO)

Fatigue Testing for Unidirectional Fiberglass
Composites
(1988) / Bhakta, S B

The University of Texas at El Paso
Anthropomorphic Hand
(1988) / Bhikha S R

Mathematical Modeling of Indirect/Direct
Heat Exchanger
(1988) / Chiang S C

Design of a Trigger Sprayers Packaging System
(1988) / Chuah K L

A Modified Inscribed Circles/Ellipses Method of
Approximating the Torsional Rigidity of Uniforms,
Non-Circular Cross Section Shafts
(1988) / Hurd S O

Interactive Computer Graphics System for
Kinematic Analysis
(1988) / Iyer S S

Applications of Extreme Value Statistics for
Maximum Wind Speed Forecast
(1988) / Joarder S H

A Customized Software Package for School
Information Management
(1988) / Kannambadi S

Performance of a Solar Pond Coupled Multistage
Flash Desalination System
(1988) / Kyathsandra J M

Analysis of a Trigger Sprayer Packaging System
(1988) / Lai M S

Vibration Frequencies of a Cantilever Wedge Beam
with Constraining Springs
(1988) / Lu Y

Design and Analysis of a Prototype Tactile Sensing
Robotic End Effector
(1988) / Magadaleno J L

Computer Aided Spur Gear Design
(1988) / Mahmoodullah S

A Robot-Vision System for Recognizing and
Manipulating Parts on a Conveyor
(1988) / Neo C K

The Theory of Kinematic Parameter Identification for
Robots with Closed-Chain Mechanisms
(1988) / Oen K T

Automatic Path Planning for a Robot Moving Among
Polyhedral Obstacles
(1988) / Pisupati R K

TEXAS, UNIVERSITY OF (EL PASO)
(continued)

Stress and Strain Distribution of Protective
A1203-HF02 Scales Protruding Into Iron
Chromium Aluminum Hydrogen Fluoride Alloys
(1988) / Ren J H

Torsion in Hollow Shafts Using Membrane Analogy
(1988) / Rhoads R P

Production and Quality Improvement of Gas Sockets
for Hospital Use
(1988) / Srinivasan V

Design and Analysis of a Prototype Multitool
End Effector
(1988) / Tjoa N T S

Computer Numerical Control Programming for
Complex Curve Machining
(1988) / Umashankar G R

TOLEDO, UNIVERSITY OF

Animated and Shaded Display of Podiatric
Pressure data
(1988) / Aryanto I

Analysis of Lubrication in Slider Bearing
with Melting Surfaces
(1988) / Atluri S

Numerical Investigation of the Unsteady
Heat Transfer from a Single Sphere
(1988) / Attar A M

Kinematics of an N-Celled Terrahedron
Tetrahedron Variable Geometry Truss
Manipulator
(1988) / Jain S

Modeling and Simulation of Servo-Controlled
Flexible Manipulators
(1988) / Konkar R R

An Electromyographic Study of Trunk Muscles
in Axial Rotation
(1988) / Rhee D

A Predictive Model of Float Heights on a
Negative Thrust Air Bearing
(1988) / Schnabel J P

Analysis and Optimization of Truncated Scarf
Nozzles Subject to External Flow Conditions
(1988) / Shyne R J

A Definition Study of the On-Orbit Assembly
Operations for the Outboard Photovoltaic
Power Modules for Space Station Freedom
(1988) / Sours T J

Cavitation in Moderately to Heavily Loaded
Noncircular Journal Bearings
(1988) / Vaidyanathan K

TULSA, UNIVERSITY OF

Computational Modeling of Single-Phase Flow
Through Pipeline Fittings with Application
to Erosion-Corrosion Damage
(1988) / Bussman W R

The Design, Construction, and Testing of a
Liquid Impingement Apparatus and a Study of
Metal Surfaces Eroded by Liquid Impingement
(1988) / Camacho R A

An Efficient Three-Dimensional Model for
Hydraulic Fracturing and Proppant Transport
(1987) / Shah G H

TUSKEGEE INSTITUTE

Effect of Moisture on Thermal Conductivity
of Insulating Materials
(1987) / Mandadi V K

Thermal Effects of Coal Dissolution
(1987) / Muchengeti P S

Residual stress Due to Machining in Steel
(AISI-4130)
(1987) / Zadoo D P

U.S. NAVAL POSTGRADUATE SCHOOL

Heat Transfer Modeling of Jet Vane Thrust
Vector Control (TVC) Systems
(1987) / Dulke M F

Modeling of Jet Vane Heat-Transfer
Characteristics and Simulation of Thermal
Response
(1988) / Hatzenbuehler M A

Experimental Investigation of the Effect of
Curvature on Heat Transfer in a Curved
Rectangular Channel of High Aspect Ratio
(1987) / Hawk J R

The Analysis of Thermal Residual Stress for
Metal Matrix Composite with Aluminum/Silicon
Carbide Particles
(1988) / Hur S H

Indirect Measurement of Local Condensing
Heat-Trannsfer Coefficient Around Horizontal
Finned Tubes
(1987) / Lester D J

A Heat Transfer Model for a Heated Helium
Airship
(1987) / Rapert R M

Comparative Controller Design for a Marine
Gas Turbine Propulsion Plant
(1988) / Stammetti V A

Two-Dimensional Computation of Heat Transfer
in Fusion Welding
(1988) / Walker L R

Condensation Heat-Transfer Measurements of
Refrigerants on Externally Enhanced Tubes
(1987) / Zebrowski D S

U.S.A.F. INSTITUTE OF TECHNOLOGY

Variable Wall Temperature Effects on
Multicellular Natural Convection in a
Horizontal Annulus
(1988) / Bennett D L

Damage Initiation in Two-Dimensional Woven
Carbon-Carbon Composites
(1988) / Dendis J C

Damage Initiation in the Short Beam Shear
Test of Composite Materials
(1988) / Dupont T D

Characterization of Delamination in Advanced
Composite Materials Under Mode III Loading
Conditions
(1988) / Lingg C L

VIRGINIA, UNIVERSITY OF
(continued)

Preliminary Design of an Intelligent Word
Processor for Use in ERICA (Eye-Gaze
Response in Computer Aid)
(1988) / Frey L A

Feasibility Study of a CAM/CAD System
(1988) / Fucci P T

Effects of Restricted Breathing on the
Pulmonary Circulation
(1988) / Hunter M A

Reverse Engineering in Design and
Manufacture of Mechanical Components Within
a CAD/CAM Environment
(1988) / Kalra V

An Expert System for Vibration Fault
Diagnosis of Turbomachinery
(1988) / Nahar R

Spatial Musculoskeletal Model and Muscular
Load Sharing for Wheelchair Propulsion
(1988) / Nayar N

Finite Element Flow Modelling with
Application to a Torque Converter and a
Centrifugal Pump
(1988) / O'Shea K M

Presynaptic Action of a Calcium Channel
Activator at the Neuromuscular Junction
(1988) / Pancrazio J J

A Research Prototype for the Integration
of CAD/CAM with Dimensional Measuring
Equipment
(1988) / Shah B H

A Minimax Indirect Boundary Element Method
(1988) / Srinivasan S

An Investigation of Parameters Affecting
Thermal Camouflage Screens Under
Simulated Outdoor Conditions
(1988) / Stott J C

Numerical Analysis of Unbalanced Forces on
an Impeller of a Centrifugal Pump Based on
Potential Theory
(1988) / Trevisan P M

Direct Three Dimensional Display of Magnetic
Resonance Images
(1988) / Watterson W K

Effect of Laparotomy and Manipulation of
the Liver on the Hepatic Circulation
(1988) / Wong J K

WASHINGTON STATE UNIVERSITY

Characterization of Acoustic Emission
Signals Generated by Degradation of
Advanced Composite Materials
(1988) / Aggarwal S M

Temperature Measurement in an End-Milling
Operation
(1988) / Arshad R

Preparation and Characterization of
Hydrolyzed Zirconia Film on Basalt Fiber
by the Sol-Gel Process
(1988) / Jung T

An Investigation of Hydrocarbon Absorption
Modeled by Molecular Structure
(1988) / Slinn A M

Heat and Mass Transfer in a Dual-Latent
Thermal Energy Storage System
(1988) / Weislogel M M

Study of the Dynamics of Flow Around a
Circular Cylinder at Subcritical Reynolds
Numbers
(1988) / Yokuda S

WASHINGTON UNIVERSITY

Three Dimensional Modeling of Human Wrist
Motion
(1987) / Bresina S J

Numerical Analysis of Stress Singularities
in Composite Materials
(1987) / Schiermeier J E

Calcium[2+] Activated Potassium[+] Channels
in B Cells: Selectivity, Gating and
Function
(1987) / Tabcharani J A

Mixed-Mode Interlaminar Fracture of
Graphite/Epoxy Composites
(1987) / White S R

WASHINGTON, UNIVERSITY OF

Swelling and Deswelling in Thermally
Reversible Hydrogels: Applications in
Delivery and Removal of Biologically Active
Agents as a Function of Temperature
(1986) / Afrassiabi A

Ultrasonic Assessment of Cardiac Ejection
Dynamics
(1987) / Albrecht E H

The Effect of Bond Imperfrections on the
Strength and Dynamic Response of Laminated
Beams
(1987) / Aliabadi B S

Assembly of Electrical Harness for Passenger
Service Unit
(1987) / Bechtolsheim F V

Microprocessor Control of an Interted
Pendulum
(1987) / Bjornsson H

Evaluation of Pulse Oximeters Using
Theoretical Models and Experimental Studies
(1987) / Brown D P

A High-Level Computer Graphics
Implementation of Three-Dimensional
B-Spline Surface Generation
(1987) / Bryant D M

The Design, Development and Preliminary
Experimental Investigation of a Coal Flow
Reactor and Fluidized Bed Feed System
(1987) / Burnett R A

Computer Analysis of a New Structural Model
of Muscle
(1987) / Chian D S

An Instrument for the Measurement of Shim
Gaps
(1987) / Cram G S

WASHINGTON, UNIVERSITY OF
(continued)

Design Considerations for Automated Assembly
of Small Disk Drives
(1987) / Dare M A

The Design, Construction, and Evaluation of
a Snow Ski Pressure Distribution Measuring
Instrument
(1987) / De Rocco A

The Design, Construction and Demonstration
of an Experimental Wood-Waste Combustion
Facility
(1987) / Ewing J B

An Experimental Investigation of an
Aero-valved (Valveless) Pulse Combustor
Using the Laser Rayleigh Scattering
Technique
(1987) / Fair M J

Two-Dimensional Vector Doppler Imaging
(1987) / FitzGerald M P

A Survey of Producibility Concepts and the
Integration of Producibility Methodology
with CAD Techniques
(1988) / Gaevert R S

Laser Doppler Monitoring of the Cutaneous
Microcirculation in Acute Burn Wounds
(1987) / Green M O

Application and Evaluation of an Imaging
Technique for 3-Dimensional Cochlear
Reconstruction
(1988) / Haq N

Modeling of Abrasive Waterjet Machining
(1987) / Hatangadi R B

High Level Computer Graphics Enhanced
Optimization Techniques
(1987) / Hergenroeder D M

Stress Distributions in Manifold Blocks
Containing Circular High Pressure Fluid
Lines
(1987) / Hippe D

Packed Bed Combustion of Wood Cylinders
(1987) / Huang C

An Investigation of the Elastic-Plastic
Behavior of Explosively Actuated Valves
(1987) / Jones B K

Surface Characterization and Blood
Interaction of Plasma Modified Vascular
Grafts
(1987) / Kiaei D

Strain Measurements in Composites Subjected
to Compressive Loading Using Moire'
Interferometry
(1987) / Klein R J

An Investigation of Brachial Plexus
Somatosensory Evoked Potentials Using an
Orthogonal Three Dimensional Electrode Array
(1987) / Law S K

Sperm Motility Measurement by Two-Angle
Detection of Laser Light Scattering:
Comparison of Spermatozoa from Normal and
Infertile Men
(1987) / Lim K

An Interferometric Analysis of a Peridontal
Model
(1987) / Loebel N G

The Development and Characterization of a
Glucose Sensing Polymer
(1986) / Miller S R

Myocardial Serotonin Exchange: Negligible
Uptake by Capillary Endothelium
(1987) / Moffett T C

A Clinical Evaluation of the Imaging Burn
Depth Indicator
(1987) / Moore M K

Testing and Calibration Refinement of the
Fringing Capacitance Profilometer
(1987) / Murphy M D

An Expert System Based Parametric Geometry
Generation Method
(1987) / Nakasone J K

Monte-Carol Simulation on Bidirectional
Reflectance
(1987) / Park H K

Measurementt and Modelling of Solar
Irradiance and Luminous Efficacy for Seattle
(1987) / Piguet J L

An Experimental Investigation of Radiative
Heat Transfer from Heptane Pool Fires to
the Fuel Surface
(1988) / Rood R C

A Computer-Assisted Morphometric Method for
Describing the Differential Geometry of the
Cochlear Spiral in Three-Dimensional Space
(1987) / Santos P M

Modal Density Measurement of a Uniform Flat
Panel and a Periodically Stiffened Cylinder
(1988) / Serati P

Immobilization of L-Asparaginase Via
Pre-Irradiation Grafting with Acrylamide
and N-Isopropyl Acrylamide
(1986) / Shoemaker S G

Configuration Dependent Robot Control
(1987) / Tschiesche T L

A Simplified Model for Thermal Radiation
Characteristics of Hypersonic Flight
(1987) / Tu P C

Forced Oscillations of a Nonlinear Single
Degree of Freedom System with Active Control
(1987) / Van Dyke D J

An Experimental Investigation of a Three
Dimensional Shock Wave-Turbulent Boundary
Layer Interaction in Supersonic Flow
(1987) / Welch J W

Experimental Studies of the Hydrodynamic
Characteistics of Four Bluff Symmetrical
Fairing Sections
(1987) / Wyszkowski M A

Robot Control with End Point Sensing
(1987) / Yang W M

WEST VIRGINIA UNIVERSITY

A Sinusoidal Low-Heat-Rejection Engine Model
(1988) / Churchill R A

WEST VIRGINIA UNIVERSITY
(continued)

Mechanical Behavior of Structural Ceramics
Failure Analysis and Design Assessment of
an Alumino-Silicate Liquid Metal Bath
Refractory Block Using the Finite Element
Method
(1988) / Crocco J

Helically Wound Composites for Performance
Enhancement of Flexible Linkages
(1988) / D'Acquisto J G

A Method for Predicting Cylinder Pressure
in the Stiller-Smith of Other Sinusoidal
Type Engines
(1988) / George A C

Vibration of Beams with Longitudinal Motion
Relative to Supports by Hip-Version Finite
Method and Bilinear Formulation
(1988) / Gokhale M R

Experimental Measurement of Surface Strains
and Deflections of a High Temperature
Split-Body Ball Valve
(1988) / Leach T E

Development of a Method for Nondestructive
Stress Wave Testing of Gluelines in Edge
Glued Wood Panels
(1988) / Mallory J C

Technique to Detect Vibration Patterns of
Excised Lungs
(1988) / Pathak M A

Isothermal Turbulent Swirling Flow Inside a
Model Combustor
(1988) / Singh S

Drag Force Measurements and Analysis for
Design Applications of Plastic
Tape-in-Channel Motion Transfer Systems
(1988) / Thompson S A

Composition and Structural Characteristics
of Eastern and Western Oil Shale
(1988) / Zhang G Q

WICHITA STATE UNIVERSITY

Determination of the Stress Concentration
Factors and the Stress Intensity Around
Notches in Graphite/Epoxy Laminate-A Finite
Element Approach
(1987) / Behzadpour F

A Study of Four-Revolute Spherical
Mechanisms
(1987) / Chakaroun S A

Simulation of Paper Machine Headboxes
(1987) / Chang T S

A Solar Heating System Performance
Prediction Method
(1987) / Christians J R

A Performance Benefit Analysis for a Tracing
Flat-Plate Solar Collector
(1987) / Mayhew D C

Effect of Environment of Static and Fatigue
of Graphite/Epoxy Composite Materials Having
Negative Poisson's Ratio
(1987) / Ziary H Z

WINDSOR, UNIVERSITY OF

Monitoring Drill Condition for Different

(1988) / Chahine J

Energy: A Computer Program for Calculating
Energy Consumption in Residential Buildings
(1988) / Grewal N

A Computer Vision System for Biological
Specimen Image Analysis
(1988) / Mantripragada S

Non-Contact Surface Flow Inspection: A
Study of Retroflection for Qualitative
and Quantitative Evaluation
(1988) / Montrose C

Development of a Residential Building
Energy Analysis Program
(1988) / Rani A S

WORCHESTER POLYTECHNIC INSTITUTE

Optimum Design of an Epicylic Gear
Mechanism
(1986) / Belon J B

Robotics in Electronic Components Assembly
(1987) / Cheng E D

Deflection and Stress Analyses of a
Butterfly Valve Using the Finite Element
Method
(1987) / Goksel B M

The Flexible Feeder Error Detection and
Recovery Hardware and software for a
Robotic PC Board Assembly System
(1986) / Lee G

An In-Vitro Durability Analysis of the
Cellular Vascular Matrix Graft
(1987) / Lipson W E

Effects of a Pulsatile Flow on Wall Shear
in a Transverse Clip Stenosis
(1987) / Nevin M

Development of a Noninvasive Method for
Determining Arterial Compliance
(1987) / Nugent J A

Application of Boundary Element Method to
Two Dimensional Steady State Potential Flow
(1986) / Tariverdian R

Automated Cardiac Mapping System for the
Localization of Ventricular Arrhythmias
(1987) / Vassallo V R

Experimental and Theoretical Modeling of the
Dynamic Behavior of a Cam Follower Arm
(1987) / Wilbur J J

The Transfer Function of the Disposable
Blood Pressure Transducer Catheter System
(1987) / Yang W

WRIGHT STATE UNIVERSITY

Agile Fighter Aircraft Simulation
(1987) / Anderson J A

Performance Prediction and Design
Optimization of a Radio Controlled Aircraft
(1987) / Rangarajan J

Evaluation and Validation of a
Phenomenological Model for Cardiac Muscle
(1987) / Shah A

WYOMING, UNIVERSITY OF

Experimental Verification of C&T Scan
Measurements of the Outer Ear Geometry of
a Cadaver Using Ear Canal Molds
(1989) / Nelson D K

Design and Fabrication of a Snow Machine
A-Frame
(1989) / Wright W B

CASE WESTERN RESERVE UNIVERSITY

Effect of Ceramic Filters on the Dross and
Fatigue Properties of Ductile Iron
(1988) / Kim H S

Study on Microstructure and Fatigue
Properties of an Oxide-Dispersion
Strengthened Nickel-Base Superalloy MA754
(1988) / Xue H

COLORADO SCHOOL OF MINES

Investigation of the Natural Hydrophobicity
of Native Gold Grains and Factors
Influencing it
(1988) / Adsoy B S

CONNECTICUT, UNIVERSITY OF

Homogenization Treatment of 3004 Aluminum
Alloy Strip
(1988) / Cevallos E E

The Effect of Composition on Marker Movement
in Ternary Alloys
(1988) / Son Y H

Laser Induced Chemical Vapor Deposition in
Synthesis of Germanium-Selenium Thin Films
(1988) / Xiao T

Near-Threshold Fatigue Crack Growth Behavior
of 304 Stainless Steel
(1988) / Zagrany W B

GEORGIA INSTITUTE OF TECHNOLOGY

Mechanisms of Creep Crack Growth in a
Copper-1 Weight Percent Antimony Alloy
(1988) / Staley J T

MARQUETTE UNIVERSITY

The Effect of Heat Treatment on the Cutting
Efficiency of Endodontic Files in Linear
Motion
(1988) / Costas-Latoni J

Precipitation Phenomena in Lead Alloys
(1988) / Crain M

Effect of Varying the Filler Content on
Diametral Tensile Strength and Fracture
Toughness of a Dental Composite Resin
(1988) / Johnson W

MASSACHUSETTS INSTITUTE OF TECHNOLOGY

Strengthening Mechanisms of Tungsten Powder
Reinforced Uranium
(1988) / Krawizcki M A

Welding Flux Binders Produced by Sol-Gel
Processing
(1988) / Kryscnski J D

Strength and Fracture Behavior of SiC Coated
Graphite Fibers and Characterization of
Interface in C Fiber/Al Matrix Composites
(1988) / Li Q

Behaviour of Oxygen and Sulphur in Mattes
Containing Nickel
(1988) / Nallicheri N V

Reduction of Metallic Iron from Fe-S-O
Melts by Gas Injection
(1988) / Tamura M

MISSOURI, UNIVERSITY OF (ROLLA)

Influence of Prior Deformation on Structure
and Properties of Heat Treated Aluminum
Casting Alloy
(1988) / Hsia C

Incorporation of Solid Charge Into a
Molten Metal Bath in a Plasma Furnace
(1988) / Rhee W H

Effects of 0.5 Weight Percent Boron on
Cobalt Aluminide
(1988) / Slattery K T

NEW MEXICO INSTITUTE OF MINING & TECH.

Effect of Dynamic Compaction on Sintering
and Properties of Aluminum Oxide-Zirconium
Oxide Ceramics
(1987) / Bengisu M

Microstructural Characterization of Impact
Welded Copper/Copper Interfaces
(1987) / Borden M J

Explosively Formed Piano-Wire-Reinforced
Aluminum Sheet
(1988) / Chang C K

Orientation Dependence of Room Temperature
Cyclic Deformation Behavior in Nickel
Germanium Single Crystals
(1987) / Chu J P

Explosive Consolidation of Titanium Alloy
Powders
(1987) / Coker H L

Biodegradation of Spodumene by Aspergillus
Niger
(1988) / Ilgar E

Development of Leadframe Alloys by
Precipitation Hardening
(1987) / Jeng N

Extraction of Gallium and Germanium from
a Complex Ore
(1988) / Jiang H

Thermal Cycling of Alumina/Magnesium Alloy
Composite
(1987) / Lee C S

The Effect of Microballoon Size on
Detonation Behavior of Emulsion Explosives
(1987) / Lcc J

Free Flight Shock Compaction of Copper
Powders
(1987) / Oh K H

Shock-Wave Consolidation of Aluminum
Lithium Alloy Powders in Cylindrical and
Plate Configurations
(1987) / Yu L H

Extraction of Gold and Silver from
Manganiferous-Silver Ores by Thiourea
Leaching
(1987) / Zegarra C R

Kinetics of Zinc Extraction from a Sulfide
Bearing Flotation Concentrate with Mercury
Chloride/Iron Chloride Solutions
(1988) / Zhang S

PENNSYLVANIA STATE UNIVERSITY

Characterization of Nickel-Chromium-Silicon
Nitride Alloys
(1987) / Bundy K A

In Situ Internal Oxidation in the Copper
Aluminum System
(1987) / Buss C

Flow and Fracture of Titanium Aluminide
Alloys
(1988) / Lukasak D A

Mass Transfer Studies of Laser-Welded
Aluminum Alloys 5083 and 7039
(1988) / Marsico T A

The Diffusivity of Hydrogen in Nickel at
25 C
(1987) / Moyle D R

Recovery of Tin from Slag
(1988) / Patankar A

Evaluation of Transport Mechanisms Occurring
in the Hot Corrosion Process
(1988) / Patton J S

Structural Investigation of Two
Aluminum-Copper-Lithium-Related Alloys
(1987) / Yuan D W

PITTSBURGH, UNIVERSITY OF

The Deformation and Annealing Textures of
Aluminum Single Crystals Deformed in Plane
Strain Compression
(1989) / Butler J F

Corrosion of Alloys in Sulfur Dioxide/
Dioxide/Sulfur Nitrate Gas Mixtures
(1989) / Cerchiara R R

Influence of Process Parameters in Weld
Deposit Quality in High Deposition Rate
Welding
(1989) / Hatzilabrou L

Multiple Phase Behavior of Carbon Dioxide
and Hydrocarbon Systems Including the
Effects of Water
(1989) / Mangone D J

An Alternative to the Surface Hardening of
Low Carbon Microalloyed Forging Steels
(1989) / Palmiere E J

The Erosion-Corrosion Behavior of Nickel in
Mixed Oxidant Atmospheres
(1989) / Rishel D M

Microchemical Investigations of Nickel
Aluminum Alloys Using Field Ion Microscopy
and Atom Probe Analysis
(1989) / Sieloff D D

QUEENS UNIVERSITY

The Effects of Minor Alloying Elements on
the Mechanical Properties of Austenitic
Maganese (Hadfield's) Steel
(1988) / Bradley D J L

Characterization of the Electrical Behaviour
and Composition of Partially Stabilized
Cubic Zirconia for Use in Furnace Elements
(1988) / Childerhose T L

STEVENS INSTITUTE OF TECHNOLOGY

Experimental Study of Some Fundamental
Aspects of the Full-Mold Casting of Grey
Iron
(1987) / Behi M

TEXAS, UNIVERSITY OF (EL PASO)

The Effect of Prior Cold Work Amount and Annealing
Temperature on Subgrain Growth in OFHC Copper
During Static Recovery
(1988) / Chang C M R

Effects of Surface Topography on the Transient
Electrochemical Response of Iron Chromium Nickel
Alloys After Depassivation
(1988) / Kang C

Physical and Mechanical Properties of an Advanced
Ceramic Composite: TIB2-ZR02
(1988) / Shiao H C

U.S.A.F. INSTITUTE OF TECHNOLOGY

Energy Dependence of the Quenching Cross
Section for the Reaction of Metastable Argon
and Water
(1988) / Novicki S W

UTAH, UNIVERSITY OF

Positron Annihilation Study of Elastic
Stresses in Metals
(1988) / Peterson D T

AKRON, UNIVERSITY OF

Numerical Studies of Modified Eden Cluster
of n=0 Axis Model
(1989) / Zhu Y

ARKANSAS, UNIVERSITY OF

Neutron and Gamma Spectral Measurements at
the Sefor Facility Using Active Detectors
(1988) / Hamblen R A

CLARKSON COLLEGE OF TECHNOLOGY

Numerical Simulation of the Onset of
Transition of a Non-Parallel Flow
(1988) / Lin L

COLORADO STATE UNIVERSITY

Transmittance of Solar Radiation Determined by
the Multiple Field of View Radiometer
(1988) / Combs C L

Observational Evaluation of Snow Cover Effects
on the Generation and Modification of
Mesoscale Circulations
(1988) / Cramer J

Cloud and Convection Frequencies Relative to
Small-Scale Geographic Features
(1988) / Gibson H M

Modeling and Observations of the Atmospheric
Response to Squall Lines
(1988) / Hertenstein R F

Microwave Remote Sensing of Cloud Liquid Water
and Surface Emittance Over Land Regions
(1988) / Jones A S

Application of the Cimms Mesoscale Model to
Operational Short-Range Forecasting
(1988) / McCoy M C

A Global Analysis and Correlation of Nimbus 7
Cloud and Longwave Radiation Data
(1988) / Ramirez D R

Instability of Horizontally and Vertically
Sheared Parallel Flow
(1988) / Ringerud M A

Surface Pressure Features Associated with a
Midlatitude Mesoscale Convective System in
O.K. Pre-Storm
(1988) / Stumpf G J

Observed Structure of a Mesovortex in the Trailing
Anvil of a Multi-Cellular Convective Complex
(1988) / Verlinde J

FLORIDA STATE UNIVERSITY

A Comparison of a Semi-Lagrangian 3-D Trajectory
Model with Hand Drawn Trajectories in Determining
the Source Region for Saharan Dust Found at an
Air Quality Testing Site in Hollywood, Florida,
on July 17, 1983
(1988) / Anderson D W

Tropical Cyclone Statistics for the State of
Florida, 1886-1986
(1988) / Artusa A M

Relations Between VAS Soundings and Their First
Guess Input
(1988) / Beven J L

Turbulent Variances and Covariances in the Marine
Atmospheric Boundary Layer Over the Fasinex Front
(1988) / Crescenti G H

The Correlation Between Rainfall and Normalized
Difference Vegetative Index in Eastern Africa
(1987) / Davenport M L

Solvent Extraction of Actinide Ions
with Phosphonated Jojoba Oils
(1987) / Gregory C M

An Objective Comparison of Vertical Motions from
the Kinematic Method and Three Forms of the
Omega Equation
(1988) / Hevrdeys G E

An Investigation Into Land Surface Forcing in
Sahelian West Africa Via Climatonomy Modeling
(1988) / Lare A R

The Relationship Between the Normalized Difference
Vegetation Index and Rainfall in West Africa
(1988) / Malo A R

A Simulation and Diagnostic Study of Water Vapor
Image Dry Bands
(1988) / Muller B M

A Case Study of Thunderstorm Development on
17 June 1986 During Cohmex
(1988) / Schudalla R L

Total Southern Hemisphere Ozone, 1978-1987:
The Quasi-Biennial Oscillation
(1988) / Schuster G S

Differential Surface Forcing of the Infrared
Cooling Profile Over the Qinghai-Xizang Plateau
(1988) / Shi L

Antecedents of a Composite Super-Typhoon
(1987) / Waters K

The Wind Driven Seasonal Circulation in the
Southern Tropical Indian Ocean
(1988) / Woodberry K E

FLORIDA, UNIVERSITY OF

The Annual Precipitation of Western Peru and
the Influence of the El-Nino-Southern
Oscillation Phenomenon
(1988) / Tapley T

HAWAII, UNIVERSITY OF

Chemistry of Cyanophytes Belonging to the
Family Stigonemataceae. Part One:
Photooxidation Studies of Hapalindole A.
Part Two: New Alkaloids from Fischerella
Muscicola
(1988) / Park A

Marine Aerosol: Their Measurement and
Influence on Cloud Base Properties
(1988) / Porter J N

An Ovservational Study of Intra-Seasonal
Low-Frequency Motions in the Equatorial
Region
(1988) / Rui H

INDIANA STATE UNIVERSITY

Surface and Upper Air Relationships in
Winter in the Central United States
(1988) / Gustin W

IOWA STATE UNIVERSITY

A Study for the Maintenance and Annual
Variation of Subtropical Jet Streams with
the NCAR Community Climate Models: Effects
of Mountain, Cumulus Convection and
Parameterization Scheme of Cumulus
Convection
(1988) / Tzeng R Y

KANSAS, UNIVERSITY OF

Spatial and Temporal Distributions of
Damaging Hail-Days in Kansas
(1988) / Schlager P H

MASSACHUSETTS INSTITUTE OF TECHNOLOGY

Momentum Transport Due to a Squall Line
System Over the Tropical Oceans
(1988) / Chang C Y

Comparison of Model Solar and Infrared
Fluxes with Aircraft-Based Radiometer
Measurements
(1988) / Stern I R

MISSOURI, UNIVERSITY OF (COLUMBIA)

Vortex Motion in Mesoscale Convective
Complexes as Revealed by Time Lapse Radar
(1988) / Akyuz F A

Relationship Between the Intertropical Convergence
Zone and Satellite Outgoing Longwave Radiation in
the Sahelian Countries
(1988) / Callis S L

Tradewind Easterlies as a Source of Extreme
Potential-Convective Instability for Mid-
Latitude Severe Storm Generation
(1988) / Hagemeyer B C

NORTH CAROLINA STATE UNIVERSITY

Precipitation Distributions Associated with
Cyclones Originating Over the Gulf of Mexico
and Surrounding Coastal Regions
(1988) / Knapp D I

Investigation on Cloud Chemistry and Acidic
Deposition at Mount Mitchell, North
Carolina, Using a Cloud Deposition Model
(1988) / Lin N H

Otolith Increment Growth Patterns in
Atlantic Menhaden (Brevoortia Tyrannus)
Larvae and Their Relation to Meteorological
and Oceanographic Variables
(1988) / Maillet G L

NORTHERN ILLINOIS UNIVERSITY

A Theoretical Analysis of the Effect of
Latitude and Season on the Nocturnal
Low-Level Jet
(1988) / Logsdon J A

OKLAHOMA, UNIVERSITY OF

An Investigation Into the Gravity Current
Aspects of a Cold Air Outbreak Using
Variational Analysis Techniques
(1988) / Allen S B

Doppler-Radar Analysis of a Dissipating
Tropical Cyclone Over Land: Hurricane
Alicia (1983) in Oklahoma
(1988) / Hazen D S

Retrieval of Horizontal Gradients of Virtual
Temperature Using Winds from the Oklahoma
Kansas Pre-Storm Profilers
(1988) / Hermes L G

Spectral and Energy Budget Analysis of the
Phoenix II Aircraft Data
(1988) / Lin J

Lightning Ground Flash Rates Relative to
Mesocyclone Evolution on 8 May, 1986
(1988) / Nielsen K E

The Specification of Precipitation Type in
Oklahoma Winter Storms
(1988) / Schichtel M L

Continental Polar Air Stream Cyclogenesis
(1988) / Smart J R

Observations of a Splitting, Low
Precipitation Severe Storm and Its
Evolution Into a Supercell
(1988) / Woodall G R

OREGON STATE UNIVERSITY

A Case Study of the May 28, 1985 Mesoscale
Convective System Observed During the O.K.
PRE-STORM Project
(1987) / Bennett S P

Simulation of Tropical Pacific Circulation Anomalies
with Linear Atmosphere and Ocean Models
(1987) / Dixit S

A Numerical Study of Sulfur and Nitrogen Scavenging
in a Mesoscale Rainband Observed in the Gale
Experiment
(1987) / Edwalds E

An Investigation of the Relation Between Total
Ozone and Synoptic Tropospheric Disturbances
(1987) / Pirlet A J

Subseasonal Variability in the Southern Hemisphere
as Simulated by a Two-Level Atmospheric General
Circulation Model
(1987) / Tomas R A

PENNSYLVANIA STATE UNIVERSITY

Waves on an Equatorial Beta-Plane in Shear
Flow
(1987) / Aberson S D

Analysis of Systematic Errors Associated
with Cold Season in the LFM-II and Spectral
Models
(1988) / Bankert R L

Analysis of Temperature and Velocity
Microturbulence Parameters from Aircraft
Data and Relationship to Atmospheric
Refractive Index Structure
(1988) / Beecher E A

Kinematic Quantities Derived from a Triangle
of VHF Doppler Wind Profiles
(1987) / Carlson C A

Composite Meteorological Fields of
Developing and Non-Developing Cold-Air
Cyclones Over the North Atlantic
(1987) / De Groodt K A

A Numerical Study of Planetary Boundary
Layer Modification in an Urban Area
(1988) / Donall E G

PENNSYLVANIA STATE UNIVERSITY
(continued)

A Micrometeorological Study of Sensible Heat
and Vapor Transfer Processes and Soil
Moisture Over a Wheat Field
(1987) / Elsalem Rabadi J K

The Influence of Soil Moisture on the Nested
Grid Model Forecasts
(1988) / Emlaw M T

The Role of a Lid in the 31 May 1985 Tornado
Outbreak
(1988) / Farrell R J

Below-Cloud Precipitation Scavenging of
Nitric Acid Vapor at a Rural Central
Pennsylvania Site
(1987) / Gibbons T H

Northwest Flow Aloft and Strong Convective
Events Within the Mid-Atlantic States
(1987) / Giordano L

Development and Testing of an Interactive
Procedure for the Improvement of Objective
Analyses
(1987) / Hamill T M

A Statistical and Structural Analysis of
Appalachian Cold-Air Damming of Various
Intensities
(1987) / Kieser E A

Kinematic Diagnoses of Frontal Structure and
Circulations Derived from Two- and Three
Station VHF Doppler Wind Profiler Networks
(1987) / Knowlton L W

Development and Evaluation of an Automated
Toxic Corridor Emergency Response System
(1987) / Messier T A

An Investigation of Aerosol Generation in
the Marine Planetary Boundary Layer
(1987) / Miller M A

A Numerical Study of the Presidents' Day
Storm of 18-19 February 1979
(1988) / Nappi A J

Wind Profiler-Derived Temperature Gradients
and Advections
(1987) / Neiman P J

The Global Atmospheric Responses to Isolated
Tropical Forcings in a National
Meteorological Center Numerical Weather
Prediction Model
(1987) / Samel A N

A Simplified MOS Technique for Specialized
Forecasting in Regions of Complex Terrain
(1987) / Shaw R M

On the Movement and Propagation of Mesoscale
Convective Systems
(1988) / Sims D L

Some Applications of 50 MHz Wind Profiler
Data: Detailed Observations of the Jet
Stream
(1987) / Syrett W J

The Structure and Evolution of a Cold Front
and Its Associated Convective System During
the 26-27 June 1985 Pre-Storm Experiment
(1987) / Trier S B

On the Improvement of Very Short-Range
Numerical Quantitative Precipitation
Forecasts: A Case Study
(1987) / Wang W

PURDUE UNIVERSITY

Precipitation Estimates in the Tropics by
the Global Heat and Moisture Budgets
Dyuring SOP-I, FGGE
(1988) / Pedigo C B

Analysis of 3DN5PH Over the North Atlantic
Ocean
(1988) / Tian L

QUEENS UNIVERSITY

Effects of Stress on Reversible Permeability
in Steel
(1988) / Makar J M

REGINA, UNIVERSITY OF

Modeling the Temperature Climate of Marmot
Creek Watershed, Alberta
(1988) / Cote M D

A Study of Possible Climatic Variation in
the Regina Area, 1953-1983: Official Data
and Newspaper Coverage
(1988) / Marsh S J

SAINT LOUIS UNIVERSITY

Role of Frontogenetical Forcing and
Conditional Symmetric Instability During
the Midwest Snowstorm of 30-31, January 1982
(1988) / Blakley P D

Intercomparison of the Atmospheric Heat
Sources Over the Bolivian and Tibetan
Plateaus in Their Respective Summers
(1988) / Erdogan S

Statistically Chosen Active Wave Lengths
Associated with Convective Events Analyzed
Through Surface Band Pass Filtering
(1988) / Hawkins D R

Case Studies of Large-Scale-Mesoscale
Interactions in the Severe Convective
Environment
(1988) / Kallas A P

Contrail Cirrus Effect on Surface
Temperature
(1988) / Karakoy S

Impact of an Increase in Atmospheric Methane
on the Temperature Field and Subsequent
Climatic Effects
(1988) / Morrison J S

Interactions of the Cold Conveyor Belt with
Jet Streaks
(1988) / Nichols W D

Sulfur Dioxide Dispersion at the Paradise
Electrical Generating Plant
(1988) / Otles Z

Quantitative Effects of Multiple Scattering
on Radiance/Transmittance in a Vertically
Inhomogeneous Hazy Atmosphere
(1988) / Pope L M

SAINT LOUIS UNIVERSITY
(continued)

Ave-Sesame I: Diabetic-Heating and Verticle
Motion
(1988) / Sammler W R

Microphysical Impact on Activity Spectra of
Radiative Cooling at the Top Convective
Clouds
(1988) / Tokay A

Structural Features of a Squall-Line
Thunderstorm Determined from SESAME
Dual-Doppler Data
(1988) / Wright P D

SOUTH DAKOTA SCHOOL OF MINES AND TECHNOL.

Development and Testing of a Triple Path
Denuder Air Sampler for Ambient Sulfate
(1988) / Hammer R J

An Objective Method of Forecasting
Precipitation in Morocco
(1988) / Madouh L

TEXAS A AND M UNIVERSITY

An Implementation and Test of the Nexrad
Transverse Wind Algorithm
(1988) / Bensinger R B

Some Statistical Associations Between Northern
Hemisphere Sea Level Pressure Patterns and
Temperatures at Selected U.S. Climate Stations
(1988) / Bryan J M

A Case Study of the Development and Organization
of Convective Rainbands in a Quasi-Barotropic
Environment
(1988) / Hadley D S

Active Modes of the Pacific ITCZ
(1988) / Hayes P M

Nightime Atmospheric Stability Changes and Their
Effects on the Temporal Intensity of a Mesoscale
Convective Complex
(1988) / Hovis J S

The Use of the Q-Vector in Operational Meteorology
(1988) / LeMay C A

Information Theory and Climate Prediction
(1988) / Leung L Y

Interdiurnal Temperature Variability Over the
Conterminous United States and Canada
(1988) / Rice P B

Synoptic-Scale East Asian Cold Surge-
Induced Phenomena
(1988) / Sautter D C

Moisture Burst Structure in Satellite Water
Vapor Imagery
(1988) / Ulsh D J

Dependence of Maximum Realizable Convective
Energy on Horizontal Scale in a One-Dimensional
Entraining Jet Model
(1988) / Wolff D B

Forces Leading to the Development of the
Low-Level Jet
(1988) / Youtsey W J

U.S. NAVAL POSTGRADUATE SCHOOL

Automated Satellite Cloud Analysis: A
Multispectral Approach to the Problem of
Snow/Cloud Discrimination
(1987) / Allen R C

Changes in the California Current System
Observed off Northern California During
July-August 1986
(1987) / Batteen M L

Changes in the California Current System
Observed off Northern California During
July-August 1986
(1987) / Beasley M E

Atmospheric Angular Momentum and Length of Day
(1988) / Benedict W L

A Study of a Rapid Cyclogenesis Event
During Gale
(1988) / Carson J L

Parameterization of Horizontal Wind
Velocity Variability
(1987) / Dowding T J

Effects of Rainfall on the Seasonal
Thermocline
(1987) / Garzon G H

A Prototype Expert System to Forecast Typhoon
Conditions at Cubi Point, Philippines
(1988) / Hagaman B M

Properties of the Atmospheric Boundary Layer
Above a Subtropical Oceanic Front
(1987) / Higgins J P

Sky Radiance Distributions for Thermal
Imaging Backgrounds
(1987) / Kotsis A

Vertical Wind Shear as a Predictor of
Tropical Cyclone Motion
(1987) / Meanor D H

Changes in the California Current System
Observed off Northern California During
July-August 1986
(1987) / Mooers C N

Utilization of Dense Packed Planar Acoustic
Echosounders to Identify Turbulence
Structure in the Lowest Levels of the
Atmosphere
(1987) / Moxcey L R

Comparison of Satellite-Derived Ocean
Velocities with Observations in the
California Coastal Region
(1987) / O'Hara J F

Three-Dimensional Analysis of Synoptic
Satellite and Conventional Meteorological
Observations
(1988) / Owen D D

A Synoptic Investigation of Maritime
Cyclogenesis During Gale
(1987) / Pertle W E

The Climatological Seasonal Response of the
Ocean Mixed Layer in the Equatorial and
Tropical Pacific Ocean
(1988) / Ries H J

U.S. NAVAL POSTGRADUATE SCHOOL
(continued)

Meteorological Features During the Marginal
Ice Zone Experiment from 20 March to 10
April 1987
(1987) / Schultz R R

Observational/Numerical Study of the Upper
Ocean Response to Hurricanes
(1987) / Shay L K

Diagnostic Study of a Genesis of Atlantic
Lows Experiment (GALE) Cyclogenesis Event
(1987) / Soper D J

Stratocummulus and Cloud-Free Reflectance
from Multi-Spectral Satellite Measurements
(1987) / Tettelbach F M

The Spatial and Temporal Variability of the
Arctic Atmospheric Boundary Layer and Its
Effect on Electromagnetic (EM) Propagation
(1987) / Willis Z S

U.S.A.F. INSTITUTE OF TECHNOLOGY

Atmospheric Corrections for Inflight
Satellite Radiometric Calibration
(1988) / Bartell R J

Analysis of Temperature and Velocity
Microturbulence Parameters from Aircraft
Data and Relationship to Atmospheric
Refractive Index Structure
(1988) / Beecher E A

Synoptic Scale Sensitivity of Tiros-N
Moisture Channels in the Tropics
(1988) / Blackwell K

Mesoscale Structure and Climatology of Rain
Snow Lines Over North Carolina
(1988) / Carey K F

Kinematic and Dynamic Studies of Microbursts
in the Subcloud Layer Derived from Jaws
Dual-Doppler Radar for a Co-Thunderstorm
(1988) / Coover J A

Some Wind Characteristics of Kahe Point,
O'Hau and Vicinity
(1988) / Cordeiro N R

On Deriving the Three-Dimensional Kinematic
Structure of Convective Storms from a Single
Radar
(1988) / Cox R B

Correction Between Rainfall and Normalized
Difference Vegetative Index in Eastern
Africa
(1988) / Davenport M L

The Role of a Lid in the 31 May 1985
Tornado Outbreak
(1988) / Farrell R

Active Modes of the Pacific Intertropical
Convergence Zone
(1988) / Hayes P M

A Modeling Perspective for Meteor Burst
Communication
(1988) / Healy B C

An Analysis: Lightning Launch Constraints
(1988) / Holland D E

An Expert System Model of Organizational
Climate and Performance
(1988) / Holt J R

Doppler Radar Analysis Coastal Marine
Atmospheric Boundary Layer Structure During
a Cold Air Outbreak
(1988) / Jones M N

Precipitation Distributions Associated
Cyclones Originating Over the Gulf of Mexico
and Surrounding Coastal Regions
(1988) / Knapp D I

An Analysis of the Error Characteristics of
Atlantic Tropical Cyclone Track Prediction
Models
(1988) / Kroll J T

The Use of the Q-Vection Operational
Meteorology
(1988) / Lemay C A

Search for Wave Induced Particle
Precipitation from Lightening and
Transmittal Sources
(1988) / Lundberg J E

Tropical Cyclone Observation and Forecasting
with and Without Aircraft Reconnaissance
(1988) / Martin J D

Origin of Tropospheric Ozone Over Central US
(1988) / McNamara D P

The Effects of Water Content on the
Predictions of the Cloud Rise Module
of DELFIC
(1988) / Minor B M

Application of a Mass-Consistent Wind Model
to Chinook Windstorms
(1988) / Muranaka N J

A Global Analysis and Correlation of Nimbus
Cloud and Longwave Radiation Data
(1988) / Ramirez D R

A Climatology of Air Parcel Trajectories
Into Mt. Michell, North Carolina
(1988) / Seabaugh S S

Tropical 200mb Divergent Circulations and
Their Relation to Outgoing Longwave
Radiation
(1988) / Sornatale F

Scattering Calculations for Various
Hailstone Models Using the Extended Boundary
Condition Method
(1988) / Stutz B J

Low-Level Oyutflow of Potential Nontornadic
Thunderstorms Inferred from Single Doppler
Radar
(1988) / Suclker G L

The Effect of Vertical Wind Shear on
Tropical Cyclone Movement
(1988) / Talbert K M

A Time-Dependent Model for the Low-Latitude
Ionosphere
(1988) / Wells G D

UTAH STATE UNIVERSITY

Middle Atmospheric Wind Measurements
Using a Medium Frequency Radar
(1987) / Coble B B

A Dendroclimatic Analysis of Fluctuations
in the Great Salt Lake
(1987) / Delehunt W J

UTAH, UNIVERSITY OF

A Linear Model Study of the Midlatitude
Response to Tropical Heating
(1988) / Lee J L

Development of Empirical Equations of Ice
Crystal Growth Microphysics for Modeling
and Analysis
(1988) / Redder C R

WASHINGTON, UNIVERSITY OF

Surface Radiative Budget Components for the
Bering Sea Marginal Ice Zone During
Mid-Winter
(1987) / Barrett M

Seasonal Variability in the 40-50 Day
Rainfall and Zonal Wind Oscillations in the
Tropics
(1987) / Gross J R

Atmospheric Structure and Momentum Balance
During a Gap Wind Event in Shelikof Strait,
Alaska
(1988) / Lackmann G M

Dual-Doppler Radar Investigation of a
Rapidly Evolving Midlatitude Mesoscale
Convective System
(1988) / Patnoe M W

Explosive Cyclogenesis in the Eastern
Pacific: A Compositing Study
(1987) / Roddy M

Analysis of Marine Stratus Cloud
Condensation Nuclei Collected with a New
Counter-Flow Virtual Impactor
(1988) / Tucker S L

On the Turbulent Fluxes of Moisture, Heat
and Stress in the Atmospheric Boundary Layer
Over the Sea in Whitecap Sea - States
(1987) / Vassiliou G D

WYOMING, UNIVERSITY OF

An Exploratory Study of Aggregate Cloud
Seeding Effect Over the Sierra Nevada of
California
(1989) / Dragomir J H

ACADIA UNIVERSITY

Petrology and Mineralization of the Deep
Cove Plauton, Gabarus Bay, Cape Breton
Island, Nova Scotia
(1988) / Dennis F

Petrology of the Black Brook Granitic Suite
and Associated Gneiss, Northeastern Cape
Breton Highland, Nova Scotia
(1988) / Yaowanoiyothin W

BRIGHAM YOUNG UNIVERSITY

Depositional Environments and Petrology of
the Middle to Upper Jurassic Carmel
Formation in the Gunlock Area, Washington
County, Utah
(1988) / Nielson D R

DALHOUSIE UNIVERSITY

The Mineralogy, Petrology, and Geochemistry
of the Halfway Cove-Queensport Pluton,
Nova Scotia, Canada
(1988) / Ham L J

The Mineralogy, Petrology and Geochemistry
of the Port Mouton Pluton, Nova Scotia,
Canada
(1988) / Woodend Douma S L

EAST CAROLINA UNIVERSITY

Mineralogy and Sedimentation of the
Clay-Sized Fraction, Pungo River Formation,
North Carolina Continental Margin
(1988) / Allison M A

Petrography and Geochemistry of the Meta-
Argillite/Metavolcanic Host Rocks and
Associated Gold Deposits at the Portis Gold
Mine, Franklin County, North Carolina
(1987) / Corbitt C L

The Distribution and Petrology of
Glauconitic Sediments Within the Pungo River
Formation, Onslow Bay, North Carolina
Continental Shelf
(1988) / Mallinson D J

Petrography of Unconformable Surfaces and
Associated Stratigraphic Units of the
Eocene Castle Hayne Formation, Southeastern
North Carolina Coastal Plain
(1988) / Moran L K

Petrology and Geochemistry of the Coronaca
Pluton, Greenwood County, South Carolina
(1988) / Schiappa C

EASTERN KENTUCKY UNIVERSITY

The Stratigraphy, Petrology, and
Depositional Environment of the Wild Member
of the Border Formation (Mississippian) in
Southeast-Central Kentucky
(1988) / Gauthier M

FORT HAYS KANSAS STATE UNIVERSITY

The Petrology and Structure of the Wet
Mountain Metamorphics and the San Isabel
Batholith Margin Along South Hardscrabble
Creek, Custer County, Colorado
(1988) / Louden R

IOWA STATE UNIVERSITY

Petrology and Geochemistry of Net-Veined
Complexes in the Austurhorn and Vesturhorn
Intrusions, S. E. Iceland
(1988) / Doden A G

Petrology and Petrogenesis of Tertiary
Ingeous Rocks, Chico Hills, New Mexico
(1988) / Potter L S

KANSAS, UNIVERSITY OF

Petrology and Geochemistry of the Volcanic
Rocks of the Nova Formation and Darwin
Plateau, Death Valley, California: The
Implications for Magmatic and Tectonic
Processes in Extensional Origens
(1988) / Coleman D S

Petrology and Petroleum Geology of the
Meramecian (Upper Mississippian) Warsaw
Formation, Kiowa and Comanche Counties,
Kansas
(1988) / Wilson M E

MASSACHUSETTS, UNIVERSITY OF

The Petrography, Mineral Chemistry and
Geochemistry of Eclogites from the Koidu
Kimberlite Complex, Sierra Leone
(1988) / Hills D V

MISSOURI, UNIVERSITY OF (COLUMBIA)

Petrology and Environments of Deposition of Limestone
Interbeds in the Womble Shale (Middle Ordovician)
Ouachita Mountains, Arkansas
(1988) / Ioos M B

MISSOURI, UNIVERSITY OF (ROLLA)

Petrology and Dolomitization of the Middle
Miocene Arun Limestone, North Sumutra,
Indonesia
(1988) / Kwasny J S

NEW YORK, STATE UNIVERSITY OF (ALBANY)

Petrology of the Kula Volcanic Field,
Western Turkey
(1987) / Dyer J M

NORTH CAROLINA STATE UNIVERSITY

Petrology of Ultramafic and Gabbroic
Inclusions in Basaltic Rocks of Kahoolawe
Island, Hawaii
(1988) / Rudek E A

NORTHEASTERN LOUISIANA UNIVERSITY

A Petrographic Analysis of the Basal
Carbonate Beds of the Ferry Lake Anhydrite,
Caddo-Pine Island Field, Caddo Parish,
Louisiana
(1988) / Kimball C E

Stratigraphy, Depositional Environments, and
Petrology of Wilcox Group Coals in North
Central Louisiana
(1988) / Wilson G D

A Subsequent Taxonomic and Distributional
Survey of the Ichthyofauna of the Bogue
Chitto River
(1988) / Wise J A

OKLAHOMA, UNIVERSITY OF

The Potential for Recovery of Residual Oil
by Microbial Enhanced Oil Recovery from the
Southeast Vassar Vertz Sand Unit
(1988) / Coughlin M K

A Study on the Ash Fusibility
Characteristics of Oklahoma Coals Blended
with Wyoming Coal
(1988) / Marnix J L

Combustion of Pulverized Coal Blends
(1988) / Prasad A

OREGON STATE UNIVERSITY

Stratigraphy and Sedimentary Petrology of the
Mascall Formation, Eastern Oregon
(1988) / Kuiper J L

PURDUE UNIVERSITY

Experimental Petrology of
Melilitites-Nephelinites
(1988) / Gee L L

REGINA, UNIVERSITY OF

Thermal Maturation and Organic Petrology of
Mesozoic Strata of Southern Saskatchewan
(1988) / Stasiuk L D

RHODE ISLAND, UNIVERSITY OF

Isotopic, Chemical, and Petrographic
Characteristics of Stratiform Manganese
Oxide Deposits in the Lake Mead Region,
Southeastern Nevada: Implications for
the Origin of the Deposits
(1988) / Bouse R M

Petrology, Geochemistry and Structure
of the Northeastern Part of the
Scituate Pluton, Rhode Island
(1988) / Hamidzada N A

STEPHAN F. AUSTIN STATE UNIVERSITY

Petrology Provenance of the Greta Sandstone,
Frio Formation (Oligocene), McFadden Field,
Victoria County, Texas
(1988) / Alford G W

Petrographic Analysis of the Gilmer
Limestone, Lovark Group (Upper Jurassic),
Box Church Field, Limestone County, Texas
(1988) / Byram K G

TENNESSEE, UNIVERSITY OF (KNOXVILLE)

Biostratigraphy, Petrology, and
Paleo-Depositional Environments of the Ross
Formation, Upper Silurian-Lower Devonian,
West-Central Tennessee
(1987) / Capaccioli D A

Geochemical, and Mineralogical
Characterization of Surficial Residuum
Formed from Lower Knox Group Dolostone, West
Chestnut Ridge, Oak Ridge, Tennessee
(1987) / Crafts A S

Petrography, Chemistry, and Origin of
Iron-Titanium Oxides in the Concord
Gabbro-Syenite, North Carolina
(1987) / Dobson M C

Petrology, Paleodepositional Environments,
Biostratigraphy and Paleontology of the
Decatur and Rockhouse Limestones (Upper
Silurian-Lower Devonian), West-Central
Tennessee
(1987) / McComb R

Quantitative Mineralogical Analysis and
Scanning Electron Microscopy Techniques for
the Study of Argillaceous Formations Which
are Potential Candidates for the Storage
of High-Level Radioactive Waste
(1987) / Mulligan P J

TEXAS, UNIVERSITY OF (ARLINGTON)

The Petrology and Geochemistry of the
Tshirege Member of the Bandelier Tuff,
Jemex Mountains Volcanic Field, New
Mexico, USA
(1988) / Balsley S D

Petrology, Physical Stratigraphy and
Biostratigraphy of the Lower Sierra Madre
Limestone, Cretaceous, West Central
Chiapas, Mexico
(1988) / Hansen D K

Petrography and Depositional Environments of
the Cretaceous (Cenomanian) Eangle Mountains
Formation, Western Trans-Pecos Texas
(1988) / Younger M A

TOLEDO, UNIVERSITY OF

A Structural Investigation of the Room
Temperature Phase Transformation in Maximum
Microcline
(1988) / Zhang H

UTAH STATE UNIVERSITY

Petrology of Pliocene Basalts in Curlew
Valley, Box Elder County, Utah
(1987) / Kerr S B

Petrology of the Middle Cambrian Langston
and Ute Formations in Southeastern Idaho
(1987) / Rogers D T

WEST TEXAS STATE UNIVERSITY

Petrology of the Contact Metamorphic Zone of
the Calumet Mine Area, Chaffee County,
Colorado
(1988) / Cleland J M

WEST VIRGINIA UNIVERSITY

Petrology and Sedimentation of the Big
Injun and Squaw Sandstones, Granny's Creek
Field, West Virginia
(1988) / Swales D L

WESTERN ONTARIO, UNIVERSITY OF

The Petrology and Geochemistry of the Lower
Zone of the Mulcahy Gabbro, Northwestern
Ontario
(1988) / Smith A R

ALABAMA, UNIVERSITY OF (UNIVERSITY)

A Two-Dimensional Two-Phase Simulation
Model for Coalbed Methane Recovery
(1988) / Chen J C

Development of a Combined Model of the Steam
Drive for a Waterflooded Reservoir
(1988) / Dindoruk B

Development of a Knowledge-Based System for
Diagnosing Electrical Faults in Mining
Machinery
(1988) / Meigs J R

Carbon Dioxide-Foam Displacement Test in
Stratified Porous Media
(1988) / Samieenasr M

Development of the Equations for a Numerical
Model of a Streamflood Applied to a
Waterflooded Reservoir
(1988) / Theodorou K

ALBERTA, UNIVERSITY OF

Column Flotation of Coal and Froth Model
Development
(1988) / Ofori P K

The Effects of Net Confining Pressure and
Temperature on the Absolute Permeability
of Consolidated Carbonate Reservoir Rocks
(1988) / Pitman N A

COLORADO SCHOOL OF MINES

Inhibition of the Water Staining Form of
Aluminum Corrosion Through the Use of
Sodium Molybdate
(1988) / Bates J

Effects of Moisture on Coal Handling
(1988) / Chu A A

The Weldability and Grain Refinement of
Aluminum-2.2 Lithium-2.7 Copper
(1988) / Dvornak M J

Open Pit Mine Production Scheduling
(1988) / Elevli B

Modified Tree Graph Algorithm for Ultimate
Pit Limit Analysis
(1988) / Huttagosol P

The Effect of Classifier Efficiency on the
Capacity of Closed-Circuit Grinding
(1988) / Llra P A

Slope Management for Economic Pit Wall
Control
(1988) / Manzanares D A

Probabilistic Risk Analysis in Evaluation
of Gold Placer Deposits in Puno-Peru
(1988) / Moises M G

Effects of Heat Input and Molybdenum Content
on the Microstructures and Mechanical
Properties of High Strength Weld Metal
(1988) / Oldland P T

Fatigue Resistance of Plasma and Gas
Carburized SAE 8719 Steel
(1988) / Pacheco J L

The Vapor Pressures of Germanium Sulfide
and Antimony Sulfide
(1988) / Shanks R

IDAHO, UNIVERSITY OF

Computer Aided Production Scheduling for
Surface Dragline Operations
(1988) / Richards D A

Dissolution of Silver with Thiourea the
Rotating Disc Study
(1988) / Seal T

KENTUCKY, UNIVERSITY OF

Air Tables: The State of Art in Pneumatic
Cleaning
(1988) / Banks R S

Observations of Rapidly Solidified Aluminum
and Aluminum-Chromium-Yttrium Alloys
(1988) / Jones J C

Influence of Moisture Fluctuations in Header
Boards on Effectiveness of Roof Bolt
Support System
(1988) / Seitz S V

MICHIGAN TECHNOLOGICAL UNIVERSITY

The Effect of Second Phase Additions on
Grain Boundary Sliding Behavior in CaO-NiO
(1988) / Brindley W J

Applications of Finite Element Method to
Open Stope Stability Analysis
(1988) / Chen C Z

Rapid Omnidirectional Compaction
(1988) / Haugh T A

Heavy Section Ductile Cast Iron: Production
and Segragation
(1988) / Hayrynen K L

Grain Size Gradient Nickel Alloys:
Processing, Kinetics, and Deformation
Behavior
(1988) / Hwang K H

The Effects of Ceramic Additions on the
Microstructure, Mechanical Properties, and
Aging Kinetics of Two Aluminum P/M Alloys
(1988) / Janowski G M

Relationships Between Plastic Strain and
Stress Corrosion Cracking
(1988) / Kasul D B

Microstructural Development and
Strengthening in Cu-Ni-Sn Composition
Gradient Materials
(1988) / Myers M J

Physical Acoustics of One Dimensional
Modulated Structures
(1988) / Ojard G C

AB Initio Intermolecular Potential Energy
Surfaces for the Hydrogen Sulfide Dimer,
Including Studies of the Third-Order
Many-Body Perturbation Theory Correction
and Nonadditive Effects
(1988) / Woon D E

MICHIGAN TECHNOLOGICAL UNIVERSITY
(continued)

The Effects of Long-Term Aging on the
Microstructure of Modified Low Carbon Single
Crystal Mar-M247 Nickel-Base Superalloys
(1988) / Yang W S

Processing/Structure Property Relationships
of Fe-Base/Yttria Stabilized Zirconia
Composites
(1988) / Yankee S J

MONTANA COLLEGE OF MINERAL SCI. & TECH.

Formation and Stability Studies of Some
Iron-Arsenic and Copper-Arsenic Compounds
(1988) / Abbas M H

Risk Qualified Pod Boundaries for a Karst
Hosted Lignite Deposit
(1988) / Fritz C J

Forming and Evaluating Ceramic Structures
with Graded Physical Properties
(1988) / Guthrie J D

Coal Quality Market Modeling for Selected
Ohio and Wisconsin Spot Markets
(1988) / Hayward T J

Hydrometallurgical Processes for Treating
Refractory Type of Silver-Bearing
Concentrates
(1988) / Ko-Lon C

Polymer Elocculation of Phosphatic Clay
(1988) / Labban M M

Conversion of Chromium Phosphate by Sodium
Carbonate Fusion
(1988) / Quinn J P

Recovery of Metal Values from Metal
Hydroxide Wastes: The Use of Ammonium
Hydroxide and Ammonia Gas at Neutralizing
Agents
(1988) / Rapkoch J M

MONTANA, UNIVERSITY OF

The Ascendancy of Surface Mining Over
Underground Mining in the United States
Coal Industry: Effects on and Issues
Relevant to the United Mine Workers of
America
(1987) / Underwood D

MONTREAL, UNIVERSITY OF

Lithogeochimie des Calcaires Superieurs
de Gaspe
(1988) / Bellehumeur C

Effets de la Redistribution de Contraintes
sur la Convergence des Excavations
Souterraines Dans le sel Gemme
(1988) / Blais M

La Carte Geotechnique de l'ile Jesus
(Laval), Quebec
(1988) / Blanchard C

Mesures des Concentrations de Phosphine et
de Phosphore a l'Aide de Piles Galvaniques
Utilisant l'Alumine Beta Comme Electrolyte
Solide
(1988) / Bouchard D

Correction de l'Effet d'une Topographie de
Surface Bi-Dimensionnelle sur les Leves
Magnetotelluriques
(1988) / Bouchard K

Petrographie et Geochimie de l'Indice
Mineralise Swanson, une Syenite Aurifere
(1988) / Bourgault G

Reacidification des Effluents Traites d'une
Parc a Residus Miniers Apres Emission Dans
le Milieu Naturel
(1988) / Cabral A

Photo-Interpretation et Analyse d'Images
des Lineaments Photo-Geologiques de l'Ile
d'Anticosti, Quebec
(1988) / Carboni S

Etude Geostatistique de Mount Leyshon Mine
(Australie)
(1988) / Chaouai N E

Effects of Additives on the Formation of
Calcium Ferrites in the Sinters Produced
from Specularite Concentrates
(1988) / Choudhury A K

Synthese Geologique des Roches Volcaniques
du Centre Nord de la Gaspesie
(1988) / Doyon M

Reductibilite de Boulettes d'Hematite a
Teneur en Silice
(1988) / Gariepy M

Gitologie de l' or a la Mine Elder, Canton
Beauchastel, Abitibi, Quebec
(1988) / Gaulin R

Reinterpretation des Donnees Aeromagnetiques
de la Region de Timgaouine (Hoggar), Algerie
(1988) / Ghanem Y

Progressive Hydrothermal Alteration
Associated with Gold Mineralization of the
Zone I Intrusion of the Callahan Property,
Val d'or Region, Quebec
(1988) / Jenkins C L

Precipitation de Niobium et Aluminum Dans
les Aciers Microallies Deformes a Chaud
(1988) / Legoux J G

Modelisation Physique de la Coulee
Continue de Brames d'Acier
(1988) / Sanchez J C

Erosion et Deformation Plastique d'Aciers
Inoxydables Austenitiques Soumis a la
Cavitation Hydraulique
(1988) / Simoneau M

Influence de la Microstructure et des
Inclusions sur la Tenacite de Depots
Soudes d'Acier Faiblement Allie
(1988) / St-Laurent S

Exploitation en Eau Souterraine a l'Ile
Perrot, Quebec
(1988) / Tremblay F M

Petrologie et Geochimie du Gisement d'or
Golden Pond East Casa-Berardi, Abitibi,
Quebec, Canada
(1988) / Verrault C

NOVA SCOTIA, TECH UNIVERSITY OF

The Effects of Fire Retardants on the
Mechanical Properties of Marine Polyester
Based Composites
(1988) / Corbin S F

PENNSYLVANIA STATE UNIVERSITY

An Analysis of the Pressure Drop Due to
Slurry Flow in the Contracting/Diverging
Geometry of a Dustpan Dredge Head
(1988) / Flegal R L

Detecting Subsided Coal Mine Entries Using
Resistivity Techniques
(1987) / Grippo A

Liquid-Liquid Separation of Ultrafine Coal
(1987) / Hsu T C

An Engineering Analysis of an Innovative
Mine Development and Ventilation Systems
Design for Large-Scale Longwall Mining
Operations
(1988) / Jain S

Quartz Levels in Airborne Dust in Continuous
Mining Sections
(1988) / Qin J

Laboratory Evaluation of Waveguides for
Acoustic Emission/Microseismic Monitoring
(1987) / Taioli F

Computer-Assisted Belt Conveyor Analysis
Utilizing User-Centered Design Criteria
(1987) / Weems R C

A Model for Thermal Permeability Enhancement
in Low-Temperature Geothermal Reservoirs
(1988) / Xiang J

PURDUE UNIVERSITY

Engineering Characteristics of Lacustrine
Deposits Involving Coal Strip Mining in
Southwest Indiana
(1988) / Lewis J C

SOUTH DAKOTA SCHOOL OF MINES AND TECHNOL.

Separation of Carbonaceous Material from
Carlin Gold Ore by Flotation
(1988) / Chen Y

Strength and Deformation of Algerie
Granite
(1988) / Simonson J R

TENNESSEE, UNIVERSITY OF (KNOXVILLE)

Effect of Niobium Concentration on the
Lambda-to-D-Type Martensitic Transformation
in Uranium Plus Niobium Alloys
(1987) / Kothandapany K

Amorphization of Iron-Boron Alloys by
Excimer Laser Processing
(1987) / McCamy J W

Microstructural Analysis of an Embrittled
422 Stainless Steel Stud Bolt After
Approximately 30 Years Service
(1987) / Zhou J P

TEXAS, UNIVERSITY OF (AUSTIN)

Engineering Economic Evaluation of Mining in
Antarctica: A Case Study of Platinum
(1988) / Beike D

UTAH, UNIVERSITY OF

An Economic Evaluation Tool for Underground
Mine Ventilation
(1988) / Gallegos A A

Maintenance of a Mine Ventilation System
(1988) / Peterson R D

WASHINGTON, UNIVERSITY OF

Potentiodynamic Anodic Polarization Studies
of Aluminum-Lithium-Copper Alloys with and
Without Magnesium
(1987) / Pak P M

WEST VIRGINIA UNIVERSITY

Batch Sieving of Cylindrical Particles by
Cascadography
(1988) / Bocoum B M

A Network Model for the Synthesis of
Regulatory Requirements Into the Mine
Design Process
(1988) / Bryan R K

The Submicrometer Fraction of Respirable
Coal Mine Dust in Diesel and Electric Mines
(1988) / Cocalis J

Economic Evaluation of the Implementation of
Longwall Mining Innovations
(1988) / Georgiana M A

An Econometric Explanation of Parameter
Variation in U. S. Copper Consumption
Models
(1988) / Labee C M

Surface Ground Movements Over Longwall
Mining in the Pittsburgh Seam
(1988) / Molesky P J

Mathematical Modeling of Coal Flotation Rate
Distributions by Gamma Function for the
Performance of Flotation Systems
(1988) / Montgomery D A

Subsurface Movement Due to Longwall Mining
(1988) / Moskal S J

Effect of Topography on Surface Ground
Movements Due to Longwall Mining
(1988) / Quinn M K

Determination of Surface Fracture Dept
Related to Subsidence Using Wave
Propagation Technique
(1988) / Ro Y S

Mine Fire Control by Explosive Fragmentation
(1988) / Shin J B

Comparison of Classifying Cyclones Utilizing
Linear Efficiency Curves
(1988) / Toney T A

Long-Run Petroleum Forecasting
(1988) / Williamson J C

WEST VIRGINIA UNIVERSITY
(continued)

Formulation of Improved Coal Mine Dust
Sampling Strategies
(1988) / Yuan S

Simulation Approach in Economic Assessment
and Risk Analysis of Carbon Dioxide
Miscible Flooding Process in West Virginia
(1988) / Zhuang Z

ALABAMA, UNIVERSITY OF (HUNTSVILLE)

The Impact of Recoverable Stage Concepts on
Manned Mars Mission Program Cost
(1988) / Montgomery E E

MASSACHUSETTS INSTITUTE OF TECHNOLOGY

Enhancing the Orbital Utilization of
Expendable Launch Vehicles
(1988) / Comstock D A

Inertia Friction Welding of Aluminum Alloys
for Space Repair Applications
(1988) / Guza D E

Effects of Force and Visual Feedback on
Space Teleoperation
Implications
(1988) / Massimino M J

Dynamics and Control of the Space Shuttle
Based Kinetic Isolation Tether Experiment
(1988) / Stephenson M W

MISSISSIPPI STATE UNIVERSITY

Development of an All Digital Control System
for an Air To Air Missile
(1986) / Kendrick J C

Optimal Guidance for Air-To-Air Missile Engagements
with Time Delays
(1986) / Killen A K

A Comparison of Three Estimation Techniques Applied
to an Air-To-Air Missile Engagement
(1986) / Kim H J

Development and Comparison of Two Iterative Kalman
Filter Techniques Applied to an Air-to-Air
Missile Engagement
(1986) / Packard M E

NEW JERSEY INSTITUTE OF TECHNOLOGY

High-Energy, Curable Polycyclopentadiene
Binder for Solid Rocket Propellants
(1988) / Ning J

OKLAHOMA, UNIVERSITY OF

A Parametric Study of Inflated Woven Space
Suit Components by the Finite Element Method
(1988) / Chengalur S N

PENNSYLVANIA STATE UNIVERSITY

Trajectory Optimization of a Single-Stage
to Orbit Composite Propulsion Vehicle
(1987) / Keith M T

Measurement of Pressure Profiles in the
Exhaust Plume of a Magnetoplasmadynamic
Thruster
(1987) / Kenney T M

A Method for Obtaining Empirical
Correlations for Predicting Crack
Propagation in a Burning Solid Propellant
Grain
(1987) / Nimis J A

Direct Measurement of the Pressure-Coupled
Response of Several Solid Propellants at
High and Intermediate Frequencies
(1988) / Taylor R D

Geosynchronous Orbit Retrieval System
Conceptual Design
(1987) / Weyandt C J

PRINCETON UNIVERSITY

Analysis and Design of Nonlinear Flight
Control Systems
(1988) / Dabundo C

The Effect of Geometrical Scale Upon MPD
Thruster Behavior
(1988) / Gilland J H

In-Flight Investigation of Longitudinal
Flying Qualities Criteria
(1988) / Glade D B
with Policy

U.S. NAVAL POSTGRADUATE SCHOOL

Satellite Signatures of Rapid Cyclogenesis
(1988) / Atangan J F

An Objective Technique for Arctic Cloud
Analysis Using Multispectral AVHRR (Advanced
Very High Resolution Radiometer) Satellite
Imagery
(1988) / Barron J P

Accuracy of Satellite Data Navigation
(1988) / Bethke W J

A Standard Library for Modeling Satellite
Orbits on a Microcomputer
(1988) / Beutel K L

Design Considerations for the Orion
Satellite: Structure, Propulsion and
Attitude Control Subsystems for a Small,
General Purpose Spacecraft
(1988) / Boyd A W

A Preliminary Study of an Omnibus
Maintenance Concept for Air Launched
Missiles
(1987) / Call K B

Design of a Three-Axis Stabilized Orion
Satellite Using an All-Thruster Attitude
Control System
(1988) / Dee S M

Comparative Analysis of Passive
Communications Satellites Employing the
SHF and HF Spectrum for Use in a Strategic
Role
(1988) / Donadio G

A Holographic Investigation of Particulates
in Metallized Solid Fuel Ramjet Combustion
(1987) / Easterling C A

Electrodynamic Driver for the Space
Thermoacoustic Refrigerator (STAR)
(1988) / Fitzpatrick M

An Analysis of Sparrow Missile Maintenance
(1987) / Gonzales P E

An Analysis of the Acquisition of the
Penguin Missile
(1987) / Hough D E

Utilization of a Kalman Observer with Large
Space Structures
(1988) / Jackson B M

U.S. NAVAL POSTGRADUATE SCHOOL
(continued)

Dynamic Analysis of the Flexible Boom in
the N-Ross Satellite
(1987) / Kang C S

Fault Tree Reliability Analysis of the
Naval Postgraduate School Mini-Satellite
(Orion)
(1987) / Keeble T G

An Analysis of the Operational Level in the
Combined Intermediate and Depot Level
Maintenance Concept for Airborne Missile
Systems
(1987) / Klimson E

Computer Simulation of a Rotational
Single-Element Flexible Spacecraft Boom
(1987) / Laufenberg R S

Relating Service Doctrine to Space System
Requirements
(1987) / Light T W

Multispectral Satellite Analysis of Marine
Stratocumulus Cloud Microphysics
(1988) / Mineart G M

Verification of a Micro-Thrusting Model to
Maintain Satellites in Low Orbit
(1987) / Noble C D

Ions Generated on or Near Satellite Surfaces
(1988) / Norwood C W

Automatic Data Retrieval from Rocket Motor
Holograms
(1987) / Orguc E S

Investigation of Design Considerations for
Telemetry, Tracking, and Command (TT&C)
Antenna System on Naval Postgraduate School
Orion Mini-Satellite
(1987) / Peters D L

Effects of Reduced Order Modeling on the
Control of a Large Space Structure
(1988) / Preson W J

Measurement of Particle Size Distribution in
a Solid Propellant Rocket Motor Using Light
Scattering
(1987) / Pruitt T E

Investigation of Anomalous Cloud Features
in 3.7 Micrometer Satellite Imagery
(1987) / Rogers J M

Design Investigation for a Microstrip Phased
Array Antenna for the Orion Satellite
(1988) / Smith M B

Thermodynamic Improvements for the Space
Thermoacoustic Refrigerator (STAR)
(1988) / Susalla M P

A Preliminary Study of an Omnibus
Maintenance Concept for Air Launched
Missiles
(1987) / Terrell A R

Use of the Tracking and Data Relay Satellite
System (TDRSS) with Low Earth Orbiting (LEO)
Satellites: A Decision Guide
(1988) / Topp A R

An Investigation Into the Feasibility of
Using a Dual-Combustion Mode Ramjet in a
High Mach Number Tactical Missile
(1987) / Vaught C B

Ion Gun Generated Electromagnetic
Interference on the Scatha Satellite
(1987) / Weddle L E

Ion Gun Operations at High Altitudes
(1988) / Werner P W

A Study of Second and Third Order Models for
the Tracking Subsystem of a Radar Guided
Missile
(1988) / Williams J W

Application of Laser Diffraction Techniques
to Particle Sizing in Solid Propellant
Rocket Motors
(1987) / Youngborg E D

U.S.A.F. INSTITUTE OF TECHNOLOGY

Continuous Low Thrust Coplanar Orbit
Transfers with Varying Eccentricity
(1988) / Beeker G L

Remote Orbital Capture Using an Orbital
Maneuvering Vehicle Equipped with Multi-Body
Grappling Arm Assembly
(1988) / Bishop J F

Active Control of a Large Space Structure as
Applied to the Phase 1 CETF Space Station
(1988) / Cope D A

Assessing Potential Benefits for Service
Repair and Retrieval of Satellites: A
Pilot Decision Analysis
(1988) / Del Pinto M A

Hypersurface Insertion Window for Long Term
Orbital Stability of Artificial Satellites
About the Planet Venus
(1988) / Dudley R L

Spacecraft Change as a Source of Electrical
Power for Spacecraft
(1988) / Gale W

Optimal Low Thrust Trajectories for
Planetary Capture
(1988) / Gaylor D E

Geosynchronous High Energy Electron (1.2
16 MeV) - Solar Wind Correlation Analysis
(1988) / Grover G P

Utility of Particle Precipitation Data as an
Input to Thermospheric Density Models for
Satellite Orbital Analysis
(1988) / Hand K J

Internal Debonding of Solid Rocket Fuel: An
Experimental Investigation
(1988) / Hanley J H

Investigation of the Validy of the
Non-Rotating Planet Assumption for
Three-Dimensional Earth Atmospheric Entry
(1988) / Karasopoulos H A

Global Positioning Satellite System
Navigation Accuracy with Updated Ephemerides
(1988) / Kirkpatrick D

U.S.A.F. INSTITUTE OF TECHNOLOGY
(continued)

Optimal Launch Time for a Discontinuous Low
Thrust Orbit Transfer
(1988) / McCann J M

A New Perspective in the Etiology,
Treatment, Prevention, and Prediction of
Space Motion Sickness
(1988) / Morales R

OFT Digital Design for Inherent
Reconfiguration of the Flight Control
System of an Unmanned Research Vehicle
(1988) / Ott P T

Configuration Comparison Analysis for the
AFIT/AAMRL Anthropomorphic Robotic
Manipulator
(1988) / Parker S L

Design of Package Cushioning for Missiles
(1988) / Pauletti S L

Light Satellites - A Dilemma for the
U. S. Army
(1988) / Pierson J R

Stability Solution to Linearized Equations
of Motion for a Symmetric Spinning Satellite
in an Elliptical Orbit Applied to the
Non-Linear Equations
(1988) / Shell D E

Control System for Maintaining Stable Orbit
Around Phobos
(1988) / Soloman M K

The Use of Chaff in Space as a Jamming
Device Between Ground Station and Satellites
(1988) / Sterns A R

Numerical Study of Orbital Trajectories
About Phobos
(1988) / Teets R

Robustness Analysis of a Moving-Bank
Multiple Model Adaptive Estimation and
Control for a Large Flexible Space Structure
(1988) / Vanderwerken D F

Estimated Satellite Cluster Elements in Near
Circular Orbit
(1988) / Ward M L

Periodic Orbits Around a Satellite Modelled
as a Triaxial Elipsoid
(1988) / Werner R A

An Analysis of the Attitude Stability of a
Space Station Subject to Parametric
Excitation of Periodic Mass Motion
(1988) / Williams T E

FLORIDA, UNIVERSITY OF

Ion Loss Characteristics of Uranium
Hexafluoride Gas Mixtures Exposed to
Ionizing Radiation
(1988) / Baumgartner M J

Inhomogeneity Delineation with Computed
Tomography for Radiation Therapy Treatment
Planning
(1988) / Leetzow M

IDAHO STATE UNIVERSITY

A Digital Controller for the Omega West
Reactor Los Alamos National Laboratory
(1988) / Smith T W

An Evaluation of the Training of Inspectors
in the U.S. Nuclear Regulatory Commission
Operational Programs Section
(1988) / Vanderniet C L

IOWA STATE UNIVERSITY

Core Physics of an Unconventional Liquid
Metal Cooled Fast Power Reactor Concept
(1988) / Schmidt R R

LOUISIANA STATE UNIVERSITY

Determination of the Temperature
Distribution Due to Neutron and Gamma
Heating in a Space-Reactor Shield
(1988) / Nadkarny S G

LOWELL, UNIVERSITY OF

Design of Yankee Fuel Assemblies with Zoned
Enrichments and Gadolinia Shims to Achieve
Longer Fuel Cycles
(1988) / Durrenberger M R

Development of a Microcomputer Model for
Dynamic Analysis of the New Hampshire
Yankee Power Plant
(1989) / Robichaud J D

MASSACHUSETTS INSTITUTE OF TECHNOLOGY

Mixed Convection Plenum-Channel Interactions
in a Vertical Heated Channel Connected to
Upper and Lower Plane
(1988) / Barakat A I

Lattice Gas Hydrodynamics with Galilean
Invariance
(1988) / Donis P A

Quantitative Lactate Imaging with Nuclear
Magnetic Resonance
(1988) / Eland B J

Vapor Condensation on a Turbulent Liquid
Interfacce, for Application in Low-Gravity
Environments
(1988) / Helmick M R

An Automated Computer Controlled Counting
System for Radionuclide Analysis of
Corrosion Products in LWR Coolant System
(1988) / Hernandez G R

Ablation of Hard Biological Tissue Using
Pulsed Hydrogen Fluoride Laser Radiation
(1988) / Izatt J A

Use of Probabilistic Risk Assessment in
Nuclear Power System Design Refinement
(1988) / Kato I

Predictive Information as an Aid to Human
Operators for Reactor Power Control
(1988) / Lau S H

Prompt Gamma Activation Analysis of Boron
10 in Blood and Dosimetric Measurements
Associated with Boron Neutron Capture
Therapy
(1988) / Lizzo N S

Atmospheric Transport of Radionuclides Due
to the Accident at the Chernobyl Unit 4
Nuclear Power Station
(1988) / McInall S G

Localized In Vivo 31p-NMR Spectroscopy for
Clinical Applications
(1988) / Moore G J

The Scattering of Gamma Radiation from
Hydrogenous Media
(1988) / Musk J H

Lattice Gas Hydrodynamics
(1988) / Myczkowski J

Alignment of Digitized Ocular Images for
Planning Proton Radiation Therapy
(1988) / Plongren S E

Ray Tracing for Millimeter/Submillimeter
Wave Plasma Diagnostics in Tokamaks
(1988) / Rhee D Y

Modular Shipbuilding and Its Relevance to
Construction of Nuclear Power Plants
(1988) / Seubert T W

Information Technology in Development:
Implications for the Energy Sector in Less
Developed Countries (LDC's)
(1988) / Srinivasan O P

Public Utility Commission Structure and
Decision Making
(1988) / Turnipseed T L

The Ramifications of a Delay in the National
High-Level Waste Repository Program
(1988) / Vance S A

Information Policy and Radiation Reporting
During the Chernobyl and Three Mile Island
Nuclear Accidents
(1988) / Woodruff M G

MISSISSIPPI STATE UNIVERSITY

Calibration of a TLD Personnel Monitoring System
at Mississippi State University
(1986) / Martin R B

Design, Construction and Demonstration of a TLD
Calibration Facility at Mississippi State University
(1986) / Pournia F

Effect of Diameter Ratio on Heat Transfer Between
Concentric Vertical Cylinders
(1986) / Rankin D B

Feasibility of Gamma-Ray Irradiation in Retarding
Shrimp Waste Decomposition
(1986) / Whitehead D W

MISSOURI, UNIVERSITY OF (COLUMBIA)

Thermal Hydraulics and Neutronics Real Time
Simulation of A Nuclear Power Reactor
(1988) / Enani M A

MISSOURI, UNIVERSITY OF (ROLLA)

Effect of an Annular Connection Between the
Plasma Source and Flux Conserver on the
Properties of CTX Spheromaks
(1988) / Hart C M

The Influence of Specimen Size on Charpy
Impact Testing of Unirradiated HT-9
(1988) / Louden B S

MONTREAL, UNIVERSITY OF

Etude des Oscillations des Mecanismes de
Reactivite du Reacteur Nucleaire de
Gentilly-2
(1988) / Paquette P

NEW MEXICO, UNIVERSITY OF

Calculation of Elastic and Inelastic Displacement
Damage in Silicon from Bombardment by Energetic
Protons
(1988) / Ek D S

Refinement of the CR-39 Range Filter Voltage
Diagnostic
(1988) / Lapin S

A 15 MWe Nuclear Turboelectric Power Supply for a
Self-Sufficient Lunar Base
(1988) / Lonz G J

Experimental Studies on the Coolability of
Stratified Dual-Layered Particle Beds
(1988) / Richardson J A

The Evaluation of a Nuclear Safeguards Physical
Security System Using the Savi Software Package
(1988) / Simpkins B E

NORTH CAROLINA STATE UNIVERSITY

Investigation of Various Techniques for Fast
Neutron Dosimetry
(1988) / Kim D

NORTHEASTERN LOUISIANA UNIVERSITY

Lanthanide-Induced Shifts in the Proton
Nuclear Magnetic Resonance Spectra of
Tropane and Tropanyl Benzoates
(1988) / Mullen D P

OREGON STATE UNIVERSITY

A Comparison of Three Coarse-Mesh Nodal Methods
for Boiling Water Reactor Analysis
(1988) / Cho B O

Computer Control of the Automatic Gamma Well
Counting System
(1987) / Ibrahim S H

PENNSYLVANIA STATE UNIVERSITY

Nuclear Analysis and Optimization of the
Molten-Salt Fusion Hybrid Reactor
(1988) / Bandini B R

Temperature Effects on Gamma Radiation
Damage in Enhanced Mode Metal Oxide Silicon
Field Effect Transistors
(1987) / Lutz R N

The Use of Ionization Chambers in Mixed
Field Dosimetry
(1987) / Meltser M S

Development of a Multicycle Scoping Code
Package for the Fuel Management in
Pressurized Water Reactors
(1988) / Moon H

Zero-Diimensional Model of a Reversed Field
Pinch Fusion Reactor
(1987) / Veerasingam R

RENSSELAER POLYTECHNIC INSTITUTE

Novel Application of High-Uranium-Density
Silicide Fuel for a University High Thermal
Flux Nuclear Reactor
(1988) / Angelo P L

A Simpel One-Dimensional, One-Fluid Divertor
Model
(1988) / Baehre M D

Wave Propagation Phenomena in Bubbly
Two-Component Two-Phase Flows
(1988) / Campanella R S

A Mechanistic Analysis of Phase Separation
in Branching Conduits Using
Three-Dimensional Two Fluid Models
(1988) / Kalkach-Navarro S

Evaluation of "Go" and "Riskman" Models for
Use in the Implementation of a Safety System
Unavailability Monitoring Program at the
Indian Point Unit 2 Nuclear Plant
(1988) / Kimball C I

Availability Analysis Improvements
(1988) / Kirchner R F

The Repair of Opaque Defects on Photomasks
Using a High Energy Focused Ion Beam System
(1988) / Liebmann L W

Three Space Reactor Studies: Thermal
Analysis of the NERVA Derivative Core
Solid Core Reactor Concept Development
Ablative Reflector/Shield System
(1988) / Snead L L

Real-Time Data Capture and Analysis of a
Moving Wood's Metal/Paraffin Interface
Utilizing Digital-Image Processing
(1988) / Valenti M E

TENNESSEE, UNIVERSITY OF (KNOXVILLE)

Development of a General Learning Algorithm
with Applications to Nuclear Reactor Systems
(1987) / Brittain C R

Design of Compact Loop Antennas for Present
and Future Fusion Heating Experiments
(1987) / Bryan W E

Extraction of Gadolinium from High Flux
Isotope Reactor Control Plates
(1987) / Kohring M W

TENNESSEE, UNIVERSITY OF (KNOXVILLE)
(continued)

High Resolution Measurement of the Uranium
Neutron Capture Yield for Incident Neutron
Energies from 1 keV to 100 keV
(1987) / Macklin R L

Interpretation of Subcriticality
Measurements with Strong Spatial Effects
(1987) / March-Leuba C

An Optimized F-Number Matching System for
the Advanced Toroidal Facility's Thomson
Scattering Plasma Diagnostic
(1987) / Painter S L

Monte Carlo Criticality Calculations Using
a Parallel Processor Computer
(1987) / Ugolini D

Development of a Nuclear Engineering
Computational System for the University
of Tennessee
(1987) / Woody N D

TEXAS A AND M UNIVERSITY

Cermet Fuel Thermal Conductivity
(1988) / Alvis J M

New Tube Bundle Heat Transfer Correlations and Flow
Regime Maps for a Once Through Steam Generator
(1988) / Blanchat T K

Analysis of a Natural Circulation Cooldown Transient
in a Westinghouse Pressurized Water Reactor Using
Trac-PF1/Mod1 and Trac-PF1/Mod2
(1988) / Breiner E M

Two-Dimensional Neutronic Analysis of the TAMU
Nuclear Science Center Reactor Using Transport
and Diffusion Theory Based Codes
(1988) / Davis J W

Numerical Modeling of the Transient Behavior of a
Thermoelectric Electromagnetic Self-Induced Pump
(1988) / Djordjevic V

Optimized Nuclear and Solar Dynamic Organic
Rankine Cycles for Space Station Applications
(1988) / Eubanks D L

Transient Thermal Analysis of a Space Reactor
Power System
(1988) / Gaeta M J

Analysis of a 4-Inch Small-Break Loss-of-Coolant
Accident in a Westinghouse Pressurized Water Reactor
Using Trac-PF1/Mod1.
(1988) / Knippel K I

U.S.A.F. INSTITUTE OF TECHNOLOGY

Neutral Particle Beam Propagation Through
a Nuclear Disturbed Atmosphere
(1988) / Lemire R A

Using ORIGEN2 to Predict Nuclear Reactor
Fuel Composition
(1988) / Lindblom B A

Direct Determination of Range from Current
Nuclear Overpressure Equations
(1988) / Wolczek R S

VIRGINIA, UNIVERSITY OF

Radioactivity in Central Virginia Sediments
and Soils
(1988) / Bose S R

Design Optimization of a Low Enrichment
University of Virginia Nuclear Reactor
(1988) / Fehr M K

Investigation of Alkali Metals Transfer to
Tobacco Smoke
(1988) / Johnson R O

The Effects of Antioxidants on Radiation
Degradation of Electric Cable Insulation
(1988) / Ray J W

Dose Rate Effects on Radiation Induced
Degradation of Electric Cable Insulation
(1988) / Terwilliger P L

WASHINGTON, UNIVERSITY OF

A Study of the Processes Controlling the
Isolation Performance of a Nuclear Waste
Package in a Deep Geologic Repository
(1988) / Baca R G

A Water Inventory Program with a Concomitant
Diagnostic Expert System for the Oak Ridge
National Laboratory High Flux Isotope
Reactor
(1987) / Curtis M M

MICRO-KINETICS: A Reactor Kinetics Code for
the Classroom
(1987) / Finfrock S H

An Analysis of Computational Difficulties
Associated with Boiling in Computer
Simulated Liquid Metal Flows
(1987) / Jaegers P J

Porous Media Resistance Formulation for
Nuclear Reactor Core Hydraulic Analysis
(1987) / Jervis A

Doppler Uncertainty in Liquid Metal Reactor
Safety Analysis
(1987) / Kahrs M C

Sensitivity of Multiple-Scattering Inverse
Transport Methods to Measurement Errors
(1987) / Oelund J C

BOSTON COLLEGE

Gamma-Ray Spectroscopy Analysis of Radon
Emanation
(1988) / Davis N M

An EPD Discharge-Flow Kinetics Study of the
F and CF3I Reaction in the Temperature
Range 195-473 K
(1988) / Donohue K

LOUISIANA STATE UNIVERSITY

Implementation and Benchmarking of Improved
Resonance Shielding Methods in a LWR Lattice
Physics Program
(1988) / Raharjo R

TOLEDO, UNIVERSITY OF

A Determination of Light Scattering
Properties of Interstellar Dust Grains
Contained in a Bok Globule
(1988) / Oliveri M V

U.S. NAVAL POSTGRADUATE SCHOOL

Emission Angles for Soft X-Ray Coherent
Transition Radiation
(1987) / Robinson R M

WASHINGTON, UNIVERSITY OF

Probabilistically Derived Concentration
Limits for Near-Surface Disposal of
Radioactive Waste
(1988) / Farris W T

CALIFORNIA, UNIVERSITY OF (IRVINE)

Reflection of Acoustic Waves from a
Discontinuity in Absorption
(1988) / Nolan E S

LOUISIANA STATE UNIVERSITY

An Evaluation of Angular Setup Errors in a
New Patient-Isolated Transvaginal Cone
System
(1988) / Haik M J

Rear-Earth Soil Horizon Markers to Determine
the Short-Term Accretion in Louisiana
Marshes
(1988) / Van Gent D L

TEMPLE UNIVERSITY

A Study of the Hydrothermal Stability of
Copper for Use as a Container Material for
Nuclear Waste at the Hanford Site, Richland,
Washington
(1988) / Lazaar P I

TEXAS A AND M UNIVERSITY

The Improvement of Methods Used to Estimate Radon
Concentration in Air Using Nuclear Track Film
(1988) / Jessick J G

Simultaneous Determination of Nickel-63 and
Nickel-59 in Radioactive Wastes by Liquid
Scintillation Spectrometry
(1988) / Kim E M

A Revised Model of the Kidney for Medical Internal
Radiation Dose Calculations
(1988) / Patel J S

Identification and Differentiation of Individual
Beta Emitters in Waste Mixtures by Liquid
Scintillation Spectrometry.
(1988) / Siskel R L

AUBURN UNIVERSITY

Removal of Common Carp (Cyprinus Carpio) Egg
Adhesiveness Using Protease Sodium Sulfite,
Salt-Urea and Tannin Solution
(1987) / Alawi H

Paddlefish Movements in the Upper Alabama River
(1987) / Brantly T B

Histopathology, Electron Microscopy and Isolation
of Channel Catfish Virus in Experimentally
Infected European Catfish
(1987) / Chumnongsittathum B

Comparison of Harvest Rates for the Marine
Recreational Fishery on the Eastern and
Western Sides of Mobile Bay, Alabama,
October 1984-September 1985
(1987) / Crone P R

Effects of the Giant Prawn Macrobrachium Rosenbergii
and Tilapia on Benthic Macroinvertebrates
(1987) / Forbin I

An Evaluation of Fixed and Demand Feeding Regimes
for Cage Culture of Tilapia Aurea
(1987) / Hargreaves J A

Status of the Niger River Fishery in Niger,
West Africa (1983-1985)
(1987) / Meredith E K

Sampling Scheme for Aerobic Heterotrophic
Bacteria in Warmwater Fishponds
(1987) / Rao S K

Effects of Dietary Protein Percentages on Yield,
Dressing Percentage and Body Composition of
Channel Catfish
(1987) / Reis L M

The Effects of Vertical Mixing on Phytoplankton
Communities and Certain Water Quality Variables
in Channel Catfish Ponds
(1987) / Self R L

Evaluating the Survey Method on the Alabama Gulf
Coast Recreational Fishery
(1987) / Stanovick J S

CALIFORNIA, UNIVERSITY OF (LA JOLLA)

Wave Induced Liquefaction Analysis Using
an Elasto-Plastic Soil Model
(1988) / DeRoos B G

Southern California Bight Sea Level Response
to Local Atmospheric Forcing
(1988) / Fu T C

The Ontogenetic Development of the Metabolic
Enzymes Citrate Synthase and Lactate
Dehydrogenase in the Swimming Muscles of
Larval and Juvenile Fishes: The Scaling of
Locomotory Power
(1987) / Kaupp S E

Pericardial and Vascular Pressures and Blood
Flow in the Albacore Tune, Thunnus Alalunga
(1987) / Lai N C

Gas Supersaturation Thresholds for Bubble
Formation in and Damage to Sea Urchin Eggs
and Embryos
(1988) / Ryan W L

CONNECTICUT, UNIVERSITY OF

The Occurrence of Glugea Stepheni in Long
Island Sound Winter Flounder, Pseudo
Pleuronectes Americanus with Reference to
Dynamics of Infection and Effects on the
Fish
(1988) / Castonguay W M

Aspects of the Life History of Pinnotheres
Maculatus and Its Effects on Mytilus Edulis
(1988) / Kelly M S

DALHOUSIE UNIVERSITY

Satellite Derived Estimates of Primary
Production on the Northeastern North
American Continental Shelf
(1988) / Kuring N

Physical Control of Chlorophyll-a in Bedford
Basin
(1988) / Mitchell M R

DELAWARE, UNIVERSITY OF

Seasonal Variation in the Hemolymph Free
Amino Acids of Crassostrea, Virginica
(1988) / Allan L B

Movement of Pre-and Post-Spawning Short Nose
Sturgeon (Acipenser Brevirostrum) in the
Delaware River
(1987) / Brundage H M

Bacterial Metabolism of Dissolved Proteins
in the Delaware Estuary
(1988) / Compton S L

Shore Erosion and Coastal Processes Along
the Eastern Shore of Chesapeake Bay,
Maryland
(1987) / Coulombe B D

The Acceptability of Stabilized Coal Waste
as a Substratum for the Colonization of
Marine Invertebrates
(1987) / Dinkins B J

The Total Deposition of PB, CD, ZN and 210
PB and Atomospheric Transport to the Western
Atlantic Using 222 RN and Air Mass
Trajectory Analysis
(1988) / Hartman M C

Utilization of Refractory Cellosic Material
Derived from Spartina Alterniflora (LOISEL)
by the Ribbed Mussel Geukensis Demissa
(DILLWYN)
(1987) / Kreeger D A

Appraisal of Economics in Fishery Management
Plans: an Application to the Mid-Atlantic
Surf Clam and Ocean Quahog Fisheries
(1987) / Leggett J J

Effects of Exogenous Androgens on Growth and
Reproduction of the Coot Clam, Mulinia
Lateralis, (say)
(1987) / Moss S M

Salinity as a Cue for Habitat Selection by
Megalopac of Two Congeneric Fiddler Crab
Species
(1987) / Rowe P M

DELAWARE, UNIVERSITY OF
(continued)

Conservation of Condominiums: a Policy
Analysis Concerning the Management of the
Development of Barrier Islands in North
Carolina, the Mid 1980's
(1987) / Smith P S

Food Habits of Dominant Piscivorous Fishes
in Delaware Bay, with Special Reference to
Predation on Juvenile Weakfish
(1988) / Taylor E T

EAST TEXAS STATE UNIVERSITY

Fish Community Structure in a Hydrologically
Variable Stream
(1988) / Capone T A

FLORIDA STATE UNIVERSITY

The Eccentricity of Warm- and Cold-Core Rings
(1988) / Shi C

FLORIDA, UNIVERSITY OF

A Two-Dimensional Finite-Difference Model
for Moving Boundary Hydrodynamic Problems
(1988) / Liu Y

Stochastic Analysis of Offshore Currents
(1988) / McMillen R

Numerical Simulation of Shallow Water Waves
(1988) / Philip R

GUAM, UNIVERSITY OF

Distribution and Production Dynamics of
Benthic Invertebrates in a Tropical Stream
on Guam
(1988) / Ellis-Neill L

HAWAII, UNIVERSITY OF

The Occurrence of Dissolved and Particulate
Adenosine-5'-Triphosphate in Antarctic
Coastal Ecosystems
(1988) / Nawrocki M P

IDAHO, UNIVERSITY OF

A Model to Assess the Effects of Hatchery
Supplementation Programs on Smolt, Recruit,
and Wild Steelhead Abundance in the
Clearwater River, Idaho
(1988) / Byrne A

Relative Survival and Growth of Steelhead
Fry Emerging Early Versus Late and
Potential Use of Dipeptidase as a Genetic
Marker in Steelhead
(1988) / Chandler G L

Nutrient Dynamics in Dingle Marsh, Bear
Lake National Wildlife Refuge, Idaho
(1988) / Myers R E

Identification of Pink Salmon Stocks in the
Fishery of Sitka Sound, Alaska
(1988) / Ralonde R L

Stress Indices in Chinook Salmon Smolts
Confined with Steelhead Trout Smolts
(1988) / Robertson C A

LOUISIANA STATE UNIVERSITY

The Age and Growth of Juvenile Spotted
Seatrout, in Coastal Louisiana
(1988) / Lorica M F

Ammonium Regeneration and Uptake in a
River-Dominated Estuary in Louisiana
(1988) / Rivera-Monroy V H

Ecological Characterization of Jean Lafitte
National Historical Park, Louisiana: Basis
for a Management Plan
(1988) / Taylor N C

MAINE, UNIVERSITY OF

Proteolytic Enyme Activity in Marine
Sediments
(1988) / Harper J C

The Feeding Biology of the Bloodworm
(1988) / Joule B J

Community Structure of Macrobenthic Found
in an Offshore Gulf of Maine Site, with
Population Structure Analysis of the
Epifaunal Ophiuroid and Its Trophic
Relationship to American Plaice
(1988) / Packer D B

Annual Production, Population Structure, and
Biomass of Some Deep-Water Benthic
Polychaetes from the Gulf of Maine
(1988) / Ward K J

MASSACHUSETTS, UNIVERSITY OF

Ecology and Range Expansion of American
Oyster-Catchers in Massachusetts
(1988) / Humphrey R C

Population Genetic Analysis of Atlantic
Salmon Transplanted to the Connecticut River
(1988) / Lindell S

MIAMI, UNIVERSITY OF

Validated Age Assessment of the Lemon Shark,
Negaprion Brevirostris, Using Tetracycline
Labeled Vertebral Centra
(1988) / Brown C A

The Descriptive Anatomy of Sound Production
and Propagation Tissues in Kogia SPP Using
Magnetic Resonance and Computed Tomography
Imaging
(1988) / Carvan M J

Magnetite and Magnetoreception in Stranded
Dwarf and Pygmy Sperm Whales, Kogia Simus
and Kogia Breviceps
(1988) / Credle V R

Induced Sex Reversal in a Lake Malawi
Cichlid, Pseudotropheus Zebra "Red Dorsal"
(1988) / DeMason L J

Turtle Brain Synaptosomes: A New Model for
the Study of the Brain
(1988) / Edwards R A

The Reproductive Anatomy of the Female
Manatee Trichechus Manatus Latirostris
Based on Gross and Histologic Observations
(1988) / Marmontel M

MIAMI, UNIVERSITY OF
(continued)

Age and Growth of Laboratory-Reared Larval
Snook, Centropomus Undecimalis, from
Otolith Microstructure
(1988) / Tellock J A

Absorption Efficiency of the Juvenile Lemon
Shark, Negaprion Brevirostris, at Varying
Rates of Energy Intake
(1988) / Wetherbee B M

MINNESOTA, UNIVERSITY OF

Production and Trophic Ecology of a Stream
Brook Trout Population
(1988) / Barstad W A

MISSOURI, UNIVERSITY OF (COLUMBIA)

Relations Among Zooplankton Abundance, Biomass,
Community Structure, and Lake Trophic State in
Selected Midwestern Waterbodies
(1988) / Canfield T J

Ecology of Migrant Shorebirds in Northeastern
Missouri
(1988) / Hands H M

Relations between Waterbird Use and the Limnological
Characteristics of Wetlands on Yukon Flats National
Wildlife Refuge, Alaska
(1988) / Heglund P J

Breeding Ecology and Habitat Use of Greater
Prairie-Chickens in Relation to Habitat Pattern
(1988) / Jones D P

NORTH CAROLINA, UNIV. OF (CHAPEL HILL)

Internal Growth Lines of the Estuarine
Bivalve, Rangia Cuneata Gray
(1987) / Cooper R B

Associational Plant Refuges: Convergent
Patterns in Marine and Terrestrial
Communities are the Result of Differing
Mechanisms
(1987) / Pfister C A

NORTHEASTERN LOUISIANA UNIVERSITY

A Subsequent Investigation of the Fishes
of Bayou Bartholomew
(1988) / Hutchins M M

OLD DOMINION UNIVERSITY

Sediment Budgets and Shoreline Dynamics
Cape Henry, Virginia
(1987) / Berman M R

Economic Considerations for Artificial Reef
Development in Coastal Virginia
(1988) / Devereaux D T

Sulfide in Shelf/Slope Waters of the Western
Atlantic
(1987) / Krahforst C F

OREGON STATE UNIVERSITY

Calibration of the Nimbus-7 Scanning Multichannel
Microwave Radiometer (SMMR), 1979-1984
(1987) / Francis E A

A Bioeconomic Analysis of Altering Instream Flows
Anadromous Fish Production and Competing Demands
for Water in the John Day River Basin, Oregon
(1987) / Johnson N S

Behavioral and Demographic Characteristics of
Northern Sea Lion Rookeries
(1987) / Merrick R L

Zooplankton Variability in the California
Current, 1951-1982
(1987) / Roesler C S

RHODE ISLAND, UNIVERSITY OF

Abundance and Catch Composition of Three
Fishing Gears (Hook and Line, Tran and
Spear) in a Coral Reef, Santiago Island,
Cape Bolinao, Philippines
(1987) / Acosta A R

Stable Isotopic Analyses of Selected
Narragansett Bay Molluscs
(1988) / Allard D J

Zooplankton Community Structure and Copepod
Population Dynamics in Mesocosms: Effects
of Benthos and Euthrophication
(1988) / Banzon P V

The Effect of Fractionated Menhaden Oil on the
Growth and Development of Coho Salmon
(1987) / Bernatonis B J

The Feeding Ecology of the Winter Flounder,
Pseudopleuronectes Americanus (Walbaum), in
Narragansett Bay, Rhode Island
(1988) / Bharadwaj A S

Subtropical Fronts in the Sargasso Sea: A Four-
Year Satellite Analysis
(1988) / Bohm E

The Retail Value of Extending Shelf Life
of Fresh Seafood
(1987) / Brooks P M

The Development of a Fluorescent In Vitro
Method to Test the Phagocytic Engulfment
and Intracellular Bactericidal Abilities
of Macrophages from Winter Flounder
(1988) / Daniels T

Performance Evaluation of Scientific
Sampling Trawls
(1987) / Fahfouhi A

A Study of the Gulf Stream Downstream of
Cape Hatteras 1975-1986
(1988) / Gilman C S

A Contribution to the Early Life History
of the Cunner in Narragansett Bay, Rhode Island
(1988) / Gleason T R

Plasma Protein Changes in Atlantic Salmon
During Parr-Smolt Transformation
(1987) / Johanning K M

The Role of the Pressure Field in the Dynamics and
Energetics of the Gulf Stream at 73 Degree W.
(1988) / Kontoyiannis H

Evaluation of Lipid and Carbohydrate Energy Sources
in the Diet of Rainbow Trout
(1987) / Lanzuela J G

RHODE ISLAND, UNIVERSITY OF
(continued)

Catchability of Estuarine Scientific Sampling
Trawls
(1987) / Liuxiong X

Application of Hepatic Macrophage Aggregates
as Health Monitors of Narragansett Bay Winter
Flounder
(1988) / Lloyd D E

The Relationship Between Perceptions of
Gentrification and Waterfront Revitalization
Policies
(1988) / Merainer N

The Biology of Juvenile Scup in Narragansett Bay,
Rhode Island: Food Habits, Metabolic Rate and
Growth Rate
(1988) / Micheleman M S

Seafloor Alteration of Hydrothermal Sulfide
Deposits: Minerology and Chemistry of Altered
Sulfide Deposits from the Endeavor Segment,
Juan de Fuca Ridge and the Explorer Ridge.
(1988) / Oakman M R

Sinking and Sedimentation Characteristics of a
Diatom Winter/Spring Bloom
(1988) / Riebesell U

Occurrence of Nitrate Reductase Along a Transect
of Narragansett Bay
(1988) / Rysza-Culver K E

Two- and Three-Dimensional Inversions of
Magnetic Anomalies in the Mark Area (Mid-
Atlantic Ridge 23 Degree N)
(1988) / Schulz N J

Application of a Visual Census Technique for
Sampling Fishes in a Heavily Fished Coral
Reef at Cape Bolinao, Philippines
(1987) / Turignan R G

Estimates of Some Population Parameters of Skipjack,
in the Waters Adjacent to Sorong, Irian Jaya,
Particularly from Length-Frequency Data
(1987) / Uktolseja J C

Secondary Circulations in the Bottom Boundary
Layer Over Sedimentary Furrows
(1988) / Viekman B E

Diel Feeding, Threshold Feeding, and Gut
Evacuation Rate in the Marine Copepod
ACARTIA HUDSONICA from Narragansett Bay,
Rhode Island
(1988) / Wlodarczyk E

SAN FRANCISCO STATE UNIVERSITY

Rocky Intertidal Patch Succession after
Disturbance: Effects of Severity, Size,
and Position Within Patch
(1988) / DeVogelaere A

SOUTH CAROLINA, UNIVERSITY OF

An Electrophoretic Survey of Geukensia sp.
in South Carolina
(1988) / Cook J P

Hydrographic Influence on Morphology in
Neogloboquadrina Pachyderma from the South
Atlantic Sector of the Southern Ocean
(1988) / Healy E A

The Geochemical Behavior of Uranium-Thorium
Series Nuclides in the South Atlantic Bight
(1988) / Killeen T P

A Four Year Study of Above- and Belowground
Decomposition of Spartina Alterniflora
(1988) / McDannell-Firth E

Prey Selection by Juvenile Spot (Leiostomus
Xanthurus) on Meiobenthic Copepods and
Nematodes
(1988) / Nelson A L

Selective Digestion of Nematode Prey by
Juvenile Spot, Leiostomus Xanthurus (Pisces)
(1988) / Salamy D K

Effects of Dietary Protein Level and Feeding
Rate on Growth of Juvenile Oreochromis
Aureus: A Quadratic Response Surface Model
(1988) / Scholz P M

SOUTHWEST TEXAS STATE UNIVERSITY

Fish Culture Production Strategy for Marine Fishes
in Saltwater Rearing Ponds
(1988) / Rutledge W P

STEPHAN F. AUSTIN STATE UNIVERSITY

Physicochemical Characteristics of Nine
First Order Streams Under Three Riparian
Management Regimes in East Texas
(1988) / Brown C E

A Comparison of the Benthic
Macroinvertebrate Communities of the Main
Tributaries of Sam Rayburn Reservoir, the
Angelina and Attoyac Rivers, Texas
(1988) / Dion E O

The Benthic Macroinvertebrate Community
Structure of Lake Houston, Texas
(1988) / Howard J G

The Zooplankton of Lone Star Reservoir,
Texas
(1988) / Ready C C

TENNESSEE, UNIVERSITY OF (KNOXVILLE)

Photosynthesis-Irradiance Relationships of
Polar Phytoplankton Populations During
Winter Conditions
(1987) / Brightman R I

TEXAS A AND M UNIVERSITY

Sub-Surface Dissolution of Evaporites in the
Eastern Mediterranean Sea
(1988) / Camerlenghi A A

Food Habits of the Mantis Shrimp, Say, in
Galveston Bay, Texas
(1988) / Cappola V A

The Distribution and Composition of Organic
Matter in Recent Deltaic and Submarine Fan
Sediments
(1988) / DeFreitas D A

Spatial and Temporal Variation of Polynuclear
Aromatic Hydrocarbons, Pesticides, and
Polychlorinated Biphenyls in Crassostrea
Virginica and Sediments from Galveston Bay, Texas
(1988) / Fox R G

TEXAS A AND M UNIVERSITY
(continued)

Temporal and Spatial Variations of Butyltin
Concentrations in Bivalve and Sediment Samples
from Some Coastal Areas of the United States
(1988) / Garcia-Romero B

Provenance and Glacial History of Very Fine Quartz
Sand from the Weddell Sea, Antarctica
(1988) / Smith C H

Geochemistry of Arsenic and Antimony in
Galveston Bay, Texas
(1988) / Tripp A R

Microstructures and Fabrics of Sediments from the
Lesser Antilles Accretionary Complex
(1988) / Wackler J D

U.S. NAVAL POSTGRADUATE SCHOOL

The Physical Oceanography of the Northern
Baffin Bay-Nares Strait Region
(1987) / Addison V C

The Effects of Time-Dependent Winds and
Ocean Eddies on Ice Motion in a Marginal
Ice Zone
(1987) / Barker J L

Underwater Acoustic Backscatter from a Model
of Arctic Ice Open Leads and Pressure Ridges
(1987) / Browne M J

A Numerical Study of Baroclinic Circulation
in Monterey Bay
(1988) / Bruner B L

A Horizontal Spatial Requirement Study of
the Gulf Stream as Modelled by the IFDPE
(Implicit Finite Difference Parabolic
Equation) Acoustic Model
(1987) / Cease K L

The Variability of the Marine Atmospheric
Boundary Layer in the Greenland Sea Marginal
Ice Zone - A Case Study
(1988) / Dinkler K L

Sea Surface Current Estimates off Central
California as Derived from Enhanced AVHRR
(Advanced Very High Resolution Radiometer)
Infrared Images
(1987) / Fang C M

Predictibility of Ice Concentration in the
High-Latitude North Atlantic from
Statistical Analysis of SST (Sea Surface
Temperature) and Ice Concentration Data
(1987) / Fleming G H

The Temporal and Spatial Variability of the
marine Atmospheric Boundary Layer and Its
Effect on Electromagnetic Propagation in
and Around the Greenland Sea Marginal Zone
(1988) / Grothers D J

Sea Surface Temperature and Salainity
Structure of Cold Upwelling Filaments Near
Point Arean as Observed Using Continuous
Underway Sampling Systems
(1988) / Snow R L

Development of a Data Analysis System for
the Detection of Lower Level Atmospheric
Turbulence with an Acoustic Sounder
(1987) / Wroblewski M R

Acoustic Backscattering from Bottom Sediment
at Normal Incidence in the Laboratory
(1987) / Yu T T

WASHINGTON, UNIVERSITY OF

Museum Education: A Framework for Creating
a Seattle Maritime Interpretive Center
(1987) / Adams R Z

Age and Growth of Concholepas (Brugiere
1789) Using Microgrowth Structure and
Spectral Analysis
(1987) / Alderstein Gonzalez S

Oceanographic Implications of SMMR Ice Cover
Observations in the Sea of Okhotsk
(1987) / Alfultis M

Communicating Imperfect Scientific
Information to Fishery Managers: Sablefish
Management off the United States West Coast
(1987) / Augerot X

Estimation of Egg Production, Spawner
Biomass, and Egg Mortality for Walleye
Pollock, in Shelikof Strait from Surveys
During 1981
(1987) / Bated R D

A Comparative Study of Short-Term Swimming
Performance in Fry of Five Salmonid Species
at Different Temperatures
(1987) / Carpenter L T

The Population Dynamics and Growth of
Dungeness Crab, in the Nearshore Coastal
Environment of Washington State and a
Comparison with an Adjacent Estuary
(1987) / Carrasco K R

Induction of Triploidy in the Pacific Oyster
Using Cytochalasin B and Hydrostatic
Pressure: Optimal Treatments Depend on
Temperature
(1987) / Downing S L

Omega - 3 Fatty Acids and the Pollock Liver:
A Potential Northwest Industry
(1988) / Emmer L

Nutritional Properties of Distillers' Dried
Grains with Solubles: Effect of Processing
on Protein Quality, Vitamin Levels and the
Pressence of Antinutritional Factors
(1987) / Gazzaz S

Comparison of Pacific Hake Stocks in Inshore
Waters of the Pacific Ocean: Puget Sound
and Strait of Georgia
(1988) / Goni R

Energy Budget of the Grays Harbor Population
of Juvenile Dungeness Crab in Summer of 1983
and 1984
(1987) / Gutermuth F B

Factors Affecting Quality of Pedah Siam
(1987) / Hanafiah T A

Economic Assessment of Projects in Salmonid
Culture at Tasu Sound, Queen Charlotte
Islands, for the Council of the Haida Nation
(1988) / Jones R R

ALBERTA, UNIVERSITY OF

Gas Evolution on Athabasca Oil Sands
(1988) / Peacock D H

CALGARY, UNIVERSITY OF

Steam-Water Buoyant Jets from a Submerged
Orifice
(1988) / Asante B

Studies in Recovering Sublimation Products
(1988) / Baah C A

Dynamic Simulation of a Reciprocating
Compressor During an Emergency Shut Down
(1988) / Beckie J K

Dynamic Modelling of a Spouted Bed Reactor
with a Draft Tube Used for Hydrocarbon
Ultrapyrolysis
(1988) / Harvey J

Recovery of Heavy Oil by Steam-Assisted
Gravity Drainage -- A Three Dimensional
Approach
(1988) / Ong T S

The Production of Saskatchewan Heavy Oils
Using Steam-Assisted Gravity Drainage
(1988) / Sugianto S

CALIFORNIA STATE UNIV. (LONG BEACH)

An Experimental Study of Confined Swirl
Flows
(1988) / Finstad R G

COLORADO SCHOOL OF MINES

Gas Flow After Cementing - A Physical Model
(1988) / Abdalbake H N

A Rule Based Expert System for Well
Completion Decisions
(1988) / Fitterer D G

Removal of a Kick with the Partition Method
(1988) / Fleckenstein W

The Effect of the Pressure Gradient in the
Water-Invaded Region on the Performance of
Water-Drive Gas Reservoirs
(1988) / Fuadi W

A Rising-Bubble Apparatus for Assessing
Minimum Miscibility Pressure
(1988) / Hedges P L

Economic Simulation of International
Petroleum Agreements
(1988) / Huang G F

Hydraulic Fracture Fluid Leakoff Analysis
(1988) / Joseph C H

Vapor-Solid Equilibrium Ratios for Structure
II Natural Gas Hydrates
(1988) / Mann S L

Carbonate Acidizing: A Waterflood Approach
for Modeling Wormhole Growth
(1988) / Voge E A
Food Habits and Daily Ration of Greenland
Halibut in the Eastern Bering Sea
(1987) / Yang M S

IDAHO STATE UNIVERSITY

Potential Geologic Hazards to the Salt Lake
City Refineries, Chevron Oil and Northwest
Natural Gas Pipelines, and Possible
Ramifications to Pocatello, Idaho's Energy
Supply and Distribution System
(1987) / Zimmerman D W

KANSAS, UNIVERSITY OF

Development of a Computer Program to Solve
Unconstrained Material Balance Problems
Using the Sequential Modular and Equation
Oriented Solution Methods
(1988) / Gutentag R S

Rehological Study of Xanthan Solutions and
Xanthan/Chromium (III) Gelling Solutions
Using Steady and Small Amplitude
Oscillatory Shear
(1988) / Hester S K

A Kinetic Study of the Reduction of Nitric
Oxide by Carbon Monoxide Over an Alumina
Supported Platinum Catalayst
(1988) / Hussain S T

Experimental Measurement and Unified
Analysis of the Phase Equilibria and Liquid
Densities for the System 1,3-Butadiene
Acetonitrile at 305 and 330 K
(1988) / Laird D G

The Potential for Carbon Dioxide Miscible
Flooding in Central Kansas
(1988) / Poyser L A

LOUISIANA STATE UNIVERSITY

Measurement and Prediction of
CO(2)/Heavy-Hydrocarbon Phase Behavior Based
on Chemical Type
(1988) / Bankston B A

An Experimental Evaluation of Containment
Properties for Shales Associated with
Deep-Well Hazardous Waste Injection Zones
(1988) / Clark D A

A Reservoir Engineering Study of the NS 3 RD
Reservoir in the University Field, East
Baton Rouge Parish, Louisiana
(1988) / McGrath F P

LOUISIANA TECH UNIVERSITY

Calculation of Laboratory Relative
Permeability Using Microcomputers
(1988) / Brumley J L

Sand Prroduction Model for the Safaniya Oil
Field in Saudi Arabia
(1988) / Desai S F

Wellbore Heat Losses in Steam Injection
Operations
(1988) / Ghafoori M

Evaluation of the Accuracy of Modified
Horner's Method in Calculation of Skin
Factor and Permeability
(1988) / Pashazadeh A

Stable Rare Earth Tracer Technique for
Determining Preferential Flow Zones in a
Petroleum Reservoir
(1988) / Pathumanun R

LOUISIANA TECH UNIVERSITY
(continued)

The Rate Normalization Method for a Variable
Flow Rate Pressure Drawdown Test
(1988) / Payedar K

A Microcomputer Mud Logging System
(1988) / Robinson S W

Interactive Data Input Functions for
Enhanced Oil Recovery Models
(1988) / Schroeder A K

MASSACHUSETTS INSTITUTE OF TECHNOLOGY

Natural Gas Marketing Strategy in the
Changing Regulatory Environment
(1988) / Agee R E

Increased Effectiveness of Petroleum
Exploration Research
(1988) / Davies E J

Optimal Profit Sharing Rules for Petroleum
Exploration and Development in Jordan
(1988) / Hampson P R

MISSISSIPPI STATE UNIVERSITY

An Investigation of Slip Flow in a Crosslinked
Fracturing Fluid
(1986) / Mann T A

MISSOURI, UNIVERSITY OF (ROLLA)

Reservoir Simulation on a Mini-Computer
(1988) / Dieckmann M D

Investigation of a New In-Situ Method to
Determine Relative Permeability
(1988) / Melvin J D

Paraffin Deposition Property Measurement
Using Differential Scanning Calaorimetry
(1988) / Thornsberry K L

Improvement of a graphical Well Test
Analysis Personal Computer Package
(1988) / Wagner P J

MONTANA COLLEGE OF MINERAL SCI. & TECH.

A Study of Pre-1960 Oil Well Abandonment and
Completion Techniques in the Cut Bank Oil
Field, Montana
(1988) / Boerner C L

A Procedure for Financial Analysis of
Petroleum Industry Firms Using Arthur
Andersen and Company's Reserve Disclosures
Database
(1988) / Meister P M

A Study of Pre-1960 Oil Well Abandonment and
Completion Techniques in the Kevin-Sunburst
Oil Field, Montana
(1988) / Walsh K D

NEW MEXICO INSTITUTE OF MINING & TECH.

Experimental Investigation of the Linearity
of Fluid Mixing During Flow Through Porous
Media
(1988) / Akram N

A Five-Spot Polymer Flood Model
(1987) / Garlough J L

Computerized Method for Simulating Sucker
Rod Pumping System
(1987) / Luo F

Simulation of Matched Viscosity Miscible
Displacement in a Heterogeneous Medium
(1987) / Sultan A J

A Fracture Matrix Model of Electric
Properties in Reservoir Rocks
(1987) / Xu X

OKLAHOMA, UNIVERSITY OF

Design of a Computer Controlled Hydraulic
System for Simulating Sucker Rod Pumping
(1988) / Cox G W

Prediction of Compressibility
[i.e. Compressibility] Factors of Gas Phase
Coal Gasification Components
(1988) / Gangadhar K

Production Interference: A Study of Its
Impact on Well Performance and Spacing in
Oil Reservoirs
(1988) / Irani R K

In-Situ Bioreclamation for Subsurface
Gasoline Cleanup
(1988) / Nanjundeswar B V

Reservoir Characterization: A Strategy for
History Matching Studies of Petroleum
Reservoirs
(1988) / Pino H

The Analysis of Four New Oil and Gas
Property Valuation Parameters-Particularly,
as Applied to the Acquisition of Oil and
Gas Properties
(1988) / Randall B L

Characterization of the Non-Darcy Flow
Coefficient in Propped Hydraulic Fractures
(1988) / Whitney D D

The Influences of Alkaline Stream Injection
on the Physical Properties of Formation
Rock and Proppants
(1988) / Whitney S W

PENNSYLVANIA STATE UNIVERSITY

The Testing of Oxy Surfactant Solutions for
Tertiary Oil Recovery
(1987) / Yeager D L

PITTSBURGH, UNIVERSITY OF

The Extension of the Generalized
Dykstra-Parsons Method to the Polymer
Flooding of Stratified Petroleum Reservoirs
(1989) / Mahfoudhi J E

SOUTHWESTERN LOUISIANA, UNIVERSITY OF

Compositional Study of Crude Oil
Vaporization Under Multicontact-Muscible
Carbon Dioxide Injection in a P.V.T. Cell
(1988) / Beladi M K

Optimization of Drillability by Prediction
of Rock Strength from Formation Evaluation
(1988) / Dowla N

SOUTHWESTERN LOUISIANA, UNIVERSITY OF
(continued)

Total Systems Approach to Performance
Prediction and Optimization of a
Conventional Linear Water Flood
(1988) / Moaveni V

The Effect of Temperature on the Size
Stability of Conventional and Ceramic
Gravel Packs
(1988) / Toups M M

TEXAS A AND M UNIVERSITY

Viscous Pill Design Using Slip Velocity Correlations
(1988) / Baxter R L

Flow Characteristics of Hydraulic Fracture Proppants
Subjected to Repeated Production Cycles
(1988) / Blakeley D M

The Development and Utilization of a High-Speed
Laboratory Rock Drilling Apparatus
(1988) / Day J D

Sucker Rod Pumping Unit Diagnostics Using an
Expert System and Pattern Recognition Technique
(1988) / Derek H J

Evaluation of Electromagnetic Stimulation of
Texas Heavy Oil Reservoirs
(1988) / Doublet L E

Simulation of Waterflood Performance in Stratified
Sandstone Reservoirs Using Dynamic Pseudo Functions
(1988) / Fischbuch D B

A Technical and Economic Evaluation of Waterflood
Infill Drilling in West Texas Clearfork Carbonate
Reservoirs
(1988) / Flores D P

Multi-Phase Decline Curve Analysis with
Normalized Rate and Time
(1988) / Fraim M L

Development of a New Model for Predicting
Sucker-Rod Pumping System Performance
(1988) / Garcia J P

An Improved Method for the Determination of the
Wellstream Gas Specific Gravity for Retrograde Gases
(1988) / Gold D K

Determination of the Pressure at the Gas-Liquid
Interface Using Acoustic Speed Measurements
(1988) / Heggelund D G

The Effects of Fracture Fluid Cleanup Upon the
Analysis of Pressure Buildup Tests in Tight
Gas Reservoirs
(1988) / Johansen A T

Well Correction Factors for Three-Dimensional
Reservoir Simulation with Nonsquare Grid Blocks
and Anisotropic Permeability
(1988) / Kim D

The Use of Pre- and Post-Stimulation Well Test
Analysis in the Evaluation of Stimulation
Effectiveness in the Devonian Shales of
the Appalachian Basin
(1988) / Lancaster D E

An Investigation of Changes in Groundwater Quality
Caused by In-Situ Gasification of East Texas Lignite
(1988) / Leach K S

Prediction of Heptanes-Plus Equilibrium
Ratios from Empirical Correlations
(1988) / McKenna M J

Effects of Stimulation/Completion Practices on
Eastern Devonian Shale Well Productivity
(1988) / Nearing T R

Hole Cleaning Requirements with Seabed Returns
(1988) / Nordt D P

A Casing String Model for the Personal Computer
(1988) / Pflucker M P

The Deconvolution of Afterflow Affected Data
(1988) / Schellenbaum S E

Analyzing Aquifers Associated with Gas Reservoirs
Using Aquifer Influence Functions
(1988) / Targac G W

Detection of Interlayer Communication Using Type
Curves
(1988) / Tiefenthal S A

An Analysis of Avery Island Rocksalt
Data Using Nonlinear Regression
(1988) / Walker S C

An Expert System Advisor for Well Log Quality Control
(1988) / Warnken D K

TEXAS, UNIVERSITY OF (AUSTIN)

Analysis of Well Test Data Using
Microcomputers
(1988) / Al-Hamadah A M

Analysis and Design of Field Tracers for
Reservoir Description
(1988) / Allison S B

An Improved Model for the Prediction of the
Critical Pressure of Complex Mixtures
(1988) / Dalle Morre J W

The Estimation of Pore Size Distribution and
Reservoir Producibility in Running Duke
Field, Houston County, Texas
(1988) / Ellis R L

The Competition for Chromium in a Berea Core
Between Xanthan Biopolymer and the Resident
Clays
(1988) / Garver F J

Computerized Tomography Applied to the
Visualization of Fluid Displacements
(1988) / Hardham W D

Development of an Improved Equation of
State
(1988) / Haynes B

Filter Cake Properties and Their
Relationship to Differential-Pressure
Sticking
(1988) / Huycke J

Analysis of Areal Permeability
Variations - San Andres Formation
(Guadalupian): Algerita Escarpment,
Otero County, New Mexico
(1988) / Kittridge M G

A Model for Clay Filter Cake Properties
(1988) / Lee Z

TEXAS, UNIVERSITY OF (AUSTIN)
(continued)

The Effects of Stress and Wettability on the
Electrical Properties of Rocks
(1988) / Lewis M G

The Alteration of Wettability Due to
Interraction with Oil-Based Drilling
Fluids and Pure Surfactants
(1988) / Menezes J L

Experimental Methods for Classification of
Clay Minerals by Their Volume Expansions,
Swelling Pressures and Surface Areas
(1988) / Mese A I

A Field Study by Use of Wireline Logs,
Southeast Bonus Field (Yegua Formation,
Eocene) Wharton County, Texas
(1988) / Mosteller W D

Compositional Simulation of Carbon Dioxide
Oil Recovery Experiments
(1988) / Ogino K

Vectorization and Microtasking of UTCHEM on
the CRAY XMP
(1988) / Pashapour Alamdary A

Development of an Image Processing
Workstation for the Quantitative Analysis
of Fluid Displacements
(1988) / Reid C A

An Experimental Investigation of Fines
Migration in Two Phase Flow
(1988) / Sarkar A K

Visual Studies on the Effect of Two-Phase
Flow on Particle Transport in Porous Media
(1988) / Shirzadi-Ghalashahi S

Design and Scaleup of Micellar-Polymer
Flooding in the Presence of Heterogeneity
(1988) / Shook G M

A Study of the Reactive Flow EDTA in Berea
Sandstone
(1988) / Simmons D M

A Computer Simulation of a Shallow Offshore
Gas Kick During Drilling Operations
(1988) / Starrett M P

A Petrophysical Model for Shaly Sands
(1988) / Stenson J D

Modeling of Gravity Override of Condensing
Steam
(1988) / Suzuki H

A Pseudofunction Approach to the Description
of Viscous Crossflow in Multi-Layered
Permeable Media
(1988) / Thiele M R

Rheological Evaluation of Invert Emulsion
Muds Under Downhole Pressure and Temperature
(1988) / Tutuncu A N

Methylene Blue and Atterberg Limits Tests on
Selected Clays and Shales
(1988) / Wahrmund E T

A Three-Dimensional Network Model for
Porous Media
(1988) / Wang Y

Modeling of Precipitation and Dissolution
Processes with Precipitate Migration
(1988) / Wu G

Automation of the Bounding Process for the
Streamline Model
(1988) / Yabrudy E

Two Phase Flow Behavior in Vertical and
Inclined Pipes
(1988) / Zavareh F

The Effect of Flow from Perforation on
Two-Phase Flow: Implication for Production
Logging
(1988) / Zhu D

TULSA, UNIVERSITY OF

Normalization of Nitrogen Loaded Gas-Lift
Valve Performance Data
(1988) / Nieberding M A

Decanting Centrifuge Performance Study
(1988) / Thurber N E

UTAH, UNIVERSITY OF

The Fluidized Bed Pyrolysis of Bitumen
Impregnated Sandstone in a Large Diameter
Reactor
(1988) / Sung S H

WEST VIRGINIA UNIVERSITY

A Preliminary Study of Carbon Dioxide
Mobility Control by In-Situ Chemical
Precipitation
(1988) / Burger R M

Carbon Dioxide Mobility Control by In-Situ
Chemical Precipitation: A Core Study
(1988) / Gosnell D K

An Economic Feasibility Study of Developing
Gas Resources in the Appalachian Basin
Through Recompletion
(1988) / Kutska H M

A Model for Gas Well Production Decline
Analysis and Prediction
(1988) / Savage M C

WYOMING, UNIVERSITY OF

Screening of Kuwait Oil Reservoirs for
Application of Different Enhanced Oil
Recovery Techniques
(1989) / Alkafeef S F

A Numerical Study of the Effects of Well
Path Trajectory and Drill Pipe to Hole
Geometry on Torque and Drag in Deviated
Wells
(1989) / Loomer G R

AUBURN UNIVERSITY

On Interpolation of Operators
(1987) / Frazier B G

On Linear Filtering Systems
(1987) / Seymour P L

On Cantor Sets and Their Topological Dynamics
(1987) / Zanati S I

CORNELL UNIVERSITY

Spectral Characterisation of Orchards with
Landsat Thematic Mapper Data
(1988) / Taberner M J

GEORGIA, UNIVERSITY OF

Map Information Content of SIR-B Image Data
(1988) / Anderson R C

An Analysis of Coastal Change Using
Photogrammetric Techniques
(1988) / Cajka J C

INDIANA STATE UNIVERSITY

Investigation of Atmospheric Particulate
Distribution and Concentration in an Urban
Plume
(1988) / Wilcoxen B

MONTREAL, UNIVERSITY OF

La Cartographie Geomorphologique au 1/20 000
de Modeles Polygeniques: un Exemple des
Basses Terres du Saint-Laurent
(1988) / Bariteau L

Cartographe Thematique d'un Milieu
Semi-Aride Sahelien: Exemple de la Basse
Vallee du Senegal
(1988) / Kane R

Cartographie des Dommages de la Tordeuse
des Bourgeons de l'Epinette Par le
Traitement Automatique des Images Landsat-TM
(1988) / St-Onge B

MURRAY STATE UNIVERSITY

Utilization of Landsat Thematic Mapper Data
for Structural Geologic Mapping in an Arid
Region, Garfield County, Utah
(1988) / Kreighbaum D W

The Use of Landsat Thematic Mapper (TM) and
Geophysical Data for Hydrocarbon Exploration
in the Illinois Basin, Western Kentucky
(1988) / Major J D

NORTHERN ILLINOIS UNIVERSITY

Raster-Mode Generalization of Land Use/Land
Cover Maps
(1988) / Rebert P

PURDUE UNIVERSITY

Quantitative Assessment of Landsat TM Data
for Detailed Soil Mapping
(1988) / Wu Y

TEXAS CHRISTIAN UNIVERSITY

Determination of Non-Point Source Pollution
Within a Rural Watershed Utilizing Landsat
TM and a GIS/USLE Model
(1988) / Hayes S A

U.S. NAVAL POSTGRADUATE SCHOOL

LO-CO-GRAF (Low Cost Graphics): Generating
Maps to Support Command and Control/Crisis
Management Using Small Computers
(1988) / Bishop R G

Optimized Observation Periods Required to
Achieve Geodetic Accuracies Using the
Global Positioning System
(1988) / Bouchard R H

Determination of Network Attributed from a
High Resolution Terrain Data Base
(1987) / Choi S C

Three-Dimensional Image Generation from an
Aerial Photograph
(1987) / Coleman L G

Tectical Use of Digitized Maps
(1987) / Daly J E

Functional Specifications to an Automated
Retinal Scanner for Use in Plotting the
Vascular Map
(1988) / Dombrowski F J

A Graphics Facility for Integration,
Editing, and Display of Slope, Curvature,
and Contours from a Digital Terrain
Elevation Database
(1988) / Felhoelter D G

Tactical Use of Digitized Maps
(1987) / Gaffney S J

Terrain Classification from Digital
Elevation Data Using Slope and Curvature
Information
(1987) / Goodpasture B K

LO-CO-GRAF (Low Cost Graphics): Generating
Maps to Support Command and Control/Crisis
Management Using Small Computers
(1988) / Sabo R P

Mesoscale Applications of High Resolution
Imagery
(1987) / Scanlon R J

U.S.A.F. INSTITUTE OF TECHNOLOGY

Land Cover Classification of Lands at
Thematic Mapper Images Using Pseudo
Invarant Feature Normalization Applied to
Change Detection
(1988) / Hawes T

WASHINGTON STATE UNIVERSITY

Monitoring Losses of Important Farmlands
Through a Landsat-Based Geographic
Information System
(1988) / Rozenbaum S J

WASHINGTON, UNIVERSITY OF

Pyramid Classification of Land Use Patterns
in Multi-Spectral Remote Sensing Images
(1987) / Cave D

AKRON, UNIVERSITY OF

Self-Diffusion of Diluents in Cellulose
Acetate Derivatives
(1988) / Doshi J

A Monte Carlp Search for a Phase Transition
in the Self Avoiding Walk Model
(1989) / Palunas P

Thermodynamical Properties of Bismuth Thin
Films
(1988) / Qi X

Chemical Potential and Carrier Density in
Bismuth Thin Films Under Three Dimensional
Quantizations
(1988) / Yao J F

ALABAMA, UNIVERSITY OF (UNIVERSITY)

Ferromagnetic Resonance Study of Several
Barium Ferrites
(1988) / Darwish A A

Metallic Mesh Mirrors for Far Infrared
Lasers
(1988) / Koppang P A

Opto-Acoustic Absorption and Infrared Radio
Frequency Double Resonance of CH3OD
(1988) / Pak H K

ALBERTA, UNIVERSITY OF

Dose Calculations for Megavoltage Photon
Beams Using Convolution
(1988) / Field G C

Beam-Foil Study of Argon (VI)
(1988) / Ge Z

Parity Violation in Nucleon-Nucleon
Scattering
(1988) / Negm K E

Non-Inertial Quantum Field Theory in Flat
Space-Time
(1988) / Ng K P

Encapsulation as a Method for Controlling
Radiation Damage in the Electron Microscope
(1988) / Rice P M

ARIZONA STATE UNIVERSITY

A Real Spinor Approach to Relativistic
Scattering of Electrons
(1988) / Bergman J W

Measurements of the Elastic Constrants in
Germanium-Arsenic-Selenium Glasses
(1988) / DeMarco C J

The Low Frequency Raman Scattering of
Homopolymer RNA Fibers
(1988) / Qi R

The Interaction of Thermoregulation with
Activity at Low Ambient Temperatures
(1988) / Zebra E

ARKANSAS, UNIVERSITY OF

The Cross Section for Quenching of Nitrogen
Dioxide Fluorescence by Xenon Gas
(1988) / Ahamid N A

Two Applications of Photothermal Deflection
Spectroscopy: The Detection of NH(2) and
Real-Time Beam Profile Measurement
(1988) / Williams K A

AUBURN UNIVERSITY

Microwave Generation Through Relativistic Electron
Beam Interaction with a Hydrogen Plasma
(1987) / Baron M H

Atomic Oxygen Reaction with Metals
(1987) / Beshears R D

Electric Field Assisted Silver-Film Diffused
Glass Waveguides
(1987) / Chen C L

Semi-Classical Statistical Model of a Small
Metallic Particle
(1987) / Cho B

Thermionic Field Emission in Annealed
Gold-Gallium Arsenide Scholtley Barriers
(1987) / Crofton J B

Design and Testing of a Model ATF Instrumented
Limiter
(1987) / Galloway M D

Investigation of the Emissive Filament Technique
of Mapping Magnetic Surfaces on the Auburn
Torsatron
(1987) / Henderson M A

The Use of a Fabry-Perot Etalon to Determine the
Temperature of a Plasma Light Source
(1987) / Hwang M T

The Time-Correlation of Emitted Light, Current and
Specimen Voltage Waveforms for the Electrical
Breakdown of Dielectrics in the Nanosecond Time
Regime
(1987) / Watkins M L

BAYLOR UNIVERSITY

Hydrodynamic Code Stimulation of Low Angles
Ejecta from Hypervelocity Impact
(1988) / Chen Y L

Non-Linear Optical Response of Four Level
Automic Systems
(1988) / Deng L

Computer Modeling of Optical Activity of an
Amino Acid in Liquids
(1988) / Dong Q

Spatial Variation of Dust in the Coma of
Comet P/Halley as Observed by the Giotto
Didsy Experiment
(1988) / Goad H S

Physical-State Effect in HE+ -Induced KLL
Auger Spectra of Carbon-Containing Molecules
(1988) / Manthani M

BOWLING GREEN STATE UNIVERSITY

Computational Method for Determining Optical
Constants in the Infrared from the Measured
Film Reflectance and Thickness
(1989) / Cai J

The Proximity Effect in High Temperature
Superconductors
(1989) / Chen J H

DELAWARE, UNIVERSITY OF

A Procedure for the Analysis of Stresses
Within Thin Films by X-Ray Diffraction
(1988) / Irvine I C P

The Search for a Roughening Transition:
Crystal Growth at High Pressure
(1988) / Johnson D R

A New Look at the Klein-Gordon Equation in
DeSitter Space
(1987) / Kinsley W N

Micellar Anisotropy in the Potassium Laurate
Regular Isotropic Phase and Re-Entrant
Isotropic Phase
(1987) / Oxborrow C A

EASTERN KENTUCKY UNIVERSITY

Proton-Induced Reactions in Scandium-45
(1988) / Englehardt R T

FLORIDA ATLANTIC UNIVERSITY

A Starting Potential for Band Theory
Calculations Generated from Thomas-Fermi
Theory
(1988) / Horvath E A

FLORIDA, UNIVERSITY OF

Study of New Polysiloxane-Based
Scintillators
(1988) / Bowen M

Tunneling in Quantum Solids: NMR Studies on
Solid Hydrogen
(1988) / Rall M

GEORGETOWN UNIVERSITY

Polarization Coupling in the Accousto-Optic
Interaction of Surface Waves
(1988) / Graver W R

Diffraction of Light by Ultrasonic Pulse
(1988) / Wolf J W

GEORGIA, UNIVERSITY OF

Phase Transition in a 2-D Lattice Gas Model
(1988) / Chen C W

Single Nondissociative Ionization of
Molecular Hydrogen by Electron Impact
(1988) / Durden A S

Monte Carlo Study of the Phase Diagrams and
Critical Behavior of the Ising Square
Lattice with Nearest, Next Nearest-Neighbor
and Three Body Interactions
(1988) / Li J

Analysis of the Two Particle Dirac Equation
(1988) / Summers J

GUELPH, UNIVERSITY OF

Model Calculations of Classical, Quantum and
Semiclassical Collision-Induced Absorption
Lineshapes
(1988) / Basile A G

Evidence for a 17 KEV Neutrino in the Beta
Spectra of Tritium and 35S
(1988) / Hime A

Charge Induced Absorption in Proton
Irradiated Solid Hydrogen Deuteride
(1988) / Miller J J

Vacuum Ultraviolet Emission from Cryogenic
Hydrogen and Helium Deuteride
(1988) / Tokaryk D W

Monte Carlo Study of the Second Virial
Coefficient for Polymers in Dilute Solution
(1988) / Van Prooyen M

Corrections to the Tritium Beta Decay
Spectrum Arising from Radiative and Atomic
Effects, and Their Relationship to Neutrino
Mass Experiments
(1988) / Weisnagel S G

IDAHO STATE UNIVERSITY

A Comparison of Thickness Measurement
Techniques
(1988) / Khosro-Shahroudi K A

Unified Theory of Gravitation, Spin, and
Isotopic Spin
(1988) / Nelson C B

The Development of Particle Induced X-Ray
Emission Analysis at Idaho State University
(1988) / Walsh D S

IOWA STATE UNIVERSITY

Performance of a Prototype Electromagnetic
Calorimeter
(1988) / Holmes R R

Cloud Fluid models for Propagating Star
Formation
(1988) / Titus T N

JOHN CARROLL UNIVERSITY

An Interferometric Fiber Optic Seismograph
(1988) / Curley J R

Fiberoptic Gallium-Arsenide Temperature
Sensor
(1988) / Das P

The Size and Shape of Sodium Dodecyl Sulfate
Micelles in Sodium-Sodium Chloride-Water
Solutions
(1988) / Mishic J R

An Ultrasonic Study of the High Temperature
Superconductor YiBa2Cu3O7-X
(1988) / Wolanski M R

KANSAS, UNIVERSITY OF

Derivation of Proton Phase Space Density
in the Magnetosphere of Earth from the
AP8 Radiation Model
(1988) / Jiang R

A Monte Carlo Simulation of Satellite
Absorption of Charged Particles in the
Uranian Magnetosphere
(1988) / McKee C M

Matching Algorithms for CCD Images: An
Application to UBVRI Photometry
(1988) / Payne T E

LOUISIANA TECH UNIVERSITY

A Programmable Interface for the IEEE-488 BUS
(1988) / Cartmill J W

LOUISVILLE, UNIVERSITY OF

Electron Beam Energy Spectrum Measurement with a
Cerenkov Detector
(1988) / Bhandare N

Preparation and Properties of Strontium Substituted
High Temperature Superconductors
(1988) / Fang K

Collision Broadening and Asymmetry in Stellar Spectra
(1988) / Gocke R K

The Physical Origin of Stellar Image Profiles
(1988) / Gorline J L

Long Term Dobson Ozone Data Trend and Its
Correlation with Latitude
(1988) / Hou J C

Radiative Collisions in Atomic Hydrogen: Satellites
and Continua on the Lyman Alpha Line
(1988) / Huang Q

Laser Spectroscopy of the OH Molecule in a Mixture
of Helium and Water Vapor
(1988) / Hunter M F

Effects of HCO(3) and CO(2) on K[+], NA[+] and
Cl[-] Pathways of Nutrient Membrane of Frog
Gastric Mucosa
(1988) / Wu J

A Study of Hadronic Production of Charmed Particles
at High Energy (530 Gev/c)
(1988) / Xiao H

LOWELL, UNIVERSITY OF

Electric Field Induced Growth of Nonlinear
Optical Materials and Their Characterization
(1988) / Francis M W

Rapid Recognition of Metallic Objects in
Ambient Lighting
(1989) / Tobin E J

Improved Small Cluster Model of Amorphous
Tetrahedrally Bonded Materials
(1988) / Union D M

Fabrication and Testing of MIS Solar Cells
(1988) / Walters F S

Measurements of Optical Phase Conjugation of
Barium Titanium Dioxide Crystals
(1988) / Yang Y

MAINE, UNIVERSITY OF

A Hall Effect Study of Silicon Doped Zinc
Selenide
(1988) / Geiger R J

A Neutron Scattering Study of the Melting
Transition of Argon-Carbon Monoxide Mixtures
on Graphite
(1988) / Jagoe R H

A Study of the Transfer of Radon-222 from
Potable Water Sources to Household Air
(1988) / Lachapelle E B

Laser Raman Spectroscopic Studies of Vapor
Deposited 4,4'-Oxydianiline Films and Vapor
Deposited 4,4'-Oxydinaline/Pyromellitic
Dianhydride Films on Polycrystalline Copper
and Polycrystalline Silver Substrates
(1988) / Mack R G

Kinetics of Thermally Induced Transition
Studied by Microcalormetry
(1988) / Yang J

MANITOBA, UNIVERSITY OF

Fluorescence Lifetime Analysis to Study
DNA Conformation
(1988) / Heller D P

Dosimetry of Technetium-99m Pyrophosphate:
A Bone Imaging Radiopharmaceutical
(1988) / Lowe D

Photomultiplier Tube Dark Current
Enhancement in Mixed Radiation Fields and
Possible Application to the Development of
a Fast Neutron Detector
(1988) / Mayer J K

Assessment of Methods of Correction for
Scatter and Attenuation in SPECT Imaging
(1988) / McFarlane Y

The Bethe Ansatz Approach to the Two-Magnon
Problem for Finite Anisotropic Ferromagnetic
Chains of Arbitrary Spin
(1988) / Scott W K

D.C. Magnetization Measurements on Two
Platinum Manganese Spin Glasses
(1988) / Yeung W T

MARQUETTE UNIVERSITY

Fission Fragment Kinetic Energy and Fission
Mass Yield Curves
(1988) / Kyrpidou A

MASSACHUSETTS INSTITUTE OF TECHNOLOGY

A Computational Physics Array Processor
Instruction Set
(1988) / Brommer K D

Magnetooptical Properties of Implanted
Manganese Cadmium Telluride
(1988) / Gonzalez H J

Time-Resolved Optical Spectroscopy of
AlGaAs/GaAs Heterostructures
(1988) / Mims V A

MASSACHUSETTS, UNIVERSITY OF

Analysis of Time Measurement Resolution of
the Brookhaven Experiment 766 PHT System
(1988) / Nordberg M M

MICHIGAN TECHNOLOGICAL UNIVERSITY

A Non-Relativistic Study of the Electron
Affinities of Bound States in Negative
Transition Metal Ions
(1988) / Cai Z

Computer Simulation of Sintering by Surface
Diffusion
(1988) / Huang K

MICHIGAN TECHNOLOGICAL UNIVERSITY
(continued)

An Ab Initio Study of the K-alpha Spectra of
Lithium Ions in Lithium Halides (LiX)
(1988) / Kaldon P E

Ab Initio Studies of the Dissociation of an
Energetic Solid (Nitromethane) in the
Presence of Neighboring Charge Defects
(1988) / Kaldon P E

Crystal Growth of Cd(1-X)Mn(X)S by Chemical
Vapor Transport
(1988) / Lu J

Study of Sintering Mechanism of Ice
(1988) / Shen W

A Potentiostatic Double-Step Method for
Measuring Diffusion and Trapping of Hydrogen
and Deuterium in Iron Single Crystal
(1988) / Shi X

An Analysis of Magnetic Field Gradient Coils
for NMR Imaging
(1988) / Wilken D E

Computer Study of the Trapping and the
Migration Behaviors of Hydrogen Atom in the
Dislocation of B.C.C. Iron
(1988) / Yan, M

Electron Spin Resonance Investigation of
Magnesium Oxide Single Crystals Containing
Mn(2+) and Fe(3+) Impurities
(1988) / Zhang Y

Simulation of Isovalent Impurites in
Magnesium Oxide Using Hartree-Fock Clusters
(1988) / Zuo J

MINNESOTA, UNIVERSITY OF

A Movpe Reactor for II-VI Compounds
(1988) / Angelo J E

A Monte Carlo Simulation of Cosmic Ray
Propagation
(1988) / Hollabaugh M

The Analysis of Magnetic Resonance Imaging
T2 Relaxation Measurements
(1988) / Koukkari M W

MISSISSIPPI STATE UNIVERSITY

The Development of an Automatic Microprocessor-Based
Sodium D-Line Reversal Instrument for Temperature
Measurement on Simulated Coal-Fired
Magnetohydrodynamic Gas Flows
(1986) / Campbell C R

The Uniqueness of Phase Retrieval from Intensity
Measurements
(1986) / Jalil M A

The Reconstruction of Bandlimited Intensities from
Their Spatially Averaged Values: The Effect of
Detector Size and Spacing
(1986) / Linder D W

Construction of a Nitrogen Laser Pumped Dye
Laser System
(1986) / Prasertwong S

The Effects of Temperature and Conductivity on
Dielectrics Under High Voltage Stress
(1986) / Sayyar H

Particle Sizing in Flames by Two Color Laser
Transmissometry
(1986) / Sribuddhachart B

MISSISSIPPI, UNIVERSITY OF

The E2 Nuclear-Resonance in 130(Te)
(1988) / Remond L A

The Effect of Ultrasound on Ion Transport
Across Biological Membranes
(1988) / Wu J

MISSOURI, UNIVERSITY OF (COLUMBIA)

Two Studies in One-Dimensional Scattering Theory:
Direct Electromagnetic Scattering from Some
Variable Index of Refraction Slabs and Inverse
Scattering from Radial Central Potentials
(1988) / Dick C R

MONTANA STATE UNIVERSITY

Magnetic Properties of the Pseudo
One-Dimensional Heisenberg Spin On-Half
Antiferromagnet Acetamidinium
Tetrachlorocuprate(II)
(1987) / Landenburger L A

MONTREAL, UNIVERSITY OF

Etats Deformes des Baryons Dans le Modele
du Sac du Mit
(1988) / Bachkhaznadji A

Un Montage Automatise Exploitant la
Technique de Photoconductivite Transitoire
et son Application a la Mesure de la
Mobilite de Derive Dans le Titanate de
Strontium
(1988) / Banville C

Mesure de Distributions Angulaires Par
Rapport a l'Axe du Spin Nucleaire,
Application Au 149Gd et a l'157Ho
(1988) / Banville F

Etude Experimentale du Bilan Energetique
d'une Decharge Entretenue Par une Onde de
Surface
(1988) / Barbeau C

Limites Dans la Perception Visuelle de
Patrons Symetriques Formes Par Radom-Dot
(1988) / Castillo Herrera C I

Le Potentiel de Bonn et Quelques Proprietes
de l'Helium 3 et du Triton
(1988) / Dufresne M

Mise au Point d'un Protocole d'Utilisation
de Dosimetres Thermoluminescents TLD-100 en
Radiotherapie
(1988) / Giroux S

l'Utilisation Simultanee de Dosimetres
Thermoluminescents et d'Emulsions
Photographiques en Dosimetrie Clinique et
la Caracterisation d'un Densitometre Optique
(1988) / Goulet M

Preparation par Pualverisation Cathodique et
Caractersation des Films d'Alliages Amorphes
Ni x Zr 1-x
(1988) / Huai Y

MONTREAL, UNIVERSITY OF
(continued)

Deconvolution par Simulation des Courbes
d'Energie Transverse, Application a la
Cible Gazeuse de l'Experience Helios
(1988) / Lussier J G

Polarisation Circulaire des Etoiles T-Tauri
(1988) / Nadeau R

Etude de la Structure Nucleaire des Noyaux
de 77BR et 79BR
(1988) / Nadon N

Analyse des Distributions d'Energie
d'Etoiles Sous-Naines de Type B et OB
(1988) / Pageau M

Etude des Processus de Piegeage et de
Depiegeage Dans le Silicium Cristallin par
Spectroscopie Transitoire des Naveaux
Profonds
(1988) / Painchaud Y

Effets de la Frequence d'Excitation d'un
Plasma d'Ondes de Surface sur la Gravure
du Polyimide
(1988) / Sauve G

MURRAY STATE UNIVERSITY

Electrical Characteristics of Frog Skin
(1988) / Starks D J

NEW MEXICO INSTITUTE OF MINING & TECH.

On Finding Charges in Thunderclouds from
Aircraft Measurements of Electric Vector
(1988) / Feng H

Faraday Rotation in Abell Clusters of
Galaxies
(1987) / Hennessy G S

A High Resolution Study of the Inne Two
Kiloparsecs of M87 at 6cm
(1987) / Hines D C

NEW MEXICO STATE UNIVERSITY

X-Ray Emissions from Cerium and Oxidized
Cerium for Incident Electron Energies Near
the M5 and M4 Ionization Energy Levels
(1988) / Mason B E

Measurement of the Absolute Population
Densities of the 4p[5]6p[3/2]2 Level of
Krypton Under Two-Photon Excitation Using
an Argon Fluoride Excimer Laser
(1988) / Oertel J A

Absolute Number Densities for Ionized
Krypton by a Three-Photon Process for Use
in Laser-Guiding of Electron Beams
(1988) / Silk R

NEW MEXICO, UNIVERSITY OF

A Method of Measure in the Thermal Characteristics
of Soil
(1988) / Bayliss S C

Effects of Thermal Instability on Intrinsic and
Extrinsic Semiconductors
(1988) / Goeller R

NEW YORK, STATE UNIVERSITY OF (ALBANY)

The Selective Low Pressure Chemical Vapor
Deposition of Tungsten on Silicon Implanted
Silicon Dioxide
(1987) / Hennessy W A

A Study of Mercuric Film Formation on
Silicon
(1987) / Jones P L

Theory of Nuclear Quadrupole Interactions in
Solid Hydrogen Fluorides
(1987) / Mohamed N S

NEW YORK, STATE UNIVERSITY OF (BUFFALO)

A Theory on Hydrogen Diffusion in Metals:
Isotope Effect
(1988) / Garcia A

Dynamic Light Scattering Study of Poly
(Ethylene Glycol)-Induced Fusion of Small
Unilamellar Vesicles of Phospholipid
Mixtures
(1988) / Guo Y

Resonance Energy Transfer Between Carbon
Monoxide and Oxygen Molecules Using Dye
Laser Intracavity Absorption Spectroscopy
(1988) / Hrinishin J

Higgs Particle Production in
Electron-Positron Collisions
(1988) / Sankar S P

NORTH CAROLINA STATE UNIVERSITY

Characterization of Mixed-Oxide
Superconductors by Resistance and Inductances
Measurements
(1988) / Desai A Y

Diamond and Diamond-Like Thin Films: A
Raman Scattering Analysis of Carbon Bonding
(1988) / Shroder R E

The Effects of Turbulent Viscosity on Mass
Entrainment and Collimation of the
Astrophysical Jets Associated with Active
Galactic Nuclei
(1988) / Weglarz R P

NORTH DAKOTA STATE UNIVERSITY

Prebreakdown Characteristics of Thin Film Coated
Molybdenum Electrodes
(1988) / He P

Ground State Energy of Nuclear Matter in Quantum
Thermodynamic Perturbation Theory
(1988) / Ren S

NORTHEASTERN LOUISIANA UNIVERSITY

Some Variational Principles in Mathematical
Physics and the Foundation of an
Approximation Method in Physics--The
Differential Equation--Functional Pair
Relations
(1988) / Liu L

NORTHERN ILLINOIS UNIVERSITY

Mossbauer Study of the Oxygen Order and
Superconductivity in Yttrium (Barium
Strontium) (Copper Iron) Oxide
(1988) / Lee H

NORTHERN ILLINOIS UNIVERSITY
(continued)

A Conversion Electron Mossbauer Study of
the Lattice Properties of Thin Film
Superconductor Niobium Tin
(1988) / Matykiewicz J L

Transverse and Longitudinal Forces of a
String on the Guitar Bridge
(1988) / Watson E T

Mossbauer Study of the Oxygen Order and
Superconductivity in Yttrium Praseodymium
Barium (Copper Iron) Oxide
(1988) / Youn P

OAKLAND UNIVERSITY

A Monte-Carlo Diffusion Approach to Quantum
Cosmology
(1988) / Kerr P L

Thermal Wave Imaging for Nondestructive
Detection of Adhesive in Composite
Structures
(1988) / Simon D L

OKLAHOMA, UNIVERSITY OF

Next Generation Ultra High Speed
Electronics: A Theoretical Analysis of
Two Quantum Effect Device Technologies
(1988) / Biegel B A

Resistive Plastic, Proportional Tube, Gas
HYadron Calorimeter
(1988) / Jaffery T S

OREGON STATE UNIVERSITY

Angular Correlations of Gamma Rays in the
Decay of [109]Pd and {111}Pd.
(1987) / Brown D E

Nuclear Magnetic Resonance in Gallium-Doped
Liquid Selenium(0.5) Telluride(0.5).
(1987) / Doll B

Monte Carto Simulation of Oxygen Mobility
in Cubic Zirconia
(1987) / Gutekunst B

Quadrupole Moments and G-Factors of Low-Lying
Bands in Odd-A-Nuclei (150<A<190)
(1987) / Rieckert V

OREGON, UNIVERSITY OF

Geometrical Branching Model of High Energy
Hadron-Hadron Collisions
(1988) / Chen W

Persistent Photocurrent Decay and Infrared
Quenching Mechanisms by Capture of
Photoelectrons in GASS/ALGAAS
Heterostructures
(1987) / He L

Fatty Acid and Lipid Monolayers on Air-Water
Interfaces
(1988) / Hifeda Y M

Radiative Weak Decay of B-Meson
(1987) / Lo P C

The Effects of the Implantation of Oxygen,
Nitrogen and Carbon Impurities on the
Density of States in the Mobility Gap of
Hydrogenated Amorphous Silicon
(1988) / Michelson C E

High Resolution Spectroscopy of Carbon
Dioxide and Sulfur Dioxide
(1987) / Milkman I W

An Experimenhtal Study of Externally
Modulated Rayleigh-Benard Convection in
Helium I
(1987) / Niemela J J

The Influence of External Modulation on the
Stability of Azimuthal Taylor-Couette
Flow: An Experimental Investigation
(1988) / Walsh T J

PENNSYLVANIA STATE UNIVERSITY

Analysis and Optimization of a Pressurized
Water Reactor Loading Pattern Using the
One-and-a-Half Dimensional Core Model
(1987) / Petrovic B

PENNSYLVANIA, UNIVERSITY OF

Calibration of Calorimeter Modules at a
Fermilab Experiment
(1988) / Gerecht J S

QUEENS UNIVERSITY

Transient Generation of Elastic Waves in
Solids and Some Applications to Materials
Characterization
(1988) / Bresse L F

A Large Hall Probel Readhead Sensor for
Magnetic Inspection of Steel Pipelines
(1988) / Cheng C W

Magnetic Properties of Pipeline Steel Under
Mechanical Stress
(1988) / Schonbachler M V

REGINA, UNIVERSITY OF

A Study of the Differential Cross-Section
of the Pion-Deuteron Breakup Reaction
(1988) / Pafilis V

RENSSELAER POLYTECHNIC INSTITUTE

The Pressure Dependence of Viscosity for a
Binary Fluid Undergoing Phase Separation
as Studied by Light Scattering
(1988) / Bilodeau L A

X-Ray Conformational Study of DNA Superhelix
(1988) / Chen S W

Solubility of Gold in Lead-Indium Alloys
(1988) / Hall C W

Computer Simulation of Electromigration
in Thin Films
(1988) / Meng P P

The Analysis of [14]N(Gamma, Pi +)[14]C
(g.s.) at E(Gamma)=226 MeV and E(Gamma)
=260 MeV
(1988) / Myers D B

RENSSELAER POLYTECHNIC INSTITUTE
(continued)

A Dynamic Study of Energetic Materials by
Rayleigh-Brillouin Scattering Techniques:
Aqueous Solutions of Hydroxyl Ammonium
Nitrate and Triethanol Ammonium Nitrate
(1988) / Patton-Hall K E

Stability in a Quantum Theory of Electron
Transport
(1988) / Radzihovsky L R

Photomodulation Spectroscopy of
Semiconductor Microcrystals
(1988) / Redwing R D

An Analysis of Various Magnetohydrodynamic
Equilibrium Configurations and the Vertical
Stability Problem for the Proposed
Rensselaer Spherical Tokamak
(1988) / Santoro R A

The Use of Numerical Methods and
Microcomputers in Undergraduate Experiments
(1988) / Stevens M R

RHODE ISLAND, UNIVERSITY OF

Structures of Large Icosahedral and Cuboctahedral
FCC Lennard-Jones Clusters
(1988) / Xie J

ROCHESTER, UNIVERSITY OF

Injection Seeding of a Q-Switched Nd:YLF
Oscillator
(1987) / Ball G

Synthesis Design for Optical Thin Films and
Automatic Construction of High/Low Index
Multilayer Systems
(1987) / Bouzid A

Interferometric Measurement of Refractive
Index
(1988) / Hopler M D

Scalar and Vector Properties of Phase
Conjugation by degenerate-Four-Wave-Mixing
(1987) / Jacobs A A

Evaporative Deposition of Aspheric Surfaces
(1988) / Martin J D

A Three Dimensional Superposition Array
(1987) / Zinter J R

ROSE-HULMAN INSTITUTE OF TECHNOLOGY

Angularly-Polished Optical Fibers and Their
Application to Right-Angle Connectors
(1989) / Caughey J P

SAN DIEGO STATE UNIVERSITY

Error Corrections for a Dual-Frequency
Interferometer
(1988) / Csanadi C J

Critical Q Vector Lamb Theory and Internal
Modulation in Helium-Neon 6328 A Lasers
(1988) / Rickards R R

The Photorefractive Effect in Bismuth
Silicon Oxide
(1987) / Roberts M W

Off Resonance Pumping Effects in the NMR of
Solids
(1988) / Sanders J P

Luminescence Studies of Gallium Arsenide
Multiple Quantum Wells in High Pulsed
Magnetic Fields
(1987) / Shen K

SASKATCHEWAN, UNIVERSITY OF (SASKATOON)

Oglow - Observations of Oxygen Emissions
from the Space Shuttle
(1988) / Gale M R

Coordinated Ground - Satellite Auroral
Studies
(1988) / Steele D P

SHERBROOKE, UNIVERSITY OF

Etude de la Reflexion d'Electrons Basse
Energie a une Interface Imparfaite
(1988) / Cobut V

Effet de l'Insuline sur le Potentiele de
Repos Depresse par l'Hypoxie et l'Absence de
Substrate Exogene Metabolisable Chez le
Coeur Isole de Lapin
(1988) / De Lorenzi F

Hamiltonien de Spin Pour un Isolant
Magnetique
(1988) / Demers J G

Proprietes Multifractales et Phenomenes
Critiqies
(1988) / Fourcade B

Photoluminescence et Absorption Infrarouge
Dans les Composes de Potassium Tantalum
Oxide et de Potassium Tantalum Niobium Oxide
(1988) / Grenier P

Interface Copolymere-Silicium
(1988) / Lebel J

Transition SPIN-PEIERLS: Etude d'un Modele
Magnetoelastique de Facture Quantigue
(1988) / Mailhot A

Etude Comparative de la Cinetique
d'Inactivation du Courant Sodium Rapide de
la Cellule Cardiaque Isolee en Utilisant la
technique de Patch-Clamp
(1988) / Morier N

Parametres Menant a la Reconnaissance des
Ondes Caracteristiques sur le Trace
Electrocardiographique
(1988) / Nguyen C L

Etudes et Applications des Phenomenes
d'Ondes de Surface a l'Aide du Microscope
Acoustique
(1988) / Rene N

Methods de Bosons et Bosons Generalises
(1988) / Savard J

SOUTH CAROLINA, UNIVERSITY OF

A Monte-Carlo Study of Photo-Breakup of the
Deuteron
(1988) / Breguet J

Bistable Motion of a Relativistic Electron
(1988) / Saunders M

SOUTH DAKOTA SCHOOL OF MINES AND TECHNOL.

Further Investigations of the Interfacial
Layer Between Ice and Surface Treated
Concrete
(1988) / Ewing A

SOUTHERN ILLINOIS UNIVERSITY

Broken Symmetry and Mass Splitting in Finite
Temperature Quantum Field Theory
(1988) / Kelly M E

Studies in Coupled Mode Theory of
Semiconductor Lasers
(1988) / Stephens E F

Piezoelectric Detection of Photoacoustic
Effect
(1988) / Tong W

SOUTHWEST TEXAS STATE UNIVERSITY

Carbon-13 Nuclear Magnetic Resonance Investigations
of Rhodium(I)-Alkene Complexes
(1988) / Ripplinger E B

SOUTHWESTERN LOUISIANA, UNIVERSITY OF

Determination of Recent Sediment Rates in
White Lake Using 210-Lead and 137-Cesium
(1988) / Lee R

STEVENS INSTITUTE OF TECHNOLOGY

High Raman Scattering Intensities in Heavy
Metal Germanate Glasses
(1987) / Miller A E

A Fundamental Study of p-n-p Gallium Indium
Arsenide/Indium Phosphide Biopolar Laser
Transistors
(1987) / Uchida T

TENNESSEE, UNIVERSITY OF (KNOXVILLE)

Laser Irradiation and Heating of Spherical
Particles
(1987) / Dobson C C

Directed Motion Doppler Shift Effects on
Infrared Band Models of Emission and
Absorption in a Hypersonic Conical Flow
(1987) / Drakes J A

Acetylene Intensity Measurements Using a
Tunable Laser Spectrometer
(1987) / Salanave J L

Optical Constants of Solid Hydrocarbon Films
Produced in a Glow Discharge
(1987) / Votaw P C

TEXAS A AND M UNIVERSITY

Measurement of the Half-Lives of [66]AS and [70]BR
High Z Test of CVC (Standard Model) Via Super-
Allowed Fermi Decay
(1988) / Burch R H

Measurements of Electron Transfer from
Carbon Dioxide and Hydrogen Atoms to Carbon
Dipositive and Carbon Tetrapositive Ions
(1988) / Chavez S C

Frustration Tuning in Regular and Random
Superconducting Networks
(1988) / Chen R L

Determination of Bound and Pseudostate Wavefunctions
of the Helium Atom for Cross Section Calculations
(1988) / Cho W

An Evaluation of the Detection System
for a Positron Emission Tamograph
(1988) / Dixon L C

Photoelectron Spectroscopic Study of the Surface
Reactivity of the High T(c) Material Yttrium Barium
Copper Oxide
(1988) / Liu H X

Two Aspects of the K500 Cyclotron Project
(1988) / Van Baalen A C

Developing Improved Nuclear Magnetic Resonance
Marginal Oscillator Spectrometers for Advanced
Teaching Laboratories
(1988) / Willingham F P

TEXAS CHRISTIAN UNIVERSITY

A Monte Carlo Simulation of a Proposed
Measurement-While-Drilling Nuclear
Well-Logging Device
(1988) / Lampley C M

TEXAS, UNIVERSITY OF (ARLINGTON)

Computer Modeling of the Vapor-Liquid
Interface of Ethane in Relation to
Positronium Decay Rate Measurrements
(1988) / Ayyalasomayajula S M

Positron Annihilation in Conducting Polymers
(1988) / Krishnamoorthy S

Positron Annihilation Induced Auger Electron
Emission from Copper and Iron
(1988) / Lei C

Positronium Localization in Carbon Dioxide
(1988) / Ward M H

TEXAS, UNIVERSITY OF (AUSTIN)

Dye Laser Frequency Stabilization Using an
External Cavity Transducer
(1987) / Boyd T L

Charge Exchange Measurements on the Texas
Experimental Tokamak
(1988) / Empson K L

Epitaxial Growth and Magnetic Properties of
Single-Crystal Thin Iron Films Grown on
Silver(100) Surface
(1988) / Hsieh T Y

An Undergraduate Electrodynamics Lecture
Series on Audio-Cassette
(1988) / Kastner C W

Dispersion Relation Check by Plasmon
Shifts Using Three Wavelengths of Light
(1988) / Koschmieder T H

A Comparison of the Experimental Results
with a Chemical Model in a
Belousov-Zhabotinskii Reaction
(1988) / Lee K G

MHD Simulation of Magnetic Reconnection in
the Earth's Polar Cusp
(1988) / Lee K Y

WASHINGTON, UNIVERSITY OF

A Monte Carlo Simulation of One-Dimensional
Electron Transport in a Thunderstorm
Environment
(1988) / Johnson J L

WEST VIRGINIA UNIVERSITY

Student Conceptions in Thermodynamics and
Kinematics
(1988) / Churchill B E

Magnetic Susceptibility of Cobalt Ions
in Magnesium Oxide
(1988) / Dean J C

Magnetic Susceptibilities of Monomers and
Dimers of Manganese Ions in Magnesium Oxide
(1988) / Gordon B L

Thermal Expansion and Magnetostriction of
Iron(z)Oxide (z=0.938) Below Room
Temperature
(1988) / Shams Q A

WESTERN MICHIGAN UNIVERSITY

Single-Electron capture and Loss Cross
Sections versus Target Z for 1 MeV/u Oxygen
Ions Incident on Gases
(1988) / Bowman S

A Study of an Integral Equation for
Computing Radial Distribution Functions
(1988) / Ismail N

Projectile Electron Capture and Loss
Accompanying Target Ionization for Highly
Charged Oxygen Ions Colliding with Helium
(1988) / Price R

Recoil Corrected Continuum Shell Model from
Factor Calculations for 4-Helium(e,e')
4-Helium (0-+)
(1988) / Yu J

WESTERN ONTARIO, UNIVERSITY OF

Multiple Scattering of Radio Waves in the
Auroral Electrojet
(1988) / Donovan E F

Externally-Applied Fracture-Fixation: The
Stability of the Original Hoffmann, AO
Tubular and RxFx Frames
(1988) / Moroz T K

WICHITA STATE UNIVERSITY

Temperature Heat Capacities of Nickel-Base
Superalloys
(1987) / Liang R C

A Green Function Model for Liquid State
Physics
(1987) / Lucas E D

The Effects of Atmospheric Opacity on
Stellar Evolution
(1987) / Novacek G R

WINDSOR, UNIVERSITY OF

Vacuum-Ultra-Violet Polarization Studies
Following Electron Impact Excitation
(1988) / Karras W L

Electron Impact Excitation Studies of Atoms
and Small Molecules Using VUV Radiation
(1988) / Wang S

WORCHESTER POLYTECHNIC INSTITUTE

Partical Size Analysis of Crosslinked
Polystyrene (Crosslinking Agent
Divinylbenzene): and a Spectroscopic Study
of the Diffusion of Sodium Polystyrene
Sulphonate in an Aqueous Solution
(1987) / Rau A

WRIGHT STATE UNIVERSITY

Cardiopulmonary and Muscular Conditioning
Effects of Isokinetic and Aerobic Exercise
on Paralyzed Patients
(1987) / Fagri P D

The Interaction of Point Defects with
Dislocations in High Purity Silver Above
Room Temperature
(1987) / Grigsby P J

DLTS Studies of High-Energy Implanted
Titanium and Hydrogen Passivated Silicon
(1987) / Locker L

Design and Testing of Microwave Plasma
Anodization Using a Magnetic Confinement
System
(1987) / Lundberg W R

A Mathematical Model of the Cardiovascular
System with Feedback Control Under Short
Term Acceleration Stress
(1987) / Provens T G

Hydrogen Passivation of Deep and Shallow
Levels in Silicon
(1987) / Robison J H

On (176, 50, 14) Abelian Difference Sets
(1987) / Stewart D L

A Study of the Histomorphometry of the
Talus Following Low Frequency Vibration
(1987) / Swenson K N

BROOKLYN, POLYTECHNIC INSTITUTE OF

Rheological Properties of Compatible Polymer
Blends
(1988) / Chung H S

The Effect of Physical Aging on the Stress
Relaxation of Poly(Methyl Methacrylate)/Poly
(Styrene-Co-Acrylonitrile) Blends
(1988) / Devine S T

Adhesion Phenomena of High Performance
Thermoplastic/Carbon Fiber Matrix Composites
(1988) / Esfandi R T

Thermal Degradation Studies of Substituted
Aromatic Polyamides
(1988) / Rodgers J

The Effect of Filler in an Epoxy/Amine
Composite on Cure Kinetics
(1988) / Yeganeh M S

CASE WESTERN RESERVE UNIVERSITY

An Investigation of the Ductile to Brittle
Transition in the Fatigue of Polycarbonate
(1988) / Benderly D

Simulation of Studies for a Pultrusion
Process of a Composite Polyester System
(1988) / Ng H

Self-Diffusion of Latex Spheres in Solution
of Semi-Flexible Polymer
(1988) / Yang T

Synthesis and Characterization of Liquid
Crystalline Monomers and Polymers
(1988) / Yen K T

LOWELL, UNIVERSITY OF

Thermoplastics Technical Data Book: A
Reference for Non-Plastics Engineers
(1989) / Arsenault P

Mixing Characteristics of Polyblends in
Processing Equipment Using PC/PET Blends
(1988) / Awojulu E A

Effect of Density on Properties of
Polystyrene Foams Made by Steam and
Injection Molding
(1988) / Basto C M

Polyblending of Styrene-Butadiene-Styrene
and Styrene-Ethylene/Butylene-Styrene with
Linear-Low-Density Polyethylene
(1989) / Burduroglu M A

Plasticization of Polymethyl Methacrylate
(1989) / Burke J T

Development of Plastics Processing
Machinery and Methods
(1988) / Chabot J P

Application of Statistical Experimental
Design Techniques for the Continuous Process
Size Reduction Studies of Thermoplastic
Films
(1988) / Chang C Y

Melt Rheology of Long Glass Fiber Reinforced
Polypropylene
(1989) / Chang S M

Preliminary Studies on Reactive Polymer
Processing for the Purpose of Preparing
Aromatic-Amine Based Polyamides from
Diisocyanates and Dicarboxylic Acids
(1988) / Chen Y P

Verification of an Injection Mold Filling
Software Package
(1988) / Clarke R E

Evaluation of Sorbital (D-Glucitol) as a
Plasticizer for Nylon 6,6 and Reactive
Processing of Nylon 6,6 with (R)PS
(1988) / Damie P

Effects of Jeffamine on Nylon-66
(1989) / Dave A K

Melt Rheology of Highly Loaded Ferrite
Filled Nylon Resins
(1989) / Do K L

Birefringence as a Function of Injection
Molding Parameters
(1988) / Gagnon M G

Adhesive and Joint Evaluation for Automative
Plastic Sealed Beam
(1988) / Gheyara V N

Effect of Conversion on Solution Properties
of Acrylamide Ionomers with Varying Ion
Content
(1988) / Goenka A

Polymer Friction and Wear
(1989) / Horan P J

Effect of Fillers on Tensile Properties of
Acrylic Latex Films
(1988) / Horrigan R

Rheological Study of Kraton Solutions
(1988) / Jeng R J

Thermal, Mechanical and Crack Propagation
Properties of Injection Molded Fibrous and
Platelet Reinforced Polypropylene Composites
(1989) / Kagle A B

Polymer Blends of Recycled Poly(Ethylene
Terephthalate) and Polypropylene with
Coupled Agent
(1988) / Kim C O

Investigation of the Effects of Fiber Blend
Ratio, Web Geometry and Bonding Process
Conditions on Tensile Properties of
Needle-Punched Thermally-Bonded Polyester
Nonwoven Fabrics
(1988) / Kim S H

Design, Material and Processing Effects on
the Tensile Properties of Filled
Polypropylene
(1988) / Kirkwood A M

Effect of Processing Variables on the
Environmental Stress Crack Resistance of
Rotationally Molded Polyethylenes
(1988) / Kohlman S D

Effect of Crystallization Conditions on
Multiple Peaks of Melt-Crystallized
Poly(Ethylene Terephthalate)
(1988) / Leu S H

PENNSYLVANIA STATE UNIVERSITY
(continued)

Anionic Initiation of Star and Telechelic
Polymers
(1987) / Clark R J

Nitro-Stabilized Proton Abstraction Doped
Model Compounds as Precursors to Conducting
Polymers
(1987) / Hilker B L

Effect of Crystalline Morphology on the
Fatigue Crack Propagation in Polyethylene
(1988) / Jacq M

Thin Film 0-3 Polymer/Piezoelectric Ceramic
Composites
(1987) / Klein K A

Improving the Natural Draw Ratio of
Polymeric Films and an Investigation of
Associated Deformation Mechanisms
(1988) / Laughner M P

SOUTHWEST TEXAS STATE UNIVERSITY

Permeation of Laminated Elastomers: Polyurethane
and Polysiloxane
(1988) / Wells L S

TENNESSEE, UNIVERSITY OF (KNOXVILLE)

Quantitative Small Angle Light Scattering of
Liquid Crystalline Copolyester Films
(1987) / Effler L J

Volume-Induced Stress Response in Amorphous
Polymers
(1987) / Ko W C

The High Speed Spinning of Nylon-66 and Its
Mathematical Modelling
(1987) / Li X

Structure Variations Through the Thickness
of Poly(Ethylene Terephthalate) Films:
Characterization and Mechanism of Formation
(1987) / Wilk R J

U.S.A.F. INSTITUTE OF TECHNOLOGY

Ultraviolet-Induced Flashover of Highly
Angled Polymeric Insulators in Vacuum
(1988) / Enloe C L

Compressive Properties of High Performance
Polymeric Fibers
(1988) / Fawaz S A

The Experimentation and Analysis Compression
Test Methods for Single High Performance
Polymeric Fibers
(1988) / Fidan S

ALBERTA, UNIVERSITY OF

Bird Watchers of Point Pelee National
Park, Ontario: Their Characteristics and
Activities, with Special Consideration to
Their Social and Resource Impacts
(1988) / Fenton G D

AUBURN UNIVERSITY

Possible Cost-Effective Forest Site Preparation
Methods for the Nonindustrial Private Landowner
(1987) / Bradley J M

Survey and Evaluation of Competing Vegetation
Impacts on Loblolly Pine Yields
(1987) / Dickens D F

BALL STATE UNIVERSITY

The Status of Fagus Grandifolia Ehrn. ifn the
Midwest: A Comparative Study of Four Virgin Forests
(1988) / Clevenger B K

Protection of Habitat Critical to the Resplendent
Quetzal, Pharomachrus Mocinno, on Private Land
Bordering the Monteverde Cloud Forest Reserve,
Costa Rica
(1988) / Guindon C F

CALGARY, UNIVERSITY OF

The Spruce Communities in the Drumheller
Area: The Relationships Between the
Communities and the Physical Properties of
the Soils
(1988) / Zhu A X

COLORADO STATE UNIVERSITY

Surface Fire Influences on Central Rocky
Mountain Lodgepole Pine Stand Structure
(1988) / Franklin T L

Economic Effects of the Tax Reform Act of 1986
on a Vertically Integrated Forest Products Firm
(1988) / Guertin D S

Fire Behavior Modeling for Burn
Prescription Specification
(1988) / Hilbruner M W

Domestic Fuelwood Consumption and Supply in
Colorado: 1986 Characteristics, Trends and
Projections
(1988) / Olsen W K

The Ecological Effects of Layering
on Spruce-Fir Forest Development
(1988) / Steigerwald D

Biotic and Abiotic Characteristics Associated
with Pure and Mixed Rocky Mountain Aspen Stands
(1988) / Wood K A

CONNECTICUT, UNIVERSITY OF

A Study of the Effects of Weather Factors on
Gypsy Moth Defoliations in Massachusetts and
Connecticut
(1988) / Mo T

DUKE UNIVERSITY

The Plant Communities of a Blackwater
River: The Waccamaw River, North Carolina
(1988) / Marty R

Social Forestry in Pakistan: An Analysis of
the Factors Limiting Project Acceptance
(1988) / Rafiq M

EMORY UNIVERSITY

Element Concentration in Red Spruce of the
Southern Appalachians, Great Smoky
Mountains and Mount Mitchell
(1988) / Chua S C

Tennessee's Copper Basin Region: Tree
Growth and Trace Metals
(1988) / Deucher R L

Fine and Coarse Root Decomposition in
Early Successional Forest Stands of the
Southern Appalachians
(1988) / Harker-Grimm A C

FLORIDA, UNIVERSITY OF

Early Genetic Evaluation of Slash Pine
(1988) / Chase C

Early Development of Loblolly and Slash Pine
Plantations Subjected to Repeated
Fertilization and Sustained Weed Control
(1988) / Colbert S

Population Structure, Density, and Biomass
of Large Herbivores in a South Indian
Tropical Forest
(1988) / Karanth U

Depth Distribution and Ecology of Two
Deep-Sea Crabs, in the Eastern Gulf of
Mexico
(1988) / Lockhart F

Canopy Characteristics and Spectral
Reflectance of Slash Pine Plantations in
North Centralo Florida Depicted by Landsat's
Multispectral Scanner
(1988) / Stanislawski L

GEORGIA, UNIVERSITY OF

A Microcomputer Timber Cruise Program for
Estimating Total Tree and Multiple Product
Stand Yield
(1988) / Burgen T M

Attractiveness of Conspecific Bladder Urine
in White-Tailed Deer
(1988) / Forster D L

Tree Rings of Shortleaf Pine (Pinus Echinata
Mill) as Indicators of Past Climatic
Variability in North Central Georgia
(1988) / Grissino-Mayer H D

The Genetic Inheritance of Interlocked Grain
in Sweetgum
(1988) / Hood K L

Reproductive Biology of White-Tailed Deer
on Cumberland Island, Georgia
(1988) / Miller S K

Foliar Nitrogen Efficiency: A Theory
Applied to Three Broadleaved, Deciduous
Species
(1988) / Moss I S

A Heuristic Procedure for the Calibration
of the U.S.D.A. Forest Service Initial
Action Assessment Fire Model
(1988) / Murray S F

GEORGIA, UNIVERSITY OF
(continued)

Effect of Site Factors on Growth Response of
9 to 15-Year-Old Slash Pine Plantations
Following Understory Vegetation Removal
(1988) / Oppenheimer M J

Water Use Efficiency of Three Hardwood
Species
(1988) / Raper S M

An Interactive Microcomputer Program for
Wilderness Travel Simulation with Graphics
(1988) / Rowell A L

Allozyme Variation, Mating Systems, and
Genetic Relatedness in Black Locust
(Robinia Pseudoacacia)
(1988) / Surles S E

Displacement and Macrophage Flow in a
Forested Piedmont Soil
(1988) / Thomas P M

HAWAII, UNIVERSITY OF

The Maui Forest Trouble: Reassessment of
an Historic Forest Dieback
(1988) / Holt R A

IDAHO, UNIVERSITY OF

Skidding and Yarding Options for Smallwood
(1987) / Carey P

Asset Valuation of Advanced Grand Fir in
Northern Idaho
(1988) / Cathcart J F

Assessment of NOAA AVHRR Data Producing
Broad Area Fire Fuel Type Maps
(1988) / Chine E P

Bracken Ecology and Management Problems on
the Selway Ranger District
(1988) / Evers L

Evaluating Alternative Target Structures
for an Uneven-Aged Ponderosa Pine Stand
Using the Stand Prognosis Model
(1988) / Hensold T S

Comparison and Development of Height Growth
and Site Index Curves for Douglas-Fir in the
Intermountain West
(1987) / Ploeg J V

Econometric Estimation of the Supply and
Demand for Softwood Stumpage in Southwest
Colorado, 1960-1986
(1988) / Powers J T

Encouraging People's Participation in
Agroforestry Using the Graap Technique
in Burkina Faso
(1988) / Sawadogo P K

Comparing Methodologies for the Valuation
of Fuelwood Benefits in Forestry Projects
in Developing Countries
(1988) / Shields M

Development and Testing of Douglas-Fir Site
Index Prediction Models from Soil and Site
Factors in the Intermountain Northwest
(1987) / Zhang L

KENTUCKY, UNIVERSITY OF

Vegetative Rehabilitation of Surfaced Mined
Land with Organic Amendments and Topsoil
Replacement
(1988) / Alderdice L

Precipitation and Streamwater Quality as
They Relate to Nutrient Cycling in a
Hardwood Forest in Eastern Kentucky
(1988) / Brooks L T

Seasonal Diets and Forging Ecology of
White-Breasted Nuthatches in the Cumberland
Plateau of Eastern Kentucky
(1988) / Brunell A M

Earthworm Species Assemblages in Undisturbed
Forest Soils of Robinson Forest
(1988) / Dotson D B

Habitat Use by White-Breasted Nutches on
the Daniel Boone National Forest: A
Radio-Telemetry Study
(1988) / Jacoby G E

Conidia from Hypovirulent Endothia
Parasitica Protect American Chestnut from
Blight
(1988) / Scibilia K

Effects of Forest Harvesting on Selected
Stormflow Parametes on Three Small Forested
Watersheds in Eastern Kentucky
(1988) / Sutardjo B

A Characterization Study of Precipitation
Events at Robinson Forest, 1972-1986
(1988) / Tiaif S

LAKE HEAD UNIVERSITY

The Effects of Wind Induced Soil Moisture
Stress on the Germination and Early Growth
of Lamb on Three Soil Texture Types
(1988) / Krygier R J

Morphological Variation in Subalpine Fir and
Its Relationship to Balsam Fir in Western
Canada and the United States
(1988) / Palmer C L

Genetic Variation in Growth and Yield
Components of Juvenile Balsam Poplar Growing
in Northwestern Ontario
(1988) / Schnekenburger F

Integrated Resource Management Planning
Through the Linking of Mathematical and
Judgement-Based Models
(1988) / Tak K

Winter Use of Upland Conifer Alternate Strip
Cuts and Clear Cuts by Moose in the Thunder
Bay District
(1988) / Todesco C J W

LOUISIANA STATE UNIVERSITY

A Growth and Yield Model for Plantation
Grown Cottonwood Along the Mississippi River
(1988) / Durand K M

Habitat Use by Mottled Duck Broods
(1988) / Esters M D

LOUISIANA STATE UNIVERSITY
(continued)

Habitat Changes and Wildlife Use of a
Pole-Sized Natural Hardwood Stand One
Year After Thinning
(1988) / Gardiner E S

Effects of Partial Overstory Removal on
Natural and Artificial Regeneration in
Bottomland Hardwood Stands
(1988) / Henkel M W

Biological and Genetic Assessment of
Eastern Bluegill, Lepomis Macrochirus
Purpurescens, Introductions in Louisiana
Waters
(1988) / Laprairie R J

Effect of Total Water Hardness and Chlorides
on Growth, Survival and Feed Conversion of
Juvenile Red Drum (Sciaenops Ocellaltus)
(1988) / Pursley M G

Use of Winter Cattle Pastures by
White-Tailed Deer, with Nutritional
Implications
(1988) / Stiegler S G

Food Habits and Nutrition of Nutria in
Selected Forested Wetland Plant Communities
(1988) / Wilsey B J

MAINE, UNIVERSITY OF

A System for Data Collection from a Single
Photographic Image
(1988) / Cole J B

Use and Effects of the Pulsation Process
for the Pressure Treatment of Northeastern
Red Spruce
(1988) / Flynn K A

Invertebrates in Spruce-Fir Stands on Mount
Moosilauke, New Hampshire and at Howland,
Maine
(1988) / Grosman D M

The Effect of Horizon Type and Nutrient
Availability on Conifer Seedling Growth
(1988) / Kraske C R

Farm Forestry in Eastern Ontario,
Canada: The Role of Marketing and
Investment Incentives
(1988) / Murphy T P

The Early Stand Development of Red Spruce
and Balsam Fir Natural Regeneration
(1988) / Strauch P J

MASSACHUSETTS, UNIVERSITY OF

Massachusetts Prime Land Classification
System: An Accuracy Assessment
(1988) / Edwards J

Timber Consumption Forecasts in
Massachusetts
(1988) / Hayden A M

The Productivity and Cost of Prebunching
with a Remote Controlled Winch in a Southern
New England Forest
(1988) / Kline B W

MICHIGAN TECHNOLOGICAL UNIVERSITY

Effects of Whole Tree Harvesting on Organic
Matter, Cation Exchange Capacity, and Water
Holding Capacity
(1988) / Brooks R H

An Economic Analysis of Northern Hardwood
Management Practices: 32-Year Results
(1988) / Erickson M D

Competition Indices for Mixed Species
Northern Hardwoods
(1988) / Holmes M J

Distribution of Armillaria Clones, Including
Models of Red Pine Seedling Mortality, on
ELF Plantation Sites in Michigan's Upper
Peninsula
(1988) / Moore J

An Evaluation of Standard Timber Harvest
Practices on Stream Dynamics in the
Pictured Rocks National Lakeshore Inland
Buffer Zone
(1988) / Mullen G

Mineralogy and Inorganic Soil Chemistry on
Three Hardwood Sites, Ontonagon County,
Upper Michigan
(1988) / Paces J B

The Inverted Pyramid: Ecosystem Dynamics of
Wolves and Moose on Isled Royale
(1988) / Page R E

MINNESOTA, UNIVERSITY OF

A System for the Collection and Analysis of
High-Resolution Spectral Reflectance
Measurement of Forest Canopies
(1988) / Dewhurst S M

Restrictions Upon Intermolecular Association
in Kraft Lignin
(1988) / Dutta S

Soil Topographic Relationships at Two
Forested Sites in Minnesota
(1988) / Hairston A B

Development of an Automated Stand
Prescription Writer for Timber Harvest
Scheduling
(1988) / Pelkki M H

Optimal Aggregation of Tree List Data for
Use with an Individual-Tree Based Growth
Projection Model
(1988) / Stegemoclle↓ K A

Engineering and Economic Analysis of Wood
Energy Systems
(1988) / Steklenski P G

White Spruce (Picea Glauca(Moench)Voss)
Bare Rooted Seedling Production at the
General Andrews Nursery, Willow River,
Minnesota
(1988) / Vaughn V M

Effects of Slope Steepness, Bulk Density,
and Surface Roughness on Interrill Soil
Erosion
(1988) / Wang Y

MISSISSIPPI STATE UNIVERSITY

Site Preparation Effects on Early Loblolly Pine Growth,
Hardwood Competition, and Soil Physical Properties:
Interior Flatwoods, Kemper, County, Mississippi
(1986) / DeWit J N

Genetic Variation in Juvenile Traits of Eastern
Cottonwood from the Southern United States
(1986) / Friend M M

Improving Reforestation of Private, Nonindustrial
Forests in North Central Mississippi-Opportunities
and Constraints
(1986) / Kinsey S E

Classification of Land Cover and Land Utilization
Potential in Central Rift Valley Province of Kenya
(1986) / Mumiukha P W

The Effect of Soil Rutting on the Growth of
Residual Loblolly Pine Trees Following a
Thinning Operation
(1986) / Nickolich M B

Genetic Variation in Wood and Bark Specific Gravity
of Five-Year-Old Sycamore in the Gulf South
(1986) / Patterson P E

A Simulation Model for Thinning Plantations
(1986) / Savelle I W

Analyzing Stump to Truck Harvesting Costs in Small
Diameter Southern Pine Plantations
(1986) / Whiting W A

The Influence of Property Value Reappraisal on
Forest Land Ad Valorem Taxes in Mississippi
(1986) / Whitman D C

TMPGENN: A Data Transformation and Analysis Method
for Landsat Multispectral Scanner Data
(1986) / Wright L R

MISSOURI, UNIVERSITY OF (COLUMBIA)

The Effects of Fire on an Oak-Hickory Forest
in the Missouri Ozarks
(1988) / Godsey D W

Stream Water Quality Protection Using Vegetated
Filter Strips: Structure and Function Related
to Sediment Control
(1988) / Gough S C

MONTANA, UNIVERSITY OF

Predicting Pinus Ponderosa Site Index from
Soil Mineralization Rates at Lubrecht
Forest, Montana
(1988) / Baker S P

The Cellular Anatomy of Penetration and
Infection of Containerized Western Larch
Seedlings by Botrytis Cinerea
(1988) / Dugan F M

Elk and Cattle Relationships on Summer Range
in Southwestern Montana
(1987) / Gniadek S

Predicting Prescribed Fire Managers'
Perceptions of Their Information Sources'
Worth
(1988) / Kaage W

Application of the SIRO-PLAN Methodology to
Decision Making in Wildland Fuel Management
Planning
(1987) / Lord B L

NEBRASKA, UNIVERSITY OF

Habitat Utilization and Movement of Adult
Channel Catfish and Flathead Catfish in
the Platte River, Nebraska
(1988) / Bunnell D

NEW BRUNSWICK, UNIVERSITY OF

Multiple-Trait Selection in Pinus Banksiana
Lamb
(1988) / Adams G W

Survival of Dispersing Spruce Grouse
(1988) / Beaudette P D

Marketing Woodlot Management
(1988) / Blouin G H

Improtance Performance Analysis of NOLS
Instructors
(1988) / Feunekes A L

Direct Current Resistance of Wood Polymer
Composites
(1988) / Hartley I D

Population Structure of Larix Laricina
(Du Roi) K. Koch in New Brunswick
(1988) / Lin Y

Interval Estimate Methods for Harvesting
Machine Evaluation
(1988) / Liu S

Correction of Snowfall Measurements on Two
Small Watersheds in New Brunswick
(1988) / Palmer D C

The Influence of a Wilderness Course on the
Self-Actualization of the Students
(1988) / Shin W S

A Geographic Information Systems Approach
for Spray Block Layout in the New Brunswick
Spruce Budworm Spray Program
(1988) / Vietinghoff L P

Learner Style, Stress and Extension Agent
(1988) / Wilson R

NEW MEXICO STATE UNIVERSITY

Field Production of Arizona Ash (Fraxinus
Velutina) Using Field-Grown Fabric
Containers, Sludge, and Three Levels of
Nitrogen Fertilizer
(1988) / Hardesty C H

Vegetative Propagation of Scots Pine by
Stem Cuttings
(1988) / Schaefer P

Vegetative Propagation of Blue Spruce by
Stem Cuttings
(1988) / Wagner A M

NORTH CAROLINA STATE UNIVERSITY

Foliar Mineral Nutrient Diagnosis with DRIS
for Identifying Nutritional Influences on
Female Cone Production in Fraser Fir
(1988) / Arnold R J

NORTH CAROLINA STATE UNIVERSITY
(continued)

Effects of Site Preparation and Cultural
Treatments on Competing Vegetation and Pine
Growth in a Piedmont Loblolly Pine
Plantation
(1988) / Fredericksen T S

Seed Germination of Fraser Fir: Influence
of Gibberellins 4+7 and and Light-Temperature
Interactions
(1988) / Henry P H

Evaluation of Loblolly Pine Pollen Quality
(1988) / Moody W R

Reduction of Available Chlorine Applied by
Using Oxygen-Enriched Alkali Extraction
(1988) / Murray J D

Effects of Spacing, Location and Height in
Bole on Engineering Properties of Loblolly
Pine Structural Lumber
(1988) / O'Quinn C N

Use of Landsat Thematic Mapper Digital Data
for Estuarine Water Quality Modeling
(1988) / Rios P

Estimating the Value of Yellow Poplar Logs
to a North Carolina Hardwood Plywood
Company Through the Use of Linear
Programming Shadow Prices
(1988) / Van Sant R L

NORTH DAKOTA STATE UNIVERSITY

The Effects of Indole-3-Butyric Acid and
2,4-Dichlorophenoxyacetic Acid on Regeneration
of Bare Root Pinus Ponderosa Dougl. ex Laws.
and Celtis Occidentalis L. Seedlings
(1988) / Ellis P L

Chemomutagenesis and in Nitro Propagation Using
Excised Ovaries of Hosta Sieboldiana
(1987) / Hill R A

Male Fertility and Tuberization of Solanum
Tuberosum Haploid-Wild Species Hybrids
(1988) / Jacobsen T L

Benzylaminopurine and Chlorflurenol as Chemical
Shearing Agents for Christmas Trees
(1988) / Poisson E J

NORTHERN ARIZONA UNIVERSITY

Response of Second-Year Ponderosa Pine
Needles to Induced Water Deficits
(1988) / Carlin K

The Influence of Induced Water Stres on
Ponderosa Pine Physiology and Pine Saw Flies
(1988) / Frantz D P

Use of Graph Theory to Analyze Constraints
on the Juxtaposition of Timber Stands
(1988) / Gross T E

Insect Defoliation, Interpretation, and
Recreation Value at Grand Canyon National
Park, Arizona
(1988) / Krisko K

A Forest Growth Simulation Model for the
Hualapai Forest
(1988) / Murphy C A

Pinyon - Juniper Seeding Establishment
Patterns in Relation to Microsite
(1988) / Sampson D A

Dominant Height and Site Index Equations for
Ponderosa Pine on the Fort Apache Indian
Reservation
(1988) / Stansfield W

Influence of Host Water Stress, Photoperiod,
Temperature and Humidity on Oviposition
Behavior and Egg Development on Ponderosa
Pine
(1988) / Tisdale R

OREGON STATE UNIVERSITY

Effects of Grazing Management and Tree Planting
Pattern in a Young Douglas-Fir Argoforest
(1987) / Carlson D H

Effects of Shrubs and Herbs on Conifer Regeneration
and Microclimate in the Rhododendron-Vaccinium-
Menziesia Community of South-Central British Columbia
(1988) / Coates K D

Shear Stress Transfer Between Roots and Soil
(1987) / Commandeur P R

Predicting the Machinability of Finger-Joints in
Ponderosa Pine Cutting Stock
(1988) / Coombs M

Relationships of Tree Growth to Nitrogen and Water
Availability in a Sheep-Tree-Pasture System in
Douglas County, Oregon Field Case Study and
Shadehouse Simulation.
(1988) / DeMeo T

Evaluation of a Shadow Price Search Heuristic as
an Alternative to Linear Programming or Binary
Search for Timber Harvest Scheduling
(1988) / Eldred P

The Influence of Distance and Floral Phenology
on Pollen Gene Flow and Mating System Patterns
in a Coastal Douglas-Fir Seed Orchard
(1987) / Erickson V J

Simulation of Storm Runoff in the Oregon Coast
Range
(1987) / Fedora M A

Models for Predicting the Effect of Factor
Changes Upon Employment in Four of Oregon's
Forest Products Industries
(1988) / Goldammer J P

Germination and First-Year Survival of Red Alder
Seedlings in the Central Coast Range of Oregon
(1987) / Haeussler S

Measurement of Within Tree Density Variations in
Douglas-Fir Using Direct Scanning X-Ray Techniques
(1988) / Hoag M L

Use of Fluorescent-Labeled Lectins for Studying
Progressive Stages of Fungal Decay in Douglas-
Fir and Ponderosa Pine
(1988) / Lin L C

Effect of Solution Nitrogen and Phosphorus on
Growth, Carbon Allocation and Nitrogen Fixation
of Red Alder Seedlings
(1987) / Lipp C C

OREGON STATE UNIVERSITY
(continued)

Synecology of the Monotropoideae within Limpy Rock
Research Natural Area, Umpqua National Forest,
Oregon
(1987) / Luoma D L

Soil Changes After Afforestation in Yellow River
Loess: A Case Study in Gansu Province, People's
Republic of China
(1988) / McCain C N

Denitrification Potential in Forest Riparian Soils
of the Western Oregon Cascades: Spatial and
Temporal Variation
(1987) / McClellan M H

Bonding Kinetics of Thermosetting Adhesive Systems
Used in Composites
(1988) / Ren S

Large Woody Debris and Channel Morphology of
Undisturbed Streams in Southeast Alaska
(1987) / Robison E G

Thinning and Urea Fertilization Effects on Emerging
Grand Fir Foliage and Growth of Western Spruce
Budworm (Choristoneura Occidentalis) Larvae
(1987) / Savage T J

Diurnal Bird Use of Snags on Clearcuts in Central
Coastal Oregon
(1987) / Schreiber B

Multitemporal Classification of Vegetation in the
Oregon Coastal Range Using Landsat Multispectral
Scanner Data
(1988) / Wang S

Impacts, Standards, and Perceived Crowding on the
Deschutes River: Extending Carrying Capacity
Research
(1987) / Whittaker D P

An Efficient Optimizing Model for Determining
Thinning Regime and Rotation Age Using the
Stand Projection System Growth Simulator
(1988) / Yoshimoto A

PENNSYLVANIA STATE UNIVERSITY

The Use of Sequential Analysis in
Controlling Final Moisture Content in
Kiln-Dried Lumber
(1988) / Hashim I B

Simulating Cable Logging in the Allegheny
Region of Pennsylvania: A Case Study
(1987) / Hodes P W

Pulp and Papermaking Properties of Gypsy
Moth-Killed Trees
(1987) / Kessler K R

Relationships Between Soil Chemistry and
Tree-Ring Chemistry for Three Tree Species
at Two Forest Sites in Central Pennsylvania
with pH 4.4 and pH 5.5 Soils
(1987) / Sayre R G

Raster-to-Vector Data Conversion with
Application to Digitally Produced Forest
Cover Maps
(1988) / Van Gorder J L

PURDUE UNIVERSITY

Tree Regeneration Response to Clearcut
Harvesting on the Hoosier National Forest
(1988) / George D W

Changes in Physical Properties of
Reconstructed Minesoil After Six Years
Exposure to Natural Processes
(1988) / Hernandez M A

Development of an Expert System for
Upholstered Furniture
(1988) / Picado J

Mechanical, Physical, and Anatomical
Properties of Short-Rotation Douglas-Fir
and White Ash
(1988) / Quanci M J

RENSSELAER POLYTECHNIC INSTITUTE

The Scientist as Technology Policy Advocate:
Charles Holmes Herty and the Establishment
of the Savannah Pulp and Paper Laboratory
(1988) / Shimazu M

RHODE ISLAND, UNIVERSITY OF

The Influence of Cuttings on Oak Forest
Composition
(1988) / Otico J B

SOUTH CAROLINA, UNIVERSITY OF

Stand Analysis, Stem Production and
Litterfall in a Mature Bottomland Hardwood
Forest
(1988) / Parks W S

STEPHAN F. AUSTIN STATE UNIVERSITY

Synthesis of High Temperature
Superconductors
(1988) / Akwani I A

A Scent-Station-Trapping Technique for
Developing Furbearer Population Indices
in East Texas
(1988) / Best M S

Effects of Thinning a Loblolly Pine
Plantation on Nongame Bird Populations in
East Texas
(1988) / Chritton C A

The Effects of Seismic Exploration on the
Woody Vegetation of the Big Thicket National
Preserve
(1988) / Duncan K L

An Ecological Study of the Crustacean
Community in a Prairie Temporary Marsh in
Central Texas
(1988) / Evans J D

Factors Affecting Visitor Selection of
Campsites on Sam Rayburn Reservoir
(1988) / Goetz L A

Seasonal Shifts in Home Range Utilization
by White-Tailed Deer
(1988) / Green W R

Woodpecker Foraging Behavior and Resource
Partitioning in an Eastern Texas Bottomland
Hardwood Forest
(1988) / Jones G D

STEPHAN F. AUSTIN STATE UNIVERSITY
(continued)

Effect of Fire Exclusion on Longleaf Pine
Savannas in the Big Thicket National
Preserve
(1988) / McClung M A

Territorial Behavior and Pair-Bonding in
the White-Breasted Nuthatch (Sitta
Caolinensis) in East-Central Texas
(1988) / Roach J

TENNESSEE, UNIVERSITY OF (KNOXVILLE)

An Analysis of Competition and Distribution
in the Tennessee Structural Composite Market
(1987) / Watson W J

TEXAS A AND M UNIVERSITY

The Use of Logistic Regression to Model the
Probability of Oak Wilt Occurrence in the
Texas Hill Country Using Forest Stand and
Site Characteristics
(1988) / Dignum D R

Using Climatic and Soils Information to Project
Loblolly Pine Growth
(1988) / Farmer D B

Modeling Distributions of Stem Characteristics
of Genetically Improved Loblolly Pine
(1988) / Janssen J E

Exogenous Application of Gibberellins to Hasten
Yaupon Seed Germination for Surface Mine Reclamation
(1988) / Whatley C M

UTAH STATE UNIVERSITY

Relative Tolerance: A Test of Some
Indirect Criteria
(1987) / Parker J N

VERMONT, UNIVERSITY OF

Variation Among Yellow Birch Population in
Developmental Photosynthesis-Growth
Relationships and Photosynthetic Response
to Varying Light Intensity and Temperature
(1988) / Shane J B

VICTORIA, UNIVERSITY OF

Development of the Crown (Nodal Diaphragm)
and Seasonal Changes in the Ultrastructure
of the Apical Meristem in Coastal Douglas
Fir Seedlings
(1988) / Krasowski M D

WASHINGTON STATE UNIVERSITY

A Study of the Forest Products Industry
in Washington
(1988) / Deffar G

Ponderosa Pine Tree Rings: Variability of
Response to Precipitation in Space and Time
(1988) / Hagan T A

Understory Production Response to Tree
Stocking Reduction in a Central Washington
Mixed-Conifer Forest
(1988) / Host L A

Grazing Behaviors of Heifers Under Short
Duration Grazing in Central Washington
(1988) / Knapik M B

Growth and Gas Exchange in Cottonwood as
Influenced by Environment, Partial
Defoliation and Nutrient Regime
(1988) / Zwier J C

WASHINGTON, UNIVERSITY OF

Study on Anthraquinone Extractives in Teak
(1987) / Aung M

The Characterization of Glass Fibers for
Papermaking
(1988) / Bailey R M

Development of DNA/DNA Hybridization
Techniques with Applications to Cactus
Systematics
(1987) / Beeman P F

Taylor Dispersion Analysis for Estimation of
Diffusion Coefficients and Molecular Weights
of Polymers and Polymeric Mixtures Such as
Lignin
(1987) / Boyle W A

Technology Creates New Veneer Product
Opportunities
(1987) / Brady T T

Foliage Respiration of Abies
Amabilis
(1987) / Brooks J R

Influence of Soil Aluminum on Armillaria
Root Disease in Washington State
(1987) / Browning J E

Galapagos National Park: The History of a
Unique System of Conservation Management
(1987) / Chamblin A W

Development of Mixed Western Hemlock Spruce
Stands on the Tongass National Forest
(1987) / Deal R L

Microbial Degradation of
Glucoiso-Saccharinic Acid
(1988) / Dykes J M

The Essential Oil of Western Red Cedar
(1987) / Frazier C E

Stand History and Host Selection of the
Mountain Pine Beetle in a Climax Lodgepole
Ecosystem in South Central Oregon
(1987) / Hobson K R

Balsam Poplar in Alaska: Ecology and Growth
Response to Climate
(1987) / Lev D J

Brown Creeper Abundance Patterns and Habitat
Use in the Southern Washington Cascades
(1987) / Mariani J M

Internal Loading of Cellulosic Pulps with
In-Situ Precipitated Insoluble Salts
(1986) / Negri A R

Soil Genesis on the 1980 Pyroclastic Flows
of Mount Saint Helens
(1987) / Nuihn W W

Analysis and Reduction of the Errors of
Predicting Prescribed Burn Emissions
(1987) / Peterson J L

WASHINGTON, UNIVERSITY OF
(continued)

Microhabitat Use of Small Mammals Revealed
by Radiotelemetry and Live-Trapping
(1987) / Reed M J

Acid Prehydrolysis of Cottonwood with Sulfur
Dioxide
(1987) / Rossi A J

The Implications of Different Spatial Scales
on Interpretation of Small Mammal Habitat
Association Patterns
(1987) / Schmidt D M

Herbicide Studies with Hybrides in the
Pacific Northwest
(1988) / Shirley D M

The Properties and Genesis of Volcanic Ash
Influenced Soils of Western Washington
(1987) / Wang L

Growth of Selected Populus Clones as
Affected by Leaf Orientation, Light
Interception and Photosynthesis
(1987) / Wiard B M

Chemistry of Phenolic Compounds in Poplar
Leaves
(1988) / Wu J

WEST VIRGINIA UNIVERSITY

Analysis of the Contents of Letters Received
at the Monongahela Forest Supervisor's
Office
(1988) / Boll C F

Growth Response to Crown Release of Pole
Size Black Cherry on an Even-Aged
Appalachian Cove Hardwood Site
(1988) / Gorman J M

Throughfall, Stemflow, and Wet Deposition
Around Single Forest Trees in the West
Virginia University Forest
(1988) / Kosuri S R

Break-Even Benefit Cost Analysis for the
Direct Control of Gypsy Moth
(1988) / Riddle K S

A Timber Trespass Evaluation Computer
Program for Appalachian Hardwoods
(1988) / Simpson B T

Herbicide Techniques for Controlling Fern
Understories in Hardwood Stands
(1988) / Smith E S

Effects of Intermediate Cuttings on Woody
Understories of Mixed Oak and Cove
Hardwood Types
(1988) / Smith S R

Home Range of Reintroduced River Otter in
West Virginia
(1988) / Tango P J

Home Range, Habitat Utilization, and
Activity of the Endangered Northern Flying
Squirrel
(1988) / Urban V K

The Competitive Position of West Virginia's
Wood Industries
(1988) / Williams M A

WESTERN MICHIGAN UNIVERSITY

The Relationship of Pore Volume and Pore
Size Distribution to the Development of
Glass and Scattering Coefficient
(1988) / Cook J

A Study of Alkaline Sodium Sulfite
Pretreatment of Spruce During
Thermomechanical Pulping
(1988) / Dendukuri N

Energy Considerations in Stock Preparation
Refining Modified by Recycling
(1988) / Deshpande R

Effects of the Addition of Binder on the
Packing Characteristics of Pigment Particles
(1988) / Garg R K

The Evaluation of Efficiency Determination
of Screening Device (Pressure Screen) by
Image Analysis
(1988) / Goyal R

A Pilot Plant Evaluation of Selected
Hardwood Pulps for Supercalendered Grade
B Paper
(1988) / Srivastava D

BAYLOR UNIVERSITY

The Management of High-Level Radioactive
Waste and the Suitability of Deaf Smith
County, Texas as a Nuclear Waste Repository
(1988) / Marcott L D

Assessment of New and Emerging Technologies
for the Treatment of Hazardous Refinery
Wastes
(1988) / Ravishankar K

GEORGETOWN UNIVERSITY

Measurements of Performance Characteristics
of Fluoroscopix X-Ray Systems in a Clinical
Setting
(1988) / Divine M P

Evaluation of the Navy Criticality Dosimeter
(1988) / Mohaupt T H

A Study of the Incineration of Radioactive
Medical-Pathological Waste at the National
Institutes of Health
(1988) / Newman N E

LOWELL, UNIVERSITY OF

A Systematic Approach to Nuclear Power Plant
Evacuation Planning
(1988) / Albright C L

Design of a Low Level Radioactive Dry Waste
Monitor to Evaluate Controlled Materials for
Free Release from a Radiologically
Controlled Area
(1989) / Callahan M A

Investigation of the Relative Surface Dose
from Lipowitz Metal Tissue Compensators for
24 and 6 mv X-Ray Beams
(1988) / Cardarelli G A

P-Channel Metal Oxide Semiconductor Field
Effect Transistor High Dose Dosimeter
(1988) / Chen F F

Investigation of Stationarity of Modulation
Transfer Function and Scatter Fraction in
Conjugate View Pre-Reconstruction Single
Photon Emission Computed Tomography
(1989) / Coleman M

Monte Carlo Simulation of a Fast Neutron
Detector
(1989) / Falo G A

Optimization of the Shielding Design for a
Germanium Based Whole Body Counting System
(1989) / Ward K D

MARQUETTE UNIVERSITY

Nuclear Fission - A Brief Survey
(1988) / Lin J

MONTANA, UNIVERSITY OF

Radioactive Fallout Monitoring Before and
After the 1963 Nuclear Weapons Test Ban
Treaty
(1988) / Tarkalson R E

U.S. NAVAL POSTGRADUATE SCHOOL

A Technique for Improving Active Network
Performance in a Radiation Environment
Without the Use of Hardened Devices
(1987) / Lohr D M

Cerenkov Radiation Field Analysis Due to a
Passing Electron Beam
(1987) / Price B K

Energy Distribution of Cerenkov Radiation
for Finite Frequency Intervals
(1987) / Wilbur T M

WASHINGTON, UNIVERSITY OF

Reduction of Fluoromisonidazole, a New
Imaging Agent for Hypoxic Cells
(1988) / Prekeges J L

WESTERN ONTARIO, UNIVERSITY OF

The Dissolution Rate and Vapour Development
Characteristics of X-Ray Irradiated PMMA
(1988) / Lehockey E M

ALBERTA, UNIVERSITY OF

Threshold Detection Values of Potential Fish
Tainting Substances from Oil Sands
Wastewaters
(1988) / Jardine C G

Demonstration of Quantitative Risk
Assessment to Some Municipal Water Treatment
Plant
(1988) / Mercer S

ARIZONA STATE UNIVERSITY

Optimization of Capacity Expansions at the
City of Chandler Surface Water Treatment
Facility
(1988) / Francis T

ARKANSAS, UNIVERSITY OF

Treatability Study of a Industrial Landfill
Leachate Containing Chlorophenols and
Chlorophenoxyacetic Acids
(1988) / Ahmed T

Modeling of Anaerobic Filter Performance
(1988) / Jahan K

Effect of Reactor Height on Anaerobic Filter
Performance
(1988) / Steinbrecher P

AUBURN UNIVERSITY

The Use of an Analog Predictor Model in an Adaptive
Control Scheme for the Waste Water Treatment System
at Auburn University's Waste Oil Reprocessing
Facility
(1987) / Bigler W J

Analysis and Design of an Aerobic, Biological
Treatment System for Oily Wastewater from a
Pilot Waste Oil Recycling Facility
(1987) / Craig B A

Water Budgets for Wastewater Lagoons Covered with
Water Hyacinth
(1987) / Ebeid M M

Upgrading Wastewater Treatment Pond Effluents
by Utilizing a Bubble Screen to Reduce Escape
of Solids
(1987) / Hendrix R L

Sequencing Batch Reactors for Nitrification of
Chicken Processing Wastewaters Pretreated by
Anaerobic Filtration
(1987) / Juang D F

A Study of an Aquatic Treatment System Using the
Water Hyacinth Anaerobic Digestion of Detrital
Material and Root Length Inhibition by Nutrient
Availability
(1987) / Odom T F

BRIGHAM YOUNG UNIVERSITY

An Analysis of Water Development in San Juan
County
(1988) / Hosler B A

CALGARY, UNIVERSITY OF

Water Conservation Capacity Building: An
Analysis of Needs and Implementing
Strategies (A Case Study of hanna, Alberta,
Canada)
(1988) / Swan B H

CALIFORNIA STATE UNIV. (SACRAMENTO)

Development of a Generalized Procedure for
the Remediation of Petroleum Hydrocarbons in
Soil and Groundwater from a Case Study
(1988) / Brenner M J

Developments in Industrial and Hazardous
Wastes Treat Technology: A Review
(1988) / Karkal S S

Ground Water Movement and Ground Water
Contamination, Southern Pacific
Transportation Company, Sacramento Railyard
(1988) / MacDonald A M

Design, Construction, and Monitoring of
Conventional On-Site Wastewater Disposal
Systems
(1988) / Rowland C A

Detention Basins for Peak Runoff Rate
Reduction in an Urbanizing Watershed: A
Sensitivity Study in Antelope Creek
Watershed, Placer County, California
(1988) / Sloat M S

CALIFORNIA, UNIVERSITY OF (DAVIS)

Evaluation of Optimization Approach for
Identifying Groundwater Pollution Sources
(1988) / Beegle J E

Seasonal Denitrification of a Secondary
Effluent Using the Overland Flow Treament
Process
(1988) / Hayashi G S

Policy Alternatives to Meet Water Quality
Objectives for Wetland Habitat in the
Grassland Water District
(1988) / McEfee C C

Stochastic Modeling of Flow and Salinity in
the Ebro River, Spain
(1988) / Quilez D

Nitrogen and Phosphorus Removal from
Secondary Effluent Using the Overland
Flow Process
(1988) / Rogalinski L M

The Effect of Soluble Soil Organic Matter on
the Adsorption of Simazine in Hanford Sandy
Loam
(1988) / Spurlock F C

Sulfate and Selenate Antagonism in the
Green Alga
(1988) / Williams M J

CARNEGIE-MELLON UNIVERSITY

Temporal Variation of Solute Concentrations
in Groundwater
(1988) / Hu H L

CLARKSON COLLEGE OF TECHNOLOGY

Resuspension and Potential for Release of
In-Place Heavy Metals from Sediments in the
Detroit River, Michigan
(1988) / Vanetti D

Feasibility of RBC Treatment of an
Industrial Wastewater Containing High
Concentrations of Ammonia and Fluoride
(1988) / Vrona M

IDAHO, UNIVERSITY OF
(continued)

Critique of a Groundwater Solute Transport
Model of Radionuclide Migration from an
Uranium Mill
(1988) / Tarr K M

Ground Water Flow Characteristics in
Fractured Basalt in a Zero Order Basin
(1987) / Winkelmaier J

IOWA STATE UNIVERSITY

Evaluation of Landfill Leachate Monitoring
Programs from Sanitary, Co-Disposal, and
Hazardous Waste Disposal Facilities
(1988) / Kroemer L R

Chemical and Geotechnical Stabilization of a
Lead-Contaminated Foundry Waste
(1988) / Megivern S J

Molecular Size Distribution of Organic
Compounds in Groundwater: Implications for
Treatment
(1988) / Spiesman A L

KANSAS, UNIVERSITY OF

Anaerobic Trickling Filter: Microbial
Adhesion and Initial Operations
(1988) / Adams C D

Effects of Water-Well Construction on
Temporal Variability of Ground-Water Quality
in Lincolnville, Marion County, Kansas
(1988) / Chaffee P K

Gasoline Biodegradation with Activated
Sludge
(1988) / Gyorog S M

Simulation and Optimization of the Operation
of the Kansas River Basin Reservoirs
(1988) / Jian X

Anaerobic Treatment of Bakery Wastewaters
(1988) / Koerfer P M

A Survey of Organic Carbon and
Trihalomethane Formation Potential in
Kansas Groundwaters
(1988) / Miller R E

An Evaluation of the Adequacy of Snowmelt
Indices for the Prediction of Flood Flow in
the Northern Plains and Upper Mississippi
Valley
(1988) / Ruhl J K

In-Situ Removal of Iron and Manganese from
Groundwater: Fundamental, Practical, and
Economical Considerations
(1988) / Rundell B D

MARQUETTE UNIVERSITY

Soil and Groundwater Contamination and
Emergency Response Guide
(1988) / Ghosh P

Hazardous Waste Site: A Case Study
(1988) / Riedl R

MASSACHUSETTS INSTITUTE OF TECHNOLOGY

Solving Boston Harbor's Combined Sewer
Overflow Problem: Is Their Room for
Innovation
(1988) / Adelman A

Repair, Replacement, and Rehabilitation
Strategies for Urban Water Supply Pipe
Networks
(1988) / Kouloumbis A

Geotechnical Engineering Aspects of
Hazardous Wastes - A Survey
(1988) / Pestana-Nascimento J M

MASSACHUSETTS, UNIVERSITY OF

Contamination of High Purity Water in the
Semi-Conductor Industry: Evaluating the
Suitability of Well Water as a Water Source
for DEC Hudson, Massachusetts
(1988) / Collagan S L

MICHIGAN TECHNOLOGICAL UNIVERSITY

Surficial Sediment Characteristics and
Sediment Phosphorus Release Rates in
Onondaga Lake, NY
(1988) / Johnson N A

MINNESOTA, UNIVERSITY OF

Selective Cyanide Recovery from Wastewater
Containing Metal Cyanide Complexes
(1988) / Chang Y W

The Removal of Oil from Oil-Water Mixtures
Using Selective Oil Filtration
(1988) / Magdich P J

Determination of Flood-Flow Frequencies for
Big Trout Creek at Pickwick, Minnesota
(1988) / Mosner J B

A Laboratory Investigation of the
Relationship Between Groundwater Velocity
and Mass Transfer from a Simulated Aromatic
Hydrocarbon Spill at the Water Table
Interface
(1988) / O'Dell C H

Acid Neutralization in Soft-Water Lakes
(1988) / Perry T E

Estimation of Transmissivities, Heads, and
Velocities in Steady-State Groundwater
Modelling
(1988) / Quinodoz H A M

The Demand for Wastewater Treatment
Facilities in Rural Minnesota Communities
(1988) / Ryan B M

Pesticide Contamination of Surface and
Groundwater in the Karst Region of
Southeastern Minnesota
(1988) / Swanson D W

Evaluation of Precipitation Samplers for
Assessing Wet-Deposition of Trace Organic
Contaminants
(1988) / Swanson M B

Ion Budgets and Cycling Processes in an
Acid-Sensitive Seepage Lake
(1988) / Tacconi J E

OREGON STATE UNIVERSITY
(continued)

The Development of a Gas-Permeable-Membrane-
Supported (GPMS) Biofilm Reactor for the Combined
Anaerobic/Aerobic Treatment of Polychlorinated
Biphenyls
(1988) / Polonsky J D

Staphylococci as Microbiological Indicators to
Estimate the Quality of Swimming Pool Waters
(1988) / Rivera J B

PENNSYLVANIA, UNIVERSITY OF

Biofilm Degradation of Cresols Under Nitrate
Respiration Conditions
(1988) / Li L

Carbon Adsorption and Biological Fluidized
Bed Treatment of Methylene Chloride
(1988) / Mitzen P G

PURDUE UNIVERSITY

Fate of Heavy Metals in Sludge-Amended Brick
(1988) / Adams R B

A Computer Analysis of Water Use in a
Foundry to Reduce Wastewater Generation
(1988) / Copeland E C

Frequency Analysis of Low Flows in Indiana
Streams
(1988) / Knox B L

An Investigation Into the Feasibility and
Applicability of Microbial In-Situ Treatment
of Ground Water Contaminated with Vinyl
Chloride in Kalamazoo, Michigan
(1988) / Oresik W L

Fixation of Lead in Resource Recovery Ash
by Solidification
(1988) / Pherson P A

RENSSELAER POLYTECHNIC INSTITUTE

Municipal Point Source Nitrogen Loads to the
Chesapeake Bay Drainage Basin
(1988) / Harrison L R

RHODE ISLAND, UNIVERSITY OF

Groundwater Modeling Analyses and Methodology
for Determining Suitable Monitoring Well
Locations at Groundwater Contamination Sites
(1988) / Eating W J

An Analysis of the Benefits of Improving Water
Quality in Narragansett Bay: An Application
of the Contingent Valuation Method
(1987) / Hayes K M

Effects of Road Salting and Deicing Practices
on the Base Flow Water Quality of Streams
(1988) / Lee Y S

A Feasibility Study of the Particle Entrainment
Simulator for Evaluating Sediment Transport in
River Systems
(1988) / Schmidt G L

SOUTH DAKOTA SCHOOL OF MINES AND TECHNOL.

Preliminary Assessment of Water Quality of
Water Wells, Pine Ridge Indian Reservation,
South Dakota
(1988) / Heintzman D L

Floodplain Modeling and Evaluation of the
HEC-2 Water Surface Profile Program
for a Mountainous Stream, Rapid Creek, South
Dakota
(1988) / Long H E

The Origin and Occurrence of Dissolved
Nitrogen and Oxygen Gas in Natural Waters
of the Black Hills Area, South Dakota
(1988) / Williams J M

SOUTH DAKOTA STATE UNIVERSITY

Evaluation of Groundwater Monitoring Data at
the Madison Infiltration-Percolation
Wastewater System
(1988) / Choi B C

Suitability of Water from a Gravel Pit Pond
in East Brookings as a Public Water Supply
(1987) / Stefanich T D

Hard Water: Occurrence, Health Effects, and
Consumer Drinking Practices in Brookings and
Sioux Falls, South Dakota
(1988) / Wong H C

SOUTHERN ILLINOIS UNIVERSITY

Implementation of the Volatile Synthetic
Organic Compound Regulation of the 1986
Amendments to the Safe Drinking Water Act
by the St. Louis Water Treatment Plants
(1988) / Armstrong S

SOUTHWEST TEXAS STATE UNIVERSITY

The Chemical and Biological Groundwater Quality of
the Edwards Aquifer in the Sink Creek Drainage
Basin, Hays County, Texas
(1988) / Kolbe C M

A Water Quality Index Based on Requirements for
Contact Recreation
(1988) / Magin D C

The Effects of a Wastewater Treatment Plant on
Water Quality in Onion Creek and Williamson Creek,
Travis County, Texas
(1988) / Palafox S D

TENNESSEE TECHNOLOGICAL UNIVERSITY

Duck River Utility Commission Water Quality
Evaluation: Seasonal Characterization of
the Source Water and Quality Assessment of
the Product Water
(1988) / Meriwether C E

An Assessment of Ground Water Contamination
by Oil and Gas Wells in Overton County,
Tennessee
(1988) / Meriwether F H

Evaluation of Surface Wastewater
Impoundments in the State of Tennessee
(1988) / Munday W W

Effects of Garton Pumps on Douglas
Reservoir Release Water Quality
(1988) / Nubbe C F

The Variation of Water Quality in the West
Tennessee Sand Aquifers
(1988) / Wilson T M A

Hydrologic Interpretation of Water Quality
Data for the Emory and French Broad Rivers
(1988) / Young J L

WASHINGTON, UNIVERSITY OF

The Seasonal Dynamics of Biologically
Available Phosphorus in Lake Sammamish
(1987) / Butkus S R

Development of Nonpoint Source Pollution
Management Plans: The Burley Minter Case
(1987) / Carter S G

Sources of Phosphorus to Phytoplankton in a
Shallow Eutrophic Lake: Importance of
Sediments Versus Macrophytes
(1987) / De Gasperi C L

Regional Administrative Centralization of
Water Management Authority in the United
States: Ideal or Impossibility
(1987) / Evans K A

Seasonal Prediction of Phosphorus
Concentration and Internal Loading in a
Shallow Lake
(1988) / Kelly T S

Guidelines for Water Quality Modeling in
Developing Countries
(1987) / Oh K D

Internal Phosphorus Loading and Water
Quality Projections in Moses Lake
(1987) / Okereke V O

Implementation Analysis and Nonpoint Source
Pollution Control: An Application in Puget
Sound
(1987) / Shigenaka G

Solid Waste Planning for a New Urban
Environment: A Systems Approach to
Developing New Institutions in Seattle, King
County, and Washington State
(1987) / Showman J W

The Rising Cost of Falling Water: The Case
for Hydropower Planning in Washington State
(1988) / Zubalik S

WINDSOR, UNIVERSITY OF

Modelling of Fate and Transport of Toxic
Chemicals in Natural Waterways
(1988) / Barycka E

Removal of Volatile Organic Compounds from
Wastewater During the Coagulation Process
(1988) / Shouli A S

WORCHESTER POLYTECHNIC INSTITUTE

Review and Analysis of a Computer Model
Designed to Simulate the Chlorination and
Disinfection Process in a Water Distribution
System
(1986) / Meader J L

WYOMING, UNIVERSITY OF

Effects of Sewage Sludge Amendment of
Abandoned Mine Spoil on Plant Growth,
Composition and Forage Quality
(1989) / Whiston K M

ALABAMA, UNIVERSITY OF (UNIVERSITY)

Computer Usage in Interior Design
Professional Practices: Implications for
Interior Design Education
(1988) / Brandon B L

BOWLING GREEN STATE UNIVERSITY

The Northwest Ohio Fur Industry: 1900 to
the Present
(1988) / Kotlarczyk L

CALIFORNIA, UNIVERSITY OF (DAVIS)

Heat-Induced Aging of Linen
(1988) / Nowak K C

CLEMSON UNIVERSITY

Determination of Spunbond Binder Content by
Near-Infrared Reflectance Analysis
(1988) / Harris S L

Predicting the Thermal Mechanical Stability
of Latex Coated Nonwovens
(1988) / Ziegler J

CORNELL UNIVERSITY

Starch as a Non-Durable Finish to Reduce
the Retention of Methyl Parathion on Cotton
and Polyester Fabrics After Laundering
(1988) / Sagan R M

Ammonia Plasma Treatment of Ultra High
Strength Polyethylene Fabric
(1988) / Sembach S A

A Strategy Analysis of Successful Textile
and Apparel Exporters in New York State
(1988) / Vanden Berghe B J

GEORGIA INSTITUTE OF TECHNOLOGY

A Study of Photopolymerized Microporous
Membranes
(1988) / Lanigan W R

Characterization of Extended Chain
Polyethylene/S-2 Glass, Interply Hybrid,
Fabric Composites
(1988) / Miller L M

Residual Property Assessment of Impacted
Carbon Fiber Composites
(1988) / Nettles A T

Effect of Processing Parameters on
Morphology and Mechanical Properties of
Carbon/PEEK (APC2) Composite
(1988) / Shukla J G

The Synthesis and Characterization of a
Regularly Alternating Copolyester
(1988) / Wusik M J

GEORGIA, UNIVERSITY OF

Half Lives of Disperse Dyes in Anaerobic
Sediment/Water Systems
(1988) / Yen C P

IOWA STATE UNIVERSITY

Development and Testing of a Computer
Simulation for Use in Apparel Design
(1988) / O'Riley A L

MANITOBA, UNIVERSITY OF

The Effectiveness of Canada Standard Sizes
for Women Aged 65 to 85
(1988) / Marshall S P

Factors Influencing Current Fabric Hood
Production in Eskimo Point, Northwest
Territories
(1988) / McBride S

MISSOURI, UNIVERSITY OF (COLUMBIA)

Textile Production in the Communalistic Colony
at Bethel, Missouri: 1844-1879
(1988) / Horn L W

Missouri Apparel Manufacturers' Attitudes Toward
Offshore Assembly, Competitiveness, and
Employment in the Missouri Apparel Industry
(1988) / Meyer M M

NEBRASKA, UNIVERSITY OF

Evaluating the Potential Synergism of
Selected Antioxidants and Ultraviolet
Absorber Combinations Applied to Textiles
(1988) / Woeppel L R

NORTH CAROLINA STATE UNIVERSITY

Evaluating the Flammability and Thermal
Degradation of Materials Used in Chemical
Protective Clothing
(1988) / An S K

Thermal Insulation Properties of Low Density
Nonwoven Battings
(1988) / Anjaria M K

The Stabilization of Silk to Light and
Heat: Screening of Stabilizers
(1988) / Becker M A

Prediction of Total Stand Ingrowth in
Southeastern Mixed Species Bottomland
Hardwood Forests
(1988) / Davidson C B

Pattern Profiles and Fabric Properties
Affecting Tailorability
(1988) / DiSanto D M

NORTH DAKOTA STATE UNIVERSITY

Factors Affecting the Transfer of Temik to Fabric
(1988) / Braaten A W

The Influence of Customer Clothing on Salesperson
Behavior
(1087) / Hatlie T L

Educational Satisfaction of Textiles and Clothing
Graduates at North Dakota State University
(1988) / Hsieh I C

RHODE ISLAND, UNIVERSITY OF

Computer Skills in Retail Education
(1987) / Huber M E

A Comparison of the Performance Properties of
Ramie, Cotton, and Flax Fabric
(1988) / Tsai X X

TENNESSEE, UNIVERSITY OF (KNOXVILLE)

Retail Apparel Buyer's Purchase Decision:
Influencing Factors
(1987) / Miller J K

A Study of Thermal Bonding of Melt Blown and
Spunbonded Laminates and Repellent Finishing
Techniques
(1987) / Salamie L M

TEXAS WOMEN'S UNIVERSITY

Interior Design in Women's Retail Apparel
Stores: An Attitude Survey
(1988) / Alexander C J

A Manual of Instruction in Vegetal Dye
(1988) / Hull P O

An Evaluation of Selected Parameters in the
Utilization of Commercially Available Flame
Retardants for Production of Carbon Fabric
from Cotton Fabric
(1988) / Iwuchukwu L O

Development of a Slide-Tape Program on the
Principles of Dart Manipulation in Flat
Pattern Making
(1988) / Jou T T

ALBERTA, UNIVERSITY OF

A Parametric Evaluation of a Mechanistic
Empirical Design Procedure for Asphalt
Concrete Pavements
(1988) / Khogali W E

Land Use Impacts of Light Rail Transit in
Edmonton's Central Business District
(1988) / Sabey W D

AUBURN UNIVERSITY

Acoustic Emission Monitoring of Concrete Pavement
Spall Repairs
(1987) / Hinds-Rico J A

Regression Analysis Correlations for Budgeting
Expenditures in Roadway Maintenance Management
Systems
(1987) / Khouri C R

Accelerated Corrosion Testing for Protective
Systems on Steel Bridges in Coastal Regions
(1987) / Yousuf S

BRIGHAM YOUNG UNIVERSITY

The Wabash River: A Transportation Barrier
to Lafayette, Indiana
(1988) / Robins D G

Developing Passenger Car Equivalencies for
Left-Turning Trucks at Diamond Interchanges
(1988) / West J E

CALGARY, UNIVERSITY OF

A Guide for Implementation and Utilization
of Pavement Management Systems
(1988) / Enslen P M

CALIFORNIA STATE UNIV. (SACRAMENTO)

Simulation of Traffic Flow on a Segment of
Twelfth Street in Sacramento Using NETSIM
Program
(1988) / Hemmati A A

CALIFORNIA, UNIVERSITY OF (IRVINE)

Analysis of Truck-Involved Freeways
Accidents
(1988) / Hsueh C W

A Conception of Coordination Between the Bus
System and the Future Mass Rapid Transit
System in the Taipei Metropolitan Area
(1988) / Wang J R

CARLETON UNIVERSITY

Towards a Policy for Establishing Multimodal
Passenger Terminals in Canada
(1988) / Bell D W

Forces at the Slab-Web Junctions in Curved
Box Girder Bridges
(1988) / Mazikins P

Inelastic Load Distribution of Composite
Concrete-Steel Slab-on-Girdger Bridges: An
Analytical Study
(1988) / Nofal M E

CARNEGIE-MELLON UNIVERSITY

Aircraft Delays in the Public
Airspace: Formulating the Problem
(1988) / Shepherd R A

CASE WESTERN RESERVE UNIVERSITY

Punching Shear Strength of Reinforced
Concrete Bridge Deck Slabs
(1988) / Bousias S N

Implementation of Reliability Based Design
and Evaluation for Highway Bridge Fatigue
Life
(1988) / Raju S K

CLARK ATLANTA UNIVERSITY

An Examination of Public Transportation Problems
Facing the Elderly and Disabled Persons in
Fulton and DeKalb Counties
(1987) / Bibb-Stevenson S E

CLARKSON COLLEGE OF TECHNOLOGY

Design of Terrorist - Resistant Passenger
Transport Vehicle
(1988) / Dicello P T

Relationships Between Design Parameters,
Operating Speeds, and Accident Rates for
Future Highway Design Consideration on
Two-Lane Rural Curved Road Sections
(1988) / Paluri A

CLEMSON UNIVERSITY

Polyester Fibers in Asphalt Paving Mixtures
(1988) / Freeman R B

Comparative Analysis of Alternate Traffic
Signal System Designs
(1988) / Harrison P A

Analytical Study of the Provision of Auto
Parking Spaces and Terminal Curb Space at
United States Airports
(1988) / Menta V K

Microcomputer Application in the Design of
Asphaltic Concrete Mixes Using the Marshall
Method
(1988) / Nune S G

An Expert System to Evaluate the Safety of
a Railroad-Highway Grade Crossing
(1988) / Prabhu S

COLORADO STATE UNIVERSITY

Automated Design and Analysis of Military
Timber Bridges
(1988) / Davis R J

Mathematical Model of a Timber
Bridge Guardrail System
(1988) / Malone M S

CONCORDIA UNIVERSITY

Analysis of Driver's Lane Change and Turning
Behavior on Urban Highways and Street
Intersections
(1988) / Verma S K

CONNECTICUT, UNIVERSITY OF

Development of a Bridge Model to Monitor
Vibrational Signatures
(1988) / Lauzon R H

DELAWARE, UNIVERSITY OF

Stresses, Displacements and the Onset of
Lift-Off Due to One and Two Axle Trucks
(1988) / Eberhardt A W

Rear-End Conflict During Traffic Signal
Vehicle Clearance Intervals
(1987) / Townsend C J

FLORIDA, UNIVERSITY OF

Analysis and Rating of Highway Bridges Using
the Finite Element Method
(1988) / Constantinides I

A Photologging System for Pavement Condition
Measurements and Analysis
(1988) / Karakadas C

Enhanced Techniques for Understanding
Concrete Highway Pavement Behavior
(1988) / Lim K

Finite Element Modeling and Field Testing
of the Eau Gallie Bridge
(1988) / Schultz D

GEORGIA INSTITUTE OF TECHNOLOGY

Reliability of System Detector Data in
Replicating Field Conditions for the
Integrated Motorist Information System
(1988) / Henson L D

HOWARD UNIVERSITY

Surface Texture Changes Due to Simulated
Traffic Under Controlled Conditions as
Indicators of Skid Resistance
(1989) / Williams D M

IDAHO, UNIVERSITY OF

The Contribution of Branchline Railroads to
the Small Town Economy in Idaho
(1987) / Corkill S A

An Asphalt Concrete Moisture Distress
Pavement Performance Model Based on
Wheelpath Permanent Deformation
(1988) / Frith D J

Factors Influencing Personnel Requirements
for Winter Maintenance of Idaho State
Highways
(1988) / Maloney M D

Relating Vehicle Delay to Traffic Volume
at Four-Way Stop-Controlled Intersections
(1988) / Marek J F

A Pavement Performance Model for Fatigue
Cracking Due to Moisture Distress in
Asphalt Concrete
(1987) / White L

IOWA STATE UNIVERSITY

Long Term Structural Movement of Bridge
Piers
(1988) / Hachicho N N

Variations in Load Distribution in Beam-Slab
Bridges Resulting from Increases in
Transverse and Longitudinal Stiffness
(1988) / Hall M J

LOUISIANA STATE UNIVERSITY

Field Investigation of Longitudinal
Movements for a Multibeam Bridge
(1988) / Levert M F

LOUISIANA TECH UNIVERSITY

Development of a Graphics Post-Processor for
Traffic Accident Reconstruction Animation
(1988) / Nguyen J S

LOUISVILLE, UNIVERSITY OF

A Knowledge-Based Approach to Constructing Urban
Transportation Models
(1988) / Inoue H

MANITOBA, UNIVERSITY OF

Overweight/Overdimension Trucking in
Manitoba and Western Canada
(1988) / Girling R R

MARQUETTE UNIVERSITY

Public Works Capital Improvement Projects
Two Examples of City of Milwaukee Bridge
Projects
(1988) / Schmalz M

MASSACHUSETTS INSTITUTE OF TECHNOLOGY

Grain Transport Pricing in the Western
United States: Application of a Bilateral
Monopoly Model
(1988) / Aeppli A E

Optimal Maintenance Frequencies for Unpaved
Roads
(1988) / Alfelor R M

Implementing a Model for Investigating the
Economics of Rail Wear
(1988) / Auzmendi A R

Competition in a Deregulated European
Airline Industry: Defining a Strategy
(1988) / Bolorinos J

Integrated Optimal Design of Feeder and
Express Bus Services
(1988) / Chan S T

Determining who Should Pay for Traffic
Congestion: Case Study of Boston's Central
Artery and Third Harbor Tunnel Project
(1988) / Cottrell W D

An Analysis of Safety Benefits and
Associated Costs of Curtailing Air Traffic
During Periods of Thunderstorm Activity
(1988) / Fisher Z J

The Downtown Seattle Transit Project: Is a
Tunnel the Appropriate Alternative?
(1988) / Gotterer E L

Effects of Junction Delays in Equilibrium
Traffic Assignment
(1988) / Habbal M B

Outsourcing Railway Engineering
(1988) / Heavin J W

MASSACHUSETTS INSTITUTE OF TECHNOLOGY
(continued)

Workforce Planning on the M.B.T.A. Rapid
Transit Lines
(1988) / Hickman M D

Detecting Defects in Steel Bridge Beams by
Remotely Sensing Dynamic Response
(1988) / Marks M H

U.S. Domestic Airline Deregulation:
Transfers of Wealth and Labor Relation
Impacts
(1988) / Maurer J E

High Speed Civil Air Transports: A Strategy
for the Future
(1988) / McCarthy J J

Seasonality in Air Transportation in Demand
(1988) / Megwinoff H N

Vehicle Routing with Open Tours
(1988) / Rotshild R

Strategies for Private Funding of Public
Highways: A Practical Approach
(1988) / Sabina E E

An Expert System for Generating Terminal
Area Flight Paths for Arriving Aircraft
(1987) / Sadoune M M

Analysis of the Reliability of Stated
Preference Data in Estimating Models of
Travel Mode Choice
(1988) / Shiroishi F

Comparison of Optimization Techniques for
Origin-Destination Seat Inventory Control
(1988) / Williamson E L

MASSACHUSETTS, UNIVERSITY OF

An Application of Strategic Planning to
Route 9 in Hadley, Massachusetts
(1988) / Annino J M

Lapsize Computer and Local Pavement
Management
(1988) / Virkud U

MIAMI, UNIVERSITY OF

Computer-Aided Visual Impact Analysis of
Elevated Mass Transit Systems
(1988) / Canete M J

Multiobjective Decision-Making Models for
the Transportation of Hazardous Waste
(1988) / Samara S M

MINNESOTA, UNIVERSITY OF

Modelling and Verification of Interrupted
Freeway Flow Dynamics
(1988) / Georgadakos G

Dynamic Interactions Between Highway
Expenditures and Income/Employment
(1988) / Zante A

Dynamic Interactions Between Highway
Expenditures and Income/Employment
(1988) / Zante A

MISSISSIPPI STATE UNIVERSITY

A Traffic Engineering Computational Package for a
Micro-Computer
(1986) / Admad S M

Feasibility of Recycling Porous Friction
Surfaced Runaways
(1986) / Vollor T W

MONTANA STATE UNIVERSITY

A Feasibility Study of the Use of Torsional
and Moment Carrying External Diaphragms on a
Double Steel Box Girder Bridge System to
Reduce Live Load Moment
(1987) / Nottingham T S

NEBRASKA, UNIVERSITY OF

Safety Effects of Left-Turn Treatments at
Intersections on Urban Four-Lane Roadways
(1988) / Malone M S

Analysis of Traffic Operations at
Coordinated Semi-Actuated Signals Under
Various Presence Mode Detection Designs
(1988) / Tobin J R

NEW BRUNSWICK, UNIVERSITY OF

Transferability of Intercity Disaggregate
Mode Choice Models
(1988) / Abdelwahab W M

Digital Topographic Data Structure
Enhancement and Road Network Simplification
(1988) / Armstrong D H

The Impact of Roadside Development on the
Safety and Capacity of the New Brunswick
Primary Highways
(1988) / Beaulieu J R

A Prototype Emergency Response Guide
Knowledge-Based System for Dangerous Goods
Movements on New Brunswick Highways
(1988) / Cantin L F

A Case Study of Truck Overweight Operations
in the Province of New Brunswick
(1988) / Gould D J

Financial Constraints to Growth in the
Airline Industry
(1988) / O'Connor S M

The Impact of Transportation Deregulation
Upon the Potato, Fishing, and Transport
Industries of New Brunswick and Prince
Edward Island
(1988) / Prescott P T

Short and Long Term Impacts of Regulatory
Reform on Air Transportation in Atlantic
Canada
(1988) / Rogers P D

An Exploration to Condense Pavement
Management
(1988) / Williams A G

NEW JERSEY INSTITUTE OF TECHNOLOGY

Reliability of Existing Bridge Inspection
and Evaluation Procedures
(1988) / Klouser L R

NEW YORK, STATE UNIVERSITY OF (BUFFALO)

An Expert System Design Approach for
Highway Bridges Undergoing Random Vibrations
(1988) / Baker D J

OKLAHOMA, UNIVERSITY OF

Vehicle-Guideway Interaction in Rigid
Airport Pavements with Discontinuities
(1988) / Ledesma R H

Simultaneous Assignments of Buses to
Journeys and Locating Bus Garages Using
Heuristics: A Methodology
(1988) / Smadi A G

Fly Ash: Aggregate Mixes for Roadway Bases
(1988) / Zenieris P E

OREGON STATE UNIVERSITY

Analysis of Traffic Accidents Involving
Radioactive Materials (RAM)
(1986) / Humphries L L

PENNSYLVANIA STATE UNIVERSITY

From Parkway to Freeway: Roadside Design
Before the Interstate, 1890-1956
(1987) / Roberts C E

Study to Determine the Effect of Tire
Parameters on Rigid Pavement Response
(1987) / Tabatabaee N

PENNSYLVANIA, UNIVERSITY OF

Application of Spasm to Route Restructuring
in Regional Rail Case Study of the
Southeastern Pennsylvania Transportation
Authority
(1988) / Calvo E R

Comparison of Lindewold Line and New Jersey
Transit Bus System in Southern New Jersey
and Philadelphia
(1988) / Olanipekun O A

Empirical Testing of Traffic Assignment
Techniques on Transportation Networks
(1988) / Patterson V

PITTSBURGH, UNIVERSITY OF

Moment-Rotation Tests of Steel Bridge
Girders for Autostress Design
(1989) / Morcos S S

PURDUE UNIVERSITY

Bond of Epoxy Coated Reinforcing Steel in
Concrete Bridge Decks
(1988) / Cleary D B

An Economic Analysis of Maintenance and
Rehabilitation Alternatives for Asphalt
Concrete Pavements in Military Installations
(1988) / Danforth L E

An Economic Analysis of Maintenance,
Rehabilitation and Reconstruction of Rigid
Roadway and Airfield Apron Pavements
(1988) / Jaspers M W

A Fuzzy Set Approach for Bridge Safety
Evaluation
(1988) / Murthy S

Development of an Expert System to Aid in
the Calibration and Use of a Stream Quality
Simulation Model
(1988) / Wood D M

Elements of Pavement Life Cycle Cost
Analysis
(1988) / Ziai D

SOUTH DAKOTA SCHOOL OF MINES AND TECHNOL.

Field Load Tests on Custer and Reptile
Gardens Bridges (Results Used to Calibrate
Finite Element Models)
(1988) / Huber M S

Analytical Parametric Study of Reinforcing
Distressed Bridge Abutments
(1988) / Schelske S D

SOUTH DAKOTA STATE UNIVERSITY

The Effect of Mineral Filler, Asphalt
Content, and Curing Time on the Properties
of Cold Mixes When Used for Low-Volume Roads
(1988) / Abdalla N A

Investigation of the Causes of Observed
Translations in a Series of Abutments of
Concrete Box Girder Bridges
(1988) / Lenning C R

SOUTH FLORIDA, UNIVERSITY OF

Collapse Load Analysis of Continuous
Prestressed Voided Slab Bridges
(1988) / Gerges A N

Ultimate Strength Analysis of Curved
Concrete Bridges Prestressed Longitudinally
and Transversely
(1988) / Hodge D S

STEVENS INSTITUTE OF TECHNOLOGY

Microcomputer Analysis of Curved Bridges
(1987) / Bilis A

TENNESSEE, UNIVERSITY OF (KNOXVILLE)

An Analysis of the Effect of a Pedestrian
Signal Upon Pedestrian Behavior at a
Given Intersection
(1987) / Fortey N J

Evaluation of Selected Pavement Management
Systems for Microcomputers
(1987) / Jackson D W

Lateral Load Distribution in Beam and Slab
Highway Bridges
(1987) / Ozbek A M

Unintended Composite Action in Beam and
Slab Bridges
(1987) / Suetoh S

TEXAS A AND M UNIVERSITY

Development of a Guardrail-to-Bridge
Rail Transition
(1988) / Bligh R P

Effects of Temperature and Moisture on
Low-Volume Roads
(1988) / Chandra D

TEXAS A AND M UNIVERSITY
(continued)

A Mechanistic Model for the Prediction of
Stresses, Strains, and Displacements in
Continuously Reinforced Concrete Pavements
(1988) / Palmer R P

A Simplified Mechanistic Rut Depth Prediction
Procedure for Low-Volume Roads
(1988) / Yapa K A

TEXAS SOUTHERN UNIVERSITY

An Evaluation of the Impact of Airport
Anti-Terrorist Measures on Passengers'
Perceptions of Safety
(1988) / Airishusdian C U

A Study of the Impact of Airport Noise on a
Neighborhood in Houston, Texas
(1988) / Ogbomon G O

Turnkey Park-and-Ride Lot Development in
Houston: Problems and Prospects
(1988) / Ohaegbulem S E

A Plan for Revitalization for Houston's
Third Ward Area
(1988) / Perkins D R

Land Use Impacts of Houston's North
Freeway (I-45N) Transitway
(1988) / Washington E J

TEXAS, UNIVERSITY OF (ARLINGTON)

Distributed High Traffic Data Elements
(1988) / Swayze L C

TEXAS, UNIVERSITY OF (AUSTIN)

Deviation Saddle Behavior and Design for
Externally Post-Tensioned Bridges
(1988) / Beaupre R J

Highway Constructability Improvement Through
Pre-Construction Planning and Design
(1988) / Brown I

A Comparative Analysis of Left-Turn Delay
Associated with Two Different Lead Lag
Phasing Arrangements
(1988) / Collins R

The Variability of Travel Times in a
Commuting Corridor During the Evening Peak
Period
(1988) / Jones E G

The Development of a Preconstruction
Management System for the Texas State
Department of Highways and Public
Transportation
(1988) / McGlamery G L

The FHWA Pavement Policy and Its Impact on
Texas
(1988) / Rogers R B

A Study of Drainage Coefficients for CRC
Pavements in Texas
(1988) / Shyam V

Monitoring and Testing of the Bonded
Concrete Overlay on Interstate Highway
610 North in Houston, Texas
(1988) / Teo K J

TOLEDO, UNIVERSITY OF

Investigation of a Truck Weight Enforcement
Program for the Toledo Area
(1988) / Brancheau R K

Evaluation Techniques Used in Measuring the
Effectiveness of the Ohio Department of
Transportation Research and Development
Process and Implementation
(1988) / Shneiker A

U.S. NAVAL POSTGRADUATE SCHOOL

A Simulation Study of a Speed Control System
for Autonomous On-Road Operation of
Automotive Vehicles
(1987) / Dolezal M J

Deregulation's Effect on Labor in the
Trucking Industry
(1987) / Dougherty M

Models for Conducting Economic Analysis of
Alternative Fuel Vehicles
(1987) / Grenier D R

An Expert System for the Diagnosis of
Vehicle Malfunctions
(1987) / Selek C

Simulation Study of Traffic Flow at a Three
Way Intersection
(1988) / Song C C

Human Factors Aspects of the Traffic Alert
and Collision Avoidance System (TCAS II)
(1988) / Tuttell R J

U.S.A.F. INSTITUTE OF TECHNOLOGY

Mathematical Model of a Timber Bridge Guard
Rail System
(1988) / Malone M S

Evaluation and Response of Aged Flexible
Airfield Pavements at Ambient Temperatures
Using the Falling Weight Deflectometer
(1988) / Manzione C

VIRGINIA, UNIVERSITY OF

Evaluation of Financing Techniques for
Secondary Roads in Virginia
(1988) / Bhargava B

Implementation of the Data Acquisition
System for the I-295 James River Bridge
(1988) / Hayes S W

Evaluation of Cracking of Polymer Concrete
Overlay on Bridge Decks
(1988) / Huq S A

Dynamic Modeling of a Cable-Stayed Bridge
During Construction
(1988) / Lissenden C J

WASHINGTON STATE UNIVERSITY

Development of Design Criteria for
Continuous Skew Slab-and-Girder Bridges
(1988) / Khaleel M A

Feasibility Study of Transportation
Alternatives Between Spokane-Pullman/Moscow
Corridor
(1988) / Ong T H

WASHINGTON, UNIVERSITY OF

The Public Port Institution in the United
States: Using Past Experiences to
Understand Modern Issues
(1987) / Bittner R S

Correlations Between Discrete and Continuous
Space Estimates of Travel Times for Three
Different Urban Highway Networks
(1987) / Chang J B

An Evaluation of Flagging Technique and
Devices on Two-Lane Highway Construction
(1987) / Ifie A O

Analysis of Travel Time and Delay on a
Concurrent-Flow Transit Lane in a Downtown
Seattle Corridor
(1988) / Lewis R J

Transportation Services as a Competitive
Strategy: Port of Seattle Case Study
(1987) / Lohaviriyasiri P

Patching of Utility Cuts in Concrete
Pavements
(1987) / Murphy S D

Special Area Management Planning: Port
Angeles Harbor Resource Managment Planning
(1987) / Rasmussen A N

End Responses on Skewed Multibeam Precast
Concrete Bridges
(1987) / Rubio Roach R M

The Journey-to-Work: Comparisons and
Analysis of Travel Flow Matrices for the
Central Puget Sound Region
(1988) / Sicko R T

Airline Complexing Operations: An
Assessment with Recommendations for
Improvement
(1987) / Wheeler C F

Application of TRANSYT-7F: A Traffic
Simulation and Signal Timing Optimization
Model
(1987) / Zandi Z

WEST TEXAS STATE UNIVERSITY

A Transportation Problem with Quantity
Discounts
(1988) / Atchison M L

WEST VIRGINIA UNIVERSITY

Computer-Based Guided Decision-Making for
Surveying Safety in Streets and Highways
(1988) / Beachy K T

Determination of Truck Tire-Open Steel Grid
Deck Contact Pressure Distribution
(1988) / Vali A

WYOMING, UNIVERSITY OF

Proposed Bridge Deck Design Specifications
Based on Recent Research Findings
(1989) / Naiknavare R D

ARKANSAS, UNIVERSITY OF

The Influence of Color Combination on Type
Legibility for Maps Viewed on Computer
Monitors
(1988) / Lockhart J J

A Cultural Perspective on Japan and Its
Rice Controversy of the 1980's
(1988) / Smith K R

COLORADO STATE UNIVERSITY

Comparison of Asbestos Fiber Distributions by TEM
and PCM from Automobile Brake Repair Operations
(1988) / Becnel C A

Utilization Patterns in Short Duration and
Continuous Grazing Systems
(1988) / Bissio J C

Calculation of Expected Final Cost for
Mined-Land Reclamation
(1988) / Law A S

Pronghorn Grazing Impacts on Winter Wheat
(1988) / Liewer J A

Estimates of Allowable Training Levels
on Fort Carson, Colorado
(1988) / Maenius-Mosley C

Demonstrating Improvement in Chair Comfort
Through Education and Proper Chair Adjustment
(1988) / Meier S J

Resource Availability and Use by Early and Late
Successional Species in a Semiarid Ecosystem
(1988) / Parachini R W

Household Hazardous Waste: Assessing
Awareness and Attitudes in Larimer County
(1988) / Scudder K

A Method to Unfold Environmental Photon Energy
Distributions from Measured Spectra
(1988) / Whicker J J

CORNELL UNIVERSITY

Uptake and Translocation of Six Organic
Chemicals in a Newly-Designed Plant Exposure
System and Evaluation of Plant Uptake
Aspects of the Pre-Biologic Screen for
Ecotoxicologic Effects
(1988) / Nolt C L

LAMAR UNIVERSITY

Occurrence of Vibrio Species in the Upper
Texas Gulf Coast
(1988) / Hughes K J

MASSACHUSETTS INSTITUTE OF TECHNOLOGY

Perceptual Correspondence of Abstract
Animation and Synthetic Sound
(1988) / Abbado A

Building Intelligent Tutors: The Macavity
Experience
(1988) / Ahuja M K

For Whom the World Stops: The Himalayan
Sadhu in a World of Constant Motion
(1988) / Ajania K

Quality Improvement for the Defense Industry
(1988) / Bartus K M

A Domain Independent Approaach to
Integrating the Technologies of Expert
Systems and NLP
(1988) / Beers T R

Computerization of the Improved Product
Development Process
(1988) / Berg J S

Taguchi Quality Improvement Methods Applied
to Product and Process Design Examples
(1988) / Burrington J S

Multiscale Coding of Images
(1988) / Butera W J

Innovations in Telecommunications: Can Lead
Users Help Guide New Services Development?
(1988) / Capell R L

The Strategic Dimension of Design
(1988) / Chiambaretta P E

Identifying the Target Market for a New
Automobile
(1988) / DeFalco S P

Joint Ventures in Related Industries:
AMC - A Case in the Automotive History
(1988) / Debry G G

Telecommunications: The Industry, the
Players, the Issues, the Future...
(1988) / Donald O C

Managing the Engineering Design Functions:
A Case Analysis of Process Redesign
(1988) / Downie E A

The Tool and Die Industry of Dayton, Ohio:
The Challenge of Change in a Traditional
Industry Sector
(1988) / Edwards D R

The Changing Television Industry: A Case
Study of the Effect of People Meters and the
Changing Television Environment Upon the
NBC Network
(1988) / Epstein D H

Global Technology Transfer in Vacuum
Equipment Industry - A Study of a Japanese
Competitor
(1988) / Erez D V

A Study of Values in the Radio Industry
(1988) / Falvey M F

Downsizing at Boston City Hospital: Its
Effects on Support Workers
(1988) / Fiske T W

Achieving Competitive Advantage in Product
and Process Design: How to Use Cost as an
Organizing Philosophy and Management Tool
in Design Processes
(1988) / Fitzpatrick D A

Perception, Aesthetics and Culture in New
Media
(1988) / Foley K A

A Case Study in Product Redesign
(1988) / Gardiner K F

MASSACHUSETTS INSTITUTE OF TECHNOLOGY
(continued)

Cost Benefit Analysis of a Flexible
Manufacturing System: The Navy's Rapid
Acquisition of Manufactured Parts (RAMP)
Project
(1988) / Gardner T A

A Study of the Effects of Acquisition Policy
on the Climate for Technological Innovation
in the Procurement of U.S. Air Force Weapon
Systems
(1988) / Gebhard C C

Preliminary Design for a Knowledge-Based
System in a Tax Auditing Environment
(1988) / Grochowski R J

Perturbative Analysis of Nonlinear Betatron
Oscillations
(1988) / Henestroza E

Support for Firing Range Placement Module
for Army Installation Planners
(1988) / Khawaja K Q

Information and Design Considerations for
Holographic Television
(1988) / Kollin J S

On the Perception of Pitch Accents
(1988) / Landel K L

An Industry Assessment and Marketing
Strategy for a New Monoclonal Antibody
Cardiovascular Diagnostic Test
(1988) / Lanning N J

Determination of Material Properties of
Ligamentous Tissues in the Human Knee
(1988) / Lanzendorf E J

The Role of Time and Time Perceptions in
the Service Industries
(1988) / Lavdas M

Integrative Bargaining in International
Cooperative Agreements and Aids of Computer
Models in Their Negotiations: Developing a
Tool for Case Analysis
(1988) / Lee B

Health Information Networks: A Preliminary
Market Analysis
(1988) / Lee D A

Firm Level Productivity Measurement and
Improving White-Collar Productivity
(1988) / Lee K S

Eastman Kodak's Implementation of
Micro-Based Manufacturing Software in
Decentralized Organizations: A Human
Resources Perspective
(1988) / Lewis B A

General Motors Marketing Competition
(1988) / Lichtenthal E T

Information Technology Planning for a
Rapidly Growing Computer Distributor
(1988) / Mankoff S M

The Relationship Between Management
Ownership and Corporate Performance: An
Empirical Analysis of Small Companies
(1988) / Mannheim D R

Children, Cybernetics, and Programmable
Turtles
(1988) / Martin F G

The Role of Strategic Planning in
Not-for-Profit Organizations Involved in
Technology Intensive Areas
(1988) / McCarthy S J

An Expert System Approach to Actuarial
Valuation
(1988) / Meltzer C S

A Study of Japanese Automotive Parts and
Component Makers Transplanting in the United
States
(1988) / Murata M

Medical Diagnostic Biosensors:
Technologies, Markets, and Strategies
(1988) / Orida N K

Software Development Management in the
Banking Industry in Japan
(1988) / Otoshi K

Survey Input Simulation for Aseptic
Packaging
(1988) / Otterbeck J

Stategic Impact of Communication Networks
for the Financial Services Industry
(1988) / Parikh A S

Exploring Controller Performance for a Large
Space Structure Via Q-Parameterization
(1988) / Paster F K

A Graphical Query Language Supporting
Flexible Database Access
(1988) / Peters N L

Volumetric Rendering for Holographic Display
of Medical Data
(1988) / Plesniak W J

The Impact of CASE Technology on Information
Systems Design Using Parallel Teams: A Case
Study
(1988) / Pumarada-O'Neill M E

Elements of Service Strategy: Make, Buy or
Borrow?
(1988) / Randig K E

A Model of the Innovation Process in
Information Technology
(1988) / Shuff R F

Hospital Information Systems
(1988) / Siderius J H

Diffusion of Technology: Expert Systems
(1988) / Siegel D L

Estimation of FM Synthesis from Sampled
Sounds
(1988) / Sullivan T M

Marketing of International
Telecommunications Services by an Asian
Developing Country in the 1980s and 1990s
(1988) / Tan K J

Programmable Forms
(1988) / Tomaino L S

Services in the American Economy
(1988) / Waldstein L A

MASSACHUSETTS INSTITUTE OF TECHNOLOGY
(continued)

Managing Key Roles and Innovation in a Small
High-Tech Firm
(1988) / Waller G W

Financing Alternatives in a High-Risk
Economy
(1988) / Weiss A

The Business Plan for Fiberspar: Graphite
Composite Masts and Booms for the
Boardsailing Market
(1988) / Weld T G

An Analysis of Patterns of Innovation
Within the Total Hip Prostheses Industry
(1988) / Wells F S

High Technology Custom Devices Marketing
Strategy in a Vertically Integrated
Company: An Internal Manufacturing Case
Study
(1988) / Willis C J

Recognizing When a Technology is Ready for
Commercilization
(1988) / Wong G E

A Business Plan for Exporting and Marketing
Apple MacIntosh Personal Computers to the
Japanese Market
(1988) / Yamauchi K

MC NEESE STATE UNIVERSITY

Immobilization of Metals by Particulate
Biomass Derived from Chlorella Vulgaris
and Scenedesmus Quadricauda]
(1988) / Harris P O

MICHIGAN TECHNOLOGICAL UNIVERSITY

Gold Price Influences in the Period 1975
to 1986
(1988) / Campbell G A

The Cost of Economic Security: Stockpiling
First Tier Strategic Materials
(1988) / Kryfka A J

MURRAY STATE UNIVERSITY

A Study of Leadership, Influence, Control
and Organizational Management of a Company
Size Military Unit
(1988) / Bechtel D W

Moral Development Scores and Rating of
Administration Policies at a Christian
Liberal Arts College
(1988) / Mullins T W

NORTHEASTERN LOUISIANA UNIVERSITY

A Comparison of the Ichthyofauna in a
Man-Made and a Natural Lotic Ecosystem in
Eastern Mississippi
(1988) / Wood R M

OREGON, UNIVERSITY OF

Simulation of Lake Evaporation with an
Energy Balance-Eddy Diffusion Model of Lake
Temperature: Model Development and
Validation, and Application to Lake-Level
Variations at Harney-Malheur Lake, Oregon
(1987) / Hostetler S W

TEXAS A AND M UNIVERSITY

Evaluation of the Performance of an Electrostatic
Coalescer
(1988) / Figueroa C E

Extremal Index Estimation
(1988) / Kampa A E

Estimating Optimal Doses in Random Coefficient
Assay Models
(1988) / Rosenthal A

Measurement of Bone Mineral Content in Caged
and Active Cats
(1988) / Tveter D E

TEXAS, UNIVERSITY OF (ARLINGTON)

A Comparison of Adenosine Triphosphatase
Activity and Morphology of Bacillus Subtilis
Grown at 37 Degrees Centigrade and 53
Degrees Centigrade
(1988) / El-Saheb M J

U.S. NAVAL POSTGRADUATE SCHOOL

A Methodology for Validation of High
Resolution Combat Models
(1988) / Coville M P

Optical Memory Cards: A Comparison with
Other Current Technologies and Potential
Military Applications
(1987) / Frink S H

Practical Applicability of Exact and
Approximate Forms of the Randomization Test
for Two Independent Samples
(1987) / Hesse D H

Assessment of Fatigue in Aviation Crews
(1987) / Hutchins M L

Strategic Allocation of Sealift: A
Gams-Based Integer Programming Approach
(1987) / Lally M J

Acquisition Streamlining Efforts Within the
Space and Naval Warfare Systems Command
(1987) / McKeever M C

A Statistical Analysis Determining Effective
and Efficient Methods of Shipboard Training
(1987) / Michealson K A

The Evolution of Naval Warfare Technology
and the Impact of Space Systems
(1987) / Sharrett P J

WASHINGTON UNIVERSITY

An Assessment of Dengue Fever in Cuba and
Puerto Rico
(1987) / Bosch A

A Technology Assessment of Environmental
Control Units as Used by High Level Spinal
Cord Injured Persons
(1987) / Schumann T L

WEST VIRGINIA UNIVERSITY

An Evaluation of Data Sources Used for
Industrial Research
(1988) / Halverson J A

Industrial Reorganization and the Local
Response to Plant Closures: A New Politics
of Manufacturing Decline
(1988) / Herod A J

WESTERN WASHINGTON UNIVERSITY

Intraspecific Variability in Selected
Morphological Characteristics of Megalopae
Among Congeneric Species of Cancer Crabs
(1988) / DeBrosse G A

Phytoplankton Composition and Temporal
Variation Among the Three Basins of Lake
Whatcom, Washington
(1988) / Ehinger W J

Home Range Size of the Northern Barred Owl
and Northern Spotted Owl in Western
Washington
(1988) / Hamer T E

WINDSOR, UNIVERSITY OF

Genome Size Variation in the Cladocera
(1988) / Beaton M J

A New Modified Newton's Method for
Minimizing Factorable Functions
(1988) / Chow K L

Riabouchinsky Flows in Magnetohydrodynamics
(1988) / Labropulu F

Hydrographic Study of Steady Transfers and
Aligned MHD Flows
(1988) / Nguyen P V